Quick Reference

connectED.mcgraw-hill.com

The icons found throughout *Math Connects* provide you with the opportunity to connect the print textbook with online interactive learning.

Investigate

 Animations present an animation of a math concept or graphic novel.

Vocabulary presents visual representations of math concepts.

Multilingual eGlossary presents key vocabulary in 13 languages.

Learn

Personal Tutor presents a teacher explaining step-by-step solutions to problems.

Virtual Manipulatives provide digital ways to explore concepts.

Graphing Calculator provides keystrokes other than the TI-83/84 Plus or TI-Nspire used in the textbook.

Audio recordings provide an opportunity to build oral and listening fluency.

Foldables provide a unique way to enhance students' study skills.

Practice

 Self-Check Practice allows students to assess their knowledge of foundational skills.

Worksheets provide additional practice and reteach opportunities.

Online Assessment checks understanding of concepts and terms.

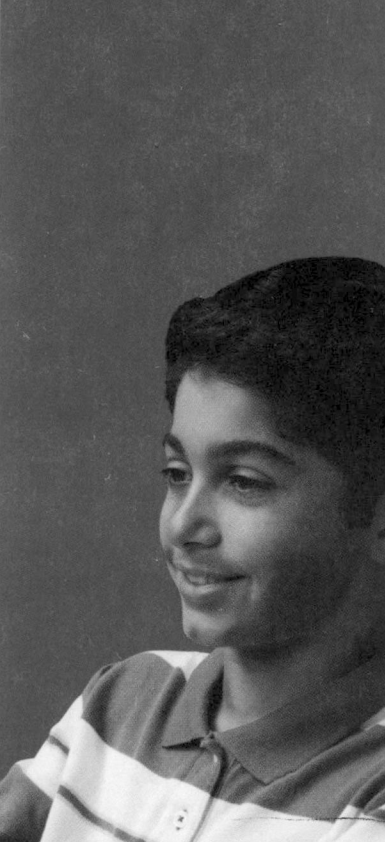

Glencoe McGraw-Hill

Math Connects

Course 2

Authors
Carter · Cuevas · Day · Malloy
Kersaint · Luchin · McClain · Molix-Bailey · Price
Reynosa · Silbey · Vielhaber · Willard

Mc Graw Hill **Glencoe**

The *McGraw-Hill* Companies

 Glencoe

Send all inquiries to:
Glencoe/McGraw-Hill
8787 Orion Place
Columbus, OH 43240-4027

ISBN: 978-0-07-895130-5
MHID: 0-07-895130-5

Math Connects, Course 2

Printed in the United States of America.

3 4 5 6 7 8 9 10 DOW 18 17 16 15 14 13 12 11

Contents in Brief

Master the Focal Points

End-of-Year Option

Problem-Solving Projects

Online Guide

"It's easy to do my assignments online and quick to find everything I need."

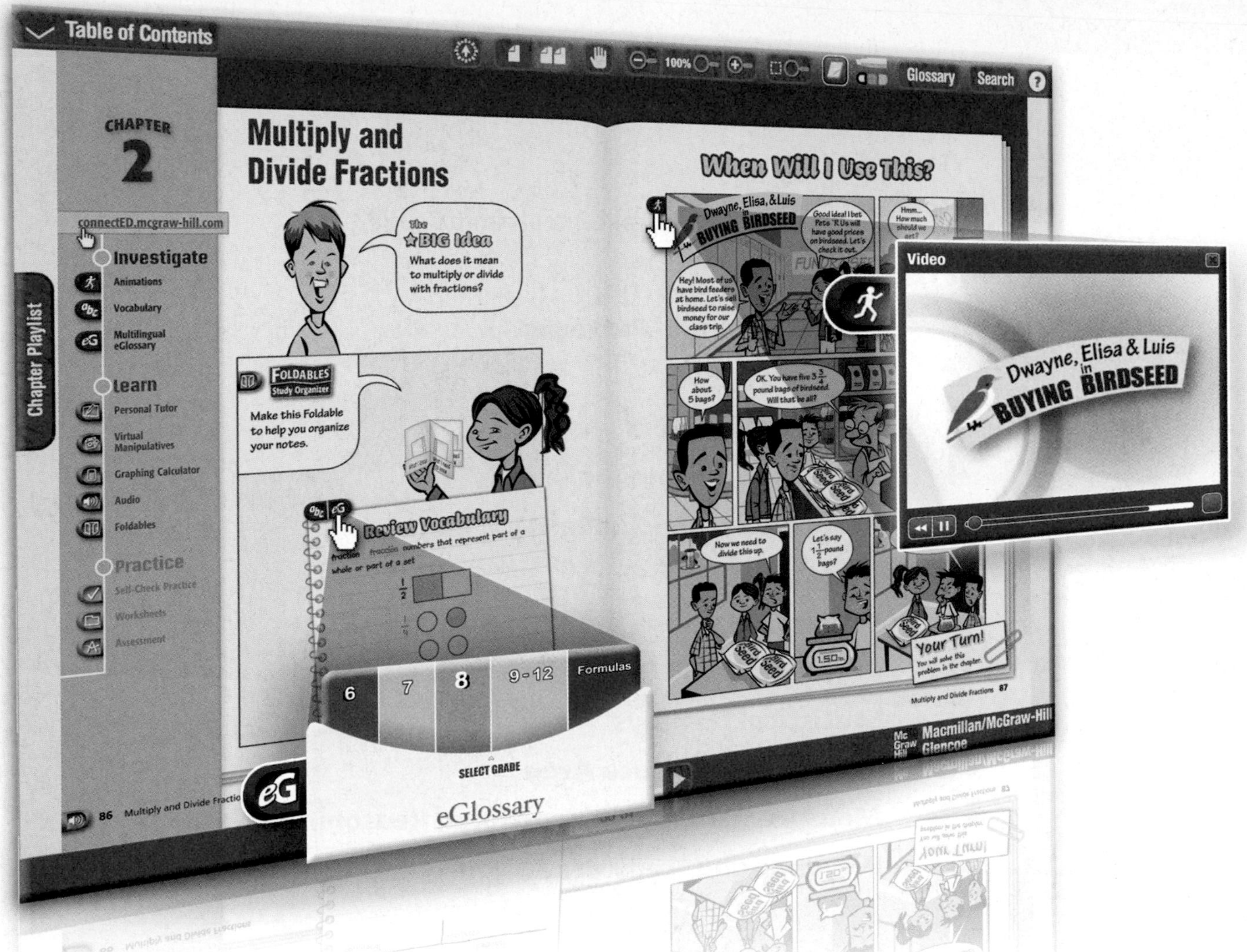

Investigate

Animations present an animation of a math concept or graphic novel.

Vocabulary presents visual representations of math concepts.

Multilingual eGlossary presents key vocabulary and formulas in 13 languages.

Digital and Print Solutions

The icons found throughout Math Connects provide you with the opportunity to connect the print textbook with online interactive learning.

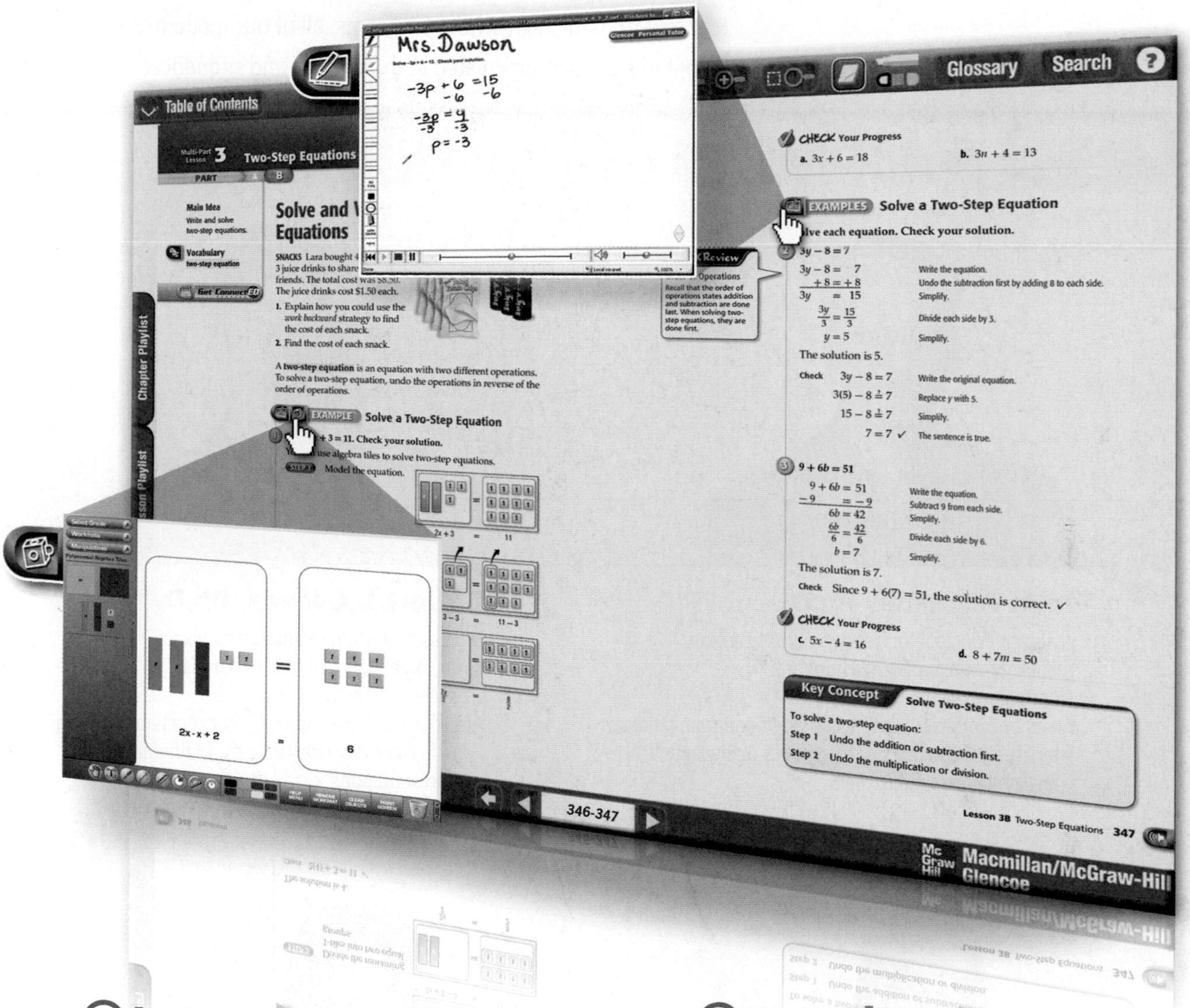

Learn

Personal Tutor presents a teacher explaining step-by-step solutions to problems.

Virtual Manipulatives provide digital ways to explore concepts.

Graphing Calculator provides keystrokes other than the TI-83/84 Plus or TI-Nspire used in the textbook.

Audio recordings provide an opportunity to build oral and listening fluency.

Foldables provide a unique way to enhance students' study skills.

Practice

Self-Check Practice allows students to assess their knowledge of foundational skills.

Worksheets provide additional practice and reteach opportunities.

Online Assessment checks understanding of concepts and terms.

Authors

Our lead authors ensure that the Macmillan/McGraw-Hill and Glencoe/McGraw-Hill mathematics programs are truly vertically aligned by beginning with the end in mind—success in Algebra 1 and beyond. By "backmapping" the content from the high school programs, all of our mathematics programs are well articulated in their scope and sequence.

Lead Authors

John A. Carter, Ph.D.
Assistant Principal for Teaching and Learning
Adlai E. Stevenson High School
Lincolnshire, Illinois
Areas of Expertise: Using technology and manipulatives to visualize concepts; Mathematics Achievement of English-Language Learners

Gilbert J. Cuevas, Ph.D.
Professor of Mathematics Education
Texas State University—San Marcos
San Marcos, Texas
Areas of Expertise: Applying concepts and skills in mathematically rich contexts; Mathematical Representations

Roger Day, Ph.D., NBCT
Mathematics Department Chairperson
Pontiac Township High School
Pontiac, Illinois
Areas of Expertise: Understanding and applying probability and statistics; Mathematics Teacher Education

Carol Malloy, Ph.D.
Associate Professor
University of North Carolina at Chapel Hill
Chapel Hill, North Carolina
Areas of Expertise: Representations and critical thinking; Student Success in Algebra 1

 **Meet the Authors.**

Program Authors

Gladis Kersaint, Ph.D.
Associate Professor of Mathematics
  Education, K–12
University of South Florida
Tampa, Florida

Kay McClain, Ed.D.
Research Professor
Arizona State University
Phoenix, Arizona

Rhonda J. Molix-Bailey
Mathematics Consultant
Mathematics by Design
DeSoto, Texas

Beatrice Moore Luchin
Mathematics Consultant
Houston, Texas

Jack Price, Ed.D.
Professor Emeritus
California State
  Polytechnic University
Pomona, California

Mary Esther Reynosa
Instructional Specialist for Elementary
  Mathematics
Northside Independent School District
San Antonio, Texas

Robyn Silbey
Math Content Coach
Montgomery County Public Schools
Gaithersburg, Maryland

Kathleen Vielhaber
Mathematics Consultant
St. Louis, Missouri

Teri Willard, Ed.D.
Associate Professor
Central Washington University
Ellensburg, Washington

Contributing Author

Dinah Zike FOLDABLES
Educational Consultant
Dinah-Might Activities, Inc.
San Antonio, Texas

Lead Consultant

Viken Hovsepian
Professor of Mathematics
Rio Hondo College
Whittier, California

Consultants and Reviewers

T hese professionals were instrumental in providing valuable input and suggestions for improving the effectiveness of the mathematics instruction.

Consultants

Mathematical Content

Dr. Michaele Chappell
Professor of Mathematics Education
Middle Tennessee State University
Murfreesboro, Tennessee

Melissa D. Young
Mathematics Specialist
Differentiated Accountability Model, Region III
Orlando, Florida

Differentiated Instruction

Jennifer Taylor-Cox, Ph.D.
Educational Consultant
Innovative Instruction: Connecting
  Research and Practice in Education
Severna Park, Maryland

Gifted and Talented

Shelbi K. Cole
Research Assistant
University of Connecticut
Storrs, Connecticut

Problem Solving

Dr. Stephen Krulik
Professor Emeritus – Math Education
Temple University
Philadelphia, Pennsylvania

Reading in the Content Areas

Sue Z. Beers
President / Consultant
Tools for Learning, Inc.
Jewell, Iowa

Reading and Vocabulary

Douglas Fisher
Professor
San Diego State University
San Diego, California

Reviewers

Sheila J. Allen
Secondary Curriculum Coordinator
Medina City Schools
Medina, OH

Kathryn Blizzard Ballin
Mathematics Supervisor
Newark Public Schools
Newark, NJ

Angelee M. Bilbao
Middle School Math Teacher
Discovery School
Glendale, AZ

Christine M. Binkley
8th Grade Math/Algebra
Celina Middle School
Celina, OH

Ronald E. Boggs
Teacher–8th Grade Math and
   Algebra 1
Lakewood Middle School
Hebron, OH

Staci Bolley
Math Teacher
Desert Sky Middle School
Glendale, AZ

Danielle Bouton
District K–12 Coordinator of
   Mathematics and Technology
Schenectady City Schools
Schenectady, NY

Matt Bowser
Math Teacher
Oil City Middle School
Oil City, PA

Thomas Brewer
6th Grade Math Teacher
West Holmes Middle School
Millersburg, OH

Lisa K. Bush
Math Curriculum Specialist–K–12
Deer Valley Unified School District
Phoenix, AZ

Janelle Chisholm Winter
Math Teacher
Palo Verde Middle School
Phoenix, AZ

S. Cox
Math Academic Facilitator
Kirksey Middle School
Rogers, AR

Mary Ellen Dierksheide
Mathematics Teacher
Elmwood Middle School
Bloomdale, OH

Dominick Galimi
Mathematics
School 5
Yonkers, NY

Jacquelyn Gawron
Liaison Instruction Department
Youngstown City Schools
Youngstown, OH

Amber Griffin
Secondary Mathematics Teacher
Kino Jr High School
Mesa, AZ

Chad D. Heuser
Math Department Chair
Elyria High School
Elyria, OH

Jerry Hicks
6th Grade Mathematics Teacher
Queensbury Middle School
Queensbury, NY

Sandra Hughes
Cross Curricular Coach
Gloucester City School District
Gloucester City, NJ

Julene M. Ippolito
Teacher
George Washington Intermediate
   School
New Castle, PA

Tracey Jaehnert
Mathematics Teacher/Dept. Chair
LaGrange Middle School
LaGrangeville, NY

Satish Jagnandan
Administrator for Mathematics and
   Science (K–12)
Mount Vernon City School District
Mount Vernon, NY

Kimberly Knisell
Teacher & Math Department
   Chairperson
Saugerties, NY

Victoria Lautsch
Middle School Math Teacher
Don Mensendick School
Glendale, AZ

Cheryl Ann Lipko
Mathematics Department Chairperson
Mount Pleasant Area Junior and Senior
   High School
Mount Pleasant, PA

Kerri L. Mahan
Middle School Math Teacher
Mount Vernon Middle School
Mount Vernon, OH

William McQuay
K–12 Director of Mathematics
Burnt Hills- Ballston Lake Central
   School District
Burnt Hills, NY

C. Vincent Pané, Ed.D.
Chair-Mathematics and Computer
   Studies
Molloy College
Rockville Centre, NY

Cheryl Peeples
Math Teacher
Tipp Middle School
Tipp City, OH

Steven M. Proehl, NBCT
Math Consultant
Chillicothe City Schools
Chillicothe, OH

Dr. Susan A. Smith
Associate Professor
Molloy College
Rockville Centre, NY

Debra S. Strayer
Middle School Mathematics
Western-Reserve Middle School
Collins, OH

Rebecca W. Sutton
Math Coach
West Jr. High
West Memphis, AR

David Thompson, M.A.
Supervisor of Mathematics
Egg Harbor Township School District
Egg Harbor Township, NJ

Kathleen M. Tucci
Math Coach
New Brighton Area School District
New Brighton, PA

William F. Wales
Director of Mathematics
Niskayuna Central Schools
Niskayuna, NY

Contents

Start Smart

Contents

Get ConnectED

connectED.mcgraw-hill.com

Investigate ▷

Learn ▷

Practice ▷

Every chapter and every lesson has a wealth of interactive learning opportunities.

Additional Lessons

Use Lesson 5 **Explore: Factor Linear Expressions** and Lesson 6 **Factor Linear Expressions** after Lesson 1-1C.

CHAPTER
2 Integers

Additional Lessons

Use Lesson 3 **Extend: Distance on the Number Line** after Lesson 2-2D.

Use Lesson 4 **Extend: Use Properties to Multiply** after Lesson 2-3C.

Contents

connectED.mcgraw-hill.com

Investigate ▷

Learn ▷

Practice ▷

Every chapter and every lesson has a wealth of interactive learning opportunities.

CHAPTER 3 Rational Numbers

CHAPTER 4
Equations and Inequalities

Additional Lessons

Use Lesson 7 **Explore: More Two-Step Equations** and Lesson 8 **More Two-Step Equations** after Lesson 4-3B.

Use Lesson 9 **Solve Two-Step Inequalities** after Lesson 4-4C.

Contents

CHAPTER 5 Proportions and Similarity

Get ConnectED

connectED.mcgraw-hill.com

Investigate ▷

Learn ▷

Practice ▷

Every chapter and every lesson has a wealth of interactive learning opportunities.

Additional Lessons

Use Lesson 1 **Complex Fractions and Unit Rates** after Lesson 5-1B.

Use Lesson 2 **Graph Proportional Relationships** after Lesson 5-1D.

Use Lesson 10 **Explore: Investigate Online Maps and Scale Drawings** before Lesson 5-2B.

CHAPTER
6 Percent

Contents

CHAPTER 7 Linear Functions

Get ConnectED

connectED.mcgraw-hill.com

Investigate ▷

Learn ▷

Practice ▷

Every chapter and every lesson has a wealth of interactive learning opportunities.

CHAPTER 8 Probability and Predictions

Additional Lessons

Use Lesson 17 **Extend: Simulate Compound Events** after Lesson 8-2B.

Use Lesson 14 **Extend: Multiple Samples of Data** after Lesson 8-3E.

Contents

connectED.mcgraw-hill.com

Investigate ▷

Learn ▷

Practice ▷

Every chapter and every lesson has a wealth of interactive learning opportunities.

CHAPTER 9
Statistical Displays

Additional Lessons

Use Lesson 15 **Explore: Visual Overlap of Data Distributions** after Lesson 9-1C.

Use Lesson 16 **Compare Populations** after Lesson 9-2B.

CHAPTER
10 Volume and Surface Area

Additional Lessons

Use Lesson 12 **Cross Sections** before Lesson 10-1A.

Use Lesson 13 **Circumference and Area of Circles** before Lesson 10-1C.

Contents

connectED.mcgraw-hill.com

CHAPTER 11 Measurement and Proportional Reasoning

Investigate ▷

Learn ▷

Practice ▷

Every chapter and every lesson has a wealth of interactive learning opportunities.

xxi

CHAPTER
12 Polygons and Transformations

Additional Lessons
Use Lesson 11 **Explore: Draw Triangles** Before Lesson 12-1B.

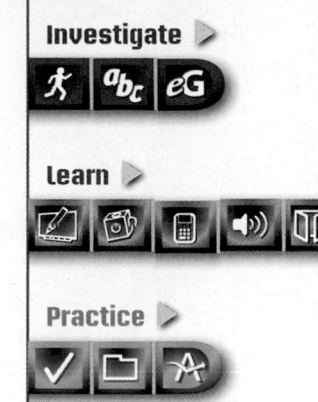

Get ConnectED

connectED.mcgraw-hill.com

Investigate

Learn

Practice

Every chapter and every lesson has a wealth of interactive learning opportunities.

Optional Projects

Problem-Solving Projects

Project 1 Are We Similar? 746

Project 2 Turn Over a New Leaf 748

Project 3 Stand Up and Be Counted! 750

Project 4 Be True to Your School 752

Project 5 Math Genes 754

Student Handbook

Additional Lessons **758**

Extra Practice **EP2**

Reference

Selected Answers and Solutions. R1

Photo Credits R43

English/Spanish Glossary. R44

Index R68

Start Smart

connectED.mcgraw-hill.com

Investigate

 Animations

 Vocabulary

 Multilingual eGlossary

Learn

 Personal Tutor

 Virtual Manipulatives

 Graphing Calculator

 Audio

Foldables

Practice

 Self-Check Practice

Worksheets

 Assessment

Here are some characters you are going to meet as you move through the book.

Jamar
"I look for bugs everywhere I go. Someday I want to work for the zoo in the entomology area."

Caitlyn
"I love to sing and play the drums. I like to play the drums to the beat of the music!"

Dario
"In my spare time, I race go-karts. I want to be a NASCAR driver when I grow up."

Divya

"My favorite subject in school is science. I'd like to be a science teacher."

Blake

"I take tennis lessons every morning before school. I want to be a tennis pro."

Theresa

"My favorite sport is soccer. My mom and I go to see professional soccer games."

Hiroshi

"My father is a professional cyclist. I enjoy riding with him on the weekends."

Marisol

"I love nature. When I visit the local park, I enjoy hearing the ranger give talks about plants and wildlife."

Seth

"I take golf lessons from a local professional at the country club. I want to be on the golf team in high school."

Aisha

"Math is my favorite subject. I love to play math games and solve problems to keep my skills sharp."

Raul

"Every year at my family reunion all the kids participate in karaoke. Last year I was the family champion."

Hannah

"I enjoy reading and spending time at the hospital reading to blind patients."

Let's Get Started!

We're going to review a little before you begin Chapter 1.

A Plan for Problem Solving

Ice Cream ▶
July is National Ice
Cream Month.

ICE CREAM Every year, each American consumes about 23.2 quarts of ice cream. However, Americans in the north-central United States consume about 18.5 quarts more. How much ice cream is consumed every year by Americans in the north-central United States?

In mathematics, there is a *four-step problem-solving plan* you can use to help you solve any problem.

Understand
- Read the problem carefully.
- What information is given?
- What do you need to find out?

Plan
- How do the facts relate to each other?
- Select a strategy for solving the problem. There may be several that you can use.
- Estimate the answer.

Solve
- Use your plan to solve the problem.
- If your plan does not work, make a new plan.
- What is the solution?

Check
- Does your answer fit the facts given in the problem?
- Is your answer reasonable?

REAL-WORLD EXAMPLE Use the Four-Step Plan

 ICE CREAM Refer to the information on the previous page. How much ice cream is consumed every year by Americans in the north-central United States?

Understand   You know that each American consumes 23.2 quarts of ice cream and that Americans in the north-central United States consume 18.5 quarts more. You need to find how much ice cream is consumed in the north-central United States.

Plan   To find the total amount, add 23.2 and 18.5.

   Estimate 20 + 20 = 40

Solve

$$\begin{array}{r} 23.2 \\ + 18.5 \\ \hline 41.7 \end{array}$$ Line up the decimal points.
Add as with whole numbers.

   Check for Reasonableness 41.7 ≈ 40 ✓

   So, 41.7 quarts of ice cream are consumed each year by Americans in the north-central United States.

Check   Check by subtracting. Since 41.7 − 23.2 = 18.5, the answer is correct. ✓

Read Math

The symbol ≈ means *is about equal to*.

 CHECK Your Progress

a. **SCHOOL** The Fort Couch Middle School cafeteria has a special lunch every Thursday. How much will you save by buying the special on Thursday instead of getting each item separately on any other day?

Entreé		Sides	
Chili	$1.49	Salad	$1.19
Chicken Fingers	$1.39	Fruit	$0.99
Hamburger	$1.99	Tortilla Chips	$1.29
		Thursday Special	
All Drinks	$0.99	Chili, Salad, and Drink	$3.49

b. **GEOMETRY** The courtyard at Eastmoor Middle School is shaped like a rectangle that is 18.8 feet long. The width of the courtyard is 4.8 feet less than the length. What is the total distance around the courtyard?

 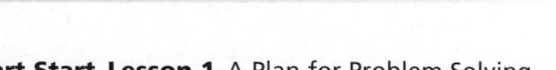

Practice and Problem Solving

Use the four-step plan to solve each problem.

1. **SCIENCE** The table shows the average lengths of the bones in a human leg. How much longer is the average femur than the average tibia?

Bones in a Human Leg	
Bone	**Length (in.)**
Femur (upper leg)	19.88
Tibia (inner lower leg)	16.94
Fibula (outer lower leg)	15.94

2. **FINANCIAL LITERACY** Terry opened a savings account in December with $150 and saved $30 each month beginning in January. Calculate the value of Terry's account at the end of July.

3. **GEOMETRY** Numbers that can be represented by a triangular arrangement of dots are called *triangular numbers*. The first five triangular numbers are shown below. Describe the pattern in the first five numbers. Then list the next three triangular numbers.

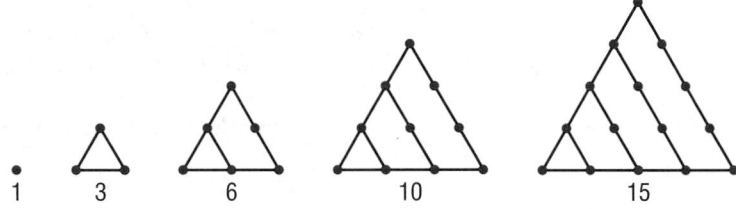

4. **SCHOOL** The Boosters expect 500 people at the annual awards banquet. If each table seats 8 people, how many tables are needed?

5. **PATTERNS** The sequence 1, 1, 2, 3, 5, 8, … is called the *Fibonacci sequence*. List the next three numbers in this sequence.

6. **TRAVEL** A commuter train departs from a train station and travels to the city each day. The schedule shows the first five departure and arrival times.

   a. How often does the commuter train arrive in the city?

   b. What is the latest time that passengers can depart from the train station if they need to arrive in the city no later than noon?

Commuter Train Schedule	
Departure	**Arrival**
6:30 A.M.	6:50 A.M.
7:15 A.M.	7:35 A.M.
8:00 A.M.	8:20 A.M.
8:45 A.M.	9:05 A.M.
9:30 A.M.	9:50 A.M.

Real-World Link The numbers in the Fibonacci sequence can be observed in a pinecone. There are usually 34 spirals of seed heads going one way and 55 going the other.

7. **MOVIES** The table shows the concession stand prices, including tax, at Mega Movies. Marlon ordered 2 medium drinks and 1 large popcorn. He gave the cashier $10. How much change should he receive?

Mega Movies Menu			
Drinks		**Popcorn**	
small	$1.99	small	$2.49
medium	$2.75	medium	$3.50
large	$3.50	large	$3.95

6 Start Smart

Multiply Decimals

KOALAS Koalas spend about 0.75 of each day sleeping. How many hours does a koala sleep each day?

To multiply decimals, multiply as with whole numbers. The product has the same number of decimal places as the sum of the decimal places of the factors.

 **REAL-WORLD EXAMPLE**   **Multiply a Decimal by a Whole Number**

 KOALAS Refer to the information above. How long does a koala sleep each day?

There are 24 hours in each day. So, multiply 0.75 by 24.

$$
\begin{array}{r}
24 \\
\times\, 0.75 \quad \longleftarrow \text{2 decimal places} \\
\hline
120 \\
1680 \\
\hline
18.00
\end{array}
$$
Count two decimal places from the right and place the decimal point.

A koala sleeps 18 hours each day.

 CHECK Your Progress

a. KOALAS A koala eats about 1.125 pounds of eucalyptus leaves each day. How many pounds of eucalyptus leaves will a koala eat in 5 days?

Vocabulary Link

Everyday Use

Annex To attach to something, as in a building that is added to another building.

Math Use

Annex To place a zero at the beginning or end of a decimal.

If there are not enough decimal places in the product, you need to annex zeros to the left.

 EXAMPLES Multiply Decimals

Multiply.

2 **1.3 × 0.9**

Estimate $1 \times 1 = 1$

$$
\begin{array}{r}
1.3 \quad \longleftarrow \text{1 decimal place} \\
\times\, 0.9 \quad \longleftarrow \text{1 decimal place} \\
\hline
1.17 \quad \longleftarrow \text{2 decimal places}
\end{array}
$$

Check for Reasonableness $1.17 \approx 1$ ✓

1 **0.054 × 1.6**

$$
\begin{array}{r}
0.054 \quad \longleftarrow \text{3 decimal places} \\
\times\, 1.6 \quad \longleftarrow \text{1 decimal place} \\
\hline
324 \\
540 \\
\hline
0.0864
\end{array}
$$
Annex a zero on the left so the answer has four decimal places.

Check for Reasonableness $0.0864 \approx 0$ ✓

 CHECK Your Progress

Multiply.

b. 15.8×11 **c.** 88×2.5 **d.** 33×0.03

Practice and Problem Solving

Place the decimal point in each product. Add zeros if necessary.

1. $1.32 \times 4 = 528$ **2.** $0.07 \times 1.1 = 77$ **3.** $0.4 \times 0.7 = 28$

4. $1.9 \times 0.6 = 114$ **5.** $1.4 \times 0.09 = 126$ **6.** $5.48 \times 3.6 = 19728$

7. INSECTS The graphic shows the speeds of three different insects.

   a. At this speed, how far can a bumblebee fly in 2 hours?

   b. How far can a dragonfly fly in 3.5 hours?

   c. How much farther can a hornet fly in 3 hours than a bumblebee?

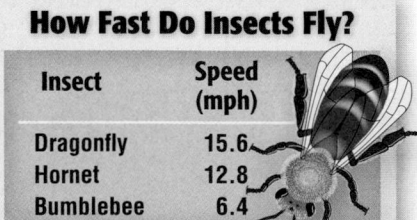

How Fast Do Insects Fly?

Insect	Speed (mph)
Dragonfly	15.6
Hornet	12.8
Bumblebee	6.4

Multiply.

8. 0.6×2 9. 0.7×18 10. 8×0.3

11. 36×0.46 12. 380×1.25 13. 42×0.17

14. 0.4×16 15. 0.23×0.2 16. 12.2×12.4

17. 0.44×5.5 18. 0.44×55 19. 44×0.55

20. **PETS** One can of dog food costs $0.56. How much will 4 cans of this dog food cost?

Real-World Link
Nearly 39% of U.S. households own a dog. Nearly 34% of U.S. households own a cat.

21. **SCHOOL SUPPLIES** Leonardo buys two 3-ring binders and a package of notebook paper. If he pays using a $20 bill, how much change will he receive?

School Supplies	
Item	**Cost**
3-ring binder	$3.95
Package of pens	$2.19
Notebook paper	$1.99
Ruler	$0.75

22. **CELL PHONES** A new brand of cell phone is 3.5 inches long. One inch is approximately equal to 2.54 centimeters. How long is the cell phone in centimeters?

23. **MEASUREMENT** Find the area of a rectangle that has a base of 7.5 meters and a height of 1.5 meters. (*Hint*: The formula for the area of a rectangle is $A = \ell w$.)

24. **EARNINGS** Cora earns $10.75 per hour. What are her total weekly earnings if she works 34.5 hours? Round to the nearest cent.

25. **SKATING** The cost of 5 youth tickets at a skating rink is $26.25. Three skaters need to rent skates for $5.50 each. What is the total cost of 5 tickets and 3 skate rentals?

26. **FIELD TRIPS** Each student who attends the field trip to the science museum must pay $6 for transportation and $5.75 for admission. There are 65 students attending the field trip. How much money is needed for transportation and admission?

27. **FIND THE ERROR** Raul is finding 2.6×2.2. Find his mistake and correct it.

$$\begin{array}{r} 2.6 \\ \times\, 2.2 \\ \hline 57.2 \end{array}$$

Divide Decimals

SPORTS Jamaican sprinter Usain Bolt ran 200 meters in 19.3 seconds at the Summer Olympics in Beijing. What was his speed in meters per second?

To find Bolt's average speed, divide 200 meters by 19.3 seconds. Dividing decimals is similar to dividing whole numbers.

 **EXAMPLE** **Divide a Decimal by a Whole Number**

1 Find 25.8 ÷ 2.   **Estimate** 26 ÷ 2 = 13

 QUICK Review

Division
Dividend → 25.8
Divisor → 2
Quotient → 12.9

$$\begin{array}{r} 12.9 \\ 2\overline{)25.8} \\ \underline{-2} \\ 05 \\ \underline{-4} \\ 18 \\ \underline{-18} \\ 0 \end{array}$$

Place the decimal directly above the decimal point in the dividend.

Divide as with whole numbers.

Check by Multiplying 12.9 × 2 = 25.8 ✓

 CHECK Your Progress

a. 8.7 ÷ 3   **b.** 613.8 ÷ 66

To divide by a decimal, multiply both the divisor and dividend by the same power of ten so that the divisor is a whole number. Then divide.

$$\text{divisor} \longrightarrow 0.5\overline{)4.5} \longleftarrow \text{dividend}$$

 EXAMPLE **Divide Decimals**

2 Find 199.68 ÷ 9.6.   **Estimate** $200 \div 10 = 20$

$$
\begin{array}{r}
20.8 \\
9.6\overline{)199.68} \\
-192 \\
\hline
768 \\
-768 \\
\hline
0
\end{array}
$$

Move each decimal point one place to the right.

Check for Reasonableness $20.8 \approx 20$ ✓

 CHECK Your Progress

c. $9.81 \div 0.3$ **d.** $5.76 \div 3.2$

In many real-world situations, the remainder is *not* zero. So, the quotient is usually rounded to a certain place-value position.

 REAL-WORLD EXAMPLE

3 **SPORTS** Refer to the information on the previous page. Divide 200 by 19.3 to find Usain Bolt's speed in meters per second. Round your answer to the nearest tenth.

Estimate $200 \div 20 = 10$

$$
\begin{array}{r}
10.36 \\
19.3\overline{)200.000} \\
-193 \\
\hline
70 \\
-0 \\
\hline
700 \\
-579 \\
\hline
1210 \\
-1158 \\
\hline
62
\end{array}
$$

Move each decimal point one place to the right.

Check for Reasonableness $10.4 \approx 10$ ✓

Usain Bolt's speed is about 10.4 meters per second.

QUICK Review

Rounding
Look at the digit to the right of the tenths place. Since 6 is greater than 5, 10.36 is rounded to 10.4.

 CHECK Your Progress

e. **FOOD** Mrs. Myers bought 2.5 pounds of hamburger for $5.20. How much did she pay in dollars per pound?

Find each quotient.

1. $812 \div 0.4$
2. $0.34 \div 0.2$
3. $2.5\overline{)14.4}$
4. $2.5\overline{)90}$
5. $4.4 \div 0.8$
6. $88.8 \div 444$
7. $6.6\overline{)5.94}$
8. $0.33\overline{)2{,}475}$
9. $20.24 \div 2.3$
10. $45 \div 0.09$
11. $36\overline{)1.8}$
12. $0.366\overline{)0.4392}$

4 for $23.96

13. **MONEY** Sachi bought four bottles of nail polish for a total of $23.96. If each bottle cost the same, how much did each cost?

14. **NUTRITION** There are 74.4 grams of sugar in 3 servings of grapes. How many grams of sugar are in a single serving of grapes?

15. **FINANCIAL LITERACY** Shiro wants to buy a pair of in-line skates that cost $140.75. So far, he has saved $56.25 and plans to save $6.50 each week. In how many weeks will he have enough money to buy the in-line skates?

16. **FOOD** Refer to the table below.

Fruit Chews		
Box	Cost	Size of Bags
Large	$4.89	13 ounces
Regular	$2.59	6.5 ounces

   a. Which package of fruit chews costs less per ounce?

   b. Would it make sense to buy one large package or two regular packages? Explain your reasoning.

17. **BASEBALL** Red Sox outfielder Jason Bay had 7 hits in 17 times at bat. To find his batting average, divide the number of hits by the number of times at bat. To the nearest thousandth, what was his batting average?

18. **OLYMPICS** Usain Bolt ran the 100-meter event in 9.69 seconds and the 200-meter event in 19.3 seconds. Was his speed in meters per second faster in the 100-meter event or the 200-meter event? How much faster? Round to the nearest hundredth.

19. **FIND THE ERROR** Caitlyn is finding $11\overline{)61.6}$. Find her mistake and correct it.

$$\begin{array}{r} 0.56 \\ 11\overline{)61.6} \end{array}$$

Ratios

START SMART 4

Little League Baseball ▶
In 2009, over 2 million little league players competed to play in the Little League World Series. The annual International Tournament is held in Williamsport, Pennsylvania.

BASEBALL The distance between bases on a Little League Baseball diamond is 60 feet. The distance between bases on a Major League Baseball diamond is 90 feet. What is the relationship between the two distances?

You can describe the relationship with a ratio. A *ratio* is a comparison of two quantities by division.

$$60 \text{ to } 90 \qquad 60 : 90 \qquad \frac{60}{90}$$

 REAL-WORLD EXAMPLE **Simplest Form**

1 **BASEBALL** Refer to the information above. Write the ratio as a fraction in simplest form. Then explain its meaning.

$$\text{Little League} \longrightarrow \quad \overset{\div 30}{\frac{60}{90}} = \frac{2}{3} \quad \longleftarrow \begin{array}{l} \text{The greatest common} \\ \text{factor, or GCF, is 30.} \end{array}$$
$$\text{Major League} \longrightarrow \quad \underset{\div 30}{}$$

The ratio of the Little League distance to the Major League distance is $\frac{2}{3}$, 2:3, or 2 to 3. That is, for every 2 feet between Little League bases, there are 3 feet between Major League bases.

 CHECK Your Progress

Write each ratio as a fraction in simplest form. Then explain its meaning.

a. 15 girls to 18 boys   **b.** 20 cats to 30 dogs

Equivalent ratios are ratios that express the same relationship between two quantities. Equivalent ratios have the same value.

 REAL-WORLD EXAMPLES Identify Equivalent Ratios

2 Determine whether the ratios 250 miles in 4 hours and 500 miles in 8 hours are equivalent.

Method 1   Compare the ratios written in simplest form.

250 miles : 4 hours $= \dfrac{250 \div 2}{4 \div 2}$ or $\dfrac{125}{2}$ Divide the numerator and denominator by the GCF, 2.

500 miles : 8 hours $= \dfrac{500 \div 4}{8 \div 4}$ or $\dfrac{125}{2}$ Divide the numerator and denominator by the GCF, 4.

The ratios simplify to the same fraction.

Method 2   Look for a common multiplier relating the two ratios.

$$\overset{\times\,2}{\overset{\frown}{\underset{\underset{\times\,2}{\smile}}{\dfrac{250}{4} = \dfrac{500}{8}}}}$$ The numerators and denominators of the ratios are related by the same multiplier, 2.

The ratios are equivalent.

3 **SCHOOL** Mrs. Garcia has 80 folders for 20 students. Mr. Walker has 75 folders for 25 students. Will a student in Mrs. Garcia's classroom receive the same number of folders as a student in Mr. Walker's classroom? Justify your answer.

80 folders : 20 students $= \dfrac{80 \div 20}{20 \div 20}$ or $\dfrac{4}{1}$ Divide the numerator and denominator by the GCF, 20.

75 folders : 25 students $= \dfrac{75 \div 25}{25 \div 25}$ or $\dfrac{3}{1}$ Divide the numerator and denominator by the GCF, 25.

No, the ratios are not equivalent. Each student in Mrs. Garcia's classroom will receive one more folder.

 CHECK Your Progress

Determine whether the ratios are equivalent.

c. 20 sandwiches for 10 people,   **d.** 100 sheets of paper in 2 notebooks,
30 sandwiches for 20 people   250 sheets of paper in 5 notebooks

e. **SWIMMING** A community pool requires there to be at least 3 lifeguards for every 20 swimmers. There are 60 swimmers and 9 lifeguards at the pool. Is this the correct number of lifeguards based on the above requirement? Justify your answer.

SOCCER Use the Madison Mavericks team statistics to write each ratio as a fraction in simplest form.

Madison Mavericks Team Statistics	
Wins	10
Losses	12
Ties	8

1. wins : losses
2. losses : ties
3. losses : games played
4. wins : games played

CARNIVALS Brighton Middle School had 6 food booths and 15 games booths at its carnival. A total of 66 adults and 165 children attended. The carnival raised a total of $1,600. Of this money, $550 came from ticket sales. Write each ratio as a fraction in simplest form.

5. children : adults
6. food booths : games booths
7. children : games booths
8. booths : money raised
9. people : children
10. non-ticket sales money : total money

Determine whether the ratios are equivalent. Explain.

11. 20 female lions to 8 male lions, 34 female lions to 10 male lions
12. $4 for every 16 ounces, $10 for every 40 ounces

13. 24 guests to 3 tables, 32 guests to 4 tables
14. 8 roses to 6 babies breath, 12 roses to 10 babies breath

15. 20 cards to 22 envelopes, 80 cards to 88 envelopes
16. 3 basketballs to 12 players, 4 basketballs to 20 players

17. **BAKING** It is recommended that ham be baked 1 hour for every 2 pounds. Latrell baked a 9-pound ham for 4.5 hours. Did he follow the recommendation? Justify your answer.

18. **FISHING** Katrina catches two similar looking fish. The larger fish is 12 inches long and 3 inches wide. The smaller fish is 6 inches long and 1 inch wide. Do these fish have equivalent length-to-width ratios? Justify your answer.

19. **MUSIC** The pitch of a musical note is measured by the number of sound waves per second, or *hertz*. If the ratio of the frequencies of two notes can be simplified to common fractions, the two notes are harmonious. Use the information below to determine if notes E and G are harmonious. Explain.

Real-World Link For Western music, the standard tuning pitch is 440 hertz, which produces the A note. Tuning forks are used to tune musical instruments.

E : 330 Hertz   G : 396 Hertz

Perimeter

PENTAGON The sides of a regular polygon are equal in length. The Pentagon in Arlington, VA, is an example of a regular pentagon. The length of each exterior wall measures 921 feet. What is the perimeter of the exterior of the Pentagon?

The distance around a geometric figure is called the *perimeter*.

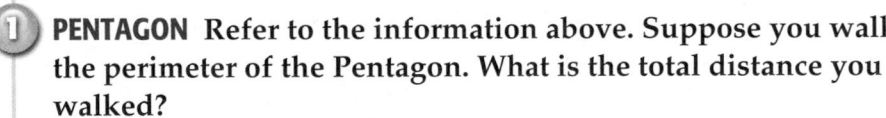 **REAL-WORLD EXAMPLE** **Find the Perimeter**

1. **PENTAGON** Refer to the information above. Suppose you walked the perimeter of the Pentagon. What is the total distance you walked?

   $P = 5s$ Perimeter of a regular pentagon

   $P = 5(921)$ Replace *s* with 921.

   $P = 4,605$ Multiply.

   You walked 4,605 feet.

CHECK Your Progress

a. **ART** A poster in the shape of a triangle has side lengths of 25 centimeters, 30 centimeters, and 35 centimeters. What is the perimeter of the poster?

b. **PHYSICAL EDUCATION** In physical education class, Mrs. Uhrlacher had the students run around the perimeter of the gym twice. The gym is 111 feet long and 107 feet wide. How far did the students run?

Sometimes you will know the perimeter of a figure and will want to determine the length of a side.

 EXAMPLE Find a Missing Side

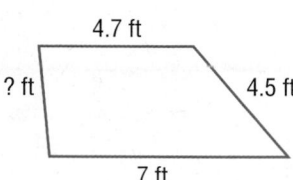

Study Tip

Perimeter is measured in linear units, such as inches, feet, centimeters, and meters.

2 The perimeter of the figure at the right is 19.7 feet. Find the missing measure.

Step 1 Add the given side lengths.

$$4.7 + 4.5 + 7 = 16.2$$

Step 2 Subtract the sum from the perimeter.

$$19.7 - 16.2 = 3.5$$

Check $4.7 + 4.5 + 7 + 3.5 = 16.2$ ✓

So, the missing measure is 3.5 feet.

 CHECK Your Progress

Find the missing measure in each figure.

c. $P = 20$ cm

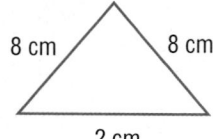

d. $P = 14$ in.

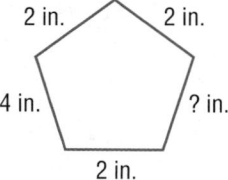

Practice and Problem Solving

Find the perimeter of each figure.

1.

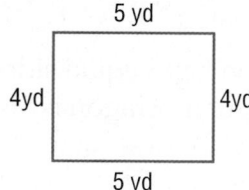

2.

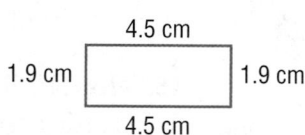

3.

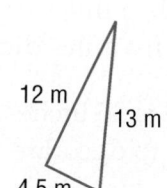

4.

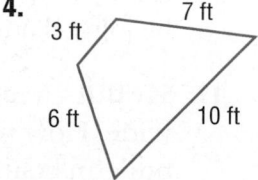

Find the perimeter of each figure.

5. rectangle with a length of 3.25 inches and a width of 2 inches

6. triangle with side lengths of 4 centimeters, 7 centimeters, and 5 centimeters

7. parallelogram with side lengths of 5 meters, 3 meters, 5 meters, and 3 meters

8. square with a side length of 12 feet

9. **PHOTOGRAPHY** The mat of a photograph is 17 inches long and 12 inches wide. What is the perimeter of the mat?

10. **HOME IMPROVEMENT** Caitlyn wants to put a wallpaper border around the ceiling of her bedroom. Find the perimeter of Caitlyn's bedroom to determine how many feet of wallpaper border she will need.

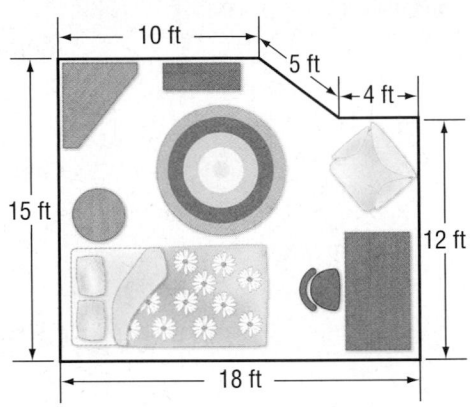

Find the missing measure in each figure.

11. $P = 14$ in.

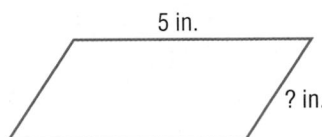

12. $P = 30$ cm

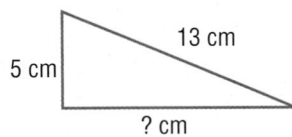

13. $P = 16$ m

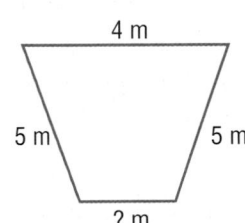

14. $P = 10$ in.

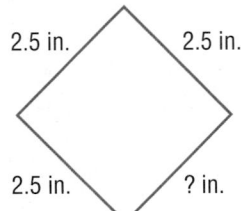

15. **TRAFFIC SIGNS** A regular octagon has 8 equal sides. The perimeter of a traffic sign in the shape of a regular octagon is 96 inches. What is the length of one side?

16. **JOGGING** Ramón jogged along the perimeter of a rectangular campground. He jogged 6 miles north, 5 miles east, and 6 miles south. If he jogged a total of 22 miles, how many miles did he jog west?

17. **SCHOOL** A rectangular bulletin board is 36 inches long and 24 inches wide. How much border does it take to decorate the perimeter of the bulletin board?

Venn Diagrams

Polar Bears ▶
Polar bears have been known to swim up to 100 miles at one time.

Mammal	Animals that Live Around Water
Giraffe	Seal
Bat	Polar Bear
Seal	Sea Turtle
Polar Bear	Swan
Walrus	Walrus

ANIMALS Polar bears are mammals. They also live around water. What other animals shown in the table are mammals that live around water?

One way to show this relationship is with a Venn diagram. A *Venn diagram* is an arrangement of overlapping circles used to show how two or more sets of data are related. In a Venn diagram, each set of data is shown separately and the items that belong to both sets are represented by the overlapping section.

Animal Characteristics

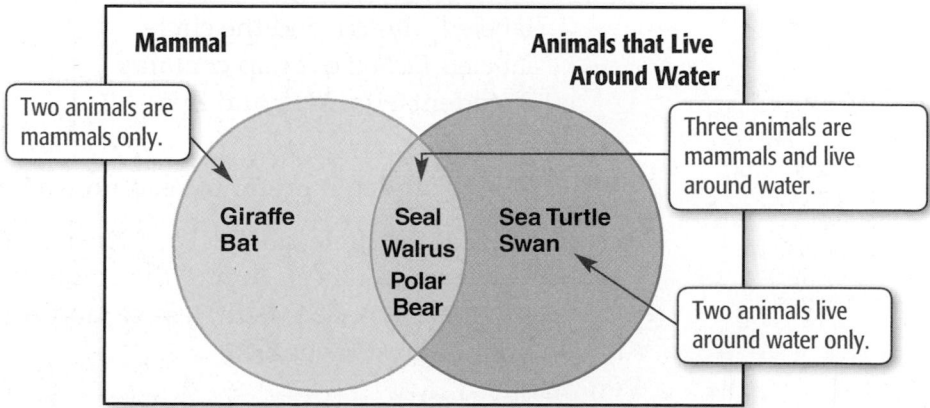

Mammal

Animals that Live Around Water

Two animals are mammals only.

Three animals are mammals and live around water.

Two animals live around water only.

Giraffe
Bat

Seal
Walrus
Polar
Bear

Sea Turtle
Swan

EXAMPLE **Display Data in a Venn Diagram**

1 **BAND** Silvia conducted a survey asking eight of her friends whether they are members of the jazz band or orchestra. Display the results shown below in a Venn diagram.

Survey Results	
Jazz Band	Trina, Jerome, Logan, Kylie, Miguel
Orchestra	Natalie, Kylie, Jerome, Malina, Tyree

First, draw two overlapping circles to represent jazz band and orchestra. Label each circle.

Since Kylie and Jerome are in both jazz band and orchestra, write their names in the overlapping region.

Write each of the other names in the appropriate region. Trina, Logan, and Miguel are in jazz band only. Natalie, Malina, and Tyree are in orchestra only.

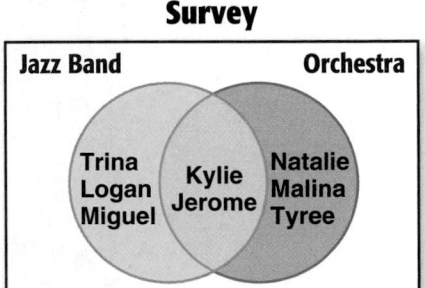

Survey

Jazz Band — Orchestra

Trina Logan Miguel | Kylie Jerome | Natalie Malina Tyree

EXAMPLES **Analyze a Venn Diagram**

BOOKS Use the Venn diagram that shows students' book preferences.

2 **Which students prefer to read mystery and fiction books?**

The region where the circle labeled *Mystery* and the circle labeled *Fiction* overlap contains 3 students: Ito, Mai, and Eva.

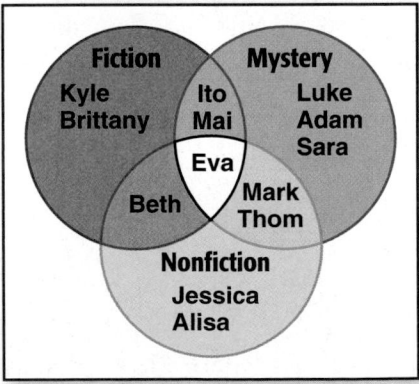

Book Preference

Fiction — Mystery
Kyle Brittany | Ito Mai | Luke Adam Sara
Eva
Beth | Mark Thom
Nonfiction
Jessica Alisa

3 **Which students prefer to read nonfiction but *not* mystery books?**

Look at the circle labeled *Nonfiction*. It contains 6 students. However, you need to find the students that do not prefer mystery books. So, Beth, Jessica, and Alisa prefer nonfiction but not mystery books.

Practice and Problem Solving

Draw a Venn diagram to show how the sets of data are related.

1.

Dessert Eaten at a Birthday Party	
Cake	Adelina, Jin, Rashid, Aisha, Santiago
Ice Cream	Rashid, Aisha, Todd, Reiko, Suzie

2.

Multiples of 4	Multiples of 6
4, 8, 12, 16, 20, 24, 28, 32, 36	6, 12, 18, 24, 30, 36

3.

Vegetable	Green in Color
Broccoli	Tree
Carrot	Grass
Corn	Frog
Pea	Broccoli
Onion	Pea

4.

Transportation to School	
Walk	Alejandro, Malcolm, Paige, Martin, Shandra, Mateo
Car/Bus	Jacob, Lisa, Ariel, Ellie, Malcolm, Diego, Kimberly

5. TRAVEL Of 50 people, 20 want to travel to Italy only, 16 want to travel to Germany only, and 14 want to travel to both Italy and Germany.

6. FOOD Of 60 people, 35 prefer pepperoni only on a pizza, 13 prefer sausage only, and 12 prefer both pepperoni and sausage.

SCHOOL Use the Venn diagram that shows students' favorite subjects.

7. How many students prefer only math?

8. How many students said they prefer science and social studies?

9. How many students prefer social studies but not math?

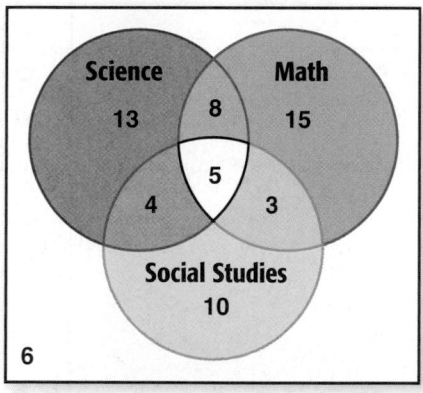

Favorite Subject

TIME Use the Venn diagram that shows students' bedtimes.

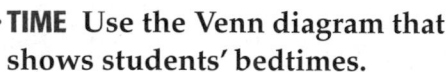

10. Which students go to bed at 10:00 P.M.?

11. How many students do not go to bed at any of these times?

12. How many students were surveyed?

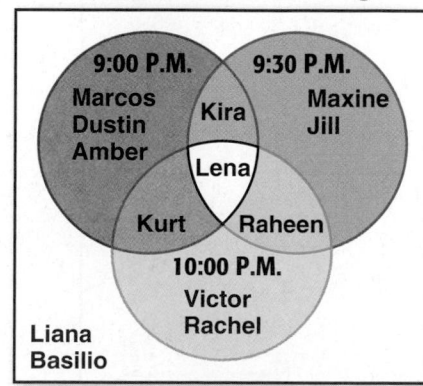

Bedtime on a School Night

Real-World Link
It is recommended that 11-18 year olds get 8 to 9 hours of sleep per night.

13. MUSIC Kristy asked 58 people to name their favorite way to listen to music, MP3 or CD. Of those surveyed, 20 said by CD only and 15 people said by both CD and MP3. Use a Venn diagram to find how many classmates prefer MP3 format only.

Smart Start Lesson 6 Venn Diagrams **21**

Expressions and Patterns

connectED.mcgraw-hill.com

Investigate

 Animations

 Vocabulary

 Multilingual eGlossary

Learn

 Personal Tutor

 Virtual Manipulatives

 Graphing Calculator

 Audio

 Foldables

Practice

 Self-Check Practice

 Worksheets

 Assessment

The ☆BIG Idea

How can algebraic expressions be written and when can they be used?

 FOLDABLES Study Organizer

Make this Foldable to help you organize your notes.

Expressions and Patterns

Review Vocabulary

product producto the result of a multiplication problem

$$5 \times 5 = 25 \leftarrow \text{product}$$

Key Vocabulary

English	Español
expression	expresión
property	propiedad
sequence	sucesión
square root	raíz cuadrada

When Will I Use This?

You have two options for checking prerequisite skills for this chapter.

Text Option Take the Quick Check below. Refer to the Quick Review for help.

QUICK Check

Add or subtract.

1. $89.3 + 16.5$
2. $7.9 + 32.4$
3. $24.6 - 13.3$
4. $9.8 - 6.7$
5. $61.14 + 21.82$
6. $30.45 + 21.39$
7. $20.57 - 18.2$
8. $69.4 - 11.53$

9. **VIDEO GAMES** Patrick bought the items shown in the table. What was the total cost, not including tax?

Item	Cost ($)
Video Game	59.99
Video Game Controller	12.95

Multiply or divide.

10. 4×7.7
11. 9.8×3
12. $37.49 \div 4.6$
13. $14.31 \div 2.7$
14. 2.7×6.3
15. 8.5×1.2
16. $6.16 \div 5.6$
17. $11.15 \div 2.5$

18. **PIZZA** Four friends decided to split the cost of a pizza evenly. How much does each friend need to pay?

Pizza $25.48

QUICK Review

EXAMPLE 1

Find $3.8 + 2.1$.

$$\begin{array}{r} 3.8 \\ + 2.1 \\ \hline 5.9 \end{array}$$ Line up the decimal points.

EXAMPLE 2

Find $37.45 - 8.52$.

$$\begin{array}{r} 37.45 \\ - 8.52 \\ \hline 28.93 \end{array}$$ Line up the decimal points.

EXAMPLE 3

Find 1.7×3.5.

$$\begin{array}{r} 1.7 \\ \times 3.5 \\ \hline 5.95 \end{array}$$

1.7 ← 1 decimal place
× 3.5 ← + 1 decimal place
5.95 ← 2 decimal places

EXAMPLE 4

Find $24.6 \div 2.5$.

$2.5\overline{)24.6} \longrightarrow 25.\overline{)246.}$ Multiply the divisor and the dividend by 10.

$$\begin{array}{r} 9.84 \\ 25\overline{)246.00} \\ -225 \\ \hline 210 \\ -200 \\ \hline 100 \\ -100 \\ \hline 0 \end{array}$$

Annex zeros.

Divide as with whole numbers.

 **Online Option** Take the Online Readiness Quiz.

Main Idea

Use powers and exponents.

 Vocabulary

factors
exponent
base
powers
squared
cubed
evaluate
standard form
exponential form

 *Get Connect ED*

Powers and Exponents

 TEXT MESSAGING
Suppose you text one of your friends. That friend then texts two friends after one minute. The pattern continues.

1. How many text messages will be sent at 4 minutes?

2. What is the relationship between the number of 2s and the number of minutes?

Minutes	Number of Text Messages	
0	1	= 1
1	1×2	= 2
2	2×2	= 4
3	$2 \times 2 \times 2$	= 8

Two or more numbers that are multiplied together to form a product are called **factors**. When the same factor is used, you may use an exponent to simplify the notation. The **exponent** tells how many times the base is used as a factor. The common factor is called the **base**.

$$16 = 2 \cdot 2 \cdot 2 \cdot 2 = 2^4 \leftarrow \text{exponent}$$
$$\uparrow$$
$$\text{base}$$

Numbers expressed using exponents are called **powers**.

Powers	Words
5^2	five to the second power or five **squared**
4^3	four to the third power or four **cubed**
2^4	two to the fourth power

 EXAMPLES Write Powers as Products

Write each power as a product of the same factor.

1 7^5

$7^5 = 7 \cdot 7 \cdot 7 \cdot 7 \cdot 7$

2 5^2

$5^2 = 5 \cdot 5$

 CHECK Your Progress

a. 6^4 **b.** 1^3 **c.** 2^5

You can **evaluate**, or find the value of, powers by multiplying the factors. Numbers written without exponents are in **standard form**.

 EXAMPLES **Write Powers in Standard Form**

Evaluate each expression.

3 2^5

$2^5 = 2 \cdot 2 \cdot 2 \cdot 2 \cdot 2$

$= 32$

4 8^3

$8^3 = 8 \cdot 8 \cdot 8$

$= 512$

CHECK Your Progress

d. 10^2 **e.** 7^3 **f.** 1^7

Numbers written with exponents are in **exponential form**.

 EXAMPLE **Write Numbers in Exponential Form**

5 Write $3 \cdot 3 \cdot 3 \cdot 3$ in exponential form.

3 is the base. It is used as a factor 4 times. So, the exponent is 4.

$3 \cdot 3 \cdot 3 \cdot 3 = 3^4$

CHECK Your Progress

Write each product in exponential form.

g. $10 \cdot 10 \cdot 10 \cdot 10$ **h.** $12 \cdot 12 \cdot 12 \cdot 12 \cdot 12 \cdot 12$

CHECK Your Understanding

Examples 1 and 2   Write each power as a product of the same factor.

1. 9^3 **2.** 8^5 **3** 5^5

Examples 3 and 4   Evaluate each expression.

4. 2^4 **5.** 10^3 **6.** 12^2

7. POPULATION There are approximately 7^6 people living in Springfield, Illinois. About how many people is this?

Example 5   Write each product in exponential form.

8. $11 \cdot 11 \cdot 11$ **9.** $1 \cdot 1 \cdot 1 \cdot 1$ **10.** $4 \cdot 4 \cdot 4 \cdot 4 \cdot 4$

Practice and Problem Solving

= **Step-by-Step Solutions** begin on page R1.
Extra Practice begins on page EP2.

Examples 1 and 2 Write each power as a product of the same factor.

11. 1^5 **12.** 4^2 **13.** 7^7

14. 8^6 **15.** 9^4 **16.** 10^4

Examples 3 and 4 Evaluate each expression.

17. 2^6 **18.** 4^3 **19.** 11^2

20. 4^6 **21.** 1^{10} **22.** 5^4

23 **BIKING** In a recent year, the number of 12- to 17-year-olds that went off-road biking was 10^6. Write this number in standard form.

24. **TRAINS** The Maglev train in China is the fastest passenger train in the world. Its average speed is 3^5 miles per hour. Write this speed in standard form.

Example 5 Write each product in exponential form.

25. $3 \cdot 3$ **26.** $2 \cdot 2 \cdot 2 \cdot 2 \cdot 2 \cdot 2$

27. $1 \cdot 1 \cdot 1 \cdot 1 \cdot 1 \cdot 1 \cdot 1 \cdot 1$ **28.** $8 \cdot 8 \cdot 8 \cdot 8 \cdot 8 \cdot 8 \cdot 8$

Write each power as a product of the same factor.

29. *four to the fifth power* **30.** *nine squared*

Evaluate each expression.

31. *six to the fourth power* **32.** *6 cubed*

33 **GEOMETRY** Refer to the puzzle cube at the right.

  a. Suppose the puzzle cube is made entirely of unit cubes. Find the number of unit cubes in the puzzle. Write using exponents.

  b. Why do you think the expression 3^3 is sometimes read as *3 cubed*?

34. **NUMBERS** Write $5 \cdot 5 \cdot 5 \cdot 5 \cdot 4 \cdot 4 \cdot 4$ in exponential form.

35. **COMPUTERS** A gigabyte is a measure of computer data storage capacity. One gigabyte stores 2^{30} bytes of data. Use a calculator to find the number in standard form that represents two gigabytes.

Order the following powers from least to greatest.

36. $6^5, 1^{14}, 4^{10}, 17^3$ **37.** $2^8, 15^2, 6^3, 3^5$ **38.** $5^3, 4^6, 2^{11}, 7^2$

39. OPEN ENDED Select a number between 1,000 and 2,000 that can be expressed as a power.

40. CHALLENGE Write and evaluate two different powers that have the same value.

41. Which One Doesn't Belong? Identify the number that does not belong with the other three. Explain your reasoning.

| 121 | 361 | 576 | 1,000 |

42. REASONING Analyze the number pattern shown at the right. Then write a convincing argument as to the value of 2^0. Based on your argument, what do you think will be the value of 2^{-1}?

$$2^4 = 16$$
$$2^3 = 8$$
$$2^2 = 4$$
$$2^1 = 2$$
$$2^0 = ?$$

43. WRITE MATH Explain the advantages of using exponents to express numerical values.

Test Practice

44. Which of the following models represents 6^3?

A.

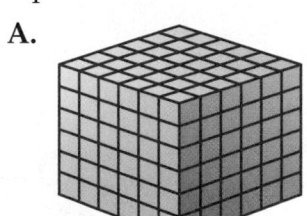

B.

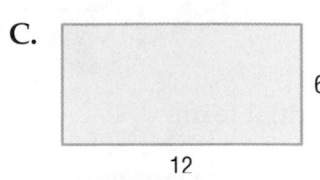

C.

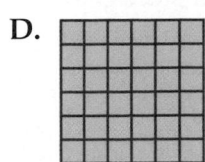

12

6

D.

45. To find the area of a square, square the length of a side. A square porch is shown.

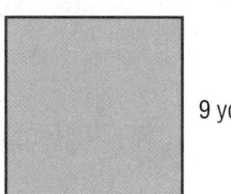

9 yd

Which of the following expressions represents the area of the porch expressed as a power?

F. 9^2 **H.** 9^3

G. 2^9 **I.** 9^9

46. Which of the following expressions when evaluated has the greatest product?

A. $5 \cdot 5 \cdot 5$

B. $4 \cdot 4 \cdot 4 \cdot 4$

C. $2 \cdot 2 \cdot 2 \cdot 2 \cdot 2 \cdot 2 \cdot 2 \cdot 2$

D. $3 \cdot 3 \cdot 3 \cdot 3 \cdot 3 \cdot 3$

Main Idea

Evaluate expressions using the order of operations.

 Vocabulary

numerical expression
order of operations

Numerical Expressions

SPORTS The Kent City football team made one 6-point touchdown and four 3-point field goals in its last game. Megan and Dexter each used an expression to find the total number of points the team scored.

Megan
$6 + 4 \cdot 3 = 6 + 12$
$= 18$
The team scored 18 points.

Dexter
$(6 + 4) \cdot 3 = 10 \cdot 3$
$= 30$
The team scored 30 points.

1. List the differences between their calculations.

2. Whose calculations are correct?

3. Make a conjecture about what should be the first step in simplifying $6 + 4 \cdot 3$.

The expression $6 + 4 \cdot 3$ is a **numerical expression**. To evaluate expressions, use the **order of operations**. These rules ensure that numerical expressions have only one value.

Key Concept **Order of Operations**

1. Evaluate the expressions inside grouping symbols.
2. Evaluate all powers.
3. Multiply and divide in order from left to right.
4. Add and subtract in order from left to right.

EXAMPLES Use Order of Operations

① **Evaluate $5 + (12 - 3)$. Justify each step.**

$$5 + (12 - 3) = 5 + 9 \qquad \text{Subtract first, since } 12 - 3 \text{ is in parentheses.}$$
$$= 14 \qquad \text{Add 5 and 9.}$$

② **Evaluate $10 - 3^2 + 7$. Justify each step.**

$$10 - 3^2 + 7 = 10 - 9 + 7 \quad \text{Find the value of } 3^2.$$
$$= 1 + 7 \qquad \text{Subtract 9 from 10.}$$
$$= 8 \qquad \text{Add 1 and 7.}$$

CHECK Your Progress

Evaluate each expression. Justify each step.

a. $39 \div (9 + 4)$ **b.** $(5 - 1)^3 \div 4$

Multiplication can be indicated by using parentheses.

 **EXAMPLE** **Use Order of Operations**

3 Evaluate $14 + 3(7 - 2)$. Justify each step.

$$14 + 3(7 - 2) = 14 + 3(5) \qquad \text{Subtract 2 from 7.}$$
$$= 14 + 15 \qquad \text{Multiply 3 and 5.}$$
$$= 29 \qquad \text{Add 14 and 15.}$$

✓ CHECK Your Progress

c. $20 - 2(4 - 1) \cdot 3$ d. $6 + 8 \div 2 + 2(3 - 1)$

REAL-WORLD EXAMPLE

4 **MONEY** Julian orders 3 rolls of crepe paper, 4 boxes of balloons, and 2 boxes of favors for the school dance. What is the total cost?

Item	Unit Cost
crepe paper	$2
favors	$7
balloons	$5

Words	cost of 3 rolls of crepe paper	+	cost of 4 boxes of balloons	+	cost of 2 boxes of favors
Expression	3×2	+	4×5	+	2×7

$$3 \times 2 + 4 \times 5 + 2 \times 7 = 6 + 20 + 14 \qquad \text{Multiply from left to right.}$$
$$= 40 \qquad \text{Add.}$$

The total cost is $40.

Real-World Link · · · ·
Crepe paper originated in the late 1700s.

✓ CHECK Your Progress

e. **MONEY** Write and evaluate an expression to find the total cost of 12 rolls of crepe paper, 3 boxes of balloons, and 3 boxes of favors.

✓ CHECK Your Understanding

Examples 1–3   Evaluate each expression. Justify each step.

1. $8 + (5 - 2)$ **2.** $25 \div (9 - 4)$

3. $14 - 2 \cdot 6 + 9$ **4.** $8 \cdot 5 - 4 \cdot 3$

5. 4×10^2 **6.** $45 \div (4 - 1)^2$

7. $17 + 2(6 - 3) - 3 \times 4$ **8.** $22 - 3(8 - 2) + 12 \div 4$

Example 4   **COINS** Roy has 3 nickels, 2 quarters, 2 dimes, and 7 pennies. Write and evaluate an expression to find how much money he has.

Practice and Problem Solving

 = **Step-by-Step Solutions** begin on page R1.
Extra Practice begins on page EP2.

Examples 1–3 Evaluate each expression. Justify each step.

10. $(1 + 8) \times 3$ **11.** $10 - (3 + 4)$ **12.** $25 \div 5 + 8$

13 $(11 - 2) \div 9$ **14.** $3 \cdot 2 + 14 \div 7$ **15.** $4 \div 2 - 1 + 7$

16. $12 + 6 \div 3 - 4$ **17.** $18 - 3 \cdot 6 + 5$ **18.** 6×10^2

19. 3×10^4 **20.** $5 \times 4^3 + 2$ **21.** $8 \times 7^2 - 6$

22. $8 \div 2 \times 6 + 6^2$ **23.** $9^2 - 14 \div 7 \cdot 3$

24. $(17 + 3) \div (4 + 1)$ **25.** $(6 + 5) \cdot (8 - 6)$

26. $6 + 2(4 - 1) + 4 \times 9$ **27.** $3(4 + 7) - 5 \cdot 4 \div 2$

Example 4 **Write an expression for each situation. Then evaluate it.**

28. MP3 PLAYERS The table shows what Reina
bought at an electronics store. What is the
total cost of her purchases?

Reina's Purchases		
Item	Quantity	Unit Cost
MP3 player	1	$200
pack of batteries	3	$4
songs	6	$2

29. BOOKS Used paperback books are $0.25,
and hardback books are $0.50. If you buy
3 paperback books and 5 hardback books,
how much money do you spend?

Evaluate each expression. Justify each step.

30. $(2 + 10)^2 \div 4$ **31.** $(3^3 + 8) - (10 - 6)^2$

32. $3 \cdot 4(5.2 + 3.8) + 2.7$ **33.** $7 \times 9 - (4 - 3.2) + 1.8$

34. MONEY Suppose you order 2 pizzas, 2 garlic
breads, and 1 order of BBQ wings. Write an
expression to find the amount of change you
would receive from $30. Then evaluate the
expression.

Item	Cost
14" pizza	$8
garlic bread	$2
BBQ wings	$4

35 **GRAPHIC NOVEL** Refer to the graphic novel frame below. Write and
evaluate an expression to find the cost of 275 text messages.

Caitlyn is helping Dario figure out what his text messaging bill will be this month.

Price Guide:

Number of Text Messages Sent	Cost
250	$5.00
252	$5.30
254	$5.60
256	$5.90

36. **OPEN ENDED** Write an expression that contains an exponent and parentheses. When evaluated it should have a value of 1.

37. **Which One Doesn't Belong?** Identify the expression that when evaluated does not belong with the other three. Explain your reasoning.

$2^4 - (12 + 4)$	$(20 - 4) - 4^2$	$2 \times 8 + 2^4$	$2^4 + 3 - (11 + 8)$

38. **CHALLENGE** Determine whether each of the following statements is *true* or *false*. If *false*, write the statement and insert parentheses to make it a true equation.

   a. $72 \div 9 + 27 - 2 = 0$

   b. $9 + 3^3 \div 6 = 6$

   c. $7 \cdot 4 - 2 \cdot 6 = 16$

   d. $32 - 10 \div 2 + 3 = 14$

   e. $15 - 2^3 + 4 = 11$

39.  **WRITE MATH** Write a real-world problem for the expression $24 \div (12 - 4)$. Then solve the problem. Explain how you solved.

✔ Test Practice

40. What is the value of $3^2 + 9 \div 3 + 3$?

   A. 3   **C.** 15

   B. 9   **D.** 18

41. The table shows the number of students Mrs. Augello has in her classes.

Mrs. Augello's Classes	
Number of classes with 24 students	2
Number of classes with 15 students	3

   Which expression *cannot* be used to find the total number of students Mrs. Augello teaches?

   F. $2(24) + 3(15)$

   G. $3 \times 15 + 2 \times 24$

   H. $5 \times (24 + 15)$

   I. $15 + 15 + 15 + 24 + 24$

42. The steps Alana took to evaluate the expression $4y + 4 \div 4$ when $y = 7$ are shown below.

> $4y + 4 \div 4$ when $y = 7$
> $4 \times 7 = 28$
> $28 + 4 = 32$
> $32 \div 4 = 8$

   What should Alana have done differently in order to evaluate the expression correctly?

   A. divided $(28 + 4)$ by (28×4)

   B. divided $(28 + 4)$ by $(28 + 4)$

   C. added $(4 \div 4)$ to 28

   D. added 4 to $(28 \div 4)$

43. **SHORT RESPONSE** Explain how you would evaluate the expression $5 \cdot 3^2 + 7$. Then evaluate the expression.

Main Idea

Evaluate simple algebraic expressions.

 Vocabulary

variable
algebra
algebraic expression
coefficient
define a variable

 7.EE.4

Algebraic Expressions

 **Explore** A pattern of squares is shown.

1. Draw the next three figures in the pattern.

2. Find the number of squares in each figure and record your data in a table like the one shown below. The first three are completed for you.

Figure	1	2	3	4	5	6
Number of Squares	3	4	5	■	■	■

3. Without drawing the figure, determine how many squares would be in the 10th figure. Check by making a drawing.

4. Find a relationship between the figure and its number of squares.

The number of squares in the figure is two more than the figure number. You can use a variable to represent the number of squares. A **variable** is a symbol that represents an unknown quantity.

$$\text{figure number} \longrightarrow \underbrace{n + 2}_{\text{number of squares}}$$

The branch of mathematics that involves expressions with variables is called **algebra**. The expression $n + 2$ is called an **algebraic expression** because it contains variables, numbers, and at least one operation.

 **EXAMPLE** **Evaluate an Algebraic Expression**

1 Evaluate $n + 3$ if $n = 4$.

$$n + 3 = 4 + 3 \qquad \text{Replace } n \text{ with 4.}$$
$$= 7 \qquad \text{Add 4 and 3.}$$

 CHECK Your Progress

Evaluate each expression if $c = 8$ and $d = 5$.

a. $c - 3$ b. $15 - c$ c. $c + d$

In algebra, the multiplication sign is often omitted.

6*d*   **9*st***   ***mn***

6 times ***d***   9 times ***s*** times ***t***   ***m*** times ***n***

The numerical factor of a multiplication expression that contains a variable is called a **coefficient**. So, 6 is the coefficient of 6*d*.

 EXAMPLES Evaluate Expressions

2 Evaluate $8w - 2v$ if $w = 5$ and $v = 3$.

$$8w - 2v = 8(5) - 2(3)$$ Replace *w* with 5 and *v* with 3.
$$= 40 - 6$$ Do all multiplications first.
$$= 34$$ Subtract 6 from 40.

3 Evaluate $4y^2 + 2$ if $y = 3$.

$$4y^2 + 2 = 4(3)^2 + 2$$ Replace *y* with 3.
$$= 4(9) + 2$$ Evaluate the power.
$$= 38$$ Multiply, then add.

CHECK Your Progress

Evaluate each expression if $a = 4$ and $b = 3$.

d. $9a - 6b$ **e.** $\dfrac{ab}{2}$ **f.** $2a^2 + 5$

The fraction bar is a grouping symbol. Evaluate the expressions in the numerator and denominator separately before dividing.

 REAL-WORLD EXAMPLE

4 **HEALTH** Use the formula at the left to find Latrina's minimum training heart rate if she is 15 years old.

$$\frac{3(220 - a)}{5} = \frac{3(220 - 15)}{5}$$ Replace *a* with 15.
$$= \frac{3(205)}{5}$$ Subtract 15 from 220.
$$= \frac{615}{5}$$ Multiply 3 and 205.
$$= 123$$ Divide 615 by 5.

Latrina's minimum training heart rate is 123 beats per minute.

CHECK Your Progress

g. **MEASUREMENT** To find the area of a triangle, use the formula $\dfrac{bh}{2}$, where *b* is the base and *h* is the height. What is the area in square inches of a triangle with a height of 6 inches and base of 8 inches?

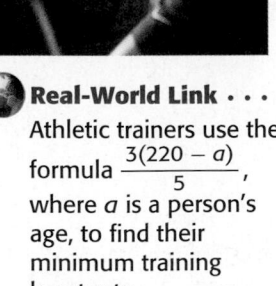

 Real-World Link · · · ·
Athletic trainers use the formula $\dfrac{3(220 - a)}{5}$, where *a* is a person's age, to find their minimum training heart rate.

Sometimes you will need to translate a verbal phrase into an algebraic expression. The first step is to define a variable. When you **define a variable**, you choose a variable to represent an unknown quantity. Follow these steps to write an algebraic expression.

①
Words
Describe the situation. Use only the most important words.

②
Variable
Choose a variable that represents the unknown quantity.

③
Expression
Write an algebraic expression that represents your verbal description.

 REAL-WORLD EXAMPLES **Write Expressions**

DVD PLAYER Marisa wants to buy the DVD player shown. She has already saved $25 and plans to save an additional $10 each week.

$150

⑤ Write an expression that represents the total amount of money Marisa has saved after any number of weeks.

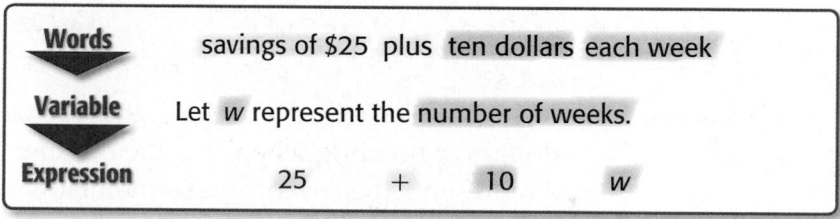

Words	savings of $25 plus ten dollars each week
Variable	Let w represent the number of weeks.
Expression	25 + 10 w

Study Tip

Defining the Variable
Although any symbol can be used, it is a good idea to use the first letter of the word you are defining as a variable. For example, w stands for the number of weeks.

⑥ Will Marisa have saved enough money to buy the DVD player in 11 weeks?

Use the expression $25 + 10w$ to find the total amount of money she has saved after 11 weeks.

$25 + 10w = 25 + 10(11)$ Replace w with 11.

$= 25 + 110$ Multiply.

$= 135$ Add.

Marisa will have saved $135 after 11 weeks. Since $135 < $150, Marisa will not have enough money to buy the DVD player.

 CHECK Your Progress

h. MUSIC An online store is having a special on music. An MP3 player costs $70 and song downloads cost $0.85 each. Write an expression that represents the cost of the MP3 player and s number of downloaded songs. Then find the total cost if 20 songs are downloaded.

Examples 1–3   Evaluate each expression if $m = 2$, $n = 6$, and $p = 4$.

1. $n - p$
2. $7m - 2$
3. $3m + 4p$
4. $n^2 + 5$
5. $\dfrac{mn}{4}$
6. $\dfrac{5n + m}{8}$

Example 4   **7. HEALTH** The standard formula for finding your maximum heart rate is $220 - a$, where a represents a person's age in years. What is your maximum heart rate?

Examples 5 and 6   **8. MUSIC** A Web site charges $0.99 to download a song and a $12.49 membership fee. Write an expression that gives the total cost in dollars to download any number of songs. Then find the cost of downloading 6 songs.

Practice and Problem Solving

● = **Step-by-Step Solutions** begin on page R1.
Extra Practice begins on page EP2.

Examples 1–3   Evaluate each expression if $d = 8$, $e = 3$, $f = 4$, and $g = 1$.

9. $d + 9$
10. $10 - e$
11. $4f + 1$
12. $8g - 3$
13. $\dfrac{d}{4}$
14. $\dfrac{16}{f}$
15. $\dfrac{5d - 25}{5}$
16. $\dfrac{(5 + g)^2}{2}$
17. $6f^2$
18. $4e^2$
19. $d^2 + 7$
20. $e^2 - 4$

Example 4   **21. BOWLING** The expression $5n + 2$ can be used to find the total cost in dollars of bowling where n is the number of games bowled and 2 represents the cost of shoe rental. How much will it cost Vincent to bowl 3 games?

22. HEALTH The expression $\dfrac{w}{30}$, where w is a person's weight in pounds, is used to find the approximate number of quarts of blood in the person's body. How many quarts of blood does a 120-pound person have?

Examples 5 and 6   **CARS** A car rental company's fees are shown.

23 Suppose you rent a car using Option 2. Write an expression that gives the total cost in dollars for driving any number of miles. Then find the cost for driving 150 miles.

Car Rental Prices	
Option 1	Option 2
• $19.99 per day	• $50 fee
• $0.17 per mi	• $0.17 per mi

24. Suppose you rent a car using Option 1. Write an expression that gives the total cost in dollars to rent a car for d days and m miles. Then find the cost for renting a car for 2 days and driving 70 miles.

Evaluate each expression if $x = 3.2$, $y = 6.1$, and $z = 0.2$.

25. $x + y - z$
26. $14.6 - (x + y + z)$
27. $xz + y^2$

28. **CHALLENGE** To find the total number of diagonals for any given polygon, you can use the expression $\frac{n(n-3)}{2}$, where n is the number of sides of the polygon.

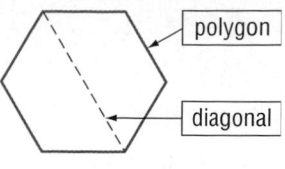

a. Determine the minimum value that n could be.

b. Make a table of four possible values of n. Then complete the table by evaluating the expression for each value of n.

29. **REASONING** Write an algebraic expression with the variable x that has a value of 3 when evaluated.

30. **CHALLENGE** Name values of x and y so that the value of $5x + 3$ is greater than the value of $2y + 14$.

31. **OPEN ENDED** Write a real-world problem that can be represented by the expression $5x + 10$.

32. **WRITE MATH** Tell whether the statement below is *sometimes*, *always*, or *never* true. Justify your reasoning.

The expressions $x - 3$ and $y - 3$ represent the same value.

Test Practice

33. Which expression could be used to find the cost of buying b books and m magazines?

School Book Fair Prices	
Item	Cost
Magazines	$4.95
Paperback books	$7.95

   A. $7.95b + 4.95m$ **C.** $12.9(b + m)$

   B. $7.95b - 4.95m$ **D.** $12.9(bm)$

34. Tonya has x quarters, y dimes, and z nickels in her pocket. Which of the following expressions gives the total amount of change she has in her pocket?

   F. $\$0.25x + \$0.05y + \$0.10z$

   G. $\$0.25x + \$0.10y + \$0.05z$

   H. $\$0.05x + \$0.25y + \$0.10z$

   I. $\$0.10x + \$0.05y + \$0.25z$

35. **CAMPING** The table shows the costs of different camping activities. Over the summer, Maura canoed 4 times and fished 3 times. Write and evaluate an expression that represents the total cost Maura spent canoeing and fishing. (Lesson 1B)

Camping Activity Costs	
Activity	Cost
Canoeing	$8
Fishing	$5

36. **PARADES** There were about 10^5 people at the parade. About how many people attended the parade? (Lesson 1A)

Main Idea

Use Commutative, Associative, Identity, and Distributive properties to solve problems.

 Vocabulary

equivalent expressions
properties

 Get ConnectED

 CCSS 7.NS.1d, 7.EE.1, 7.EE.2

Properties

MUSEUMS The admission costs are shown.

Science Center Admission	
Admission	$12
Movie	$8

1. Find the total cost of admission and a movie ticket for a 4-person family.

2. Explain how you solved.

Here are two ways to find the total cost.

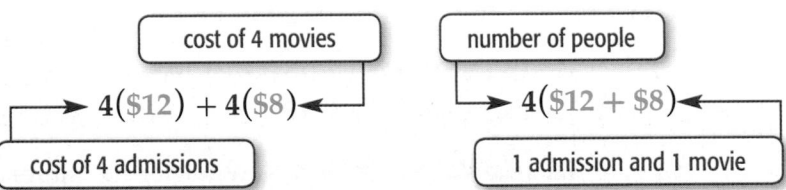

The expressions 4($12) + 4($8) and 4($12 + $8) are **equivalent expressions** because they have the same value, $80. This shows how the **Distributive Property** combines addition and multiplication.

🏃 **Key Concept**   **Distributive Property**

Words   To multiply a sum by a number, multiply each addend of the sum by the number outside the parentheses.

Examples   **Numbers**   **Algebra**

$3(4 + 6) = 3(4) + 3(6)$ $a(b + c) = a(b) + a(c)$

$5(7) + 5(3) = 5(7 + 3)$ $a(b) + a(c) = a(b + c)$

📝 **EXAMPLES** **Use the Distributive Property**

Use the Distributive Property to rewrite each expression. Then evaluate it.

① $5(3 + 2)$

$5(3 + 2) = 5(3) + 5(2)$

$\quad\quad = 15 + 10$ Multiply.

$\quad\quad = 25$ Add.

② $3(7) + 3(4)$

$3(7) + 3(4) = 3(7 + 4)$

$\quad\quad = 3(11)$ Add.

$\quad\quad = 33$ Multiply.

✓ **CHECK Your Progress**

a. $6(1 + 4)$ **b.** $6(9) + 6(3)$

 REAL-WORLD EXAMPLE

 TOUR DE FRANCE The Tour de France is a cycling race through France that lasts 22 days. If a cyclist averages 90 miles per day, about how far does he travel?

Use the Distributive Property to multiply 90×22 mentally.

$90(22) = 90(20 + 2)$	Rewrite 22 as $20 + 2$.
$= 90(20) + 90(2)$	Distributive Property
$= 1,800 + 180$	Multiply.
$= 1,980$	Add.

The cyclist travels about 1,980 miles.

CHECK Your Progress

c. MONEY Jennifer saved $120 each month for five months. How much did she save in all? Explain your reasoning.

Properties are statements that are true for all numbers.

 Key Concept   **Real Number Properties**

Commutative Properties
The order in which two numbers are added or multiplied does not change their sum or product.
$$a + b = b + a \qquad a \cdot b = b \cdot a$$

Associative Properties
The way in which three numbers are grouped when they are added or multiplied does not change their sum or product.
$$a + (b + c) = (a + b) + c \qquad a \cdot (b \cdot c) = (a \cdot b) \cdot c$$

Identity Properties
The sum of an addend and 0 is the addend. The product of a factor and 1 is the factor.
$$a + 0 = a \qquad a \cdot 1 = a$$

 EXAMPLE   **Use Properties to Evaluate Expressions**

 Evaluate $4 \cdot 12 \cdot 25$ mentally. Justify each step.

$4 \cdot 12 \cdot 25 = 4 \cdot 25 \cdot 12$	Commutative Property of Multiplication
$= (4 \cdot 25) \cdot 12$	Associative Property of Multiplication
$= 100 \cdot 12$ or $1,200$	Multiply 100 and 12 mentally.

CHECK Your Progress

Evaluate each expression. Justify each step.

d. $40 \cdot (7 \cdot 5)$ **e.** $(89 + 15) + 1$

Examples 1, 2 Use the Distributive Property to rewrite each expression. Then evaluate it.

1. $7(4 + 3)$ **2.** $5(6 + 2)$ **3.** $3(9) + 3(6)$ **4.** $6(17) + 6(3)$

Example 3 **5** **MENTAL MATH** Admission to a baseball game is \$12, and a hot dog costs \$5. Use the Distributive Property to mentally find the total cost for 4 tickets and 4 hot dogs. Explain your reasoning.

6. **MENTAL MATH** A cheetah can run 65 miles per hour at maximum speed. At this rate, how far could a cheetah run in 2 hours? Use the Distributive Property to multiply mentally. Explain your reasoning.

Example 4 Evaluate each expression mentally. Justify each step.

7. $44 + (23 + 16)$ **8.** $50 \cdot (33 \cdot 2)$

Practice and Problem Solving

 = **Step-by-Step Solutions** begin on page R1.
Extra Practice begins on page EP2.

Examples 1, 2 Use the Distributive Property to rewrite each expression. Then evaluate it.

9. $2(6 + 7)$ **10.** $5(8 + 9)$ **11.** $4(3) + 4(8)$ **12.** $7(3) + 7(6)$

Example 3 Evaluate each expression mentally. Justify each step.

13. $(8 + 27) + 52$ **14.** $(13 + 31) + 17$

15. $91 + (15 + 9)$ **16.** $85 + (46 + 15)$

17. $(4 \cdot 18) \cdot 25$ **18.** $(5 \cdot 3) \cdot 8$

19 $15 \cdot (8 \cdot 2)$ **20.** $2 \cdot (16 \cdot 50)$

21. $5 \cdot (30 \cdot 12)$ **22.** $20 \cdot (48 \cdot 5)$

Example 4 **MENTAL MATH** Use the Distributive Property to multiply mentally. Explain your reasoning.

23. **TRAVEL** Each year about 27 million people visit Paris, France. About how many people will visit Paris over a five-year period?

24. **ROLLER COASTERS** One ride on a roller coaster lasts 108 seconds. How long will it take to ride this coaster three times?

The Distributive Property also can be applied to subtraction. Use the Distributive Property to rewrite each expression. Then evaluate it.

25. $7(9) - 7(3)$ **26.** $12(8) - 12(6)$ **27.** $9(7) - 9(3)$ **28.** $6(12) - 6(5)$

ALGEBRA Use one or more properties to rewrite each expression as an equivalent expression that does not use parentheses.

29. $(y + 1) + 4$ **30.** $2 + (x + 4)$ **31.** $4(8b)$ **32.** $(3a)2$

33. $2(x + 3)$ **34.** $4(2 + b)$ **35.** $6(c + 1)$ **36.** $3(f + 4) + 2f$

H.O.T. Problems

37. OPEN ENDED The Multiplication Property of Zero states that when any number is multiplied by zero, the product is zero. Write an equation that supports this property.

38. FIND THE ERROR Dario rewrote the expression shown. Find his mistake and correct it.

$$2(3x - 2) = 6x - 2$$

39. NUMBER SENSE Analyze the statement $(18 + 35) \times 4 = 18 + 35 \times 4$. Then tell whether the statement is *true* or *false*. Explain your reasoning.

40. CHALLENGE A *counterexample* is an example showing that a statement is not true. Provide a counterexample to the following statement.

Division of whole numbers is associative.

41. ![icon] **WRITE MATH** Explain the importance of applying properties of operations when simplifying expressions.

✓ Test Practice

42. Which expression can be written as $6(9 + 8)$?

    A. $8 \cdot 6 + 8 \cdot 9$ **C.** $6 \cdot 9 \cdot 6 \cdot 8$

    B. $6 \cdot 9 + 6 \cdot 8$ **D.** $6 + 9 \cdot 6 + 8$

43. The model below represents the area of patio and a new addition to the patio.

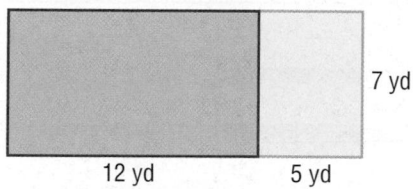

Which represents the area in square yards of the entire patio?

    F. $12(7 + 5)$ **H.** $5(19)$

    G. $7(12 + 5)$ **I.** $5(12) + 5(7)$

44. THINK SOLVE EXPLAIN **EXTENDED RESPONSE** The table shows the driving distance from certain cities.

From	To	Driving Distance (mi)
Pittsburgh	Johnstown	55
Johnstown	Allentown	184

Part A Write a number sentence that compares the mileage from Pittsburgh to Johnstown to Allentown, and the mileage from Allentown to Johnstown to Pittsburgh.

Part B Name the property that is illustrated by this sentence.

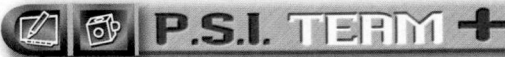
Problem-Solving Investigation

Main Idea Solve problems by looking for a pattern.

☑ 📷 P.S.I. TERM ➕

Look for a Pattern

HOSHI: Last year, members of Student Council wrote a blog for the school Web page. We used a rule to determine the number of posts we would publish each month. We published 2 posts in August, 4 posts in September, and 8 posts in October.

YOUR MISSION: Look for a pattern to find the number of posts that Student Council members published in January.

Understand	Student Council members published 2 posts in August, 4 posts in September, and 8 posts in October. You need to find the number of posts that they published in January.
Plan	Look for a pattern. Then extend the pattern to find the solution.
Solve	In the first three months, Student Council members published 2 posts, 4 posts, and 8 posts. Extend the pattern.

Months	A	S	O	N	D	J
Number of Posts	2	4	8	16	32	64

×2 ×2 ×2 ×2 ×2

The Student Council published 64 posts in January.

Check	Use counters or cubes to model the monthly pattern of published posts. Count the number of objects used to represent published posts in January. ✔

Analyze the Strategy

1. Suppose Student Council members published 2 posts in August, 6 posts in September, and 18 posts in October. Find the number of posts that Student Council members would publish in January.

= **Step-by-Step Solutions** begin on page R1.
Extra Practice begins on page EP2.

- Look for a pattern.
- Choose an operation.
- Draw a diagram.

Use the *look for a pattern* strategy to solve
Exercises 2–4.

2. FINANCIAL LITERACY Latoya is saving
money to buy a camera. After 1 month, she
has $75. After 2 months, she has $120.
After 3 months, she has $165. She plans to
keep saving in this pattern. How long will
it take Latoya to save enough money to
buy a camera that costs $300?

3. WORK The table shows the number of
hours that Omar worked at his job.

Week 1	Week 2	Week 3	Week 4
10	11.5	13	14.5

If the pattern continues, how many hours
will Omar work in Week 7?

4. DISPLAYS Cans of soup are arranged in a
triangular display. The top row has 1 can.
The second row has 2 cans. The third row
has 3 cans. The bottom row has 10 cans.
How many total cans are there?

Use any strategy to solve Exercises 5–12.

5. BRIDGES The total length of wire used in
the cables supporting the Golden Gate
Bridge in San Francisco is about 80,000
miles. This is 5,300 miles longer than three
times the distance around Earth at the
equator. What is the distance around Earth
at the equator?

6. GEOMETRY What are the next two figures
in the pattern?

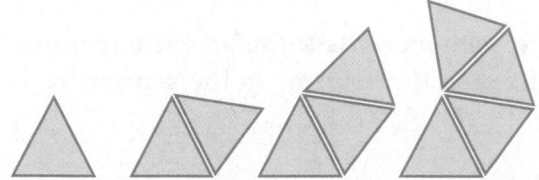

7. COOKIES Angelo places 4 chocolate chip
cookies and 3 sugar cookies into each tin.
He has used 24 chocolate chip cookies and
18 sugar cookies. How many tins did
Angelo make?

8. ALGEBRA What are the next two numbers
in the pattern?

16, 32, 64, 128, 256, ____, ____

9 ROLLER COASTERS The Jackrabbit roller
coaster can handle 1,056 passengers per
hour. The coaster has 8 vehicles. If each
vehicle carries 4 passengers, how many
runs are made in one hour?

10. ANALYZE TABLES The table gives the
average snowfall for Valdez, Alaska.

Month	Snowfall (in.)
October	11.6
November	40.3
December	73.0
January	65.8
February	59.4
March	52.0
April	22.7

How many total inches of snowfall could
a resident of Valdez expect to receive from
October to April?

11. HOMEWORK Angel has guitar practice at
7:00 P.M. He has homework in math,
science, and history that will take him
30 minutes each to complete. He also has
to allow 20 minutes for dinner. What is
the latest time Angel can start his
homework?

12. WRITE MATH Write a real-world
problem that could be solved by looking
for a pattern. Then write the steps you
would take to find the solution to your
problem.

Main Idea

Describe the relationships and extend terms in arithmetic sequences.

 Vocabulary

sequence
term
arithmetic sequence
geometric sequence

Get ConnectED

CCSS 7.EE.4

Sequences

 **Explore** Use centimeter cubes to make the figures shown.

Figure 1   Figure 2   Figure 3

1. How many centimeter cubes are used in each figure?

2. What pattern do you see? Describe it in words.

3. Suppose this pattern continues. Copy and complete the table to find the number of cubes needed to make each figure.

Figure	1	2	3	4	5	6	7	8
Cubes Needed	4	8	12					

4. How many cubes are needed to make the 10th figure? Explain.

A **sequence** is an ordered list of numbers. Each number in a sequence is called a **term**. In an **arithmetic sequence**, each term is found by adding the same number to the previous term. An example of an arithmetic sequence is shown.

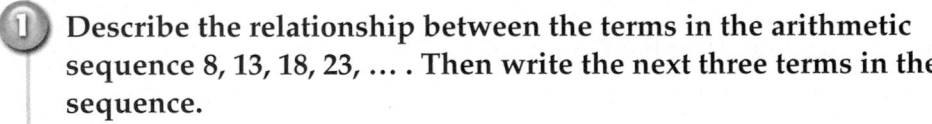

8, 11, 14, 17, 20, …
  +3 +3 +3 +3

Each term is found by adding 3 to the previous term.

EXAMPLE **Describe and Extend Sequences**

1. Describe the relationship between the terms in the arithmetic sequence 8, 13, 18, 23, … . Then write the next three terms in the sequence.

8, 13, 18, 23, …
  +5 +5 +5

Each term is found by adding 5 to the previous term.

Continue the pattern to find the next three terms.

$$23 + 5 = 28 \qquad 28 + 5 = 33 \qquad 33 + 5 = 38$$

The next three terms are 28, 33, and 38.

CHECK Your Progress

Describe the relationship between the terms in each arithmetic sequence. Then write the next three terms in the sequence.

a. 0, 13, 26, 39, …   **b.** 4, 7, 10, 13 …

Arithmetic sequences can also involve decimals.

 EXAMPLE **Describe and Extend Sequences**

2 Describe the relationship between the terms in the arithmetic sequence 0.4, 0.6, 0.8, 1.0, ... Then write the next three terms in the sequence.

0.4, 0.6, 0.8, 1.0, ...
+0.2 +0.2 +0.2

> Each term is found by adding 0.2 to the previous term.

Continue the pattern to find the next three terms.

$$1.0 + 0.2 = 1.2 \qquad 1.2 + 0.2 = 1.4 \qquad 1.4 + 0.2 = 1.6$$

The next three terms are 1.2, 1.4, and 1.6.

 CHECK Your Progress

Describe the relationship between the terms in each arithmetic sequence. Then write the next three terms in the sequence.

c. 1.0, 1.3, 1.6, 1.9, …   **d.** 2.5, 3.0, 3.5, 4.0, …

e. 1.75, 2.5, 3.25, 4.0 …   **f.** 0.01, 0.02, 0.03, 0.04 …

In a sequence, each term has a specific position within the sequence. Consider the sequence 2, 4, 6, 8, 10, … .

2nd position   4th position

2, 4, 6, 8, 10, …

1st position   3rd position   5th position

Study Tip

Arithmetic Sequences When looking for a pattern between the position number and each term in the sequence, it is often helpful to make a table.

The table below shows the position of each term in this sequence. Notice that as the position number increases by 1, the value of the term increases by 2.

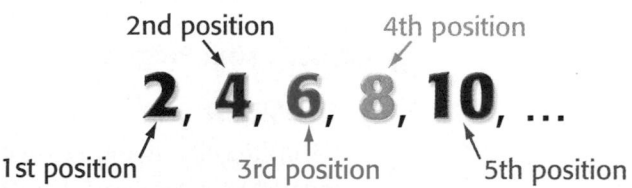

Position	Operation	Value of Term
1	$1 \cdot 2 = 2$	2
2	$2 \cdot 2 = 4$	4
3	$3 \cdot 2 = 6$	6
4	$4 \cdot 2 = 8$	8
5	$5 \cdot 2 = 10$	10

You can also write an algebraic expression to represent the relationship between any term in a sequence and its position in the sequence. In this case, if n represents the position in the sequence, the value of the term is $2n$.

REAL-WORLD EXAMPLE

3) GREETING CARDS The homemade greeting cards that Meredith makes are sold in boxes at a local gift store. Each week, the store sells five more boxes.

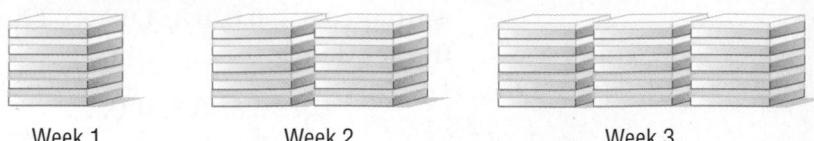

Week 1   Week 2   Week 3

If this pattern continues, what algebraic expression can be used to help her find the total number of boxes sold at the end of the 100th week? Then use the expression to find the total.

Make a table to display the sequence.

Position	Operation	Value of Term
1	1 • 5	5
2	2 • 5	10
3	3 • 5	15
n	n • 5	$5n$

Each term is 5 times its position number. So, the expression is $5n$.

$5n$ Write the expression.

$5(100) = 500$ Replace n with 100.

At the end of 100 weeks, 500 boxes will have been sold.

✓ CHECK Your Progress

g. GEOMETRY If the pattern continues, what algebraic expression can be used to find the number of circles used in any figure? How many circles will be in the 50th figure?

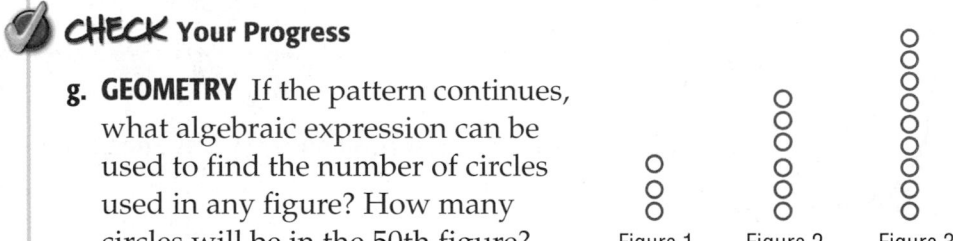

Figure 1   Figure 2   Figure 3

✓ CHECK Your Understanding

Examples 1, 2   Describe the relationship between the terms in each arithmetic sequence. Then write the next three terms in each sequence.

1. 0, 9, 18, 27, …   **2.** 4, 9, 14, 19, …

3. 1, 1.1, 1.2, 1.3, …   **4.** 5, 5.4, 5.8, 6.2, …

Example 3   **5) PLANTS** Refer to the table shown. If the pattern continues, what algebraic expression can be used to find the plant's height for any month? What will be the plant's height at 12 months?

Month	Height (in.)
1	3
2	6
3	9

= **Step-by-Step Solutions** begin on page R1.
Extra Practice begins on page EP2.

Examples 1–2   Describe the relationship between the terms in each arithmetic sequence. Then write the next three terms in each sequence.

6. 0, 7, 14, 21, …   **7.** 1, 7, 13, 19, …   **8.** 26, 34, 42, 50, …

9. 19, 31, 43, 55, …   **10.** 6, 16, 26, 36, …   **11.** 33, 38, 43, 48, …

12. 0.1, 0.4, 0.7, 1.0, … **13.** 2.4, 3.2, 4.0, 4.8, … **14.** 2.0, 3.1, 4.2, 5.3, …

15. 4.5, 6.0, 7.5, 9.0, … **16.** 1.2, 3.2, 5.2, 7.2, … **17.** 4.6, 8.6, 12.6, 16.6, …

Example 3   **18. COLLECTIONS** Hannah is starting a doll collection. Each year, she buys 6 dolls. Suppose she continues this pattern. What algebraic expression can be used to find the number of dolls in her collection after any number of years? How many dolls will Hannah have after 25 years?

19. EXERCISE The table shows the number of laps that Jorge swims each week. Jorge's goal is to continue this pace. What algebraic expression can be used to find the total number of laps he will swim after any given number of weeks? How many laps will Jorge swim after 6 weeks?

Week	Number of Laps
1	7
2	14
3	21
4	28

Describe the relationship between the terms in each arithmetic sequence. Then write the next three terms in each sequence.

20. 18, 33, 48, 63, …   **21** 20, 45, 70, 95, …   **22.** 38, 61, 84, 107, …

The terms of an arithmetic sequence can be related by subtraction. Write the next three terms of each sequence.

23. 32, 27, 22, 17, …   **24.** 45, 42, 39, 36, …   **25.** 10.5, 10, 9.5, 9, …

26. 🧩 **MULTIPLE REPRESENTATIONS** Kendra is stacking boxes of tissues for a store display. Each minute, she stacks another layer of boxes. Suppose the pattern continues for parts a–d.

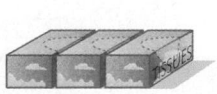
1 Minute

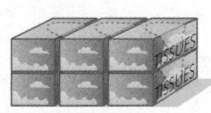
2 Minutes

3 Minutes

a. SYMBOLS Write an expression to find the *n*th term in the sequence.

b. TABLE Make a table of values for 1, 2, 3, 4, and 5 minutes.

c. GRAPH Graph the table of values from part **b** on the coordinate plane. Let *x* represent the number of minutes and *y* represent the number of boxes. Then describe the graph.

d. NUMBERS How many boxes will be displayed after 45 minutes?

QUICK Review

Coordinate Plane

The *x*-coordinate corresponds to the horizontal number line, or *x*-axis. The *y*-coordinate corresponds to the vertical number line, or *y*-axis.

27. GRAPHIC NOVEL Refer to the graphic novel frame below for parts **a–c**.

a. Explain how the number of text messages Dario sent and the cost form an arithmetic sequence.

b. Write an expression to find Dario's text messaging bill if he sends n number of text messages over 250.

c. Use your expression from part **b** to find Dario's text messaging bill if n equals 25, 50, and 75.

NUMBER SENSE Find the 100th number in each sequence.

28. 12, 24, 36, 48, …

29. 14, 28, 42, 56, …

30. 0.25, 0.5, 0.75, 1, …

31. 0.2, 0.4, 0.6, 0.8, …

32. 0, 50, 100, 150, …

33. 0, 75, 150, 225, …

34. MATH IN THE MEDIA The Fibonacci sequence is a well-known sequence in mathematics. Use the Internet or another source to find information about the Fibonacci sequence. List the first 10 terms in the sequence and explain its pattern.

35 GEOMETRY Refer to the figures for parts **a** and **b**. The pattern below was formed using rectangles.

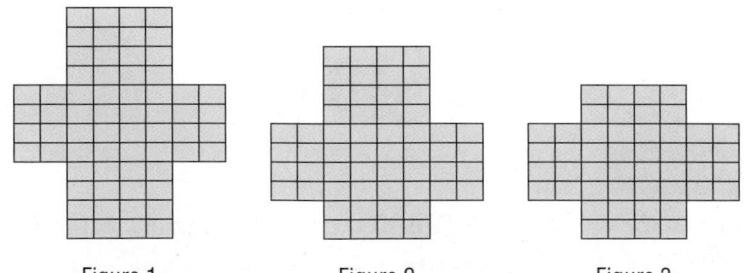

Figure 1   Figure 2   Figure 3

a. Describe the relationship between the figures and the number of rectangles shown.

b. If the pattern continues, how many rectangles will be in the next 2 figures?

H.O.T. Problems

CHALLENGE Not all sequences are arithmetic. But, there is still a pattern. Describe the relationship between the terms in each sequence. Then write the next three terms in the sequence.

36. 1, 2, 4, 7, 11, …

37. 0, 2, 6, 12, 20, …

38. **OPEN ENDED** Write five terms of an arithmetic sequence and describe the rule for finding the terms.

39. **WRITE MATH** Janice earns $6.50 per hour running errands for her neighbor. Explain how her hourly earnings form an arithmetic sequence.

✔ Test Practice

40. Which sequence follows the rule $3n - 2$, where n represents the position of a term in the sequence?

   A. 21, 18, 15, 12, 9, …

   B. 3, 6, 9, 12, 15, …

   C. 1, 7, 10, 13, 16, …

   D. 1, 4, 7, 10, 13, …

41. Which expression can be used to find the nth term in this sequence?

Position	1	2	3	4	5	n
Value of Term	2	5	10	17	26	

   F. $n^2 + 1$ **H.** $n + 1$

   G. $2n + 1$ **I.** $2n^2 + 2$

More About Sequences

In a **geometric sequence**, each term is found by multiplying the previous term by the same number.

EXAMPLE

Describe the relationship between the terms in the geometric sequence shown in the table below. Then write the next 3 terms in the sequence.

Term	1	2	3	4
Value of term	2	4	8	16

×2 ×2 ×2

Each term is found by multiplying the previous term by 2.

So, the next 3 terms are 32, 64, and 128.

Describe the relationship between the terms in the geometric sequence. Then write the next three terms of each geometric sequence.

42. 1, 4, 16, 64, …   **43.** 2, 6, 18, 54, …   **44.** 5, 25, 125, 625, …

Extend Sequences

Main Idea

Explore patterns in sequences of geometric figures.

Get ConnectED

ACTIVITY

STEP 1 Use toothpicks to build the figures below.

Figure 1   Figure 2   Figure 3

STEP 2 Make a table like the one shown and record the figure number and number of toothpicks used in each figure.

Figure Number	Number of Toothpicks
1	4
2	
3	

STEP 3 Construct the next figure in this pattern. Record your results.

STEP 4 Repeat Step 3 until you have found the next four figures in the pattern.

Analyze the Results

1. How many additional toothpicks were used each time to form the next figure in the pattern? Where is this pattern found in the table?

2. Based on your answer to Exercise 1, how many toothpicks would be in Figure 0 of this pattern?

3. Remove one toothpick from your pattern so that Figure 1 is made up of just three toothpicks as shown. Then create a table showing the number of toothpicks that would be in the first 7 figures by using the same pattern as above.

   Figure 1

4. How many toothpicks would there be in Figure *n* of this new pattern?

5. How could you adapt the expression you wrote in Exercise 4 to find the number of toothpicks in Figure *n* of the original pattern?

6. **MAKE A PREDICTION** How many toothpicks would there be in Figure 10 of the original pattern? Explain. Then check your answer by constructing the figure.

Mid-Chapter Check

Write each power as a product of the same factor. (Lesson 1A)

1. 4^5

2. 9^6

3. **ZOOS** The Lincoln Park Zoo in Illinois is $2 \cdot 2 \cdot 2 \cdot 2 \cdot 2 \cdot 2 \cdot 2$ years old. Write this age in exponential form. (Lesson 1A)

Evaluate each expression. Justify each step. (Lesson 1B)

4. $25 - (3^2 + 2 \times 5)$

5. $(8 + 4) \div (6 - 3)$

6. **MULTIPLE CHOICE** A coach spent $201 on baseball bats and baseball gloves. Let b represent the number of bats and g represent the number of gloves. Which expression represents the number of items that she bought? (Lesson 1C)

   A. $b + g$

   B. $b \cdot g$

   C. $2b + 3g$

   D. $3b + 2g$

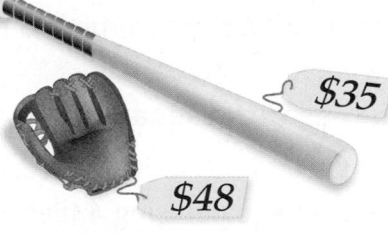

$35

$48

Use the Distributive Property to rewrite each expression. (Lesson 1D)

7. $4(6 + 9)$

8. $2(7 + 5)$

9. $3(4) + 3(8)$

10. $9(20) + 9(7)$

11. **SCHOOL SUPPLIES** Mrs. Martinez has three children. She purchased five folders and six pens for each child. Use the Distributive Property to write two expressions that represent the total number of folders and pens purchased. (Lesson 1D)

12. **FOOD** At a local bakery, 6 bagels cost $0.75 each, 12 bagels cost $0.70 each, and 18 bagels cost $0.65 each. Look for a pattern to find the cost of each bagel when a customer buys 30 bagels. (Lesson 2A)

Describe the relationship between the terms in each arithmetic sequence. Then write the next three terms in each sequence. (Lesson 2B)

13. 5, 8, 11, 14, …

14. 4, 11, 18, 25, …

15. 5.8, 10.8, 15.8, 20.8, …

16. 2.1, 2.3, 2.5, 2.7, …

17. **MULTIPLE CHOICE** Which of the following statements describes the relationship between the terms in the sequence 2, 6, 10, …? (Lesson 2B)

   F. Add 4 to each previous term.

   G. Subtract 4 from each previous term.

   H. Multiply each previous term by 3.

   I. Divide each previous term by 3.

18. **MONEY** Antoine is saving money for a laptop computer. The table shows his savings for the first four months. (Lesson 2B)

Month	Savings
1	$30
2	$60
3	$90
4	$120

Write an algebraic expression to find each month's savings. Then use the expression to find Antoine's savings after 12 months.

19. **TOOTHPICKS** Find the number of toothpicks in Figure 8 of the pattern below. (Lesson 2B)

Figure 1   Figure 2   Figure 3

Explore  Perfect Squares

Main Idea

Identify perfect squares.

Get ConnectED

🏃 📷 **AREA** The rug in Michaela's bedroom is in the shape of a square and has an area of 36 square feet. How long is the length of one side of the rug?

ACTIVITIES

1 **What do you need to find?** the side length of the rug

STEP 1 Construct a square using 36 tiles.

STEP 2 Count the total number of tiles for one side.

So, the length of each side of Michaela's rug is 6 feet.

2 **Determine if you can construct a square using 4, 9, 12, 16, 18, and 20 tiles.**

STEP 1 Try to construct a square using 4 tiles, 9 tiles, and 16 tiles.

STEP 2 Try to construct a square using 12 tiles, 18 tiles, and 20 tiles.

Analyze the Results

1. Refer to Activity 2. Which of the tile groups form squares?

2. Construct a square using 100 tiles. What are the lengths of the sides of the square?

3. Name two other tile groups that can be used to construct a square.

4. **MAKE A CONJECTURE** What is the relationship between the lengths of the sides and the areas of the squares?

Main Idea
Find squares of numbers and square roots of perfect squares.

 Vocabulary
square
perfect square
square root
radical sign

 Get Connect**ED**

Squares and Square Roots

WATERMELON Farmers are now growing square watermelons. Yancy cut a piece of square watermelon with a side length of 5 inches. What is the area of the piece of watermelon?

Recall the area of a square can be found by multiplying the lengths of two sides. The area of the square at the right is 5 • 5 or 25 square units. The product of a number and itself is the **square** of that number. So, the square of 5 is 25.

5 units   25 units²

5 units

So, the area of Yancy's piece of watermelon is 25 square inches.

 **EXAMPLES** **Find Squares of Numbers**

① **Find the square of 3.**

$3 \cdot 3 = 9$ Multiply 3 by itself.

So, the square of 3 is 9.

9 units² 3 units

3 units

② **Find the square of 28.**

Method 1 Use paper and pencil.	**Method 2 Use a calculator.**
28 Multiply 28 by itself. × 28 —— 224 + 560 —— 784	28 $\boxed{x^2}$ $\boxed{\text{ENTER}}$ 784

So, the square of 28 is 784.

 **CHOOSE** Your Method

Find the square of each number.

a. 8   **b.** 12   **c.** 23

Numbers like 9, 16, and 225 are called square numbers or **perfect squares** because they are squares of whole numbers.

The factors multiplied to form perfect squares are called **square roots**. A **radical sign**, $\sqrt{}$, is the symbol used to indicate a square root.

Read Math

Square Roots
Read $\sqrt{16} = 4$ as *the square root of 16 is 4*.

Key Concept   Square Root

Words	A square root of a number is one of its two equal factors.

Examples

        Numbers   **Algebra**

$4 \cdot 4 = 16$, so $\sqrt{16} = 4$.   If $x \cdot x$ or $x^2 = y$, then $\sqrt{y} = x$.

 EXAMPLES   Find Square Roots

3 Find $\sqrt{81}$.

$9 \cdot 9 = 81$, so $\sqrt{81} = 9$.   What number times itself is 81?

4 Find $\sqrt{225}$.

[2nd] [$\sqrt{}$] 225 [ENTER] 15

So, $\sqrt{225} = 15$.

CHECK Your Progress

Find each square root.

d. $\sqrt{64}$ **e.** $\sqrt{289}$

 Real-World Link · · · ·
The average number of pitches for a major league baseball is 7 pitches.

 REAL-WORLD EXAMPLE

5 **SPORTS** The infield of a baseball field is a square with an area of 8,100 square feet. What are the dimensions of the infield?

The infield is a square. By finding the square root of the area, 8,100, you find the length of one side of the infield.

$90 \cdot 90 = 8,100$, so $\sqrt{8,100} = 90$.

The dimensions of the infield are 90 feet by 90 feet.

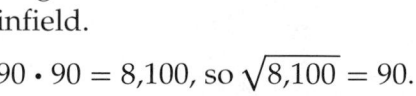

2nd base

Pitcher's Mound

3rd base   1st base

Home plate

 CHECK Your Progress

f. **SPORTS** The largest ring in amateur boxing is a square with an area of 400 square feet. What are the dimensions of the ring?

CHECK Your Understanding

Examples 1 and 2   Find the square of each number.

     1. 6   **2.** 10   **3** 17   **4.** 30

Examples 3 and 4   Find each square root.

     5. $\sqrt{9}$ **6.** $\sqrt{36}$ **7.** $\sqrt{121}$ **8.** $\sqrt{169}$

Example 5   **9. ROAD SIGNS** Historic Route 66 from Chicago to Los Angeles is known as the Main Street of America. If the area of a Route 66 sign measures 576 square inches and the sign is a square, what are the dimensions of the sign?

Practice and Problem Solving

> ● = **Step-by-Step Solutions** begin on page R1.
> **Extra Practice** begins on page EP2.

Examples 1 and 2   Find the square of each number.

     10. 4   **11.** 1   **12.** 7   **13.** 11

     14. 16   **15.** 20   **16.** 18   **17.** 34

Examples 3 and 4   Find each square root.

     18. $\sqrt{4}$ **19.** $\sqrt{16}$ **20.** $\sqrt{49}$ **21.** $\sqrt{100}$

     22. $\sqrt{144}$ **23.** $\sqrt{256}$ **24.** $\sqrt{529}$ **25.** $\sqrt{625}$

Example 5   **26. MEASUREMENT** Roslyn's bedroom is shaped like a square. What are the dimensions of the room if the area of the floor is 196 square feet?

27. SPORTS For the floor exercise, gymnasts perform their tumbling skills on a mat that has an area of 1,600 square feet. How much room does a gymnast have to run along one side of the mat?

28. GEOGRAPHY Refer to the squares in the diagram. They represent the approximate areas of Florida, North Carolina, and Pennsylvania.

     a. What is the area of North Carolina in square miles?

     b. How much larger is Florida than Pennsylvania?

     c. The water areas of Florida, North Carolina, and Pennsylvania are approximately 11,881 square miles, 5,041 square miles, and 1,225 square miles, respectively. Make a similar diagram comparing the water areas of these states.

29 MEASUREMENT A chessboard has an area of 324 square inches. There is a 1-inch border around the 64 squares on the board. What is the length of one side of the region containing the small squares?

30. **OPEN ENDED** Write a number whose square is between 100 and 150.

31. **CHALLENGE** Use the diagram that shows the area of a dog's pen.

    a. Could the area of the dog's pen be made larger using the same amount of fencing? Explain.

6 ft

14 ft

    b. Describe the largest rectangular pen area possible using the same amount of fencing. How do the perimeter and area compare to the original pen?

32. **REASONING** What is the value of $(\sqrt{n})^2$? Support your answer with an example.

33. **CHALLENGE** Which expression is $\sqrt{25 \cdot 4}$ equal to: $\sqrt{25} \cdot \sqrt{4}$ or $\sqrt{25} + \sqrt{4}$? Explain.

34. **WRITE MATH** Explain why raising a number to the second power is called *squaring* the number. Then explain how taking a square root and squaring a number are different.

Test Practice

35. Which model represents the square of 4?

    A.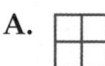

    B.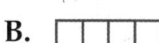

    C.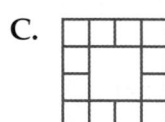

    D.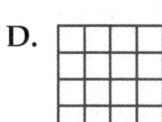

36. Which measure can be the area of a square if the measure of the side length is a whole number?

    F. 836 sq ft   H. 1,100 sq ft

    G. 949 sq ft   I. 1,225 sq ft

37. Theresa is making a quilt with the letter T as shown. Each square of the letter T has an area of 25 square inches.

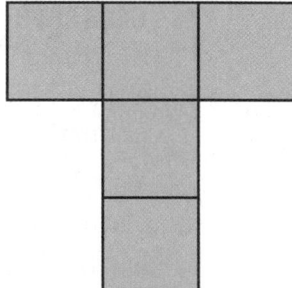

    What is the perimeter of the figure T?

    A. 20 inches   C. 60 inches

    B. 50 inches   D. 100 inches

38. Which of the following numbers is NOT a perfect square?

    F. 121   H. 182

    G. 144   I. 225

Main Idea

Estimate square roots.

 Vocabulary

irrational number
cube root

Get Connect**ED**

Estimate Square Roots

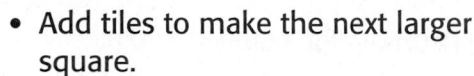 **Explore** Estimate the square root of 27.

- Arrange 27 tiles into the largest square possible.

- Add tiles to make the next larger square.

- The square root of 27 is between 5 and 6. Since 27 is much closer to 25 than 36, the square root of 27 is closer to 5 than 6.

Use algebra tiles to estimate the square root of each number to the nearest whole number.

1. 40   **2.** 28   **3.** 85   **4.** 62

5. Describe another method that you could use to estimate the square root of a number.

The square root of a perfect square is an integer. You can estimate the square root of a number that is *not* a perfect square.

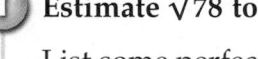 **EXAMPLE** **Estimate a Square Root**

① Estimate $\sqrt{78}$ to the nearest whole number.

List some perfect squares.   1, 4, 9, 16, 25, 36, 49, 64, 81, …

The first perfect square less than 78 is 64. $\sqrt{64} = 8$

The first perfect square more than 78 is 81. $\sqrt{81} = 9$

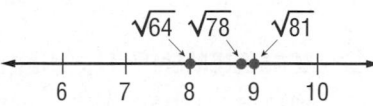

So, $\sqrt{78}$ is between 8 and 9. Since 78 is much closer to 81 than to 64, the best whole number estimate is 9.   Verify with a calculator.

 CHECK Your Progress

a. Estimate $\sqrt{50}$ to the nearest whole number.

Numbers like $\sqrt{2}$, $\sqrt{50}$, and $\sqrt{78}$ are *not* perfect squares. These numbers have square roots that are irrational numbers. A number that cannot be expressed as the ratio of two integers is an **irrational number**.

REAL-WORLD EXAMPLE

2 **SCIENCE** Physicist and mathematician Galileo performed an experiment by dropping stones from the edge of the Leaning Tower of Pisa. The formula $t = \dfrac{\sqrt{h}}{4}$ represents the time t in seconds that it takes an object to fall from a height of h feet. If the tower is 180 feet tall, estimate the time it took the stones to hit the ground.

First estimate the value of $\sqrt{180}$.

$169 < 180 < 196$ 169 and 196 are the closest perfect squares.

$13^2 < 180 < 14^2$ $169 = 13^2$ and $196 = 14^2$

$\sqrt{13^2} < \sqrt{180} < \sqrt{14^2}$ Find the square root of each number.

$13 < \sqrt{180} < 14$ Simplify.

Since 180 is closer to 169 than 196, the best whole number estimate for $\sqrt{180}$ is 13. Use this value to evaluate the expression.

$t = \dfrac{\sqrt{h}}{4} \approx \dfrac{13}{4}$ or 3.25

So, stones dropped from the Leaning Tower of Pisa would take about 3.25 seconds to hit the ground.

CHECK Your Progress

b. Estimate the time for an object to fall if dropped from the observation deck of the Eiffel Tower at 311 feet.

✓ CHECK Your Understanding

Example 1   Estimate each square root to the nearest whole number.

1. $\sqrt{39}$ **2.** $\sqrt{106}$ $\sqrt{90}$ **4.** $\sqrt{140}$

5. MEASUREMENT The diagram shows the floor plan of a square kitchen. What is the approximate length of one side of the kitchen floor? Round to the nearest whole number.

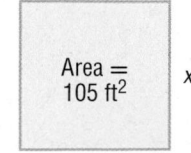
Area = 105 ft² | x

x

Example 2   **6. SPEED** A vehicle's speed can be estimated based on the length of the skid marks made by the car tires using the formula $s = \sqrt{24m}$, where s represents the speed in miles per hour and m represents the length of the skid in feet. A car leaves a mark that is 10 feet long. Estimate the speed of the vehicle.

Practice and Problem Solving

= **Step-by-Step Solutions** begin on page R1.
Extra Practice begins on page EP2.

Example 1 **Estimate each square root to the nearest whole number.**

   7. $\sqrt{11}$ **8.** $\sqrt{20}$ **9.** $\sqrt{35}$ **10.** $\sqrt{65}$

   11 $\sqrt{89}$ **12.** $\sqrt{116}$ **13.** $\sqrt{137}$ **14.** $\sqrt{409}$

15. MEASUREMENT The bottom of a square baking pan has an area of 67 square inches. What is the length of one side of the pan to the nearest whole number?

16. ALGEBRA What whole number is closest to $\sqrt{m-n}$ if $m = 45$ and $n = 8$?

Example 2   **17. ARCHITECTURE** The Parthenon in Athens, Greece, contains the *golden rectangle* proportion repeatedly. The length of the longer side divided by the length of the shorter side is equal to $\dfrac{1 + \sqrt{5}}{2}$. Estimate the value of the ratio.

18. BASEBALL In Little League, the field is a square with sides of 60 feet. The expression $\sqrt{(s^2 + s^2)}$ represents the diagonal distance *across* a square of side length s. Second base and home plate are in opposite corners across the field. Estimate the distance the catcher at home plate would have to throw the ball to reach the second baseman at second base.

Estimate each square root to the nearest whole number.

  19. $\sqrt{925}$ **20.** $\sqrt{2{,}480}$ **21.** $\sqrt{1{,}610}$ **22.** $\sqrt{6{,}500}$

ALGEBRA Estimate each expression to the nearest whole number if $a = 5$, $b = 10$, and $c = 20$.

  23. $\sqrt{a+b}$ **24.** $\sqrt{6b-a}$ **25.** $\sqrt{a^2 + b^2}$

  26. $\sqrt{b^2 - a^2}$ **27.** $\sqrt{2(b+c)}$ **28.** $\sqrt{3abc}$

29 STAMPS The Special Olympics commemorative stamp is square in shape with an area of 1,008 square millimeters.

  a. Find the length of one side of the postage stamp to the nearest whole number.

  b. What is the length of one side in centimeters? Round to the nearest whole number.

30. ALGEBRA The formula $D = 1.23 \times \sqrt{h}$ can be used to estimate the distance D in miles you can see from a point h feet above Earth's surface. Use the formula to find the distance D in miles that you can see from the top of a 120-foot hill. Round to the nearest tenth.

31. Which One Doesn't Belong? Identify the number that does not have the same characteristic as the other three. Explain your reasoning.

| $\sqrt{5}$ | $\sqrt{18}$ | $\sqrt{81}$ | $\sqrt{1,000}$ |

32. OPEN ENDED Select three numbers with square roots between 4 and 5.

33. NUMBER SENSE Explain why 8 is the best whole number estimate for $\sqrt{71}$.

CHALLENGE Estimate the square root of each number to the nearest tenth.

34. $\sqrt{0.26}$ **35.** $\sqrt{0.51}$ **36.** $\sqrt{1.99}$ **37.** $\sqrt{3.04}$

38. FIND THE ERROR Caitlyn is estimating $\sqrt{125}$. Find her mistake and correct it.

$$\sqrt{125} \approx 12$$

39. **WRITE MATH** Apply what you know about numbers to explain why $\sqrt{30}$ is an irrational number.

Test Practice

40. Reina wrote four numbers on a piece of paper. She then asked Tyron to select the number closest to 5. Which number should he select?

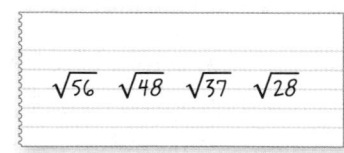

$\sqrt{56}$ $\sqrt{48}$ $\sqrt{37}$ $\sqrt{28}$

A. $\sqrt{56}$

B. $\sqrt{48}$

C. $\sqrt{37}$

D. $\sqrt{28}$

41. Which of the following is an irrational number?

F. $\sqrt{25}$

G. $\sqrt{7}$

H. -13

I. $\frac{4}{5}$

42. **THINK SOLVE EXPLAIN** **EXTENDED RESPONSE** The area of a square is 169 square inches.

Part A What is the length of a side?

Part B What is the perimeter of the square? Explain how you solved.

More About Estimating Roots

Recall the square root of a number is one of its two equal factors. Similarly, a **cube root** of a number is one of its *three* equal factors.

> ### Key Concept Cube Root
>
> **Words**   A cube root of a number is one of its three equal factors.
>
> **Examples**   **Numbers**   **Algebra**
>
> $2 \cdot 2 \cdot 2 = 8$, so $\sqrt[3]{8}$ If $x^3 = y$, then $\sqrt[3]{y} = x$.

✎ EXAMPLES

① Find $\sqrt[3]{27}$.

Since $3 \times 3 \times 3 = 27$,
the cube root of 27 is 3.

You can also estimate cube roots by using perfect cubes. The first ten perfect cubes are shown at the right.

② Find $\sqrt[3]{125}$.

Since $5 \times 5 \times 5 = 125$,
the cube root of 125 is 5.

$1 = 1^3$	$216 = 6^3$
$8 = 2^3$	$343 = 7^3$
$27 = 3^3$	$512 = 8^3$
$64 = 4^3$	$729 = 9^3$
$125 = 5^3$	$1{,}000 = 10^3$

✎ EXAMPLE

③ Estimate $\sqrt[3]{90}$ to the nearest whole number.

The first perfect cube less than 90 is 64. $\sqrt[3]{64} = 4$

The first perfect cube more than 90 is 125. $\sqrt[3]{125} = 5$

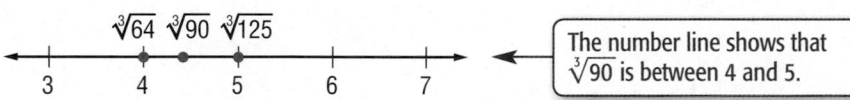

> The number line shows that $\sqrt[3]{90}$ is between 4 and 5.

Since 90 is closer to 64 than 125, $\sqrt[3]{90}$ is closer to 4 than 5.

So, $\sqrt[3]{90} \approx 4$.

Check Use a calculator. ⎣MATH⎦ 4 90 ⎣ENTER⎦ 4.4814047

4.4814047 rounds to 4. ✓

Find each cube root.

43. $\sqrt[3]{216}$ **44.** $\sqrt[3]{1{,}000}$ **45.** $\sqrt[3]{729}$

Estimate each cube root to the nearest whole number. Check using a calculator.

46. $\sqrt[3]{35}$ **47.** $\sqrt[3]{100}$ **48.** $\sqrt[3]{75}$

Problem Solving in Animal Conservation

Tag, You're It!

Are you fascinated by sharks, especially those that are found around the coasts of the United States? If so, you should consider a career as a shark scientist. Shark scientists use satellite-tracking devices, called *tags*, to study and track the movements of sharks. By analyzing the data transmitted by the tags, scientists are able to learn more about the biology and ecology of sharks. Their research is helpful in protecting shark populations around the world.

21st Century Careers

Are you interested in a career as a shark scientist? Take some of the following courses in high school.

- Algebra
- Physics
- Calculus
- Statistics

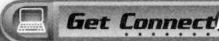
Get Connect**ED**

Hammerhead Shark
The total length of a hammerhead shark is about 1.3 times the fork length.

Tiger Shark
A study found that the average fork length of a tiger shark is 55 centimeters less than twice the average fork length of a sandbar shark.

White Shark
The weight in pounds of a white shark can be estimated using the expression $18x^2 - 134x + 304$, where x is the fork length in feet.

Real-World Math

The *fork length* of a shark is the length from the tip of the snout to the fork of the tail. Use the information on the note cards to solve each problem.

1. Write an expression to represent the total length of a hammerhead shark that has a fork length of f feet.

2. Use the expression from Exercise 1 to find the total length of a hammerhead shark that has a fork length of 11.6 feet.

3. Write an expression to represent the average fork length of a tiger shark, given the average fork length s of a sandbar shark.

4. Use the expression from Exercise 3 to find the average fork length of a tiger shark if the average fork length of a sandbar shark is 129 centimeters.

5. About how much would a white shark with a fork length of 13.8 feet weigh?

6. The largest white shark in a recent study had a fork length of 16.2 feet. How much more did this shark weigh than the shark described in Exercise 5?

FOLDABLES Study Organizer

Be sure the following Key Concepts are noted in your Foldable.

Key Concepts

Expressions (Lesson 1)
To evaluate expressions, use the order of operations and properties.

Order of Operations
- Do all operations within grouping symbols first.
- Evaluate all powers before other operations.
- Multiply and divide in order from left to right.
- Add and subtract in order from left to right.

Sequences (Lesson 2)
- A sequence is an ordered list of numbers.
- Each number in a sequence is called a term.
- In an arithmetic sequence, each term is found by adding the same number to the previous term.

8, 13, 18, 23, ... ⟵ An arithmetic sequence
  +5 +5 +5

Squares and Square Roots (Lesson 3)
- The square of a number is the product of a number and itself.
- A square root of a number is one of its two equal factors.

radical sign ⟶ $\sqrt{25} = 5$

Key Vocabulary

algebra	numerical expression
algebraic expression	order of operations
arithmetic sequence	perfect square
base	powers
coefficient	radical sign
cubed	sequence
equivalent expressions	squared
evaluate	square root
exponent	standard form
exponential form	term
factors	variable

Vocabulary Check

State whether each sentence is *true* or *false*. If *false*, replace the underlined word or number to make a true sentence.

1. <u>Numerical expressions</u> have the same value.

2. The <u>coefficient</u> of the expression $4n$ is 4.

3. Two or more numbers that are multiplied together are called <u>powers</u>.

4. A <u>sequence</u> is an ordered list of numbers.

5. The product of a number and itself is the <u>square root</u> of the number.

6. The number 3^3 is written in <u>standard form</u>.

7. A square root of a number is one of its <u>three</u> equal factors.

8. The expression 9^2 is read as nine <u>squared</u>.

Multi-Part Lesson Review

Lesson 1 **Expressions**

Powers and Exponents (Lesson 1A)

Evaluate each expression.

9. 3^5 **10.** 5^1 **11.** 18^2

12. PATHS At the edge of a forest, there are two paths. At the end of each path, there are two more paths. At the end of each of those paths, there are two more paths. How many paths are there at the end?

> **EXAMPLE 1**   Evaluate 4^5.
>
> The base is 4. The exponent 5 means that 4 is used as a factor 5 times.
>
> $4^5 = 4 \cdot 4 \cdot 4 \cdot 4 \cdot 4$
>
> $ = 1{,}024$

Numerical and Algebraic Expressions (Lessons 1B and 1C)

Evaluate each expression.

13. $48 \div 6 + 2 \cdot 5$ **14.** $9 + 3(7 - 5)^3$

Evaluate each expression if $a = 10$ and $b = 4$.

15. $(a - b)^2$ **16.** $(b + a)^2 \div 2$

17. CLOTHING Write an expression that represents the cost in dollars of buying any number of hats and any number of shirts. Then find the cost of buying 3 hats and 5 shirts.

$8.95

$5.75

> **EXAMPLE 2**   Evaluate $24 - (8 \div 4)^4$.
>
> $24 - (8 \div 4)^4 = 24 - 2^4$ Divide 8 by 4.
>
> $ = 24 - 16$ Find the value of 2^4.
>
> $ = 8$ Subtract.

> **EXAMPLE 3**   Evaluate $2m^2 - 5n$ if $m = 4$ and $n = 3$.
>
> $2m^2 - 5n = 2(4)^2 - 5(3)$ Replace m with 4 and n with 3.
>
> $ = 2(16) - 5(3)$ Find the value of 4^2.
>
> $ = 32 - 15$ Multiply.
>
> $ = 17$ Subtract.

Properties (Lesson 1D)

Evaluate each expression mentally. Justify each step.

18. $(25 \times 15) \times 4$ **19.** $14 + (38 + 16)$

20. ROSES Wesley sold roses for $2 a rose. He sold 15 roses on Monday and 12 roses on Tuesday. Use the Distributive Property to mentally find the total amount Wesley earned. Explain.

> **EXAMPLE 4**   Evaluate $8 + (17 + 22)$ mentally. Justify each step.
>
> $8 + (17 + 22)$
>
> $= 8 + (22 + 17)$ Commutative Property of Addition
>
> $= (8 + 22) + 17$ Associative Property of Addition
>
> $= 30 + 17$ or 47 Add 30 and 17 mentally.

Patterns

PSI: Look for a Pattern (Lesson 2A)

Use the *look for a pattern* strategy to solve.

21. GEOMETRY What are the next two figures in the pattern shown below?

22. FOOTBALL Nick is practicing kicking field goals for football tryouts. He makes 2 out of 3 attempts. How many field goals should Nick make after 15 attempts?

23. AGES The table shows how Jarred's age is related to his sister Emily's age. How old will Emily be when Jarred is 25?

Emily's Age (yr)	3	6	9	12
Jarred's age (yr)	7	10	13	16

EXAMPLE 5 The table shows how to convert cups to ounces. If you had 56 ounces, how many cups would you have? Use the *look for a pattern* strategy.

Convert Cups to Ounces	
Number of Cups	Number of Ounces
1	8
2	16
3	24

Based on the table, there are 8 ounces for every cup. So, 56 ÷ 8 is 7.

You would have 7 cups.

Sequences (Lesson 2B)

Describe the relationship between the terms in each arithmetic sequence. Then find the next three terms in each sequence.

24. 3, 9, 15, 21, 27, …

25. 2.6, 3.4, 4.2, 5, 5.8, …

26. 0, 7, 14, 21, 28, …

MONEY Tanya collected $4.50 for the first car washed at a band fundraiser. After the second and third cars were washed, the donations totaled $9 and $13.50, respectively.

27. If this donation pattern continues, what algebraic expression can be used to find the amount of money earned for any number of cars washed?

28. How much money will be collected after a total of 8 cars have been washed?

EXAMPLE 6 At the end of day 1, Sierra read 25 pages of a novel. By the end of days 2 and 3, she read a total of 50 and 75 pages, respectively. If the pattern continues, what expression will give the total number of pages read after any number of days?

Make a table to display the sequence.

Position	Operation	Value of Term
1	$1 \cdot 25$	25
2	$2 \cdot 25$	50
3	$3 \cdot 25$	75
n	$n \cdot 25$	$25n$

Each term is 25 times its position number. So, the expression is $25n$.

Square Roots

Squares and Square Roots (Lesson 3B)

Find each square root.

29. $\sqrt{81}$ **30.** $\sqrt{324}$

31. FARMING The area of a farm field is shown. What is the length of one side of the square field?

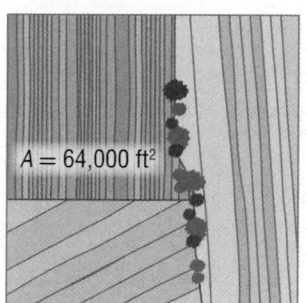

$A = 64{,}000 \text{ ft}^2$

EXAMPLE 7   Find the square root of 441.

$21 \cdot 21 = 441$, so $\sqrt{441} = 21$.

EXAMPLE 8   The area of a square ceramic tile is 25 square inches. What is the length of one side?

To find the side length, find $\sqrt{25}$.

$5 \cdot 5 = 25$, so $\sqrt{25} = 5$.

So, the side length of the square tile is 5 inches.

Estimate Square Roots (Lesson 3C)

Estimate each square root to the nearest whole number.

32. $\sqrt{26}$ **33.** $\sqrt{99}$ **34.** $\sqrt{48}$

35. $\sqrt{76}$ **36.** $\sqrt{19}$ **37.** $\sqrt{52}$

ALGEBRA Estimate each expression to the nearest whole number if $x = 6$, $y = 12$, and $z = 24$.

38. $\sqrt{y - x}$ **39.** $\sqrt{3z - x}$

40. $\sqrt{3(y + z)}$ **41.** $\sqrt{2xyz}$

42. GAMES The area of a popular square game puzzle is 77 square inches. What is the approximate length of one of the sides of the game board to the nearest whole number?

EXAMPLE 9   Estimate $\sqrt{29}$ to the nearest whole number.

$25 < 29 < 36$ 29 is between the perfect squares 25 and 36.

$\sqrt{25} < \sqrt{29} < \sqrt{36}$ Find the square root of each number.

$5 < \sqrt{29} < 6$ $\sqrt{25} = 5$ and $\sqrt{36} = 6$

So, $\sqrt{29}$ is between 5 and 6.

Check Use a number line.

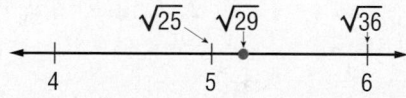

Since 29 is closer to 25 than to 36, the best whole number estimate is 5. ✓

Write each power as a product of the same factor. Then evaluate the expression.

1. 3^5

2. 15^4

3. Write $2 \cdot 2 \cdot 2 \cdot 2 \cdot 2 \cdot 2$ in exponential form.

4. MEASUREMENT Gregory wants to stain his 15-foot-by-15-foot deck. Is one can of the stain shown enough to cover his entire deck? Explain your reasoning.

Deck Stain
One can covers 200 square feet!

5. MULTIPLE CHOICE Which of the following is the value of $8 + (12 \div 3)^3 - 5 \times 9$?

A. 603

B. 135

C. 27

D. 19

Evaluate each expression if $x = 12$, $y = 5$, and $z = 3$.

6. $x - 9$

7. $8y$

8. $(y - z)^3$

9. $(x + z) \div y$

10. SAVINGS Shannon is saving $24 per month to buy a new digital camera. Use the Distributive Property to mentally find how much she will have saved after 7 months. Justify each step.

Evaluate each expression mentally. Justify each step.

11. $13 + (34 + 17)$

12. $50 \cdot (17 \cdot 2)$

13. SCHOOL The table shows the number of hours Teodoro spent studying for his biology test over four days.

Day	Number of Hours Spent Studying
Monday	0.5
Tuesday	0.75
Wednesday	1.0
Thursday	1.25

If the pattern continues, how many hours will Teodoro study on Sunday?

Describe the relationship between the terms in each sequence. Then write the next three terms in each sequence.

14. 7, 16, 25, 34, …

15. 59, 72, 85, 98, …

16. PHOTOGRAPHY Suzi cut a photograph in the shape of a square to place in her scrapbook. If the photograph has an area of 121 square inches, what are the dimensions of the photograph?

Estimate each square root to the nearest whole number.

17. $\sqrt{7}$

18. $\sqrt{84}$

19. $\sqrt{570}$

20. THINK SOLVE EXPLAIN **EXTENDED RESPONSE** Jacob had 10 points on his game card. An additional 5 points is added to his game card every time he plays a game.

Part A Write an expression that represents the total number of points Jacob has on his card for playing any number of games.

Part B How many points would Jacob have on his card if he has played an additional 12 games? Explain how you solved.

Preparing for Standardized Tests

Multiple Choice: Making Tables

One strategy for solving word problems is to organize the data in a table. It allows you to see patterns and sequences more easily.

TEST EXAMPLE

A restaurant offers discounts based on the number of sandwiches ordered up to a maximum of five sandwiches. One sandwich costs $6.00. Two sandwiches cost $5.75 each. Three sandwiches cost $5.50 each. Based on this pattern, how much would a sandwich cost if five sandwiches were ordered?

A. $5.25

B. $5.00

C. $4.75

D. $4.50

Make a table to organize the data. Then look for a pattern.

The cost of each sandwich is decreasing by $0.25. When 5 sandwiches are ordered, a sandwich would cost $5.00.

The correct answer is B.

Number of Sandwiches	Cost per Sandwich
1	$6.00
2	$5.75
3	$5.50
4	$5.25
5	$5.00

−0.25
−0.25
−0.25
−0.25

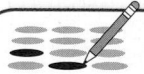 **Work on It**

At a movie theater, popcorn is sold in five sizes of containers: child-size, small, medium, large, and x-large. The price of a child-size container of popcorn is $2.00. The cost of a small container of popcorn is $2.75. The cost of a medium container is $3.50. If the pattern continues, what is the cost of the x-large container of popcorn?

F. $6.50   **H.** $5.00

G. $5.75   **I.** $4.25

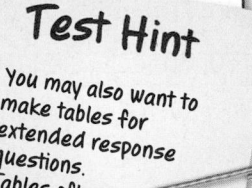

Test Hint

You may also want to make tables for extended response questions. Tables often help identify and extend patterns.

Test Practice

Read each question. Then fill in the correct answer on the answer document provided by your teacher or on a sheet of paper.

1. The world's largest ocean, the Pacific Ocean, covers approximately 2^6 million square miles. What is this area in standard form?

   A. 12 square miles

   B. 64 square miles

   C. 12,000,000 square miles

   D. 64,000,000 square miles

2. Roberto is training for the cross country team. The table shows the number of minutes he ran the first five days.

Day	Number of Minutes
Day 1	30
Day 2	30
Day 3	40
Day 4	40
Day 5	50

   If the pattern continues, which of the following shows the number of minutes he will run the next three days?

   F. 50, 50, 60

   G. 50, 60, 60

   H. 60, 60, 70

   I. 60, 70, 80

3. ≡≡✍ **GRIDDED RESPONSE** What is the value of the expression below if $x = 6$ and $y = 4$?

   $$(x + y)^2 + 5$$

4. Which of the following describes the relationship between the value of a term and n, its position in the sequence?

Position	1	2	3	4	5	n
Value of Term	3	6	9	12	15	■

   A. Add 2 to n.

   B. Divide n by 3.

   C. Multiply n by 3.

   D. Subtract n from 2.

5. ≡≡✍ **GRIDDED RESPONSE** Parker and Paige baked cookies for a bake sale. They baked 4^3 cookies. How many cookies did they bake?

6. Which does NOT have a value of 256?

   F. 16^2 H. 4^4

   G. 8^3 I. 2^8

7. What is the first step in evaluating the expression $3 \times (5 + 4) - 27 \div 9$?

   A. multiplying 3 and 5

   B. adding 5 and 4

   C. subtracting 27

   D. dividing 27 and 9

8. ≡≡✍ **GRIDDED RESPONSE** A square-shaped bulletin board is shown.

   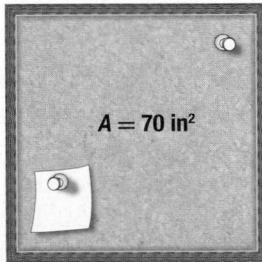

   $A = 70 \text{ in}^2$

   What is the approximate length in inches of a side of the bulletin board? Round to the nearest whole number.

9. What is the perimeter of the square garden?

A = 25 ft²

F. 5 feet   H. 50 feet

G. 20 feet   I. 100 feet

10. Sachi collects stamps. Each year, the number of stamps in her collection is ten times n, the number's position in the sequence. Which sequence represents Sachi's number of stamps?

A. 1, 11, 21, 31   C. 10, 11, 12, 13

B. 1, 10, 100, 1,000   D. 10, 20, 30, 40

11. Which of the following expressions when evaluated has a value of 22?

F. $2^5 - 3$ H. $5^2 - 3$

G. $2^5 + 3$ I. $5^2 + 3$

12. **THINK SOLVE EXPLAIN** **SHORT RESPONSE** Lemisha drove an average of 50 miles per hour on Sunday, 55 miles per hour on Monday, and 53 miles per hour on Tuesday. Let s represent the number of hours she drove on Sunday, m represent the number of hours she drove on Monday, and t represent the number of hours she drove on Tuesday. Write an expression that represents the total distance Lemisha drove.

13. Which of the following expressions can be written as $5(3 + 4)$?

A. $4 \cdot 5 + 4 \cdot 3$ C. $5 \cdot 3 + 4$

B. $5 \cdot 3 + 5 \cdot 4$ D. $3 + 5 \cdot 4$

14. **THINK SOLVE EXPLAIN** **SHORT RESPONSE** Use the Distributive Property to rewrite $4(12) + 4(8)$. Then evaluate the expression.

15. Which statement below is an example of the Associative Property of Addition?

F. $7 + (3 + 5) = 7 + (5 + 3)$

G. $9 + (11 + 6) = (9 + 11) + 6$

H. $3(6 + 5) = 3 \cdot 6 + 3 \cdot 5$

I. $12(8 + 4) = 12(8) + 12(4)$

16. **THINK SOLVE EXPLAIN** **EXTENDED RESPONSE** The first and fifth terms of a sequence are shown.

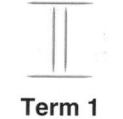

Term 1

Term 5

Part A   What might the third term look like?

Part B   Describe the relationship between the term number and the sequence.

Part C   Write a rule that connects the term number and the number of toothpicks in the sequence.

NEED EXTRA HELP?																
If You Missed Question...	1	2	3	4	5	6	7	8	9	10	11	12	13	14	15	16
Go to Chapter-Lesson...	1-1A	1-2A	1-1C	1-2B	1-1A	1-1A	1-1B	1-3B	1-3B	1-2B	1-1B	1-1C	1-1D	1-1D	1-1D	1-2B

Integers

connectED.mcgraw-hill.com

Investigate

 Animations

 Vocabulary

 Multilingual eGlossary

Learn

 Personal Tutor

 Virtual Manipulatives

 Graphing Calculator

 Audio

 Foldables

Practice

 Self-Check Practice

 Worksheets

 Assessment

The ☆BIG Idea

How can you determine if the sums, differences, products, and quotients of integers are either positive or negative?

FOLDABLES®
Study Organizer

Make this Foldable to help you organize your notes.

Lesson 1
Integers and the Coordinate Plane

Review Vocabulary

numerical expression expression numerica a combination of numbers and operations

$$3 + (4 \cdot 5)$$

Key Vocabulary

English	Español
integer	entero
negative integer	entero negative
positive integer	entero positivo
absolute value	valor absoluto
coordinate plane	plano de coordenadas

When Will I Use This?

You have two options for checking prerequisite skills for this chapter.

Text Option Take the Quick Check below. Refer to the Quick Review for help.

QUICK Check	QUICK Review

Evaluate.

1. $54 \div (6 + 3)$

2. $29 + 46 - 34$

3. $7 + 50 \div 5$

4. $20 + 30 - 38$

5. $18 + 2(4 - 1)$

6. $(4 + 13 \times 2) \div 3$

7. $(30 \div 6) + (3 - 1)$

8. $5 + (91 \div 7)$

EXAMPLE 1

Evaluate $48 \div 6 + 2 \cdot 5$.

Follow the order of operations.

$48 \div 6 + 2 \cdot 5$
$= 8 + 2 \cdot 5$ Divide 48 by 6.
$= 8 + 10$ Multiply 2 by 5.
$= 18$ Add.

Use the coordinate plane to name the ordered pair for each point.

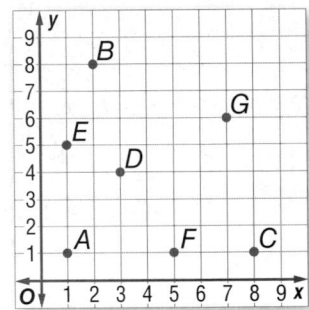

9. A

10. B

11. C

12. D

13. E

14. F

15. G

EXAMPLE 2

Use the coordinate plane to name the ordered pair for point M.

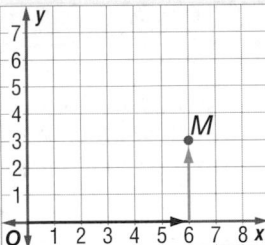

Start from the origin. Point M is located 6 units to the right and 3 units up.

So, point M corresponds to the ordered pair (6, 3).

Online Option Take the Online Readiness Quiz.

Writing Math

Compare and Contrast

When you compare, you notice how things are alike. When you contrast, you notice how they are different.

I know *compare* means to tell how things are alike. But what does *contrast* mean? Can you help?

The table below shows two monthly plans a cell phone company offers.

Plan A	Plan B
$34.99	$34.99
200 anytime minutes	300 anytime minutes
200 text messages	100 text messages
Free weekend minutes	Free weekend minutes

To compare and contrast the monthly plans, make a list of how they are alike and how they are different.

Alike/Compare

- The monthly cost is the same.
- They both have free weekend minutes.

Different/Contrast

- Plan B has more anytime minutes.
- Plan A has more text messages.

Practice

1. Write a few sentences that compare and contrast the figures below.

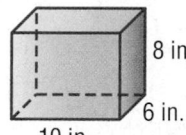 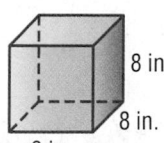

2. Write a few sentences that compare and contrast the numbers below.

$$30\% \qquad 0.3 \qquad \frac{3}{10}$$

3. Refer to Lesson 1B of this chapter. Compare and contrast negative and positive integers.

4. Compare and contrast the units below.

inch   foot   meter

5. Compare and contrast a rectangle and a square.

Explore Absolute Value

Main Idea

Use models to find the absolute value of an integer.

Get ConnectED

SCUBA DIVING A scuba diver is diving at −130 feet. At the same time, a hiker is at the top of a cliff 130 feet *above* the surface of the water. Compare and contrast −130 feet and 130 feet.

ACTIVITY

What do you need to find? how −130 feet and 130 feet are the same and how they are different

130 ft

−130 ft

Both the hiker and the diver are the same distance from the surface of the water. The hiker is *above* and the diver is *below*.

Analyze the Results

1. Write about a real-world situation with two values that can be represented by −10 and 10. Compare and contrast them.

2. Refer to the street map at the right. Compare and contrast walking from the house to the school and walking from the house to the park.

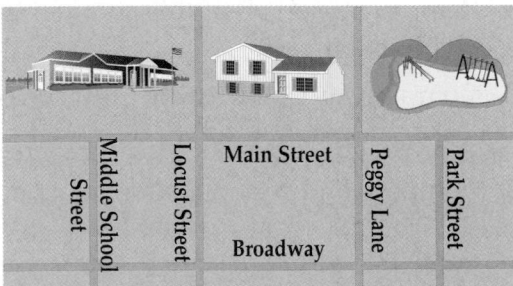

Middle School Street · Locust Street · Main Street · Broadway · Peggy Lane · Park Street

3. **WRITE MATH** Find the definition of *absolute* in a dictionary. Explain how the definition could be applied to math.

Main Idea

Read and write integers, and find the absolute value of an integer.

 Vocabulary

integer
negative integer
positive integer
graph
absolute value

Integers and Absolute Value

SKATEBOARDING The bottom of a skateboarding ramp is 8 feet below street level. A value of −8 represents 8 feet *below* street level.

1. For this situation, what would a value of −10 represent?

2. The top deck of the ramp is 5 feet *above* street level. How can you represent 5 feet *above* street level?

Numbers like 5 and −8 are called integers. An **integer** is any number from the set {…, −4, −3, −2, −1, 0, 1, 2, 3, 4, …}, where … means *continues without end*.

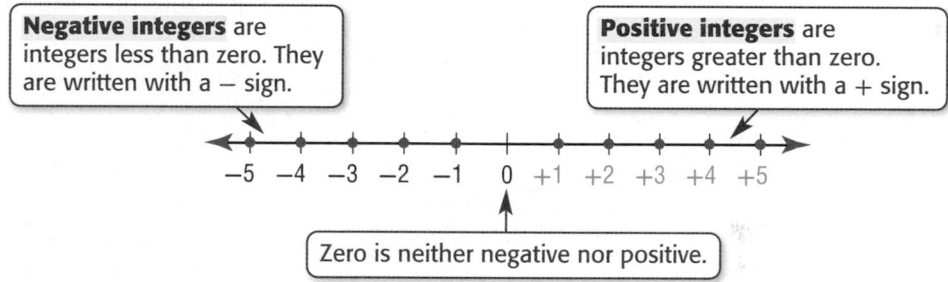

Negative integers are integers less than zero. They are written with a − sign.

Positive integers are integers greater than zero. They are written with a + sign.

−5 −4 −3 −2 −1   0   +1 +2 +3 +4 +5

Zero is neither negative nor positive.

 REAL-WORLD EXAMPLE

WEATHER Write an integer for each situation.

1 an average temperature of 5 degrees below normal

Because it represents *below* normal, the integer is −5.

2 an average rainfall of 5 inches above normal

Because it represents *above* normal, the integer is +5 or 5.

CHECK Your Progress

WEATHER Write an integer for each situation.

    a. 6 degrees above normal   **b.** 2 inches below normal

Integers can be graphed on a number line. To **graph** an integer on the number line, draw a dot on the line at its location.

 **EXAMPLE** Graph Integers

(3) Graph the set of integers {4, −6, 0} on a number line.

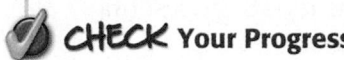

CHECK Your Progress

Graph each set of integers on a number line.

c. {−2, 8, −7}   **d.** {−4, 10, −3, 7}

Numbers that are the same distance from zero on a number line have the same **absolute value**.

Read Math

Absolute Value
|−5| *absolute value of negative five*

Key Concept   Absolute Value

Words   The absolute value of a number is the distance between the number and zero on a number line.

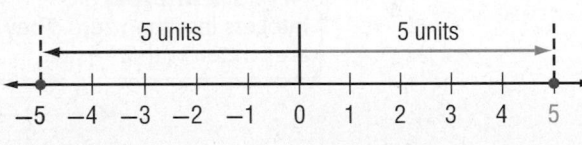

Examples $|-5| = 5$ $|5| = 5$

 **EXAMPLES** Evaluate Expressions

Evaluate each expression.

(4) $|-4|$

The graph of −4 is 4 units from 0.

So, $|-4| = 4$.

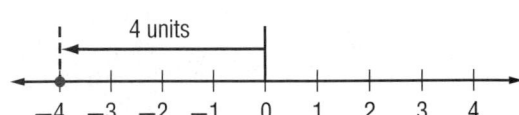

Study Tip

Order of Operations
The absolute value bars are considered to be a grouping symbol. When evaluating |−5| − |2|, evaluate the absolute values before subtracting.

(5) $|-5| - |2|$

$|-5| - |2| = 5 - 2$ $|-5| = 5, |2| = 2$
So, $|-5| - |2| = 3$.

CHECK Your Progress

e. $|8|$ **f.** $2 + |-3|$ **g.** $|-6| - 5$

✓ CHECK Your Understanding

Examples 1 and 2 Write an integer for each situation.

1. a loss of 11 yards   **2.** 6°F below zero

3. a deposit of $16   **4.** 250 meters above sea level

5. FOOTBALL The quarterback lost 15 yards on one play. Write an integer to represent the number of yards lost.

Example 3 Graph each set of integers on a number line.

6. {11, −5, −8}   **7.** {2, −1, −9, 1}

Examples 4 and 5 Evaluate each expression.

8. |−9| **9.** 1 + |7| **10.** |−11| − |−6|

Practice and Problem Solving

 = **Step-by-Step Solutions** begin on page R1.
Extra Practice begins on page EP2.

Examples 1 and 2 Write an integer for each situation.

11. a profit of $9   **12.** a bank withdrawal of $50

13. 53°C below zero   **14.** 7 inches more than normal

15. 2 feet below flood level   **16.** 160 feet above sea level

17. an elevator goes up 12 floors   **18.** no gains or losses on first down

19. GOLF In golf, scores are often written in relationship to *par*, the average score for a round at a certain course. Write an integer to represent a score that is 7 under par.

20. PETS Jasmine's pet guinea pig gained 8 ounces in one month. Write an integer to describe the amount of weight her pet gained.

Example 3 Graph each set of integers on a number line.

21. {0, 1, −3}   **22.** {3, −7, 6}

23. {−5, −1, 10, −9}   **24.** {−2, −4, −6, −8}

Examples 4 and 5 Evaluate each expression.

25. |10| **26.** |−12| **27.** |−7| − 5

28. 7 + |4| **29.** |−9| + |−5| **30.** |18| − |−10|

31. |−10| ÷ 2 × |5| **32.** 12 − |−8| + 7   **33.** |27| ÷ 3 − |−4|

34. SCUBA DIVING One diver descended 10 feet and another ascended 8 feet. Which situation has the greater absolute value? Explain.

35. MATH IN THE MEDIA Find an example of a positive integer and a negative integer in a newspaper or magazine, on television, or on the Internet. Explain what each integer means.

36. **REASONING** If $|x| = 3$, what is the value of x?

37. **CHALLENGE** Two numbers A and B are graphed on a number line. Is it *always*, *sometimes*, or *never* true that $A - |B| \leq A + B$? Explain.

38. **NUMBER SENSE** Explain why -18 is less than -10. Which one has the greater absolute value? Justify your reasoning.

39. **Which One Doesn't Belong?** Identify the expression that is not equal to the other three. Explain your reasoning.

| $|15 - |-5||$ | $|-4| + 6$ | $-|7 + 3|$ | $|-10|$ |

40. **WRITE MATH** Why is the absolute value of a number positive? Explain your reasoning.

Test Practice

41. Which point has a coordinate with the greatest absolute value?

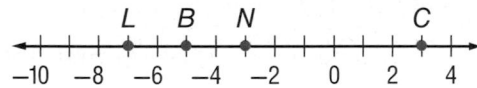

A. Point B

B. Point C

C. Point L

D. Point N

42. Which of the following statements about these real-world situations is NOT true?

F. A $100 check deposited in a bank can be represented by $+100$.

G. A loss of 15 yards in a football game can be represented by -15.

H. A temperature of 20 below zero can be represented by -20.

I. A submarine diving 300 feet under water can be represented by $+300$.

43. Rachel recorded the low temperatures for one week in the table.

Low Temperatures	
Day	Temperature (°F)
Sunday	2
Monday	−6
Tuesday	4
Wednesday	−8
Thursday	2
Friday	0
Saturday	−1

On which day was the low temperature the farthest from 0°F?

A. Monday

B. Tuesday

C. Wednesday

D. Friday

44. Write an integer to represent the temperature shown on the thermometer.

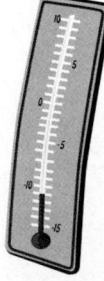

F. −11°F

G. −10°F

H. 10°F

I. 11°F

Main Idea

Graph points on a coordinate plane.

Vocabulary

coordinate plane
origin
y-axis
x-axis
quadrant
ordered pair
x-coordinate
y-coordinate

The Coordinate Plane

GPS A GPS, or global positioning system, is a satellite-based navigation system. A GPS map of Tallahassee, Florida, is shown.

1. Suppose Mr. Diaz starts at Carter Howell Strong Park and drives 2 blocks east. Name the street he will cross.

2. Using the words *north*, *south*, *east*, and *west*, write directions to go from Old City Cemetery to Carter Howell Strong Park.

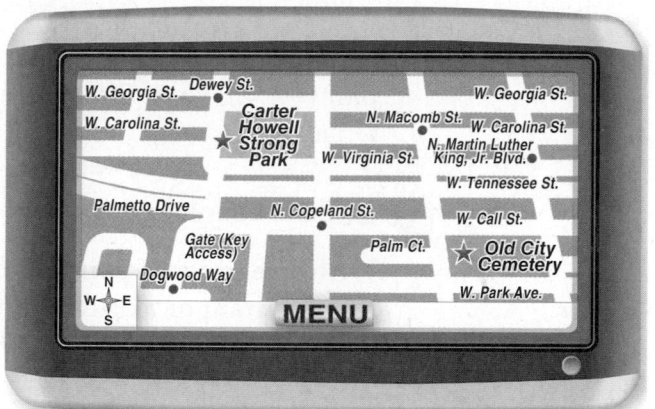

A **coordinate plane** is formed when two number lines intersect. The number lines separate the coordinate plane into four regions called quadrants.

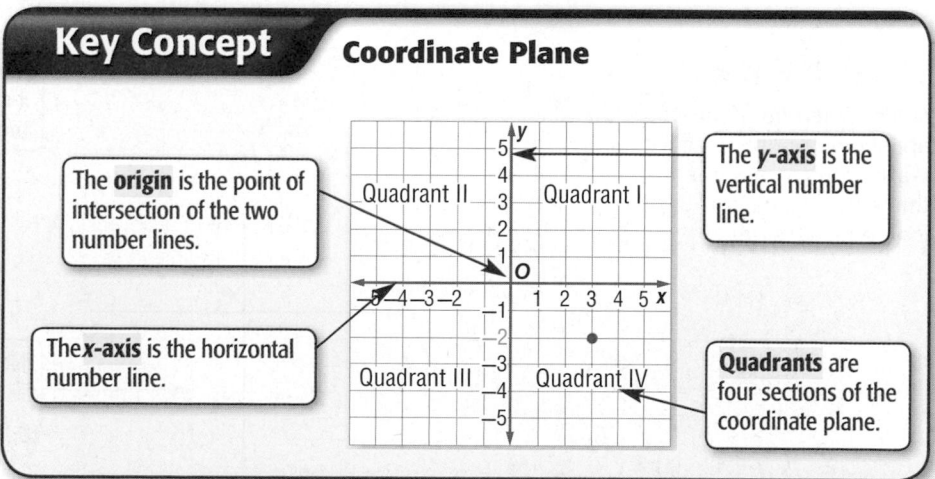

Key Concept — Coordinate Plane

The **origin** is the point of intersection of the two number lines.

The **y-axis** is the vertical number line.

The **x-axis** is the horizontal number line.

Quadrants are four sections of the coordinate plane.

An **ordered pair** is a pair of numbers, such as (3, −2), used to locate a point in the coordinate plane.

The **x-coordinate** corresponds to a number on the x-axis.

$(3, -2)$

The **y-coordinate** corresponds to a number on the y-axis.

When locating an ordered pair, moving *right* or *up* on a coordinate plane is in the *positive* direction. Moving *left* or *down* is in the *negative* direction.

EXAMPLE   Naming Points Using Ordered Pairs

1 Write the ordered pair that corresponds to point *D*. Then state the quadrant or axis on which the point is located.

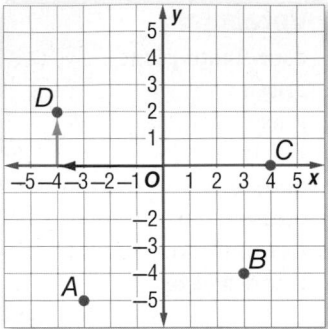

- Start at the origin.
- Move left on the *x*-axis to find the *x*-coordinate of point *D*, which is −4.
- Move up to find the *y*-coordinate, which is 2.

So, point *D* corresponds to the ordered pair (−4, 2). Point *D* is located in Quadrant II.

CHECK Your Progress

Write the ordered pair that corresponds to each point. Then state the quadrant or axis on which the point is located.

a. *A*   **b.** *B*   **c.** *C*

EXAMPLE   Graph an Ordered Pair

2 Graph and label point *K* at (2, −5) on a coordinate plane.

Read Math

Scale When no numbers are shown on the *x*- or *y*-axis, you can assume that each square is 1 unit long on each side.

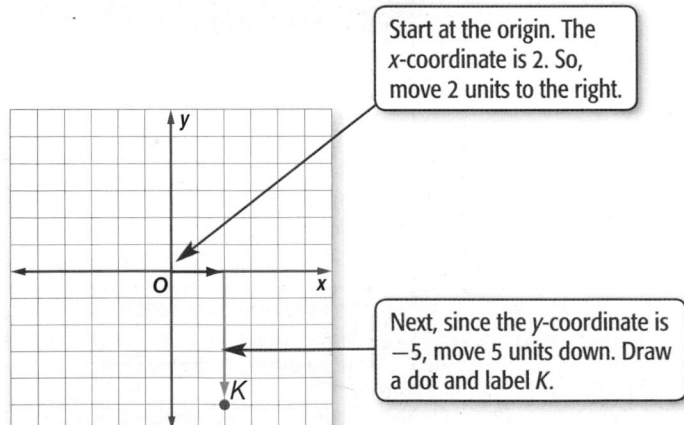

Start at the origin. The *x*-coordinate is 2. So, move 2 units to the right.

Next, since the *y*-coordinate is −5, move 5 units down. Draw a dot and label *K*.

CHECK Your Progress

Graph and label each point on a coordinate plane.

d. *L*(−4, 2)   **e.** *M*(−5, −3)   **f.** *N*(0, 1)

3 **AQUARIUMS** A map can be divided into a coordinate plane where the x-coordinate represents how far to move right or left and the y-coordinate represents how far to move up or down. What exhibit is located at (6, 5)?

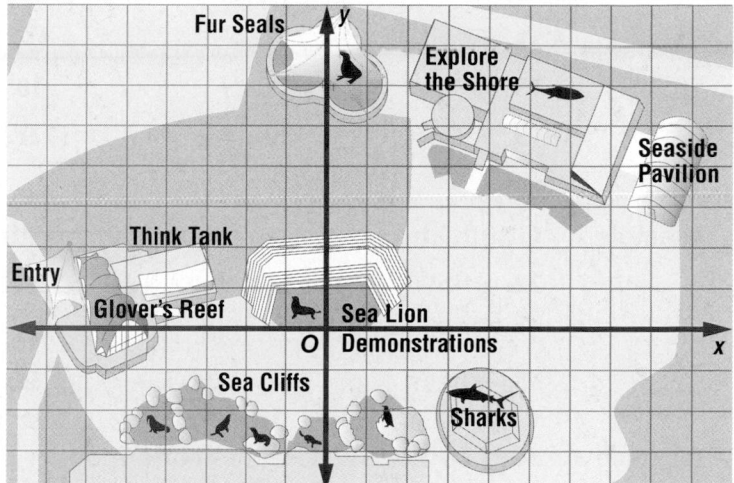

Real-World Link · · · · ·
The leatherback turtle is the world's largest turtle, weighing as much as 1,100 pounds. Its shell can measure over 5 feet long.

Start at the origin. Move 6 units to the right and then 5 units up. Explore the Shore is located at (6, 5).

4 In which quadrant is the Shark Exhibit located?

The Shark Exhibit is located in Quadrant IV.

CHECK Your Progress

g. What ordered pair represents the location of the Think Tank?

h. What exhibit is located at the origin?

CHECK Your Understanding

Example 1 Write the ordered pair corresponding to each point graphed at the right. Then state the quadrant or axis on which each point is located.

1. P 2. Q
3 R 4. S

Example 2 Graph and label each point on a coordinate plane.

5. $T(2, 3)$ 6. $U(-4, 6)$ 7. $V(-5, 0)$ 8. $W(1, -2)$

Examples 3 and 4 9. **GEOGRAPHY** Use the map in Example 3 above.

a. What exhibit is located at (0, −3)?

b. In which quadrant is the Seaside Pavilion located?

Practice and Problem Solving

● = **Step-by-Step Solutions** begin on page R1.
Extra Practice begins on page EP2.

Example 1 Write the ordered pair corresponding to each point graphed at the right. Then state the quadrant or axis on which each point is located.

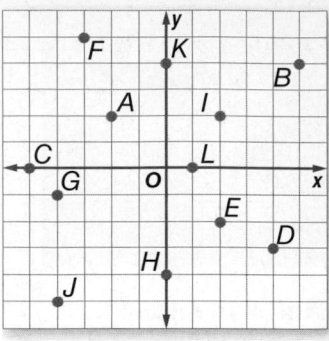

10. A	**11.** B	**12.** C
13. D	**14.** E	**⑮** F
16. G	**17.** H	**18.** I
19. J	**20.** K	**21.** L

Example 2 Graph and label each point on a coordinate plane.

22. $M(5, 6)$	**23.** $N(-2, 10)$	**24.** $P(7, -8)$	**25.** $Q(3, 0)$
26. $R(-1, -7)$	**27.** $S(8, 1)$	**28.** $T(-3, 7)$	**29.** $U(5, -2)$
30. $V(0, 6)$	**31.** $W(-5, -7)$	**32.** $X(-4, 0)$	**33.** $Y(0, -5)$

Example 3 **34. GEOGRAPHY** Use the world map.

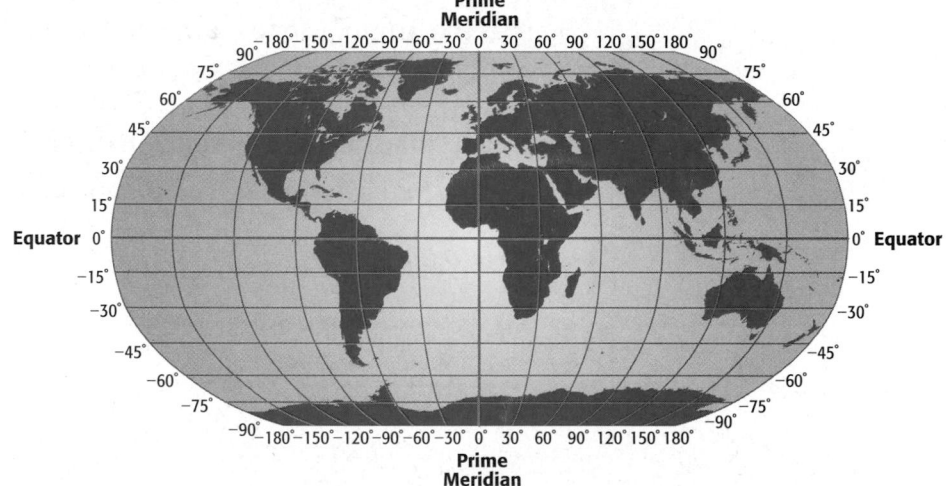

a. The world map can be divided into a coordinate plane where (x, y) represents (degrees longitude, degrees latitude). In what continent is the point (30° longitude, −15° latitude) located?

b. Which of the continents is located entirely in Quadrant II?

c. In what continent is the point (−60° longitude, 0° latitude) located?

d. Name a continent on the map that is located in Quadrant I.

Graph and label each point on a coordinate plane.

㉟ $X(1.5, 3.5)$ **36.** $Y\left(3\frac{1}{4}, 2\frac{1}{2}\right)$ **37.** $Z\left(2, 1\frac{2}{3}\right)$

38. GEOMETRY Graph four points on a coordinate plane so that they form a square when connected. Identify the ordered pairs.

39. RESEARCH Use the Internet or other resources to explain why the coordinate plane is sometimes called the Cartesian plane.

H.O.T. Problems

40. **OPEN ENDED** Create a display that shows how to determine in what quadrant a point is located without graphing. Then provide an example that demonstrates how your graphic is used.

41. **FIND THE ERROR** Hannah is plotting the point $(-3, 4)$. Find her mistake and correct it.

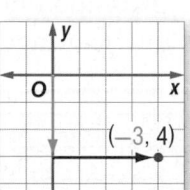

Move 3 units down and then 4 units to the right.

42. **CHALLENGE** Find the possible locations for any ordered pair with x- and y-coordinates always having the same sign. Explain.

43. **WRITE MATH** Explain why the location of point $A(1, -2)$ is different than the location of point $B(-2, 1)$.

Test Practice

44. Which of the following points lie within the triangle?

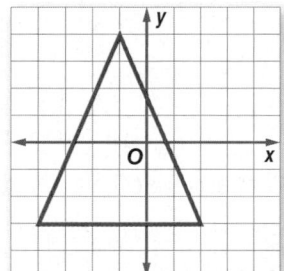

   A. $A(-4, -1)$ **C.** $C(-1, 2)$

   B. $B(1, 3)$ **D.** $D(2, -2)$

45. The location of a given point is 3 units below the origin on the y-axis. What is the ordered pair for this point?

   F. $(3, 0)$ **H.** $(0, -3)$

   G. $(0, 3)$ **I.** $(-3, 0)$

46. Which point **best** represents the location of the lunch room?

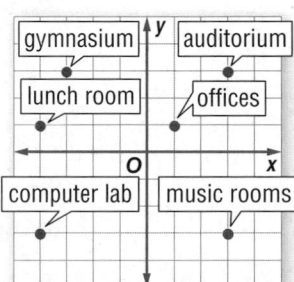

   A. $(4, -1)$ **C.** $(1, 4)$

   B. $(-4, 1)$ **D.** $(1, -4)$

47. Point A has an x-coordinate of -2. If point A's y-coordinate is positive, in which quadrant is point A located?

   F. I   **H.** III

   G. II   **I.** IV

Chapter 2 Lesson 1C Integers and the Coordinate Plane   **85**

Explore Add Integers

Main Idea

Model addition of integers.

Vocabulary

zero pair

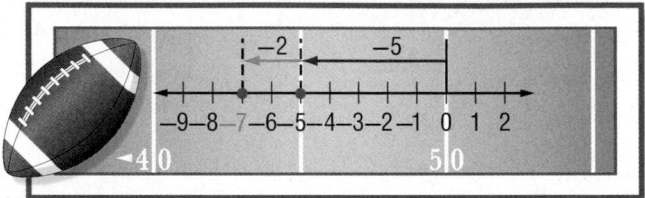 Get ConnectED

CCSS 7.NS.1, 7.NS.3

  FOOTBALL In football, forward progress is represented by a positive integer. Losing yardage is represented by a negative integer. On the first play a team loses 5 yards and on the second play they lose 2 yards. What is the team's total yardage on the two plays?

ACTIVITIES

(1) **What do you need to find?** the total yardage on the two plays

Use counters to find the total yardage.

STEP 1 Combine a set of 5 negative counters and a set of 2 negative counters.

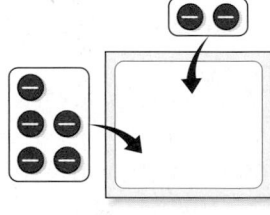

STEP 2 Find the total number of counters.

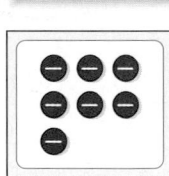

So, $-5 + (-2) = -7$. The football team lost 7 yards on the two plays.

The following two properties are important when modeling operations with integers.

- When one positive counter is paired with one negative counter, the result is called a **zero pair**. The value of a zero pair is 0.

- You can add or remove zero pairs from a mat because adding or removing zero does not change the value of the counters on the mat.

2 **Use counters to find −4 + 2.**

STEP 1 Combine 4 negative counters with 2 positive counters.

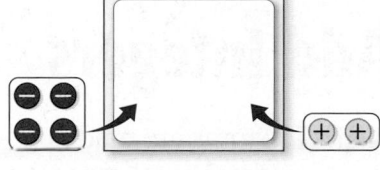

STEP 2 Remove all zero pairs.

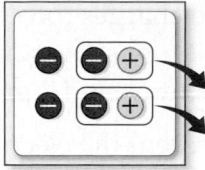

STEP 3 Find the number of counters remaining.

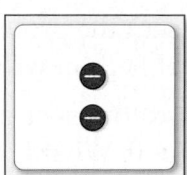

So, −4 + 2 = −2.

Practice and Apply

Use counters or a drawing to find each sum.

1. 5 + 6	**2.** −3 + (−5)	**3.** −5 + (−4)
4. 7 + 3	**5.** −2 + (−5)	**6.** −8 + (−6)
7. −6 + 5	**8.** 3 + (−6)	**9.** −2 + 7
10. 8 + (−3)	**11.** −9 + 1	**12.** −4 + 10

Analyze the Results

13. Write two addition sentences where the sum is positive. In each sentence, one addend should be positive and the other negative.

14. Write two addition sentences where the sum is negative. In each sentence, one addend should be positive and the other negative.

15. Write two addition sentences where the sum is zero. Describe the numbers.

16. GAME SHOWS A contestant has −350 points after the first two rounds of questions. What is his point standing after earning 500 points in the third round of questions?

17. MAKE A CONJECTURE What is a rule you can use to find the sum of two integers with the same sign? two integers with different signs?

Main Idea
Add integers.

 Vocabulary
opposites
additive inverse

 7.NS.1, 7.NS.1a, 7.NS.1b, 7.NS.1d, 7.NS.3, 7.EE.3

Add Integers

SCIENCE Atoms are made of negative charges (electrons) and positive charges (protons). The helium atom shown has a total of 2 electrons and 2 protons.

1. Represent the electrons in an atom of helium with an integer.

2. Represent the protons in an atom of helium with an integer.

3. Each proton-electron pair has a value of 0. What is the total charge of an atom of helium?

Combining protons and electrons in an atom is similar to adding integers.

 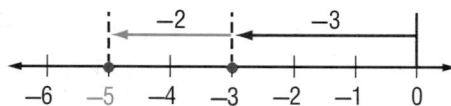 **EXAMPLE** **Add Integers with the Same Sign**

① **Find $-3 + (-2)$.**

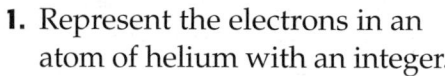

Start at 0. Move 3 units left to show -3.

From there, move 2 units left to show -2.

So, $-3 + (-2) = -5$.

 CHECK Your Progress

a. $-5 + (-7)$

b. $-10 + (-4)$

Key Concept **Add Integers with the Same Sign**

Words To add integers with the same sign, add their absolute values. The sum is:

• positive if both integers are positive.

• negative if both integers are negative.

Examples $7 + 4 = 11$ $-7 + (-4) = -11$

EXAMPLE ~ Add Integers with the Same Sign

2 Find $-26 + (-17)$.

$-26 + (-17) = -43$ Both integers are negative, so the sum is negative.

✓ CHECK Your Progress

c. $-14 + (-16)$ **d.** $23 + 38$

Vocabulary Link

Everyday Use

Opposite something that is across from or is facing the other way, as in running the opposite way

Math Use

Opposite two numbers that are the same distance from 0 on a number line

The integers 5 and -5 are called **opposites** because they are the same distance from 0, but on opposite sides of 0. Two integers that are opposites are also called **additive inverses**.

Key Concept ~ Additive Inverse Property

Words	The sum of any number and its additive inverse is 0.
Example	$5 + (-5) = 0$

When you add integers with different signs, start at zero. Move right for positive integers. Move left for negative integers.

EXAMPLES ~ Add Integers with Different Signs

3 Find $5 + (-3)$.

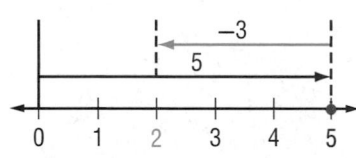

So, $5 + (-3) = 2$.

4 Find $-3 + 2$.

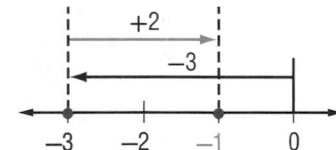

So, $-3 + 2 = -1$.

✓ CHECK Your Progress

e. $6 + (-7)$ **f.** $-15 + 19$

Key Concept ~ Add Integers with Different Signs

Words	To add integers with different signs, subtract their absolute values. The sum is: • positive if the positive integer's absolute value is greater. • negative if the negative integer's absolute value is greater.
Examples	$9 + (-4) = 5$ $-9 + 4 = -5$

Add Integers with Different Signs

5 Find $7 + (-1)$.

$7 + (-1) = 6$ Subtract absolute values; $7 - 1 = 6$. Since 7 has the greater absolute value, the sum is positive.

6 Find $-8 + 3$.

$-8 + 3 = -5$ Subtract absolute values; $8 - 3 = 5$. Since -8 has the greater absolute value, the sum is negative.

QUICKReview

Commutative Properties
$a + b = b + a$
$a \cdot b = b \cdot a$
Associative Properties
$a + (b + c) = (a + b) + c$
$a \cdot (b \cdot c) = (a \cdot b) \cdot c$
Identity Properties
$a + 0 = a$
$a \cdot 1 = a$

7 Find $2 + (-15) + (-2)$.

$$2 + (-15) + (-2) = 2 + (-2) + (-15)$$ Commutative Property (+)
$$= [2 + (-2)] + (-15)$$ Associative Property (+)
$$= 0 + (-15)$$ Additive Inverse Property
$$= -15$$ Additive Identity Property

CHECK Your Progress

g. $10 + (-12)$ **h.** $-13 + 18$ **i.** $(-14) + (-6) + 6$

REAL-WORLD EXAMPLE

8 **ROLLER COASTERS** The graphic shows the change in height at several points on a roller coaster. Write an addition sentence to find the height at point D in relation to point A.

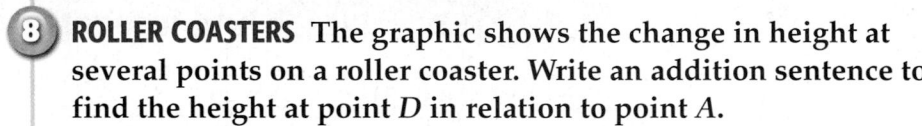
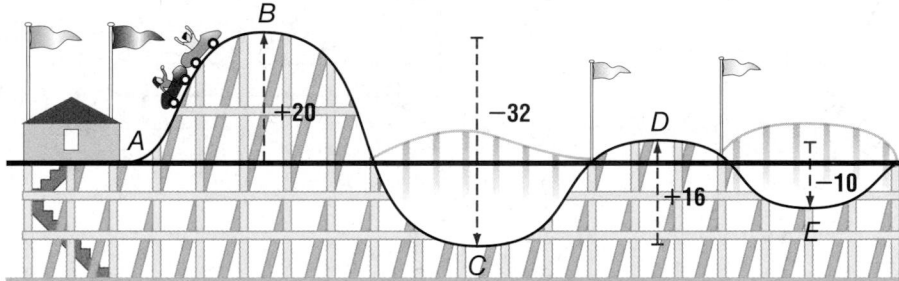

$$20 + (-32) + 16 = 20 + 16 + (-32)$$ Commutative Property (+)
$$= 36 + (-32)$$ $20 + 16 = 36$
$$= 4$$ Subtract absolute values. Since 36 has the greater absolute value, the sum is positive.

Point D is 4 feet higher than point A.

CHECK Your Progress

j. **WEATHER** The temperature is $-3°$F. An hour later, it drops $6°$ and 2 hours later, it rises $4°$. Write an addition sentence to describe this situation. Then find the sum and explain its meaning.

Examples 1–7 Add.

1. $-6 + (-8)$ 2. $4 + 5$ 3. $-3 + 10$

4. $-15 + 8$ 5. $7 + (-11)$ 6. $14 + (-6)$

7 $-17 + 20 + (-3)$ 8. $15 + 9 + (-9)$ 9. $-4 + 12 + (-9)$

Example 8   10. **MONEY** Sofia owes her brother $25. She gives her brother the $18 she earned dog-sitting. Write an addition expression to describe this situation. Then find the sum and explain its meaning.

Practice and Problem Solving

 = **Step-by-Step Solutions** begin on page R1.
Extra Practice begins on page EP2.

Examples 1–7 Add.

11. $-22 + (-16)$ 12. $-10 + (-15)$ 13. $6 + 10$ 14. $17 + 11$

15. $18 + (-5)$ 16. $13 + (-19)$ 17. $-19 + 24$ 18. $-12 + 10$

19. $21 + (-21) + (-4)$ 20. $-8 + (-4) + 12$ 21. $-34 + 25 + (-25)$

22. $-16 + 16 + 22$ 23. $25 + 3 + (-25)$ 24. $7 + (-19) + (-7)$

Example 8   Write an addition expression to describe each situation. Then find each sum and explain its meaning.

25. **SCUBA DIVING** Lena was scuba diving 14 meters below the surface of the water. She saw a nurse shark 3 meters above her.

26. **PELICANS** A pelican starts at 60 feet above sea level. It descends 60 feet to catch a fish.

27 **FINANCIAL LITERACY** Stephanie has $152 in the bank. She withdraws $20. Then she deposits $84.

28. **FOOTBALL** A quarterback is sacked for a loss of 5 yards. On the next play, his team loses 15 yards. Then the team gains 12 yards on the third play.

29. **GRAPHIC NOVEL** Find the total profit or loss for each color of T-shirt.

We are creating T-shirts to be sold for homecoming!

Green T-shirt:
Short-sleeve shirt: $8.00
Printing: $6.00
Selling price: $15.00
White T-shirt:
Long-sleeve shirt: $10.00
Printing: $7.00
Selling price: $20.00
Black T-shirt:
Short-sleeve shirt: $8.00
Printing on Front: $4.00
Printing on Back: $3.00
Selling price: $18.00

30. **OPEN ENDED** Give an example of an addition sentence containing at least four integers with a sum of zero.

31. **CHALLENGE** Name the property illustrated by the following.

    a. $x + (-x) = 0$ **b.** $x + (-y) = -y + x$

CHALLENGE Simplify.

32. $8 + (-8) + a$ 33. $x + (-5) + 1$ 34. $-9 + m + (-6)$ 35. $-1 + n + 7$

36. 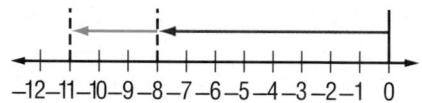 **WRITE MATH** Explain how you know whether a sum is positive, negative, or zero without actually adding.

Test Practice

37. **SHORT RESPONSE** Write an addition sentence to represent the number line below.

    -12 -11 -10 -9 -8 -7 -6 -5 -4 -3 -2 -1 0

38. At 8 A.M., the temperature was 3°F below zero. By 1 P.M., the temperature rose 14°F and by 10 P.M., dropped 12°F. What was the temperature at 10 P.M.?

    A. 5°F above zero

    B. 5°F below zero

    C. 1°F above zero

    D. 1°F below zero

39. What is the value of $-8 + 7 + (-3)$?

    F. -18

    G. -4

    H. 2

    I. 18

40. Which of the following expressions is represented by the number line below?

    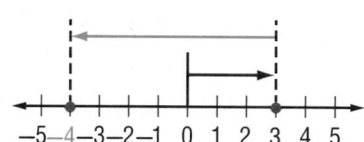
    -5 -4 -3 -2 -1 0 1 2 3 4 5

    A. $-4 + 3$ **C.** $3 + (-7)$

    B. $-4 + 7$ **D.** $0 + (-7)$

Spiral Review

Write the ordered pair corresponding to each point graphed at the right. Then state the quadrant or axis on which each point is located. (Lesson 1C)

41. J 42. K 43. L 44. M

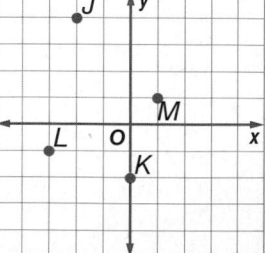

Write an integer for each situation. (Lesson 1B)

45. a bank deposit of $75   46. a loss of 8 pounds

47. 13° below zero   48. a gain of 4 yards

49. spending $12   50. a gain of 5 hours

Explore Subtract Integers

Main Idea

Model subtraction of integers.

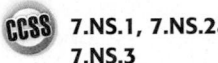

CCSS 7.NS.1, 7.NS.2a, 7.NS.3

DOLPHINS At a local aquarium, a popular attraction features dolphins. Dolphins jump through rings that are 5 meters above the surface of the water. To prepare, they start from 6 meters below the surface of the water. What is the difference between the two distances?

ACTIVITY

1 **What do you need to find?** the difference between the height of the rings and the depth at which the dolphins start

Use counters to find 5 − (−6), the difference between the two distances.

STEP 1 Place 5 positive counters on the mat. Remove 6 negative counters. However, there are 0 negative counters.

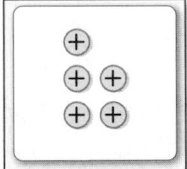

STEP 2 Add 6 zero pairs to the mat.

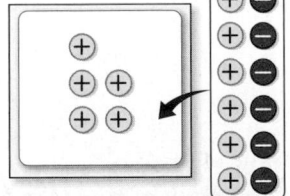

STEP 3 Now you can remove 6 negative counters. Find the remaining number of counters.

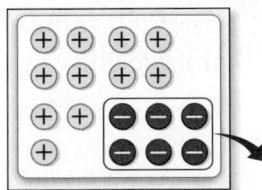

So, 5 − (−6) = 11. The difference between the two distances is 11 meters.

Use counters to find each difference.

2 −6 − (−3)

> **STEP 1** Place 6 negative counters on the mat.
>
> **STEP 2** Remove 3 negative counters.

So, −6 − (−3) = −3.

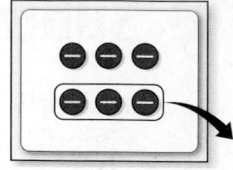

3 −5 − 1

> **STEP 1** Place 5 negative counters on the mat. Remove 1 positive counter. However, there are 0 positive counters.

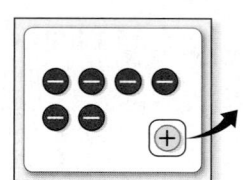

> **STEP 2** Add 1 zero pair to the mat.

> **STEP 3** Now you can remove 1 positive counter. Find the remaining number of counters.

So, −5 − 1 = −6.

Practice and Apply

Use counters or a drawing to find each difference.

1. 7 − 6   **2.** 5 − (−3)   **3.** 6 − (−3)   **4.** 5 − 8

5. −7 − (−2)   **6.** −7 − 3   **7.** −5 − (−7)   **8.** −2 − 9

Analyze the Results

9. Write two subtraction sentences where the difference is positive. Use a combination of positive and negative integers.

10. Write two subtraction sentences where the difference is negative. Use a combination of positive and negative integers.

11. **E⌧ WRITE MATH** Jake owes his sister $3. She decides to "take away" his debt. That is, he does not have to pay her back. Write a subtraction sentence for this situation. Explain why subtracting a negative integer is the same as adding a positive integer.

Main Idea

Subtract integers.

 7.NS.1, 7.NS.1c, 7.NS.3, 7.EE.3

Subtract Integers

Explore You can use a number line to model a subtraction sentence.

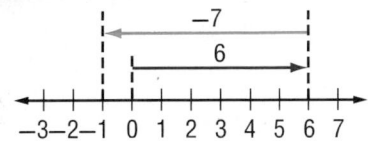

1. Write a related addition sentence for the subtraction sentence.

Use a number line to find each difference. Write an equivalent addition sentence for each.

2. $1 - 5$ **3.** $-2 - 1$ **4.** $-3 - 4$ **5.** $0 - 5$

When you subtract 7, the result is the same as adding its opposite, -7.

opposite

$$6 - 7 = -1 \qquad 6 + (-7) = -1$$

same result

Key Concept   **Subtract Integers**

Words   To subtract an integer, add its opposite.

Examples $4 - 9 = 4 + (-9) = -5$ $7 - (-10) = 7 + (10) = 17$

EXAMPLES **Subtract Positive Integers**

① Find $8 - 13$.

$8 - 13 = 8 + (-13)$ To subtract 13, add -13.

$\qquad = -5$ Simplify.

② Find $-10 - 7$.

$-10 - 7 = -10 + (-7)$ To subtract 7, add -7.

$\qquad = -17$ Simplify.

 CHECK Your Progress

a. $6 - 12$ **b.** $-20 - 15$ **c.** $-22 - 26$

EXAMPLES **Subtract Negative Integers**

3 Find $1 - (-2)$.

$1 - (-2) = 1 + 2$ To subtract −2, add 2.

$\quad\quad\quad = 3$ Simplify.

4 Find $-10 - (-7)$.

$-10 - (-7) = -10 + 7$ To subtract −7, add 7.

$\quad\quad\quad\quad = -3$ Simplify.

CHECK Your Progress

d. $4 - (-12)$ **e.** $-15 - (-5)$ **f.** $18 - (-6)$

EXAMPLE **Evaluate an Expression**

5 **ALGEBRA** Evaluate $x - y$ if $x = -6$ and $y = -5$.

$x - y = -6 - (-5)$ Replace x with −6 and y with −5.

$\quad\quad = -6 + (5)$ To subtract −5, add 5.

$\quad\quad = -1$ Simplify.

CHECK Your Progress

Evaluate each expression if $a = 5$, $b = -8$, and $c = -9$.

g. $b - 10$ **h.** $a - b$ **i.** $c - a$

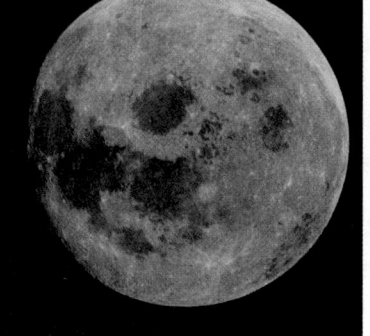

Real-World Link · · · ·
The mean surface
temperature on the
Moon during the day
is 107°C.

REAL-WORLD EXAMPLE

6 **SPACE** The temperatures on the Moon vary from −173°C to
127°C. Find the difference between the maximum and minimum
temperatures.

Subtract the lower temperature from the higher temperature.

Estimate $100 + 200 = 300$

$127 - (-173) = 127 + 173$ To subtract −173, add 173.

$\quad\quad\quad\quad = 300$ Simplify.

So, the difference between the temperatures is 300°C.

CHECK Your Progress

j. **MONEY** Brenda had a balance of −$52 in her account. The bank
charged her a fee of $10 for having a negative balance. What is
her new balance?

✓ CHECK Your Understanding

Examples 1–4   Subtract.

1. $14 - 17$

2. $10 - 30$

3 $-4 - 8$

4. $-2 - 23$

5. $14 - (-10)$

6. $5 - (-16)$

7. $-3 - (-1)$

8. $-11 - (-9)$

Example 5   **ALGEBRA** Evaluate each expression if $p = 8$, $q = -14$, and $r = -6$.

9. $r - 15$

10. $q - r$

11. $p - q$

Example 6   **12. EARTH SCIENCE** The sea surface temperatures range from $-2°C$ to $31°C$. Find the difference between the maximum and minimum temperatures.

Practice and Problem Solving

 = **Step-by-Step Solutions** begin on page R1.
Extra Practice begins on page EP2.

Examples 1–4   Subtract.

13. $0 - 10$

14. $13 - 17$

15. $-9 - 5$

16. $-8 - 9$

17. $12 - 26$

18. $31 - 48$

19. $-25 - 5$

20. $-44 - 41$

21. $4 - (-19)$

22. $27 - (-8)$

23. $-11 - (-42)$

24. $-27 - (-19)$

25. $52 - (-52)$

26. $15 - (-14)$

27. $-27 - (-33)$

28. $-18 - (-20)$

Example 5   **ALGEBRA** Evaluate each expression if $f = -6$, $g = 7$, and $h = 9$.

29. $g - 7$

30. $f - 6$

31. $-h - (-9)$

32. $f - g$

33. $h - f$

34. $g - h$

35. $5 - f$

36. $4 - (-g)$

Example 6   **37 ANALYZE TABLES** Use the information below.

State	Alabama	California	Florida	Louisiana	New Mexico
Lowest Elevation (ft)	0	−282	0	−8	2,842
Highest Elevation (ft)	2,407	14,494	345	535	13,161

a. What is the difference between the highest elevation in Alabama and the lowest elevation in Louisiana?

b. Find the difference between the lowest elevation in New Mexico and the lowest elevation in California.

c. Find the difference between the highest elevation in Florida and the lowest elevation in California.

d. What is the difference between the lowest elevation in Alabama and in Louisiana?

ALGEBRA Evaluate each expression if $h = -12$, $j = 4$, and $k = 15$.

38. $-j + h - k$

39. $|h - j|$

40. $k - j - h$

41. OPEN ENDED Write a subtraction sentence using integers. Then, write the equivalent addition sentence and explain how to find the sum.

42. FIND THE ERROR Hiroshi is finding $-15 - (-18)$. Find his mistake and correct it.

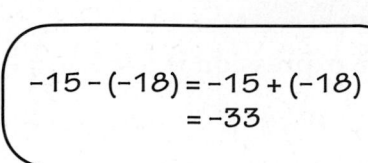

$$-15 - (-18) = -15 + (-18)$$
$$= -33$$

43. CHALLENGE *True* or *False*? When n is a negative integer, $n - n = 0$.

44.  **WRITE MATH** If x and y are positive integers, is $x - y$ always positive? Explain.

Test Practice

45. Which sentence about integers is NOT always true?

   A. positive − positive = positive

   B. positive + positive = positive

   C. negative + negative = negative

   D. positive − negative = positive

46. Morgan drove from Los Angeles (elevation 330 feet) to Death Valley (elevation −282 feet). What is the difference in elevation between Los Angeles and Death Valley?

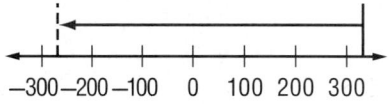

   F. 48 feet   **H.** 582 feet

   G. 148 feet   **I.** 612 feet

Spiral Review

Add. (Lesson 2B)

47. $10 + (-3)$ **48.** $-2 + (-9)$ **49.** $-7 + (-6)$ **50.** $-18 + 4$

51. State the quadrant in which the graphed point is located.
(Lesson 1C)

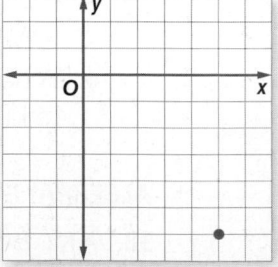

Evaluate each expression. (Lesson 1B)

52. $|-12|$

53. $|-3| + |-5|$

54. $|-25| \div 5 - |-3|$

Mid-Chapter Check

Write an integer for each situation. (Lesson 1B)

1. dropped 45 feet

2. a bank deposit of $100

3. gained 8 pounds

4. lost a $5 bill

5. **OCEANS** The deepest point in the world is the Mariana Trench in the Western Pacific Ocean at a depth of 35,840 feet below sea level. Write this depth as an integer. (Lesson 1B)

Evaluate each expression. (Lesson 1B)

6. $|-16|$

7. $|24|$

8. $|-9| - |3|$

9. $|-13| + |-1|$

Graph and label each point on a coordinate plane. (Lesson 1C)

10. $D(4, -3)$

11. $E(-1, 2)$

12. $F(0, -5)$

13. $G(-3, 0)$

14. **MULTIPLE CHOICE** Which line contains the ordered pair $(-1, 4)$? (Lesson 1C)

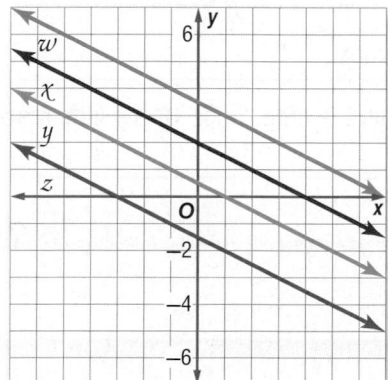

    A. line w **C.** line y

    B. line x **D.** line z

Add. (Lesson 2B)

15. $3 + 4 + (-3)$

16. $7 + (-11)$

17. $-5 + (-6)$

18. $8 + (-1) + 1$

19. **MULTIPLE CHOICE** Kendra deposited $78 into her savings account. Two weeks later, she deposited a check for $50 into her account and withdrew $27. Which of the following expressions represents the amount of money left in her account? (Lesson 2B)

    F. $78 + (-$50) + (-$27)

    G. $78 + (-$50) + $27

    H. $78 + $50 + (-$27)

    I. $78 + $50 + $27

Subtract. (Lesson 2D)

20. $9 - 15$

21. $-3 - 10$

22. $8 - (-12)$

23. $-4 - (-13)$

24. **ANALYZE TABLES** The table shows the record low temperatures for Des Moines, Iowa.

Month	Temperature (°F)
January	−24
May	30

What is the difference between the low temperature for January and the low temperature for May? (Lesson 2D)

25. **CHEMISTRY** The melting point of mercury is $-36°F$ and its boiling point is $672°F$. What is the difference between these two temperatures? (Lesson 2D)

26. **MULTIPLE CHOICE** During two plays of a football game, the Valley Tigers had a loss of 4 yards and a gain of 16 yards. Which of the following expressions represents the difference between the number of yards? (Lesson 2D)

    A. $-4 + 16$ **C.** $4 + 16$

    B. $-4 + (-16)$ **D.** $4 - 16$

Problem-Solving Investigation

Main Idea Solve problems by looking for a pattern.

🏃 📝 P.S.I. TERM ✛

Look for a Pattern

LAURA: I've been practicing free throws every day after school to get ready for basketball tryouts. Now, I can make 3 free throws out of every 5 attempts.

YOUR MISSION: Look for a pattern to find the number of free throws Laura can make after 30 attempts.

Understand	Laura can make an average of 3 free throws out of every 5 attempts. You need to find the number of free throws she can make after 30 attempts.
Plan	Look for a pattern. Then extend the pattern to find the solution.
Solve	Laura can make 3 free throws out of every 5 she attempts. Extend the pattern.

+3   +3   +3   +3   +3

Free throws	3	6	9	12	15	18
Attempts	5	10	15	20	25	30

+5   +5   +5   +5   +5

She can make 18 free throws out of 30 attempts.

Check	She makes free throws a little more than half the time. Since 18 is a little more than 15, the answer is reasonable. ✓

Analyze the Strategy

1. Suppose Laura can make 4 out of 5 attempts. Find the number of free throws she can make after 30 attempts.

- Look for a pattern.
- Guess, check, and revise.
- Make a list.
- Choose an operation.

 = **Step-by-Step Solutions** begin on page R1.
Extra Practice begins on page EP2.

Use the *look for a pattern* strategy to solve Exercises 2–4.

2. **DISPLAYS** A display of cereal boxes is stacked as shown below.

If the display contains 7 rows of boxes and the top three rows are shown, how many boxes are in the display?

3. **FINANCIAL LITERACY** Peter is saving money to buy an MP3 player. After one month, he has $50. After 2 months, he has $85. After 3 months, he has $120. After 4 months, he has $155. He plans to keep saving at the same rate. How long will it take Peter to save enough money to buy an MP3 player that costs $295?

4. **INSECTS** The table shows how many times a cricket chirps at different temperatures. If this pattern continues, about how many times will a cricket chirp when the temperature is 60°F?

Outside Temperature (°F)	Chirps per Minute
70	120
75	140
80	160
85	180

Use any strategy to solve Exercises 5–10.

5. **COINS** Adelina has exactly six coins that total $0.86. What are the coins?

6. **DIVING** The table shows a diver's position after several minutes. If she keeps descending at this rate, find the diver's position after ten minutes.

Time (min)	Position (ft)
1	−15
2	−30
3	−45

7. **GEOMETRY** The pattern below is made from toothpicks. How many toothpicks would be needed for the sixth term in the pattern?

  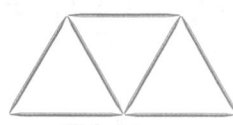

First term　　**Second term**　　**Third term**

8. **GOLF** Allie's golf scores for the first five holes are given in the table. What is her total score after the first five holes?

Hole	Score
1	0
2	1
3	−1
4	−2
5	3

9. **NATURE** A sunflower usually has two different spirals of seeds, one with 34 seeds and the other with 55 seeds.

The numbers 34 and 55 are part of the Fibonacci sequence: 1, 1, 2, 3, 5, 8, 13, 21, 34, 55, … . Find the pattern in the Fibonacci sequence and identify the next two terms.

10. **WRITE MATH** Write a real-world problem that could be solved by looking for a pattern. Then solve.

Explore **Multiply and Divide Integers**

Main Idea

Model multiplication and division of integers.

7.NS.2, 7.NS.3

SCHOOL The number of students who bring their lunch to Phoenix Middle School has been decreasing at a rate of 4 students per month. What integer represents the total change after three months?

ACTIVITY

1 **What do you need to find?** the total change in the number of students who bring their lunch to school

The integer −4 represents a decrease of 4 students each month. After three months, the total change will be $3 \times (-4)$. Use counters to model 3 groups of 4 negative counters.

STEP 1 Place 3 sets of 4 negative counters on the mat.

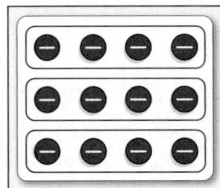

STEP 2 Count the number of negative counters.

The integer −12 represents the total change in the number of students who bring their lunch to school.

If the first factor is *negative*, you will need to remove counters from the mat.

ACTIVITY

2 **Use counters to find −2 × (−4).**

The expression $-2 \times (-4)$ means to *remove* 2 sets of 4 negative counters.

STEP 1 Place 2 sets of 4 zero pairs on the mat.

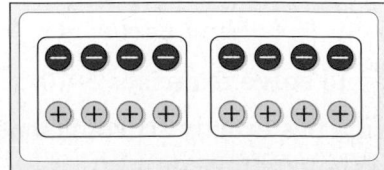

STEP 2 Then remove 2 sets of 4 negative counters from the mat. There are 8 positive counters remaining.

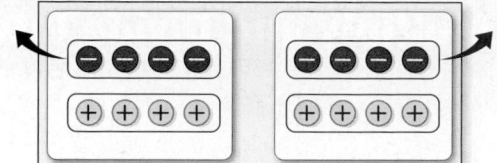

So, $-2 \times (-4) = 8$.

You can model division by separating algebra counters into equal-size groups.

ACTIVITY

3 **Model $-9 \div 3$ using algebra counters.**

STEP 1 Place 9 negative counters on the mat to represent -9.

STEP 2 Separate the counters into 3 equal-size groups. There are 3 negative counters in each of the three groups.

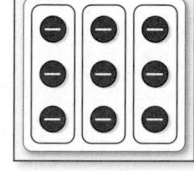

So, $-9 \div 3 = -3$.

Practice and Apply

Find each product. Use models if needed.

1. $7 \times (-2)$ **2.** $2 \times (-3)$ **3.** $4 \times (-4)$ **4.** $8 \times (-1)$

5. $-5 \times (-1)$ **6.** $-2 \times (-2)$ **7.** $-4 \times (-3)$ **8.** $-6 \times (-2)$

Find each quotient. Use models if needed.

9. $-12 \div 4$ **10.** $-18 \div 9$ **11.** $-20 \div 5$ **12.** $-10 \div 2$

13. $-6 \div 6$ **14.** $-14 \div 7$ **15.** $-16 \div 4$ **16.** $-8 \div 2$

Analyze the Results

17. How are the operations -5×4 and $4 \times (-5)$ the same? How do they differ?

18. MAKE A CONJECTURE Write a rule you can use to find the sign of the product of two integers given the sign of both factors. Justify your rule.

19. When the dividend is negative and the divisor is positive, is the quotient positive or negative? How does this compare to a multiplication problem when one factor is positive and one is negative?

20. MAKE A CONJECTURE Write a rule you can use to find the sign of the quotient of two integers. Justify your rule.

Main Idea

Multiply integers.

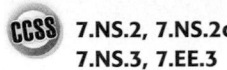

CCSS 7.NS.2, 7.NS.2c, 7.NS.3, 7.EE.3

Multiply Integers

COOKING Genevieve's freezer decreases the temperature of a piece of pie by 3° every minute. What is the change in degrees after 4 minutes?

1. Write a multiplication sentence to represent the situation above.

2. How could you model the multiplication sentence above with counters?

Remember that multiplication is the same as repeated addition.

$4(-3) = (-3) + (-3) + (-3) + (-3)$ −3 is used as an addend four times.

$ = -12$

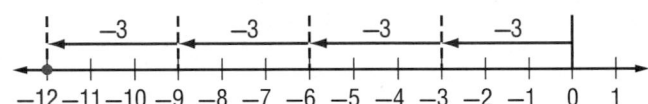

The Commutative Property of Multiplication states that you can multiply in any order. So, $4(-3) = -3(4)$.

Key Concept | **Multiply Integers with Different Signs**

Words   The product of two integers with different signs is negative.

Examples $6(-4) = -24$ $-5(7) = -35$

EXAMPLES **Multiply Integers with Different Signs**

1 Find $3(-5)$.

   $3(-5) = -15$ The integers have different signs. The product is negative.

2 Find $-6(8)$.

   $-6(8) = -48$ The integers have different signs. The product is negative.

CHECK Your Progress

   a. $9(-2)$ **b.** $-7(4)$

The product of two positive integers is positive. You can use a pattern to find the sign of the product of two negative integers.

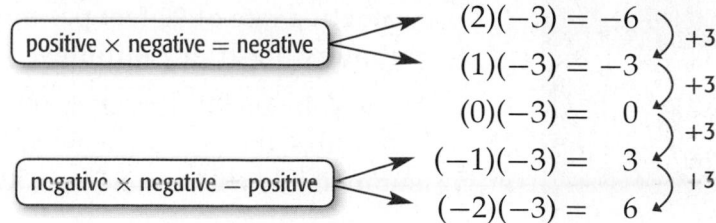

Study Tip

Multiplying by Zero
The Multiplicative Property
of Zero states that when
any number is multiplied by
zero, the product is zero.

Each product is 3 more than the previous product. This pattern can also be shown on a number line.

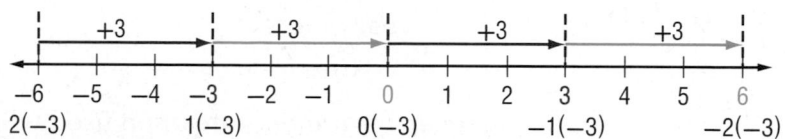

Key Concept   **Multiply Integers with the Same Sign**

Words   The product of two integers with the same sign is positive.

Examples $2(6) = 12$ $-10(-6) = 60$

📝 **EXAMPLES** **Multiply Integers with the Same Sign**

3 **Find $-11(-9)$.**

$-11(-9) = 99$ The integers have the same sign. The product is positive.

QUICK Review

Exponents

When expressing a
number using exponents,
the base is the factor and
the exponent tells how
many times to use the
base as a factor.
$a^3 = a \cdot a \cdot a$

4 **Find $(-4)^2$.**

$(-4)^2 = (-4)(-4)$ There are two factors of -4.

$\quad\quad = 16$ The product is positive.

5 **Find $-3(-4)(-2)$.**

$-3(-4)(-2) = [\mathbf{-3(-4)}](-2)$ Associative Property

$\quad\quad\quad = \mathbf{12}(-2)$ $-3(-4) = 12$

$\quad\quad\quad = -24$ $12(-2) = -24$

✏️ **CHECK Your Progress**

c. $-12(-4)$ **d.** $(-5)^2$ **e.** $-7(-5)(-3)$

REAL-WORLD EXAMPLE

6 **SUBMERSIBLES** A submersible is diving from the surface of the water at a rate of 90 feet per minute. What is the depth of the submersible after 7 minutes?

If the submersible descends 90 feet per minute, then after 7 minutes, the vessel will be at 7(−90) or −630 feet. The submersible will descend to 630 feet below the surface.

CHECK Your Progress

f. FINANCIAL LITERACY Mr. Simon's bank automatically deducts a $4 monthly maintenance fee from his savings account. What integer represents the change in his savings account from one year of fees?

Negative numbers are often used when evaluating algebraic expressions.

EXAMPLE **Evaluate Expressions**

7 **ALGEBRA** Evaluate pqr if $p = -3$, $q = 4$, and $r = -1$.

$$pqr = -3(4)(-1) \qquad \text{Replace } p \text{ with } -3, q \text{ with } 4, \text{ and } r \text{ with } -1.$$
$$= (-12)(-1) \qquad \text{Multiply } -3 \text{ and } 4.$$
$$= 12 \qquad \text{Multiply } -12 \text{ and } -1.$$

CHECK Your Progress

g. Evaluate xyz if $x = -7$, $y = -4$, and $z = 2$.

CHECK Your Understanding

Examples 1 and 2   Multiply.

1. $6(-10)$ 2. $11(-4)$ 3. $-2(14)$ 4. $-8(5)$

Examples 3–5   Multiply.

5. $-15(-3)$ 6. $-7(-9)$ 7. $(-8)^2$

8. $(-3)^3$ 9. $-1(-3)(-4)$ 10. $2(4)(5)$

Example 6   **11. MONEY** Tamera owns 100 shares of a certain stock. Suppose the price of the stock drops by $3 per share. Write a multiplication expression to find the change in Tamera's investment. Explain your answer.

Example 7   **ALGEBRA** Evaluate each expression if $f = -1$, $g = 7$, and $h = -10$.

12. $5f$ **13** fgh

● = **Step-by-Step Solutions** begin on page R1.
Extra Practice begins on page EP2.

Examples 1–5 Multiply.

14. $8(-12)$ 15. $11(-20)$ 16. $-15(4)$ 17. $-7(10)$

18. $-7(11)$ 19. $25(-2)$ 20. $-20(-8)$ 21. $-16(-5)$

22. $(-6)^2$ 23. $(-5)^3$ 24. $(-4)^3$ 25. $(-9)^2$

26. $-4(-2)(-8)$ 27. $-3(-2)(1)$ 28. $-9(-1)(-5)$ 29. $-4(3)(-2)$

Example 6 Write a multiplication expression to represent each situation. Then find each product and explain its meaning.

30. **ECOLOGY** Wave erosion causes a certain coastline to recede at a rate of 3 centimeters each year. This occurs uninterrupted for a period of 8 years.

31 **EXERCISE** Ethan burns 650 Calories when he runs for 1 hour. Suppose he runs 5 hours in one week.

Example 7 **ALGEBRA** Evaluate each expression if $w = 4$, $x = -8$, $y = 5$, and $z = -3$.

32. $-4w$ 33. $7wz$ 34. $-2wx$ 35. wyx

36. **🔀 MULTIPLE REPRESENTATIONS** When a movie is rented it has a due date. If the movie is not returned on time, a late fee is assessed. Kaitlyn is charged \$5 each day for a movie that is 4 days late.

  a. **WORDS** Explain why $4 \times (-5) = -20$ describes the situation.

  b. **ALGEBRA** Write an expression to represent the fee when the movie is x days late.

  c. **WORDS** Kaitlyn resolved the problem of the late movie and did not have to pay the late fees. Explain why $(-4) \times (-5) = 20$ represents this situation. Why is the product a positive integer?

ALGEBRA Evaluate each expression if $a = -6$, $b = -4$, $c = 3$, and $d = 9$.

37. $-3a^2$ 38. $-cd^2$ **39** $-2a + b$ 40. $b^2 - 4ac$

41. **GRAPHIC NOVEL** Refer to the graphic novel frame below. How many T-shirts would Hannah and Dario need to sell to make up the loss in profit?

Refer to the graphic novel at the start of the chapter.

It looks like we spend \$15 on every black T-shirt that is printed on both back and front. We should give one to the mascot for free.

42. OPEN ENDED Write a multiplication sentence with a product of −18.

43. NUMBER SENSE Explain how to evaluate each expression as simply as possible.

    a. $(-9)(-6)(15)(-7+7)$ **b.** $(-15)(-26)+(-15)(25)$

44. CHALLENGE Evaluate $(-1)^{50}$. Explain your reasoning.

45. WRITE MATH Explain when the product of three integers is positive.

Test Practice

46. The temperature drops 2 degrees per hour for 3 hours. Which expression does NOT describe the change in temperature?

    A. $-2(3)$

    B. $-2+(-2)+(-2)$

    C. $-2-2-2$

    D. $2(3)$

47. GRIDDED RESPONSE Which number is the seventh number in the sequence shown?

Position	1	2	3	4	5	6	7
Number	1	−2	4	−8	16	?	?

48. ANT FARMS The table below shows the number of ants in an ant farm on different days. The number of ants doubles every ten days. (Lesson 3A)

Day	51	61	71
Number of Ants	320	640	1,280

    a. How many ants were in the farm on Day 1?

    b. How many ants will be in the farm on Day 91?

49. VIDEO GAMES Nieves and her three friends are playing a video game. The table shows their scores at the end of the first round. (Lesson 2D)

    a. What is the difference between the highest and lowest scores?

    b. By how many points is Nieves losing to Polly?

Player	Score
Nieves	−189
Polly	−142
Saul	230
Harry	−48

Add. (Lesson 2B)

50. $-12+(-19)$ **51.** $-8+(-11)$ **52.** $-5+6$ **53.** $7+(-11)$

Main Idea
Divide integers.

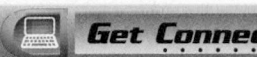

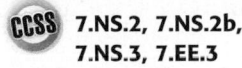

CCSS 7.NS.2, 7.NS.2b, 7.NS.3, 7.EE.3

Divide Integers

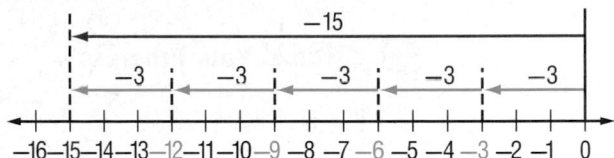 **Explore** You can find the product of $5 \times (-3)$ on a number line. Start at zero and then move -3 units five times.

1. What is the product of $5 \times (-3)$?

2. What division sentence is also shown on the line?

3. Draw a number line to find the product of $4 \times (-2)$. Then find the related division sentence.

Division of numbers is related to multiplication. When finding the quotient of two integers, you can use a related multiplication sentence.

The factor in the multiplication sentence ...		... is the quotient in the division sentence.
$2(6) = 12 \longrightarrow$	$12 \div 2 = 6$	
$4(5) = 20 \longrightarrow$	$20 \div 4 = 5$	

Since multiplication and division sentences are related, you can use them to find the quotient of integers with different signs.

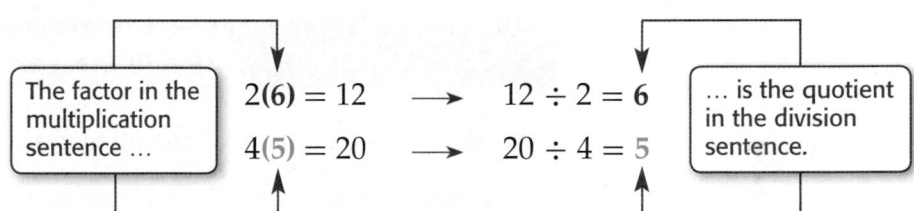

different signs

$2(-6) = -12 \longrightarrow -12 \div 2 = -6$
$-2(-6) = 12 \longrightarrow 12 \div (-2) = -6$ } negative quotient

---

Key Concept	**Divide Integers with Different Signs**

Words   The quotient of two integers with different signs is negative.

Examples $33 \div (-11) = -3$ $-64 \div 8 = -8$

EXAMPLES **Divide Integers with Different Signs**

1 Find $80 \div (-10)$. The integers have different signs.

$80 \div (-10) = -8$ The quotient is negative.

2 Find $\dfrac{-55}{11}$. The integers have different signs.

$\dfrac{-55}{11} = -5$ The quotient is negative.

CHECK Your Progress

a. $20 \div (-4)$ **b.** $\dfrac{-81}{9}$ **c.** $-45 \div 9$

Study Tip

Dividing Integers
Division of integers with same or different signs follows the same rules as the ones for multiplication.

You can also use multiplication and division sentences to find the quotient of integers with the same sign.

same signs	$4(5) = 20$ $\longrightarrow$	$20 \div 4 = 5$
	$-4(5) = -20$ $\longrightarrow$	$-20 \div (-4) = 5$

positive quotient

Key Concept **Divide Integers with the Same Sign**

Words The quotient of two integers with the same sign is positive.

Examples $15 \div 5 = 3$ $-64 \div (-8) = 8$

EXAMPLES **Divide Integers with the Same Sign**

3 Find $-14 \div (-7)$. The integers have the same sign.

$-14 \div (-7) = 2$ The quotient is positive.

4 **ALGEBRA** Evaluate $-16 \div x$ if $x = -4$.

$-16 \div x = -16 \div (-4)$ Replace x with -4.

$ = 4$ Divide. The quotient is positive.

CHECK Your Progress

d. $-24 \div (-4)$ **e.** $-9 \div (-3)$ **f.** $\dfrac{-28}{-7}$

g. **ALGEBRA** Evaluate $a \div b$ if $a = -33$ and $b = -3$.

 REAL-WORLD EXAMPLE

5 **ANIMALS** One year, the estimated Australian koala population was 1,000,000. After 10 years, there were about 100,000 koalas. Find the average change in the koala population per year.

$$\frac{N - P}{10} = \frac{100,000 - 1,000,000}{10}$$ *N* is the new population, 100,000. *P* is the previous population, 1,000,000.

$$= \frac{-900,000}{10} \text{ or } -90,000$$ Divide.

The koala population has changed by −90,000 per year.

 CHECK Your Progress

h. WEATHER The average temperature in January for North Pole, Alaska, is −24°C. Use the expression $\frac{9C + 160}{5}$ to find this temperature in degrees Fahrenheit. Round to the nearest degree.

Key Concept **Operations with Integers**

Operation	Rule
Add	**Same Signs:** Add absolute values. The sum has the same sign as the integers. **Different Signs:** Subtract absolute values. The sum has the sign of the integer with the greater absolute value.
Subtract	To subtract an integer, add its opposite.
Multiply and Divide	**Same Signs:** The product or quotient is positive. **Different Signs:** The product or quotient is negative.

 CHECK Your Understanding

Examples 1–3 Divide.

1. $32 \div (-8)$ **2.** $-16 \div 2$ **3.** $\frac{42}{-7}$

4. $-30 \div (-5)$ **5.** $55 \div 11$ **6.** $\frac{-16}{-4}$

Example 4 **ALGEBRA** Evaluate each expression if $x = 8$ and $y = -5$.

7 $15 \div y$ **8.** $xy \div (-10)$

Example 5 **9. TEMPERATURE** The lowest recorded temperature in Wisconsin is −55°F on February 4, 1996. Use the expression $\frac{5(F - 32)}{9}$ to find this temperature in degrees Celsius. Round to the nearest tenth.

Practice and Problem Solving

● = **Step-by-Step Solutions** begin on page R1.
Extra Practice begins on page EP2.

Examples 1–3 Divide.

10. $50 \div (-5)$ **11.** $56 \div (-8)$ **12.** $-18 \div 9$ **13.** $-36 \div 4$

14. $-15 \div (-3)$ **15.** $-100 \div (-10)$ **16.** $\dfrac{22}{-2}$ **17.** $\dfrac{84}{-12}$

18. $\dfrac{-26}{13}$ **19.** $\dfrac{-27}{3}$ **20.** $\dfrac{-21}{-7}$ **㉑** $\dfrac{-54}{-6}$

22. Divide -200 by -100.   **23.** Find the quotient of -65 and -13.

Example 4 **ALGEBRA** Evaluate each expression if $r = 12$, $s = -4$, and $t = -6$.

24. $-12 \div r$ **25.** $72 \div t$ **26.** $r \div s$ **27.** $rs \div 16$

28. $\dfrac{t - r}{3}$ **29.** $\dfrac{8 - r}{-2}$ **30.** $\dfrac{s + t}{5}$ **31.** $\dfrac{t + 9}{-3}$

Example 5 **32. FINANCIAL LITERACY** Last year, Mr. Engle's total income was $52,000, while his total expenses were $53,800. Use the expression $\dfrac{I - E}{12}$, where I represents total income and E represents total expenses, to find the average difference between his income and expenses each month.

33. SCIENCE The boiling point of water is affected by changes in elevation. Use the expression $\dfrac{-2A}{1,000}$, where A represents the altitude in feet, to find the number of degrees Fahrenheit at which the boiling point of water changes at an altitude of 5,000 feet.

ALGEBRA Evaluate each expression if $d = -9$, $f = 36$, and $g = -6$.

34. $\dfrac{-f}{d}$ **35.** $\dfrac{12 - (-f)}{-g}$ **36.** $\dfrac{f^2}{d^2}$ **37.** $g^2 \div f$

38. PLANETS The temperature on Mars ranges widely from $-207°$F at the winter pole to almost $80°$F on the dayside during the summer. Use the expression $\dfrac{-207 + 80}{2}$ to find the average of the temperature extremes on Mars.

㊴ ANALYZE GRAPHS The *mean* of a set of data is the sum of the data divided by the number of items in the data set. The graph shows the approximate depths at which certain fish are found in the Caribbean. What is the mean depth of the fish shown?

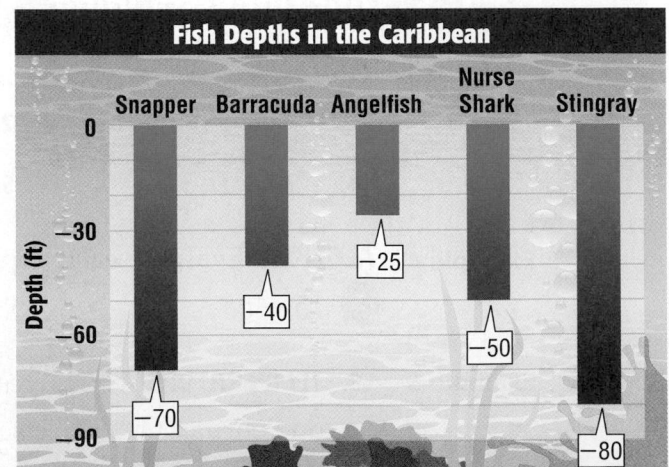

Fish Depths in the Caribbean

H.O.T. Problems

40. OPEN ENDED Write a division sentence with a quotient of -12.

41. Which One Doesn't Belong? Identify the expression that does not belong with the other three. Explain your reasoning.

| $-66 \div 11$ | $-32 \div (-4)$ | $16 \div (-4)$ | $-48 \div 4$ |

42. CHALLENGE Find values for x, y, and z so that all of the following statements are true.

- $y > x$, $z < y$, and $x < 0$
- $x \div z = -z$

- $z \div 2$ and $z \div 3$ are integers
- $x \div y = z$

43. ✍ **WRITE MATH** Evaluate $-2 \cdot (2^2 + 2) \div 2^2$. Justify each step.

✔ Test Practice

44. THINK SOLVE EXPLAIN **SHORT RESPONSE** The table shows the points that each student lost on the first math test. Each question on the test was worth an equal number of points.

Student	Points
Christopher	-24
Nythia	-16
Raul	-4

If Christopher answered 6 questions incorrectly, how many questions did Nythia answer incorrectly? Explain.

45. On December 24, 1924, the temperature in Fairfield, Montana, fell from 63°F at noon to -21°F at midnight. What was the average temperature change per hour?

A. -3.5°F

B. -7°F

C. -42°F

D. -84°F

Spiral Review

Multiply. (Lesson 3C)

46. $14(-2)$ **47.** $-20(-3)$ **48.** $-5(7)$ **49.** $(-9)^2$

50. DISPLAYS A display of cereal boxes has one box in the top row, two boxes in the second row, three boxes in the third row, and so on, as shown. How many rows of boxes will there be in a display of 45 boxes? (Lesson 3A)

51. Find $6 - (-12)$. (Lesson 2D)

52. Name the quadrant in which the point $(-4, -3)$ could be found on the coordinate plane. (Lesson 1C)

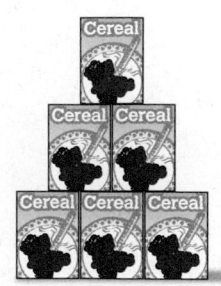

FOLDABLES®
Study Organizer

Be sure the following Key Concepts are noted in your Foldable.

Lesson 1
Integers and the
Coordinate Plane

Key Concepts

Absolute Value (Lesson 1)
• The absolute value of a number is the distance of the number from zero on a number line.

Integer Operations (Lessons 2 and 3)
• To add integers with the same sign, add their absolute value. The sum is positive if both integers are positive and negative if both integers are negative.

$$-2 + (-3) = -5$$

• To add integers with different signs, subtract their absolute values. The sum has the sign of the integer with the larger absolute value.

$$-7 + 1 = -6 \qquad -3 + 5 = 2$$

• To subtract an integer, add its opposite.

$$-2 - 3 = -2 + (-3) \text{ or } -5$$

• The product or quotient of two integers with different signs is negative.

$$6(-2) = -12 \qquad -10 \div 2 = -5$$

• The product or quotient of two integers with the same sign is positive.

$$-4(-5) = 20 \qquad -12 \div (-4) = 3$$

Key Vocabulary

absolute value	origin
additive inverse	positive integer
coordinate plane	quadrant
graph	*x*-axis
integer	*x*-coordinate
negative integer	*y*-axis
opposites	*y*-coordinate
ordered pair	

Vocabulary Check

State whether each sentence is *true* or *false*. If *false*, replace the underlined word or number to make a true sentence.

1. Integers less than zero are <u>positive</u> integers.

2. The <u>origin</u> is the point at which the *x*-axis and *y*-axis intersect.

3. The <u>absolute value</u> of 7 is −7.

4. The sum of two negative integers is <u>positive</u>.

5. The <u>*x*-coordinate</u> of the ordered pair (2, −3) is −3.

6. Two integers that are opposites are also called <u>additive inverses</u>.

7. The product of a positive and a negative integer is <u>negative</u>.

8. The *x*-axis and the *y*-axis separate the plane into four <u>coordinates</u>.

9. The quotient of two negative integers is <u>negative</u>.

Multi-Part Lesson Review

Lesson 1 Integers and the Coordinate Plane

Integers and Absolute Value (Lesson 1B)

Write an integer for each situation.

10. a loss of $150

11. 350 feet above sea level

12. a loss of 8 yards

13. **JUICE** Mavis drank 48 milliliters of apple juice before replacing the carton in the refrigerator. Write an integer that shows the change in the volume of juice in the carton.

Evaluate each expression.

14. $|100|$

15. $|-32|$

16. $|-16| + |9|$

17. $|7 + 12| + |-14|$

18. **PLANETS** The average temperature of Saturn is 218°F below zero. What is the absolute value of the integer that represents this temperature?

EXAMPLE 1 Write an integer for an altitude of 8 feet.

Since this situation represents an elevation *above* the ground, 8 represents the situation.

EXAMPLE 2 Write an integer for a loss of 5 pounds.

This situation represents a decrease in weight, so −5 represents this situation.

EXAMPLE 3 Evaluate $|-10|$.

On the number line, the graph of −10 is 10 units from 0.

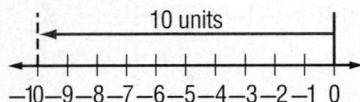

So, $|-10| = 10$.

The Coordinate Plane (Lesson 1C)

Graph and label each point on a coordinate plane.

19. $F(1, -4)$ 20. $G(-4, 2)$

21. $H(-2, -3)$ 22. $I(4, 0)$

23. **ROUTES** Starting at the school, Pilar walked 1 block east and 3 blocks south. From there, she walked 5 blocks west and 4 blocks north to the park. If the school represents the origin, what is the ordered pair for the park?

EXAMPLE 4 Graph and label the point $S(3, -1)$.

Draw a coordinate plane. Move 3 units to the right. Then move 1 unit down. Draw a dot and label it $S(3, -1)$.

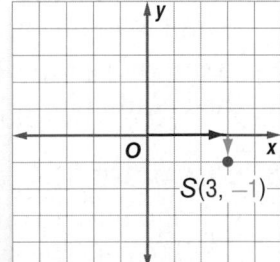

Lesson 2 Add and Subtract Integers

Add Integers (Lesson 2B)

Add.

24. $-6 + 8$

25. $-4 + (-9)$

26. $7 + (-12)$

27. $-18 + 18$

28. $-2 + 9$

29. $-10 + (-5)$

30. $6 + (-9)$

31. $17 + -17$

32. **HIKING** Alicia hiked 75 feet up a mountain. She then hiked 22 feet higher. Then, she descended 8 feet, and finally climbed up another 34 feet. What is the final change in Alicia's elevation?

33. **HIBERNATION** A black bear can weigh as much as 900 pounds when it starts hibernating. It loses 350 pounds during hibernation and gains 50 pounds two months after hibernation ends. What is the bear's final weight?

EXAMPLE 5 Find $-4 + 3$.

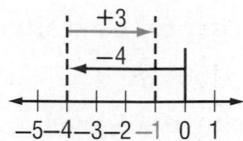

So, $-4 + 3 = -1$.

EXAMPLE 6 The population of Ankeny grew by 2,394. Then it decreased 1,459. Then it decreased 3,490 more. What is the net change in population?

The changes in population can be represented by the addition sentence $2{,}394 + (-1{,}459) + (-3{,}490)$.

$2{,}394 + (-1{,}459) + (-3{,}490) = -2{,}555$

So, the net change in population is $-2{,}555$.

Subtract Integers (Lesson 2D)

Subtract.

34. $-5 - 8$

35. $3 - 6$

36. $5 - (-2)$

37. $-4 - (-8)$

38. $-3 - 7$

39. $12 - 17$

40. $4 - (-7)$

41. $-10 - (-9)$

42. **GOLF** Owen shot 2 under par while his friend Nathan shot 3 over par. By how many shots was Owen's score better than Nathan's?

43. **VOLCANOES** The summit of an oceanic volcano once had an elevation of $-3{,}780$ feet. The summit now has an elevation of -800 feet. What was the change in elevation?

EXAMPLE 7 Find $-3 - 9$.

$-3 - 9 = -3 + (-9)$ To subtract 9, add -9.

$\qquad = -12$ Simplify.

EXAMPLE 8 On Wednesday, a company's stock closed at $13.54. On Thursday, it closed at $12.28. What was the change in the closing price?

To find the change in closing price, subtract $13.54 from $12.28.

$\$12.28 - \$13.54 = -\$1.26$

The change in closing price is $-\$1.26$.

PSI: Look for a Pattern (Lesson 3A)

Solve. Look for a pattern.

44. **HEALTH** The average person blinks 12 times per minute. At this rate, how many times does the average person blink in 12 hours?

45. **SALARY** Koko gets a job that pays $31,000 per year. She is promised a $2,200 raise after each year. At this rate, what will her salary be after 7 years?

46. **DOGS** A kennel determined that they need 144 feet of fencing to board 2 dogs, 216 feet to board 3 dogs, and 288 feet to board 4 dogs. If this pattern continues, how many feet of fencing are needed to board 8 dogs?

EXAMPLE 9 A theater has 18 seats in the first row, 24 seats in the second row, 30 seats in the third row, and so on. If this pattern continues, how many seats are in the sixth row?

Begin with 18 seats and add 6 seats for each additional row.

So, there are 48 seats in the sixth row.

Row	Number of Seats
1	18
2	24
3	30
4	36
5	42
6	48

Multiply Integers (Lesson 3C)

Multiply.

47. $-4(3)$
48. $8(-6)$
49. $-5(-7)$
50. $-2(40)$

ALGEBRA Evaluate each expression if $a = -4$, $b = -7$, and $c = 5$.

51. ab
52. $-3c$
53. bc
54. abc

EXAMPLE 10 Find $-4(3)$.

$-4(3) = -12$ The integers have different signs. The product is negative.

EXAMPLE 11 Evaluate xyz if $x = -6$, $y = 11$, and $z = -10$.

$$xyz = (-6)(11)(-10)$$
$$= (-66)(-10) \quad \text{Multiply } -6 \text{ and } 11.$$
$$= 660 \quad \text{Multiply } -66 \text{ and } -10.$$

Divide Integers (Lesson 3D)

Divide.

55. $-45 \div (-9)$
56. $36 \div (-12)$
57. $-12 \div 6$
58. $-81 \div (-9)$

EXAMPLE 12 Find $-72 \div (-9)$.

$-72 \div (-9) = 8$ The integers have the same sign. The quotient is positive.

1. **WEATHER** Adam is recording the change in the outside air temperature for a science project. At 8:00 A.M., the high temperature was 42°F. By noon, the outside temperature had fallen 11°F. By mid-afternoon, the outside air temperature had fallen another 12°F and by evening, it had fallen an additional 5°F. Write an integer that describes the final change in temperature.

Evaluate each expression.

2. $|-3|$

3. $|-18| - |6|$

4. **STOCKS** A certain stock dropped 9 points one day and gained 13 points the following day. What was the net change in the stock's worth?

5. **FREEZING POINTS** The table shows the freezing points of various chemicals.

Chemical	Freezing Point (°F)
Carbon dioxide (dry ice)	−109
Water	32
Hydrogen	−435

a. What is the difference between the freezing point of dry ice and the freezing point of hydrogen?

b. What is the difference between the freezing points of water and dry ice?

6. **DEBT** Amanda owes her brother $24. If she plans to pay him back an equal amount from her piggy bank each day for six days, describe the change in the amount of money in her piggy bank each day.

Write the ordered pair corresponding to each point graphed. Then state the quadrant in which each point is located.

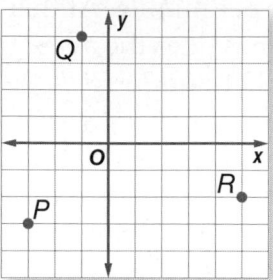

7. P

8. Q

9. R

10. **MULTIPLE CHOICE** Kendrick created a six-week schedule for practicing the piano. The table shows the number of hours he practiced in the first three weeks. If the pattern continues, how many hours will he practice during the sixth week?

Week	1	2	3
Hours	4	7	10

A. 15 hours   C. 19 hours

B. 18 hours   D. 22 hours

Add, subtract, multiply, or divide.

11. $12 + (-9)$

12. $-3 - 4$

13. $-7 - (-20)$

14. $-7(-3)$

15. $5(-11)$

16. $-36 \div (-9)$

17. $-15 + (-7)$

18. $8 + (-6) + (-4)$

19. $-9 - 7$

20. $-13 + 7$

21. **THINK SOLVE EXPLAIN** **EXTENDED RESPONSE** Rectangle $ABCD$ has vertices $A(-4, -2)$, $B(-4, 5)$, and $C(6, 5)$.

Part A Graph points A, B, and C on a coordinate plane.

Part B Find the coordinates of point D.

Part C Explain how you found the coordinates for point D.

Preparing for Standardized Tests

Gridded Response: Integers

Some standardized tests have gridded-response questions. To answer a gridded-response question, first write your answer in the boxes at the top of the answer grid. Then fill in the bubble under each box to match your answer.

TEST EXAMPLE

Firefly squid are typically found 365 meters below the ocean's surface, or at a depth of −365 meters. Some are found at a depth 183 meters higher than this. At what depth can the shallower firefly squid be found?

$-365 + 183 = -182$

The shallower squid are at a depth of −182 meters. So, the answer is −182. Fill in −182.

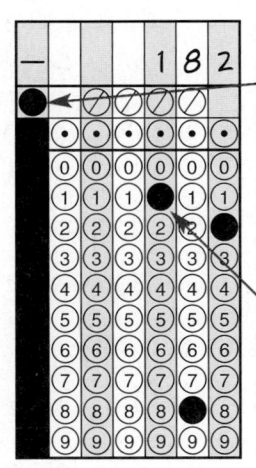

or

Write the (−) in the answer box, then fill in the ⊖ bubble.

Write 1, 8, and 2 in separate answer boxes. Fill in the 1 bubble under the 1, the 8 bubble under the 8, and the 2 bubble under the 2.

Work on It

The lowest point in Egypt is the Qattara Depression, which has an elevation of −133 meters. The highest point is Mount Catherine, which has an elevation of 2,629 meters. What is the difference in meters between the highest and lowest points? Fill in your answer on an answer grid.

Test Hint

The bubbles must be filled in completely and accurately for you to receive credit for your answer.

✓ Test Practice

Read each question. Then fill in the correct answer on the answer document provided by your teacher or on a sheet of paper.

1. The table shows the daily low temperatures for Cleveland, Ohio, over five days.

Day	Temperature
1	15°F
2	−2°F
3	8°F
4	−6°F
5	5°F

Which expression can be used to find the average daily low temperature during the five days?

A. $(15 + 2 + 8 + 6 + 5) \div 5$

B. $15 + 2 + 8 + 6 + 5 \div 5$

C. $[15 + (-2) + 8 + (-6) + 5] \div 5$

D. $15 + (-2) + 8 + (-6) + 5 \div 5$

2. Three vertices of a parallelogram are given as coordinates $(-4, 2)$, $(-2, 4)$, and $(1, -3)$ in the graph. Which coordinates best represent the location of the fourth vertex of the parallelogram?

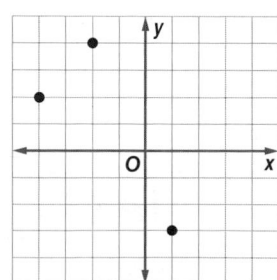

F. $(-3, 1)$

G. $(3, -1)$

H. $(1, -3)$

I. $(-1, 3)$

3. ✎ **GRIDDED RESPONSE** The lowest point in Japan is Hachiro-gata (elevation −4 m), and the highest point is Mount Fuji (elevation 3,776 m). What is the difference in elevation, in meters, between Mount Fuji and Hachiro-gata?

4. ✎ **GRIDDED RESPONSE** A submarine is cruising 8 meters below the surface. The captain orders a dive of another 17 meters. What is the new cruising depth of the submarine in meters?

5. In what quadrant is point P located?

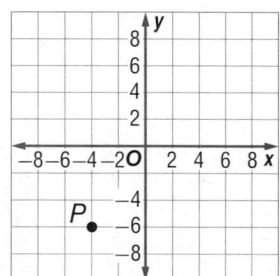

A. Quadrant I

B. Quadrant II

C. Quadrant III

D. Quadrant IV

6. What integer added to −9 gives a sum of 3?

F. 12

G. 6

H. 3

I. −12

7. By the end of the third quarter of a football game, Ricky had gained 112 yards and had lost 12 yards. If Ricky lost an additional 8 yards and gained 22 yards in the fourth quarter, which equation could be used to represent his total yardage for the game?

A. $112 + 12 + 8 + 22 = 154$

B. $112 + (-12) + (-8) + 22 = 114$

C. $112 + 12 + (-8) + (-22) = 94$

D. $(-112) + (-12) + 8 + 22 = -94$

8. 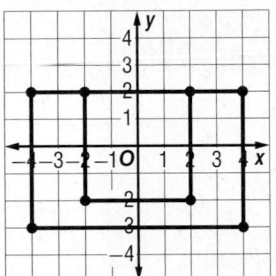 **GRIDDED RESPONSE** Bobby is diving 50 feet below sea level at the beach. His sister is at the swimming pool deck, which is 15 feet above sea level. What is the difference, in feet, between the pool deck and Bobby's position?

9. **THINK SOLVE EXPLAIN** **SHORT RESPONSE** Larry borrowed $12,000 from his grandfather to buy a car. He bought a used car, so he returned $4,411 to his grandfather. Write and solve an equation using integers that shows the total amount that Larry owes his grandfather.

10. Pablo and three of his friends are playing paintball. The table shows their scores at the end of one round. By how many points is Winston beating Pablo?

Player	Score
Pablo	−189
Winston	−124
Nevin	130
Marsella	48

   F. 65
   G. 135
   H. 178
   I. 313

11. Each of the first 4 pit stops a race car driver makes loses ten seconds off the leader. The pit crew makes adjustments, and at each of the next two pit stops he gains 7 seconds on the leader. How much time is the driver off the leader?

   A. 40 seconds
   B. 14 seconds
   C. −26 seconds
   D. −54 seconds

12. **THINK SOLVE EXPLAIN** **SHORT RESPONSE** A rectangle and a square are graphed on a coordinate plane. Name an ordered pair that is inside the rectangle but outside the square.

13. **THINK SOLVE EXPLAIN** **EXTENDED RESPONSE** The use of computers to download music has decreased the sales of music CDs. Use the following table to answer the questions.

Year	Estimated Number of New Music CD Releases
2000	36,000
2001	32,000
2002	34,000
2003	8,000
2004	14,000
2005	10,000
2006	12,000

   Part A   During which year was there the greatest decrease in CD releases from the previous year? What was the decrease?

   Part B   Write and evaluate an expression that shows the change in CD releases from 2004 to 2005.

NEED EXTRA HELP?

If You Missed Question...	1	2	3	4	5	6	7	8	9	10	11	12	13
Go to Chapter-Lesson...	2-3D	2-1C	2-2D	2-2D	2-1C	2-2B	2-2B	2-2D	2-2B	2-2D	2-3C	2-1C	2-2D

Rational Numbers

The
☆BIG Idea
How are fractions, decimals, and percents related?

connectED.mcgraw-hill.com

Investigate

 Animations

 Vocabulary

 Multilingual eGlossary

Learn

 Personal Tutor

 Virtual Manipulatives

 Graphing Calculator

 Audio

 Foldables

Practice

 Self-Check Practice

 Worksheets

 Assessment

FOLDABLES
Study Organizer

Make this Foldable to help you organize your notes.

Review Vocabulary

percent *por ciento* a ratio that compares a number to 100

$$50\% = 50 \text{ out of } 100 \text{ or } \frac{50}{100}$$

Key Vocabulary

English	Español
exponent	exponente
like fractions	fracciones semejantes
rational number	número racional
unlike fractions	fracciones con distinto denominador

When Will I Use This?

Are You Ready
for the Chapter?

You have two options for checking prerequisite skills for this chapter.

Text Option Take the Quick Check below. Refer to the Quick Review for help.

QUICK Check	QUICK Review
Write each percent as a fraction in simplest form.	**EXAMPLE 1**

Write each percent as a fraction in simplest form.

1. 75%   **2.** 90%

3. 47%   **4.** 14%

5. FRIENDS Of Max's friends, 35% have a birthday in the summer months. Write 35% as a fraction in simplest form.

EXAMPLE 1

Write 25% as a fraction in simplest form.

$25\% = \dfrac{25}{100}$ Definition of percent.

$= \dfrac{\overset{1}{\cancel{25}}}{\underset{4}{\cancel{100}}}$ or $\dfrac{1}{4}$ Simplify. Divide the numerator and denominator by the GCF, 25.

Graph on a number line.

6. $2\frac{1}{4}$ **7.** $5\frac{1}{2}$

8. $\frac{4}{5}$ **9.** $1\frac{1}{3}$

10. MEASUREMENT Jenna is baking a cake that requires $2\frac{3}{4}$ cups of flour and $1\frac{1}{2}$ cups of sugar. Graph the amounts of flour and sugar on a number line.

EXAMPLE 2

Graph $3\frac{2}{3}$ on a number line.

Find the two whole numbers that $3\frac{2}{3}$ lies between.

$3 < 3\frac{2}{3} < 4$

Since the denominator is 3, divide each space into 3 sections.

Place a dot at $3\frac{2}{3}$.

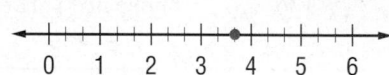

Write each percent as a decimal.

11. 57%   **12.** 38%

13. 94%   **14.** 6%

15. SOCCER Last season, Chris made 72% of his penalty kicks. Write 72% as a decimal.

EXAMPLE 3

Write 21% as a decimal.

$21\% = \dfrac{21}{100}$ Rewrite the percent as a fraction with a denominator of 100.

$= 0.21$ Write *21 hundredths* as a decimal.

 **Online Option** Take the Online Readiness Quiz.

Explore The Number Line

Main Idea

Graph rational numbers on the number line.

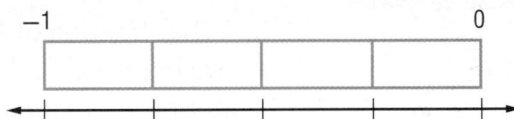 You have already graphed integers and positive fractions on a number line. In this Activity, you will graph negative fractions.

ACTIVITY

Use models to graph $-\dfrac{3}{4}$ on a number line.

STEP 1 Draw a number line with arrows on each end. Place a fraction strip divided in fourths above the number line. Mark a 0 on the right side and a -1 on the left side.

STEP 2 Starting from the right, shade three fourths. Label the number line with $-\dfrac{1}{4}$, $-\dfrac{2}{4}$, and $-\dfrac{3}{4}$.

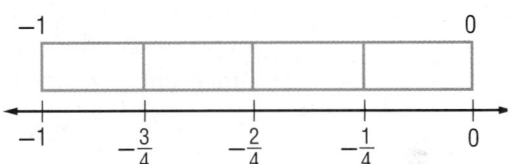

STEP 3 Draw a dot on the number line above the $-\dfrac{3}{4}$ mark.

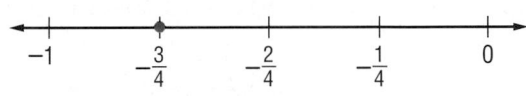

Practice and Apply

Graph each fraction on a number line. Use a fraction strip.

1. $-\dfrac{3}{8}$ **2.** $-\dfrac{1}{3}$ **3.** $-1\dfrac{2}{5}$ **4.** $-2\dfrac{5}{6}$

Graph each pair of numbers on a number line. Then write which number is less.

5. $-\dfrac{7}{8}, -\dfrac{3}{8}$ **6.** $-\dfrac{5}{8}, -1\dfrac{1}{8}$ **7.** $-\dfrac{13}{8}, -\dfrac{3}{8}$ **8.** $-1\dfrac{7}{8}, -1\dfrac{5}{8}$

9. **WRITE MATH** How does graphing $-\dfrac{3}{4}$ differ from graphing $\dfrac{3}{4}$?

Main Idea

Write fractions as terminating or repeating decimals and write decimals as fractions.

 Vocabulary

terminating decimal
repeating decimal
bar notation

 Get ConnectED

 CCSS 7.NS.2d

Terminating and Repeating Decimals

NASCAR The table shows the winning speeds for a 10-year period at the Daytona 500.

1. What fraction of the speeds are between 130 and 145 miles per hour?

2. Express this fraction using words and then as a decimal.

3. What fraction of the speeds are between 145 and 165 miles per hour? Express this fraction using words and then as a decimal.

Daytona 500		
Year	Winner	Speed (mph)
1999	J. Gordon	148.295
2000	D. Jarrett	155.669
2001	M. Waltrip	161.783
2002	W. Burton	142.971
2003	M. Waltrip	133.870
2004	D. Earnhardt Jr.	156.345
2005	J. Gordon	135.173
2006	J. Johnson	142.667
2007	K. Harvick	149.335
2008	R. Newman	152.672

Our decimal system is based on powers of 10 such as 10, 100, and 1,000. If the denominator of a fraction is a power of 10, you can use place value to write the fraction as a decimal.

Words	Fraction	Decimal
seven tenths	$\frac{7}{10}$	0.7

If the denominator of a fraction is a *factor* of 10, 100, 1,000, or any greater power of ten, you can use mental math and place value.

 EXAMPLES **Use Mental Math**

Write each fraction or mixed number as a decimal.

1 $\frac{7}{20}$

Think $\overset{\times 5}{\frac{7}{20} = \frac{35}{100}}\underset{\times 5}{}$

So, $\frac{7}{20} = 0.35$.

2 $5\frac{3}{4}$

$5\frac{3}{4} = 5 + \frac{3}{4}$ Think of it as a sum.

$= 5 + 0.75$ You know that $\frac{3}{4} = 0.75$.

$= 5.75$ Add mentally.

So, $5\frac{3}{4} = 5.75$.

 CHECK Your Progress

a. $\frac{3}{10}$ b. $\frac{3}{25}$ c. $-6\frac{1}{2}$

Any fraction can be written as a decimal by dividing its numerator by its denominator. Division ends when the remainder is zero.

 EXAMPLES Use Division

3 Write $\frac{3}{8}$ as a decimal.

$$
\begin{array}{r}
0.375 \\
8\overline{)3.000} \\
-24 \\
\hline
60 \\
-56 \\
\hline
40 \\
-40 \\
\hline
0
\end{array}
$$
Divide 3 by 8.

Division ends when the remainder is 0.

So, $\frac{3}{8} = 0.375$.

4 Write $-\frac{1}{40}$ as a decimal.

$$
\begin{array}{r}
0.025 \\
40\overline{)1.000} \\
-80 \\
\hline
200 \\
-200 \\
\hline
0
\end{array}
$$
Divide 1 by 40.

So, $-\frac{1}{40} = -0.025$.

CHECK Your Progress

Write each fraction or mixed number as a decimal.

d. $-\frac{7}{8}$ **e.** $2\frac{1}{8}$ **f.** $7\frac{9}{20}$

Vocabulary Link

Everyday Use

Terminate coming to an end, as in terminate a game

Math Use

Terminate a decimal whose digits end

In Examples 1–4, the decimals 0.35, 5.75, 0.375, and −0.025 are called terminating decimals. A **terminating decimal** is a decimal with digits that end. **Repeating decimals** have a pattern in their digit(s) that repeats forever. Consider $\frac{1}{3}$.

$$
\begin{array}{r}
0.333\ldots \\
3\overline{)1.000} \\
-9 \\
\hline
10 \\
-9 \\
\hline
10 \\
-9 \\
\hline
1
\end{array}
$$

> The number 3 repeats. The repetition of 3 is represented by three dots.

You can use **bar notation** to indicate that a number pattern repeats indefinitely. A bar is written over only the digit(s) that repeat.

$0.33333\ldots = 0.\overline{3}$ $0.121212\ldots = 0.\overline{12}$ $11.3858585\ldots = 11.3\overline{85}$

EXAMPLE Write Fractions as Repeating Decimals

⑤ Write $\frac{7}{9}$ as a decimal.

$$
\begin{array}{r}
0.777\ldots \\
9\overline{)7.000} \\
-63 \\
\hline
70 \\
-63 \\
\hline
70 \\
-63 \\
\hline
7
\end{array}
$$

Divide 7 by 9.

> Notice that the remainder will never be zero. That is, the division never ends.

So, $\frac{7}{9} = 0.777\ldots$ or $0.\overline{7}$.

✓ CHECK Your Progress

Write each fraction or mixed number as a decimal. Use bar notation if the decimal is a repeating decimal.

g. $\frac{2}{3}$ **h.** $-\frac{3}{11}$ **i.** $8\frac{1}{3}$

Every terminating decimal can be written as a fraction with a denominator of 10, 100, 1,000, or a greater power of ten. Use the place value of the final digit as the denominator.

numerator

$$
0.25 = \frac{25}{100}
$$

hundredths place

Real-World Link · · · ·
The recommended water temperature for goldfish is 65–72°F.

REAL-WORLD EXAMPLE

⑥ **FISH** Use the table to find what fraction of the fish in an aquarium are goldfish. Write in simplest form.

$0.15 = \frac{15}{100}$ The final digit, 5, is in the hundredths place.

$= \frac{3}{20}$ Simplify.

So, $\frac{3}{20}$ of the fish are goldfish.

Fish	Amount
Guppy	0.25
Angelfish	0.4
Goldfish	0.15
Molly	0.2

✓ CHECK Your Progress

Determine the fraction of the aquarium made up by each fish. Write the answer in simplest form.

j. molly   **k.** guppy   **l.** angelfish

✓ CHECK Your Understanding

Examples 1–5 Write each fraction or mixed number as a decimal. Use bar notation if the decimal is a repeating decimal.

1. $\frac{2}{5}$

2. $-\frac{9}{10}$

3. $7\frac{1}{2}$

4. $-4\frac{3}{20}$

5 $\frac{1}{8}$

6. $-3\frac{5}{8}$

7. $\frac{5}{9}$

8. $1\frac{5}{6}$

Example 6 Write each decimal as a fraction or mixed number in simplest form.

9. -0.22

10. 0.1

11. 4.6

12. **HOCKEY** During a hockey game, an ice resurfacer travels 0.75 mile during each ice resurfacing. What fraction represents this distance?

Practice and Problem Solving

 = **Step-by-Step Solutions** begin on page R1.
Extra Practice begins on page EP2.

Examples 1–5 Write each fraction or mixed number as a decimal. Use bar notation if the decimal is a repeating decimal.

13. $\frac{4}{5}$

14. $\frac{1}{2}$

15. $-4\frac{4}{25}$

16. $-7\frac{1}{20}$

17. $\frac{5}{16}$

18. $\frac{3}{16}$

19. $-\frac{33}{50}$

20. $-\frac{17}{40}$

21. $5\frac{7}{8}$

22. $9\frac{3}{8}$

23. $-\frac{4}{9}$

24. $-\frac{8}{9}$

25. $-\frac{1}{6}$

26. $-\frac{8}{11}$

27. $5\frac{1}{3}$

28. $2\frac{6}{11}$

Example 6 Write each decimal as a fraction or mixed number in simplest form.

29. -0.2

30. -0.9

31. 0.55

32. 0.34

33. 5.96

34. 2.66

35 **INSECTS** A praying mantis can be 30.5 centimeters long. What mixed number represents this length?

36. **GROCERIES** Suppose you buy a 1.25-pound package of ham for $4.99. What fraction of a pound did you buy?

37. **MATH IN THE MEDIA** Find examples of negative fractions or decimals in a newspaper or magazine or on the Internet. Write a real-world problem in which you use negative fractions or decimals.

Write each of the following as an improper fraction.

38. -13

39. $7\frac{1}{3}$

40. -1.028

41. -3.2

42. **MUSIC** Nicolás practiced playing the cello for 2 hours and 18 minutes. Write the time Nicolás spent practicing as a decimal.

**H.O.T. Problems**

43. OPEN ENDED Write a fraction that is equivalent to a terminating decimal between 0.5 and 0.75.

44. CHALLENGE Fractions in simplest form that have denominators of 2, 4, 8, 16, and 32 produce terminating decimals. Fractions with denominators of 6, 12, 18, and 24 produce repeating decimals. What causes the difference? Explain.

45. 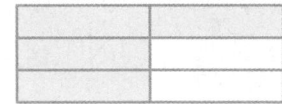 **WRITE MATH** The value of pi (π) is 3.1415926… . The mathematician Archimedes believed that π was between $3\frac{1}{7}$ and $3\frac{10}{71}$. Was Archimedes correct? Explain your reasoning.

✔ Test Practice

46. Tanya drew a model for the fraction $\frac{4}{6}$.

Which of the following decimals is equal to $\frac{4}{6}$?

A. 0.666

B. $0.\overline{6}$

C. 0.667

D. $0.66\overline{7}$

47. Use the table that shows decimal and fraction equivalents.

Decimal	Fraction
$0.\overline{3}$	$\frac{3}{9}$
$0.\overline{4}$	$\frac{4}{9}$
$0.\overline{5}$	$\frac{5}{9}$
$0.\overline{6}$	$\frac{6}{9}$

Which fraction represents $0.\overline{8}$?

F. $\frac{4}{5}$

G. $\frac{80}{99}$

H. $\frac{5}{6}$

I. $\frac{8}{9}$

48. The sign shows the lengths of four hiking trails.

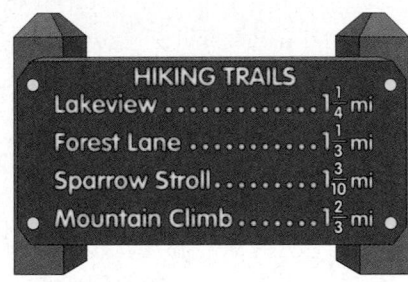

HIKING TRAILS
Lakeview .... $1\frac{1}{4}$ mi
Forest Lane .... $1\frac{1}{3}$ mi
Sparrow Stroll........ $1\frac{3}{10}$ mi
Mountain Climb ...... $1\frac{2}{3}$ mi

Which trail length is equivalent to $1.\overline{3}$?

A. Forest Lane

B. Lakeview

C. Mountain Climb

D. Sparrow Stroll

49. Zoe went to lunch with a friend. After tax, her bill was $12.05. Which mixed number represents this amount in simplest form?

F. $12\frac{1}{2}$

G. $12\frac{1}{20}$

H. $12\frac{5}{10}$

I. $12\frac{5}{100}$

Main Idea

Compare and order rational numbers.

 Vocabulary

rational numbers
common denominator
least common
denominator (LCD)

 Get Connect ED

Compare and Order Rational Numbers

SOFTBALL The batting average of a softball player is found by comparing the number of hits to the number of times at bat. Felisa had 50 hits in 175 at bats, and Harmony had 42 hits in 160 at bats.

1. Write the two batting averages as fractions.

2. Which girl had the better batting average? Explain.

3. Describe two methods you could use to compare the batting averages.

A **rational number** is a number that can be expressed as a ratio of two integers expressed as a fraction, in which the denominator is not zero. Common fractions, terminating and repeating decimals, percents, and integers are all rational numbers.

Rational Numbers

0.8   Integers $\frac{1}{2}$

20%   −3   Whole Numbers   −1

2.$\overline{2}$ 2   1 $1\frac{2}{3}$

−1.4444...

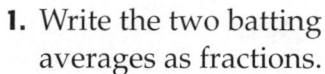 **EXAMPLE** Compare Rational Numbers

 Replace the ● with <, >, or = to make $-1\frac{5}{6}$ ● $-1\frac{1}{6}$ a true sentence.

Graph each rational number on a number line. Mark off equal-size increments of $\frac{1}{6}$ between −2 and −1.

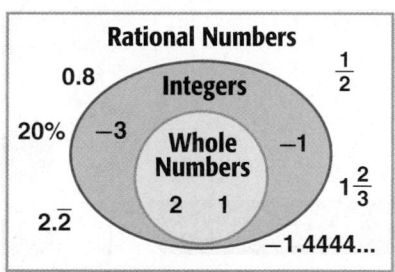

$$-2 \quad -1\frac{5}{6} \quad -1\frac{4}{6} \quad -1\frac{3}{6} \quad -1\frac{2}{6} \quad -1\frac{1}{6} \quad -1$$

The number line shows that $-1\frac{5}{6} < -1\frac{1}{6}$.

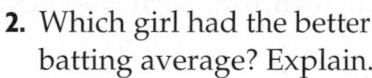 **CHECK** Your Progress

a. Replace the ● with <, >, or = to make $-5\frac{5}{9}$ ● $-5\frac{1}{9}$ a true sentence.

A **common denominator** is a common multiple of the denominators of two or more fractions. The **least common denominator** or **LCD** is the LCM or least common multiple of the denominators. You can use the LCD to compare fractions.

 **EXAMPLE** Compare Rational Numbers

2 Replace the ⬤ with $<$, $>$, or $=$ to make $\frac{7}{12}$ ⬤ $\frac{8}{18}$ a true sentence.

The LCD of the denominators 2 and 3 is 6.

$$\frac{7}{12} = \frac{7 \times 3}{12 \times 3} \qquad\qquad \frac{8}{18} = \frac{8 \times 2}{18 \times 2}$$

$$= \frac{21}{36} \qquad\qquad\qquad = \frac{16}{36}$$

Since $\frac{21}{36} > \frac{16}{36}$, $\frac{7}{12} > \frac{8}{18}$.

QUICK Review

LCD

To find the least common denominator for $\frac{7}{12}$ and $\frac{8}{18}$, find the LCM of 12 and 18.

$12 = 2 \times 2 \times 3$
$18 = 2 \times 3 \times 3$
$LCM = 2 \times 2 \times 3 \times 3$
$\qquad = 36$

✓ **CHECK Your Progress**

Replace each ⬤ with $<$, $>$, or $=$ to make a true sentence.

b. $\frac{5}{6}$ ⬤ $\frac{7}{9}$ **c.** $\frac{1}{5}$ ⬤ $\frac{7}{50}$ **d.** $-\frac{9}{16}$ ⬤ $-\frac{7}{10}$

You can also compare fractions by writing each fraction as a decimal and then comparing the decimals.

 REAL-WORLD EXAMPLE

3 **ROLLER SHOES** In Mr. Huang's math class, 20% of students own roller shoes. In Mrs. Trevino's math class, 5 out of 29 students own roller shoes. In which class does a greater fraction of students own roller shoes?

Express each number as a decimal and then compare.

$$20\% = 0.2 \qquad \frac{5}{29} = 5 \div 29$$

$$\approx 0.1724$$

Since $0.2 > 0.1724$, $20\% > \frac{5}{29}$.

A greater fraction of students in Mr. Huang's class owns roller shoes.

QUICK Review

Percents as Decimals

To write a percent as a decimal, remove the percent sign and then move the decimal point two places to the left. Add zeros if necessary.

$20\% = 20\%$
$\qquad = 0.20$

✓ **CHECK Your Progress**

e. **BOWLING** In a second period class, 37.5% of students like to bowl. In a fifth period class, 12 out of 29 students like to bowl. In which class does a greater fraction of the students like to bowl?

Not all numbers are rational numbers. The Greek letter π (pi) represents the nonterminating and nonrepeating number whose first few digits are 3.1415926… . This number is an *irrational* number.

EXAMPLE Order Rational Numbers

4 **List the numbers 3.44, π, 3.14, and $3.\overline{4}$ in order from least to greatest.**

Line up the decimal points and compare using place value.

3.140	Annex a zero.	3.440	Annex a zero.
3.1415926…	π ≈ 3.1415926…	3.444…	$3.\overline{4} = 3.444…$
Since 0 < 1, 3.14 < π.		Since 0 < 4, 3.44 < $3.\overline{4}$.	

So, the order of the numbers from least to greatest is 3.14, π, 3.44, and $3.\overline{4}$.

Check Graph each number on a number line.

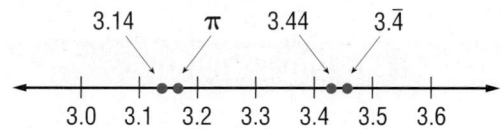

The number line confirms the order of the numbers is correct.

CHECK Your Progress

f. **SCORES** Class average scores on the last four science quizzes were 0.82, $\frac{4}{5}$, 83%, and $\frac{3}{4}$. List the scores from least to greatest.

These common equivalents are used frequently. It will be useful to memorize them.

Key Concept Common Equivalents

$\frac{1}{4} = 0.25 = 25\%$	$\frac{1}{5} = 0.2 = 20\%$	$\frac{1}{8} = 0.125 = 12.5\%$	$\frac{1}{10} = 0.1 = 10\%$
$\frac{1}{2} = 0.5 = 50\%$	$\frac{2}{5} = 0.4 = 40\%$	$\frac{3}{8} = 0.375 = 37.5\%$	$\frac{3}{10} = 0.3 = 30\%$
$\frac{3}{4} = 0.75 = 75\%$	$\frac{3}{5} = 0.6 = 60\%$	$\frac{1}{3} = 0.\overline{3} = 33.\overline{3}\%$	$\frac{7}{10} = 0.7 = 70\%$
$1 = 1.00 = 100\%$	$\frac{4}{5} = 0.8 = 80\%$	$\frac{2}{3} = 0.\overline{6} = 66.\overline{6}\%$	$\frac{9}{10} = 0.9 = 90\%$

Examples 1 and 2   Replace each ● with <, >, or = to make a true sentence. Use a number line if necessary.

1. $-\dfrac{4}{9}$ ● $-\dfrac{7}{9}$

2. $-1\dfrac{3}{4}$ ● $-1\dfrac{6}{8}$

3. $\dfrac{3}{8}$ ● $\dfrac{6}{15}$

4. $2\dfrac{4}{5}$ ● $2\dfrac{7}{8}$

Example 3   5. **SOCCER** The table shows the average saves for two soccer goalies. Who has the better average, Elliot or Shanna? Explain.

Name	Average
Elliot	3 saves out of 4
Shanna	7 saves out of 11

6. **SCHOOL** On her first quiz in social studies, Meg answered 92% of the questions correctly. On her second quiz, she answered 27 out of 30 questions correctly. On which quiz did Meg have the better score?

Example 4   7. **INSECTS** The lengths of four insects are 0.02 inch, $\dfrac{1}{8}$ inch, 0.1 inch, and $\dfrac{2}{3}$ inch. List the lengths in inches from least to greatest.

Practice and Problem Solving

● = **Step-by-Step Solutions** begin on page R1.
Extra Practice begins on page EP2.

Examples 1 and 2   Replace each ● with <, >, or = to make a true sentence. Use a number line if necessary.

8. $-\dfrac{3}{5}$ ● $-\dfrac{4}{5}$

9. $-\dfrac{5}{7}$ ● $-\dfrac{2}{7}$

10. $-7\dfrac{5}{8}$ ● $-7\dfrac{1}{8}$

11. $-3\dfrac{2}{3}$ ● $-3\dfrac{4}{6}$

12. $\dfrac{7}{10}$ ● $\dfrac{2}{3}$

13 $\dfrac{4}{7}$ ● $\dfrac{5}{8}$

14. $\dfrac{2}{3}$ ● $\dfrac{10}{15}$

15. $-\dfrac{17}{24}$ ● $-\dfrac{11}{12}$

16. $2\dfrac{3}{4}$ ● $2\dfrac{2}{3}$

17. $6\dfrac{2}{3}$ ● $6\dfrac{1}{2}$

18. $5\dfrac{5}{7}$ ● $5\dfrac{11}{14}$

19. $3\dfrac{11}{16}$ ● $3\dfrac{7}{8}$

Example 3   20. **MONEY** The table shows how much copper is in each type of coin. Which coin contains the greatest amount of copper?

Coin	Amount of Copper
Dime	$\dfrac{12}{16}$
Nickel	$\dfrac{3}{4}$
Penny	$\dfrac{1}{400}$
Quarter	$\dfrac{23}{25}$

21. **BASKETBALL** Gracia and Jim were shooting free throws. Gracia made 4 out of 15 free throws. Jim *missed* 6 out of 16 free throws. Who made the free throw a greater fraction of the time?

Example 4   Order each set of numbers from least to greatest.

22. $0.23, 19\%, \dfrac{1}{5}$

23. $\dfrac{8}{10}, 0.81, 0.805$

24. $-0.615, -\dfrac{5}{8}, -0.62$

25. $-1.4, -1\dfrac{1}{25}, -1.25$

26. $7.49, 7\dfrac{49}{50}, 7.5\%$

27. $3\dfrac{4}{7}, 3\dfrac{3}{5}, 3.47$

Replace each ● with <, >, or = to make a true sentence. Use a number line if necessary.

28. 40% ● 112 out of 250

29. 3 out of 5 ● 59%

30. 82% ● 5 out of 6

31. 9 out of 20 ● 45%

MEASUREMENT Replace each ● with <, >, or = to make a true sentence.

32. $\frac{5}{8}$ yard ● $\frac{1}{16}$ yard

33. 0.25 pound ● $\frac{2}{9}$ pound

34. $2\frac{5}{6}$ hours ● 2.8 hours

35 $1\frac{7}{12}$ gallons ● $1\frac{5}{8}$ gallons

MEASUREMENT Order each of the following from least to greatest.

36. 4.4 miles, $4\frac{3}{8}$ miles, $4\frac{5}{12}$ miles

37. 6.5 cups, $6\frac{1}{3}$ cups, 6 cups

38. 1.2 laps, 2 laps, $\frac{1}{2}$ lap

39. $\frac{1}{5}$ gram, 5 grams, 1.5 grams

40. ANIMALS Use the table that shows the lengths of small mammals.

Animal	Length (ft)
Eastern Chipmunk	$\frac{1}{3}$
Kitti's Hog-Nosed Bat	$0.8\overline{3}$
European Mole	$\frac{5}{12}$
Masked Shrew	$\frac{1}{6}$
Spiny Pocket Mouse	0.25

a. Which animal is the smallest mammal?

b. Which animal is smaller than the European Mole but larger than the Spiny Pocket Mouse?

c. Order the animals from greatest to least size.

41. GRAPHIC NOVEL Refer to the graphic novel frame below for Exercises a–b.

a. Rewrite the organizer dimensions so they all have a common denominator.

b. If the closet organizer has a total width of $69\frac{1}{8}$ inches and the closet is $69\frac{3}{4}$ inches wide, will the organizer fit?

42. Which One Doesn't Belong? Identify the ratio that does not have the same value as the other three. Explain your reasoning.

| 12 out of 15 | 0.08 | 80% | $\frac{4}{5}$ |

43. CHALLENGE Explain how you know which number, $1\frac{15}{16}$, $\frac{17}{8}$, or $\frac{63}{32}$, is closest to 2.

44. ✏ WRITE MATH Write a word problem about a real-world situation in which you would compare rational numbers. Then solve the problem.

✔ Test Practice

45. Which point shows the location of $\frac{7}{2}$ on the number line?

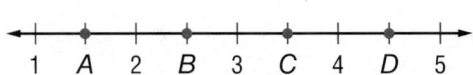

    1 *A* 2 *B* 3 *C* 4 *D* 5

A. point *A*   C. point *C*

B. point *B*   D. point *D*

46. Tabitha surveyed high school students to find how many students have a pet.

Grade	Students Who Have Pets Out of Total Students
Freshmen	45 out of 56
Sophomores	40 out of 50
Juniors	35 out of 42
Seniors	38 out of 43

Which list shows the results in order from least to greatest?

F. freshmen, sophomores, juniors, seniors

G. sophomores, freshmen, juniors, seniors

H. seniors, juniors, sophomores, freshmen

I. juniors, seniors, freshmen, sophomores

47. Which of the following fractions is closest to 0?

A. $-\frac{3}{4}$ C. $\frac{7}{12}$

B. $-\frac{2}{3}$ D. $\frac{5}{8}$

48. Which list of numbers is ordered from least to greatest?

F. $\frac{1}{4}$, $4\frac{1}{4}$, 0.4, 4%

G. 4%, 0.4, $4\frac{1}{4}$, $\frac{1}{4}$

H. 4%, $\frac{1}{4}$, 0.4, $4\frac{1}{4}$

I. 0.4, $\frac{1}{4}$, 4%, $4\frac{1}{4}$

49. Which of the following fractions is the least?

A. $-\frac{7}{8}$ C. $-\frac{7}{10}$

B. $-\frac{7}{9}$ D. $-\frac{7}{11}$

50. The daily price changes for a stock are shown in the table.

Day	Price Change
Monday	−0.21
Tuesday	−1.05
Wednesday	−0.23
Thursday	+0.42
Friday	−1.15

On which day did the price decrease by the greatest amount?

F. Monday   H. Wednesday

G. Tuesday   I. Friday

Main Idea

Add and subtract fractions with like denominators.

 **Vocabulary**

like fractions

 **Get ConnectED**

 7.NS.1, 7.NS.1b, 7.NS.1c, 7.NS.3, 7.EE.3

Add and Subtract Like Fractions

SHOES Sean surveyed ten classmates to find which type of tennis shoe they like to wear.

1. What fraction liked cross trainers? high tops?

2. What fraction likes either cross trainers or high tops?

Shoe Type	Number
Cross Trainer	5
Running	3
High Top	2

Fractions that have the same denominators are called **like fractions**.

Key Concept Add and Subtract Like Fractions

Words To add or subtract like fractions, add or subtract the numerators and write the result over the denominator.

Examples

Numbers

$$\frac{5}{10} + \frac{2}{10} = \frac{5+2}{10} \text{ or } \frac{7}{10}$$

$$\frac{11}{12} - \frac{4}{12} = \frac{11-4}{12} \text{ or } \frac{7}{12}$$

Algebra

$$\frac{a}{c} + \frac{b}{c} = \frac{a+b}{c}, \text{ where } c \neq 0$$

$$\frac{a}{c} - \frac{b}{c} = \frac{a-b}{c}, \text{ where } c \neq 0$$

EXAMPLES Add Like Fractions

Add. Write in simplest form.

1 $\dfrac{5}{9} + \dfrac{2}{9}$

$\dfrac{5}{9} + \dfrac{2}{9} = \dfrac{5+2}{9}$ Add the numerators.

$= \dfrac{7}{9}$ Simplify.

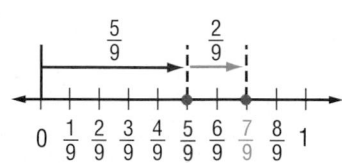

2 $-\dfrac{3}{5} + \left(-\dfrac{1}{5}\right)$

$-\dfrac{3}{5} + \left(-\dfrac{1}{5}\right) = -\dfrac{3}{5} + -\dfrac{1}{5}$

$= \dfrac{-3 + (-1)}{5}$ Add the numerators.

$= \dfrac{-4}{5} \text{ or } -\dfrac{4}{5}$ Use the rules for adding integers.

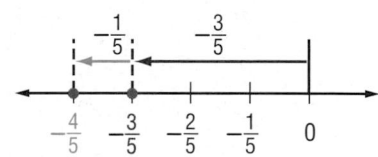

CHECK Your Progress

a. $\dfrac{1}{3} + \dfrac{2}{3}$ b. $-\dfrac{3}{7} + \dfrac{1}{7}$ c. $-\dfrac{2}{5} + \left(-\dfrac{2}{5}\right)$

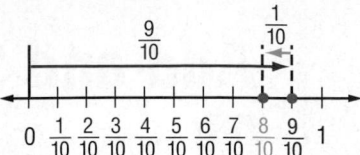 **EXAMPLES** **Subtract Like Fractions**

Subtract. Write in simplest form.

3 $\dfrac{9}{10} - \dfrac{1}{10}$

$\dfrac{9}{10} - \dfrac{1}{10} = \dfrac{9-1}{10}$ Subtract the numerators.

$= \dfrac{8}{10}$ Simplify.

$= \dfrac{4}{5}$ Simplify.

4 $-\dfrac{5}{8} - \dfrac{3}{8}$

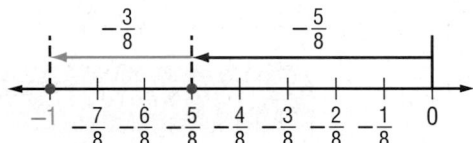

$-\dfrac{5}{8} - \dfrac{3}{8} = -\dfrac{5}{8} + \left(-\dfrac{3}{8}\right)$ To subtract $\dfrac{3}{8}$, add $-\dfrac{3}{8}$.

$= \dfrac{-5 + (-3)}{8}$ Add the numerators.

$= -\dfrac{8}{8}$ Simplify.

$= -1$ Simplify.

5 Find $\dfrac{5}{8} - \dfrac{7}{8}$.

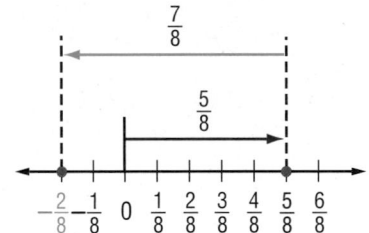

$\dfrac{5}{8} - \dfrac{7}{8} = \dfrac{5-7}{8}$ Subtract the numerators.

$= -\dfrac{2}{8}$ Simplify.

$= -\dfrac{1}{4}$ Simplify.

Check $-\dfrac{2}{8} + \dfrac{7}{8} = \dfrac{5}{8}$ ✓

✓ CHECK Your Progress

d. $-\dfrac{5}{9} - \dfrac{2}{9}$ **e.** $-\dfrac{11}{12} - \left(-\dfrac{5}{12}\right)$ **f.** $-\dfrac{3}{4} - \dfrac{1}{4}$

g. $\dfrac{5}{9} - \dfrac{2}{9}$ **h.** $\dfrac{11}{12} - \dfrac{5}{12}$ **i.** $\dfrac{7}{10} - \dfrac{3}{10}$

REAL-WORLD EXAMPLE

6 **POPULATION** About $\frac{6}{100}$ of the population of the United States lives in Florida. Another $\frac{4}{100}$ lives in Ohio. How much more of the U.S. population lives in Florida than in Ohio?

$$\frac{6}{100} - \frac{4}{100} = \frac{6-4}{100} \qquad \text{Subtract the numerators.}$$

$$= \frac{2}{100} \text{ or } \frac{1}{50} \qquad \text{Simplify.}$$

About $\frac{1}{50}$ more of the U.S. population lives in Florida than in Ohio.

Check 6 hundredths minus 4 hundredths equals 2 hundredths. ✓

 CHECK Your Progress

j. **JUICE** Two-fifths quart of pineapple juice was added to a bowl containing $\frac{3}{5}$ quart of orange juice. How many total quarts of pineapple juice and orange juice are in the bowl?

✓ CHECK Your Understanding

Examples 1–5 **Add or subtract. Write in simplest form.**

1. $\frac{3}{5} + \frac{1}{5}$ **2.** $\frac{2}{7} + \frac{1}{7}$ **3** $-\frac{3}{4} + \left(-\frac{3}{4}\right)$

4. $\frac{3}{8} - \frac{1}{8}$ **5.** $-\frac{4}{5} - \left(-\frac{1}{5}\right)$ **6.** $\frac{2}{7} - \frac{6}{7}$

Example 6

7. **STATES** Of the 50 states in the United States, 14 have an Atlantic Ocean coastline and 5 have a Pacific Ocean coastline. What fraction of U.S. states have either an Atlantic Ocean or Pacific Ocean coastline?

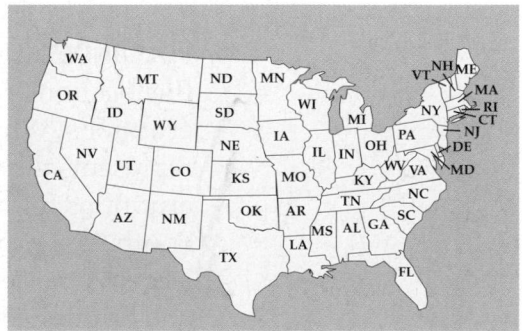

8. **LUNCH** The table shows the lunch count for three seventh-grade classes. Write the fraction in simplest form that represents how much more of the seventh-grade class is buying lunch rather than packing.

Lunch Count		
Class	Buying	Packing
Mrs. Savage	15	10
Mr. LeGault	19	8
Miss Cappizella	12	12

● = **Step-by-Step Solutions** begin on page R1.
Extra Practice begins on page EP2.

Examples 1–5 Add or subtract. Write in simplest form.

9 $\frac{4}{5}+\frac{3}{5}$ **10.** $\frac{5}{7}+\frac{6}{7}$ **11.** $\frac{3}{8}+\left(-\frac{7}{8}\right)$ **12.** $-\frac{1}{9}+\left(-\frac{5}{9}\right)$

13. $-\frac{5}{6}+\left(-\frac{5}{6}\right)$ **14.** $-\frac{15}{16}+\left(-\frac{7}{16}\right)$ **15.** $\frac{9}{10}-\frac{3}{10}$ **16.** $\frac{5}{8}-\frac{3}{8}$

17. $\frac{5}{14}-\left(-\frac{1}{14}\right)$ **18.** $-\frac{5}{9}-\frac{2}{9}$ **19.** $\frac{7}{12}-\frac{2}{12}$ **20.** $\frac{15}{18}-\frac{13}{18}$

Example 6 **21** **GRADES** In Mr. Navarro's first period class, $\frac{17}{28}$ of the students got an A on their math test. In his second period class, $\frac{11}{28}$ of the students got an A. What fraction more of the students got an A in Mr. Navarro's first period class than in his second period class? Write in simplest form.

22. COOKING A recipe for Michigan blueberry pancakes calls for $\frac{3}{4}$ cup flour, $\frac{1}{4}$ cup milk, and $\frac{1}{4}$ cup blueberries. How much more flour is needed than milk? Write in simplest form.

23. INSTANT MESSENGER The table shows the Instant Messenger abbreviations students at Hillside Middle School use the most.

Instant Messenger Abbreviations	
L8R (Later)	$\frac{48}{100}$
LOL (Laughing out loud)	$\frac{26}{100}$
BRB (Be right back)	$\frac{19}{100}$
CUL8R (See you later)	$\frac{7}{100}$

  a. What fraction of these students uses LOL or CUL8R when using Instant Messenger?

  b. What fraction of these students uses L8R or BRB when using Instant Messenger?

Use the order of operations to add or subtract. Write in simplest form.

24. $\frac{4}{5}+\frac{1}{5}+\frac{3}{5}$ **25.** $\frac{7}{8}+\left(-\frac{5}{8}\right)-\frac{1}{8}$ **26.** $\frac{13}{14}-\frac{5}{14}+\frac{6}{14}$

27. $\frac{2}{3}+\frac{2}{3}+\frac{2}{3}$ **28.** $\frac{4}{15}+\left(-\frac{9}{15}\right)+\frac{1}{15}$ **29.** $\frac{5}{7}-\frac{3}{7}+\frac{6}{7}$

30. VOLCANOES The graph shows the location of volcanic eruptions. What fraction represents the volcanic eruptions for both North and South America? How much larger is the section for Asia and South Pacific than for Europe? Write in simplest form.

31 **MEASUREMENT** How much longer than $\frac{5}{16}$ inch is $\frac{13}{16}$ inch?

32. MEASUREMENT What is the total of $1\frac{3}{4}$ cups and $\frac{3}{4}$ cup in simplest form?

Worldwide Volcano Eruptions

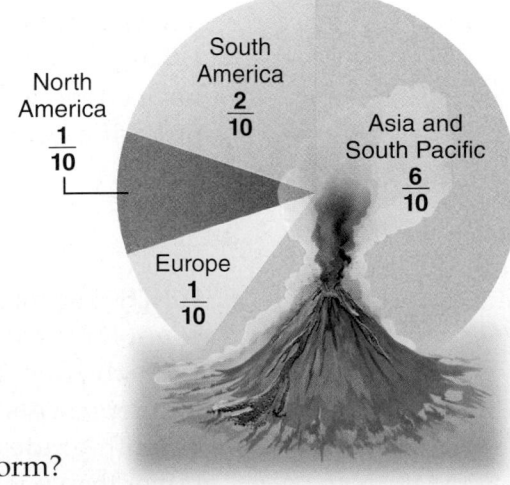

H.O.T. Problems

33. **OPEN ENDED** Select two like fractions with a difference of $\frac{1}{3}$ and with denominators that are *not* 3. Justify your selection.

34. **CHALLENGE** Simplify the following expression.
$$\frac{14}{15} + \frac{13}{15} - \frac{12}{15} + \frac{11}{15} - \frac{10}{15} + \cdots - \frac{4}{15} + \frac{3}{15} - \frac{2}{15} + \frac{1}{15}$$

35. ✏️ **WRITE MATH** Write a simple rule for adding and subtracting like fractions.

✔️ Test Practice

36. A group of friends bought two large pizzas and ate only part of each pizza. The pictures show how much of the pizzas were left.

First Pizza   Second Pizza

How many pizzas did they eat?

A. $\frac{3}{8}$ C. $1\frac{1}{4}$

B. $\frac{5}{8}$ D. $1\frac{3}{8}$

37. At a school carnival, homemade pies were cut into 8 equal-size pieces. Eric sold 13 pieces, Elena sold 7 pieces, and Tanya sold 10 pieces. Which expression can be used to find the total number of pies sold by Eric, Elena, and Tanya?

F. $13 + 7 + 10$ H. $\frac{13}{8} \times \frac{7}{8} \times \frac{10}{8}$

G. $8(13 + 7 + 10)$ I. $\frac{13}{8} + \frac{7}{8} + \frac{10}{8}$

38. A male Jumping Spider is shown below.

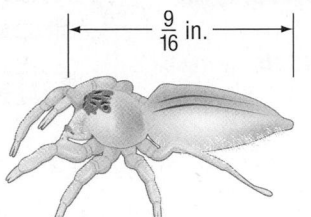

$\frac{9}{16}$ in.

The length of a female Jumping Spider is about $\frac{5}{16}$ inch. How much longer is a male Jumping Spider than a female Jumping Spider?

A. $\frac{7}{8}$ inch   C. $\frac{1}{4}$ inch

B. $\frac{1}{2}$ inch   D. $\frac{3}{16}$ inch

39. ✏️ **GRIDDED RESPONSE** What is the value of x that makes the statement below true?
$$\frac{7}{9} - \frac{x}{9} = \frac{1}{3}$$

Spiral Review

Replace each ● with <, >, or = to make a true sentence. (Lesson 1C)

40. $2\frac{7}{8}$ ● 2.75 41. $-\frac{1}{3}$ ● $-\frac{7}{3}$ 42. $\frac{5}{7}$ ● $\frac{4}{5}$ 43. $3\frac{6}{11}$ ● $3\frac{9}{14}$

Write each decimal as a fraction. (Lesson 1B)

44. 0.56 45. 0.375 46. 0.07 47. 0.019

Explore Unlike Fractions With Models

Main Idea
Use models to add and subtract fractions with unlike denominators.

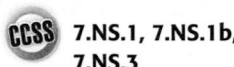

CCSS 7.NS.1, 7.NS.1b, 7.NS.3

TELEVISION Half of Brandon's class likes reality television best, and $\frac{1}{5}$ of the class likes sports television best. What fraction of the class likes either reality or sports television?

 **ACTIVITY**

1 **What do you need to find?** what fraction of the class likes reality or sports television

STEP 1   Model each fraction with fraction tiles.

$\frac{1}{2}$ [$\frac{1}{2}$]

$\frac{1}{5}$ [$\frac{1}{5}$]

STEP 2   To add, line up the end of the shaded part of the first tile with the beginning of the second tile.

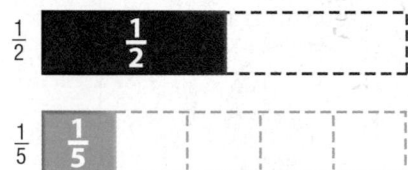

STEP 3   Test different fraction tiles below the model, lining up each with the beginning of the first tile. Does the last tile line up with the end of the second tile? If not, try another tile.

Once the correct tile is found, shade the sections between the beginning of the tile to the point where they line up.

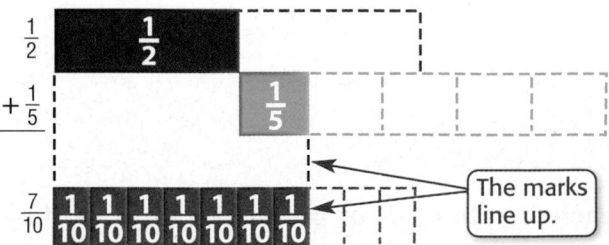

The marks line up.

$$\frac{1}{2} + \frac{1}{5} = \frac{7}{10}$$

So, $\frac{7}{10}$ of the class prefer reality or sports television.

ACTIVITY

2 Use fraction strips to find $\frac{7}{8} - \frac{3}{4}$.

STEP 1 Model each fraction.

$\frac{7}{8}$ | 1/8 | 1/8 | 1/8 | 1/8 | 1/8 | 1/8 | 1/8 |

$\frac{3}{4}$ | 1/4 | 1/4 | 1/4 |

STEP 2 To subtract, line up the ends of the shaded parts of each strip.

$\frac{7}{8}$ | 1/8 | 1/8 | 1/8 | 1/8 | 1/8 | 1/8 | 1/8 |

$-\frac{3}{4}$ | 1/4 | 1/4 | 1/4 |

STEP 3 Test different fraction strips below the model, checking to see if the marks line up. Then shade the sections between the beginning of the strip and the point where they line up.

$\frac{7}{8}$ | 1/8 | 1/8 | 1/8 | 1/8 | 1/8 | 1/8 | 1/8 |

$-\frac{3}{4}$ | 1/4 | 1/4 | 1/4 |

$\frac{1}{8}$ | 1/8 |

The marks line up.

So, $\frac{7}{8} - \frac{3}{4} = \frac{1}{8}$.

Practice and Apply

Use fraction strips to add or subtract.

1. $\frac{1}{10} + \frac{2}{5}$ **2.** $\frac{1}{6} + \frac{1}{2}$ **3.** $\frac{1}{2} + \frac{3}{4}$

4. $\frac{3}{8} - \frac{1}{4}$ **5.** $\frac{8}{9} - \frac{1}{3}$ **6.** $\frac{2}{3} - \frac{1}{4}$

Analyze the Results

Use the models from Activities 1 and 2 to complete the following.

7. $\frac{1}{2} + \frac{1}{5} = \frac{\blacksquare}{10} + \frac{\blacksquare}{10}$ **8.** $\frac{7}{8} - \frac{3}{4} = \frac{\blacksquare}{8} - \frac{\blacksquare}{8}$

9. MAKE A CONJECTURE What is the relationship between the number of separations on the answer fraction strip and the denominators of the fractions added or subtracted?

10. **WRITE MATH** Two thirds of the baseball team at Park Street Middle School lives within walking distance of the practice field. Another $\frac{1}{6}$ of the team lives close enough to ride their bike. What fraction of the team lives close enough to the practice field to walk or ride their bike there? Explain your reasoning.

Main Idea
Add and subtract fractions with unlike denominators.

 Vocabulary
unlike fractions

 7.NS.1, 7.NS.1b, 7.NS.1c, 7.NS.3, 7.EE.3

Add and Subtract Unlike Fractions

 MEASUREMENT The table shows the fractions of one hour for different minutes.

Number of Minutes	Fraction of One Hour
1	$\frac{1}{60}$
5	$\frac{5}{60}$
10	$\frac{10}{60}$
15	$\frac{15}{60}$
20	$\frac{20}{60}$
30	$\frac{30}{60}$

1. Write each fraction in simplest form.

2. What fraction of one hour is equal to the sum of 15 minutes and 20 minutes? Write in simplest form.

3. Explain why $\frac{1}{6}$ hour + $\frac{1}{3}$ hour = $\frac{1}{2}$ hour.

4. Explain why $\frac{1}{12}$ hour + $\frac{1}{2}$ hour = $\frac{7}{12}$ hour.

Before you can add two **unlike fractions**, or fractions with different denominators, one or both of the fractions must be renamed so that they have a common denominator.

Key Concept   **Add or Subtract Unlike Fractions**

To add or subtract fractions with different denominators,
- Rename the fractions using the least common denominator (LCD).
- Add or subtract as with like fractions.
- If necessary, simplify the sum or difference.

 EXAMPLE   **Add Unlike Fractions**

1 Find $\frac{1}{2} + \frac{1}{4}$.

Method 1   Use a model.

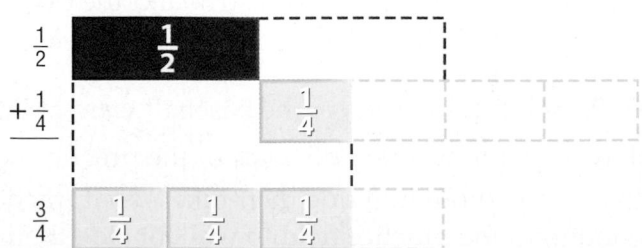

Method 2   Use the LCD.

The least common denominator of $\frac{1}{2}$ and $\frac{1}{4}$ is 4.

$$\frac{1}{2} + \frac{1}{4} = \frac{1 \times 2}{2 \times 2} + \frac{1 \times 1}{4 \times 1} \qquad \text{Rename using the LCD, 4.}$$
$$= \frac{2}{4} + \frac{1}{4} \qquad \text{Add the fractions.}$$
$$= \frac{3}{4} \qquad \text{Simplify.}$$

Using either method, $\frac{1}{2} + \frac{1}{4} = \frac{3}{4}$.

 CHOOSE Your Method

Add. Write in simplest form.

  a. $\frac{1}{6} + \frac{2}{3}$ **b.** $\frac{9}{10} + \left(-\frac{1}{2}\right)$ **c.** $\frac{1}{4} + \frac{3}{8}$

EXAMPLE   Subtract Unlike Fractions

2 Find $\frac{2}{3} - \frac{1}{2}$.

Method 1   Use a model.

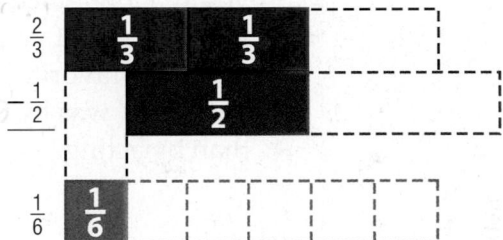

Study Tip

Check for Reasonableness
Estimate the difference in
Example 2.
$\frac{2}{3} - \frac{1}{2} \approx \frac{1}{2} - \frac{1}{2}$ or 0.
Compare $\frac{1}{6}$ to the estimate.
$\frac{1}{6} \approx 0$. So, the answer is
reasonable.

Method 2   Use the LCD.

The least common denominator of $\frac{2}{3}$ and $\frac{1}{2}$ is 6.

$$\frac{2}{3} - \frac{1}{2} = \frac{2 \times 2}{3 \times 2} - \frac{1 \times 3}{2 \times 3} \qquad \text{Rename using the LCD, 6.}$$
$$= \frac{4}{6} - \frac{3}{6} \qquad \text{Subtract the fractions.}$$
$$= \frac{1}{6} \qquad \text{Simplify.}$$

Check by adding $\frac{1}{6} + \frac{1}{2} = \frac{1}{6} + \frac{3}{6} = \frac{4}{6}$ or $\frac{2}{3}$ ✓

Using either method, $\frac{2}{3} - \frac{1}{2} = \frac{1}{6}$.

 CHOOSE Your Method

Subtract. Write in simplest form.

  d. $\frac{5}{8} - \frac{1}{4}$ **e.** $\frac{3}{4} - \frac{1}{3}$ **f.** $\frac{1}{2} - \left(-\frac{2}{5}\right)$

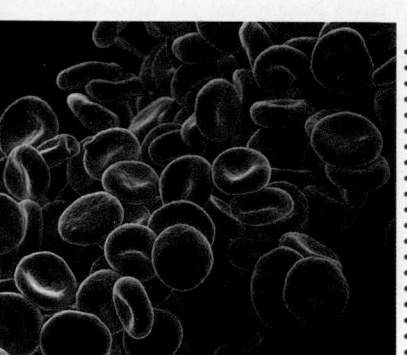

 HEALTH Use the table to find the fraction of the total population that has type A or type B blood.

Blood Type Frequencies				
ABO Type	O	A	B	AB
Fraction	$\frac{11}{25}$	$\frac{21}{50}$	$\frac{1}{10}$	$\frac{1}{25}$

To find the fraction of the total population, add $\frac{21}{50}$ and $\frac{1}{10}$.

$$\frac{21}{50} + \frac{1}{10} = \frac{21 \times 1}{50 \times 1} + \frac{1 \times 5}{10 \times 5}$$ Rename using the LCD, 50.

$$= \frac{21}{50} + \frac{5}{50}$$ Add the fractions.

$$= \frac{26}{50}$$ Simplify.

$$= \frac{13}{25}$$ Simplify.

So, $\frac{13}{25}$ of the population has type A or type B blood.

CHECK Your Progress

g. SURVEY The graphic shows the results of an online survey of over 36,000 youth. How much greater was the part of youth that said their favorite way to be "artsy" was by drawing than by acting?

What is your favorite way to be artsy?

Drawing	$\frac{8}{25}$
Acting	$\frac{7}{50}$
Making music	$\frac{7}{50}$
Taking pictures	$\frac{11}{100}$
Writing	$\frac{3}{50}$

✓ CHECK Your Understanding

Examples 1 and 2 Add or subtract. Write in simplest form.

1. $\frac{4}{9} + \frac{1}{3}$ **2.** $-\frac{5}{6} + \left(-\frac{4}{9}\right)$ **3.** $\frac{3}{8} - \left(-\frac{1}{4}\right)$ **4.** $\frac{4}{5} - \frac{3}{10}$

5 $\frac{1}{6} + \frac{3}{8}$ **6.** $\frac{2}{3} + \frac{5}{6}$ **7.** $\frac{7}{12} - \frac{5}{6}$ **8.** $\frac{3}{4} - \frac{1}{3}$

Example 3 Choose an operation to solve each problem. Explain your reasoning. Then solve the problem. Write in simplest form.

9. MEASUREMENT Cassandra cuts $\frac{5}{16}$ inch off the top of a photo and $\frac{3}{8}$ inch off the bottom. How much shorter is the total height of the photo now?

10. CHORES A bucket was $\frac{7}{8}$ full with soapy water. After washing the car, the bucket was only $\frac{1}{4}$ full. What part of the water was used?

Practice and Problem Solving

 = **Step-by-Step Solutions** begin on page R1.
Extra Practice begins on page EP2.

Examples 1 and 2   Add or subtract. Write in simplest form.

11. $\frac{3}{5} + \frac{1}{10}$ 12. $\frac{5}{8} + \frac{1}{4}$ **13** $\frac{5}{6} - \left(-\frac{2}{3}\right)$ 14. $\left(-\frac{7}{10}\right) - \frac{2}{5}$

15. $-\frac{1}{15} + \left(-\frac{3}{5}\right)$ 16. $-\frac{7}{12} + \frac{7}{10}$ 17. $\frac{5}{8} + \frac{11}{12}$ 18. $\frac{7}{9} + \frac{5}{6}$

19. $\frac{7}{9} - \frac{1}{3}$ 20. $\frac{4}{5} - \frac{1}{6}$ 21. $-\frac{4}{9} - \frac{2}{15}$ 22. $\frac{3}{10} - \left(-\frac{1}{4}\right)$

Example 3   Choose an operation to solve each problem. Explain your reasoning. Then solve the problem. Write in simplest form.

23. **MEASUREMENT** Ebony is building a shelf to hold the two boxes shown. What is the least width she should make the shelf?

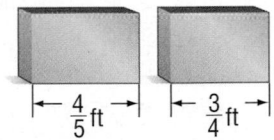

24. **MEASUREMENT** Mrs. Escalante was riding a bicycle on a bike path. After riding $\frac{2}{3}$ of a mile, she discovered that she still needed to travel $\frac{3}{4}$ of a mile to reach the end of the path. How long is the bike path?

25. **MEASUREMENT** Makayla bought $\frac{1}{4}$ pound of ham and $\frac{5}{8}$ pound of turkey. How much more turkey did she buy?

26. **ANIMALS** The three-toed sloth can travel $\frac{3}{20}$ mile per hour while a giant tortoise can travel $\frac{17}{100}$ mile per hour. How much faster, in miles per hour, is the giant tortoise?

Simplify.

27. $\frac{1}{7} + \frac{1}{2} + \frac{5}{28}$ 28. $\frac{1}{4} + \frac{5}{6} + \frac{7}{12}$ 29. $\frac{1}{6} + \left(\frac{2}{3} - \frac{1}{4}\right)$ 30. $\frac{5}{6} - \left(\frac{1}{2} + \frac{1}{3}\right)$

31. $1 + \frac{1}{4}$ 32. $1 - \frac{5}{8}$ 33. $2 + \frac{2}{3}$ 34. $3 - \frac{1}{6}$

35. **MONEY** Chellise saves $\frac{1}{5}$ of her allowance and spends $\frac{2}{3}$ of her allowance at the mall. What fraction of her allowance remains?

36. **ANALYZE TABLES** Pepita and Francisco each spend an equal amount of time on homework. The table shows the fraction of time they spend on each subject. Determine the missing fraction for each student.

| Homework | Fraction of Time | |
	Pepita	Francisco
Math	■	$\frac{1}{2}$
English	$\frac{2}{3}$	■
Science	$\frac{1}{6}$	$\frac{3}{8}$

ALGEBRA Evaluate each expression in simplest form if $a = \frac{3}{4}$ and $b = \frac{5}{6}$.

37 $\frac{1}{2} + a$ 38. $b - \frac{7}{10}$ 39. $b - a$ 40. $a + b$

41 **BOOK REPORTS** Four students were scheduled to give book reports in a 1-hour class period. After the first report, $\frac{2}{3}$ hour remained. If the next two students' reports took $\frac{1}{6}$ hour and $\frac{1}{4}$ hour, respectively, what fraction of the hour remained for the final student's report? Justify your answer.

42. **CELL PHONES** One hundred sixty cell phone owners were surveyed. What fraction of owners prefers using their cell phone for text messaging or taking pictures?

How do you use a cell phone?

Taking pictures $\frac{3}{8}$

Playing games $\frac{1}{4}$

$\frac{3}{8}$ Text messaging

43. **MEASUREMENT** LaTasha and Colin are jogging on a track. LaTasha jogs $\frac{1}{4}$ of a mile and then stops. Colin jogs $\frac{5}{8}$ of a mile, stops, and then turns around and jogs $\frac{1}{2}$ of a mile. Who is farther ahead on the track? How much farther? Explain.

H.O.T. Problems

44. **CHALLENGE** Fractions whose numerators are 1, such as $\frac{1}{2}$ or $\frac{1}{3}$, are called *unit fractions*. Describe a method you can use to add two unit fractions mentally. Explain your reasoning and use your method to find $\frac{1}{99} + \frac{1}{100}$.

45. **OPEN ENDED** Provide a counterexample to the following statement.

The sum of three fractions with odd numerators is never $\frac{1}{2}$.

46. **FIND THE ERROR** Theresa is finding $\frac{1}{4} + \frac{3}{5}$. Find her mistake and correct it.

$$\frac{1}{4} + \frac{3}{5} = \frac{1+3}{4+5}$$

47. **WRITE MATH** To make a cake, Felicia needs 1 cup of flour, but she only has a $\frac{2}{3}$-cup measure and a $\frac{3}{4}$-cup measure. Which method will bring her closest to having the amount of flour she needs? Explain.

a. Fill the $\frac{2}{3}$-cup measure twice.

c. Fill the $\frac{2}{3}$-cup measure once.

b. Fill the $\frac{3}{4}$-cup measure twice.

d. Fill the $\frac{3}{4}$-cup measure once.

48. The table gives the number of hours Orlando spent at football practice for one week.

Day	Time (h)
Monday	$\frac{1}{2}$
Tuesday	2
Wednesday	$\frac{1}{3}$
Thursday	$\frac{5}{6}$
Friday	$\frac{1}{2}$
Saturday	$\frac{3}{4}$

How many more hours did he practice on Thursday than on Saturday?

A. $\frac{2}{3}$ h

B. $\frac{1}{3}$ h

C. $\frac{1}{4}$ h

D. $\frac{1}{12}$ h

49. Which of the following is the prime factorization of the lowest common denominator of $\frac{7}{12} + \frac{11}{18}$?

F. 2×3

G. 2×3^2

H. $2^2 \times 3^2$

I. $2^3 \times 3$

50. Brett has $\frac{5}{6}$ of his weekly allowance left to spend. He has budgeted $\frac{1}{8}$ of his allowance to save for a new video game. How much of his weekly allowance will he have left after putting the savings away?

A. $\frac{4}{7}$

B. $\frac{3}{8}$

C. $\frac{7}{12}$

D. $\frac{17}{24}$

Spiral Review

Add or subtract. Write in simplest form. (Lesson 2A)

51. $\frac{7}{10} + \frac{1}{10}$ **52.** $\frac{3}{8} - \frac{1}{8}$ **53.** $\frac{5}{18} + \frac{7}{18}$ **54.** $\frac{11}{20} - \frac{3}{20}$

55. $\frac{5}{14} + \frac{3}{14}$ **56.** $\frac{20}{21} - \frac{5}{21}$ **57.** $\frac{11}{19} + \frac{8}{19}$ **58.** $\frac{29}{40} - \frac{14}{40}$

Replace each ● with >, <, or = to make a true sentence. (Lesson 1C)

59. $\frac{3}{8}$ ● $\frac{4}{9}$ **60.** $\frac{2}{11}$ ● $\frac{1}{12}$ **61.** 80% ● $\frac{12}{15}$

62. SCHOOL This week, Tia took quizzes in her math and science classes. On which quiz did Tia have the better score? (Lesson 1C)

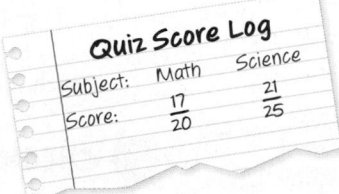

Quiz Score Log

Subject:	Math	Science
Score:	$\frac{17}{20}$	$\frac{21}{25}$

63. SPORTS To practice for a track meet, Lucinda ran $\frac{3}{8}$ mile and Vijay ran $\frac{4}{7}$ mile. Who ran the greater distance? Explain your reasoning. (Lesson 1C)

Main Idea

Add and subtract mixed numbers.

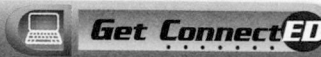

 7.NS.1, 7.NS.1b, 7.NS.1c, 7.NS.3, 7.EE.3

Add and Subtract Mixed Numbers

BABIES The birth weights of several babies in the hospital nursery are shown.

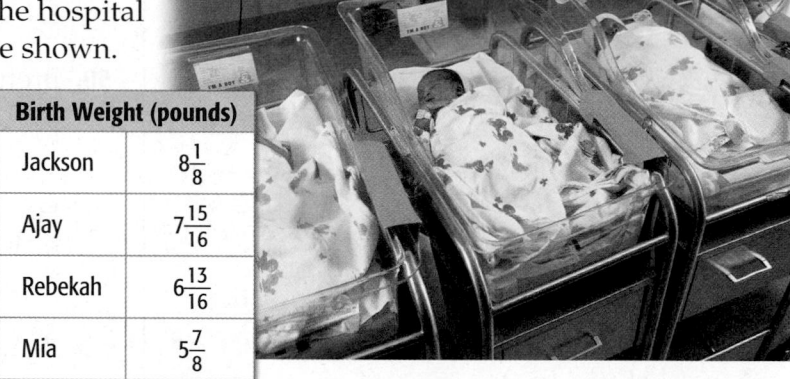

Birth Weight (pounds)	
Jackson	$8\frac{1}{8}$
Ajay	$7\frac{15}{16}$
Rebekah	$6\frac{13}{16}$
Mia	$5\frac{7}{8}$

1. Write an expression to find how much more Ajay weighs than Mia.

2. Rename the fractions using the LCD.

3. Find the difference of the fractional parts of the mixed numbers.

4. Find the difference of the whole numbers.

5. **MAKE A CONJECTURE** Explain how to find $7\frac{15}{16} - 5\frac{7}{8}$. Then use your conjecture to find the difference.

To add or subtract mixed numbers, first add or subtract the fractions. If necessary, rename them using the LCD. Then add or subtract the whole numbers and simplify if necessary.

 EXAMPLES Add and Subtract Mixed Numbers

1 **Find $7\frac{4}{9} + 10\frac{2}{9}$. Write in simplest form.**

Estimate $7 + 10 = 17$

$$
\begin{array}{r}
7\frac{4}{9} \\
+\ 10\frac{2}{9} \\
\hline
17\frac{6}{9} \text{ or } 17\frac{2}{3}
\end{array}
$$
Add the whole numbers and fractions separately.

Simplify.

Check for Reasonableness $17\frac{2}{3} \approx 17$ ✓

 CHECK Your Progress

a. $6\frac{1}{8} + 2\frac{5}{8}$ b. $5\frac{1}{5} + 2\frac{3}{10}$ c. $1\frac{5}{9} + 4\frac{1}{6}$

2 Find $8\frac{5}{6} - 2\frac{1}{3}$. Write in simplest form.

Estimate $9 - 2 = 7$

$$8\frac{5}{6} \rightarrow 8\frac{5}{6}$$

$$-2\frac{1}{3} \rightarrow -2\frac{2}{6} \qquad \text{Rename the fraction using the LCD. Then subtract.}$$

$$6\frac{3}{6} \text{ or } 6\frac{1}{2} \qquad \text{Simplify.}$$

Check for Reasonableness $6\frac{1}{2} \approx 7$ ✓

 CHECK Your Progress

Subtract. Write in simplest form.

d. $5\frac{4}{5} - 1\frac{3}{10}$ **e.** $13\frac{7}{8} - 9\frac{3}{4}$ **f.** $8\frac{2}{3} - 2\frac{1}{2}$

g. $7\frac{3}{4} - 4\frac{1}{3}$ **h.** $11\frac{5}{6} - 3\frac{1}{8}$ **i.** $9\frac{4}{7} - 5\frac{1}{2}$

> ## Study Tip
>
> **Fractions Greater Than One** An improper fraction has a numerator that is greater than or equal to the denominator. Examples of improper fractions are $\frac{5}{4}$ and $2\frac{6}{5}$.

Sometimes when you subtract mixed numbers, the fraction in the first mixed number is less than the fraction in the second mixed number. In this case, rename the first fraction as a fraction greater than or equal to one in order to subtract.

EXAMPLES **Rename Mixed Numbers to Subtract**

3 Find $2\frac{1}{3} - 1\frac{2}{3}$.

Estimate $2 - 1\frac{1}{2} = \frac{1}{2}$

Since $\frac{1}{3}$ is less than $\frac{2}{3}$, rename $2\frac{1}{3}$ before subtracting.

Change 1 to $\frac{3}{3}$.

$$2\frac{1}{3} \qquad = \qquad 1\frac{3}{3} + \frac{1}{3} \text{ or } 1\frac{4}{3}$$

$$2\frac{1}{3} \rightarrow 1\frac{4}{3} \qquad \text{Rename } 2\frac{1}{3} \text{ as } 1\frac{4}{3}.$$

$$-1\frac{2}{3} \rightarrow -1\frac{2}{3} \qquad \text{Subtract the whole numbers and then the fractions.}$$

$$\frac{2}{3}$$

Check for Reasonableness $\frac{2}{3} \approx \frac{1}{2}$ ✓

Find $8 - 3\frac{3}{4}$.   **Estimate** $8 - 4 = 4$

Using the denominator of the fraction in the subtrahend, $8 = 8\frac{0}{4}$.
Since $\frac{0}{4}$ is less than $\frac{3}{4}$, rename 8 before subtracting.

$$
\begin{array}{rll}
8 & \rightarrow & 7\frac{4}{4} \qquad \text{Rename 8 as } 7 + \frac{4}{4} \text{ or } 7\frac{4}{4}. \\
-3\frac{3}{4} & \rightarrow & -3\frac{3}{4} \qquad \text{Subtract.} \\
\hline
& & 4\frac{1}{4} \qquad \text{Check for Reasonableness } 4\frac{1}{4} \approx 4 \checkmark
\end{array}
$$

CHECK Your Progress

j. $11\frac{2}{5} - 2\frac{3}{5}$ **k.** $5\frac{3}{8} - 4\frac{11}{12}$ **l.** $7 - 1\frac{1}{2}$

REAL-WORLD EXAMPLE

 MEASUREMENT An urban planner is designing a skateboard park. What will be the length of the park and the parking lot combined?

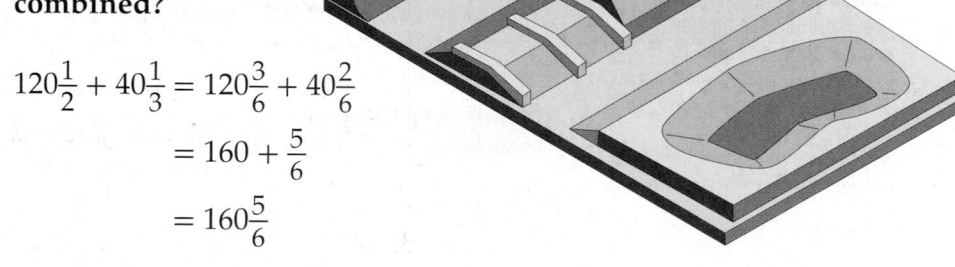

$40\frac{1}{3}$ ft

$120\frac{1}{2}$ ft

$$120\frac{1}{2} + 40\frac{1}{3} = 120\frac{3}{6} + 40\frac{2}{6}$$
$$= 160 + \frac{5}{6}$$
$$= 160\frac{5}{6}$$

The total length is $160\frac{5}{6}$ feet.

CHECK Your Progress

m. MEASUREMENT Jermaine walked $1\frac{5}{8}$ miles on Saturday and $2\frac{1}{2}$ miles on Sunday. How many more miles did he walk on Sunday?

✓ **CHECK Your Understanding**

Examples 1–4   **Add or subtract. Write in simplest form.**

1. $1\frac{5}{7} + 8\frac{1}{7}$ **2.** $8\frac{1}{2} + 3\frac{4}{5}$ **3.** $7\frac{5}{6} - 3\frac{1}{6}$ **4.** $9\frac{4}{5} - 2\frac{3}{4}$

 $3\frac{1}{4} - 1\frac{3}{4}$ **6.** $5\frac{2}{3} - 2\frac{3}{5}$ **7.** $11 - 6\frac{3}{8}$ **8.** $16 - 5\frac{5}{6}$

Example 5   **9. CARS** A hybrid car's gas tank can hold $11\frac{9}{10}$ gallons of gasoline. It contains $8\frac{3}{4}$ gallons of gasoline. How much more gasoline is needed to fill the tank?

Practice and Problem Solving

 = **Step-by-Step Solutions** begin on page R1.
Extra Practice begins on page EP2.

Examples 1–4 **Add or subtract. Write in simplest form.**

10. $2\frac{1}{9} + 7\frac{4}{9}$ **11.** $3\frac{2}{7} + 4\frac{3}{7}$ **12.** $10\frac{4}{5} - 2\frac{1}{5}$ **13.** $8\frac{6}{7} - 6\frac{5}{7}$

14. $9\frac{4}{5} - 2\frac{3}{10}$ **15** $11\frac{3}{4} - 4\frac{1}{3}$ **16.** $8\frac{5}{12} + 11\frac{1}{4}$ **17.** $8\frac{3}{8} + 10\frac{1}{3}$

18. $9\frac{1}{5} - 2\frac{3}{5}$ **19.** $6\frac{1}{4} - 2\frac{3}{4}$ **20.** $6\frac{3}{5} - 1\frac{2}{3}$ **21.** $4\frac{3}{10} - 1\frac{3}{4}$

22. $14\frac{1}{6} - 7\frac{1}{3}$ **23.** $12\frac{1}{2} - 6\frac{5}{8}$ **24.** $8 - 3\frac{2}{3}$ **25.** $13 - 5\frac{5}{6}$

Example 5 **Choose an operation to solve each problem. Explain your reasoning. Then solve the problem. Write your answer in simplest form.**

26. HIKING If Sara and Maggie hiked both of the trails listed in the table, how far did they hike altogether?

Trail	Length (mi)
Woodland Park	$3\frac{2}{3}$
Mill Creek Way	$2\frac{5}{6}$

27. JEWELRY Margarite made the jewelry shown. If the necklace is $10\frac{5}{8}$ inches longer than the bracelet, how long is the necklace that Margarite made?

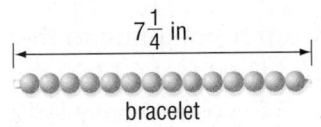

$7\frac{1}{4}$ in.

bracelet

necklace

28. GARDENS The length of Kasey's garden is $4\frac{5}{8}$ feet. Find the width of Kasey's garden if it is $2\frac{7}{8}$ feet shorter than the length.

29. HAIRSTYLES Before Alameda got her haircut, the length of her hair was $9\frac{3}{4}$ inches. After her haircut, the length was $6\frac{1}{2}$ inches. How many inches did she have cut?

Add or subtract. Write in simplest form.

30. $10 - 3\frac{5}{11}$ **31.** $24 - 8\frac{3}{4}$ **32.** $6\frac{1}{6} + 1\frac{2}{3} + 5\frac{5}{9}$ **33.** $3\frac{1}{4} + 2\frac{5}{6} - 4\frac{1}{3}$

34. TIME Karen wakes up at 6:00 A.M. It takes her $1\frac{1}{4}$ hours to shower, get dressed, and comb her hair. It takes her $\frac{1}{2}$ hour to eat breakfast, brush her teeth, and make her bed. At what time will she be ready for school?

MEASUREMENT Find the perimeter of each figure. Write in simplest form.

35

$2\frac{3}{8}$ yd $2\frac{3}{8}$ yd

$2\frac{3}{8}$ yd

36.

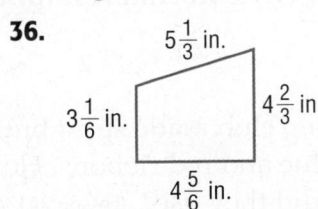

$5\frac{1}{3}$ in.

$3\frac{1}{6}$ in. $4\frac{2}{3}$ in.

$4\frac{5}{6}$ in.

37. **OPEN ENDED** A board with a length of $3\frac{7}{8}$ feet needs to be cut from a $5\frac{1}{2}$-foot existing board. Write and solve a subtraction problem to find the amount of wood that would be left after the cut is made.

38. **CHALLENGE** A string is cut in half. One of the halves is thrown away. One fifth of the remaining half is cut away and the piece left is 8 feet long. How long was the string initially? Justify your answer.

39. **WRITE MATH** The fence of a rectangular garden is constructed from 12 feet of fencing wire. Suppose that one side of the garden is $2\frac{5}{12}$ feet long. Explain how to find the length of the other side.

Test Practice

40. The distance from home plate to the pitcher's mound is 60 feet 6 inches and from home plate to second base is 127 feet $3\frac{3}{8}$ inches. Find the distance from the pitcher's mound to second base.

 A. 68 ft $3\frac{1}{4}$ in.

 B. 67 ft $8\frac{3}{4}$ in.

 C. 67 ft $2\frac{5}{8}$ in.

 D. 66 ft $9\frac{3}{8}$ in.

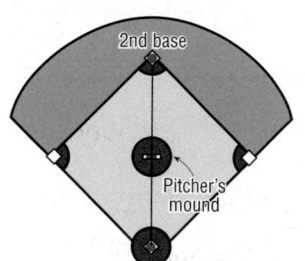

2nd base

Pitcher's mound

Home plate

41. A recipe for party mix calls for $4\frac{3}{4}$ cups of cereal. The amount of peanuts needed is $1\frac{2}{3}$ cups less than the amount of cereal needed. How many cups of peanuts and cereal are needed?

 F. $3\frac{1}{12}$ cups

 G. $6\frac{1}{2}$ cups

 H. $7\frac{5}{6}$ cups

 I. $8\frac{1}{2}$ cups

Spiral Review

42. **SCHOOL** Kai did $\frac{1}{5}$ of her homework in class and $\frac{1}{3}$ more on the bus. What fraction of homework does she still need to do? (Lesson 2C)

Add or subtract. Write in simplest form. (Lesson 2C)

43. $\frac{1}{5} + \frac{1}{4}$

44. $\frac{1}{3} - \frac{1}{6}$

45. $\frac{4}{9} + \frac{2}{7}$

46. $\frac{11}{15} - \frac{3}{20}$

47. **CLUBS** The pep club made spirit buttons. They used blue and red ribbons. How much total ribbon did they use? (Lesson 2A)

Blue	Red
$\frac{3}{8}$ ft	$\frac{3}{8}$ ft

Mid-Chapter Check

Write each fraction or mixed number as a decimal. Use bar notation if the decimal is a repeating decimal. (Lesson 1B)

1. $\frac{7}{8}$ **2.** $-\frac{2}{9}$ **3.** $3\frac{13}{20}$

Write each decimal as a fraction in simplest form. (Lesson 1B)

4. 0.6 **5.** 0.48 **6.** -7.02

7. ANIMALS The maximum height of an Asian elephant is 9.8 feet. What mixed number represents this height? (Lesson 1B)

Replace each ● with <, >, or = to make a true sentence. (Lesson 1C)

8. $-\frac{3}{5}$ ● $-\frac{5}{9}$ **9.** $4\frac{7}{12}$ ● 4.75 **10.** $\frac{13}{20}$ ● 0.65

11. WATER The table at the right shows the fraction of each state that is water. Order the states from least to greatest fraction of water. (Lesson 1C)

What Part is Water?	
State	**Fraction**
Alaska	$\frac{3}{41}$
Michigan	$\frac{40}{97}$
Wisconsin	$\frac{1}{6}$

Add or subtract. Write in simplest form.
(Lesson 2A)

12. $-\frac{11}{15} - \frac{1}{15}$ **13.** $\frac{4}{14} - \frac{3}{14}$

14. $-\frac{1}{9} + \frac{2}{9}$ **15.** $\frac{5}{8} + \frac{3}{8}$

16. SCIENCE $\frac{39}{50}$ of Earth's atmosphere is made up of nitrogen while only $\frac{21}{100}$ is made up of oxygen. What fraction of Earth's atmosphere is either nitrogen or oxygen? (Lesson 2C)

Add or subtract. Write in simplest form.
(Lesson 2D)

17. $8\frac{3}{4} - 2\frac{5}{12}$ **18.** $5\frac{1}{6} - 1\frac{1}{3}$

19. $2\frac{5}{9} + 1\frac{2}{3}$ **20.** $2\frac{3}{5} + 6\frac{13}{15}$

21. MULTIPLE CHOICE The table shows the weight of a newborn infant for its first year. (Lesson 2D)

Month	Weight (lb)
0	$7\frac{1}{4}$
3	$12\frac{1}{2}$
6	$16\frac{5}{8}$
9	$19\frac{4}{5}$
12	$23\frac{3}{20}$

During which three-month period was the infant's weight gain the greatest?

A. 0–3 months **C.** 6–9 months

B. 3–6 months **D.** 9–12 months

22. MEASUREMENT How much does a $50\frac{1}{4}$-pound suitcase weigh in simplest form after $3\frac{7}{8}$ pounds are removed? (Lesson 2D)

23. MULTIPLE CHOICE The table gives the average annual snowfall for several U.S. cities. (Lesson 2D)

City	Average Snowfall (in.)
Anchorage, AK	$70\frac{4}{5}$
Mount Washington, NH	$259\frac{9}{10}$
Buffalo, NY	$93\frac{3}{5}$
Birmingham, AL	$1\frac{1}{2}$

On average, how many more inches of snow does Mount Washington, New Hampshire, receive than Anchorage, Alaska?

F. $330\frac{7}{10}$ in. **H.** $166\frac{3}{10}$ in.

G. $189\frac{1}{10}$ in. **I.** $92\frac{1}{10}$ in.

PART ⟩ A ⟩ B ⟩ C ⟩ D

Explore Multiply Fractions with Models

Main Idea

Model multiplication of fractions and mixed numbers.

7.NS.2, 7.NS.3

Just as the product 3×4 is the number of square units in a rectangle, the product of two fractions can be shown using area models.

ACTIVITY

1 Find $\frac{3}{4} \times \frac{2}{3}$ using a geoboard.

The first factor is 3 *fourths* and the second factor is 2 *thirds*.

STEP 1 Use one geoband to show fourths and another to show thirds on the geoboard.

STEP 2 Use geobands to form a rectangle. Place one geoband on the peg to show 3 fourths and another on the peg to show 2 thirds.

STEP 3 Connect the geobands to show a small rectangle.

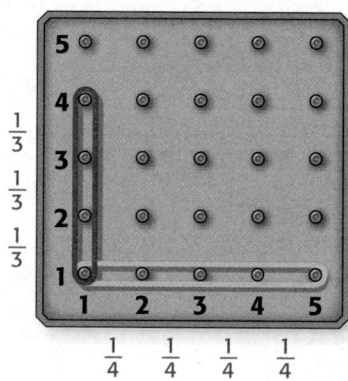

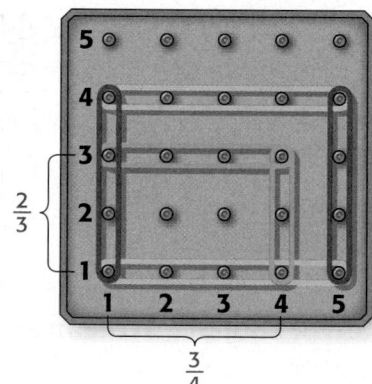

The area of the small rectangle is 6 square units. The area of the large rectangle is 12 square units. So, $\frac{3}{4} \times \frac{2}{3} = \frac{6}{12}$ or $\frac{1}{2}$.

ACTIVITY

2 Find $2 \times \frac{1}{4}$ using an area model.

STEP 1 Draw two squares side by side. Shade a rectangle that is 2 units long and $\frac{1}{4}$ unit wide.

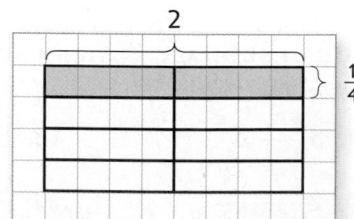

STEP 2   Rearrange the shaded parts into a unit square.

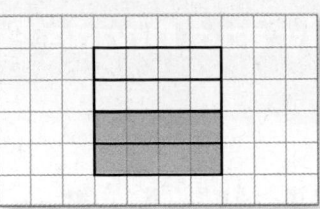

Since $\frac{2}{4}$ or $\frac{1}{2}$ of the unit square is shaded, $2 \times \frac{1}{4} = \frac{1}{2}$.

ACTIVITY

3   Find $1\frac{2}{3} \times \frac{1}{2}$ using a model.

  STEP 1   Draw two squares side by side. Shade a rectangle that is $1\frac{2}{3}$ units long and $\frac{1}{2}$ unit wide.

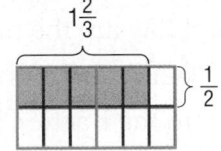

  STEP 2   Rearrange the shaded parts into a unit square.

Since $\frac{5}{6}$ of the unit square is shaded, $1\frac{2}{3} \times \frac{1}{2} = \frac{5}{6}$.

Practice and Apply

Write a multiplication sentence for the product shown in each model.

1.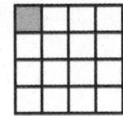

2.

Find each product using a geoboard or an area model.

3. $\frac{1}{4} \times \frac{1}{3}$ **4.** $\frac{1}{2} \times \frac{1}{2}$ **5.** $\frac{3}{4} \times \frac{1}{2}$ **6.** $\frac{2}{3} \times \frac{1}{4}$

7. $3 \times \frac{2}{3}$ **8.** $2 \times \frac{2}{5}$ **9.** $1\frac{1}{4} \times \frac{1}{5}$ **10.** $2\frac{1}{2} \times \frac{3}{4}$

Analyze the Results

11. MAKE A CONJECTURE Refer to Activities 1 and 2. How are the numerators and denominators of the factors related to each product?

12. WRITE MATH Write a rule you can use to multiply two fractions.

Main Idea

Multiply fractions and mixed numbers.

 **7.NS.2, 7.NS.2a, 7.NS.2c, 7.NS.3, 7.EE.3**

Multiply Fractions

LUNCH Two thirds of the students at the lunch table ordered a hamburger for lunch. One half of those students ordered cheeseburgers.

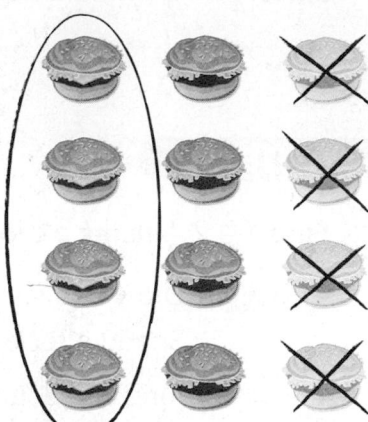

1. What fraction of the students at the lunch table ordered a cheeseburger?

2. How are the numerators and denominators of $\frac{2}{3}$ and $\frac{1}{2}$ related to the fraction in Exercise 1?

Key Concept   Multiply Fractions

Words   To multiply fractions, multiply the numerators and multiply the denominators.

Examples   **Numbers**   **Algebra**

$$\frac{1}{2} \times \frac{2}{3} = \frac{1 \times 2}{2 \times 3} \text{ or } \frac{2}{6} \qquad \frac{a}{b} \cdot \frac{c}{d} = \frac{a \cdot c}{b \cdot d} \text{ or } \frac{ac}{bd}, \text{ where } b, d \neq 0$$

EXAMPLES   Multiply Fractions

Multiply. Write in simplest form.

1 $\frac{1}{2} \times \frac{1}{3}$

$\frac{1}{2} \times \frac{1}{3} = \frac{1 \times 1}{2 \times 3}$ ← Multiply the numerators.
              ← Multiply the denominators.

      $= \frac{1}{6}$ Simplify.

2 $2 \times \left(-\frac{3}{4}\right)$

$2 \times \left(-\frac{3}{4}\right) = \frac{2}{1} \times \left(\frac{-3}{4}\right)$ Write 2 as $\frac{2}{1}$ and $-\frac{3}{4}$ as $\frac{-3}{4}$.

          $= \frac{2 \times (-3)}{1 \times 4}$ ← Multiply the numerators.
                  ← Multiply the denominators.

          $= \frac{-6}{4} \text{ or } -1\frac{1}{2}$ Simplify.

✓ **CHECK Your Progress**

a. $\frac{3}{5} \times \frac{1}{2}$ **b.** $\frac{1}{3} \times \frac{3}{4}$ **c.** $\frac{2}{3} \times (-4)$

---

 160   Rational Numbers

If the numerator and denominator of either fraction have common factors, you can simplify before multiplying.

 EXAMPLE **Simplify Before Multiplying**

(3) Find $\frac{2}{7} \times \frac{3}{8}$. Write in simplest form.

$$\frac{2}{7} \times \frac{3}{8} = \frac{\overset{1}{\cancel{2}}}{7} \times \frac{3}{\underset{4}{\cancel{8}}} \qquad \text{Divide 2 and 8 by their GCF, 2.}$$

$$= \frac{1 \times 3}{7 \times 4} \text{ or } \frac{3}{28} \qquad \text{Multiply.}$$

✓ **CHECK Your Progress**

Multiply. Write in simplest form.

d. $-\frac{1}{3} \times \left(-\frac{3}{7}\right)$ **e.** $\frac{4}{9} \times \frac{1}{8}$ **f.** $\frac{5}{6} \times \frac{3}{5}$

 EXAMPLE **Multiply Mixed Numbers**

(4) Find $\frac{1}{2} \times 4\frac{2}{5}$. Write in simplest form. **Estimate** $\frac{1}{2} \times 4 = 2$

Study Tip

Simplifying If you forget to simplify before multiplying, you can always simplify the final answer. However, it is usually easier to simplify before multiplying.

Method 1 **Rename the mixed number.**

$$\frac{1}{2} \times 4\frac{2}{5} = \frac{1}{\underset{1}{\cancel{2}}} \times \frac{\overset{11}{\cancel{22}}}{5} \qquad \begin{array}{l}\text{Rename } 4\frac{2}{5} \text{ as an improper fraction, } \frac{22}{5}. \\ \text{Divide 2 and 22 by their GCF, 2.}\end{array}$$

$$= \frac{1 \times 11}{1 \times 5} \qquad \text{Multiply.}$$

$$= \frac{11}{5} \text{ or } 2\frac{1}{5} \qquad \text{Simplify.}$$

Method 2 **Use mental math.**

The mixed number $4\frac{2}{5}$ is equal to $4 + \frac{2}{5}$.

So, $\frac{1}{2} \times 4\frac{2}{5} = \frac{1}{2}\left(4 + \frac{2}{5}\right)$. Use the Distributive Property to multiply, then add mentally.

$$\frac{1}{2}\left(4 + \frac{2}{5}\right) = 2 + \frac{1}{5} \qquad \text{\textbf{Think} Half of 4 is 2 and half of 2 fifths is 1 fifth.}$$

$$= 2\frac{1}{5} \qquad \text{Rewrite the sum as a mixed number.}$$

So, $\frac{1}{2} \times 4\frac{2}{5} = 2\frac{1}{5}$. **Check for Reasonableness** $2\frac{1}{5} \approx 2$ ✓

✓ **CHOOSE Your Method**

Multiply. Write in simplest form.

g. $\frac{1}{4} \times 8\frac{4}{9}$ **h.** $5\frac{1}{3} \times 3$ **i.** $-1\frac{7}{8} \times \left(-2\frac{2}{5}\right)$

REAL-WORLD EXAMPLES

⑤ **SLEEP** Humans sleep about $\frac{1}{3}$ of each day. If each year is equal to $365\frac{1}{4}$ days, determine the number of days in a year the average human sleeps.

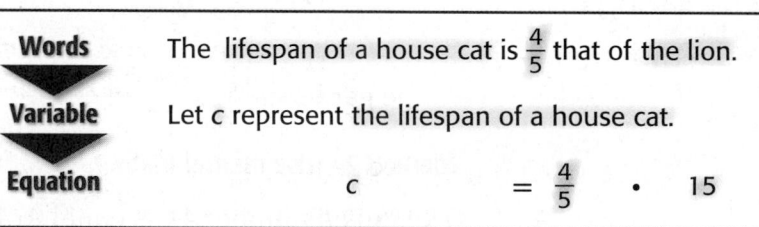

Words	Humans sleep about $\frac{1}{3}$ of $365\frac{1}{4}$ days.
Variable	Let d represent the number of days a human sleeps.
Equation	$d = \frac{1}{3} \cdot 365\frac{1}{4}$

$d = \frac{1}{3} \cdot 365\frac{1}{4}$ Write the equation.

$d = \frac{1}{3} \cdot \frac{1,461}{4}$ Rename the mixed number as an improper fraction.

$d = \frac{1}{\overset{}{\underset{1}{3}}} \cdot \frac{\overset{487}{\cancel{1,461}}}{4}$ Divide 3 and 1,461 by their GCF, 3.

$d = \frac{487}{4}$ or $121\frac{3}{4}$ Multiply. Then rename as a mixed number.

The average human sleeps $121\frac{3}{4}$ days each year.

⑥ **ANIMALS** The house cat has an average lifespan that is $\frac{4}{5}$ of a lion's lifespan. If a lion's average lifespan is 15 years, find the average lifespan of a house cat.

Words	The lifespan of a house cat is $\frac{4}{5}$ that of the lion.
Variable	Let c represent the lifespan of a house cat.
Equation	$c = \frac{4}{5} \cdot 15$

$c = \frac{4}{5} \cdot 15$ Write the equation.

$c = \frac{4}{5} \cdot \frac{15}{1}$ Write the whole number 15 as an improper fraction.

$c = \frac{4}{\underset{1}{\cancel{5}}} \cdot \frac{\overset{3}{\cancel{15}}}{1}$ Divide 5 and 15 by their GCF, 5.

$c = \frac{12}{1}$ or 12 Multiply, then simplify.

The average lifespan of a house cat is 12 years.

 CHECK Your Progress

j. COOKING Lloyd wishes to make $\frac{1}{2}$ of a recipe. If the original recipe calls for $3\frac{3}{4}$ cups of flour, how many cups should he use?

CHECK Your Understanding

Examples 1–4   **Multiply. Write in simplest form.**

1. $\frac{2}{3} \times \frac{1}{3}$ 2. $-2 \times \frac{2}{5}$ 3. $\frac{1}{6} \times 4$

4. $-\frac{1}{4} \times \left(-\frac{8}{9}\right)$ 5. $2\frac{1}{4} \times \frac{2}{3}$ 6. $-1\frac{5}{6} \times \left(-3\frac{3}{5}\right)$

Examples 5 and 6   7. **WEIGHT** The weight of an object on Mars is about $\frac{2}{5}$ its weight on Earth. How much would an 80-pound dog weigh on Mars?

Practice and Problem Solving

> ● = **Step-by-Step Solutions** begin on page R1.
> **Extra Practice** begins on page EP2.

Examples 1–4   **Multiply. Write in simplest form.**

8. $\frac{3}{4} \times \frac{1}{8}$ 9. $\frac{2}{5} \times \frac{2}{3}$ 10. $-9 \times \frac{1}{2}$ 11. $\frac{4}{5} \times (-6)$

12. $-\frac{1}{5} \times \left(-\frac{5}{6}\right)$ 13. $-\frac{4}{9} \times \left(-\frac{1}{4}\right)$ 14. $\frac{2}{3} \times \frac{1}{4}$ 15. $\frac{1}{12} \times \frac{3}{5}$

16. $\frac{4}{7} \times \frac{7}{8}$ $\frac{2}{5} \times \frac{15}{16}$ 18. $\left(-1\frac{1}{2}\right) \times \frac{2}{3}$ 19. $3\frac{1}{3} \times \left(-\frac{1}{5}\right)$

Examples 5 and 6   20. **DVDs** Each DVD storage case is about $\frac{1}{5}$ inch thick. What will be the height in simplest form of 12 cases sold together?

21. **PIZZA** Mark left $\frac{3}{8}$ of a pizza in the refrigerator. On Friday, he ate $\frac{1}{2}$ of what was left of the pizza. What fraction of the entire pizza did he eat on Friday?

22. **MEASUREMENT** The width of a vegetable garden is $\frac{1}{3}$ times its length. If the length of the garden is $7\frac{3}{4}$ feet, what is the width in simplest form?

23. **RECIPES** A recipe to make one batch of blueberry muffins calls for $4\frac{2}{3}$ cups of flour. How many cups of flour are needed to make 3 batches of blueberry muffins? Write in simplest form.

Multiply. Write in simplest form.

24. $-14 \times 1\frac{1}{7}$ 25. $3\frac{3}{4} \times 8$ 26. $9 \times 4\frac{2}{3}$ 27. $4 \times \frac{75}{6}$

28. $-3\frac{1}{4} \times \left(-2\frac{2}{3}\right)$ 29. $\left(\frac{1}{4}\right)^2$ 30. $\left(\frac{3}{10}\right)^4$ 31. $\left(-\frac{2}{3}\right)^3$

32. **MEASUREMENT** The width of the fish tank is $\frac{2}{5}$ of its length. What is the width of the fish tank in simplest form?

33. **BICYCLING** Philip rode his bicycle at $9\frac{2}{5}$ miles per hour. If he rode for $\frac{3}{4}$ of an hour, how many miles in simplest form did he cover?

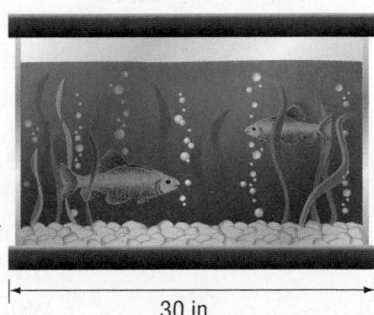

30 in.

Evaluate each verbal expression.

34. one half of five eighths

35. four sevenths of two thirds

36. nine tenths of one fourth

37. one third of eleven sixteenths

38. GRAPHIC NOVEL Refer to the graphic novel frame below for Exercises a–b.

a. If the height of the closet is 96 inches and Aisha would like to have 4 rows of cube organizers, what is the most the height of each cube organizer can be?

b. Aisha would like to stack 3 shoe boxes on top of each other at the bottom of the closet. If the height of each shoe box is $4\frac{1}{2}$ inches, what is the total height of the 3 boxes?

MEASUREMENT For Exercises 39–42, use measurement conversions.

39. Find $\frac{1}{2}$ of $\frac{1}{4}$ of a gallon.

40. What is $\frac{1}{60}$ of $\frac{1}{24}$ of a day?

41. Find $\frac{1}{100}$ of $\frac{1}{1,000}$ of a kilometer.

42. What is $\frac{1}{12}$ of $\frac{1}{3}$ of a yard?

ALGEBRA Evaluate each expression if $a = 4$, $b = 2\frac{1}{2}$, and $c = 5\frac{3}{4}$.

43. $a \times b + c$

44. $b \times c - a$

45. $2bc$

46. TELEVISION One evening, $\frac{2}{3}$ of the students in Rick's class watched television. Of those students, $\frac{3}{8}$ watched a reality show. Of the students that watched the show, $\frac{1}{4}$ of them recorded the show. What fraction of the students in Rick's class watched and recorded a reality TV show?

47 FOOD Alano wants to make one and a half batches of the pasta salad recipe shown at the right. How much of each ingredient will Alano need? Explain how you solved the problem.

Pasta Salad Recipe	
Ingredient	**Amount**
Broccoli	$1\frac{1}{4}$ c
Cooked pasta	$3\frac{3}{4}$ c
Salad dressing	$\frac{2}{3}$ c
Cheese	$1\frac{1}{3}$ c

48. MATH IN THE MEDIA Find examples of fractions in a newspaper or magazine, on television, or on the Internet. Write a real-world problem in which you would multiply fractions.

H.O.T. Problems

49. **CHALLENGE** Two positive improper fractions are multiplied. Is the product *sometimes*, *always*, or *never* less than 1? Explain your reasoning.

50. **OPEN ENDED** Write a real-world problem that involves finding the product of $\frac{3}{4}$ and $\frac{1}{8}$.

51. **WRITE MATH** Explain the difference in the processes of addition and multiplication of fractions.

Test Practice

52. Of the dolls in Marjorie's doll collection, $\frac{1}{5}$ have red hair. Of these, $\frac{3}{4}$ have green eyes. What fraction of Marjorie's doll collection has both red hair and green eyes?

 A. $\frac{2}{9}$

 B. $\frac{3}{20}$

 C. $\frac{4}{9}$

 D. $\frac{19}{20}$

53. Which description gives the relationship between a term and n, its position in the sequence?

Position	1	2	3	4	5	n
Value of Term	$\frac{1}{4}$	$\frac{1}{2}$	$\frac{3}{4}$	1	$1\frac{1}{4}$	

 F. Subtract 4 from n.

 G. Add $\frac{1}{4}$ to n.

 H. Multiply n by $\frac{1}{4}$.

 I. Divide n by $\frac{1}{4}$.

Spiral Review

54. **MEASUREMENT** How much longer is a $2\frac{1}{2}$-inch-long piece of string than a $\frac{2}{5}$-inch-long piece of string? (Lesson 2D)

55. **TRAILS** The table shows lengths of trails at Sharon Woods Park. (Lesson 2C)

 a. How much longer is Oak Trail than Willow Trail? Write in simplest form.

 b. If you walked Maple Trail and Oak Trail, what is the total distance you walked in simplest form?

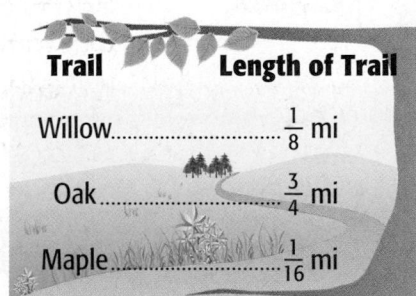

Trail	Length of Trail
Willow	$\frac{1}{8}$ mi
Oak	$\frac{3}{4}$ mi
Maple	$\frac{1}{16}$ mi

Replace each ● with <, >, or = to make a true sentence. (Lesson 1C)

56. $\frac{5}{12}$ ● 0.4

57. $\frac{3}{16}$ ● 12.5%

58. $3\frac{7}{6}$ ● $3\frac{6}{5}$

Problem-Solving Investigation

Main Idea Solve problems by drawing a diagram.

P.S.I. TEAM +

Draw a Diagram

CACEY: I drop a ball from a height of 12 feet. It hits the ground and bounces up half as high as it fell. This is true for each successive bounce.

YOUR MISSION: Draw a diagram to find the height the ball reaches after the fourth bounce.

Understand	You know the ball is dropped from a height of 12 feet. It bounces up half as high.
Plan	Draw a diagram to show the height of the ball after each bounce.
Solve	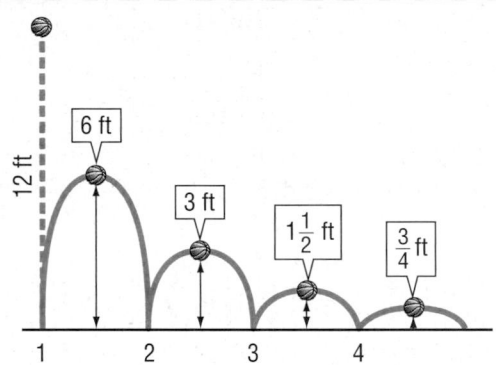 The ball reaches a height of $\frac{3}{4}$ foot after the fourth bounce.
Check	Start at 12 feet. Multiply by $\frac{1}{2}$ for each bounce: $12 \cdot \frac{1}{2} \cdot \frac{1}{2} \cdot \frac{1}{2} \cdot \frac{1}{2} = \frac{12}{16}$ or $\frac{3}{4}$. ✓

Analyze the Strategy

1. Determine what height a ball would reach after the fourth bounce if it is dropped from 12 feet and bounces up $\frac{2}{3}$ as high on each successive bounce.

• Draw a diagram.
• Look for a pattern.
• Choose an operation.

 = **Step-by-Step Solutions** begin on page R1.
Extra Practice begins on page EP2.

Use the *draw a diagram* strategy to solve Exercises 2–4.

2. **TRAVEL** Mr. Garcia has driven 60 miles, which is $\frac{2}{3}$ of the way to his sister's house. How much farther does he have to drive to get to his sister's house?

```
|------ 60 miles -----|
|   30    |    30     |          |
```

3. **DISTANCE** Alejandro and Pedro are riding their bikes to school. After 1 mile, they are $\frac{4}{5}$ of the way there. How much farther do they have to go?

4. **VOLUME** A swimming pool is being filled with water. After 25 minutes, $\frac{1}{6}$ of the swimming pool is filled. How much longer will it take to completely fill the pool, assuming the water rate is constant?

Use any strategy to solve Exercises 5–11.

5. **BASEBALL** Of Lee's baseball cards, $\frac{1}{5}$ show California players. Of these, $\frac{3}{8}$ show San Diego Padres players. Is the fraction of Lee's collection that shows Padres players $\frac{23}{40}$, $\frac{4}{13}$, or $\frac{3}{40}$? Justify your answer.

6. **FRACTIONS** Marta ate a quarter of a whole pie. Edwin ate $\frac{1}{4}$ of what was left. Cristina then ate $\frac{1}{3}$ of what was left. What fraction of the pie remains?

7. **GAMES** Eight members of a chess club are having a tournament. In the first round, every player will play a chess game against every other player. How many games will be in the first round of the tournament?

8. **MEASUREMENT** Kiaya is adding a 2-inch border to the length and width of a photograph as shown.

Find the area of the border to be added to the original photograph.

9. **RACES** Anna, Isabela, Mary, and Rachana ran a race. Anna is just ahead of Rachana. Rachana is two places behind Isabela. Isabela is a few seconds behind the leader, Mary. Place the girls in order from first to last.

10. **SEATS** The number of seats in the first row of a concert hall is 6. The second row has 9 seats, the third row has 12 seats, and the fourth row has 15 seats. How many seats will be in the eighth row?

11. **WRITE MATH** Write a real-world problem that could be solved by drawing a diagram. Exchange your problem with a classmate and solve.

Main Idea

Divide fractions and mixed numbers.

 7.NS.2, 7.NS.2b, 7.NS.3, 7.EE.3

Divide Fractions

Explore Cut two paper plates into four equal pieces each to show $2 \div \frac{1}{4}$.

1. How many $\frac{1}{4}$s are in 2 plates?

2. How would you model $3 \div \frac{1}{2}$?

3. What is true about $3 \div \frac{1}{2}$ and 3×2?

Dividing 2 by $\frac{1}{4}$ is the same as multiplying 2 by the reciprocal of $\frac{1}{4}$, which is 4.

reciprocals

$$2 \div \frac{1}{4} = 8 \qquad 2 \cdot 4 = 8$$

same result

Is this pattern true for any division expression?

Consider $\frac{7}{8} \div \frac{3}{4}$, which can be rewritten as $\dfrac{\frac{7}{8}}{\frac{3}{4}}$.

$$\frac{\frac{7}{8}}{\frac{3}{4}} = \frac{\frac{7}{8} \times \frac{4}{3}}{\frac{3}{4} \times \frac{4}{3}}$$ Multiply the numerator and denominator by the reciprocal of $\frac{3}{4}$, which is $\frac{4}{3}$.

$$= \frac{\frac{7}{8} \times \frac{4}{3}}{1} \qquad \frac{3}{4} \times \frac{4}{3} = 1$$

$$= \frac{7}{8} \times \frac{4}{3}$$

So, $\frac{7}{8} \div \frac{3}{4} = \frac{7}{8} \times \frac{4}{3}$.

Key Concept   **Divide by Fractions**

Words   To divide by a fraction, multiply by its multiplicative inverse, or reciprocal.

Examples   **Numbers**   **Algebra**

$$\frac{7}{8} \div \frac{3}{4} = \frac{7}{8} \cdot \frac{4}{3} \qquad \qquad \frac{a}{b} \div \frac{c}{d} = \frac{a}{b} \cdot \frac{d}{c}, \text{ where } b, c, d \neq 0$$

 **EXAMPLE** **Divide by Fractions**

① Find $\frac{3}{4} \div \left(-\frac{1}{2}\right)$. Write in simplest form.

Estimate $1 \div \left(-\frac{1}{2}\right) = \blacksquare$

Think How many groups of $\frac{1}{2}$ are in 1? $1 \div \frac{1}{2} = 2$, so $1 \div \left(-\frac{1}{2}\right) = -2$.

$\frac{3}{4} \div \left(-\frac{1}{2}\right) = \frac{3}{4} \cdot \left(-\frac{2}{1}\right)$ Multiply by the reciprocal of $-\frac{1}{2}$, which is $-\frac{2}{1}$.

$= \frac{3}{\overset{}{\underset{2}{4}}} \cdot \left(-\frac{\overset{1}{2}}{1}\right)$ Divide 4 and 2 by their GCF, 2.

$= -\frac{3}{2}$ or $-1\frac{1}{2}$ Multiply.

Check for Reasonableness $-1\frac{1}{2} \approx -2$ ✓

QUICK Review

Reciprocal
The reciprocal of a number is its multiplicative inverse. For example, the reciprocal of $\frac{5}{9}$ is $\frac{9}{5}$. The reciprocal of 8 is $\frac{1}{8}$.

CHECK Your Progress

Divide. Write in simplest form.

a. $\frac{3}{4} \div \frac{1}{4}$ **b.** $-\frac{4}{5} \div \frac{8}{9}$ **c.** $-\frac{5}{6} \div \left(-\frac{2}{3}\right)$

To divide by a mixed number, first rename the mixed number as a fraction greater than one. Then multiply the first fraction by the reciprocal, or multiplicative inverse, of the second fraction.

EXAMPLE **Divide by Mixed Numbers**

② Find $\frac{2}{3} \div 3\frac{1}{3}$. Write in simplest form.

Estimate $\frac{1}{2} \div 3 = \frac{1}{2} \cdot \frac{1}{3}$ or $\frac{1}{6}$

$\frac{2}{3} \div 3\frac{1}{3} = \frac{2}{3} \div \frac{10}{3}$ Rename $3\frac{1}{3}$ a fraction greater than one.

$= \frac{2}{3} \cdot \frac{3}{10}$ Multiply by the reciprocal of $\frac{10}{3}$, which is $\frac{3}{10}$.

$= \frac{\overset{1}{2}}{\underset{1}{3}} \cdot \frac{\overset{1}{3}}{\underset{5}{10}}$ Divide out common factors.

$= \frac{1}{5}$ Multiply.

Check for Reasonableness $\frac{1}{5}$ is close to $\frac{1}{6}$. ✓

CHECK Your Progress

Divide. Write in simplest form.

d. $5 \div 1\frac{1}{3}$ **e.** $-\frac{3}{4} \div 1\frac{1}{2}$ **f.** $2\frac{1}{3} \div 5$

g. NUTS In planning for a party, $5\frac{1}{4}$ pounds of cashews will be divided into $\frac{3}{4}$-pound bags. How many such bags can be made?

Study Tip

Dividing by a Whole Number Remember that a whole number can be written as a fraction with a 1 in the denominator. So, $2\frac{1}{3} \div 5$ can be rewritten as $2\frac{1}{3} \div \frac{5}{1}$.

3 **WOODWORKING** Students in a woodworking class are making butterfly houses. The side pieces of the house need to be $8\frac{1}{4}$ inches long. How many side pieces can be cut from a board measuring $49\frac{1}{2}$ inches long?

$8\frac{1}{4}$ in.

To find how many side pieces can be cut, divide $49\frac{1}{2}$ by $8\frac{1}{4}$.

Estimate Use compatible numbers. $48 \div 8 = 6$

$$49\frac{1}{2} \div 8\frac{1}{4} = \frac{99}{2} \div \frac{33}{4}$$ Rename the mixed numbers as fractions greater than one.

$$= \frac{99}{2} \cdot \frac{4}{33}$$ Multiply by the reciprocal of $\frac{33}{4}$, which is $\frac{4}{33}$.

$$= \frac{\overset{3}{\cancel{99}}}{\underset{1}{\cancel{2}}} \cdot \frac{\overset{2}{\cancel{4}}}{\underset{1}{\cancel{33}}}$$ Divide out common factors.

$$= \frac{6}{1} \text{ or } 6$$ Multiply.

So, 6 side pieces can be cut.

Check for Reasonableness Compare to the estimate. $6 = 6$ ✓

 CHECK Your Progress

h. FOOD Suppose a small box of cereal contains $12\frac{2}{3}$ cups of cereal. How many $1\frac{1}{3}$-cup servings are in the box? Write in simplest form.

i. MEASUREMENT The area of a rectangular bedroom is $146\frac{7}{8}$ square feet. If the width is $11\frac{3}{4}$ feet, find the length in simplest form.

✓ CHECK Your Understanding

Examples 1 and 2 — Divide. Write in simplest form.

1. $\frac{1}{8} \div \frac{1}{3}$ 2. $-\frac{3}{5} \div \left(-\frac{1}{4}\right)$ 3. $-3 \div \left(-\frac{6}{7}\right)$ 4. $\frac{3}{4} \div 6$

5. $\frac{1}{2} \div 7\frac{1}{2}$ 6. $\frac{4}{7} \div \left(-1\frac{2}{7}\right)$ 7. $5\frac{3}{5} \div 4\frac{2}{3}$ 8. $6\frac{1}{2} \div 3\frac{5}{7}$

Example 3 — **9. FOOD** Deandre has 7 apples and each apple is divided evenly into eighths. How many apple slices does Deandre have?

10. WALKING On Saturday, Lindsay walked $3\frac{1}{2}$ miles in $1\frac{2}{5}$ hours. What was her walking pace in miles per hour? Write in simplest form.

Practice and Problem Solving

  = **Step-by-Step Solutions** begin on page R1.
Extra Practice begins on page EP2.

Examples 1 and 2 **Divide. Write in simplest form.**

11. $\frac{3}{8} \div \frac{6}{7}$ **12.** $\frac{5}{9} \div \frac{5}{6}$ **13.** $-\frac{2}{3} \div \left(-\frac{1}{2}\right)$ **14.** $-\frac{7}{8} \div \frac{3}{4}$

15. $6 \div \left(-\frac{1}{2}\right)$ **16.** $-\frac{4}{9} \div (-2)$ **17.** $2\frac{2}{3} \div 4$ **18.** $5 \div \frac{1}{3}$

Example 3 **19. FOOD** Mason has 8 cups of popcorn kernels to divide into $\frac{2}{3}$-cup portions. How many portions will there be?

20. MOVIES Cheryl is organizing her movie collection. If each movie case is $\frac{3}{4}$ inch wide, how many movies can fit on a shelf 5 feet wide?

Divide. Write in simplest form.

21. $\frac{2}{3} \div 2\frac{1}{2}$ **22.** $\frac{8}{9} \div 5\frac{1}{3}$ **23.** $-4\frac{1}{2} \div 6\frac{3}{4}$ **24.** $-5\frac{2}{7} \div \left(-2\frac{1}{7}\right)$

25 $3\frac{4}{5} \div 1\frac{1}{3}$ **26.** $9\frac{1}{2} \div 2\frac{5}{6}$ **27.** $-5\frac{1}{5} \div \frac{2}{3}$ **28.** $-6\frac{7}{8} \div \left(-\frac{3}{4}\right)$

29. ICE CREAM Vinh bought $4\frac{1}{2}$ gallons of ice cream to serve at his birthday party. If a pint is $\frac{1}{8}$ of a gallon, how many pint-sized servings can be made?

30. BEVERAGES William has $8\frac{1}{4}$ cups of fruit juice. If he divides the juice into $\frac{3}{4}$-cup servings, how many servings will he have?

31. BIRDS Use the table that gives information about several types of birds of prey. Write your answers in simplest form.

a. How many times as heavy is the Golden Eagle as the Red-Tailed Hawk?

b. How many times as heavy is the Golden Eagle as the Northern Bald Eagle?

Bird	Maximun Weight (lb)
Golden Eagle	$13\frac{9}{10}$
Northern Bald Eagle	$9\frac{9}{10}$
Red-Tailed Hawk	$3\frac{1}{2}$

Real-World Link
Red-tailed hawks are large, stocky birds. Females are larger than males and can weigh up to $3\frac{1}{2}$ pounds.

Draw a model of each verbal expression and then evaluate the expression. Explain how the model shows the division process.

32. one half divided by two fifths

33. five eighths divided by one fourth

34. one and three eighths divided by one half

35. two and one sixth divided by two thirds

36. PIZZA A concession stand sells three types of pizza. The diagram shows how much pizza of each type was left when the concession stand closed. If the pizza is sold in slices that are $\frac{1}{8}$ of a whole pizza, how many more slices can be sold?

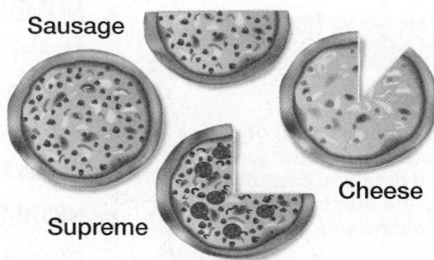

Sausage

Cheese

Supreme

ALGEBRA Evaluate each expression if $g = -\frac{1}{6}$, $h = \frac{1}{2}$, and $j = 3\frac{2}{3}$. Write in simplest form.

37. $j \div h$ **38.** $g \div j$ **39.** $3g \div h$ **40.** $h \div \left(\frac{1}{2}j\right)$

For each of the following, write your answers in simplest form.

41 **SHOPPING** A supermarket sells pretzels in $\frac{3}{4}$-ounce snack-sized bags or $12\frac{1}{2}$-ounce regular-sized bags. How many times larger is the regular-sized bag than the snack-sized bag?

42. MEASUREMENT A recipe calls for $2\frac{2}{3}$ cups of brown sugar and $\frac{2}{3}$ cup of confectioner's sugar. How many times greater is the number of cups of brown sugar in the recipe than of confectioner's sugar?

43. SCHOOL The table shows the number of hours students spend studying each week.

a. How many times greater was the number of students who spent over 10 hours each week studying than those who spent only 1–2 hours each week studying?

b. How many times greater was the number of students who spent 3 or more hours each week studying than those who spent less than 3 hours each week studying?

Weekly Study Hours	
Hours	**Fraction of Students**
None	$\frac{1}{50}$
1–2	$\frac{2}{25}$
3–5	$\frac{11}{50}$
6–7	$\frac{17}{100}$
8–10	$\frac{1}{5}$
Over 10	$\frac{19}{100}$
Not sure	$\frac{3}{25}$

44. SCHOOL SUPPLIES Tara bought a dozen folders. She took $\frac{1}{3}$ of the dozen and then divided the remaining folders equally among her four friends. What fraction of the dozen did each of her four friends receive and how many folders was this per person?

45. WEATHER A meteorologist has issued a thunderstorm warning. So far, the storm has traveled 35 miles in $\frac{1}{2}$ hour. If it is currently 5:00 P.M. and the storm is 105 miles away from you, at what time will the storm reach you? Explain how you solved the problem.

46. MULTIPLE REPRESENTATIONS Jorge recorded the distance that five of his friends live from his house on the chart that is shown.

a. **NUMBERS** Thuy lives about how many times farther away than Jamal?

b. **ALGEBRA** Write and solve an equation to find the mean number of miles that Jorge's friends live from his house. Write your answer in simplest form.

c. **MODEL** Draw a bar diagram that can be used to find how many more miles Lon travels than Lucia to get to Jorge's house.

Student	Miles
Lucia	$5\frac{1}{2}$
Lon	$8\frac{2}{3}$
Sam	$12\frac{5}{6}$
Jamal	$2\frac{7}{9}$
Thuy	$17\frac{13}{18}$

QUICK Review

Mean
To find the mean of a data set, find the sum of the data and then divide by the number of items in the data set.

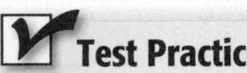

H.O.T. Problems

47. CHALLENGE If $\frac{5}{6}$ is divided by a certain fraction $\frac{a}{b}$, the result is $\frac{1}{4}$. What is the fraction $\frac{a}{b}$?

48. FIND THE ERROR Blake is finding $\frac{4}{5} \div \frac{6}{7}$. Find his mistake and correct it.

$$\frac{4}{5} \div \frac{6}{7} = \frac{5}{4} \cdot \frac{6}{7}$$
$$= \frac{30}{28} \text{ or } 1\frac{1}{14}$$

49. WRITE MATH If you divide a proper fraction by another proper fraction, is it possible to get a mixed number as an answer? Explain your reasoning.

 **Test Practice**

50. Which expression represents the **least** value?

A. $298 + \frac{1}{2}$

B. $298 - \frac{1}{2}$

C. $298 \times \frac{1}{2}$

D. $298 \div \frac{1}{2}$

51. How many small boxes of peanuts shown can be filled from the large box of peanuts?

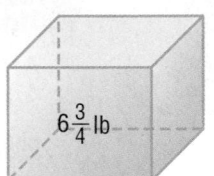 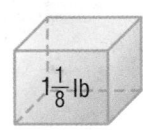

F. 4   H. 6

G. 5   I. 7

Spiral Review

52. ROCK CLIMBING A rock climber stops to rest at a ledge 150 feet above the ground. If this represents 75% of the total climb, how high above the ground is the top of the rock? (Lesson 3C)

53. Find $\frac{1}{10} \times \frac{5}{8}$. Write in simplest form. (Lesson 3B)

54. EXERCISE The table shows the number of miles Tonya jogged each week. What is the total number of miles that Tonya jogged? (Lesson 3B)

Week	1	2	3	4	5
Distance Jogged (mi)	$7\frac{1}{6}$	$7\frac{1}{6}$	$5\frac{2}{3}$	$5\frac{2}{3}$	$5\frac{2}{3}$

Chapter 3 Lesson 3D Multiply and Divide Fractions   **173**

Problem Solving in Fashion Design

A flair for *Fashion*

Do you enjoy reading fashion magazines, keeping up with the latest trends, and creating your own unique sense of style? You might want to consider a career in fashion design. Fashion designers create new designs for clothing, accessories, and shoes. In addition to being creative and knowledgeable about current fashion trends, fashion designers need to be able to take accurate measurements and calculate fit by adding, subtracting, and dividing measurements.

21ˢᵗ Century Careers

Are you interested in a career as a fashion designer? Take some of the following courses in high school.

- Algebra
- Art
- Digital Design
- Geometry

 Get Connect**ED**

Amount of Fabric Needed (yards)				
Dress Style	Size 8	Size 10	Size 12	Size 14
A	$3\frac{3}{8}$	$3\frac{1}{2}$	$3\frac{3}{4}$	$3\frac{7}{8}$
B	$3\frac{1}{4}$	$3\frac{1}{2}$	$3\frac{7}{8}$	4

Real-World Math

Use the information in the table to solve each problem. Write in simplest form.

1. For size 8, does Dress Style A or B require more fabric? Explain.

2. How many yards of fabric are needed to make Style A in sizes 8 and 14?

3. Estimate how many yards of fabric are needed to make Style B in each of the sizes shown. Then find the actual amount of fabric.

4. For Style B, how much more fabric is required for size 14 than for size 12?

5. A designer has half the amount of fabric needed to make Style A in size 10. How much fabric does she have?

6. A bolt has $12\frac{1}{8}$ yards of fabric left on it. How many dresses in Style B size 12 could be made? How much fabric is left over?

Main Idea
Multiply and divide monomials.

 **Vocabulary**
monomial

Multiply and Divide Monomials

EARTHQUAKES For each increase on the Richter scale, an earthquake's vibrations, or *seismic waves*, are 10 times greater. So, an earthquake of magnitude 4 has seismic waves that are 10 times greater than that of a magnitude 3 earthquake.

Richter Scale	Times Greater than Magnitude 3 Earthquake	Written Using Powers
4	$10 \times 1 = 10$	10^1
5	$10 \times 10 = 100$	$10^1 \times 10^1 = 10^2$
6	$10 \times 100 = 1,000$	$10^1 \times 10^2 = 10^3$
7	$10 \times 1,000 = 10,000$	$10^1 \times 10^3 = 10^4$
8	$10 \times 10,000 = 100,000$	$10^1 \times 10^4 = 10^5$

1. Examine the exponents of the powers in the last column. What do you observe?

2. MAKE A CONJECTURE Write a rule for determining the exponent of the product when you multiply powers with the same base. Test your rule by multiplying 2^2 and 2^4 using a calculator.

Recall that exponents are used to show repeated multiplication. You can use the definition of an exponent to find a rule for multiplying powers with the same base.

$$2^3 \cdot 2^4 = \underbrace{(2\cdot2\cdot2)}_{\text{3 factors}} \cdot \underbrace{(2\cdot2\cdot2\cdot2)}_{\text{4 factors}}$$

$$\underbrace{}_{\text{7 factors}}$$

$$= 2^7$$

Notice the sum of the original exponents is the exponent in the final product.

🗂 Key Concept **Product of Powers**

Words To multiply powers with the same base, add their exponents.

Symbols $a^m \cdot a^n = a^{m+n}$

Example $3^2 \cdot 3^4 = 3^{2+4}$ or 3^6

 **EXAMPLE** **Multiply Powers**

 Find $7^3 \cdot 7$. Express using exponents.

$7^3 \cdot 7 = 7^3 \times 7^1$ $7 = 7^1$ **Check** $7^3 \cdot 7 = (7 \cdot 7 \cdot 7)(7)$

$ = 7^{3+1}$ The common base is 7. $ = 7 \cdot 7 \cdot 7 \cdot 7$ or 7^4 ✓

$ = 7^4$ Add the exponents.

> **Study Tip**
>
> **Common Misconception**
> When multiplying powers, do not multiply the bases. $3^2 \cdot 3^4 = 3^6$, not 9^6.

✓ **CHECK Your Progress**

Simplify. Express using exponents.

a. $5^3 \cdot 5^4$ **b.** $\left(\frac{1}{2}\right)^2 \cdot \left(\frac{1}{2}\right)^9$

A **monomial** is a number, variable, or product of a number and one or more variables. Monomials can also be multiplied using the rule for the product of powers.

 EXAMPLES **Multiply Monomials**

Find each product. Express using exponents.

② $x^5 \cdot x^2$

$x^5 \cdot x^2 = x^{5+2}$ The common base is x.

$ = x^7$ Add the exponents.

③ $(-4n^3)(2n^6)$

$(-4n^3)(2n^6) = (-4 \cdot 2)(n^3 \cdot n^6)$ Use the Commutative and Associative Properties.

$ = (-8)(n^{3+6})$ The common base is n.

$ = -8n^9$ Add the exponents.

> **QUICK Review**
>
> **Properties**
> Commutative Property of Multiplication
> $a \cdot b = b \cdot a$
> Associative Property of Multiplication
> $(ab)c = a(bc)$

✓ **CHECK Your Progress**

Simplify. Express using exponents.

c. $-3m(-8m^4)$ **d.** $5^2x^2y^4 \cdot 5^3xy^3$

You can also write a rule for finding quotients of powers.

$$\frac{2^6}{2^1} = \frac{2 \cdot 2 \cdot 2 \cdot 2 \cdot 2 \cdot 2}{2} \quad \longleftarrow \boxed{\text{6 factors}}$$
$$\quad \longleftarrow \boxed{\text{1 factor}}$$

$$= \frac{2 \cdot 2 \cdot 2 \cdot 2 \cdot 2 \cdot \overset{1}{\cancel{2}}}{\underset{1}{\cancel{2}}} \quad \text{Divide out common factors.}$$

$$= 2^5 \quad \longleftarrow \boxed{\text{5 factors}} \quad \text{Simplify.}$$

Compare the difference between the original exponents and the exponent in the final quotient. This relationship is stated in the following Key Concept box.

Key Concept · Quotient of Powers

Words   To divide powers with the same base, subtract their exponents.

Symbols $\dfrac{a^m}{a^n} = a^{m-n}$, where $a \neq 0$

Example $\dfrac{4^5}{4^2} = 4^{5-2}$ or 4^3

 EXAMPLES Divide Powers

Find each quotient. Express using exponents.

4 $\dfrac{5^7}{5^4}$

$\dfrac{5^7}{5^4} = 5^{7-4}$ The common base is 5.

$\qquad = 5^3$ Subtract the exponents.

5 $\dfrac{y^5}{y^3}$

$\dfrac{y^5}{y^3} = y^{5-3}$ The common base is y.

$\qquad = y^2$ Subtract the exponents.

✔ CHECK Your Progress

Simplify. Express using exponents.

e. $\dfrac{7^6}{7^2}$ **f.** $\dfrac{9c^7}{3c^2}$

 REAL-WORLD EXAMPLE

6 **COMPUTERS** The table compares the processing speeds of a specific type of computer in 1999 and in 2008. Find how many times faster the computer was in 2008 than in 1999.

Year	Processing Speed (instructions per second)
1999	10^3
2008	10^9

Write a division expression to compare the speeds.

$\dfrac{10^9}{10^3} = 10^{9-3}$ The common base is 10.

$\qquad = 10^6$ Subtract the exponents.

So, the computer was 10^6 or one million times faster in 2008 than in 1999.

✔ CHECK Your Progress

g. **FOOD** A candy store owner has 3^8 chocolate bars and 3^6 packages of sour candy. How many times as many chocolate bars does the candy store owner have than packages of sour candy?

Examples 1–5   Simplify. Express using exponents.

1. $6^5 \cdot 6^4$

2. $3^3 \cdot 3^2$

3. $\left(\frac{2}{3}\right)^4\left(\frac{2}{3}\right)^5$

4. $m^4 \cdot m^8$

5. $-4c^2(3c^6)$

6. $4^3x^4y^6 \cdot 4^2x^3y$

7 $\dfrac{4^7}{4^3}$

8. $\dfrac{x^8}{x^6}$

9. $\dfrac{6t^4}{3t}$

Example 6   **10. SURVEY** A school surveyed people who like fruit and vegetables. Of the students surveyed, 2^6 like vegetables and 2^4 like fruit. How many times as many students prefer vegetables over fruit?

Practice and Problem Solving

● = **Step-by-Step Solutions** begin on page R1.
Extra Practice begins on page EP2.

Examples 1–5   Simplify. Express using exponents.

11. $5^6 \cdot 5^4$

12. $3^7 \cdot 3^8$

13. $\left(\frac{1}{5}\right)^3\left(\frac{1}{5}\right)^2$

14. $4^3 \cdot 4^4$

15. $7 \cdot 7^7$

16. $\left(\frac{3}{4}\right)\left(\frac{3}{4}\right)^4$

17. $n^2 \cdot n$

18. $4b^3 \cdot 5b^4$

19. $3j^3k^4 \cdot 6jk^9$

20. $t^6 \cdot t^3$

21. $(7h^3)(2h^8)$

22. $(-7m^3n^7p^9)(-8m^4n^5p^4)$

23. $(6x^5)(4x^7)$

24. $(-5f)(8f^6)$

25. $(-6p^9)(-7p^7)$

26. $\dfrac{5^4}{5}$

27. $\dfrac{8^8}{8^3}$

28. $\dfrac{6^7}{6^4}$

29. $\dfrac{c^6}{c^3}$

30. $\dfrac{h^8}{h^4}$

31. $\dfrac{x^8}{x^7}$

32. $\dfrac{36g^8}{4g^3}$

33. $\dfrac{50n^7}{5n^2}$

34. $\dfrac{x^3 \cdot y^6 \cdot z^9}{x \cdot y^4 \cdot z^3}$

Example 6   **35. FISH** The number of fish in a school of fish is 4^3. If the number of fish in the school increased by 4^2 times the original number of fish, how many fish are now in the school?

36. MUSIC Jon has 5^4 songs on his computer. His friend has 5^2 times the number of songs he has. How many songs does his friend have?

37 **POWERS OF TEN** Use the information in the table.

a. How many times greater is one trillion than one million?

b. How many times greater is one quintillion than one billion?

Power of Ten	U.S. Name
10^3	One thousand
10^6	One million
10^9	One billion
10^{12}	One trillion
10^{15}	One quadrillion
10^{18}	One quintillion

 **Problems**

38. **CHALLENGE** What is twice 2^{20}? Write using exponents.

39. **NUMBER SENSE** Is $\frac{4^{200}}{4^{199}}$ greater than, less than, or equal to 4? Explain your reasoning.

40. **WRITE MATH** Does $3^2 + 3^5$ have the same value as $3^2 \cdot 3^5$? Explain.

✓ Test Practice

41. Multiply $7xy$ and $x^{14}z$.
   A. $7x^{15}yz$
   B. $7x^{15}y$
   C. $7x^{13}yz$
   D. $x^{15}yz$

42. Find the quotient of $a^5 \div a$.
   F. a^5
   G. a^4
   H. a^6
   I. a

43. THINK SOLVE EXPLAIN **SHORT RESPONSE** What is the area of the rectangle?

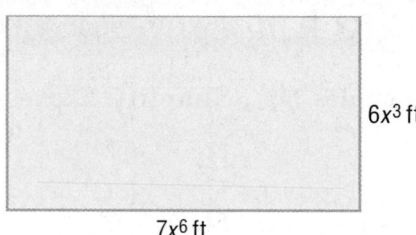

$6x^3$ ft

$7x^6$ ft

Spiral Review

Divide. Write in simplest form. (Lesson 3D)

44. $1\frac{1}{2} \div \frac{3}{4}$

45. $5 \div \frac{3}{4}$

46. $\frac{5}{6} \div \left(-\frac{2}{3}\right)$

47. $-\frac{4}{5} \div \left(-\frac{1}{5}\right)$

48. $6\frac{3}{4} \div 2\frac{1}{2}$

49. $-3\frac{2}{9} \div \frac{1}{3}$

50. **BAKING** The table shows the amount of walnuts needed for making pumpkin bread. John made pumpkin bread using Mom's recipe and used 6 cups of walnuts. How many batches of pumpkin bread did John make? (Lesson 3D)

Pumpkin Bread	
Recipe	Amount of Walnuts
Mom's	$\frac{3}{4}$ c
Cookbook's	$\frac{1}{4}$ c

51. **MEASUREMENT** Will doubled just the length of the rectangular ice skating rink in his backyard from 16 to 32 feet. That increased the area from 128 square feet to 256 square feet. Find the width of both rinks. Use the *draw a diagram* strategy. (Lesson 3C)

Order each set of numbers from least to greatest. (Lesson 1C)

52. $\frac{2}{25}$, 9%, 0.089

53. $\frac{7}{16}$, 44%, 0.38

54. $\frac{2}{3}$, 65%, 0.69

Main Idea

Write expressions using negative exponents.

 Vocabulary

negative exponent

 **Get Connect ED**

Negative Exponents

 Explore Copy the table at the right.

1. Describe the pattern of the powers in the first column. Continue the pattern by writing the next two values in the table.

2. Describe the pattern of values in the second column. Then complete the second column.

3. Determine how 3^{-1} should be defined.

Power	Value
2^6	64
2^5	32
2^4	16
2^3	8
2^2	4
2^1	2
2^0	■
2^{-1}	■

Key Concept — Negative Exponent

Words Any nonzero number to the negative n power is the multiplicative inverse of its nth power.

Symbols $a^{-n} = \dfrac{1}{a^n}$, for $a \neq 0$ and any integer n

Example $5^{-4} = \dfrac{1}{5^4}$

 EXAMPLES Use Positive Exponents

Write each expression using a positive exponent.

1 6^{-2}

$6^{-2} = \dfrac{1}{6^2}$ Definition of negative exponent

2 x^{-5}

$x^{-5} = \dfrac{1}{x^5}$ Definition of negative exponent

CHECK Your Progress

a. 5^{-6}

b. t^{-4}

 EXAMPLES Use Negative Exponents

Write each expression using a negative exponent other than -1.

3 $\dfrac{1}{9} = \dfrac{1}{3^2}$ Definition of exponent

$= 3^{-2}$ Definition of negative exponent

4 $\dfrac{1}{d^5} = d^{-5}$ Definition of negative exponent

 CHECK Your Progress

c. $\dfrac{1}{16}$

d. $\dfrac{1}{t^6}$

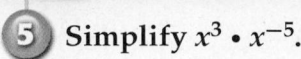 **EXAMPLES** **Perform Operations with Exponents**

⑤ Simplify $x^3 \cdot x^{-5}$.

Method 1   Product of Powers

$x^3 \cdot x^{-5} = x^{3+(-5)}$

$\qquad = x^{-2}$

Method 2   Definition of Power

$x^3 \cdot x^{-5}$

$= x \cdot x \cdot x \cdot \dfrac{1}{x \cdot x \cdot x \cdot x \cdot x}$

$= \cancel{x}^1 \cdot \cancel{x}^1 \cdot \cancel{x}^1 \cdot \dfrac{1}{\cancel{x}_1 \cdot \cancel{x}_1 \cdot \cancel{x}_1 \cdot x \cdot x}$

$= \dfrac{1}{x \cdot x}$ or x^{-2}

⑥ Simplify $\dfrac{g^5}{g^2}$.

Method 1   Quotient of Powers

$\dfrac{g^5}{g^2} = g^{5-2}$

$\qquad = g^3$

Method 2   Definition of Power

$\dfrac{g^5}{g^2} = \dfrac{\cancel{g}^1 \cdot \cancel{g}^1 \cdot g \cdot g \cdot g}{\cancel{g}^1 \cdot \cancel{g}^1}$

$\qquad = g^3$

✔ **CHOOSE** Your Method

e. $\dfrac{x^{-2}}{x^{-1}}$

f. $c^{-4} \cdot c^3$

Real-World Link · · · ·

A single drop of water contains about 10^{20} molecules.

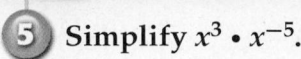 **REAL-WORLD EXAMPLE**

⑦ **WATER** A molecule of water contains two hydrogen atoms and one oxygen atom. A hydrogen atom is only 0.00000001 centimeter in diameter. Write the decimal as a power of ten.

The digit 1 is in the 100-millionths place.

$0.00000001 = \dfrac{1}{100,000,000}$ Write the decimal as a fraction.

$\qquad\qquad = \dfrac{1}{10^8}$ $100,000,000 = 10^8$

$\qquad\qquad = 10^{-8}$ Definition of negative exponent

A hydrogen atom is 10^{-8} centimeter in diameter.

✔ **CHECK** Your Progress

g. **MEASUREMENT** A unit of measure called a *micron* equals 0.001 millimeter. Write this number using a negative exponent.

✓ CHECK Your Understanding

Examples 1 and 2   Write each expression using a positive exponent.

    1. 5^{-2} **2.** $(-7)^{-1}$ **3** t^{-10} **4.** n^{-2}

Examples 3 and 4   Write each fraction as an expression using a negative exponent other than −1.

    5. $\dfrac{1}{3^4}$ **6.** $\dfrac{1}{x^2}$ **7.** $\dfrac{1}{49}$ **8.** $\dfrac{1}{8}$

Examples 5 and 6   Simplify each expression.

    9. $h^4 \cdot h^{-2}$ **10.** $n^{-6} \cdot n^{-1}$ **11.** $\dfrac{r^5}{r^4}$ **12.** $\dfrac{s^{-8}}{s^{-3}}$

Example 7   **13. MEASUREMENTS** A unit of measure called a *microgram* equals 0.000001 gram. Write this number using a negative exponent.

Practice and Problem Solving

> ● = **Step-by-Step Solutions** begin on page R1.
> **Extra Practice** begins on page EP2.

Examples 1 and 2   Write each expression using a positive exponent.

    14. 4^{-1} **15.** 5^{-3} **16.** $(-6)^{-2}$ **17.** $(-3)^{-3}$

    18. 3^{-5} **19.** 10^{-4} **20.** p^{-1} **21.** a^{-10}

    22. d^{-3} **23.** q^{-4} **24.** $2s^{-5}$ **25.** x^{-2}

Examples 3 and 4   Write each fraction as an expression using a negative exponent other than −1.

    26. $\dfrac{1}{9^4}$ **27.** $\dfrac{1}{5^5}$ **28.** $\dfrac{1}{b^3}$ **29.** $\dfrac{1}{k^2}$

    30. $\dfrac{1}{1,000}$ **31.** $\dfrac{1}{81}$ **32.** $\dfrac{1}{27}$ **33.** $\dfrac{1}{16}$

Examples 5 and 6   Simplify each expression.

    34. $d^3 \cdot d^4$ **35.** $g^{-4} \cdot g^2$ **36.** $3m^7 \cdot 2m^{-2}$ **37.** $5v^{-2} \cdot 3v^{-3}$

    38. $\dfrac{f^8}{f^2}$ **39.** $\dfrac{k^{-9}}{k^{-8}}$ **40.** $\dfrac{36b^5}{6b^3}$ **41.** $\dfrac{81c^{-6}}{9c^{-2}}$

Example 7   **42. ANIMALS** A common flea is $\dfrac{1}{2^4}$ inch long. Write this using a negative exponent.

43 MEASUREMENTS What fraction of a meter does one centimeter represent? Write this using a negative exponent other than −1.

Meter Equivalents
10 decimeters
100 centimeters
1,000 millimeters

Write each decimal using a negative exponent.

    44. 0.1   **45.** 0.01   **46.** 0.0001   **47.** 0.00001

48. OPEN ENDED Write a convincing argument that $3^0 = 1$ using the fact that $3^4 = 81$, $3^3 = 27$, $3^2 = 9$, and $3^1 = 3$.

49. REASONING Order 8^{-8}, 8^3, and 8^0 from greatest to least. Explain your reasoning.

50. CHALLENGE Compare and contrast x^{-n} and x^n. Then give a numerical example to show the relationship.

51. ✍ WRITE MATH Explain the meaning of a negative exponent when the base is 10.

✔ Test Practice

52. Which is 15^{-5} written as a fraction?

   A. $\dfrac{1}{5^5}$ **C.** $\dfrac{1}{15}$

   B. $\dfrac{1}{15^5}$ **D.** $-\dfrac{1}{15^5}$

53. The table shows the length of a cell and virus.

Structure	Length (m)
Cell	10^{-4}
Virus	10^{-7}

Which of the following represents the length of a virus using a positive exponent?

   F. 10^7 **H.** $\dfrac{1}{10^7}$

   G. 7^{10} **I.** $-\dfrac{1}{10^7}$

54. THINK SOLVE EXPLAIN SHORT RESPONSE Rewrite the numbers with negative exponents so they have positive exponents.

Negative Exponents	m^{-5}	$b^{-3} \cdot b^{-1}$	$\dfrac{x^{-5}}{x^{-3}}$
Positive Exponents	■	■	■

55. The prefix *centi* equals 0.01. Which of the following represents the decimal written with a negative exponent?

   A. 10^{-1}

   B. 10^{-2}

   C. 10^1

   D. 10^2

Spiral Review

ALGEBRA Simplify each expression. (Lesson 4A)

56. $3^6 \cdot 3$ **57.** $x^2 \cdot x^4$ **58.** $\dfrac{5^5}{5^2}$ **59.** $(n^4)(-2n^3)$

60. SPORTS The length of an athletic field is twice its width. Write an expression in simplest form that represents the area of the field. (Lesson 4A)

61. WHALES During the first year, a baby whale gains about $27\frac{3}{5}$ tons. What is the average weight gain in simplest form per month? (Lesson 3D)

62. PETS Hannah made the dog bed shown. If the area of the dog bed is $2\frac{2}{3}$ square feet, what is the length of the bed? (Lesson 3D)

x ft $1\frac{1}{3}$ ft

Main Idea

Express numbers in scientific notation and in standard form.

 Vocabulary

scientific notation

Scientific Notation

More than 425 million pounds of gold have been discovered in the world. If all this gold were in one place, it would form a cube seven stories on each side.

1. Write 425 million in standard form.

2. Complete: $4.25 \times$ ___?___ $= 425$ million.

When you deal with very large numbers like 425,000,000, it can be difficult to keep track of the zeros. You can express numbers such as this in **scientific notation** by writing the number as the product of a factor and a power of 10.

Key Concept   Scientific Notation

Words   A number is expressed in scientific notation when it is written as the product of a factor and a power of 10. The factor must be greater than or equal to 1 and less than 10.

Symbols $a \times 10^n$, where $1 \le a < 10$ and n is an integer

Example $425{,}000{,}000 = 4.25 \times 10^8$

EXAMPLE   Express Large Numbers in Standard Form

1. Express 2.16×10^5 in standard form.

$$2.16 \times 10^5 = 2.16 \times 100{,}000 \qquad 10^5 = 100{,}000$$

$$= 216{,}000 \qquad\qquad \text{Move the decimal point 5 places.}$$

 **CHECK Your Progress**

Express each number in standard form.

   a. 7.6×10^6 **b.** 3.201×10^4

Scientific notation is also used to express very small numbers. Study the pattern of products at the right. Notice that multiplying by a negative power of 10 moves the decimal point to the left the same number of places as the absolute value of the exponent.

$1.25 \times 10^2 = 125$
$1.25 \times 10^1 = 12.5$
$1.25 \times 10^0 = 1.25$
$1.25 \times 10^{-1} = 0.125$
$1.25 \times 10^{-2} = 0.0125$
$1.25 \times 10^{-3} = 0.00125$

 EXAMPLE **Express Small Numbers in Standard Form**

 Express 5.8×10^{-3} in standard form.

$5.8 \times 10^{-3} = 5.8 \times 0.001$ $10^{-3} = 0.001$

$= 0.0058$ Move the decimal point 3 places to the left.

✓ **CHECK Your Progress**

c. 4.7×10^{-5} **d.** 9×10^{-4}

To write a number in scientific notation, place the decimal point after the first nonzero digit. Then find the power of 10.

 EXAMPLES **Express Numbers in Scientific Notation**

Express each number in scientific notation.

 1,457,000

$1,457,000 = 1.457 \times 1,000,000$ The decimal point moves 6 places to the left.

$= 1.457 \times 10^6$ The exponent is positive.

 0.00063

$0.00063 = 6.3 \times 0.0001$ The decimal point moves 4 places to the right.

$= 6.3 \times 10^{-4}$ The exponent is negative.

✓ **CHECK Your Progress**

e. 35,000   **f.** 0.00722

Compare the exponents to compare numbers in scientific notation. With positive numbers, any number with a greater exponent is greater. If the exponents are the same, compare the factors.

 **REAL-WORLD EXAMPLE** **Compare Numbers in Scientific Notation**

 OCEANS The Atlantic Ocean has an area of 3.18×10^7 square miles. The Pacific Ocean has an area of 6.4×10^7 square miles. Which ocean has the greater area?

$3.18 < 6.4 \longrightarrow 3.18 \times 10^7 < 6.4 \times 10^7$

So, the Pacific Ocean has the greater area.

✓ **CHECK Your Progress**

g. Replace ● with <, >, or = to make 4.13×10^{-2} ● 5.0×10^{-3} a true sentence.

🌐 **Real-World Link** · · · ·
At the deepest point in the ocean, the pressure is greater than 8 tons per square inch and the temperature is only a few degrees above freezing.

Examples 1 and 2 — Express each number in standard form.

1. 3.754×10^5 **2.** 8.34×10^6

3. 1.5×10^{-4} **4.** 2.68×10^{-3}

Examples 3 and 4 — Express each number in scientific notation.

5 4,510,000 **6.** 0.00673

7. 0.000092 **8.** 11,620,000

9. PHYSICAL SCIENCE Light travels 300,000 kilometers per second. Write this number in scientific notation.

Example 5 **10. TECHNOLOGY** The distance between tracks on a CD and DVD are shown in the table. Which disc has the greater distance between tracks?

Disc	Distance (mm)
CD	1.6×10^{-3}
DVD	7.4×10^{-4}

Replace each ● with <, >, or = to make a true sentence.

11. 2.3×10^5 ● 1.7×10^5 **12.** 0.012 ● 1.4×10^{-1}

Practice and Problem Solving

 = **Step-by-Step Solutions** begin on page R1.
Extra Practice begins on page EP2.

Examples 1 and 2 — Express each number in standard form.

13. 6.1×10^4 **14.** 5.72×10^6 **15.** 3.3×10^{-1} **16.** 5.68×10^{-3}

17. 9.014×10^{-2} **18.** 1.399×10^5 **19.** 2.505×10^3 **20.** 7.4×10^{-5}

21. SPIDERS The diameter of a spider's thread is 1×10^{-3} inch. Write this number in standard form.

22. DINOSAURS The *Giganotosaurus* dinosaur weighed about 1.4×10^4 pounds. Write this number in standard form.

Examples 3 and 4 — Express each number in scientific notation.

23. 499,000 **24.** 2,000,000 **25.** 0.006 **26.** 0.0125

27. 50,000,000 **28.** 39,560 **29.** 0.000078 **30.** 0.000425

31. CHESS The number of possible ways that a player can play the first four moves in a chess game is 3 billion. Write this number in scientific notation.

32. SCIENCE A particular parasite is approximately 0.025 inch long. Write this number in scientific notation.

Example 5 **33. SPORTS** Use the table. Determine which category in each pair had a greater amount of sales.

a. golf or tennis

b. camping or golf

Category	Sales ($)
Camping	1.547×10^9
Golf	3.243×10^9
Tennis	3.73×10^8

Replace each ● with <, >, or = to make a true sentence.

34. 1.8×10^3 ● 1.9×10^{-1}

35. 5.2×10^2 ● $5{,}000$

36. 0.00701 ● 7.1×10^{-3}

37. 6.49×10^4 ● 649×10^2

38. **MEASUREMENT** The table at the right shows the values of different prefixes that are used in the metric system. Write the units attometer, gigameter, kilometer, nanometer, petameter, and picometer in order from greatest to least measure.

Metric Measures	
Prefix	**Meaning**
atto	10^{-18}
giga	10^9
kilo	10^3
nano	10^{-9}
peta	10^{15}
pico	10^{-12}

39. **NUMBER SENSE** Write the product of 0.00004 and 0.0008 in scientific notation.

40. **NUMBER SENSE** Order 6.1×10^4, 6,100, 6.1×10^{-5}, 0.0061, and 6.1×10^{-2} from least to greatest.

41 **PHYSICAL SCIENCE** The table shows the maximum amounts of lava in cubic meters per second that erupted from four volcanoes.

Volcanic Eruptions	
Volcano, Year	**Eruption Rate (m³/s)**
Mount St. Helens, 1980	2×10^4
Ngauruhoe, 1975	2×10^3
Hekla, 1970	4×10^3
Agung, 1963	3×10^4

a. How many times greater was the Mount St. Helens eruption than the Ngauruhoe eruption?

b. How many times greater was the Hekla eruption than the Ngauruhoe eruption?

Write each number in standard form.

42. $(8 \times 10^0) + (4 \times 10^{-3}) + (3 \times 10^{-5})$

43. $(4 \times 10^4) + (8 \times 10^3) + (3 \times 10^2) + (9 \times 10^1) + (6 \times 10^0)$

44. ⬜ **MULTIPLE REPRESENTATIONS** A square piece of property has a side of 3,250 feet.

a. **ALGEBRA** Write an equation to represent the area of the property in square feet.

b. **ALGEBRA** Solve the equation in part a to find the area of the property.

c. **NUMBERS** Express the area in scientific notation.

3,250 ft

3,250 ft

🌐 **Real-World Link · · · ·**
Mount St. Helens is located in the state of Washington about 100 miles south of Seattle. The height of Mount St. Helens was 9,677 feet before the 1980 eruption and 8,363 feet after the eruption.

H.O.T. Problems

45. **CHALLENGE** Convert the numbers in each expression to scientific notation. Then evaluate the expression. Express in scientific notation and in decimal notation.

a. $\dfrac{(420,000)(0.015)}{0.025}$

b. $\dfrac{(0.078)(8.5)}{0.16(250,000)}$

46. **REASONING** Which is a better estimate for the number of times per year that a person blinks: 6.25×10^{-2} times or 6.25×10^6 times? Explain your reasoning.

47. **OPEN ENDED** Describe a real-world value or measure using numbers in scientific notation and in standard form.

48. **WRITE MATH** Explain when scientific notation should be used.

Test Practice

49. Which shows 0.00000029 in scientific notation?

   A. 2.9×10^7

   B. 2.9×10^6

   C. 2.9×10^{-6}

   D. 2.9×10^{-7}

50. The average width of a strand of thread is 2.2×10^{-4} meter. Which expression represents this number in standard form?

   F. 22,000 meters

   G. 220,000 meters

   H. 0.00022 meter

   I. 0.000022 meter

More About Scientific Notation

You have used scientific notation to express numbers as the products of a factor and a power of 10. For example, 487,000,000,000 can be written as 4.87×10^{11}.

Numbers expressed in scientific notation appear differently on a calculator screen. The screen at the right shows a calculator display for 4.87×10^{11}. The number following E indicates the exponent that should be used for the power of 10.

Write each calculator screen expression as it would appear in scientific notation. Then write the value in standard form.

51. 5.21E4

52. 8.53E8

53. 9.2E−2

54. 6.53E−4

Chapter Study Guide and Review

FOLDABLES® Study Organizer

Be sure the following Key Concepts are noted in your Foldable.

2-1 Rational Numbers
2-2 Add and Subtract Fractions
2-3 Multiply and Divide Fractions

Key Concepts

Terminating and Repeating Decimals
(Lesson 1)
• A terminating decimal is a decimal whose digits end. Repeating decimals have a pattern in their digits that repeats forever.

Adding and Subtracting Fractions (Lesson 2)
• To add or subtract fractions, rename the fractions using the LCD. Then add or subtract the numerators and write the result over the denominator.

Multiplying and Dividing Fractions (Lesson 3)
• To multiply fractions, multiply the numerators and multiply the denominators.
• To divide by a fraction, multiply by its multiplicative inverse, or reciprocal.

Scientific Notation (Lesson 4)
• A number is expressed in scientific notation when it is written as the product of a factor and a power of 10. The factor must be greater than or equal to 1 and less than 10.

Key Vocabulary

bar notation

common denominator

least common denominator (LCD)

like fractions

monomial

negative exponent

rational numbers

repeating decimal

scientific notation

standard form

terminating decimal

unlike fractions

Vocabulary Check

Choose the correct term or number to complete each sentence.

1. 1.875 is an example of a (terminating, repeating) decimal.

2. A common denominator for the fractions $\frac{2}{3}$ and $\frac{1}{4}$ is (7, 12).

3. To add like fractions, add the (numerators, denominators).

4. When dividing by a fraction, multiply by its (value, reciprocal).

5. Fractions with different denominators are called (like, unlike) fractions.

6. The mixed number $2\frac{4}{7}$ can be renamed as $\left(2\frac{7}{7}, 1\frac{11}{7}\right)$.

7. When multiplying fractions, multiply the numerators and (multiply, keep) the denominators.

8. Fractions, terminating decimals, and repeating decimals are (integers, rational numbers).

9. 3.16×10^3 is expressed in (scientific notation, standard form).

10. The least common denominator for $\frac{5}{8}$ and $\frac{7}{12}$ is (4, 24).

Multi-Part Lesson Review

Lesson 1   Rational Numbers

Terminating and Repeating Decimals (Lessons 1A and 1B)

Write each decimal as a fraction in simplest form.

11. 0.7   **12.** 0.44   **13.** 0.05

14. RUNNING Jeremy ran a mile in 5 minutes and 8 seconds. Write this time in minutes as a decimal.

> **EXAMPLE 1**   Write $\frac{2}{3}$ as a decimal.
>
> $$\begin{array}{r} 0.66... \\ 3\overline{)2.00} \\ -18 \\ \hline 20 \\ -18 \\ \hline 2 \end{array}$$
>
> Since the pattern continues to repeat, a bar is placed over the 6 to indicate it repeats infinitely. So, $\frac{2}{3} = 0.\overline{6}$.

Compare and Order Rational Numbers (Lesson 1C)

Replace each ● with <, >, or = to make a true sentence.

15. 37.5% ● $\frac{2}{3}$ **16.** -0.45 ● $-\frac{9}{20}$

17. SCHOOL Michael received a $\frac{26}{30}$ on his English test and an 81% on his biology test. On which test did he receive the higher score?

> **EXAMPLE 2**   Replace each ● with <, >, or = to make 60% ● $\frac{5}{8}$ a true sentence.
>
> Write each as a decimal.
>
> $60\% = 0.6$ $\frac{5}{8} = 0.625$
>
> Since $0.6 < 0.625$, then $60\% < \frac{5}{8}$.

Lesson 2   Add and Subtract Fractions

Add and Subtract Like Fractions (Lesson 2A)

Add or subtract. Write in simplest form.

18. $-\frac{5}{8} + \frac{1}{8}$ **19.** $\frac{7}{12} + \frac{1}{12}$

20. $\frac{11}{12} - \frac{7}{12}$ **21.** $\frac{7}{9} - \left(-\frac{4}{9}\right)$

> **EXAMPLE 3**   Find $\frac{7}{12} - \frac{5}{12}$. **Estimate** $\frac{1}{2} - \frac{1}{2} = 0$
>
> $\frac{7}{12} - \frac{5}{12} = \frac{7-5}{12}$ Subtract the numerators.
>
> $= \frac{2}{12}$ or $\frac{1}{6}$ Simplify.

Add and Subtract Unlike Fractions (Lessons 2B and 2C)

Add or subtract. Write in simplest form.

22. $\frac{7}{9} - \frac{1}{6}$ **23.** $\frac{4}{5} + \frac{2}{10}$

24. RUNNING Teresa ran $\frac{5}{6}$ mile while Yolanda ran $\frac{1}{4}$ mile. By what fraction did Teresa run more than Yolanda?

> **EXAMPLE 4**   Find $\frac{3}{8} + \frac{2}{3}$. **Estimate** $\frac{1}{2} + \frac{1}{2} = 1$
>
> The LCD of $\frac{3}{8}$ and $\frac{2}{3}$ is 24.
>
> $$\begin{array}{ccccc} \frac{3}{8} & \rightarrow & \frac{3 \times 3}{8 \times 3} & \rightarrow & \frac{9}{24} \\ +\frac{2}{3} & \rightarrow & \frac{2 \times 8}{3 \times 8} & \rightarrow & +\frac{16}{24} \\ \hline & & & & \frac{25}{24} \text{ or } 1\frac{1}{24} \end{array}$$

Chapter Study Guide and Review

Lesson 2 Add and Subtract Fractions (continued)

Add and Subtract Mixed Numbers (Lesson 2D)

Add or subtract. Write in simplest form.

25. $-3\frac{2}{15} + 6\frac{9}{15}$ **26.** $4\frac{1}{3} - 2\frac{2}{3}$

27. $7\frac{3}{5} - 5\frac{1}{3}$ **28.** $5\frac{3}{4} - \left(-1\frac{1}{6}\right)$

29. BABYSITTING Lucas watched his little sister for $2\frac{1}{2}$ hours on Friday, $3\frac{2}{3}$ hours on Saturday, and $1\frac{3}{4}$ hours on Sunday. For how many hours did Lucas watch his little sister?

EXAMPLE 5   Find $4\frac{1}{5} - 2\frac{3}{5}$.

$$4\frac{1}{5} - 2\frac{3}{5} = 3\frac{6}{5} - 2\frac{3}{5}$$ Rename $4\frac{1}{5}$ as $3\frac{6}{5}$.

$$= 1\frac{3}{5}$$ Subtract the whole numbers and subtract the fractions.

Lesson 3 **Multiply and Divide Fractions**

Multiply Fractions (Lessons 3A and 3B)

Multiply. Write in simplest form.

30. $\frac{3}{5} \times \frac{10}{21}$ **31.** $4 \times \left(-\frac{13}{20}\right)$

32. $2\frac{1}{3} \times \frac{3}{4}$ **33.** $4\frac{1}{2} \times 2\frac{1}{12}$

34. FOOD An average slice of American cheese is about $\frac{1}{8}$-inch thick. What is the height in simplest form of a package containing 20 slices?

EXAMPLE 6   Find $3\frac{1}{2} \times 2\frac{3}{4}$.

$$3\frac{1}{2} \times 2\frac{3}{4} = \frac{7}{2} \times \frac{11}{4}$$ Rename $3\frac{1}{2}$ and $2\frac{3}{4}$.

$$= \frac{7 \times 11}{2 \times 4}$$ Multiply the numerators and multiply the denominators.

$$= \frac{77}{8} \text{ or } 9\frac{5}{8}$$ Simplify.

PSI: Draw A Diagram (Lesson 3C)

35. COOKIES A cookie jar contains three types of cookies: oatmeal, chocolate chip, and sugar. Sixty percent are chocolate chip. Half of the remaining cookies are oatmeal. If there are 9 oatmeal cookies, how many cookies are in the jar? Use the *draw a diagram* strategy.

EXAMPLE 7   **Marian is painting a fence that is 72 feet long. She has already painted $\frac{5}{8}$ of the fence. How many feet of fence does she have left to paint?**

Make a bar diagram.

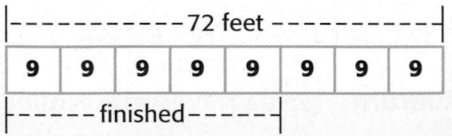

So, she has 3×9 or 27 feet of the fence left to paint.

Multiply and Divide Fractions (continued)

Divide Fractions (Lesson 3D)

Divide. Write in simplest form.

36. $\frac{3}{5} \div \frac{6}{7}$ **37.** $4 \div \left(-\frac{2}{3}\right)$

38. $2\frac{3}{4} \div \frac{5}{6}$ **39.** $-\frac{2}{7} \div \left(-\frac{8}{21}\right)$

40. MEASUREMENT How many $\frac{1}{8}$-inch lengths are in $6\frac{3}{4}$ inches?

> **EXAMPLE 8** Find $2\frac{4}{5} \div \frac{7}{10}$.
>
> $2\frac{4}{5} \div \frac{7}{10} = \frac{14}{5} \div \frac{7}{10}$ Rename $2\frac{4}{5}$.
>
> $= \frac{\overset{2}{\cancel{14}}}{\underset{1}{\cancel{5}}} \cdot \frac{\overset{2}{\cancel{10}}}{\underset{1}{\cancel{7}}}$ Multiply by the reciprocal of $\frac{7}{10}$.
>
> $= \frac{4}{1}$ or 4 Simplify.

Monomials

Multiply and Divide Monomials (Lesson 4A)

41. PHONE CALLS Mrs. Jones, as a telemarketer, made 3^3 phone calls on Monday. Suppose on Tuesday she made 3^4 phone calls. How many times as many phone calls did she make on Tuesday than on Monday?

> **EXAMPLE 9** Find $3 \cdot 3^5$. Express using exponents.
>
> $3 \cdot 3^5 = 3^1 \cdot 3^5$ $3 = 3^1$
>
> $\quad\quad\ = 3^{1+5}$ The common base is 3.
>
> $\quad\quad\ = 3^6$ Add the exponents.

Negative Exponents (Lesson 4B)

Write each expression using a positive exponent.

42. 6^{-3} **43.** 3^{-5} **44.** b^{-4}

Write each fraction as an expression using a negative exponent other than −1.

45. $\frac{1}{5^2}$ **46.** $\frac{1}{64}$ **47.** $\frac{1}{x^4}$

> **EXAMPLE 10** Write m^{-5} using a positive exponent.
>
> $m^{-5} = \frac{1}{m^5}$ Definition of negative exponent

Scientific Notation (Lesson 4C)

Express each number in scientific notation.

48. 0.00027 **49.** 0.0000196

50. INSECTS The giant weta is the largest known insect, having a mass of about 7.1×10^1 grams. Write this mass in standard form.

> **EXAMPLE 11** Write 3.06×10^6 in standard form.
>
> $3.06 \times 10^6 = 3,060,000$ Move the decimal point 6 places.
>
> **EXAMPLE 12** Write 0.00016 in scientific notation.
>
> $0.00016 = 16 \times 0.0001$ Move the decimal point 4 places.
>
> $\quad\quad\quad\ = 1.6 \times 10^{-4}$ Since $0 < 0.00016 < 1$, the exponent is negative.

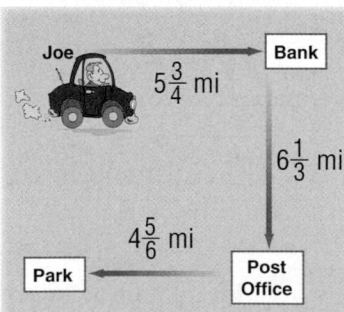

Write each fraction or mixed number as a decimal. Use bar notation if the decimal is a repeating decimal.

1. $\frac{7}{9}$ **2.** $4\frac{5}{8}$ **3.** $\frac{91}{100}$

Write each decimal as a fraction in simplest form.

4. 0.84 **5.** 0.006 **6.** 0.42

Replace each ● with <, >, or = to make a true sentence.

7. $-\frac{2}{3} ● \frac{4}{7}$ **8.** $3\frac{5}{8} ● 3\frac{3}{5}$ **9.** $\frac{11}{20} ● 0.55$

Add, subtract, multiply, or divide. Write in simplest form.

10. $-\frac{4}{15} + \frac{8}{15}$ **11.** $\frac{7}{10} - \frac{1}{6}$

12. $\frac{5}{8} + \frac{2}{5}$ **13.** $6 + \frac{8}{21}$

14. $-4\frac{5}{12} - 2\frac{1}{12}$ **15.** $6\frac{7}{9} + 3\frac{5}{12}$

16. $8\frac{4}{7} - 1\frac{5}{14}$ **17.** $4\frac{5}{6} - \left(-1\frac{2}{3}\right)$

18. MULTIPLE CHOICE Use the diagram to find the total distance Joe drove.

A. $15\frac{9}{13}$ miles   **C.** $\frac{11}{12}$ mile

B. $\frac{7}{12}$ mile   **D.** $16\frac{11}{12}$ miles

19. SPORTS Tyler's football practice lasted $2\frac{1}{2}$ hours. If $\frac{1}{4}$ of the time was spent catching passes, what fraction of time was spent catching passes? Write in simplest form.

20. FINANCIAL LITERACY For his birthday, Keith received a check from his grandmother. Of this amount, the table shows how he spent or saved the money. Two weeks later, he withdrew $\frac{2}{3}$ of the amount he had deposited into his savings account. What fraction in simplest form of the original check did he withdraw from his savings account?

Fraction of Check	How Spent or Saved
$\frac{2}{5}$	Spent on baseball cards
$\frac{1}{4}$	Spent on a CD
$\frac{7}{20}$	Deposited into savings account

21. MEASUREMENT An ounce is $\frac{1}{16}$ of a pound. How many ounces are in $8\frac{3}{4}$ pounds?

Simplify each expression.

22. $c^7 \times c^{-2}$ **23.** $m^{-5} \times m^8$

24. $4g^{10} \times 3g^{-5}$ **25.** $8v^{-4} \times 5v^6$

26. SOLAR SYSTEM The distance from Earth to the Sun is 9.3×10^7 miles. What is this distance in standard form?

27. THINK SOLVE EXPLAIN **EXTENDED RESPONSE** For a measurement experiment in class, students measured the distance around their heads. The chart shows the results for part of the class.

Student	Distance Around Head (in.)
Max	17.5
Juan	$17\frac{3}{8}$
Hue	$16\frac{7}{8}$
Kobe	17.25
Martha	16.8

Part A Which student had the greatest head size?

Part B Which student had the least head size?

Part C Arrange the head sizes in order from least to greatest.

Gridded Response: Gridding Mixed Numbers

A mixed number cannot be written in the answer grid of a gridded-response question. Write the answer as an improper fraction and then fill in the grid.

TEST EXAMPLE

The tail of a salamander is $\frac{2}{3}$ of its total length. The salamander is 7 inches long. How many inches long is its tail?

$\frac{2}{3} \times 7 = \frac{2}{3} \times \frac{7}{1}$ Rename 7 as an improper fraction.

$\qquad = \frac{14}{3}$ or $4\frac{2}{3}$ Multiply.

The tail of the salamander is $\frac{14}{3}$ or $4\frac{2}{3}$ inches long. Since $4\frac{2}{3}$ cannot be written in the answer grid, fill in $\frac{14}{3}$.

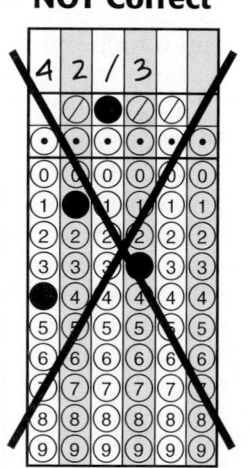

Correct

NOT Correct

or

Work on It

Juliette played the same song twice on her MP3 player. It took $9\frac{1}{2}$ minutes. How many minutes long is the song? Fill in your answer on a grid like the one shown above.

Test Hint

Don't forget to include the fraction bar in the answer box.

Test Practice

Read each question. Then fill in the correct answer on the answer sheet provided by your teacher or on a sheet of paper.

1. **THINK SOLVE EXPLAIN** **SHORT RESPONSE** Mrs. Brown needs to make two different desserts for a party. The first recipe requires $2\frac{1}{4}$ cups of flour and the second recipe requires $\frac{3}{4}$ cup less than the first. Write an equation that can be used to find the number of cups of flour needed for the second recipe.

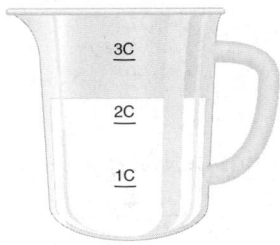

2. The fraction $\frac{5}{6}$ is found between which pair of fractions on a number line?

   A. $\frac{1}{4}$ and $\frac{5}{8}$

   B. $\frac{1}{3}$ and $\frac{4}{9}$

   C. $\frac{11}{12}$ and $\frac{31}{36}$

   D. $\frac{7}{12}$ and $\frac{17}{18}$

3. At 7 A.M., the temperature was 15°F below zero. By 2 P.M. the temperature rose 32°F and by 5 P.M. it dropped 10°F. What was the temperature at 5 P.M.?

   F. 10°F   H. 7°F

   G. 9°F   I. 11°F

4. **GRIDDED RESPONSE** A diver is swimming 11 meters below the surface. The diver sees a shark 19 meters below him. How many meters from the surface is the shark?

5. **GRIDDED RESPONSE** Maria had $240 in her savings account. The table shows the change in her account for four consecutive weeks.

Week	Change
1	Deposit of $25
2	Withdrawal of $45
3	Withdrawal of $10
4	Deposit of $60

   How much money, in dollars, did Maria have in her account at the end of the four weeks?

6. The table shows the distance Kelly swam over a four-day period. What was the total distance, in miles, that Kelly swam?

Kelly's Swimming	
Day	**Distance (mi)**
Monday	1.5
Tuesday	$2\frac{3}{4}$
Wednesday	2.3
Thursday	$3\frac{1}{2}$

   A. 10.5 miles   C. $10\frac{1}{20}$ miles

   B. $10\frac{1}{4}$ miles   D. 9 miles

7. Which of the following gives the correct meaning of the expression $\frac{5}{8} \div \frac{1}{3}$?

   F. $\frac{5}{8} \div \frac{1}{3} = \frac{8}{5} \times \frac{3}{1}$

   G. $\frac{5}{8} \div \frac{1}{3} = \frac{5+1}{8+3}$

   H. $\frac{5}{8} \div \frac{1}{3} = \frac{5}{8} \times \frac{3}{1}$

   I. $\frac{5}{8} \div \frac{1}{3} = \frac{5}{8} \times \frac{1}{3}$

8. The table shows the lowest temperature readings to the nearest degree recorded for four countries.

City	Temperature (°F)
Finland	−61°
France	−42°
India	−27°
United States	−80°

Which of the countries has the lowest recorded temperature?

A. Finland   **C.** France

B. India   **D.** United States

9. **GRIDDED RESPONSE** Nate had 25 action figures. He gave away 10 to his brother. He then got 3 new action figures as a gift. How many action figures does Nate have now?

10. Which expression represents the least value?

F. $678 \div \frac{1}{3}$ **H.** $678 \times \frac{1}{3}$

G. $678 + \frac{1}{3}$ **I.** $678 - \frac{1}{3}$

11. **GRIDDED RESPONSE** Jacob had $25 for back-to-school shopping. He bought a shirt for $15 and then returned a shirt he bought a week ago and got $20 in return. How much money in dollars does Jacob have now?

12. **GRIDDED RESPONSE** Evan runs $2\frac{3}{8}$ miles each week. He runs $\frac{3}{4}$ mile on Mondays and $\frac{3}{4}$ mile on Tuesdays. How far does he run, in miles, on Thursday if it is the only other day he runs?

13. **THINK SOLVE EXPLAIN** **SHORT RESPONSE** A recipe for a batch of cookies calls for $2\frac{1}{3}$ cups of flour for 24 cookies. Manuel wants to make 72 cookies. How many cups of flour will he need?

14. **GRIDDED RESPONSE** The diameter of Saturn measures 7.46×10^4 miles. What is this distance in standard notation?

15. **THINK SOLVE EXPLAIN** **EXTENDED RESPONSE** A box of laundry detergent contains 35 cups. It takes $1\frac{1}{4}$ cups per load of laundry.

Part A Write an equation to represent how many loads ℓ you can wash with one box.

Part B How many loads can you wash with one box?

Part C How many loads can you wash with 3 boxes?

NEED EXTRA HELP?															
If You Missed Question...	1	2	3	4	5	6	7	8	9	10	11	12	13	14	15
Go to Chapter-Lesson...	3-2D	3-1C	2-2B	2-2D	2-2D	3-2D	3-3D	2-1B	2-2D	3-3D	2-2D	3-2D	3-3B	3-4C	3-3D

Equations and Inequalities

connectED.mcgraw-hill.com

Investigate

 Animations

 Vocabulary

 Multilingual eGlossary

Learn

 Personal Tutor

 Virtual Manipulatives

 Graphing Calculator

 Audio

 Foldables

Practice

 Self-Check Practice

 Worksheets

 Assessment

The ☆BIG Idea

How are the procedures for solving equations and inequalities the same? How are they different?

 FOLDABLES®
Study Organizer

Make this Foldable to help you organize your notes.

Review Vocabulary

solution *solución* a value for the variable that makes an equation true

The solution of $12 = x + 7$ is **5**.

Key Vocabulary

English	Español
coefficient	coeficiente
equivalent equation	ecuaciones equivalentes
formula	formula
two-step equation	ecuación de dos pasos

When Will I Use This?

Your Turn!
You will solve this problem in the chapter.

Are You Ready for the Chapter?

You have two options for checking prerequisite skills for this chapter.

Text Option Take the Quick Check below. Refer to the Quick Review for help.

QUICK Check

Write the phrase as an algebraic expression.

1. 3 more runs than the Pirates scored

2. a number increased by eight

3. ten dollars more than Grace has

Identify the solution of each equation from the list given.

4. $8 + w = 17$; 7, 8, 9

5. $d - 12 = 5$; 16, 17, 18

6. $6 = 3y$; 2, 3, 4

7. $7 \div c = 7$; 0, 1, 2

8. $a + 8 = 23$; 13, 14, 15

9. $25 + d = 54$; 28, 29, 30

10. $10 = 45 - n$; 35, 36, 37

Solve each equation mentally.

11. $p + 3 = 9$
12. $5 + y = 20$
13. $40 = 25 + m$
14. $14 - n = 10$
15. $47 = x - 3$
16. $18 = 25 - h$

17. **AGE** The equation $13 + a = 51$ describes the sum of the ages of Elizabeth and her mother. If a is Elizabeth's mother's age, how old is Elizabeth's mother?

QUICK Review

EXAMPLE 1

Write the phrase as an algebraic expression.

Words: five dollars more than Jennifer earned

Variable: Let d represent the number of dollars Jennifer earned.

Expression: $d + 5$

EXAMPLE 2

Is 3, 4, or 5 the solution of the equation $x + 8 = 12$?

Value of x	$x + 8 = 12$	Are both sides equal?
3	$3 + 8 = 12$ $11 \neq 12$	no
4	$4 + 8 = 12$ $12 = 12$	yes ✔
5	$5 + 8 = 12$ $13 \neq 12$	no

The solution is 4 since replacing x with 4 results in a true sentence.

EXAMPLE 3

Solve $18 = 3h$ mentally.

$18 = 3h$ What number times 3 is 18?

$18 = 3 \cdot 6$ You know that 3 • 6 is 18.

$h = 6$

The solution is 6.

 **Online Option** Take the Online Readiness Quiz.

Reading Math

Identify Key Information

> Have you ever tried to solve a long word problem and didn't know where to start? Start by reading the problem carefully.

STEP 1 Look for key words.

> During a recent Super Bowl, an estimated 12.4 million pounds of potato chips were consumed. This was 3.1 million pounds **more than** the number of pounds of tortilla chips consumed. How many pounds of tortilla chips were consumed?

> The word **this** refers to the number of pounds of potato chips.

> The potato chips were 3.1 million pounds **more than** the tortilla chips.

STEP 2 Write the important information in one sentence.

> The number of pounds of potato chips was 3.1 million pounds more than the number of pounds of tortilla chips.

STEP 3 Replace any phrases with numbers that you know.

> 12.4 million was 3.1 million more than the number of pounds of tortilla chips.

Practice

Write the important information in one sentence. Replace any phrases with numbers that you know. Do not solve the problem.

1. **BASEBALL** Last year, Scott attended 13 Minnesota Twins baseball games. This year, he attended 24. How many more games did he attend this year?

2. **HOT AIR BALLOONS** Miyoki paid $140 for a four-hour hot air balloon ride. The cost of each hour was the same. Find the cost per hour of the ride.

Problem-Solving Investigation

Main Idea Solve problems using the *work backward* strategy.

⚡ ✐ P.S.I. TEAM ＋

Work Backward

MIGUEL: Yesterday, I earned extra money by doing yard work for my neighbor. Then I spent $5.50 at the convenience store and four times that amount at the bookstore. Now I have $7.75 left.

YOUR MISSION: Work backward to find how much money Miguel had before he went to the convenience store and the bookstore.

Understand	You know he has $7.75 left. You need to find the amount he started with.
Plan	Start with the end result and work backward.
Solve	He has $7.75 left. **Undo** the four times $5.50 spent at the bookstore. Since $5.50 × 4 is $22, add $7.75 and $22. **Undo** the $5.50 spent at the convenience store. Add $5.50 and $29.75. So, Miguel had $35.25 to start with.
Check	Assume Miguel started with $35.25. He spent $5.50 and $22. He had $35 − $5.50 − $22 or $7.75 left. So, $35.25 is correct. ✓

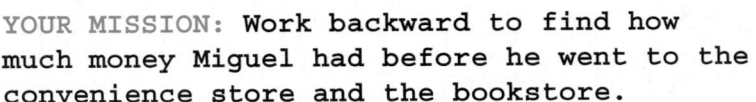

$$\begin{array}{r} \$7.75 \\ +\ \$22.00 \\ \hline \$29.75 \\ +\ \ \$5.50 \\ \hline \$35.25 \end{array}$$

Analyze the Strategy

1. Explain when you would use the *work backward* strategy to solve a problem.

2. Describe how to solve a problem by working backward.

3. **▣ Write Math** Write a problem that could be solved by working backward. Then write the steps you would take to find the solution to your problem.

• Work backward.
• Look for a pattern.
• Draw a diagram.
• Choose an operation.

 = **Step-by-Step Solutions** begin on page R1.
Extra Practice begins on page EP2.

Use the *work backward* strategy to solve Exercises 4–7.

4. MONEY Marisa spent $8 on a movie ticket. Then she spent $5 on popcorn and one half of what was left on a drink. She has $2 left. How much did she have initially?

5 NUMBER THEORY A number is multiplied by −3. Then 6 is subtracted from the product. After adding −7, the result is −25. What is the number?

6. TIME Timothy's morning schedule is shown. At what time does Timothy wake up if he arrives at school at 9:00 A.M.?

Timothy's Schedule	
Activity	**Time**
Wake up	■
Get ready for school – 45 min	■
Walk to school – 25 min	9:00 A.M.

7. LOGIC A small box contains 4 tennis balls. Six of these boxes are inside a medium box. Eight medium boxes are inside each large box, and 100 large boxes are shipped in a large truck. How many tennis balls are on the truck?

Use any strategy to solve Exercises 8–15.

8. GEOGRAPHY The land area of North Dakota is 68,976 square miles. This is about 7 times the land area of Vermont. Estimate the land area of Vermont.

9. AGE Brie is two years older than her sister Kiana. Kiana is 4 years older than their brother Jeron, who is 8 years younger than their brother Percy. Percy is 16 years old. How old is Brie?

10. ELEVATION New Orleans, Lousiana, has an elevation of −8 feet related to sea level. Death Valley, California, is 274 feet lower than New Orleans. What is the elevation of Death Valley?

11. GEOMETRY Draw the sixth figure in the pattern shown.

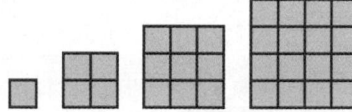

12. WATERFALLS Angel Falls in Venezuela is 3,212 feet high. It is 87 feet higher than 2.5 times the height of the Empire State Building. Find the height of the Empire State Building.

13. AIRCRAFT An aircraft carrier travels about 6 inches per gallon of fuel. Raquel's car travels about 28 miles per gallon of fuel. There are 5,280 feet in one mile. How many more inches per gallon does Raquel's car get than an aircraft carrier?

14. SCHOOL SUPPLIES Alexandra wishes to buy 5 pens, 1 ruler, and 7 folders.

Item	Cost
Pens	$2.09
Ruler	$0.99
Folder	$1.19

If there is no tax, is $20 enough to pay for Alexandra's school supplies? Explain your reasoning.

15. MONEY Antonio has saved $27 in cash to spend at the arcade. If he has 10 bills, how many of each kind of bill does he have?

Solve Addition and Subtraction Equations with Bar Diagrams

Main Idea

Write and solve addition and subtraction equations using bar diagrams.

 Get ConnectED

CCSS 7.EE.4

CELL PHONES In a recent year, 15 of the 50 states had a law banning the use of handheld cell phones while driving a school bus. How many states did *not* have this law?

ACTIVITY

① **What do you need to find?** how many states did not have a cell phone law for school bus drivers

You can represent this situation with an equation.

STEP 1 Draw a bar diagram that represents the total number of states and how many have passed a law.

⊢ ─ ─ ─ ─ ─ ─ ─ ─ ─ ─ ─ ─ 50 states ─ ─ ─ ─ ─ ─ ─ ─ ─ ─ ─ ─ ⊣
states with a law	**states that do not have a law**
⊢ ─ ─ ─ 15 ─ ─ ─ ⊣ ─ ─ ─ ─ ─ ─ ─ ? ─ ─ ─ ─ ─ ─ ─ ⊣

STEP 2 Write an equation from the bar diagram. Let *x* represent the states that do not have a cell phone law for school bus drivers.

$$15 + x = 50$$

STEP 3 Use the *work backward* strategy to solve the equation. Since $15 + x = 50$, $x = 50 - 15$. So, $x = 35$.

Check $15 + 35 = 50$ ✓

So, 35 states did *not* have a law banning the use of cell phones by bus drivers.

Analyze the Results

1. Suppose nine more states adopt similar laws. How would the equation change?

2. How would the diagram change if the U.S. Virgin Islands and Puerto Rico were counted with the United States?

ACTIVITY

② **BASEBALL CARDS** Jack had some baseball cards. He sold 7 of them. Then he had 29 left. Write and solve a subtraction equation to find the number he had at the beginning.

STEP 1 Draw a bar diagram representing the sentence. Let n represent the number.

⊢------------------- n baseball cards ----------------⊣	
sold	**number left**
⊢----7----⊣	⊢----------------- 29 ---------------⊣

STEP 2 Write a subtraction equation from the bar diagram.

$$n - 7 = 29$$

STEP 3 Solve the equation by working backward. Since $n - 7 = 29$, then $n = 29 + 7$. So, $n = 36$.

Check $36 - 7 = 29$ ✓

If 7 less than a number is 29, the number is 36.

Practice and Apply

BAR | DIAGRAM | **Draw a bar diagram and write an equation for each situation. Then solve the equation.**

3. The sum of a number and four is equal to 18.

4. Two more than the number of frogs is 4.

5. A number of students increased by 10 is 28.

6. A number decreased by 15 is 22.

7. **BABYSITTING** Katie earned $15 babysitting this week. Last week she earned $46. How much more did she earn last week than this week?

8. **AGES** The median age of people living in Arizona is 1 year older than the median age of people living in the United States. If the median age in Arizona is 34, what is the median age of people living in the United States?

9. ✎ **WRITE MATH** Write a real-world problem that can be represented by the bar diagram below.

⊢------------------------ n ----------------------⊣	
money spent for DVD	**money left**
⊢------------------- $19------------------⊣	⊢------$8------⊣

Explore

Solve Equations Using Algebra Tiles

Main Idea

Solve one-step equations using models.

Get Connect ED

On the balance at the right, the paper bag contains a certain number of orange blocks. (Assume that the paper bag weighs nothing.) Without looking in the bag, how can you determine the number of blocks it contains?

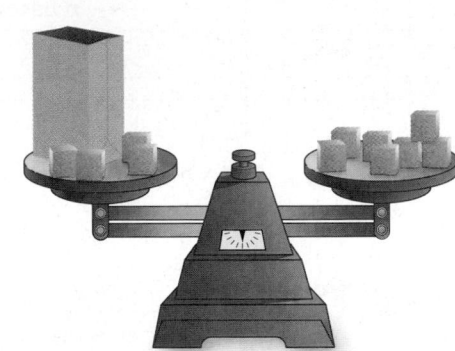

One way is to remove 4 blocks from each side of the balance. So, the bag contains 3 blocks.

In algebra, an *equation* is like a balance. The weights on each side of the scale are equal. The expressions on each side of an equation are equal. You can use algebra tiles to model and solve equations.

ACTIVITY

① **Solve $x + 4 = 7$ using algebra tiles.**

STEP 1 Model the equation.

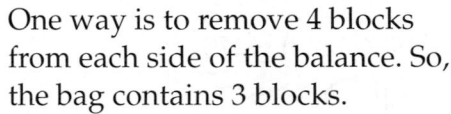

$$x + 4 \quad = \quad 7$$

STEP 2 Remove the same number of 1-tiles from each side of the mat so that the variable is by itself on one side.

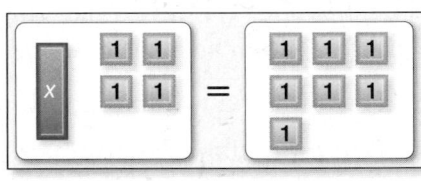

$$x + 4 - 4 \quad = \quad 7 - 4$$

STEP 3 The number of 1-tiles remaining on the right side of the mat represents the value of x.

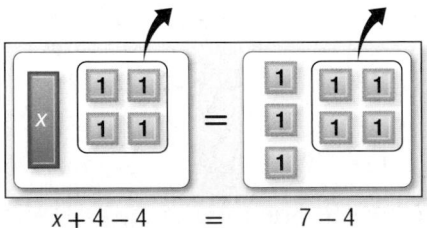

$$x \quad = \quad 3$$

Therefore, $x = 3$.

Check Since $3 + 4 = 7$, the solution is correct. ✔

A *zero pair* is a number paired with its opposite, like 2 and −2. You can add or subtract a zero pair from either side of an equation without changing its value, because the value of a zero pair is zero.

ACTIVITY

2 **Solve $x + 2 = -1$ using models.**

STEP 1 Model the equation.

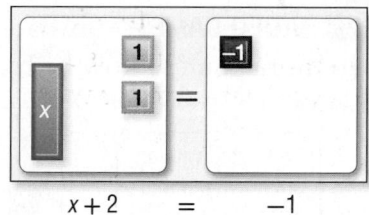

$$x + 2 = -1$$

STEP 2 Add 2 negative tiles to the left side of the mat and add 2 negative tiles to the right side of the mat to form zero pairs on the left.

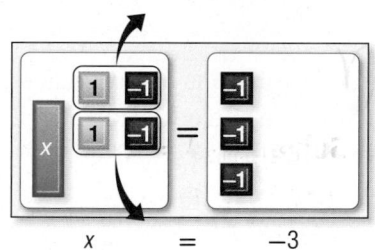

$$x + 2 + (-2) = -1 + (-2)$$

STEP 3 Remove all of the zero pairs from the left side. There are 3 negative tiles on the right side of the mat.

$$x = -3$$

Therefore, $x = -3$.

Check Since $-3 + 2 = -1$, the solution is correct. ✓

Practice and Apply

Solve each equation using algebra tiles.

1. $x + 4 = 4$
2. $5 = x + 4$
3. $4 = 1 + x$
4. $2 = 2 + x$
5. $-2 = x + 1$
6. $x - 3 = -2$
7. $x - 1 = -3$
8. $4 = x - 2$

9. Explain how the *work backward* strategy is used to solve equations.

10. How is solving an equation similar to keeping a scale in balance?

11. **MAKE A CONJECTURE** Write a rule that you can use to solve an equation like $x + 3 = 2$ without using models or a drawing.

12. **WRITE MATH** Write a real-world problem that can be solved using algebra tiles.

Main Idea

Solve addition and subtraction equations.

 Vocabulary

equation
equivalent equations

Get ConnectED

CCSS   7.EE.4

Solve One-Step Addition and Subtraction Equations

VIDEO GAMES Max had some video games, and then he bought two more games. Now he has six games.

He started with an unknown number of games.

He bought two more.

Now he has six games.

1. What does x represent in the figure?

2. What addition equation is shown in the figure?

3. How many games did Max have in the beginning?

An **equation** is a sentence stating that two quantities are equal. You can solve the equation $x + 2 = 6$ by subtracting 2 from each side of the equation.

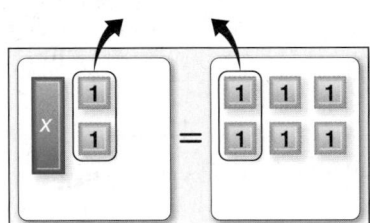

$$\begin{aligned} x + 2 &= 6 \\ -2 &= -2 \\ \hline x &= 4 \end{aligned}$$

The equations $x + 2 = 6$ and $x = 4$ are **equivalent equations** because they have the same solution, 4.

Key Concept   **Subtraction Property of Equality**

Words   If you subtract the same number from each side of an equation, the two sides remain equal.

Symbols   If $a = b$, then $a - c = b - c$.

Examples   **Numbers**   **Algebra**

$$\begin{aligned} 6 &= 6 \\ -2 &= -2 \\ \hline 4 &= 4 \end{aligned}$$ $$\begin{aligned} x + 2 &= 6 \\ -2 &= -2 \\ \hline x &= 4 \end{aligned}$$

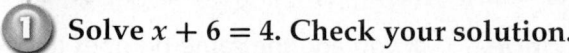 **EXAMPLE** Solve Addition Equations

① Solve $x + 6 = 4$. Check your solution.

$$
\begin{array}{ll}
x + 6 = 4 & \text{Write the equation.} \\
\underline{-6 = -6} & \text{Subtraction Property of Equality} \\
x = -2 & \text{Simplify.}
\end{array}
$$

Check
$$
\begin{array}{ll}
x + 6 = 4 & \text{Write the original equation.} \\
-2 + 6 \stackrel{?}{=} 4 & \text{Replace } x \text{ with } -2. \\
4 = 4 \checkmark & \text{The sentence is true.}
\end{array}
$$

The solution is −2.

 CHECK Your Progress

Solve each equation. Check your solution.

a. $y + 6 = 9$ **b.** $x + 3 = 1$ **c.** $-3 = a + 4$

Study Tip

Solutions Notice that your new equation, $x = -2$, has the same solution as the original equation, $x + 6 = 4$.

 **REAL-WORLD EXAMPLE**

② **MARINE BIOLOGY** Clown fish and angelfish are popular tropical fish. An angelfish can grow to be 12 inches long. If an angelfish is 8.5 inches longer than a clown fish, how long is a clown fish?

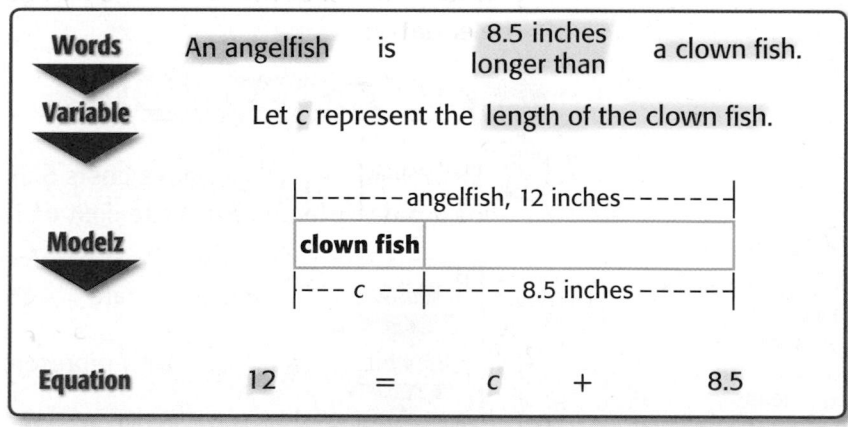

| Words | An angelfish | is | 8.5 inches longer than | a clown fish. |

Variable   Let c represent the length of the clown fish.

Modelz

angelfish, 12 inches
clown fish
c ---- 8.5 inches

| Equation | 12 | = | c | + | 8.5 |

$$
\begin{array}{ll}
12 = c + 8.5 & \text{Write the equation.} \\
\underline{-8.5 = -8.5} & \text{Subtraction Property of Equality} \\
3.5 = c & \text{Simplify.}
\end{array}
$$

A clown fish is 3.5 inches long.

 CHECK Your Progress

d. WEATHER The highest recorded temperature in Warsaw, Missouri, is 118°F. This is 158° greater than the lowest recorded temperature. Write and solve an equation to find the lowest recorded temperature.

 **Real-World Link** · · · ·

The deeper the descent of a dive, the quicker the diver consumes air. Most divers can spend 45 minutes to an hour at 40 feet below the surface.

Key Concept   Addition Property of Equality

Words   If you add the same number to each side of an equation, the two sides remain equal.

Symbols   If $a = b$, then $a + c = b + c$.

Examples

Numbers	Algebra

$$
\begin{array}{r}
5 = 5 \\
+3 = +3 \\
\hline
8 = 8
\end{array}
$$

$$
\begin{array}{r}
x - 2 = 4 \\
+2 = +2 \\
\hline
x = 6
\end{array}
$$

 EXAMPLE   Solve a Subtraction Equation

3   Solve $x - 2 = 1$. Check your solution.

$$
\begin{array}{ll}
x - 2 = 1 & \text{Write the equation.} \\
\underline{+2 = +2} & \text{Addition Property of Equality} \\
x = 3 & \text{Simplify.}
\end{array}
$$

The solution is 3.   **Check** $3 - 2 = 1$ ✓

CHECK Your Progress

Solve each equation. Check your solution.

e. $y - 3 = 4$ **f.** $r - 4 = -2$ **g.** $q - 8 = -9$

 REAL-WORLD EXAMPLE

4   **SHOPPING** A pair of shoes costs $25. This is $14 less than the cost of a pair of jeans. Find the cost of the jeans.

Study Tip

Models A bar diagram can be used to represent this situation.

```
|------ jeans, j ------|
| shoes     |          |
|----$25----|--$14--|
```

Words	Shoes	are	$14 less than	jeans.
Variable		Let j represent the cost of jeans.		
Equation	25	=	j − 14	

$$
\begin{array}{ll}
25 = j - 14 & \text{Write the equation.} \\
\underline{+14 = +14} & \text{Addition Property of Equality} \\
39 = j & \text{Simplify.}
\end{array}
$$

The jeans cost $39.

CHECK Your Progress

h. **ANIMALS** The average lifespan of a tiger is 17 years. This is 3 years less than the average lifespan of a lion. Write and solve an equation to find the average lifespan of a lion.

Example 1   Solve each equation. Check your solution.

1. $n + 6 = 8$ 2. $7 = y + 2$

3. $m + 5 = 3$ 4. $-2 = a + 6$

Example 2   5. **BAR DIAGRAM** Orville and Wilbur Wright made the first airplane flights in 1903. Wilbur's flight was 364 feet. This was 120 feet longer than Orville's flight. Draw a bar diagram to represent the flights. Then write and solve an equation to find the length of Orville's flight.

Example 3   Solve each equation. Check your solution.

6. $x - 5 = 6$ 7. $-1 = c - 6$

Example 4   8. **PRESIDENTS** John F. Kennedy was the youngest president to be inaugurated. He was 43 years old. This was 26 years younger than the oldest president to be inaugurated—Ronald Reagan. Write and solve an equation to find how old Reagan was when he was inaugurated.

Practice and Problem Solving

 = **Step-by-Step Solutions** begin on page R1.
Extra Practice begins on page EP2.

Examples 1 and 3   Solve each equation. Check your solution.

9. $a + 3 = 10$ 10. $y + 5 = 11$ 11. $9 = r + 2$

12. $14 = s + 7$ 13. $x + 8 = 5$ 14. $y + 15 = 11$

15. $r + 6 = -3$ 16. $k + 3 = -9$ 17. $s - 8 = 9$

18. $w - 7 = 11$ 19. $-1 = q - 8$ 20. $-2 = p - 13$

Examples 2 and 4   **BAR DIAGRAM** Write an equation. Use a bar diagram if needed. Then solve the equation.

21. **MUSIC** Last week Tiffany practiced her bassoon a total of 7 hours. This was 2 hours more than she practiced the previous week. How many hours did Tiffany practice the previous week?

22. **CIVICS** In a recent presidential election, Ohio had 20 electoral votes. This is 14 votes less than Texas had. How many electoral votes did Texas have?

23. **AGES** Zack is 15 years old. This is 3 years younger than his brother Louis. How old is Louis?

24. **BASKETBALL** The Miami Heat scored 79 points in a recent game. This was 13 points less than the Chicago Bulls scored. How many points did the Chicago Bulls score?

Solve each equation. Check your solution.

25. $a - 3.5 = 14.9$

26. $x - 2.8 = 9.5$

27. $r - 8.5 = -2.1$

28. $z - 9.4 = -3.6$

29. $n + 1.4 = 0.72$

30. $b + 2.25 = 1$

31. $m + \dfrac{5}{6} = \dfrac{11}{12}$

32. $y + \dfrac{1}{2} = \dfrac{3}{4}$

33. $s - \dfrac{1}{9} = \dfrac{5}{18}$

34. $-\dfrac{1}{3} = r - \dfrac{3}{4}$

35. $-\dfrac{5}{6} + c = -\dfrac{11}{12}$

36. $-\dfrac{33}{34} = t - \dfrac{13}{17}$

Write an equation. Then solve the equation.

37. MONEY Suppose you have d dollars. After you pay your sister the $5 you owe her, you have $18 left. How much money did you have?

38. FINANCIAL LITERACY Suppose you have saved $38. How much more do you need to save to buy a small television that costs $65?

39. GEOMETRY The sum of the measures of the angles of a triangle is 180°. Find the missing measure.

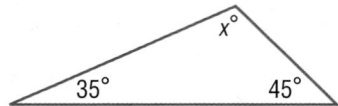

40. VOLCANOES Alaska, Hawaii, and Washington have active volcanoes. Alaska has 43, Hawaii has 5, and Washington has v. If they have 52 active volcanoes in all, how many volcanoes does Washington have?

41 GOLF The table shows Cristie Kerr's scores for four rounds of a recent U.S. Women's Open. Her total score was even with par. What was her score for the third round?

Round	Score
First	−1
Second	−3
Third	s
Fourth	+2

42. BUSINESS At the end of the day, the closing price of XYZ Stock was $62.87 per share. This was $0.62 less than the opening price. Find the opening price.

43. MULTIPLE REPRESENTATIONS The table shows information about the tallest wooden roller coasters.

Tallest Wooden Roller Coasters	Height (feet)	Drop (feet)	Speed (mph)
Son of Beast	218	214	s
El Toro	181	176	70
The Rattler	180	d	65
Voyage	173	154	67
Colossos	h	159	75

a. ALGEBRA The difference in speeds of Son of Beast and The Rattler is 13 miles per hour. If Son of Beast has the greater speed, write and solve a subtraction equation to find its speed.

b. BAR DIAGRAM The Rattler has a drop that is 52 feet less than El Toro. Draw a bar diagram and write an equation to find the height of The Rattler.

c. ALGEBRA Let h represent the height of the Colossos roller coaster. Explain why $h + 10 = 180$ and $h + 48 = 218$ are equivalent equations. Then explain the meaning of the solution.

44. FIND THE ERROR Aisha is finding $b + 5 = -8$. Find her mistake and correct it.

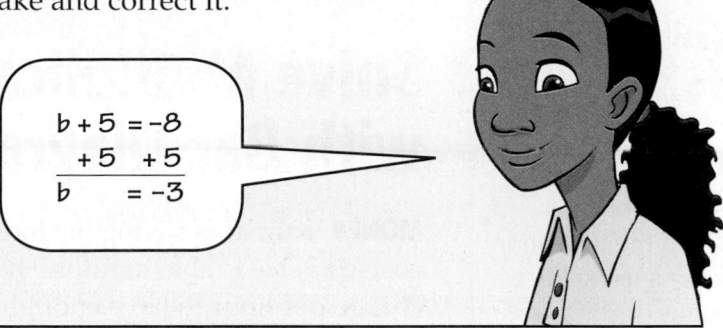

$$b + 5 = -8$$
$$\underline{+5 \quad +5}$$
$$b \quad\quad = -3$$

45. CHALLENGE Suppose $x + y = 11$ and the value of x increases by 2. If their sum remains the same, what must happen to the value of y?

46. OPEN ENDED Write an addition equation and a subtraction equation that have 10 as a solution.

47. ✎ **WRITE MATH** Write a problem about a real-world situation that can be represented by the equation $p - 25 = 50$.

Test Practice

48. The Oriental Pearl Tower in China is 1,535 feet tall. It is 280 feet shorter than the Canadian National Tower in Canada. Which equation can be used to find the height of the Canadian National Tower?

A. $1{,}535 + h = 280$

B. $h = 1{,}535 - 280$

C. $1{,}535 = h - 280$

D. $280 - h = 1{,}535$

49. Which of the following statements is true concerning the equation $x + 3 = 7$?

F. To find the value of x, add 3 to each side.

G. To find the value of x, add 7 to each side.

H. To find the value of x, find the sum of 3 and 7.

I. To find the value of x, subtract 3 from each side.

50. The model represents the equation $x - 2 = 5$.

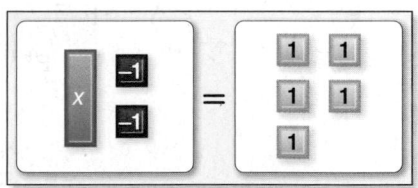

How could you find the value of x?

A. Add two positive counters to each side of the equation mat.

B. Add two negative counters to each side of the equation mat.

C. Add five positive counters to each side of the equation mat.

D. Add five negative counters to each side of the equation mat.

51. ✎ **GRIDDED RESPONSE** What is the solution to the equation below?

$$-8 = x - 15$$

Explore

Solve Multiplication Equations with Bar Diagrams

Main Idea

Write and solve one-step multiplication equations using bar diagrams.

Get ConnectED

CCSS 7.EE.4

MONEY Kumar is saving his money to buy a new DVD player that costs $63. He is able to tutor 9 hours in a week. How much should he charge per hour to have enough money by the end of the week?

ACTIVITY

What do you need to find? how much he should charge per hour

STEP 1 Draw a bar diagram that represents the money he needs to earn and the number of hours he is available to tutor that week.

$63								
hour 1	hour 2	hour 3	hour 4	hour 5	hour 6	hour 7	hour 8	hour 9
--?--	--?--	--?--	--?--	--?--	--?--	--?--	--?--	--?--

STEP 2 Write an equation from the bar diagram. Let x represent the amount he should charge each hour.

$$9x = 63$$

STEP 3 Use the *work backward* strategy to solve the equation. Since $9x = 63$, $x = 63 \div 9$. So, $x = 7$.

Check $9 \times 7 = 63$ ✓

Kumar should charge $7 an hour.

Practice and Apply

| BAR | DIAGRAM | Draw a bar diagram and write an equation for each situation. Then solve the equation.

1. **DANCE CLASS** Keyani spent $70 for 4 hours of dance classes. How much did she spend per hour of dance class?

2. **TEXT MESSAGING** The screen on Lin's cell phone allows for 8 lines of text per message. The maximum number of characters for each message is 160. How many characters can each line hold?

Main Idea

Solve one-step multiplication and division equations.

 Vocabulary

coefficient
formula

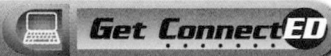

 7.EE.4

Solve One-Step Multiplication and Division Equations

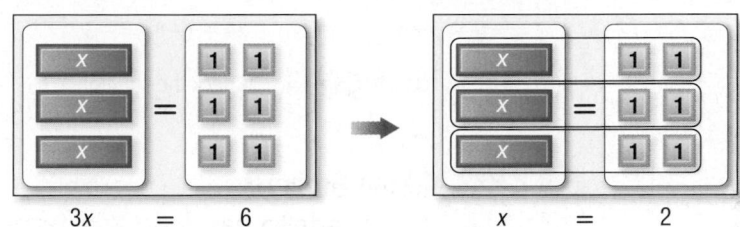 **Explore** Suppose three friends order an appetizer of nachos that costs \$6. They agree to split the cost equally. The figure below illustrates the multiplication equation $3x = 6$, where x represents the amount each friend pays.

| $3x$ | $=$ | 6 | | x | $=$ | 2 |

Since there are 3 xs, each x is matched with \$2.

The solution of $3x = 6$ is 2. So, each friend pays \$2.

Solve each equation using models or a drawing.

1. $3x = 12$ 3. $8 = 2x$

2. $2x = -8$ 4. $3x = -9$

5. What operation did you use to find each solution?

6. How can you use the 8 being multiplied by x to solve $8x = 40$?

The expression $3x$ means *3 times the value of* x. The numerical factor of a multiplication expression like $3x$ is called a **coefficient**. So, 3 is the coefficient of $3x$.

Key Concept   Division Property of Equality

Words	If you divide each side of an equation by the same nonzero number, the two sides remain equal.

Symbols   If $a = b$ and $c \neq 0$, then $\dfrac{a}{c} = \dfrac{b}{c}$.

Examples	**Numbers**	**Algebra**
	$8 = 8$	$2x = -6$
	$\dfrac{8}{2} = \dfrac{8}{2}$	$\dfrac{2x}{2} = \dfrac{-6}{2}$
	$4 = 4$	$x = -3$

EXAMPLES **Solve Multiplication Equations**

(1) **Solve $20 = 4x$. Check your solution.**

$20 = 4x$	Write the equation.
$\dfrac{20}{4} = \dfrac{4x}{4}$	Division Property of Equality
$5 = x$	$20 \div 4 = 5$

The solution is 5.   Check the solution.

(2) **Solve $-8y = 24$. Check your solution.**

$-8y = 24$	Write the equation.
$\dfrac{-8y}{-8} = \dfrac{24}{-8}$	Division Property of Equality
$y = -3$	$24 \div (-8) = -3$

The solution is -3.   Check the solution.

 CHECK Your Progress

Solve each equation. Check your solution.

   a. $30 = 6x$ **b.** $-6a = 36$ **c.** $-9d = -72$

 **REAL-WORLD EXAMPLE**

(3) **TEXT MESSAGING** Lelah sent 574 text messages last week. On average, how many messages did she send each day?

Words	Total	is equal to	number of days	times	number of messages.
Variable	\multicolumn{5}{l}{Let m represent the number of messages Lelah sent.}
Equation	574	=	7	•	m

$574 = 7m$	Write the equation.
$\dfrac{574}{7} = \dfrac{7m}{7}$	Division Property of Equality
$82 = m$	$574 \div 7 = 82$

Lelah sent 82 messages on average each day.

Real-World Link · · · ·
Over 60% of teenagers' text messages are sent from their homes— even when a landline is available.

CHECK Your Progress

d. TRAVEL Mrs. Acosta's car can travel an average of 24 miles on each gallon of gasoline. Write and solve an equation to find how many gallons of gasoline she will need for a trip of 348 miles.

Key Concept — Multiplication Property of Equality

Words   If you multiply each side of an equation by the same number, the two sides remain equal.

Examples

Numbers	Algebra
$3 = 3$	$\dfrac{x}{4} = 7$
$3(-6) = 3(-6)$	$\dfrac{x}{4}(4) = 7(4)$
$-18 = -18$	$x = 28$

 EXAMPLE Solve Division Equations

④ Solve $\dfrac{a}{-4} = -9$.

$\dfrac{a}{-4} = -9$ Write the equation.

$\dfrac{a}{-4}(-4) = -9(-4)$ Multiplication Property of Equality

$a = 36$ $-9 \cdot (-4) = 36$

 CHECK Your Progress

e. $\dfrac{y}{-3} = -8$ **f.** $\dfrac{m}{5} = -7$ **g.** $30 = \dfrac{b}{-6}$

Vocabulary Link

Everyday Use

Formula A plan or method for doing something.

Math Use

Formula An equation that shows the relationship among certain quantities.

A **formula** is an equation that shows the relationship among certain quantities. One of the most common formulas is the equation $d = rt$, which gives the relationship among distance d, rate r, and time t. The formula can also be written as $r = \dfrac{d}{t}$ or $t = \dfrac{d}{r}$.

 REAL-WORLD EXAMPLE

⑤ **DISTANCE** The distance d Tina travels in her car while driving 60 miles per hour for 3 hours is given by the equation $\dfrac{d}{3} = 60$. How far did she travel?

$\dfrac{d}{3} = 60$ Write the equation.

$\dfrac{d}{3}(3) = 60(3)$ Multiplication Property of Equality

$d = 180$ $60 \cdot 3 = 180$

Tina traveled 180 miles.

 CHECK Your Progress

h. ANIMALS A Fitch ferret has a mass of about 2 kilograms. To find its weight in pounds p, you can use the equation $\dfrac{p}{2} = 2.2$. How many pounds does a Fitch ferret weigh?

Examples 1, 2, and 4   Solve each equation. Check your solution.

1. $6c = 18$ **2.** $15 = 3z$ **3.** $-8x = 24$

4. $-9r = -36$ **5.** $\dfrac{p}{9} = 9$ **6.** $\dfrac{a}{12} = -3$

Example 3   **7. WORK** Antonia earns $6 per hour helping her grandmother. Write and solve an equation to find how many hours she needs to work to earn $48.

Example 5   **8. SHARKS** A shark can swim at an average speed of 25 miles per hour. At this rate, how far can a shark swim in 2.4 hours? Use $r = \dfrac{d}{t}$.

Practice and Problem Solving

 = **Step-by-Step Solutions** begin on page R1.
Extra Practice begins on page EP2.

Examples 1, 2, and 4   Solve each equation. Check your solution.

9. $7a = 49$ **10.** $9e = 27$ **11.** $2x = -6$

12. $-4j = 36$ **13.** $-12y = 60$ **14.** $-4s = -16$

(15) $\dfrac{m}{10} = 7$ **16.** $\dfrac{u}{6} = 9$ **17.** $\dfrac{h}{-3} = 12$

18. $-30 = \dfrac{q}{-5}$ **19.** $-8 = \dfrac{c}{-10}$ **20.** $\dfrac{r}{20} = -2$

| BAR | DIAGRAM | Write an equation. Use a bar diagram if needed. Then solve the equation.

Example 3   **21. MONEY** Brandy wants to buy a digital camera that costs $300. If she saves $15 each week, in how many weeks will she have enough money for the camera?

22. COMPUTERS The width of a computer monitor is 1.25 times its height. Find the height of the computer monitor at the right.

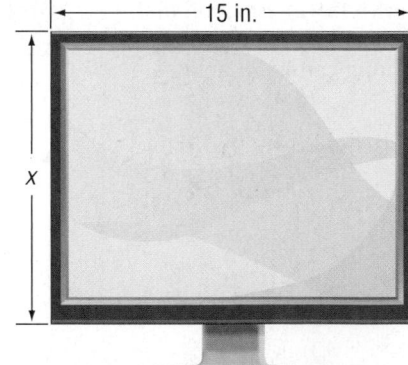

Example 5   **(23) SPEED** A race car can travel at a rate of 205 miles per hour. At this rate, how far would it travel in 3 hours? Use $r = \dfrac{d}{t}$.

24. INSECTS A dragonfly, the fastest insect, can fly a distance of 50 feet at a speed of 25 feet per second. Find the time in seconds. Use $d = rt$.

25. HURRICANES A certain hurricane travels at 20.88 kilometers per hour. The distance from Cuba to Key West is 145 kilometers. Write and solve a multiplication equation to find how long it would take the hurricane to travel from Cuba to Key West.

26. **OPEN ENDED** Describe a real-world situation in which you would use a division equation to solve a problem. Write your equation and then solve your problem.

27. **FIND THE ERROR** Raul is solving $-6x = 72$. Find his mistake and correct it.

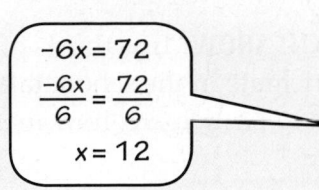
$$-6x = 72$$
$$\frac{-6x}{6} = \frac{72}{6}$$
$$x = 12$$

28. **CHALLENGE** Solve $3|x| = 12$. Explain your reasoning.

29. **REASONING** *True* or *false*. To solve the equation $5x = 20$ you can use the Multiplication Property of Equality. Explain your reasoning.

 **WRITE MATH** **Write a real-world problem that could be represented by each equation.**

30. $2x = 16$ 31. $3x = 75$ 32. $4x = -8$

![checkmark] **Test Practice**

33. A football player can run 20 yards in 3.4 seconds. Which equation could be used to find y, the number of yards the football player can run in a second?

   A. $20y = 3.4$

   B. $3.4 - y = 20$

   C. $3.4y = 20$

   D. $20 + y = 3.4$

34. **SHORT RESPONSE** Use the formula $A = bh$ to find the base in inches of a rhombus with a height of 7 inches and an area of 56 square inches.

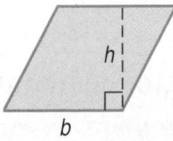

35. **EXTENDED RESPONSE** The table shows the prices of different online DVD rental plans. Mrs. Freedman prepaid for x number of months of the unlimited plan.

DVD Rental Plans	
Plan	**Cost per Month ($)**
Unlimited	16.50
3 DVDs at a time	14.35
2 DVDs at a time	11.99

Part A Mrs. Freedman paid $99 for x number of months of the unlimited plan. Write an equation to find the number of prepaid months.

Part B Solve your equation. Explain how you solved.

Solve Equations with Rational Coefficients

Main Idea

Write and solve equations with rational coefficients using bar diagrams.

TALENT SHOW Two thirds of Chen's homeroom class plan to participate in the school talent show. If 16 students from the class plan to participate, how many students are in the homeroom class?

ACTIVITY

What do you need to find? how many students are in Chen's homeroom class

You can represent this situation with an equation.

STEP 1 Draw a bar diagram that represents the total number of students in the class and how many plan to participate.

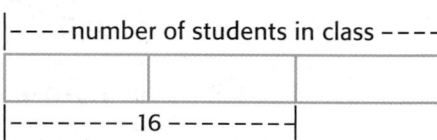

STEP 2 Write an equation from the bar diagram. Let c represent the total number of students in the class.

$$\frac{2}{3}c = 16$$

STEP 3 Find the number of students represented by the sections of the bar.

--number of students in class, c--		
8	8	8

Since each section represents 8 students, there are 8 × 3 or 24 students in the class. **Check** $\frac{2}{3} \times 24 = 16$ ✓

Practice and Apply

1. Suppose $\frac{3}{4}$ of the class planned to participate. How would the diagram and equation be different?

2. BAR DIAGRAM Eliana is spending $\frac{3}{5}$ of her monthly allowance on a costume for the talent show. She plans to spend $24. Draw a bar diagram to represent the situation. Then write and solve an equation to find the amount of Eliana's monthly allowance.

PART A B C **D**

Main Idea

Solve equations with rational coefficients.

 Vocabulary

multiplicative inverse
reciprocal

 Get ConnectED

 CCSS 7.EE.4

Solve Equations with Rational Coefficients

HOMEWORK Shawnda spends $\frac{1}{2}$ hour doing homework after school. Then she spends another $\frac{1}{2}$ hour doing homework before bed. In all, she spends $2 \cdot \frac{1}{2}$ or 1 hour doing homework.

1. Copy and complete the table below.

$\frac{3}{2} \times \frac{2}{3} = \blacksquare$	$\frac{1}{5} \times \blacksquare = 1$	$\frac{5}{6} \times \frac{6}{5} = \blacksquare$	$\frac{7}{8} \times \frac{8}{7} = \blacksquare$
$\blacksquare \times \frac{5}{7} = 1$	$\frac{2}{6} \times \frac{6}{2} = \blacksquare$	$\frac{7}{1} \times \blacksquare = 1$	$\blacksquare \times 8 = 1$

2. What is true about the products in Exercise 1?

Two numbers with a product of 1 are called **multiplicative inverses**, or **reciprocals**.

Key Concept **Inverse Property of Multiplication**

Words The product of a number and its multiplicative inverse is 1.

Examples **Numbers** **Algebra**

$$\frac{3}{4} \times \frac{4}{3} = 1 \qquad \frac{a}{b} \cdot \frac{b}{a} = 1, \text{ for } a, b \neq 0$$

 **EXAMPLES** Find Multiplicative Inverses

1 Find the multiplicative inverse of $\frac{2}{5}$.

Since $\frac{2}{5} \cdot \frac{5}{2} = 1$, the multiplicative inverse of $\frac{2}{5}$ is $\frac{5}{2}$, or $2\frac{1}{2}$.

2 Find the multiplicative inverse of $2\frac{1}{3}$.

$2\frac{1}{3} = \frac{7}{3}$ Rename the mixed number as an improper fraction.

Since $\frac{7}{3} \cdot \frac{3}{7} = 1$, the multiplicative inverse of $2\frac{1}{3}$ is $\frac{3}{7}$.

CHECK Your Progress

Find the muliplicative inverse of each number.

a. $\frac{5}{6}$ **b.** $1\frac{1}{2}$ **c.** 8 **d.** $\frac{4}{3}$

Sometimes the coefficient of a term in a multiplication equation is a rational number. If the coefficient is a decimal, divide each side by the coefficient. If the coefficient is a fraction, multiply each side by the reciprocal of the coefficient.

$$0.75x = 3.75$$
$$\frac{0.75x}{0.75} = \frac{3.75}{0.75}$$
$$x = 5$$

$$\frac{2}{3}x = 4$$
$$\frac{3}{2} \cdot \frac{2}{3}x = \frac{3}{2} \cdot 4$$
$$x = 6$$

 **EXAMPLE** **Decimal Coefficients**

3 **Solve $16 = 0.25n$. Check your solution.**

$16 = 0.25n$	Write the equation.
$\dfrac{16}{0.25} = \dfrac{0.25n}{0.25}$	Division Property of Equality
$64 = n$	Simplify.

Check $16 = 0.25n$ Write the original equation.

$16 \overset{?}{=} 0.25 \cdot 64$ Replace n with 64.

$16 = 16$ ✓   This sentence is true.

The solution is 64.

 CHECK Your Progress

e. $6.4 = 0.8m$ **f.** $-2.8p = 4.2$ **g.** $-4.7k = -10.81$

 REAL-WORLD EXAMPLE

4 **ICE CREAM** Jaya's softball coach agreed to buy ice cream cones for all of the team members. Ice cream cones are $2.40 each. How many ice cream cones can the coach buy with $30?

Let n represent the number of ice cream cones the coach can buy.

$2.4n = 30$	Write the equation; $2.40 = 2.4$.
$\dfrac{2.4n}{2.4} = \dfrac{30}{2.4}$	Division Property of Equality
$n = 12.5$	Simplify.

 Real-World Link · · · · ·

Each year, over 1.5 billion gallons of ice cream are produced in the United States.

Since the number of ice cream cones must be a whole number, there is enough money for 12 ice cream cones.

 CHECK Your Progress

h. ICE CREAM Suppose the ice cream cones cost $2.80 each. How many ice cream cones could the coach buy with $42?

Study Tip

Fractions as Coefficients

The expression $\frac{3}{4}x$ can be read as $\frac{3}{4}$ of x, $\frac{3}{4}$ multiplied by x, $3x$ divided by 4, or $\frac{x}{4}$ multiplied by 3.

5 Solve $\frac{3}{4}x = \frac{12}{20}$. Check your solution.

$$\frac{3}{4}x = \frac{12}{20}$$ Write the equation.

$$\left(\frac{4}{3}\right) \cdot \frac{3}{4}x = \left(\frac{4}{3}\right) \cdot \frac{12}{20}$$ Multiply each side by the reciprocal of $\frac{3}{4}$, $\frac{4}{3}$.

$$\overset{1}{\underset{1}{\cancel{\frac{4}{3}}}} \cdot \overset{1}{\underset{1}{\cancel{\frac{3}{4}}}}x = \overset{1}{\underset{1}{\cancel{\frac{4}{3}}}} \cdot \overset{4}{\underset{5}{\cancel{\frac{12}{20}}}}$$ Divide by common factors.

$$x = \frac{4}{5}$$ Simplify.

Check $\frac{3}{4}x = \frac{12}{20}$ Write the original equation.

$$\frac{3}{4}\left(\frac{4}{5}\right) \overset{?}{=} \frac{12}{20}$$ Replace x with $\frac{4}{5}$.

$$\frac{12}{20} = \frac{12}{20} ✓$$ This sentence is true.

CHECK Your Progress

i. $\frac{1}{2}x = 8$ **j.** $-\frac{3}{4}x = 9$ **k.** $-\frac{7}{8}x = -\frac{21}{64}$

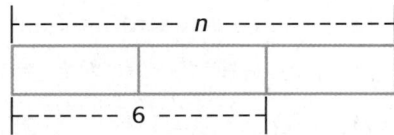 **REAL-WORLD EXAMPLE**

6 **SCHOOL PLAY** Valerie needs $\frac{2}{3}$ yard of fabric to make each hat for the school play. Write and solve an equation to find how many hats can she make with 6 yards of fabric.

Each hat needs $\frac{2}{3}$ yard of fabric. So, $\frac{2}{3}$ of some number n is 6. Draw a bar diagram.

```
+---------- n ----------+
|       |       |       |
+------ 6 ------+
```

Write and solve a multiplication equation.

$$\frac{2}{3}n = 6$$ Write the equation.

$$\left(\frac{3}{2}\right) \cdot \frac{2}{3}n = \left(\frac{3}{2}\right) \cdot 6$$ Multiply each side by $\frac{3}{2}$.

$$n = 9$$ Simplify.

Valerie can make 9 hats.

CHECK Your Progress

l. **TRAIL MIX** Wilson has 9 pounds of trail mix. How many $\frac{3}{4}$-pound bags of trail mix can he make?

Examples 1 and 2   Find the multiplicative inverse of each number.

   1. $\dfrac{8}{5}$ **2.** $\dfrac{2}{9}$ **3.** $5\dfrac{4}{5}$ **4.** 9

Examples 3 and 5   Solve each equation. Check your solution.

   5. $1.6k = 3.2$ **6.** $3.9 = 1.3y$ **7.** $-2.5b = 20.5$

   8. $-\dfrac{1}{2} = -\dfrac{5}{18}h$ **9.** $\dfrac{3}{8}a = \dfrac{12}{40}$ **10.** $-6 = \dfrac{4}{7}x$

Examples 4 and 6   **11. FRUIT** Three fourths of the fruit in a refrigerator are apples. There are 24 apples in the refrigerator. Write and solve an equation to find how many pieces of fruit are in the refrigerator.

   12. FINANCIAL LITERACY Dillon deposited $\dfrac{3}{4}$ of his paycheck into the bank. The deposit slip shows how much he deposited. Write and solve an equation to find the amount of his paycheck.

DEPOSIT		
	CHECKS	4 6 . 5 0
Name: *Dillon Gates*		
Date: *9/22*		
🏛 Great Savings Bank		
Transaction # •543345890•3221•8755P	DEPOSIT	$ 4 6 . 5 0

Practice and Problem Solving

    = **Step-by-Step Solutions** begin on page R1.
         Extra Practice begins on page EP2.

Examples 1 and 2   Find the multiplicative inverse of each number.

   13. $\dfrac{5}{6}$ **14.** $\dfrac{11}{2}$ **15.** $\dfrac{1}{6}$ **16.** $\dfrac{1}{10}$

   17. 3 **18.** 14 **19.** $5\dfrac{1}{8}$ **20.** $6\dfrac{2}{3}$

Examples 3 and 5   Solve each equation. Check your solution.

   21. $1.2x = 6$ **22.** $2.8 = 0.4d$ **23.** $-2.4b = 14.4$

   24. $-5w = -24.5$ **25.** $3.6h = -10.8$ **26.** $2.8m = 12.88$

   27. $\dfrac{2}{5}t = \dfrac{12}{25}$ **28.** $\dfrac{24}{16} = \dfrac{3}{4}a$ **㉙** $\dfrac{7}{8}k = \dfrac{5}{6}$

   30. $\dfrac{2}{3} = \dfrac{8}{3}b$ **31.** $-\dfrac{1}{2}g = -3\dfrac{1}{3}$ **32.** $\dfrac{3}{5}c = -6\dfrac{1}{4}$

Examples 4 and 6   **33. LIFE SCIENCE** The average growth of human hair per month is 0.5 inch. Write and solve an equation to find how long it takes a human hair to grow 3 inches.

   34. SEWING Jocelyn has nine yards of fabric to make table napkins for a senior citizens' center. She needs $\dfrac{3}{8}$ yard for each napkin. Write and solve an equation to find the number of napkins that she can make with this amount of fabric.

For Exercises 35–38, write an equation. Then solve.

35. **CAVES** The self-guided Mammoth Cave Discovery Tour includes an elevation change of 140 feet. This is $\frac{7}{15}$ of the elevation change on the Wild Cave Tour. What is the elevation change on the Wild Cave Tour?

36. **MUSEUMS** Twenty-four students brought their permission slips to attend the class field trip to the local art museum. If this represented $\frac{4}{5}$ of the class, how many students are in the class?

37. **MEASUREMENT** If one serving of cooked rice is $\frac{3}{4}$ cup, how many servings will $16\frac{1}{2}$ cups of rice yield?

38. **HIKING** After Alana hiked $2\frac{5}{8}$ miles along a hiking trail, she realized that she was only $\frac{3}{4}$ of the way to the end of the trail. How long is the trail?

39. **GRAPHIC NOVEL** Refer to the graphic novel frame below. Write and solve an equation to find how many movies they have time to show.

H.O.T. Problems

40. **REASONING** Complete the statement: If $8 = \frac{m}{4}$, then $m - 12 = \blacksquare$. Explain.

41. **Which One Doesn't Belong?** Identify the pair of numbers that does not belong with the other three. Explain.

| $\frac{9}{6}, \frac{6}{9}$ | $4, \frac{1}{4}$ | $\frac{3}{5}, 5$ | $\frac{2}{7}, \frac{7}{2}$ |

42. **CHALLENGE** The formula for the area of a trapezoid is $A = \frac{1}{2}h(b_1 + b_2)$, where b_1 and b_2 are both bases and h is the height. Find the value of h in terms of A, b_1, and b_2. Justify your answer.

43. **WRITE MATH** Explain the Multiplication Property of Equality. Then give an example of an equation in which you would use this property.

44. Audrey drove 200 miles in 3.5 hours. Which equation can you use to find the rate r at which Audrey was traveling?

   A. $200 = 3.5r$

   B. $200 \cdot 3.5 = r$

   C. $\dfrac{r}{3.5} = 200$

   D. $200r = 3.5$

45. A high-speed train travels 100 miles in $\dfrac{2}{3}$ hour. Which speed represents the rate of the train?

   F. 50 mph

   G. 75 mph

   H. 100 mph

   I. 150 mph

46. The table shows the results of a survey.

Music Preference	
Type	Fraction of Students
Pop	$\dfrac{5}{8}$
Jazz	$\dfrac{1}{8}$
Rap	$\dfrac{1}{4}$

If there are 420 students surveyed, which equation can be used to find the number of students s who prefer rap?

   A. $\dfrac{1}{4}s = 420$ **C.** $s + \dfrac{1}{4} = 420$

   B. $s = \dfrac{1}{4} \cdot 420$ **D.** $420 + s = \dfrac{1}{4}$

47. THINK SOLVE EXPLAIN **SHORT RESPONSE** Nithia earns $6.25 per hour at work. She wants to earn $100 for a class camping trip. Use the equation below to find h, the number of hours she will have to work to earn the money.

$$6.25h = 100$$

ALGEBRA Solve each equation. Check your solution. (Lesson 2B)

48. $4f = 28$

49. $-3y = -15$

50. $\dfrac{p}{14} = 3$

51. $\dfrac{x}{30} = 15$

52. $-40.5 = -4.5a$

53. $-58 = 7.25c$

54. $6.8 = \dfrac{b}{-7}$

55. $2.25 = \dfrac{9}{m}$

56. BAKING Carlota made one batch of brownies. She used the recipe shown. If 8 eggs remain after making the brownies, how many eggs did Carlota have originally? Write and solve an equation. (Lesson 1D)

Brownies (makes 1 batch)

2 cups white sugar
1 cup butter
$\frac{1}{2}$ cup cocoa powder
1 teaspoon vanilla extract
4 eggs
$1\frac{1}{2}$ cups all-purpose flour
$\frac{1}{2}$ teaspoon baking powder
$\frac{1}{2}$ teaspoon salt

57. NUMBER THEORY Molly is thinking of a number. Suppose you divide the number by 5 and then add −2. After subtracting 10, the result is −25. What is Molly's number? Use the *work backward* strategy. (Lesson 1A)

Mid-Chapter Check

1. **GIFT** Tom and Angela shared the cost of a gift for Jael. Angela contributed two dollars more than twice the amount that Tom contributed, who spent $6.00 on the gift. Use the *work backward* strategy to determine how much Angela spent on the gift. (Lesson 1A)

Solve each equation. Check your solution.
(Lesson 1D)

2. $21 + m = 33$

3. $a - 5 = -12$

4. $p + 1.7 = -9.8$

5. $56 = k - (-33)$

6. **GEOMETRY** The sum of the measures of the angles of a triangle is 180°. Write and solve an equation to find the missing measure m. (Lesson 1D)

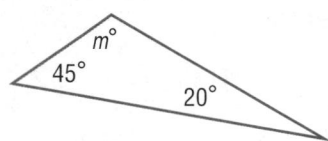

7. **MULTIPLE CHOICE** Trevor's test score was 5 points lower than Ursalina's test score. If Ursalina scored 85 on the test, which equation would give Trevor's score d when solved? (Lesson 1D)

A. $85 = d + 5$

B. $d - 5 = 85$

C. $80 = d + 5$

D. $d - 5 = 80$

8. **PETS** Cameron has 11 adult Fantail goldfish. This is 7 fewer Fantail goldfish than his friend Julia has. Write and solve a subtraction equation to determine the number of Fantail goldfish g that Julia has. (Lesson 1D)

9. **MEASUREMENT** The Grand Canyon has a maximum depth of almost 5,280 feet. An average four-story apartment building has a height of 66 feet. Write and solve a multiplication equation to determine the number of apartment buildings b, stacked on top of each other, that would fill the depth of the Grand Canyon. (Lesson 2B)

Solve each equation. Check your solution.
(Lesson 2B)

10. $5f = -75$

11. $\frac{w}{3} = 16$

12. $63 = 7y$

13. $-28 = -2d$

14. $\frac{g}{12} = 6$

15. $15 = \frac{b}{15}$

16. **MULTIPLE CHOICE** Michelann drove 44 miles per hour and covered a distance of 154 miles. Which equation accurately describes this situation if h represents the number of hours Michelann drove? (Lesson 2B)

F. $154 = 44 + h$

G. $44h = 154$

H. $154 = 44 \div h$

I. $h - 44 = 154$

17. **LAWN SERVICE** Trey estimates he will earn $470 next summer cutting lawns in his neighborhood. This amount is 2.5 times the amount a he earned this summer. Write and solve a multiplication equation to find how much Trey earned this summer. (Lesson 2B)

Solve each equation. Check your solution.
(Lesson 2D)

18. $-1.3x = 3.9$

19. $3.7k = -4.44$

20. $2.56 = 1.6c$

21. $\frac{3}{4}z = 12$

22. $\frac{2}{5}n = 8$

23. $\frac{7}{8} = \frac{1}{4}p$

Explore

Main Idea

Write and solve two-step equations using bar diagrams.

 Get Connect**ED**

 CCSS 7.EE.4, 7.EE.4a

Solve Two-Step Equations with Bar Diagrams

SPORTS Two identical basketballs and five identical tennis balls weigh a total of 52 ounces. Each tennis ball weighs 2 ounces. What is the weight of a basketball?

ACTIVITY

1. **What do you need to find?** the weight of a basketball

   STEP 1 Draw a bar diagram that represents the two basketballs and five tennis balls. Label the parts.

basketball	basketball	tennis	tennis	tennis	tennis	tennis

   $\vdash$ ------?------ $+$ ------?------ $+$ 2 oz $+$ 2 oz $+$ 2 oz $+$ 2 oz $+$ 2 oz $\dashv$

   STEP 2 Write an equation from the bar diagram. Let x represent the weight of a basketball.

   $$2x \quad + \quad 10 \quad = \quad 52$$

   STEP 3 Use the *work backward* strategy to find the weight of the basketballs. Since $2x + 10 = 52$, $2x = 52 - 10$ or 42, and $42 \div 2 = 21$.

   Check $2 \cdot 21 + 10 = 52$ ✓

   The weight of one basketball is 21 ounces.

Analyze the Results

1. Suppose there were 10 identical tennis balls and 3 identical soccer balls that had a total weight of 53 ounces. You want to find the weight of one soccer ball. How would the above equation change?

2. **BAR DIAGRAM** Ryan is saving money to buy a skateboard that costs $85. He has already saved $40 and plans to save the same amount each week for three weeks. Draw a bar diagram and write an equation to find how much he should save each week. Then solve the equation.

ACTIVITY

2 **COMPUTERS** Adriana bought a computer and three pieces of software. She spent a total of $1,220. The computer costs $995 and each piece of software costs the same. Write and solve an equation to find the cost of one piece of software.

What do you need to find? the cost of one piece of software

STEP 1 Draw a bar diagram to represent the situation.

$1,220			
computer	software	software	software
$995	?	?	?

STEP 2 Write an equation from the bar diagram. Let x represent the cost of one piece of software.

$$3x \quad + \quad 995 \quad = \quad 1,220$$

STEP 3 Use the *work backward* strategy to find the cost of one piece of software. Since $3x + 995 = 1,220$, $3x = 225$ and $x = 225 \div 3$ or 75.

Each piece of software costs $75.   **Check** $3 \cdot 75 + 995 = 1,220$ ✓

Analyze the Results

3. How much more does the computer cost than the three pieces of software?

4. How would the equation change if she purchased the computer and three pieces of software during a sale in which she received one piece of software free with the purchase of two?

Practice and Apply

BAR DIAGRAM Draw a bar diagram and write an equation for each situation. Then solve the equation.

5. **GROCERY** Lindsey is buying a pound of cheese and 2 pounds of ham at the deli counter. The cheese is $3.99 per pound and her total cost at the deli counter is $14.97. What is the cost per pound of the ham?

6. **MAIL** It costs $0.42 to mail a letter. Jacob needs to mail a letter and 3 packages having equal size and weight. The post office charges a total of $4.92. What is the cost of mailing each package?

7. **WRITE MATH** Write a real-world problem that could be solved by the bar diagram.

Main Idea
Solve two-step equations.

 **Vocabulary**
two-step equation

 Get ConnectED

CCSS 7.EE.4, 7.EE.4a

Solve Two-Step Equations

⚙ Explore

A florist charges \$2 for each balloon in an arrangement and a \$3 delivery fee. You have \$9 to spend. The model illustrates the equation $2x + 3 = 9$, where x represents the number of balloons.

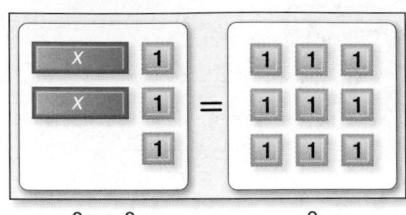

$2x + 3 \quad = \quad 9$

To solve $2x + 3 = 9$, remove three 1-tiles from each side of the mat. Then divide the remaining tiles into two equal groups. The solution of $2x + 3 = 9$ is 3.

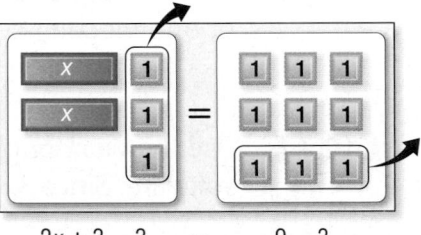

$2x + 3 - 3 \quad = \quad 9 - 3$

Solve each equation by using algebra tiles or a drawing.

1. $2x + 1 = 5$

2. $3x + 2 = 8$

3. $2 = 5x + 2$

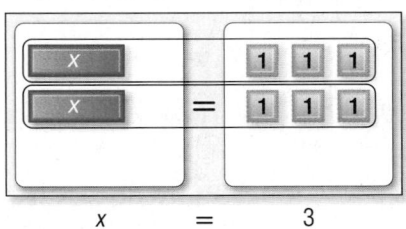

$x \quad = \quad 3$

A **two-step equation** has two different operations. To solve a two-step equation, undo the operations in reverse order of the order of operations.

📝 EXAMPLE   Solve a Two-Step Equation

① Solve $2x + 3 = 9$. Check your solution.

$$2x + 3 = 9 \qquad \text{Write the equation.}$$
$$\underline{ - 3 = -3} \qquad \text{Undo the addition first by subtracting 3 from each side.}$$
$$2x = 6$$
$$\frac{2x}{2} = \frac{6}{2} \qquad \text{Next, undo the multiplication by dividing each side by 2.}$$
$$x = 3 \qquad \text{Simplify.}$$

Check Since $2(3) + 3 = 9$, the solution is 3. ✔

✍ CHECK Your Progress

Solve each equation. Check your solution.

a. $2x + 4 = 10$ **b.** $3x + 5 = 14$ **c.** $5 = 2 + 3x$

Solve Two-Step Equations

② Solve $3x + 2 = 23$. Check your solution.

$$
\begin{aligned}
3x + 2 &= 23 && \text{Write the equation.} \\
\underline{-2} &= \underline{-2} && \text{Undo the addition first by subtracting 2 from each side.} \\
3x &= 21 \\
\frac{3x}{3} &= \frac{21}{3} && \text{Division Property of Equality} \\
x &= 7 && \text{Simplify.}
\end{aligned}
$$

Check $3x + 2 = 23$ Write the original equation.

$3(7) + 2 \stackrel{?}{=} 23$ Replace x with 7.

$23 = 23$ ✓   The sentence is true.

The solution is 7.

③ Solve $-2y - 7 = 3$. Check your solution.

$$
\begin{aligned}
-2y - 7 &= 3 && \text{Write the equation.} \\
\underline{+7} &= \underline{+7} && \text{Undo the subtraction first by adding 7 to each side.} \\
-2y &= 10 \\
\frac{-2y}{-2} &= \frac{10}{-2} && \text{Division Property of Equality} \\
y &= -5 && \text{Simplify.}
\end{aligned}
$$

The solution is -5.   Check the solution.

④ Solve $4 + \frac{1}{5}r = -1$. Check your solution.

$$
\begin{aligned}
4 + \frac{1}{5}r &= -1 && \text{Write the equation.} \\
\underline{-4} &= \underline{-4} && \text{Undo the addition first by subtracting 4 from each side.} \\
\frac{1}{5}r &= -5 \\
5 \cdot \frac{1}{5}r &= 5 \cdot (-5) && \text{Multiplication Property of Equality} \\
r &= -25 && \text{Simplify.}
\end{aligned}
$$

The solution is -25.   Check the solution.

✓ CHECK Your Progress

d. $4x + 5 = 13$ **e.** $-5s + 8 = -2$ **f.** $-2 + \frac{2}{3}w = 10$

Key Concept   **Solving Two-Step Equations**

Step 1   Undo the addition or subtraction first.

Step 2   Then undo the multiplication or division.

⑤ **MOVIES** Toya wants to have her birthday party at the movies. It costs $27 for pizza and $8.50 per friend for the movie tickets. How many friends can Toya have at her party if she has $78 to spend?

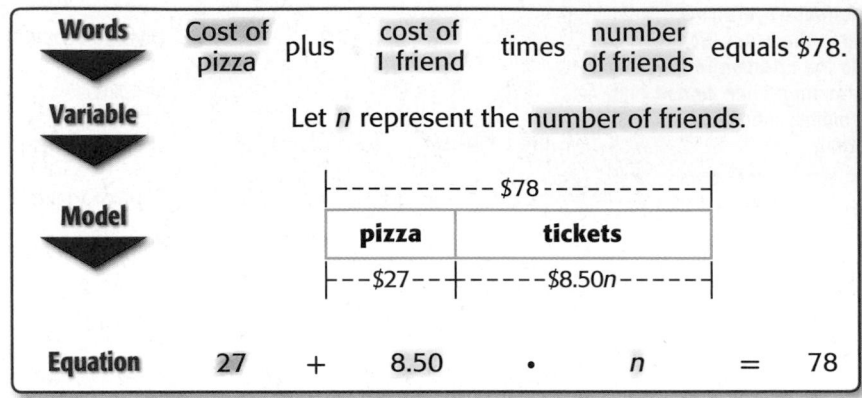

Words	Cost of pizza	plus	cost of 1 friend	times	number of friends	equals $78.

Variable   Let n represent the number of friends.

Model

$$\overset{\text{\quad\quad\quad\quad\quad \$78 \quad\quad\quad\quad\quad}}{\begin{array}{|c|c|}\hline \text{pizza} & \text{tickets} \\ \hline \end{array}}$$
$$\underset{\text{\$27}\quad\quad\quad\text{\$8.50}n}{}$$

Equation	27	+	8.50	·	n	=	78

 Real-World Link · · · ·
Most teenagers see more than seven movies a year.

$$27 + 8.50n = 78 \qquad \text{Write the equation.}$$
$$\underline{-27 \qquad\qquad = -27} \qquad \text{Subtract 27 from each side.}$$
$$8.50n = \quad 51$$
$$\frac{8.50n}{8.50} = \frac{51}{8.50} \qquad \text{Division Property of Equality}$$
$$n = 6 \qquad\qquad 51 \div 8.50 = 6$$

Check $27 + 8.50n = 78$ Write the original equation.
$$27 + 8.50(\mathbf{6}) \overset{?}{=} 78 \qquad \text{Replace } n \text{ with 6.}$$
$$27 + 51 \overset{?}{=} 78 \qquad \text{Simplify.}$$
$$78 = 78 \checkmark \qquad \text{The sentence is true.}$$

Toya can have 6 friends at her party.

✍ **CHECK Your Progress**

g. FITNESS A fitness club is having a special offer where you pay $22 to join plus a $16 monthly fee. You have $150 to spend. Write and solve an equation to find how many months you can use the fitness club.

✓ **CHECK Your Understanding**

Examples 1–4   Solve each equation. Check your solution.

     1. $3x + 1 = 7$ **2.** $4h - 6 = 22$ ③ $-6r + 1 = -17$

     4. $-3y - 5 = 10$ **5.** $13 = 1 + 4s$ **6.** $-7 = 1 + \frac{2}{3}n$

Example 5   **7. MONEY** Syreeta wants to buy some CDs, each costing $14, and a DVD that costs $23. She has $65 to spend. Write and solve an equation to find how many CDs she can buy.

Practice and Problem Solving

 = **Step-by-Step Solutions** begin on page R1.
Extra Practice begins on page EP2.

Examples 1–4 Solve each equation. Check your solution.

8. $3x + 1 = 10$ 9. $5x + 4 = 19$ 10. $2t + 7 = -1$

11. $6m + 1 = -23$ 12. $-4w - 4 = 8$ 13. $-7y + 3 = -25$

14. $-8s + 1 = 33$ 15. $-2x + 5 = -13$ 16. $-3 + 8n = -5$

17. $5 + 4d = 37$ 18. $14 + \frac{2}{3}p = 8$ 19. $25 + \frac{11}{12}y = 47$

Example 5 BAR | DIAGRAM Write an equation. Use a bar diagram if needed. Then solve the equation.

20. **BICYCLES** Cristiano is saving money to buy a new bike that costs $189. He has saved $99 so far. He plans on saving $10 each week. In how many weeks will Cristiano have enough money to buy the new bike?

21. **PETTING ZOOS** It costs $10 to enter a petting zoo. Each cup of food to feed the animals is $2. If you have $14, how many cups of food can you buy?

Solve each equation. Check your solution.

22. $2r - 3.1 = 1.7$ 23. $4t + 3.5 = 12.5$ 24. $10 = b(2 \div 3)$

25. $5w + 9.2 = 19.7$ 26. $16 = 0.5r - 8$ 27. $n + 9 \div 3 = 14$

28. **GRAPHIC NOVEL** Refer to the graphic novel frame below. Jamar figured that they need to spend $39 for popcorn. Write and solve an equation to find how many movies they can purchase.

29 **TEMPERATURE** Temperature is usually measured on the Fahrenheit scale (°F) or the Celsius scale (°C). Use the formula $F = 1.8C + 32$ to convert from one scale to the other.

Alaska Record Low Temperatures (°F) by Month	
January	−80
April	−50
July	16
October	−48

a. Convert the temperature for Alaska's record low in July to Celsius. Round to the nearest degree.

b. Hawaii's record low temperature is −11°C. Find the difference in degrees Fahrenheit between Hawaii's record low temperature and the record low temperature for Alaska in January.

30. **CHALLENGE** Refer to Exercise 29. Is there a temperature in the table at which the number of degrees Celsius is the same as the number of degrees Fahrenheit? If so, find it. If not, explain why not.

31. **CHALLENGE** Suppose your school is selling magazine subscriptions. Each subscription costs $20. The company pays the school half of the total sales in dollars. The school must also pay a one-time fee of $18. What is the fewest number of subscriptions that can be sold to earn a profit of $200?

32. ✏️ **WRITE MATH** When solving an equation, explain why it is important to perform identical operations on each side of the equals sign.

✓ Test Practice

33. A rental car company charges $30 a day plus $0.05 a mile. This is represented by the equation below, where m is the number of miles and c is the total cost of the rental.

$$c = 30 + 0.05m$$

If the Boggs family paid $49.75 for their car rental, how many miles did they travel?

A. 95 miles   C. 295 miles

B. 195 miles   D. 395 miles

34. ✏️ **GRIDDED RESPONSE** The Rodriguez family went on a vacation. They started with $1,875. They spent $140 each day and have $895 left for the rest of their trip. Use the equation below to find d, the number of days they have been on their vacation so far.

$$1{,}875 - 140d = 895$$

How many days have they vacationed?

ALGEBRA Solve each equation. Check your solution. (Lesson 2D)

35. $\frac{1}{2}f = \frac{3}{4}$

36. $-\frac{3}{4}y = \frac{-15}{16}$

37. $\frac{3}{5}p = 12$

38. $\frac{11}{12} = \frac{11}{13}n$

39. **E-CARDS** The prices for E-cards are shown. Seth's grandmother spent $22.75 on song E-cards. How many song E-cards did she send? Write and solve an equation. (Lesson 2B)

e cards

HELLO FROM ALL OF US

There's Music in the Air

Animations: $2.50   Song: $3.25

Solve each equation. Check your solution. (Lesson 1D)

40. $15 + x = 19$

41. $x + 5 = -24$

42. $-13 = x - 8$

43. $-4 = x + 20$

Explore

Main Idea

Use algebra tiles to solve equations with variables on each side of the equation.

Equations with Variables on Each Side

Some equations have variables on each side of the equals sign.

 **ACTIVITY**

Solve $x + 5 = 2x + 2$ using algebra tiles.

STEP 1 Model the equation.

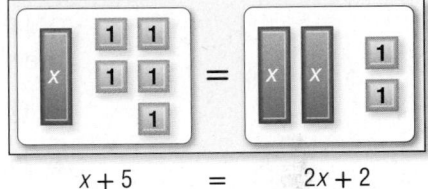

$$x + 5 = 2x + 2$$

STEP 2 Remove one x-tile from each side of the mat. All of the x-tiles are on one side of the mat.

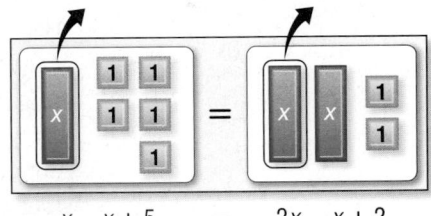

$$x - x + 5 = 2x - x + 2$$

STEP 3 Now you can remove two 1-tiles from each side. There are 3 tiles on the left side of the mat.

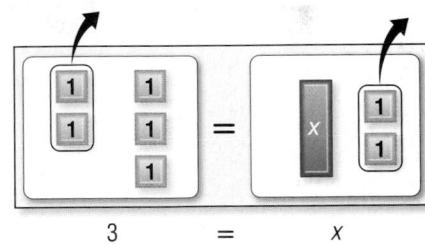

$$3 = x$$

The solution is 3.

Analyze the Results

Use algebra tiles to model and solve each equation.

1. $3x - 2 = 2x + 1$

2. $2x + 6 = 4x - 2$

3. $x + 1 = 3x - 3$

4. $2x - 9 = 5x + 3$

5. Does it matter whether you remove x-tiles or 1-tiles first? Explain.

6. What property of equality allows you to remove an x-tile from each side of the mat?

Main Idea

Use the properties of equality to simplify and solve equations with variables on each side.

Solve Equations with Variables on Each Side

FOOD Julian takes two orders at the fast food restaurant where he works. One order is for two hamburgers and three orders of fries. The other order is for three hamburgers and one order of fries. The two orders cost the same.

1. If an order of fries costs $2, write expressions for each order.

2. How could you show the expressions are equal?

You can use the properties of equality to solve equations with variables on each side. The new equation is equivalent to the original equation.

EXAMPLE **Writing Equivalent Equations**

1 **Express $2x + 6 = 3x + 2$ as an equivalent equation.**

Method 1   Subtract a Variable

$$
\begin{array}{ll}
2x + 6 = \quad 3x + 2 & \text{Write the equation.} \\
\underline{-2x \quad\quad = -2x} & \text{Subtract } 2x \text{ from each side of the equation.} \\
\quad\quad 6 = \quad x + 2 & \text{Simplify.}
\end{array}
$$

Method 2   Subtract a Number

$$
\begin{array}{ll}
2x + 6 = 3x + 2 & \text{Write the equation.} \\
\underline{\quad -2 = \quad\quad -2} & \text{Subtract 2 from each side of the equation.} \\
2x + 4 = 3x & \text{Simplify.}
\end{array}
$$

So, $2x + 6 = 3x + 2$ is equivalent to $6 = x + 2$ and $2x + 4 = 3x$.

 CHECK Your Progress

Express each equation as another equivalent equation using properties of equality. Justify your answer.

a. $2x + 4 = x + 6$ **b.** $3x - 4 = x - 6$

To solve equations with variables on each side, use the Properties of Equality to write an equivalent equation with the variable on one side. Then solve the equation.

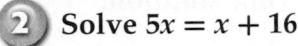 **EXAMPLE** **Equations with Variables on Each Side**

2 Solve $5x = x + 16$.

$5x =$	$x + 16$	Write the equation.
$-x =$	$-x$	Subtraction Property of Equality
$4x =$	16	Simplify.

$$\frac{4x}{4} = \frac{16}{4}$$ Division Property of Equality

$$x = 4$$ Simplify. Check your solution.

 CHECK Your Progress

Solve each equation. Check your solution.

c. $7x = 5x + 6$ **d.** $3x - 2 = x + 10$

Real-World Link · · · · ·
The Memphis Redbirds is a minor league baseball team for the St. Louis Cardinals. Their attendance record was set in 2000. A total of 902,110 fans attended that season.

REAL-WORLD EXAMPLE

3 **TICKETS** If you pay a one-time fee of $20, you can purchase reserve tickets for your local minor league baseball team for only $4 a game. Regular tickets sell at the stadium for $6. Write and solve an equation to find how many reserve tickets you would need to buy to equal the cost of regular tickets. Let x represent the number of tickets purchased.

$20 + 4x =$	$6x$	Write the equation.
$-4x =$	$-4x$	Subtraction Property of Equality
$20 =$	$2x$	Simplify.

$$\frac{20}{2} = \frac{2x}{2}$$ Division Property of Equality

$$10 = x$$ Simplify.

Check $20 + 4(10) = 60, 6(10) = 60.$ ✔

So, ten reserve tickets would have to be purchased to equal the cost of the regular tickets.

 CHECK Your Progress

e. **BASKETBALL** Bill averages 10 points a game and has 110 points for the season. Aaron averages 12 points a game and has 96 points for the season. Write and solve an equation to find how many games it will take until they tie in points scored if they continue at the same rate.

Example 1   Express each equation as another equivalent equation. Justify your answer.

1. $4x + 8 = 2x + 40$ 2. $9x - 2 = 34 + 3x$ 3. $6x - 7 = 43 + x$

Example 2   Solve each equation. Check your solution.

4. $\frac{x}{3} - 15 = 12 + x$ 5. $\frac{1}{4}x - 8 = 5$ 6. $11 + 4x = 7 + 5x$

7. $3 - x = 4 - 3x$ 8. $-7 + x = -8 - x$ 9. $-x + 4 = -9.8 + x$

Example 3   10. **RENTAL CARS** Refer to the table. Write and solve an equation to find the number of miles a rental car must be driven in one day for each company to cost the same.

	Per Day Charge	Per Mile Charge
Rentals R Us	$25	$0.45
EZ Rental	$40	$0.25

Practice and Problem Solving

● = **Step-by-Step Solutions** begin on page R1.
Extra Practice begins on page EP2.

Example 1   Express each equation as another equivalent equation. Justify your answer.

11. $6x + 14 = 20 + 4x$ 12. $15x + 6 = 71 + 10x$ 13. $6x - 3 = 33 + 3x$

14. $16 - 2x = 2x + 4$ 15. $4x - 15 = 5 + 2x$ 16. $4x - 13 = 11 - 2x$

Example 2   Solve each equation. Check your solution.

17. $5x + 1 = 7x - 11$ 18. $5x - 12 = 3x - 2$ 19. $2x - 27 = 45 - x$

20. $9 - 4x = 2x + 6$ 21. $18.6 - 2x = 3x - 2.4$ 22. $7.5 + x = 4.5 + 4x$

23. $-7 + x = \frac{x}{10} - 16$ 24. $-73 + 3x = 15 + 11x$ 25. $-1 - 5x = 15x - 6$

26. $3x - 11 = 34 + 2x$ 27. $-20 - 2x = 2x + 4$ 28. $143 + 2x = -10x - 1$

Example 3   29. **SHOPPING** Manny bought car wash supplies for $48 and 3 buckets for his car wash team. Jin did not buy any car wash supplies but bought 7 buckets. All the buckets cost the same amount, and they both spent the same amount of money. Write and solve an equation to find the cost of one bucket.

30. **BABYSITTING** Catie charges an initial amount of $5 for babysitting and then $3 per hour. Jolisa does not charge an initial amount but charges $4 per hour. Write and solve an equation to find how many hours each girl will have to babysit to earn the same amount.

31. **ALGEBRA** Explain how the two given equations are equivalent.

$$-2b - 11 = -7b + 14$$
$$-25 = -5b$$

32. **OPEN ENDED** Write a real-world problem that could be solved by using the equation $4x + 8 = 2x + 32$. Then solve the equation.

33. **CHALLENGE** If $4x + 2 = y$ and $3x - 1 = y$, find the value of y.

34. **WRITE MATH** Explain how to solve $2 - 4x = 6x - 8$.

Test Practice

35. Rena spent $18 including tax on ribbon she bought at a craft store. The price of the ribbon, including tax, was $3 for two spools. In the equation $18 = r(3 \div 2)$, r represents the number of spools of ribbon Rena bought. How many spools of ribbon did Rena buy?

   A. 6   C. 18

   B. 12   D. 36

36. **SHORT RESPONSE** Gary bought a box of 230 tennis balls for $110. He made a profit of $97 after selling all the balls. Each ball was sold at the same price. Use the equation below to find b, the selling price of one ball.

$$230b - 110 = 97$$

37. **EXTENDED RESPONSE** Damian and Sergio bought the baseball cards shown in the table. Sergio also bought gum that cost $0.99. All of the cards cost the same and the boys spent the same amount of money.

Boy	Number of Cards	Packs of Gum
Sergio	12	1
Damian	15	0

Part A Write an equation to represent the situation.

Part B Solve the equation from Part **A**. Explain the solution.

Part C How many cards could Sergio buy if he bought 2 packs of gum but spent the same amount of money?

Solve each equation. Check your solution. (Lesson 3B)

38. $6 + 2x = 16$ 39. $5 + 3n = -4$ 40. $-3 = 4k + 9$ 41. $9 = 5y + 12$

42. **BOOKS** Of the books on a shelf, $\frac{2}{3}$ are mysteries. Write and solve an equation to find how many books are on the shelf when there are 10 mystery books. (Lesson 2D)

43. **BAR DIAGRAMS** Use the bar diagram to find n the number of guests who still have not replied about attending the party. How many guests have still not replied? (Lesson 1D)

```
|-------- 42 guests invited --------|
|        guests replied        |    |
|---------- 34 ----------|- n -|
```

Problem Solving in Veterinary Medicine

Vet Techs
Don't Monkey Around

If you love being around animals, enjoy working with your hands, and are good at analyzing problems, a challenging career in veterinary medicine might be a perfect fit for you. **Veterinary technicians** help veterinarians by helping to diagnose and treat medical conditions. They may work in private clinics, animal hospitals, zoos, aquariums, or wildlife rehabilitation centers.

21st Century Careers

Are you interested in a career as a veterinary technician? Take some of the following courses in high school.

- Algebra
- Animal Science
- Biology
- Chemistry
- Veterinary Assisting

Get ConnectED

GOLDEN LION TAMARIN MONKEYS

Measure	Minimum	Maximum
Body length	7.9 in.	ℓ
Tail length	t	15.7 in.
Weight	12.7 oz	28 oz

EMPEROR TAMARIN MONKEYS

Measure	Minimum	Maximum
Body length	9.2 in.	b
Tail length	14 in.	16.6 in.
Weight	10.7 oz	w

Real-World Math

For each problem, use the information in the tables to write an equation. Then solve the equation.

1. The minimum tail length of an emperor tamarin is 1.6 inches greater than that of a golden lion tamarin. What is the minimum tail length of a golden lion tamarin?

2. The minimum body length of a golden lion tamarin is 5.3 inches less than the maximum body length. What is the maximum body length?

3. Tamarins live an average of 15 years. This is 13 years less than the years that one tamarin in captivity lived. How long did the tamarin in captivity live?

4. The maximum weight of a golden lion tamarin is about 1.97 times the maximum weight of an emperor tamarin. What is the maximum weight of an emperor tamarin? Round to the nearest tenth.

5. For an emperor tamarin, the maximum total length, including the body and tail, is 27 inches. What is the maximum body length of an emperor tamarin?

Explore Solve Inequalities

Main Idea

Use models to solve problems involving inequalities.

 **Vocabulary**

inequality

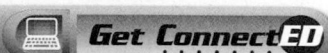 Get Connect**ED**

LUGGAGE Airlines charge a fee for checked luggage that weighs greater than 50 pounds. Mia's suitcase currently weighs 35 pounds and she still needs to pack her shoes. What is the maximum amount her shoes can weigh so Mia will not be charged a fee?

An **inequality** is a mathematical sentence that compares quantities.

Words	Symbols
x is less than two	$x < 2$
x is greater than or equal to four	$x \geq 4$

To solve an inequality means to find values for the variable that make the sentence true. You can use bar diagrams to solve inequalities.

ACTIVITY

1. **What do you need to find?** the maximum amount the shoes can weigh so Mia is not charged a fee

   STEP 1 Draw a bar above a number line that represents the maximum luggage weight.

   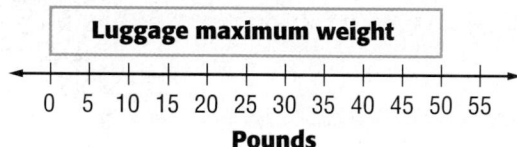

   STEP 2 Draw a bar above that bar for Mia's luggage.

   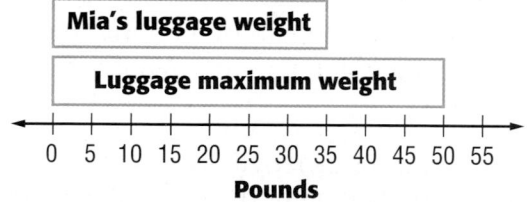

   STEP 3 Draw a bar connected to the bar that represents the weight of Mia's luggage. Label this bar *x*.

   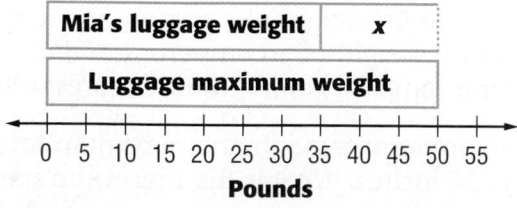

Analyze the Results

1. What does the bar labeled x represent?

2. What is the value of x?

3. If x represents the possible weight of Mia's shoes, could there be more than one value of x? If so, give an example.

4. Refer to Step 3. Write an inequality to represent the situation.

5. Refer to the bar diagram in Activity 1. Explain why the bar diagram is considered an inequality.

You can also use a balance to model and solve inequalities.

ACTIVITY

2 **Model $x + 2 < 5$.**

> **STEP 1** On one side of a balance place a paper bag and 2 cubes.
>
> **STEP 2** On the other side of the balance place 5 cubes.

The side with the bag and 2 blocks weighs less than the side with 5 blocks.

The model shows the inequality $x + 2 < 5$.

Analyze the Results

6. Assume the paper bag is weightless. How many blocks would be in the bag if the left side balanced the right side?

7. Refer to Exercise 6. Explain how you determined your answer.

8. What numbers of blocks can be in the bag so that the left side weighs less than the right side? Write an inequality to represent your answer.

9. Refer to the model in Activity 2. Explain why the blocks and the bag on the balance is considered an inequality.

10. 📝 **WRITE MATH** Compare and contrast the solution of an equation and an inequality. Include an example in your explanation.

Main Idea

Solve inequalities by using the Addition and Subtraction Properties of Inequality.

 7.EE.4

Solve Inequalities by Addition or Subtraction

MAIL A first class stamp can be used for letters and packages weighing thirteen ounces or less. Fisher is mailing pictures to his grandmother, and only has a first class stamp. His envelope weighs 2 ounces. How much can the pictures weigh so that Fisher can use the stamp?

1. Let x represent the weight of the pictures. Write and solve an equation to find the maximum weight of the pictures.

2. Replace the equal sign in your equation with the less than or equal to symbol, $\leq$.

3. Refer to Exercise 2. Name three possible values of x.

An inequality is a mathematical sentence that compares quantities. Solving an inequality means finding values for the variable that make the inequality true. In the situation above, any number less than or equal to 11 is a solution. The solution is written as the inequality $x \leq 11$.

You can solve inequalities by using the Properties of Inequality.

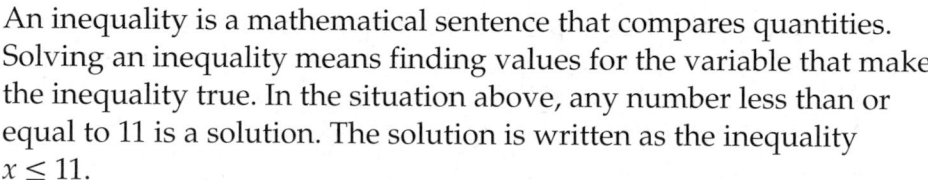

Key Concept **Addition and Subtraction Properties of Inequality**

Words When you add or subtract the same number from each side of an inequality, the inequality remains true.

Symbols For all numbers a, b, and c,
1. if $a > b$, then $a + c > b + c$ and $a - c > b - c$.
2. if $a < b$, then $a + c < b + c$ and $a - c < b - c$.

Examples

$$\begin{array}{r} 2 < 4 \\ +3 \quad +3 \\ \hline 5 < 7 \end{array} \qquad \begin{array}{r} 6 > 3 \\ -4 \quad -4 \\ \hline 2 > -1 \end{array}$$

These properties are also true for $a \geq b$ and $a \leq b$.

Inequalities				
Words	• is less than • is fewer than	• is greater than • is more than • exceeds	• is less than or equal to • is no more than • is at most	• is greater than or equal to • is no less than • is at least
Symbols	$<$	$>$	$\leq$	$\geq$

✎ EXAMPLES Solve Inequalities

① Solve $x + 3 > 10$.

$$x + 3 > 10 \quad \text{Write the inequality.}$$
$$\underline{-3 \quad -3} \quad \text{Subtract 3 from each side.}$$
$$x > 7 \quad \text{Simplify.}$$

Therefore, the solution is $x > 7$.

② Solve $-6 \geq n - 5$.

$$-6 \geq n - 5 \quad \text{Write the inequality.}$$
$$\underline{+5 \qquad +5} \quad \text{Add 5 to each side.}$$
$$-1 \geq n \qquad \text{Simplify.}$$

The solution is $-1 \geq n$ or $n \leq -1$.

> **Study Tip**
>
> **Checking Solutions**
> To check Example 1, write the inequality, replace x with a value in the solution set, and check to see if the result is a true statement.
>
> $x + 3 > 10$
> $8 + 3 \overset{?}{>} 10$
> $11 > 10$ ✓

✔ CHECK Your Progress

Solve each inequality.

a. $a - 3 < 8$ **b.** $14 + y \geq 7$

✎ EXAMPLE Graph Solutions of Inequalities

③ Solve $a + \frac{1}{2} < 2$. Graph the solution set on a number line.

$$a + \frac{1}{2} < \quad 2 \qquad \text{Write the inequality.}$$
$$\underline{-\frac{1}{2} \quad -\frac{1}{2}} \qquad \text{Subtract } \tfrac{1}{2} \text{ from each side.}$$
$$a < 1\tfrac{1}{2} \qquad \text{Simplify.}$$

The solution is $a < 1\frac{1}{2}$.   Check your solution.

Graph the solution.

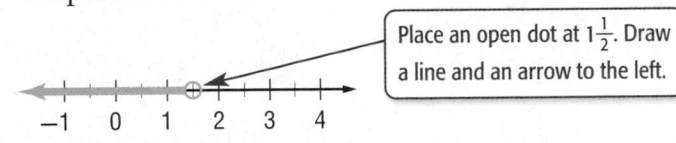

> Place an open dot at $1\frac{1}{2}$. Draw a line and an arrow to the left.

> **Study Tip**
>
> **Open and Closed Dots**
> When graphing inequalities, an open dot is used when the value should not be included in the solution, as with $>$ and $<$ inequalities. A closed dot indicates the value is included in the solution, as with $\leq$ and $\geq$ inequalities.

✔ CHECK Your Progress

Solve each inequality. Graph the solution set on a number line.

c. $h + 4 > 4$ **d.** $x - 6 \leq 4$

 **REAL-WORLD EXAMPLE** **Write an Inequality**

 STATE FAIRS Dylan has $18 to ride go-karts and play games at the state fair. If the go-karts cost $5.50, what is the most he can spend on games?

We need to find the greatest amount of money Dylan can spend on games.

Let x represent the amount Dylan can spend on games. Write an inequality to represent the problem.

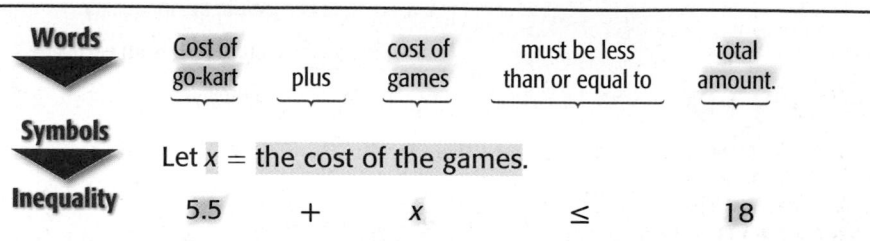

Words	Cost of go-kart	plus	cost of games	must be less than or equal to	total amount.
Symbols	Let x = the cost of the games.				
Inequality	5.5	+	x	$\leq$	18

$$5.5 + x \leq 18 \qquad \text{Write the inequality. } (5.50 = 5.5)$$
$$\underline{-5.5 \qquad\quad -5.5} \qquad \text{Subtract 5.5 from each side.}$$
$$x \leq 12.5 \qquad \text{Simplify.}$$

Check by choosing an amount less than or equal to $12.50, such as $10. Then Dylan would spend $5.50 + $10 or $15.50 in all. Since $15.50 < $18, the answer is reasonable.

So, the most Dylan can spend on games is $12.50.

 CHECK Your Progress

e. **SAVINGS** Shane is saving money for a ski trip. He has $62.50, but his goal is to save at least $100. Write and solve an inequality to determine the least amount Shane needs to save to reach his goal.

 **Real-World Link** · · · ·
State Fairs

According to a recent survey, corn dogs are one of the favorite foods of fairgoers. They were first invented at the Texas State Fair about 70 years ago.

✓ CHECK Your Understanding

Examples 1 and 2 **Solve each inequality.**

1. $c + 4 < 8$ **2.** $14 + t \geq 5$ **3.** $y - 9 < 11$

4. $10 > x + 5$ **5** $c + 4 \geq 17$ **6.** $t - 7 < 25$

Example 3 **Solve each inequality. Graph the solution set on a number line.**

7. $6 + h \geq 12$ **8.** $15 - y < 8$ **9.** $-7 \leq n + 9$

Example 4 **10. ELEVATORS** An elevator can hold 2,800 pounds or less. Write and solve an inequality that describes how much more weight the elevator can hold if it is currently holding 2,375 pounds.

● = **Step-by-Step Solutions** begin on page R1.
Extra Practice begins on page EP2.

Examples 1 and 2 Solve each inequality.

11. $-3 < n - 8$ **12.** $h - 16 \leq -24$ **13.** $y - 6 \geq -13$

14. $3 \leq m + 1.4$ **15.** $x + 0.7 > -0.3$ **16.** $w - 8 \geq 5.6$

Example 3 Solve each inequality. Graph the solution set on a number line.

17. $-11 > t + 7$ **18.** $m + 5 \geq -1$ **19.** $-21 < a - 16$

20. $t - 6.2 < 4$ **21.** $n - \frac{1}{5} \leq \frac{3}{10}$ **22.** $6 > x + 3\frac{1}{3}$

Example 4 Write an inequality and solve each problem.

23. Four more than a number is more than 13.

24. The sum of a number and 19 is at least 8.

25. Eight less than a number is less than 10.

26. The difference between a number and 21 is no more than 14.

27. SOCCER The high school soccer team can have no more than 26 players. Write and solve an inequality to determine how many more players can make the team if the coach has already chosen 17 players.

28. CARS There were a total of 125 cars at a car dealership. A salesperson sold 68 of the cars in one month. Write and solve an inequality that describes how many more cars, at most, the salesman has left to sell.

29 **CELL PHONES** Lalo has 1,500 minutes per month on his cell phone plan. How many more minutes can he use if he has already talked for 785 minutes?

30. MATH IN THE MEDIA Find real-world data using a newspaper or magazine, the television, or the Internet. Then write and solve a real-world problem that involves an inequality.

31. WEATHER Refer to the diagram below.

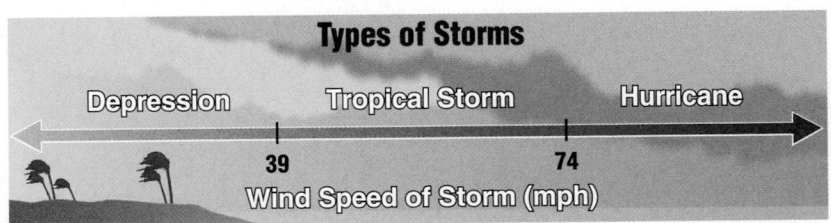

Types of Storms

Depression Tropical Storm Hurricane

39 74

Wind Speed of Storm (mph)

a. A hurricane has winds that are at least 74 miles per hour. Suppose a tropical storm has winds that are 42 miles per hour. Write and solve an inequality to find how much the winds must increase before the storm becomes a hurricane.

b. A *major storm* has wind speeds that are at least 110 miles per hour. Write and solve an inequality that describes how much greater these wind speeds are than the slowest hurricane.

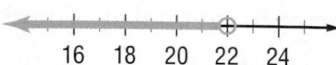
H.O.T. Problems

32. **REASONING** Compare and contrast the solutions of $a - 3 = 15$ and $a - 3 \geq 15$.

33. **OPEN ENDED** Write an addition inequality with the solution set graphed below.

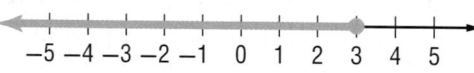

34. **WRITE MATH** Explain when you would use addition and when you would use subtraction to solve an inequality.

Test Practice

35. Which inequality represents a temperature that is equal to or less than 42°?

   A. $t \geq 42$

   B. $t > 42$

   C. $t \leq 42$

   D. $t < 42$

36. Which inequality represents the graph below?

$$-5\ -4\ -3\ -2\ -1\ \ 0\ \ 1\ \ 2\ \ 3\ \ 4\ \ 5$$

   F. $x > 3$

   G. $x \geq 3$

   H. $x < 3$

   I. $x \leq 3$

37. Arlo has $25 to spend on a T-shirt and shorts for gym class. The shorts cost $14. Use the inequality $14 + t \leq 25$, where t represents the cost of the T-shirt. What is the most Arlo can spend on the T-shirt?

   A. $9

   B. $10.99

   C. $11

   D. $11.50

38. **SHORT RESPONSE** Write and solve an inequality for the following sentence: Jan has $50 in her savings account and needs to save at least $268 for camp this summer. What is the least amount that she can save to meet her goal?

Spiral Review

39. **CATERING** The prices that two companies charge for an order of appetizers is shown. Food Inc. charges an additional fee of $21. Write and solve an equation to find how many orders of appetizers you would have to order for the costs to be the same. (Lesson 3D)

Company	Cost
Catering Co.	$25
Food Inc.	$18

Solve each equation. Check your solution. (Lesson 3B)

40. $7x + 21 = 56$

41. $-1 = 6x + 11$

42. $3x - 4 = 14$

43. $-2 = 9x - 65$

Main Idea

Solve inequalities by using the Multiplication or Division Properties of Inequality.

 7.EE.4

Solve Inequalities by Multiplication or Division

SCIENCE An astronaut in a space suit weighs about 300 pounds on Earth, but only 50 pounds on the Moon.

weight on Earth   weight on Moon
300   >   50

If the astronaut and space suit each weighed half as much, would the inequality still be true?

Location	Weight of Astronaut (lb)
Earth	300
Moon	50
Pluto	67
Mars	113
Neptune	407
Jupiter	796

1. Divide each side of the inequality $300 > 50$ by 2. Is the inequality still true? Explain by using an inequality.

2. Would the weight of 5 astronauts be greater on Pluto or on Earth? Explain by using an inequality.

The examples above demonstrate how you can solve inequalities by using the Multiplication and Division Properties of Inequality.

Key Concept — Multiplication and Division Properties of Inequality, Positive Number

Words When you multiply or divide each side of an inequality by a positive number, the inequality remains true.

Symbols For all numbers a, b, and c, where $c > 0$,
1. if $a > b$, then $ac > bc$ and $\frac{a}{c} > \frac{b}{c}$.
2. if $a < b$, then $ac < bc$ and $\frac{a}{c} < \frac{b}{c}$.

These properties are also true for $a \geq b$ and $a \leq b$.

Study Tip

Reading Math The inequality $c > 0$ means that c is a positive number.

EXAMPLES Multiply or Divide by a Positive Number

1. Solve $8x \leq 40$.

$8x \leq 40$ Write the inequality.
$\frac{8x}{8} \leq \frac{40}{8}$ Divide each side by 8.
$x \leq 5$ Simplify.

The solution is $x \leq 5$. You can check this solution by substituting 5 or a number less than 5 into the inequality.

2 Solve $\frac{d}{2} > 7$.

$$\frac{d}{2} > 7 \qquad \text{Write the inequality.}$$

$$2\left(\frac{d}{2}\right) > 2(7) \qquad \text{Multiply each side by 2.}$$

$$d > 14 \qquad \text{Simplify.}$$

The solution is $d > 14$. You can check this solution by substituting a number greater than 14 into the inequality.

CHECK Your Progress

a. $4x < 40$

b. $6 \geq \frac{x}{7}$

 Key Concept — **Multiplication and Division Properties of Inequality, Negative Number**

Words When you multiply or divide each side of an inequality by a negative number, the inequality symbol must be reversed for the inequality to remain true.

Symbols For all numbers a, b, and c, where $c < 0$,

1. if $a > b$, then $ac < bc$ and $\frac{a}{c} < \frac{b}{c}$.

2. if $a < b$, then $ac > bc$ and $\frac{a}{c} > \frac{b}{c}$.

Examples

$7 > 1$

$-2(7) < -2(1)$ Reverse the symbols.

$-14 < -2$

$-4 < 16$

$\frac{-4}{-4} > \frac{16}{-4}$

$1 > -4$

These properties are also true for $a \geq b$ and $a \leq b$.

right **Study Tip**

Reading Math The inequality $c < 0$ means that c is a negative number.

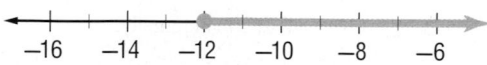 **EXAMPLE** **Multiply or Divide by a Negative Number**

3 Solve $\frac{x}{-3} \leq 4$. Graph the solution set on a number line.

$$\frac{x}{-3} \leq 4 \qquad \text{Write the inequality.}$$

$$-3\left(\frac{x}{-3}\right) \geq -3(4) \qquad \text{Multiply each side by } -3 \text{ and reverse the symbol.}$$

$$x \geq -12 \qquad \text{Simplify.}$$

Graph the solution, $x \geq -12$.

 CHECK Your Progress

c. $\frac{k}{-2} < 9$

d. $-6a \geq -78$

250 Equations and Inequalities

Some inequalities involve more than one operation. To solve the inequality, work backward to undo the operations, just as you did to solve multi-step equations.

EXAMPLE   Solve a Multi-Step Inequality

④ Solve $\frac{6}{7}x + 15 > 9$. Graph the solution set on a number line.

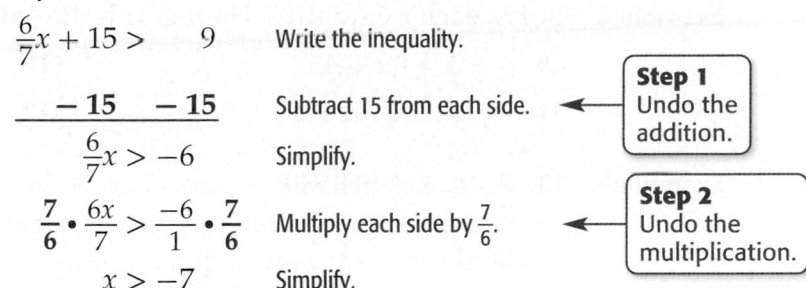

$$\frac{6}{7}x + 15 > \quad 9 \qquad \text{Write the inequality.}$$

$$\underline{\quad -15 \qquad -15\quad} \qquad \text{Subtract 15 from each side.} \quad \leftarrow \quad \boxed{\begin{array}{l}\textbf{Step 1}\\ \text{Undo the}\\ \text{addition.}\end{array}}$$

$$\frac{6}{7}x > -6 \qquad \text{Simplify.}$$

$$\frac{7}{6} \cdot \frac{6x}{7} > \frac{-6}{1} \cdot \frac{7}{6} \qquad \text{Multiply each side by } \frac{7}{6}. \quad \leftarrow \quad \boxed{\begin{array}{l}\textbf{Step 2}\\ \text{Undo the}\\ \text{multiplication.}\end{array}}$$

$$x > -7 \qquad \text{Simplify.}$$

Graph the solution, $x > -7$.

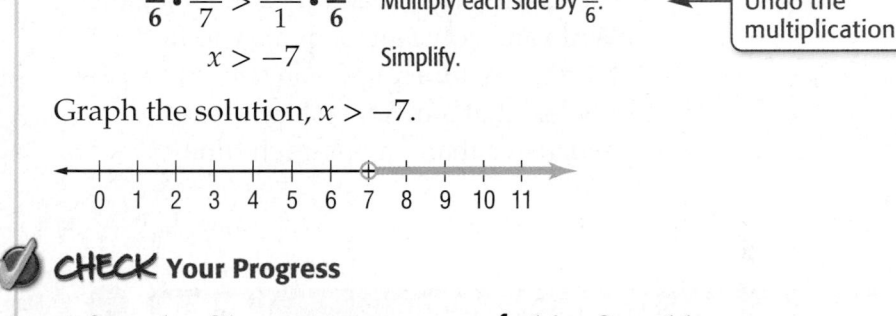

✓ CHECK Your Progress

e. $3x + 4 \leq 31$ **f.** $16 - 3c > 14$

REAL-WORLD EXAMPLE   Write an Inequality

⑤ **JOBS** Ling earns $8 per hour working at the zoo. Write and solve an inequality that can be used to find how many hours she must work in a week to earn at least $120.

Let x represent the number of hours worked.

Words	Amount earned per hour	times	number of hours	is at least	amount earned each week.
Variable	Let x represent the number of hours.				
Inequality	8	•	x	$\geq$	120

$$8x \geq 120$$

$$\frac{8x}{8} \geq \frac{120}{8} \qquad \text{Divide each side by 8.}$$

$$x \geq 15 \qquad \text{Simplify.}$$

✓ CHECK Your Progress

g. EARNINGS Elisa delivers pizzas. Her average tip is $1.50 for each pizza that she delivers. Write and solve an inequality to represent how many pizzas she must deliver to earn at least $21 in tips.

Examples 1–3   Solve each inequality.

1. $5x > 15$ 2. $\frac{2}{3} < \frac{4}{5}y$ 3. $9h \geq 63$ 4. $\frac{h}{6} \leq 8$

5. $-7y > 28$ 6. $-3n \leq -21$ **7** $\frac{t}{-4} < -11$ 8. $\frac{x}{-5} \geq -6$

Example 4   Solve each inequality. Then graph the solution set on a number line.

9. $y + 1 \geq 4y + 4$ 10. $16 - 2c < 14$

11. $-6.1n \geq 3.9n + 5$ 12. $-4 \leq \frac{x}{4} - 6$

Example 5   **13. POOL MEMBERSHIP** A pool charges $4 each visit, or you can buy a membership. Write and solve an inequality to find how many times a person should use the pool so that a membership is less expensive than paying each time.

Practice and Problem Solving

● = **Step-by-Step Solutions** begin on page R1.
Extra Practice begins on page EP2.

Examples 1–3   Solve each inequality. Check your solution.

14. $6y < 18$ 15. $4x \geq 36$ 16. $12n \leq 48$

17. $20 < 5t$ 18. $60 \leq \frac{m}{3}$ 19. $\frac{h}{9} > 9$

20. $-3s \geq 33$ 21. $-7y < 35$ 22. $-56 \leq -8x$

23. $-10n > -20$ 24. $\frac{w}{-5} \geq 9$ 25. $\frac{t}{-2} < 6$

26. $\frac{m}{-14} \leq -4$ 27. $\frac{s}{-6} > -16$ 28. $\frac{x}{-4} \geq -8$

Example 4   Solve each inequality. Graph the solution set on a number line.

29. $4x + 3 < 19$ 30. $7h + 1 \geq -6$ 31. $-6y + 6 \leq -8 + y$

32. $9 + \frac{n}{-2} > 5$ 33. $\frac{t}{5} - 6 \leq -11$ 34. $44 + 4x < 11 + x$

Example 5   **35. CARNIVAL** Each game at a carnival costs $0.50, or you can pay $15 and play an unlimited amount of games. Write and solve an inequality to find how many times a person should play a game so that the unlimited game play for $15 is less expensive than paying each time.

36. BASEBALL At a baseball game you can get a single hot dog for $2. You have $10 to spend. Write and solve an inequality to find the number of hot dogs you can buy.

Write an inequality for each sentence. Then solve the inequality.

37 Five times a number decreased by seven is less than −52.

38. The product of a number and 4 minus three is at least −15.

H.O.T. Problems

39. OPEN ENDED Write two different inequalities that have the solution $y > 6$. One inequality should be solved using multiplication properties, and the other should be solved using division properties.

40. CHALLENGE You score 15, 16, 17, 14, and 19 points out of 20 possible points on five tests. What must you score on the sixth test to have an average of at least 16 points?

41. 📝 **WRITE MATH** Explain when you should not reverse the inequality symbol when solving an inequality.

✓ Test Practice

42. Which inequality represents *five more than twice a number is less than ten*?

A. $(5 + 2)n < 10$

B. $2n - 5 < 10$

C. $10 < 2n + 5$

D. $5 + 2n < 10$

43. Which sentence represents the following inequality?

$$\frac{x}{5} - 3 \leq 8$$

F. The difference of a number and 5 increased by 3 is at most 8.

G. The quotient of a number and 5 decreased by 3 is at most 8.

H. The quotient of a number and 5 decreased by 3 is 8.

I. The quotient of a number and 5 decreased by 3 is at least 8.

44. 📄 **EXTENDED RESPONSE** Karl's scores on the first five science tests are shown in the table.

Test	1	2	3	4	5
Score	85	84	90	95	88

Part A Write an inequality that represents how to find the score he must receive on the sixth test to have an average score of more than 88.

Part B Solve the inequality to determine the lowest score he can receive in order to have an average score of more than 88.

Solve each inequality. Check your solution. (Lesson 4B)

45. $20 < -9 + k$

46. $22 \leq -15 + y$

47. $6 + x < -27$

48. $n - 4 \leq -11$

49. DOG WALKING Doug charges an initial fee of $12 and then $5 for each time he walks a dog. Kip does not charge an initial fee but charges $8 each time he walks a dog. How many times will a dog have to be walked for the boys to earn the same amount? (Lesson 3D)

Chapter Study Guide and Review

FOLDABLES
Study Organizer

Be sure the following Key Concepts are noted in your Foldable.

Key Concepts

Addition and Subtraction Equations
(Lesson 1)
• If you add or subtract the same number from each side of an equation, the two sides remain equal.

Multiplication and Division Equations
(Lesson 2)
• If you multiply each side of an equation by the same number, the two sides remain equal.
• If you divide each side of an equation by the same nonzero number, the two sides remain equal.

Multi-Step Equations (Lesson 3)
• A multi-step equation has variables on each side. For example, $3x + 3 = x + 5$.

Inequalities (Lesson 4)
• When you multiply or divide each side of an inequality by a negative number, the direction of the symbol must be reversed for the inequality to be true.

Key Vocabulary

coefficient   inequality
equation   multiplicative inverse
equivalent equations   reciprocal
formula   two-step equation

Vocabulary Check

State whether each sentence is *true* or *false*. If *false*, replace the underlined word or number to make a true sentence.

1. The expression $\frac{1}{3}y$ means <u>one third of y</u>.
2. Another term for multiplicative inverse is <u>reciprocal</u>.
3. The formula <u>$d = rt$</u> gives the distance d traveled at a rate of r for t units of time.
4. The algebraic expression representing the words *six less than m* is <u>$6 - m$</u>.
5. The symbol $<$ means <u>greater than</u>.
6. The word *each* sometimes suggests the operation of <u>division</u>.
7. In solving the equation $4x + 3 = 15$, first <u>divide each side by 4</u>.
8. A solution of the inequality $p + 4.4 < 11.6$ is <u>7.2</u>.
9. The process of solving a <u>two-step equation</u> uses the *work backward* strategy.
10. The coefficient in the term $15x$ is <u>x</u>.
11. The reciprocal of $\frac{2}{3}$ is <u>$-\frac{2}{3}$</u>.
12. The word *per* sometimes suggests the operation of <u>subtraction</u>.

Multi-Part Lesson Review

Lesson 1   Addition and Subtraction Equations

PSI: Work Backward (Lesson 1A)

13. BASEBALL Last baseball season, Nelson had four less than twice the number of hits Marcus had. Nelson had 48 hits. How many hits did Marcus have last season?

14. BANKING Trina had $320 in her savings account after a withdrawal of $75 and a deposit of $120. How much did she have in her account originally?

EXAMPLE 1   A number is divided by 2. Then 4 is added to the quotient. After subtracting 3, the result is 18. What is the number?

Start with the final value and work backward with each resulting value until you arrive at the starting value.

$18 + 3 = 21$ Undo subtracting 3.

$21 - 4 = 17$ Undo adding 4.

$17 \cdot 2 = 34$ Undo dividing by 2.

Solve One-Step Addition and Subtraction Equations (Lesson 1D)

Solve each equation. Check your solution.

15. $x + 5 = 8$ **16.** $r + 8 = 2$

17. $p + 9 = -4$ **18.** $s - 8 = 15$

19. $n - 1 = -3$ **20.** $w - 9 = 28$

21. $b + \dfrac{1}{2} = \dfrac{3}{4}$ **22.** $t - \dfrac{5}{7} = \dfrac{20}{21}$

23. COOKIES Hector baked some chocolate chip cookies for himself and his sister. His sister ate 6 of these cookies. If there were 18 cookies left, write and solve an equation to find how many cookies c Hector baked.

24. ANIMALS A giraffe is 3.5 meters taller than a camel. If a giraffe is 5.5 meters tall, how tall is a camel?

EXAMPLE 2   Solve $x + 6 = 4$.

$$x + 6 = 4$$
$$\underline{-6 = -6} \quad \text{Subtraction Property of Equality}$$
$$x = -2$$

EXAMPLE 3   Solve $y - 3 = -2$.

$$y - 3 = -2$$
$$\underline{+3 = +3} \quad \text{Addition Property of Equality}$$
$$y = 1$$

EXAMPLE 4

Admission to a popular amusement park is $8.75 more than the previous year's admission price. If this year's admission price is $20, write and solve an equation to find the previous year's admission.

$$A + 8.75 = 20 \quad \text{Write the equation.}$$
$$\underline{- 8.75 = - 8.75} \quad \text{Subtraction Property of Equality}$$
$$A = 11.25$$

So, the previous year's admission was $11.25.

Multiplication and Division Equations

Solve One-Step Multiplication and Division Equations (Lesson 2B)

Solve each equation. Check your solution.

25. $7c = 28$

26. $-8w = 72$

27. $10y = -90$

28. $-12r = -36$

29. $\frac{a}{3} = 4$

30. $\frac{x}{9} = 6$

31. MONEY Matt borrowed $98 from his father. He plans to repay his father at $14 per week. Write and solve an equation to find the number of weeks w required to pay back his father.

EXAMPLE 5 Solve $-4b = 32$.

$-4b = 32$

$\dfrac{-4b}{-4} = \dfrac{32}{-4}$ Division Property of Equality

$b = -8$

EXAMPLE 6 Solve $\frac{b}{7} = -6$.

$\dfrac{b}{7} = -6$

$\dfrac{b}{7} \cdot 7 = -6 \cdot 7$ Multiplication Property of Equality

$b = -42$

Solve Equations with Rational Coefficients (Lesson 2D)

Solve each equation. Check your solution.

32. $-3.4 = 1.7d$

33. $0.5x = 0.75$

34. $0.42y = 1.26$

35. $1.5t = 30$

36. $\frac{4}{5}r = 1$

37. $-2 = \frac{2}{5}m$

38. BLIMP A blimp travels 300 miles in 7.5 hours. Assuming the blimp travels at a constant speed, write and solve an equation to find the speed of the blimp. Use the formula $d = rt$.

EXAMPLE 7 Solve $2.5r = 30$.

$2.5r = 30$

$\dfrac{2.5r}{2.5} = \dfrac{30}{2.5}$ Division Property of Equality

$r = 12$

EXAMPLE 8 Solve $-4 = \frac{2}{3}s$.

$-4 = \dfrac{2}{3}s$

$\left(\dfrac{3}{2}\right)-4 = \left(\dfrac{3}{2}\right)\dfrac{2}{3}s$ Multiplication Property of Equality

$-6 = s$

Multi-Step Equations

Solve Two-Step Equations (Lesson 3B)

Solve each equation. Check your solution.

39. $3y - 12 = 6$

40. $6x - 4 = 20$

41. $2x + 5 = 3$

42. $\frac{3}{5}m + 6 = -4$

43. DVDS Blake had 6 times the amount of DVDs that Daniel had. Blake just bought 5 more. Write and solve an equation to find how many DVDs Daniel had if Blake now has 155.

EXAMPLE 9 Solve $3p - 4 = 8$.

$$3p - 4 = 8$$
$$\underline{+4 = +4} \quad \text{Addition Property of Equality}$$
$$3p = 12$$
$$\dfrac{3p}{3} = \dfrac{12}{3} \quad \text{Division Property of Equality}$$
$$p = 4$$

Multi-Step Equations (continued)

Solve Equations with Variables on Each Side (Lesson 3D)

Solve each equation. Check your solution.

44. $m + 1 = 2m + 7$

45. $12b = 7b + 5$

46. $3.21 - 7y = 10y - 1.89$

47. **VIDEOS** A video store has two membership plans. Under plan A, a yearly membership costs $30 plus $1.50 for each rental. Under plan B, the yearly membership costs $12 plus $3 for each rental. What number of rentals results in the same yearly cost?

EXAMPLE 10 Solve $5x + 4 = 3x - 2$.

$$
\begin{array}{ll}
5x + 4 = \quad 3x - 2 & \\
\underline{-3x \qquad = -3x} & \text{Subtraction Property of Equality} \\
2x + 4 = -2 & \text{Simplify.} \\
\underline{\quad -4 = -4} & \text{Subtraction Property of Equality} \\
2x = -6 & \text{Simplify.} \\
\dfrac{2x}{2} = \dfrac{-6}{2} & \text{Division Property of Equality} \\
x = -3 & \text{Simplify.}
\end{array}
$$

Inequalities

Solve Inequalities by Addition or Subtraction (Lesson 4B)

Solve each inequality. Check your solution.

48. $m + 3 < 9$

49. $4 \geq 6 + n$

50. $x + \dfrac{3}{4} \leq 5$

51. $-1\dfrac{1}{3} < g - 5$

52. **LIFTING** Ben is training for football and is lifting 120 pounds on the bench press. He can lift a maximum of 180 pounds. Write and solve an inequality to determine how much additional weight Ben can lift.

EXAMPLE 11 Solve $x + 5 \geq 7$. Check your solution.

$$
\begin{array}{ll}
x + 5 \geq \quad 7 & \text{Write the inequality.} \\
\underline{-5 \quad -5} & \text{Subtract 5 from each side.} \\
x \geq 2 & \text{Simplify.}
\end{array}
$$

Check
$$
\begin{array}{ll}
x + 5 \geq 7 & \text{Write the inequality.} \\
2 + 5 \overset{?}{\geq} 7 & \text{Replace } x \text{ with a number greater than or equal to 2.} \\
7 \geq 7 \checkmark & \text{This statement is true.}
\end{array}
$$

Solve Inequalities by Multiplication or Division (Lesson 4C)

Solve each inequality.

53. $\dfrac{p}{3} < 6$

54. $\dfrac{w}{2.4} \leq 3$

55. $-0.9d > 6.3$

56. $-42 \leq 6y$

57. **SHOPPING** Jess wants to spend less than $18.75 on new socks. Each pack costs $6. Write and solve an inequality to find the maximum number of packs she can buy.

EXAMPLE 12 Solve $-3k \geq 33$.

$$
\begin{array}{ll}
-3k \geq 33 & \text{Write the inequality.} \\
\dfrac{-3k}{-3} \leq \dfrac{33}{-3} & \text{Divide each side by } -3 \text{ and reverse the symbol.} \\
k \leq -11 & \text{Simplify.}
\end{array}
$$

The solution is $k \leq -11$.

Practice Chapter Test

Solve each equation. Check your solution.

1. $x + 5 = -8$ **2.** $y - 11 = 15$

3. $12 = z + 14$ **4.** $13 = t - 13$

5. $s + 1.5 = 2.7$ **6.** $\frac{1}{3} + r = \frac{5}{6}$

7. FLOWERS The number of tulips in Paula's garden is 8 less than the number of marigolds. If there are 16 tulips, write and solve an equation to determine the number of marigolds m.

8. PIZZA Chris and Heladio shared a pizza. Chris ate two more than twice as many pieces as Heladio, who ate 3 pieces. If there were 3 pieces left, how many pieces were there initially? Use the *work backward* strategy.

Solve each equation. Check your solution.

9. $9z = -81$ **10.** $-6k = -72$

11. $4 = 8n$ **12.** $120 = \frac{a}{2}$

13. $27 = \frac{b}{3}$ **14.** $3.3 = \frac{c}{3}$

15. PHONE Susie's phone service is $0.15 per minute of use. Write and solve an equation to find how long she can talk for $5.00.

I-Call Wireless
15¢ per minute

16. MULTIPLE CHOICE Which of the following equations does NOT have a solution of 3?

A. $3x = 9$

B. $\frac{x}{5} = 15$

C. $21 = 7x$

D. $\frac{1}{3}x = 1$

Solve each equation. Check your solution.

17. $-0.5m = -10$ **18.** $-14.2 = -7.1t$

19. $-\frac{1}{9}t = 7$ **20.** $6 = -\frac{3}{4}x$

21. $-6k + 4 = -38$ **22.** $3z - 7 = 17$

23. $4x + 9 = 7x$ **24.** $2y - 6 = 7y + 24$

25. ROLLER COASTERS The track length of a popular roller coaster is 5,106 feet. The roller coaster has an average speed of about 2,000 feet per minute. At that speed, how long will it take to travel its length of 5,106 feet? Round to the nearest tenth. Use the formula $d = rt$.

Solve each inequality. Check your solution.

26. $-6 < \frac{r}{5}$ **27.** $3x \geq -27$

28. $6h - 6 < 30$ **29.** $9 - 2c \leq 55$

30. MONEY Keiko has $17 in savings. She receives an $11 weekly allowance. Write and solve an inequality to find the least amount of weeks it will take her to save $72.

31. **EXTENDED RESPONSE** A taxicab company charges $2 plus $1.25 for each mile of a trip.

Part A Write the expression for a trip of m miles.

Part B Write an equation to find the number of miles if the trip is $12.

Part C How many miles was the trip?

Part D Another taxicab company advertises $1.50 plus $1.60 per mile. If a taxicab ride is 2 miles long, is this company less expensive than the first company? Explain.

Preparing for Standardized Tests

Multiple Choice: Using the Answer Choices

Sometimes you can find the correct answer more quickly by substituting each answer choice into the problem.

TEST EXAMPLE

Renata earns $7.50 per hour babysitting. She spent $16 of the total amount that she earned this week on a new CD and had $29 left over. Use the equation below to find h, the number of hours she babysat this week.

$$7.5h - 16 = 29$$

A. 1.7 hours

C. 6 hours

B. 3.9 hours

D. 13 hours

Substitute each answer choice into the equation until you find the correct solution.

$7.5(\textbf{1.7}) - 16 = 29$ ← Substitute 1.7 for h.
$12.75 - 16 = 29$
$-3.25 \neq 29$ ✗

$7.5(\textbf{3.9}) - 16 = 29$ ← Substitute 3.9 for h.
$29.25 - 16 = 29$
$13.25 \neq 29$ ✗

$7.5(\textbf{6}) - 16 = 29$ ← Substitute 6 for h.
$45 - 16 = 29$
$29 = 29$ ✓

Since this is a true sentence, 6 is the solution of the equation. So, the correct answer is C.

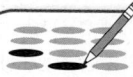

Work on It

The Berk family camped for 6 nights and paid a total of $89. Their total cost, which includes an admission fee of $5 and the cost per night d, can be represented by $6d + 5 = 89$. What was the cost per night?

F. $14.00

H. $15.67

G. $14.83

I. $16.60

> **Test Hint**
>
> Once you find the solution, you do not need to substitute the remaining answer choices into the equation.

Read each question. Then fill in the correct answer on the answer sheet provided by your teacher or on a sheet of paper.

1. A sports store sells two different field hockey kits shown in the table.

Hockey Kits	
Beginner	**Basic**
hockey stick ball shin guards	hockey stick ball

The beginner's field hockey kit costs $150. It is $15 more than three times the cost of the basic kit. What is the cost of the basic kit?

A. $35.00   **C.** $45.00

B. $40.00   **D.** $50.00

2. Which line contains the ordered pair $(-2, 4)$?

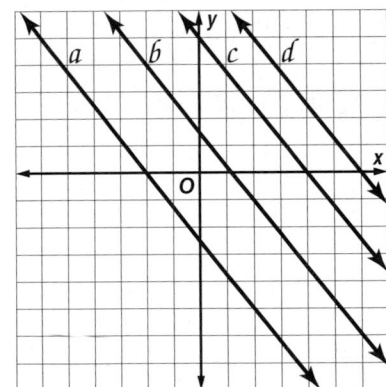

F. line a **H.** line b

G. line c **I.** line d

3. Which integer added to 12 gives a sum of -14?

A. -18 **C.** -24

B. -20 **D.** -26

4. ✎ **GRIDDED RESPONSE** Mrs. McDowell is making a big batch of cookies for her son's birthday. The price of the chocolate chips is 2 bags for $4.00. Use the table to determine the number of bags of chocolate chips r that Mrs. McDowell bought if the cost c was $12.

r	$r(4 \div 2)$	c
1	$1(4 \div 2)$	$2
2	$2(4 \div 2)$	$4
3	$3(4 \div 2)$	$6

5. ✎ **GRIDDED RESPONSE** Aida bought a costume box containing 50 costumes for $300. She sold all of the costumes and made a $250 profit. She sold all of the costumes for the same price. Use the equation $50c - 300 = 250$, where c is the selling price of each costume. What was the selling price of one costume in dollars?

6. Which of the following problems can be solved using the equation $x - 9 = 15$?

F. Allison is 9 years younger than her sister Pam. Allison is 15 years old. What is x, Pam's age?

G. David's portion of the bill is $9 more than Jaleel's portion of the bill. If Jaleel pays $9, find x, the amount in dollars that David pays.

H. The sum of two numbers is 15. If one of the numbers is 9, what is x, the other number?

I. Calvin owns 15 CDs. If he gave 9 of them to a friend, what is x, the number of CDs he has left?

7. What value of x makes this equation true?

$$4x + 7 = 43$$

    A. 12

    B. 10

    C. 9

    D. 8

8. Joshua spends $0.25 for every song he downloads to his cell phone. Which of the following represents the number of songs he can download if he has at least $3?

    F.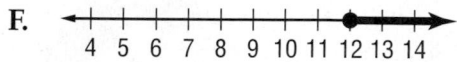
4 5 6 7 8 9 10 11 12 13 14

    G.
4 5 6 7 8 9 10 11 12 13 14

    H.
4 5 6 7 8 9 10 11 12 13 14

    I. Not enough information is given.

9. THINK SOLVE EXPLAIN **SHORT RESPONSE** Rico, Carolina, and Gloria have pizza that they are going to be sharing with other people. Rico gave away $\frac{1}{3}$ of his cheese pizza to Carolina and she gave him $\frac{3}{7}$ of her pepperoni. Rico then gave Gloria $\frac{1}{7}$ of his cheese pizza. How much pizza, pepperoni and cheese, does Rico have now?

10. For a warm up, Samuel runs 200 yards less than half the maximum distance he can run. This is represented by the equation $r = \frac{1}{2}x - 200$, where x represents the maximum distance he can run and r represents the distance run during his warm up. If Samuel ran 1,600 yards during his warm up, what is the maximum distance he can run?

    A. 3,600 yards

    B. 2,400 yards

    C. 1,800 yards

    D. 1,600 yards

11. What is the value of $20 \div (-4)$?

    F. -5 **H.** 5

    G. -7 **I.** 4

12. ✏️ **GRIDDED RESPONSE** Ines is in a hot air balloon 89 feet above the ground. A bird is flying 15 feet above the hot air balloon. How high off the ground is the bird in feet?

13. THINK SOLVE EXPLAIN **EXTENDED RESPONSE** A first-time bungee jumper is about to make his first jump. When the bungee jumper jumps, he will fall 5 feet every 0.5 second.

    Part A Let s be the total number of seconds in a jump and h be the height of the jump. Write an equation that can be used to find s.

    Part B Use your equation to calculate the total seconds for a 150-foot jump. Show your work.

NEED EXTRA HELP?													
If You Missed Question...	1	2	3	4	5	6	7	8	9	10	11	12	13
Go to Chapter-Lesson...	4-3B	2-1C	2-2B	4-2B	4-3B	4-1D	4-3B	4-4C	3-2C	4-3B	2-3D	2-2B	4-3B

Proportions and Similarity

connectED.mcgraw-hill.com

Investigate

 Animations

 Vocabulary

 Multilingual eGlossary

Learn

 Personal Tutor

 Virtual Manipulatives

 Graphing Calculator

 Audio

 Foldables

Practice

 Self-Check Practice

 Worksheets

 Assessment

The ☆ BIG Idea

What makes two quantities proportional?

 FOLDABLES Study Organizer

Make this Foldable to help you organize your notes.

Review Vocabulary

ratio *razón* a comparison of two numbers by division; the ratio of 2 to 3 can be written as 2 out of 3, 2 to 3, 2 : 3, or $\frac{2}{3}$

The ratio of squares to triangles is 2 : 3.

Key Vocabulary

English	Español
proportion	proporción
rate	tasa
scale factor	factor de escala
similar figures	figuras semejantes

When Will I Use This?

Raul, Caitlyn, and Jamar in
Campfire Song

Wow! Kayaking sure is fun! Camp is turning out to be a blast!

If Caitlyn hadn't rescued me, I'd be going in circles for hours! It took me forever to get it straight!

Now I'm so hungry I could eat a kayak!

Well, you can do that if you want to, Jamar! As for me, I need to get back to the cabin and get my camera before dinner.

Gotcha!

After that, Raul and I need to get to the campfire early to practice our song.

Song?

Yes! We have been preparing a song for the campfire tonight!

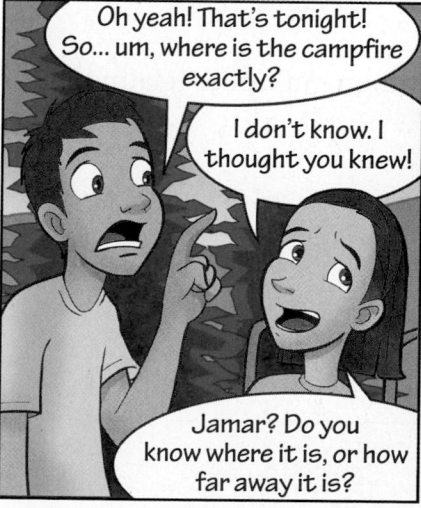

Oh yeah! That's tonight! So... um, where is the campfire exactly?

I don't know. I thought you knew!

Jamar? Do you know where it is, or how far away it is?

And Jamar saves the day! I brought a map! We can figure out how far it is from here.

CABIN
LAKE
ROPES COURSE
1 SQUARE = 75 YARDS
MESS HALL
PICNIC AREA
CAMPFIRE

Awesome! Jamar, you are the greatest!

Maybe you can use your greatness to get us to dinner quickly! I'm getting hungry!

Your Turn!
You will solve this problem in the chapter.

Are You Ready for the Chapter?

You have two options for checking prerequisite skills for this chapter.

Text Option Take the Quick Check below. Refer to the Quick Review for help.

QUICK Check

FIELD TRIPS Write each ratio as a fraction in simplest form.

1. adults : students
2. students : buses
3. buses : people
4. adults : people
5. students : people

Seventh-Grade Field Trip	
Students	180
Adults	24
Buses	4

Determine whether the ratios are equivalent. Explain.

6. 20 nails for every 5 shingles, 12 nails for every 3 shingles

7. 2 cups of flour to 8 cups of sugar, 8 cups of flour to 14 cups of sugar

8. 12 out of 20 doctors agree, 6 out of 10 doctors agree

9. 2 DVDs to 7 CDs, 10 DVDs to 15 CDs

10. 27 students to 6 microscopes, 18 students to 5 microscopes

QUICK Review

EXAMPLE 1

SOCCER Write the ratio of wins to losses as a fraction in simplest form.

Madison Mavericks Team Statistics	
Wins	10
Losses	12
Ties	8

$$\text{wins} \rightarrow \frac{10}{12} = \frac{5}{6} \leftarrow \text{losses}$$

The ratio of wins to losses is $\frac{5}{6}$.

EXAMPLE 2

Determine whether the ratios 250 miles in 4 hours and 500 miles in 8 hours are equivalent.

Method 1

Compare the ratios written in simplest form.

$$250 \text{ miles} : 4 \text{ hours} = \frac{250}{4} \text{ or } \frac{125}{2}$$

$$500 \text{ miles} : 8 \text{ hours} = \frac{500}{8} \text{ or } \frac{125}{2}$$

The ratios simplify to the same fraction.

Method 2

Look for a common multiplier relating the two ratios.

$$\frac{250}{4} \xlongequal{\times 2} \frac{500}{8}$$

The numerator and denominator of the ratios are related by the same multiplier, 2.

The ratios are equivalent.

Online Option Take the Online Readiness Quiz.

Unit Rates

Main Idea
Model proportions using bar diagrams.

CCSS 7.RP.1

MONEY When Jeremy gets his allowance, he agrees to save part of it. His savings and expenses are in the ratio 7 : 5. If his daily allowance is $3, how much does he save each day?

ACTIVITY

What do you need to find? how much money Jeremy saves each day

STEP 1 The ratio of savings to expenses is 7 to 5.

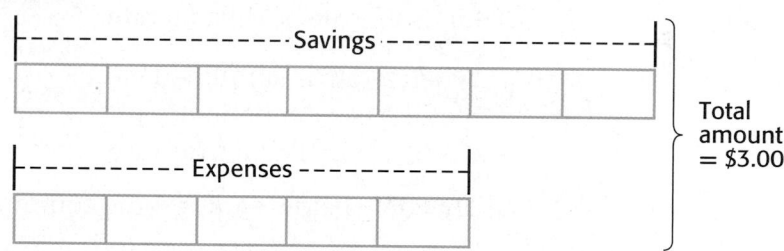

STEP 2 Let x represent each part of a bar. Write an equation.

$7x + 5x = 3$	Write the equation.
$12x = 3$	There are 12 parts in all.
$\dfrac{12x}{12} = \dfrac{3}{12}$	Division Property of Equality
$x = \dfrac{1}{4}$ or 0.25	Simplify.

STEP 3 Each part of the bar represents $0.25. Jeremy's savings are represented by 7 parts. 7 × $0.25 or $1.75

So, Jeremy saves $1.75 each day.

Analyze the Results

Use a bar diagram to solve.

1. **ATHLETES** The ratio of the number of boys to the number of girls on the swim team is 7 : 6. If there are 39 athletes on the swim team, how many more boys than girls are there?

2. **DOGS** In a dog kennel, the number of long-haired dogs and the number of short-haired dogs are in the ratio of 5 : 3. If there are 20 long-haired dogs, how many dogs are there altogether?

Main Idea
Determine unit rates.

 Vocabulary
rate
unit rate

CCSS 7.RP.1

Rates

Explore Choose a partner and take each other's pulse for 2 minutes.

1. Count the number of beats for each of you.

2. Write the ratio *beats* to *minutes* as a fraction.

A ratio that compares two quantities with different kinds of units is called a **rate**.

$$\frac{160 \text{ beats}}{2 \text{ minutes}}$$

> The units *beats* and *minutes* are different.

When a rate is simplified so that it has a denominator of 1 unit, it is called a **unit rate**.

$$\frac{80 \text{ beats}}{1 \text{ minute}}$$

> The denominator is 1 unit.

The table below shows some common unit rates.

Rate	Unit Rate	Abbreviation	Name
$\frac{\text{number of miles}}{1 \text{ hour}}$	miles per hour	mi/h or mph	average speed
$\frac{\text{number of miles}}{1 \text{ gallon}}$	miles per gallon	mi/gal or mpg	gas mileage
$\frac{\text{number of dollars}}{1 \text{ pound}}$	price per pound	dollars/lb	unit price

 REAL-WORLD EXAMPLE **Find a Unit Rate**

1 **BIKING** Adrienne biked 24 miles in 4 hours. If she biked at a constant speed, how many miles did she ride in one hour?

24 miles in 4 hours $= \dfrac{24 \text{ mi}}{4 \text{ h}}$ Write the rate as a fraction.

$= \dfrac{24 \text{ mi} \div 4}{4 \text{ h} \div 4}$ Divide the numerator and the denominator by 4.

$= \dfrac{6 \text{ mi}}{1 \text{ h}}$ Simplify.

Adrienne biked 6 miles in one hour.

 CHECK Your Progress

Find each unit rate. Round to the nearest hundredth if necessary.

a. $300 for 6 hours   **b.** 220 miles on 8 gallons

REAL-WORLD EXAMPLE **Find a Unit Rate**

② **JUICE** Find the unit price if it costs $2 for eight juice boxes. Round to the nearest cent if necessary.

QUICK Review

Dividing by Whole Numbers

Place the decimal point in the quotient directly above the decimal point in the dividend.

$$
\begin{array}{r}
0.25 \\
8\overline{)2.00} \\
-16 \\
\hline
40 \\
-40 \\
\hline
0
\end{array}
$$

$$\$2 \text{ for eight boxes} = \frac{\$2}{8 \text{ boxes}} \qquad \text{Write the rate as a fraction.}$$

$$= \frac{\$2 \div 8}{8 \text{ boxes} \div 8} \qquad \text{Divide the numerator and the denominator by 8.}$$

$$= \frac{\$0.25}{1 \text{ box}} \qquad \text{Simplify.}$$

The unit price is $0.25 per juice box.

 CHECK Your Progress

c. ESTIMATION Find the unit price if a 4-pack of mixed fruit sells for $2.12.

 REAL-WORLD EXAMPLE **Compare Using Unit Rates**

③ **DOG FOOD** The prices of 3 different bags of dog food are given in the table. Which size bag has the lowest price per pound?

Dog Food Prices	
Bag Size (lb)	**Price ($)**
40	49.00
20	23.44
8	9.88

- 40-pound bag
  $49.00 ÷ 40 pounds = $1.225 per pound

- 20-pound bag
  $23.44 ÷ 20 pounds = $1.172 per pound

- 8-pound bag
  $9.88 ÷ 8 pounds = $1.235 per pound

At $1.172 per pound, the 20-pound bag sells for the lowest price per pound.

Study Tip

Alternative Method
One 40-lb bag is equivalent to two 20-lb bags or five 8-lb bags. The cost for one 40-lb bag is $49, the cost for two 20-lb bags is about 2 x $23 or $46, and the cost for five 8-lb bags is about 5 x $10 or $50. So, the 20-lb bag has the lowest price per pound.

 CHECK Your Progress

d. FOOD Tito wants to buy some peanut butter to donate to the local food pantry. If Tito wants to save as much money as possible, which brand should he buy?

Peanut Butter Sales	
Brand	**Sale Price**
Nutty	12 ounces for $2.19
Grandma's	18 ounces for $2.79
Bee's	28 ounces for $4.69
Save-A-Lot	40 ounces for $6.60

REAL-WORLD EXAMPLE ▸ **Use a Unit Rate**

④ **FACE PAINTING** Lexi painted 3 faces in 12 minutes at the Crafts Fair. At this rate, how many faces can she paint in 40 minutes?

Method 1   Draw a Bar Diagram

```
|------------- 12 min -------------|
| time to paint | time to paint | time to paint |
|   one face    |   one face    |   one face    |
|--- 4 min ---|--- 4 min ---|--- 4 min ---|
```

It takes 4 minutes to paint one face. In 40 minutes, Lexi can paint 40 ÷ 4 or 10 faces.

Method 2   Find a Unit Rate

Find the unit rate. Then multiply this unit rate by 40 to find the number of faces she can paint in 40 minutes.

3 faces in 12 minutes $= \dfrac{3 \text{ faces} \div 12}{12 \text{ min} \div 12} = \dfrac{0.25 \text{ face}}{1 \text{ min}}$ Find the unit rate.

$\dfrac{0.25 \text{ face}}{1 \text{ min}} \cdot 40 \text{ min} = 10$ faces   Divide out the common units.

Using either method, Lexi can paint 10 faces in 40 minutes.

 CHOOSE Your Method

e. **SCHOOL SUPPLIES** Kimbel bought 4 notebooks for $6.32. At this same unit price, how much would he pay for 5 notebooks?

Real-World Link
Face paint can be made from 1 teaspoon cornstarch and $\frac{1}{2}$ teaspoon each of water and cold cream.

✓ **CHECK Your Understanding**

Examples 1 and 2   Find each unit rate. Round to the nearest hundredth if necessary.

1. 90 miles on 15 gallons
2. 1,680 kilobytes in 4 minutes
3. 5 pounds for $2.49
4. 152 feet in 16 seconds

Example 3   5. **CDs** Four stores offer customers bulk CD rates. Which store offers the best buy?

Example 4   6. **TRAVEL** After 3.5 hours, Pasha had traveled 217 miles. If she travels at a constant speed, how far will she have traveled after 4 hours?

Bulk CD Offers	
Store	**Offer**
CD Express	4 CDs for $60
Music Place	6 CDs for $75
CD Rack	5 CDs for $70
Music Shop	3 CDs for $40

Practice and Problem Solving

= **Step-by-Step Solutions** begin on page R1.
Extra Practice begins on page EP2.

Examples 1 and 2   **Find each unit rate. Round to the nearest hundredth if necessary.**

7. 360 miles in 6 hours

8. 6,840 customers in 45 days

9. 150 people for 5 classes

10. 815 Calories in 4 servings

11 45.5 meters in 13 seconds

12. $7.40 for 5 pounds

13. $1.12 for 8.2 ounces

14. 144 miles on 4.5 gallons

15. ESTIMATION Estimate the unit rate if 12 pairs of socks sell for $5.79.

16. ESTIMATION Estimate the unit rate if a 26-mile marathon was completed in 5 hours.

Example 3   **17. SPORTS** The results of a swim meet are shown. Who swam the fastest? Explain your reasoning.

Name	Event	Time (s)
Tawni	50-m Freestyle	40.8
Pepita	100-m Butterfly	60.2
Susana	200-m Medley	112.4

18. MONEY A grocery store sells three different packages of bottled water. Which package costs the least per bottle? Explain your reasoning.

6-pack
for $3.79

9-pack
for $4.50

12-pack
for $6.89

19. NUTRITION Use the table at the right.

a. Which soft drink has about twice the amount of sodium per ounce than the other two? Explain.

b. Which soft drink has the least amount of sugar per ounce? Explain.

Soft Drink Nutritional Information			
Soft Drink	Serving Size (oz)	Sodium (mg)	Sugar (g)
A	12	40	22
B	8	24	15
C	7	42	30

Example 4   **20. WORD PROCESSING** Ben can type 153 words in 3 minutes. At this rate, how many words can he type in 10 minutes?

21. FABRIC Kenji buys 3 yards of fabric for $7.47. Later he realizes that he needs 2 more yards. How much will he pay for the extra fabric?

22. ESTIMATION A player scores 87 points in 6 games. At this rate, about how many points would she score in the next 4 games?

23. JOBS Dalila earns $108.75 for working 15 hours as a holiday helper wrapping gifts. If she works 18 hours the next week, how much money will she earn?

24. POPULATION Florida has approximately 18.1 million people living in 65,795 square miles. What is the *population density* or number of people per square mile in Florida?

ESTIMATION Estimate the unit price for each item. Justify your answers.

25.
$2.49

26.
$1.89

27.
$1.13

28. **RECIPES** A recipe that makes 10 mini-loaves of banana bread calls for $1\frac{1}{4}$ cups flour. How much flour is needed to make 2 dozen mini-loaves using this recipe?

29. **SPORTS** Use the information at the left.

   a. The Boston Marathon is 26.2 miles long. What was the average speed of the record winner of the wheelchair division? Round to the nearest hundredth.

   b. At this rate, about how long would it take this competitor to complete a 30-mile race?

30. **MONEY** Suppose that 1 euro is worth $1.25. In Europe, a book costs 19 euros. In Los Angeles, the same book costs $22.50. In which location is the book less expensive?

31. **ANIMALS** Use the graph that shows the average number of heartbeats for an active adult brown bear and a hibernating brown bear.

   a. What does the point (2, 120) represent on the graph?

   b. What does the point (1.5, 18) represent on the graph?

   c. What does the ratio of the y-coordinate to the x-coordinate for each pair of points on the graph represent?

   d. Use the graph to find the bear's average heart rate when it is active and when it is hibernating.

   e. When is the bear's heart rate greater, when it is active or when it is hibernating? How can you tell this from the graph?

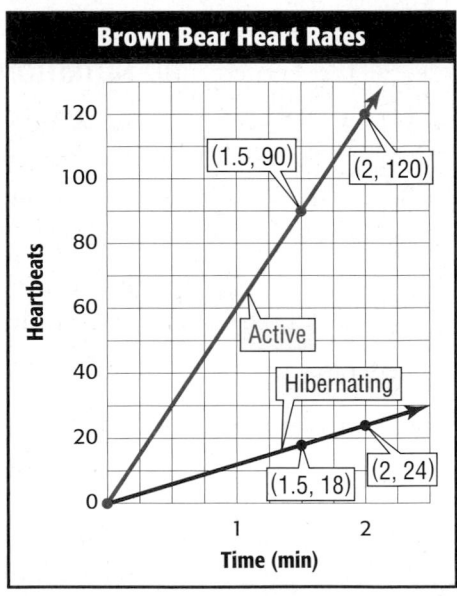

32. **TIRES** At Tire Depot, a pair of new tires sells for $216. The manager's special advertises the same tires selling at a rate of $380 for 4 tires. How much do you save per tire if you purchase the manager's special?

33. **MATH IN THE MEDIA** Find examples of grocery item prices in a newspaper, on television, or on the Internet. Compare unit prices of two different brands of the same item. Explain which item is the better buy.

34. **FIND THE ERROR** Seth is trying to find the unit price for a package of blank compact discs on sale at 10 for $5.49. Find his mistake and correct it.

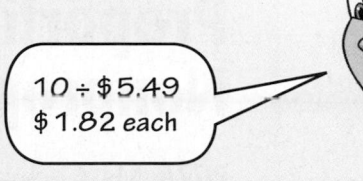

$10 \div \$5.49$
$\$1.82$ each

CHALLENGE Determine whether each statement is *sometimes*, *always*, or *never* true. Give an example or a counterexample.

35. A ratio is a rate.

36. A rate is a ratio.

37. **WRITE MATH** Describe, using an example, how a *rate* is a measure of one quantity per unit of another quantity.

✔ Test Practice

38. Mrs. Ross needs to buy dish soap. There are four differently sized containers.

Dish Soap Prices	
Brand	**Price**
Lots of Suds	$0.98 for 8 ounces
Bright Wash	$1.29 for 12 ounces
Spotless Soap	$3.14 for 30 ounces
Lemon Bright	$3.45 for 32 ounces

Which brand costs the least per ounce?

A. Lots of Suds   C. Spotless Soap

B. Bright Wash   D. Lemon Bright

39. Bonita spent $2.00 for 20 pencils, Jamal spent $1.50 for 10 pencils, and Hasina spent $2.10 for 15 pencils. Which of the following lists the students from least to greatest according to unit price paid?

F. Bonita, Jamal, Hasina

G. Hasina, Jamal, Bonita

H. Bonita, Hasina, Jamal

I. Jamal, Hasina, Bonita

40. The Jimenez family took a four-day road trip. They traveled 300 miles in 5 hours on Sunday, 200 miles in 3 hours on Monday, 150 miles in 2.5 hours on Tuesday, and 250 miles in 6 hours on Wednesday. On which day did they average the greatest miles per hour?

A. Sunday   C. Tuesday

B. Monday   D. Wednesday

41. The table shows the total distance traveled by a car driving at a constant rate of speed.

Time (h)	Distance (mi)
2	130
3.5	227.5
4	260
7	455

How far will the car have traveled after 10 hours?

F. 520 miles   H. 650 miles

G. 585 miles   I. 715 miles

Main Idea

Identify proportional and nonproportional relationships.

 Vocabulary

proportional
nonproportional

 *Get ConnectED*

 CCSS **7.RP.2, 7.RP.2a**

Proportional and Nonproportional Relationships

PIZZA Ms. Cochran is planning a year-end pizza party for her students. Ace Pizza offers free delivery and charges $8 for each medium pizza.

1. Copy and complete the table to determine the cost for different numbers of pizzas ordered.

Cost ($)	8			
Pizzas Ordered	1	2	3	4

2. For each number of pizzas, write the relationship of the cost and number of pizzas as a ratio in simplest form. What do you notice?

Two quantities are **proportional** if they have a constant ratio. For relationships in which this ratio is not constant, the two quantities are **nonproportional**.

$$\frac{\text{cost of order}}{\text{pizzas ordered}} = \frac{8}{1} = \frac{16}{2} = \frac{24}{3} = \frac{32}{4} \text{ or } \$8 \text{ per pizza}$$

The cost of an order is *proportional* to the number of pizzas ordered.

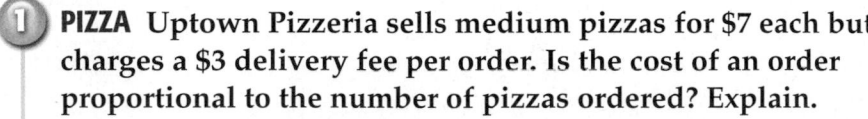 **EXAMPLES** Identify Proportional Relationships

1 **PIZZA** Uptown Pizzeria sells medium pizzas for $7 each but charges a $3 delivery fee per order. Is the cost of an order proportional to the number of pizzas ordered? Explain.

Cost ($)	10	17	24	31
Pizzas Ordered	1	2	3	4

For each number of pizzas, write the relationship of the cost and number of pizzas as a ratio in simplest form.

$$\frac{\text{cost of order}}{\text{pizzas ordered}} \longrightarrow \quad \frac{10}{1} \text{ or } 10 \qquad \frac{17}{2} \text{ or } 8.5 \qquad \frac{24}{3} \text{ or } 8 \qquad \frac{31}{4} \text{ or } 7.75$$

Since the ratios of the two quantities are not the same, the cost of an order is *not* proportional to the number of pizzas ordered.

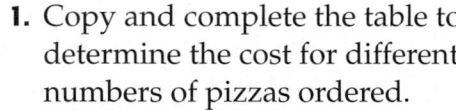

② **BEVERAGES** You can use the recipe shown to make a healthier version of a popular beverage. Is the amount of mix used proportional to the amount of sugar used? Explain.

Fruit Punch
½ cup sugar
1 envelope of mix
2 quarts of water

Find the amount of mix and sugar needed for different numbers of batches and make a table to show these mix and sugar measures.

Cups of Sugar	$\frac{1}{2}$	1	$1\frac{1}{2}$	2
Envelopes of Mix	1	2	3	4
Quarts of Water	2	4	6	8

For each number of cups of sugar, write the relationship of the cups and number of envelopes of mix as a ratio in simplest form.

$$\frac{\text{cups of sugar}}{\text{envelopes of mix}} \rightarrow \frac{\frac{1}{2}}{1}, \frac{1}{2}, \frac{1\frac{1}{2}}{3}, \frac{2}{4}$$

All of the ratios between the two quantities can be simplified to 0.5. The amount of mix used is proportional to the amount of sugar used.

 CHECK Your Progress

a. BEVERAGES In Example 2, is the amount of sugar used proportional to the amount of water used?

b. MONEY At the beginning of the year, Isabel had $120 in the bank. Each week, she deposits another $20. Is her account balance proportional to the number of weeks of deposits?

✓ CHECK Your Understanding

For Exercises 1–4, explain your reasoning.

Examples 1 and 2   **①** **ELEPHANTS** An adult elephant drinks about 225 liters of water each day. Is the number of days that an elephant's water supply lasts proportional to the number of liters of water the elephant drinks?

2. PACKAGES A package shipping company charges $5.25 to deliver a package. In addition, they charge $0.45 for each pound over one pound. Is the cost to ship a package proportional to the weight of the package?

3. SCHOOL At a certain middle school, every homeroom teacher is assigned 28 students. There are 3 teachers who do not have a homeroom. Is the number of students at this school proportional to the number of teachers?

4. JOBS Andrew earns $18 per hour for mowing lawns. Is the amount of money he earns proportional to the number of hours he spends mowing?

Practice and Problem Solving

 = **Step-by-Step Solutions** begin on page R1.
Extra Practice begins on page EP2.

For Exercises 5–12, explain your reasoning.

Examples 1 and 2   **5. RECREATION** The Vista Marina rents boats for $25 per hour. In addition to the rental fee, there is a $12 charge for fuel. Is the number of hours you can rent the boat proportional to the total cost? Explain.

6. ELEVATORS An elevator *ascends,* or goes up, at a rate of 750 feet per minute. Is the height to which the elevator ascends proportional to the number of minutes it takes to get there? Justify your response.

7. PLANTS A vine grows 7.5 feet every 5 days. Is the length of the vine on the last day proportional to the number of days of growth? Explain.

8. TEMPERATURE To convert a temperature in degrees Celsius to degrees Fahrenheit, multiply the Celsius temperature by $\frac{9}{5}$ and then add 32°. Is a temperature in degrees Celsius proportional to its equivalent temperature in degrees Fahrenheit? Explain your reasoning.

9. ADVERTISING On Saturday, Querida gave away 416 coupons for a free appetizer at a local restaurant. The next day, she gave away about 52 coupons an hour.

   a. Is the number of coupons Querida gave away on Sunday proportional to the number of hours she worked that day? Explain.

   b. Is the total number of coupons Querida gave away on Saturday and Sunday proportional to the number of hours she worked on Sunday? Justify your response.

10. TAXES MegaMart collects a sales tax equal to $\frac{1}{16}$ of the retail price of each purchase and sends this money to the state government.

   a. Is the amount of tax collected proportional to the cost of an item before tax is added?

   b. Is the amount of tax collected proportional to the cost of an item after tax has been added?

11 **MEASUREMENT** Determine whether the measures for the figure shown are proportional.

   a. the length of a side and the perimeter

   b. the length of a side and the area

s

12. RIDES The table shows the fee for ride tickets at a carnival.

   a. Is the fee for ride tickets proportional to the number of tickets? Explain your reasoning.

   b. Can you determine the fee for 30 ride tickets?

Number of Ride Tickets	5	10	15	20	25
Fee ($)	5	9.50	13.50	16	18.75

Real-World Link · · ·
Ascending at a speed of 1,000 feet per minute, the five outside elevators of the Westin St. Francis are the fastest glass elevators in San Francisco.

13. FIND THE ERROR Blake ran laps around the gym. His times are shown in the table. Blake is trying to decide whether the number of laps is proportional to the time. Find his mistake and correct it.

Time (min)	1	2	3	4
Laps	4	6	8	10

It is proportional because the number of laps always increases by 2.

CHALLENGE For each of the following, determine whether each situation is *sometimes*, *always*, or *never* proportional. Explain your reasoning.

14. the cost for ordering multiple items that will be delivered

15. the number of cars on a highway and the number of tires

16. WRITE MATH Explain what makes two quantities proportional.

Test Practice

17. Mr. Martinez is comparing the price of oranges from several different markets. Which market's pricing guide is based on a constant unit price?

A.

Number of Oranges	5	10	15	20
Total Cost ($)	3.50	6.00	8.50	11.00

B.

Number of Oranges	5	10	15	20
Total Cost ($)	3.50	6.50	9.50	12.50

C.

Number of Oranges	5	10	15	20
Total Cost ($)	3.00	5.00	7.00	9.00

D.

Number of Oranges	5	10	15	20
Total Cost ($)	3.00	6.00	9.00	12.00

18. GRIDDED RESPONSE The middle school is planning a family movie night where popcorn will be served. The constant relationship between the number of people n and p, the number of cups of popcorn, is shown in the table. How many people can be served with 519 cups of popcorn?

n	30	60	120	?
p	90	180	360	519

19. Which relationship has a unit rate of 60 miles per hour?

F. 300 miles in 6 hours

G. 300 miles in 5 hours

H. 240 miles in 6 hours

I. 240 miles in 5 hours

Main Idea
Use proportions to solve problems.

 Vocabulary
equivalent ratios
proportion
cross products

CCSS 7.RP.1, 7.RP.2, 7.RP.2b, 7.RP.2c, 7.RP.3

Solve Proportions

SHOPPING A local department store advertised a sale as shown at the right.

1. Write a ratio in simplest form that compares the cost to the number of bottles of nail polish.

2. Suppose Kate and some friends wanted to buy 6 bottles of polish. Write a ratio comparing the cost to the number of bottles of polish.

3. Is the cost proportional to the number of bottles of polish purchased? Explain.

The ratios of the cost to the number of bottles of polish for two and six bottles are both equal to $\frac{5}{2}$. **Equivalent ratios** have the same value.

$$\frac{\$5}{2 \text{ bottles of polish}} = \frac{\$15}{6 \text{ bottles of polish}}$$

Key Concept   **Proportion**

Words   A **proportion** is an equation stating that two ratios or rates are equivalent.

Symbols   **Numbers**   **Algebra**

$$\frac{6}{8} = \frac{3}{4} \qquad \frac{a}{b} = \frac{c}{d}, b \neq 0, d \neq 0$$

Consider the following proportion.

$$\frac{a}{b} = \frac{c}{d}$$

$$\frac{a}{\cancel{b}} \cdot \cancel{b}d \overset{1}{=} \frac{c}{\cancel{d}} \cdot b\cancel{d} \qquad$$ Multiply each side by bd and divide out common factors.

$$ad = bc \qquad$$ Simplify.

The products ad and bc are called the **cross products** of this proportion. The cross products of any proportion are equal.

$$\frac{6}{8} = \frac{3}{4} \quad \longrightarrow \quad \begin{array}{l} 8 \cdot 3 = 24 \\ 6 \cdot 4 = 24 \end{array}$$

EXAMPLE Write and Solve a Proportion

① **TEMPERATURE** After 2 hours, the air temperature had risen 7°F. Write and solve a proportion to find the amount of time it will take at this rate for the temperature to rise an additional 13°F.

Write a proportion. Let t represent the time in hours.

temperature → $\dfrac{7}{2} = \dfrac{13}{t}$ ← temperature
time → ← time

$7 \cdot t = 2 \cdot 13$ Find the cross products.

$7t = 26$ Multiply.

$\dfrac{7t}{7} = \dfrac{26}{7}$ Divide each side by 7.

$t \approx 3.7$ Simplify.

It will take about 3.7 hours to rise an additional 13°F.

CHECK Your Progress

Solve each proportion.

a. $\dfrac{x}{4} = \dfrac{9}{10}$ **b.** $\dfrac{2}{34} = \dfrac{5}{y}$ **c.** $\dfrac{7}{3} = \dfrac{n}{2.1}$

REAL-WORLD EXAMPLE Make Predictions

② **BLOOD** During a blood drive, the ratio of Type O donors to non-Type O donors was 37 : 43. About how many Type O donors would you expect in a group of 300 donors?

Write the ratio for the given information.

Type O donors → $\dfrac{37}{37 + 43}$ or $\dfrac{37}{80}$
total donors →

Write a proportion. Let t represent the number of Type O donors.

Type O donors → $\dfrac{37}{80} = \dfrac{t}{300}$ ← Type O donors
total donors → ← total donors

$37 \cdot 300 = 80t$ Find the cross products.

$11{,}100 = 80t$ Multiply.

$\dfrac{11{,}100}{80} = \dfrac{80t}{80}$ Divide each side by 80.

$t = 138.75$ Simplify.

There would be about 139 Type O donors.

Real-World Link · · · ·
There are four different blood types: A, B, AB, and O. People with Type O blood are considered *universal donors.* Their blood can be transfused into people with any blood type.

CHECK Your Progress

d. RECYCLING Recycling 2,000 pounds of paper saves about 17 trees. Write and solve a proportion to determine how many trees you would save by recycling 5,000 pounds of paper.

You can also use the unit rate to write an equation expressing the relationship between two proportional quantities.

 EXAMPLE **Write and Use an Equation**

3 **GASOLINE** Jaycee bought 8 gallons of gasoline for $31.12. Write an equation relating the cost to the number of gallons of gasoline. How much would Jaycee pay for 11 gallons at this same rate?

Find the unit rate between cost and gallons.

$$\frac{\text{cost in dollars}}{\text{gasoline in gallons}} = \frac{31.12}{8} \text{ or } \$3.89 \text{ per gallon}$$

Words	The cost is $3.89 times the number of gallons.
Variable	Let c represent the cost. Let g represent the number of gallons.
Equation	$c = 3.89 \cdot g$

Find the cost for 11 gallons sold at the same rate.

$c = 3.89g$	Write the equation.
$c = 3.89(11)$	Replace g with the number of gallons.
$c = 42.79$	Multiply.

The cost for 11 gallons is $42.79.

Study Tip

Checking Your Equation
You can check to see if the equation you wrote is accurate by testing the two known quantities.

$c = 3.89g$

$31.12 = 3.89(8)$

$31.12 = 31.12$

 CHECK Your Progress

e. TYPING Olivia typed 2 pages in 15 minutes. Write an equation relating the number of minutes m to the number of pages p typed. If she continues typing at this rate, how many minutes will it take her to type 10 pages? to type 25 pages?

✓ **CHECK Your Understanding**

Example 1   Solve each proportion.

1 $\frac{1.5}{6} = \frac{10}{p}$

2. $\frac{3.2}{9} = \frac{n}{36}$

3. $\frac{41}{x} = \frac{5}{2}$

For Exercises 4 and 5, assume all situations are proportional.

Example 2   **4. PAINT** Sheila mixed 3 ounces of blue paint with 2 ounces of yellow paint. She decided to create 20 ounces of the same mixture. Write and solve a proportion to determine the number of ounces of yellow paint Sheila needed for the new mixture.

Example 3   **5. TUTORING** Trina earns $28.50 tutoring for 3 hours. Write an equation relating her earnings m to the number of hours h she tutors. How much would Trina earn tutoring for 2 hours? for 4.5 hours?

Practice and Problem Solving

 = **Step-by-Step Solutions** begin on page R1.
Extra Practice begins on page EP2.

Example 1 Solve each proportion.

6. $\dfrac{k}{7} = \dfrac{32}{56}$ 7. $\dfrac{x}{13} = \dfrac{18}{39}$ 8. $\dfrac{44}{p} = \dfrac{11}{5}$ 9. $\dfrac{2}{w} = \dfrac{0.4}{0.7}$

10. $\dfrac{6}{25} = \dfrac{d}{30}$ 11. $\dfrac{2.5}{6} = \dfrac{h}{9}$ 12. $\dfrac{3.5}{8} = \dfrac{a}{3.2}$ 13. $\dfrac{48}{9} = \dfrac{72}{n}$

Example 2 Assume the situations are proportional. Write and solve a proportion.

14. **COOKING** Evarado paid $1.12 for a dozen eggs. Determine the cost of 3 eggs.

15. **TRAVEL** A certain vehicle can travel 483 miles on 14 gallons of gasoline. Determine how many gallons of gasoline this vehicle will need to travel 600 miles. Round to the nearest tenth.

16. **ILLNESS** For every person who actually has the flu, there are 6 people who have flu-like symptoms resulting from a cold. If a doctor sees 40 patients, determine how many of these you would expect to have a cold.

17. **LIFE SCIENCE** For every left-handed person, there are about 4 right-handed people. If there are 30 students in a class, predict the number of students who are right-handed.

18. **TRAVEL** A speed limit of 100 kilometers per hour (kph) is approximately equal to 62 miles per hour (mph). Predict the following measures. Round your answers to the nearest whole number.

   a. a speed limit in mph for a speed limit of 75 kph

   b. a speed limit in kph for a speed limit of 20 mph

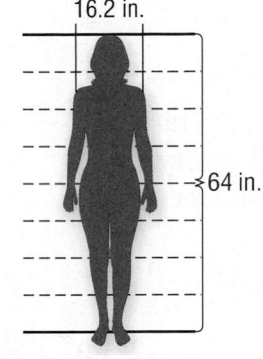

16.2 in.

64 in.

Real-World Link . . . .
Although people vary in size and shape, in general, people do not vary in proportion.

Example 3 19. **MEASUREMENT** The width of a woman's shoulders is proportional to her height. A woman who is 64 inches tall has a shoulder width of 16.2 inches. Find the height of a woman who has a shoulder width of 18.5 inches.

20. **COOKING** A recipe calls for $1\frac{1}{2}$ tablespoons of cinnamon for every 3 tablespoons of sugar. If you want to increase the recipe and use 6 tablespoons of sugar, how many tablespoons of cinnamon would you need? Explain your reasoning.

21. **PLANETS** Use the table to write a proportion relating the weights on two planets. Then find the missing weight. Round to the nearest tenth.

   a. Earth: 90 pounds; Venus: ▓ pounds

   b. Mercury: 55 pounds; Earth: ▓ pounds

   c. Jupiter: 350 pounds; Uranus: ▓ pounds

   d. Venus: 115 pounds; Mercury: ▓ pounds

Weights on Different Planets Earth Weight = 120 pounds	
Mercury	45.6 pounds
Venus	109.2 pounds
Uranus	96 pounds
Jupiter	304.8 pounds

22. **MATH IN THE MEDIA** Find an example of a ratio in a newspaper, on television, or on the Internet. Write an equivalent ratio. Then use a proportion to verify that the ratios are equivalent.

23. REASONING A powdered drink mix calls for a ratio of powder to water of 1 : 8. If there are 32 cups of powder, how many total cups of water are needed? Explain your reasoning.

CHALLENGE Solve each equation.

24. $\dfrac{2}{3} = \dfrac{18}{x + 5}$

25. $\dfrac{x - 4}{10} = \dfrac{7}{5}$

26. $\dfrac{4.5}{17 - x} = \dfrac{3}{8}$

27. WRITE MATH Explain why it might be easier to write an equation to represent a proportional relationship rather than using a proportion.

Test Practice

28. BAKING A recipe for making 3 dozen muffins requires 1.5 cups of flour. At this rate, how many cups of flour are required to make 5 dozen muffins?

   A. 2 cups   **C.** 3 cups

   B. 2.5 cups   **D.** 3.5 cups

29. An amusement park line is moving about 4 feet every 15 minutes. At this rate, approximately how long will it take for a person at the back of the 50-foot line to reach the front of the line?

   F. 1 hour   **H.** 5 hours

   G. 3 hours   **I.** 13 hours

30. THINK SOLVE EXPLAIN EXTENDED RESPONSE Crystal's mother kept a record of Crystal's height at different ages. She recorded the information in a table.

Age (y)	Height (in.)
0 (birth)	19
1	25
2	30
5	42
10	55
12	60

Part A Write a ratio for each age and the corresponding height.

Part B Is the relationship between Crystal's age and her height proportional? Explain.

Spiral Review

31. BABYSITTING Brenna charges $15, $30, $45, and $60 for babysitting 1, 2, 3, and 4 hours, respectively. Is the relationship between the number of hours and the amount charged proportional? If so, find the unit rate. If not, explain why not. (Lesson 1C)

Find each unit rate. Round to the nearest hundredth if necessary. (Lesson 1B)

32. 50 miles on 2.5 gallons

33. 2,500 kilobytes in 5 minutes

34. 5 peppers for $6.45

35. 64.8 meters in 9 seconds

Extend

Wildlife Sampling

Main Idea

Use proportions to estimate populations.

CCSS 7.RP.2, 7.RP.3

Naturalists can estimate the population in a wildlife preserve by using the capture-recapture technique. You will model this technique using dried beans in a bowl to represent bears in a forest.

ACTIVITY

STEP 1 Fill a small bowl with dried beans. Scoop out some of the beans. These represent the original *captured* bears. Count and record the number of beans. Mark each bean with an × on both sides. Then return these beans to the bowl and mix well.

STEP 2 Scoop another cup of beans from the bowl and count them. This is the *sample* for Trial A. Count the beans with the ×s. These are the *recaptured* bears. Record both numbers in a table.

Trial	Sample	Recaptured	P
A			
B			
⋮			
Total			

STEP 3 Use the proportion below to estimate the total number of beans in the bowl. This represents the total population *P*. Record the value of *P* in the table.

$$\frac{\text{captured}}{\text{total population } (P)} = \frac{\text{recaptured}}{\text{sample}}$$

STEP 4 Return all of the beans to the bowl.

STEP 5 Repeat Steps 2–4 nine times.

Analyze the Results

1. **ESTIMATION** Find the average of the estimates in column *P*. Is this a good estimate of the number of beans in the bowl? Explain your reasoning.

2. Count the actual number of beans in the bowl. How does this number compare to your estimate?

Problem-Solving Investigation

CCSS 7.RP.3

Main Idea Solve problems by drawing a diagram.

P.S.I. TEAM +

Draw a Diagram

GRACE: I am making trail mix for my sister's birthday party. On Saturday, I bought sunflower seeds and peanuts in a ratio of 3:4. On Sunday, I bought sunflower seeds and peanuts in a ratio of 2:3. I bought 18 ounces of sunflower seeds and peanuts altogether. How many ounces of peanuts did I buy?

YOUR MISSION: Draw a diagram to determine how many ounces of peanuts Grace bought.

Understand	You know that she bought two mixtures of sunflower seeds and peanuts, one in the ratio of 3 : 4 and the other in the ratio of 2 : 3. She bought 18 ounces of mixture.
Plan	Draw a diagram to represent the two mixtures.
Solve	Start with a bar diagram representing the mixtures. 3 : 4 ratio of sunflower seeds to peanuts 2 : 3 ratio of sunflower seeds to peanuts 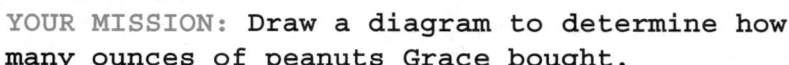 There is a total of 12 parts. Each part represents 18 ÷ 12 or 1.5 ounces. So, she bought 7 · 1.5 or 10.5 ounces of peanuts.
Check	For each mixture, there are more peanuts than sunflower seeds. So, more than half of the total should be peanuts. The answer is reasonable. ✓

Analyze the Strategy

1. **WRITE MATH** How can drawing a diagram be useful when solving a real-world problem?

= **Step-by-Step Solutions** begin on page R1.
Extra Practice begins on page EP2.

- Draw a diagram.
- Look for a pattern.
- Work backward.
- Choose an operation.

Use the *draw a diagram* strategy to solve Exercises 2–5.

2. A string is cut into 3 pieces in the ratio of 1 : 2 : 4. If the longest piece is 16 inches longer than the medium piece, find the length of the original string.

Piece 1 [ ]

Piece 2 [   ]

16 in.

Piece 3 [   ]

3. TILES A kitchen is 10 feet long and 8 feet wide. If kitchen floor tiles are $2\frac{1}{2}$ inches by 3 inches, how many tiles are needed for the kitchen?

4. TRAVEL There are four seats in Clarence's car: two in the front and two in the back. If Benny, Carlita, and Juanita are all in the car with Clarence, how many ways can they be seated in the car if Clarence is driving?

5 **GARDENING** Mr. Sanchez has a flower bed with a length of 10 meters and a width of 5 meters. If he can only change the width of the flower bed, describe what he can do to increase the perimeter by 12 meters.

Use any strategy to solve Exercises 6–12.

6. SPORTS Every 12 times at bat, Shim hits the ball an average of 3 times. About how many times will he hit the ball after 20 times at bat? 40? 84?

Times at Bat	Number of Hits
12	3
20	?
40	?
84	?

7. NUMBERS A number is halved. Then three is subtracted from the quotient, and 5 is multiplied by the difference. Finally, 1 is added to the product. If the ending number is 26, what was the beginning number?

8. PRIZES By reaching into a bag that has letters A, B, and C, George will select three letters without replacing them. In how many possible ways could the letters be drawn?

9. HEALTH The human body is about $\frac{7}{10}$ water. About how much would a person weigh if he or she had 70 pounds of water weight?

10. FINANCIAL LITERACY The ratio of Julio's money to Marcus's money was 7 : 10. After Julio bought 4 books at $4.50 each, the ratio of Julio's money to Marcus's money was 1 : 4. Use a bar diagram to find how much money Julio and Marcus have left.

11. GEOMETRY What is the next figure in the pattern below?

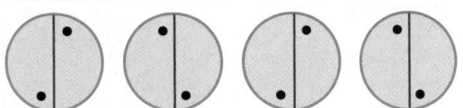

12. BASKETBALL Last basketball season, Simon had four less than twice the number of free throws Amit had. Simon had 48 free throws. How many free throws did Amit have last season?

13. **WRITE MATH** Write a real-world problem that can be solved by drawing a diagram. Then solve the problem.

Main Idea
Solve problems involving scale drawings.

 Vocabulary
scale drawing
scale model
scale
scale factor

 **Get ConnectED**

 **CCSS** 7.G.1

Scale Drawings

🏃 Explore

- Measure the length of each item in a room, such as a gymnasium.

- Record each length to the nearest $\frac{1}{2}$ foot.

1. Let 1 unit on the grid paper represent 2 feet. So, 4 units = 8 feet. Convert all your measurements to units.

2. On grid paper, make a drawing of your room like the one shown.

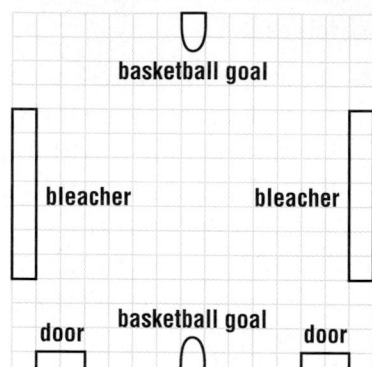

Scale drawings and **scale models** are used to represent objects that are too large or too small to be drawn or built at actual size. The **scale** gives the ratio that compares the measurements of the drawing or model to the measurements of the real object. The measurements on a drawing or model are proportional to the measurements on the actual object.

📝 EXAMPLE   Use a Map Scale

① **MAPS** What is the actual distance between Hagerstown and Annapolis?

Step 1 Use a centimeter ruler to find the map distance between the two cities. The map distance is about 4 centimeters.

Step 2 Write and solve a proportion using the scale. Let d represent the actual distance between the cities.

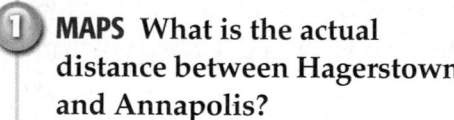

$$1 \times d = 24 \times 4 \qquad \text{Cross products}$$
$$d = 96 \qquad \text{Simplify.}$$

The distance between the cities is about 96 miles.

CHECK Your Progress

a. MAPS On the map of Arkansas shown, find the actual distance between Clarksville and Little Rock. Use a ruler to measure.

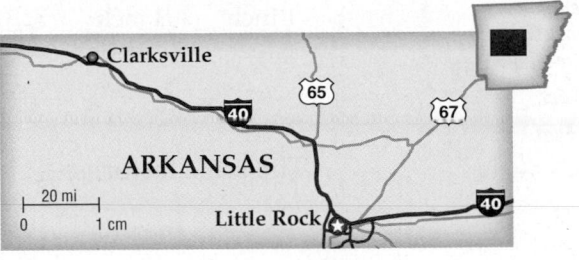

EXAMPLE Use a Scale Model

4 in.

② PHONES A graphic artist is creating an advertisement for a new cell phone. If she uses a scale of 5 inches = 1 inch, what is the length of the cell phone on the advertisement?

Write a proportion using the scale. Let a represent the length of the advertisement cell phone.

	Scale		Length	
advertisement →	$\dfrac{5 \text{ inches}}{1 \text{ inch}}$	=	$\dfrac{a \text{ inches}}{4 \text{ inches}}$	← advertisement
actual →				← actual

$5 \cdot 4 = 1 \cdot a$ Cross products

$20 = a$ Multiply.

The length of the cell phone on the advertisement is 20 inches long.

CHECK Your Progress

b. SCOOTERS A scooter is $3\dfrac{1}{2}$ feet long. Find the length of a scale model of the scooter if the scale is 1 inch $= \dfrac{3}{4}$ feet.

A scale written as a ratio without units in simplest form is called the **scale factor**.

scale → $\dfrac{\dfrac{1}{4} \text{ inch}}{2 \text{ feet}} = \dfrac{\dfrac{1}{4} \text{ inch}}{24 \text{ inches}}$ Convert 2 feet to 24 inches.

$= \dfrac{4}{4} \cdot \dfrac{\dfrac{1}{4} \text{ inch}}{24 \text{ inches}}$ Multiply by $\dfrac{4}{4}$ to eliminate the fraction in the numerator. Divide out the common units.

$= \dfrac{1}{96}$ ← scale factor

EXAMPLE Find a Scale Factor

3 SAILBOATS Find the scale factor of a model sailboat if the scale is 1 inch = 6 feet.

$$\frac{1 \text{ inch}}{6 \text{ feet}} = \frac{1 \text{ inch}}{72 \text{ inches}} \qquad \text{Convert 6 feet to inches.}$$

$$= \frac{1}{72} \qquad \text{Divide out the common units.}$$

The scale factor is $\frac{1}{72}$.

 CHECK Your Progress

c. CARS What is the scale factor of a model car if the scale is 1 inch = 2 feet?

REAL-WORLD EXAMPLE Construct a Scale Model

4 FERRIS WHEELS Penny is making a model of a Ferris wheel that is 60 feet tall. The model is 15 inches tall. Penny also is making a model of the sky needle ride that is 100 feet tall using the same scale. How tall is the model?

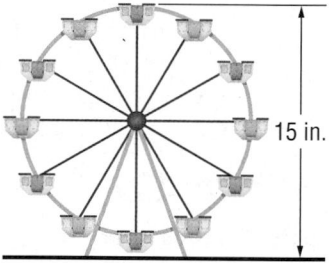

15 in.

$$\begin{array}{c}\text{model} \rightarrow \\ \text{actual} \rightarrow\end{array} \frac{15 \text{ in.}}{60 \text{ ft}} = \frac{1 \text{ in.}}{4 \text{ ft}} \qquad \begin{array}{l}\text{Divide the numerator and denominator by 15 so the}\\\text{numerator equals 1.}\end{array}$$

The scale is 1 inch = 4 feet. Using this scale, find the height of the sky needle ride.

$$\begin{array}{c}\text{model} \rightarrow \\ \text{actual} \rightarrow\end{array} \frac{1 \text{ in.}}{4 \text{ ft}} = \frac{x \text{ in.}}{100 \text{ ft}}$$

$$1 \cdot 100 = 4 \cdot x \qquad \text{Find the cross products.}$$

$$100 = 4x \qquad \text{Multiply.}$$

$$25 = x \qquad \text{Divide each side by 4.}$$

The height of the sky needle model is 25 inches.

 CHECK Your Progress

d. SOLAR SYSTEM Leah is constructing a scale model of the solar system for science class. She knows the diameter of the Sun is 1,391,900 kilometers, but would like to make it 10 inches. If Mercury's orbit radius is 57,950,000 kilometers, how many feet away from the Sun will Mercury be in the model?

Example 1 **GEOGRAPHY** Find the actual distance between each pair of cities in New Mexico. Use a ruler to measure.

1 Carlsbad and Artesia

2. Hobbs and Eunice

3. Artesia and Eunice

4. Lovington and Carlsbad

Example 2 **5. BRIDGES** An engineer makes a model of a bridge using a scale of 1 inch = 3 yards.

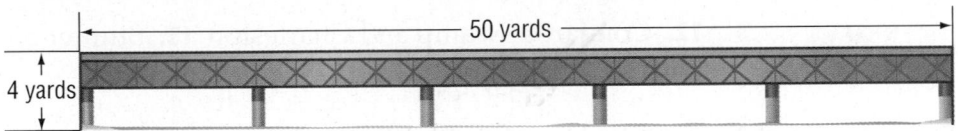

50 yards

4 yards

   a. What is the length of the model?

   b. What is the height of the model?

Example 3 Find the scale factor of each scale drawing or model.

6.

1 inch = 4 feet

7.

1 centimeter = 15 millimeters

8. CITY PLANNING In the aerial view of a city block at the right, the length of Main Street is 2 inches. If Main Street's actual length is 2 miles, find the scale factor of the drawing.

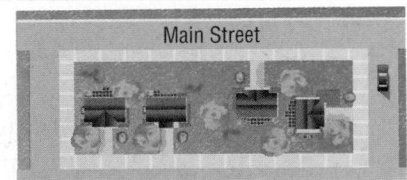

Main Street

Example 4 **9. DECORATING** Julianne is constructing a scale model of her family room to decide how to redecorate it. The room is 14 feet long by 18 feet wide. If she wants the model to be 8 inches long, about how wide will it be?

8 ″

● = **Step-by-Step Solutions** begin on page R1.
Extra Practice begins on page EP2.

Practice and Problem Solving

Example 1 **GEOGRAPHY** Find the actual distance between each pair of locations in South Carolina. Use a ruler to measure.

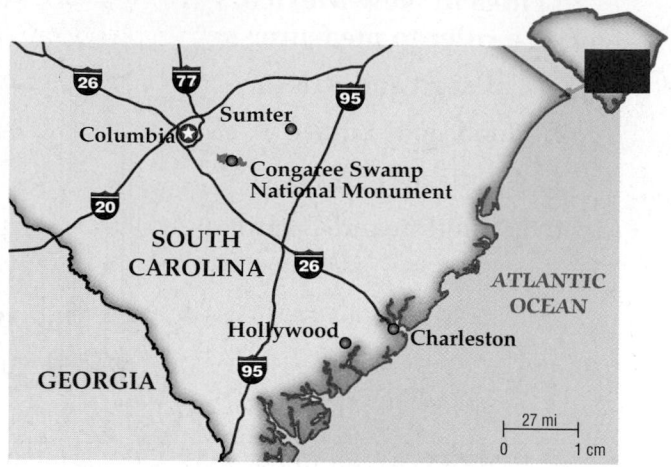

10. Columbia and Charleston   **11.** Hollywood and Sumter

12. Congaree Swamp and Charleston   **13.** Sumter and Columbia

Examples 2 and 3 Find the length of each model. Then find the scale factor.

14.

|←——— 87 ft ———→|

2 in. = 15 ft

15.

36 m

0.5 cm = 1.5 m

16.

|←——— 120 yd ———→|

1 in. = 20 yd

17.

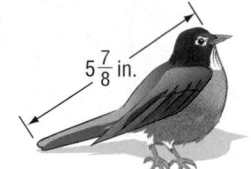

$5\frac{7}{8}$ in.

1 in. = 0.5 in.

Example 4 **18.** A model of an apartment is shown where $\frac{1}{4}$ inch represents 3 feet in the actual apartment.

   a. What is the actual length of the living room?

   b. Find the actual dimensions of the master bedroom.

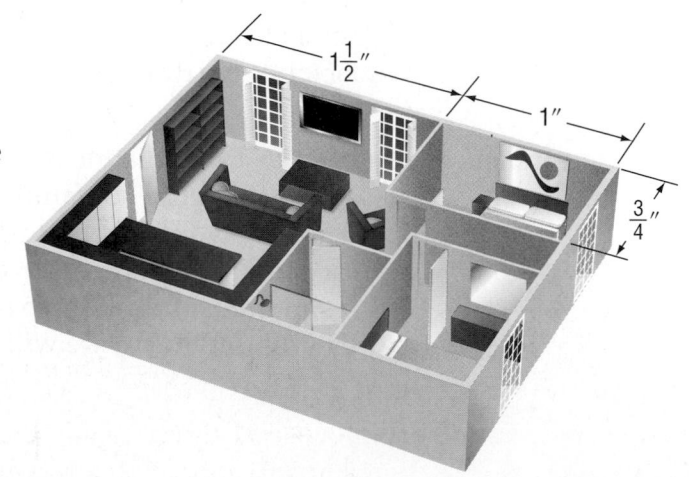

$1\frac{1}{2}''$

1″

$\frac{3}{4}''$

19. **TREES** A model of a tree is made using a scale of 1 inch : 25 feet. What is the height of the actual tree if the height of the model is $4\frac{3}{8}$ inches?

20. **GEOGRAPHY** A map of Bakersfield, California, has a scale of 1 inch to 5 miles. If the city is $5\frac{1}{5}$ inches across on the map, what is the actual distance across the actual city? Use estimation to check your answer.

21. 🔲 **MULTIPLE REPRESENTATIONS** Refer to the information at the left.

   a. **NUMBERS** Find the scale factor between the actual height of Thomas Jefferson and the statue.

   b. **TABLE** Make a table showing the height of the statue for every foot in height of Thomas Jefferson.

   c. **ALGEBRA** Write an expression to represent the height of the statue if Thomas Jefferson is x feet in height.

   d. **NUMBERS** Find the actual height of Thomas Jefferson if the height of the statue is 19 feet.

22. **BUILDINGS** If you are making a model of your bedroom, which would be an appropriate scale: 1 inch = 2 feet, or 1 inch = 12 feet? Explain.

23 **GRAPHIC NOVEL** Refer to the graphic novel frame below for Exercises a and b.

a. How many units on the map will Raul, Caitlyn, and Jamar go from the lake to the cabin?

b. What is the actual distance?

(H.O.T. Problems)

24. **OPEN ENDED** On grid paper, create a scale drawing of a room in your home. Include the scale that you used.

25. **REASONING** Compare and contrast the terms *scale* and *scale factor*. Include an example in your comparison.

26. 📝 **WRITE MATH** Explain how you could use a map to estimate the actual distance between Miami, Florida, and Atlanta, Georgia.

27. A landscape designer created the scale drawing below showing the bench that will be in the garden area.

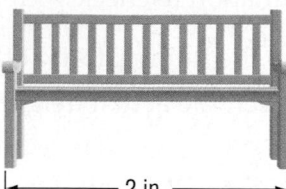

|← 2 in. →|

Which of these was the scale used for the drawing if the actual width of the bench is 6 feet?

A. $\frac{1}{4}$ inch = 1 foot

B. 3 inches = 1 foot

C. $\frac{2}{3}$ inch = 1 foot

D. 1 inch = 3 feet

28. ✏️ **GRIDDED RESPONSE** Ernesto drew a map of his school. He used a scale of 1 inch : 50 feet. What distance in inches on Ernesto's map should represent the 625 feet between the cafeteria and the science lab?

29. How many miles are represented by 4 inches on this map?

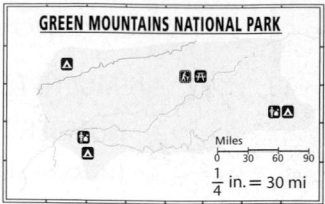

GREEN MOUNTAINS NATIONAL PARK

Miles
0   30   60   90
$\frac{1}{4}$ in. = 30 mi

F. 480 miles   **H.** 30 miles

G. 120 miles   **I.** 16 miles

30. A scale drawing of a doctor's office is shown.

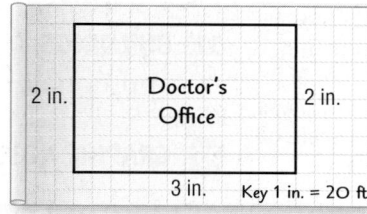

2 in.   Doctor's Office   2 in.

3 in.   Key 1 in. = 20 ft

What are the actual dimensions of the doctor's office?

A. 24 feet × 48 feet

B. 30 feet × 52 feet

C. 40 feet × 60 feet

D. 37.5 feet × 65 feet

Spiral Review

31. CARPENTRY A carpenter sawed a piece of wood into 3 pieces. The ratio of wood pieces is 1 : 3 : 6. The longest piece is 2.5 feet longer than the shortest piece. Use the *draw a diagram* strategy to find the length of the original piece. (Lesson 2A)

32. BAKING Write and solve a proportion to find the number of cups of cookie mix needed to make 72 cookies. (Lesson 1D)

Old-fashioned
Sugar
Cookie
Mix

3 cups makes 48 cookies

Solve each proportion. (Lesson 1D)

33. $\frac{5}{7} = \frac{a}{35}$

34. $\frac{12}{p} = \frac{36}{45}$

35. $\frac{3}{9} = \frac{21}{k}$

36. $\frac{n}{15} = \frac{17}{34}$

37. $\frac{-7}{10} = \frac{3.5}{j}$

38. $\frac{12}{18} = \frac{-40}{x}$

Main Idea

Use a spreadsheet to calculate measurements for scale drawings.

 7.G.1

Spreadsheet Lab: Scale Drawings

A computer spreadsheet is a useful tool for calculating measures for scale drawings. You can change the scale factors and the dimensions, and the spreadsheet will automatically calculate the new values. Suppose you want to make a scale drawing of your school.

ACTIVITY

STEP 1 Measure the rooms for your scale drawing.

STEP 2 Set up a spreadsheet like the one shown below. In this spreadsheet, the actual measures are in feet and the scale drawing measures are in inches.

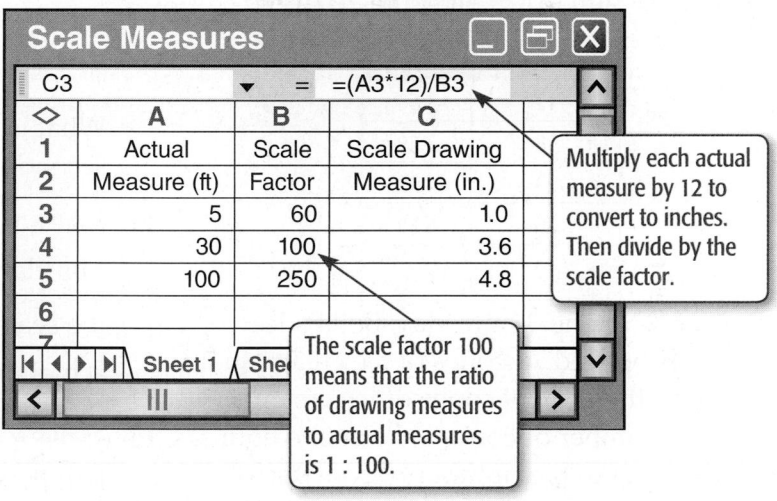

Analyze the Results

1. The length of one side of the school building is 100 feet. If you use a scale factor of 1 : 250, what is the length on your scale drawing?

2. The length of a classroom is 30 feet. What is the scale factor if the length of the classroom on a scale drawing is 3.6 inches?

3. **MAKE A SCALE DRAWING** Choose three rooms in your home and use a spreadsheet to make scale drawings. First, choose an appropriate scale and calculate the scale factor. Include a sketch of the furniture drawn to scale in each room.

Mid-Chapter Check

Express each rate as a unit rate. (Lesson 1B)

1. 750 yards in 25 minutes

2. $420 for 15 tickets

3. 42 laps in 6 races

4. **MULTIPLE CHOICE** In her last race, Bergen swam 1,500 meters in 30 minutes. On average, how many meters did she swim per minute? (Lesson 1B)

   A. 25   **C.** 40

   B. 30   **D.** 50

5. Which amount of nuts shown in the table has the best unit price? (Lesson 1B)

Weight (oz)	Cost ($)
12	2.50
18	3.69
24	4.95
30	6.25

6. **ICE CREAM** In one 8-hour day, Bella's Ice Cream Shop sold 72 cones of ice cream. In one hour, they sold 9 cones of ice cream. Is the total number of cones sold in one hour proportional to the number of cones sold during the day? Explain. (Lesson 1C)

7. **DISHES** Alan washed 60 plates in 30 minutes. It took him 3 minutes to wash 6 plates. Is the number of plates washed in 3 minutes proportional to the total number of plates he washed in 30 minutes? Explain. (Lesson 1C)

Solve each proportion. (Lesson 1D)

8. $\dfrac{33}{r} = \dfrac{11}{2}$

9. $\dfrac{x}{36} = \dfrac{15}{24}$

10. $\dfrac{5}{9} = \dfrac{4.5}{a}$

11. **MULTIPLE CHOICE** A bread recipe uses 4 cups of flour and $2\frac{1}{2}$ cups of water. If a baker puts 24 cups of flour into the mixer, how many cups of water will he need? (Lesson 1D)

    F. 15   **G.** 12   **H.** 8   **I.** 6

12. The diagram shows a rectangular patio. The length and width of another patio is 4 times the length and width of the first.

    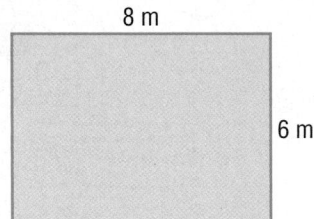

    What is the perimeter, in meters, of the second patio? (Lesson 2A)

13. **MAPS** Washington, D.C., and Baltimore, Maryland, are $2\frac{7}{8}$ inches apart on a map. If the scale is $\frac{1}{2}$ inch : 6 miles, what is the actual distance between the cities? (Lesson 2B)

14. **SCALE DRAWING** Each square has a side length of $\frac{1}{4}$ inch. (Lesson 2B)

    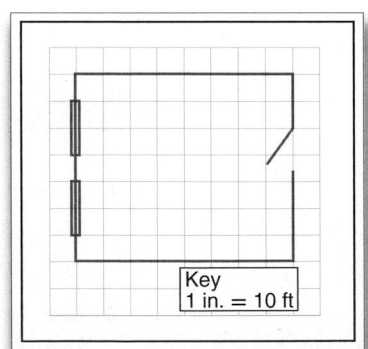

    a. What is the actual length of each window?

    b. What are the actual dimensions of the office space?

Main Idea

Solve problems involving similar figures.

Vocabulary

similar figures
corresponding sides
corresponding angles
indirect measurement
Side-Side-Side Similarity (SSS)
Angle-Angle Similarity (AA)
Side-Angle-Side Similarity (SAS)

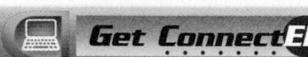

 7.RP.2

Similar Figures

Explore Each pair of figures below have the same shape but different sizes. Copy each pair onto centimeter dot paper. Measure each side using a centimeter ruler.

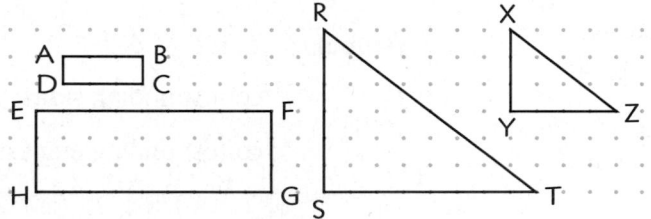

1. $\overline{AB}$ is in the same relative position as $\overline{EF}$. Name all such pairs of sides in each pair of figures.

   The notation $\overline{AB}$ means the segment with endpoints at A and B.

2. Write each ratio in simplest form.

   The notation AB means the *measure* of segment AB.

   a. $\dfrac{AB}{EF}, \dfrac{BC}{FG}, \dfrac{DC}{HG}, \dfrac{AD}{EH}$

   b. $\dfrac{RS}{XY}, \dfrac{ST}{YZ}, \dfrac{RT}{XZ}$

3. What do you notice about the ratios of these sides?

4. Name all pairs of angles in the same relative position. Cut out the figures. Compare the angles by matching them up. What do you notice about the measure of these angles?

5. **MAKE A CONJECTURE** Write a sentence about the sides and angles that are in the same position.

Figures that have the same shape but not necessarily the same size are **similar figures**. Triangle RST *is similar to* triangle XYZ.

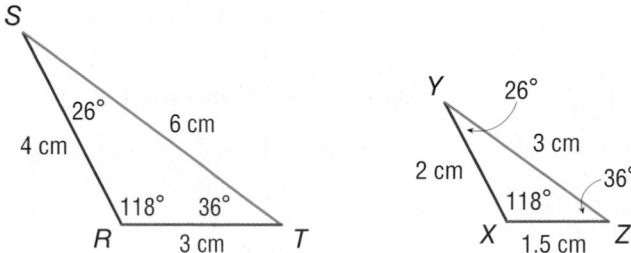

Corresponding sides — $\overline{ST}$ and $\overline{YZ}$, $\overline{SR}$ and $\overline{YX}$, and $\overline{RT}$ and $\overline{XZ}$ — are sides that are in the same relative position.

Corresponding angles — $\angle S$ and $\angle Y$, $\angle R$ and $\angle X$, and $\angle T$ and $\angle Z$ — are angles that are in the same relative position.

Words Two figures are similar if
  • the corresponding sides are proportional, and
  • the corresponding angles are congruent.

Models

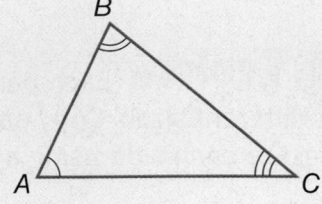
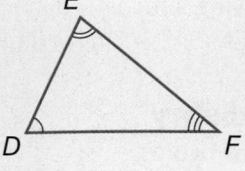

Symbols $\triangle ABC \sim \triangle DEF$

corresponding sides: $\dfrac{AB}{DE} = \dfrac{BC}{EF} = \dfrac{AC}{DF}$

corresponding angles: $\angle A \cong \angle D$; $\angle B \cong \angle E$; $\angle C \cong \angle F$

 EXAMPLE **Identify Similar Figures**

① **Which trapezoid below is similar to trapezoid** *DEFG*?

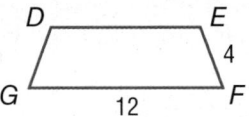

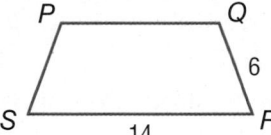
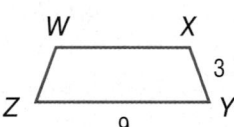
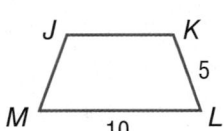

Find the ratios of the corresponding sides to see if they are the same.

Trapezoid *PQRS*	**Trapezoid** *WXYZ*	**Trapezoid** *JKLM*
$\dfrac{EF}{QR} = \dfrac{4}{6}$ or $\dfrac{2}{3}$	$\dfrac{EF}{XY} = \dfrac{4}{3}$	$\dfrac{EF}{KL} = \dfrac{4}{5}$
$\dfrac{FG}{RS} = \dfrac{12}{14}$ or $\dfrac{6}{7}$	$\dfrac{FG}{YZ} = \dfrac{12}{9}$ or $\dfrac{4}{3}$	$\dfrac{FG}{LM} = \dfrac{12}{10}$ or $\dfrac{6}{5}$
Not similar	Similar	Not similar

So, trapezoid *WXYZ* is similar to trapezoid *DEFG*.

 CHECK Your Progress

a. Which triangle below is similar to triangle *DEF*?

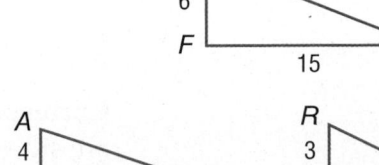

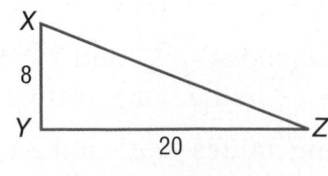

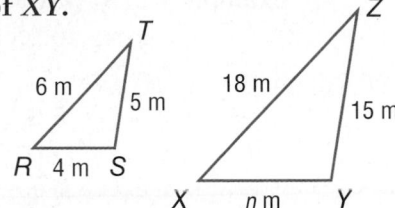

EXAMPLE Find Missing Measures

2 If $\triangle RST \sim \triangle XYZ$, find the length of $\overline{XY}$.

Since the two triangles are similar, the ratios of their corresponding sides are equal. Write and solve a proportion to find XY.

$\dfrac{RT}{XZ} = \dfrac{RS}{XY}$ Write a proportion.

$\dfrac{6}{18} = \dfrac{4}{n}$ Let n represent the length of $\overline{XY}$. Then substitute.

$6n = 18(4)$ Find the cross products.

$6n = 72$ Simplify.

$n = 12$ Divide each side by 6. The length of $\overline{XY}$ is 12 meters.

Read Math

Similarity Corresponding angles are written in the same order.

$\triangle RST \sim \triangle XYZ$

✔ CHECK Your Progress

b. If $\triangle ABC \sim \triangle EFD$, find the length of $\overline{AC}$.

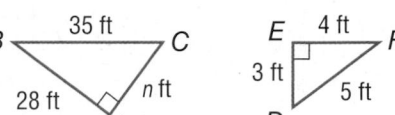

Indirect measurement uses similar figures to find the length, width, or height of objects that are too difficult to measure directly.

🏃 ✎ REAL-WORLD EXAMPLE

3 **GEYSERS** Old Faithful in Yellowstone National Park shoots water 60 feet into the air and casts a shadow of 42 feet. What is the height of a nearby tree that casts a shadow 63 feet long? Assume the triangles are similar.

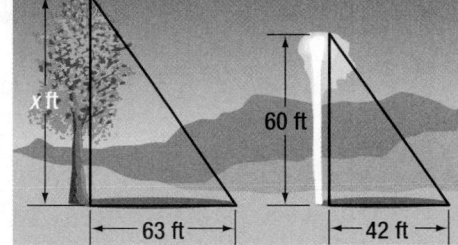

$\dfrac{x}{60} = \dfrac{63}{42}$ Write a proportion.

$42x = 60(63)$ Find the cross products.

$42x = 3{,}780$ Simplify.

$x = 90$ Divide each side by 42.

The height of the tree is 90 feet.

✔ CHECK Your Progress

c. **PHOTOGRAPHY** Destiny wants to resize a 4-inch-wide by 5-inch-long photograph so that it will fit in a space that is 2 inches wide. What is the new length?

Chapter 5 Lesson 3A Similarity and Proportional Reasoning **295**

Example 1 **1.** Which rectangle below is similar to rectangle *ABCD*?

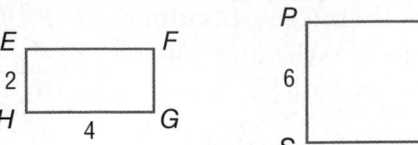

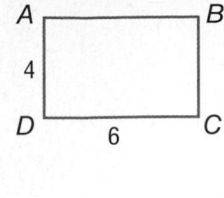

Example 2 **ALGEBRA** Find the value of *x* in each pair of similar figures.

2.

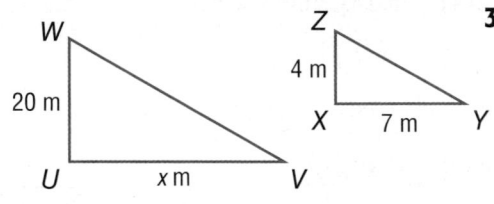

3.

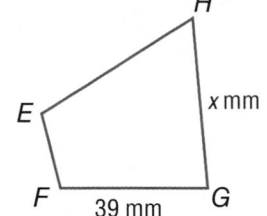

Example 3 **4. SHADOWS** A flagpole casts a 20-foot shadow. At the same time, Humberto, who is 6 feet tall, casts a 5-foot shadow. What is the height of the flagpole? Assume the triangles are similar.

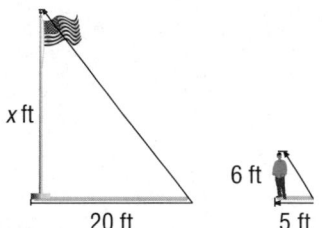

Practice and Problem Solving

● = **Step-by-Step Solutions** begin on page R1.
Extra Practice begins on page EP2.

Example 1 **⑤** Which triangle below is similar to △ *FGH*?

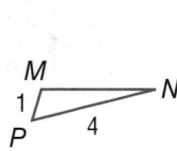

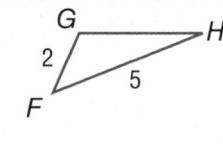

6. Which parallelogram below is similar to ▱ *HJKM*?

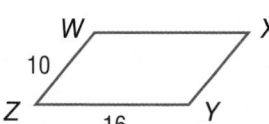

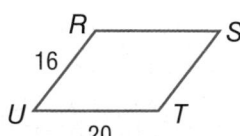

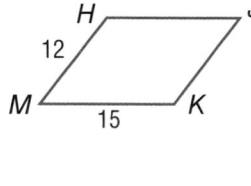

Example 2 **ALGEBRA** Find the value of *x* in each pair of similar figures.

7.

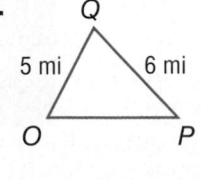

8.

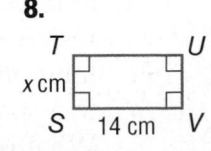

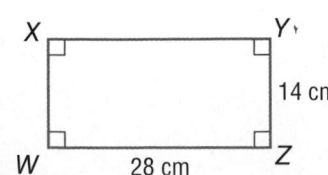

Example 3   The triangles are similar. Write a proportion and solve the problem.

9. **BASKETBALL** What is the height of the basketball hoop?

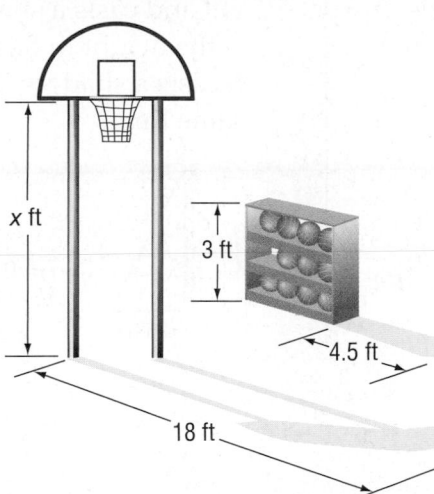

10. **TREES** How tall is the tree?

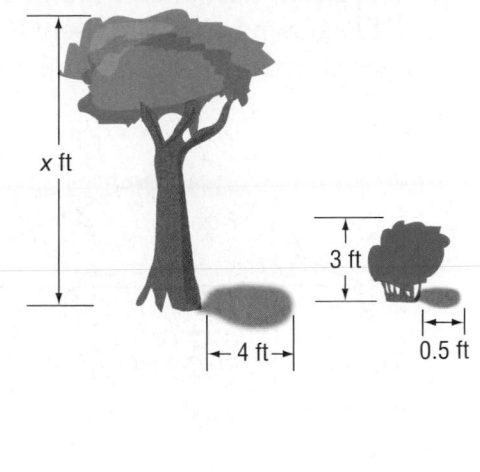

⓫ **GRAPHIC NOVEL** Refer to the graphic novel frame below for Exercises **a** and **b**.

  a. If the red line represents the path they took, how far have Raul, Caitlyn, and Jamar traveled since they left the lake?

  b. What will be their total distance by the time they return to the cabin?

12. **FURNITURE** A child's desk is made so that it is a replica of a full-size adult desk. Suppose the top of the full-size desk measures 54 inches long by 36 inches wide. If the top of the child's desk is 24 inches wide and is similar to the full-size desk, what is the length?

H.O.T. Problems

13. **CHALLENGE** Determine whether each statement is *true* or *false*. Explain.

  a. Any two squares are similar.   **b.** Any two rectangles are similar.

14. ✏️ **WRITE MATH** Write a real-world problem that could be solved using proportions and the concept of similarity. Solve the problem.

15. **SHORT RESPONSE** Triangles *ABC* and *DEF* shown below represent 2 puzzle pieces. Triangle *ABC* is similar to triangle *DEF*.

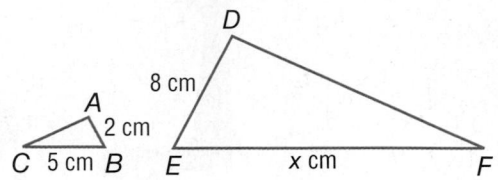

What is the length of $\overline{EF}$?

16. **GRIDDED RESPONSE** Horatio is 6 feet tall and casts a shadow 3 feet long. What is the height in feet of a nearby tower if it casts a shadow 25 feet long at the same time?

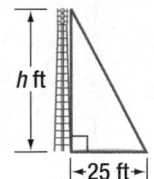 

Not to scale

More About Similar Figures

Two triangles are similar if their corresponding sides are proportional. This is called **Side-Side-Side Similarity (SSS)**. They are also similar if two of their corresponding angles are congruent. This is called **Angle-Angle Similarity (AA)**.

Are there any other ways to test for similarity? Try the following activity.

ACTIVITY

STEP 1 Draw a 30° angle that is formed by a segment measuring 3 centimeters and a segment measuring 5 centimeters.

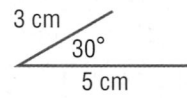

STEP 2 Draw a 30° angle that is formed by a segment measuring 4.5 centimeters and a segment measuring 7.5 centimeters.

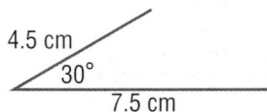

STEP 3 Form two triangles by connecting the endpoints of both pairs of segments. Measure the lengths of the formed segments.

STEP 4 Show that corresponding sides are proportional.

17. Are the two triangles similar? Justify your response.

Repeat Steps 1–3 using the following angle and segment measurements.

18. 60°; 1 inch, 1.5 inches
60°; 2 inches, 3 inches

19. 45°; 3 inches, 5 inches
45°; 9 inches, 15 inches

20. Are the triangles similar? Justify your response.

21. **MAKE A CONJECTURE** The activities suggest another way to test triangles for similarity. This is called **Side-Angle-Side Similarity (SAS)**. Write a sentence or two that explains Side-Angle-Side Similarity.

Main Idea

Find the relationship between perimeters and areas of similar figures.

Perimeter and Area of Similar Figures

Explore Suppose you double each dimension of the rectangle at the right. The new rectangle is similar to the original rectangle with a scale factor of 2.

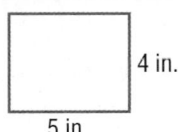

4 in.

5 in.

1. What is the perimeter of the original rectangle?

2. What is the perimeter of the new rectangle?

3. How is the perimeter of the new rectangle related to the perimeter of the original rectangle and the scale factor?

In similar figures, the perimeters are related by the scale factor. What about area? Consider the rectangles from the Explore activity.

Original Rectangle

$A = \ell w$

$A = 4 \cdot 5$

New Rectangle

$A = \ell w$

$A = (2 \cdot 4)(2 \cdot 5)$

$\quad = (2 \cdot 2)(4 \cdot 5)$

$\quad = 2^2(4 \cdot 5)$

> The scale factor, 2, is used as a factor twice.

The area of the new rectangle is equal to the area of the original rectangle times the *square* of the scale factor.

Key Concept — Perimeter and Area of Similar Figures

Perimeter		**Models**
Words	If figure B is similar to figure A by a scale factor, then the perimeter of B is equal to the perimeter of A times the scale factor.	*a* **Figure A**
Symbols	$\dfrac{\text{perimeter of}}{\text{figure } B} = \dfrac{\text{perimeter of}}{\text{figure } A} \cdot \text{scale factor}$	
Area		*b* **Figure B**
Words	If figure B is similar to figure A by a scale factor, then the area of B is equal to the area of A times the square of the scale factor.	
Symbols	$\dfrac{\text{area of}}{\text{figure } B} = \dfrac{\text{area of}}{\text{figure } A} \cdot (\text{scale factor})^2$	

Vocabulary Link

Everyday Use

Perimeter A line or strip protecting an area.

Math Use

Perimeter The distance around the outside of a closed figure.

 EXAMPLE Determine Perimeter

① Two rectangles are similar. One has a length of 6 inches and a perimeter of 24 inches. The other has a length of 7 inches. What is the perimeter of this rectangle?

The scale factor is $\frac{7}{6}$. The perimeter of the original is 24 inches.

$$x = 24\left(\frac{7}{6}\right) \qquad \text{Multiply by the scale factor.}$$

$$x = \frac{\overset{4}{24}}{1}\left(\frac{7}{\underset{1}{6}}\right) \qquad \text{Divide out common factors.}$$

$$x = 28 \qquad \text{Simplify.}$$

So, the perimeter of the new rectangle is 28 inches.

 CHECK Your Progress

a. Triangle *LMN* is similar to triangle *PQR*. If the perimeter of △*LMN* is 64 meters, what is the perimeter of △*PQR*?

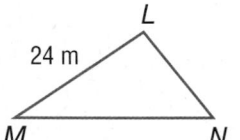
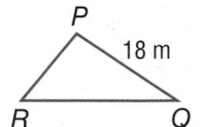

REAL-WORLD EXAMPLE Determine Area

② **CONSTRUCTION** The Eddingtons have a 5-foot by 8-foot porch on the front of their house. They are building a similar porch on the back with double the dimensions. Find the area of the back porch.

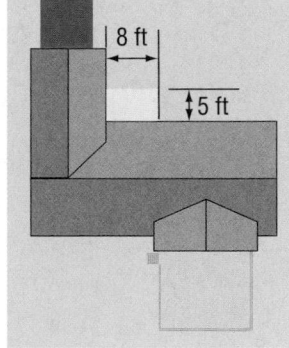

The scale factor is 2.

The area of the front porch is (5)(8) or 40 square feet.

$$x = 40(2)^2 \qquad \text{Multiply by the square of the scale factor.}$$

$$x = 40(4) \text{ or } 160 \qquad \text{Evaluate the power.}$$

The back porch will have an area of 160 square feet.

QUICK Review

Squaring Fractions

$$\left(\frac{1}{20}\right)^2 = \frac{1}{20} \cdot \frac{1}{20}$$

$$= \frac{1}{400}$$

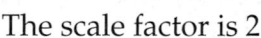 **CHECK Your Progress**

b. MURALS Malia is painting a mural on her bedroom wall. The image she is reproducing is $\frac{1}{20}$ of her wall and has an area of 36 square inches. Find the area of the mural.

✓ CHECK Your Understanding

Example 1   **For each pair of similar figures, find the perimeter of the second figure.**

1

12 mm   18 mm

$P = 38$ mm $P = ?$ mm

2.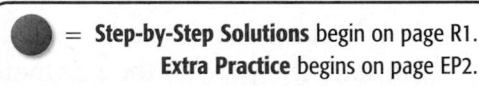

$P = 21$ ft

5 ft $P = ?$ ft

2 ft

Example 2   **3. DIGITAL PHOTOGRAPHY** Julie is enlarging a digital photograph on her computer. The original photograph is 5 inches by 7 inches. If she enlarges it 1.5 times, what will be the area, in square inches, of the new image?

Practice and Problem Solving

● = **Step-by-Step Solutions** begin on page R1.
Extra Practice begins on page EP2.

Example 1   **For each pair of similar figures, find the perimeter of the second figure.**

4.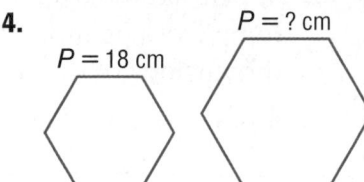

$P = ?$ cm

$P = 18$ cm

3 cm   4 cm

5.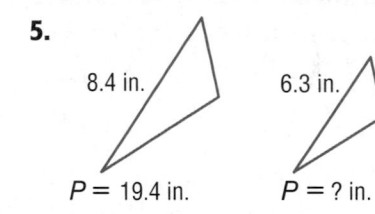

8.4 in.   6.3 in.

$P = 19.4$ in. $P = ?$ in.

6. INVITATIONS For your birthday party, you make a map to your house on a 3-inch-wide by 5-inch-long index card. What will be the perimeter of your map if you use a copier to enlarge it so it is 8 inches long?

Example 2   **7. MODEL TRAINS** Craig is making a model version of his neighborhood that uses model trains. The ratio of the model train to the actual train is 1 : 64. His neighborhood covers an area of 200,704 square feet. What will be the area of the model neighborhood?

8. ADVERTISING A company wants to reduce the size of its logo by one fourth to use on business cards. If the area of the original logo is 4 square inches, what is the area of the logo that will be used on the business cards?

9 GOLF Theo is constructing a miniature putting green in his backyard. He wants it to be similar to a putting green at the local golf course, but one third the size. The area of the putting green at the golf course is 1,134 square feet. What will be the area of the putting green Theo constructs?

10. LOGOS Mr. James is enlarging a logo for printing on the back of a T-shirt. He wants to enlarge a logo that is 3 inches by 5 inches so that the dimensions are 3 times larger than the original. How many times as large as the original logo will the area of the printing be?

H.O.T. Problems

11. **OPEN ENDED** Draw and label two similar figures with a scale factor of $\frac{5}{6}$. Find the perimeter and area of each figure.

12. **CHALLENGE** Two circles have circumferences of π and 3π. What is the ratio of the area of the circles? the diameters? the radii?

13. **WRITE MATH** A company wants to reduce the dimensions of its logo from 6 inches by 4 inches to 3 inches by 2 inches to use on business cards. Robert thinks that the new logo is $\frac{1}{4}$ the size of the original logo. Denise thinks that is $\frac{1}{2}$ of the original size. Explain their thinking.

Test Practice

14. Two rectangular pieces of wood are similar. The ratio of the perimeters of the two pieces is $2:3$. If the area of the smaller piece is 12 square inches, what is the area of the larger piece?

    A. 8 in²   **C.** 27 in²

    B. 18 in²   **D.** 36 in²

15. A photograph is enlarged to three times the size of the original. Which of the following statements is true?

    F. The area of the enlargement is three times the area of the original.

    G. The area of the enlargement is six times the area of the original.

    H. The area of the enlargement is nine times the area of the original.

    I. The area of the enlargement is twelve times the area of the original.

16. **THINK SOLVE EXPLAIN** **SHORT RESPONSE** A smaller copy of the 3-foot by 5-foot school flag at Brook Park Middle School is being made to appear on the front of the students' homework agenda books. The dimensions of the copy are to be $\frac{1}{6}$ of the school flag.

How many times larger is the area of the flag than the area of the copy?

Spiral Review

Find the value of x in each pair of similar figures. (Lesson 3A)

17.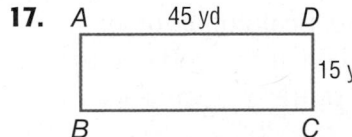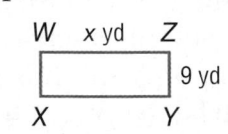

18.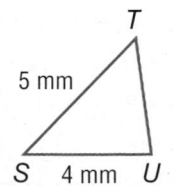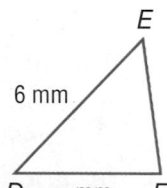

19. **GEOGRAPHY** Omaha, Nebraska, and Sioux City, Iowa, are 90 miles apart. If the distance on the map is $1\frac{1}{2}$ inches, find the scale of the map. (Lesson 2B)

Main Idea

Find the value of the golden ratio.

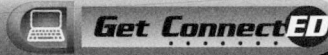

The Golden Rectangle

ACTIVITY

STEP 1   Cut out a rectangle that measures 34 units by 21 units. Find the ratio of the length to the width. Express it as a decimal to the nearest hundredth. Record your data in a table.

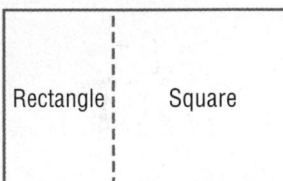

length	34	21			
width	21	13			
ratio					
decimal					

STEP 2   Cut this rectangle into two parts. One part is the largest possible square, and the other part is a rectangle. Record the rectangle's length and width. Write the ratio of length to width. Express it as a decimal to the nearest hundredth and record in the table.

Rectangle | Square

STEP 3   Repeat the procedure described in Step 2 until the remaining rectangle measures 3 units by 5 units.

Analyze the Results

1. Describe the pattern in the ratios you recorded.

2. **MAKE A CONJECTURE** If the rectangles you cut out are described as *golden rectangles*, what is the value of the *golden ratio*?

3. Write a definition for a golden rectangle. Use the word *ratio* in your definition. Then describe the shape of a golden rectangle.

4. Determine whether all golden rectangles are similar. Explain.

5. **MATH IN THE MEDIA** There are many examples of the golden rectangle in architecture. One is shown at the right. Use the Internet or another resource to find three places where the golden rectangle is used in architecture.

Problem Solving in Design Engineering

A Thrilling Ride!

If you have a passion for amusement parks, a great imagination, and enjoy building things, you might want to consider a career in roller coaster design. Roller coaster designers combine creativity, engineering, mathematics, and physics to develop rides that are both exciting and safe. In order to analyze data and make precise calculations, a roller coaster designer must have a solid background in high school math and science.

21st Century Careers

Are you interested in a career as a roller coaster designer? Take some of the following courses in high school.

- Algebra
- Calculus
- Geometry
- Physics
- Trigonometry

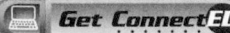 Get ConnectED

Real-World Math

Use the information in the table to solve each problem.

1. In a scale drawing of SheiKra, a designer uses a scale of 1 inch = 16 feet. What is the height of the roller coaster in the drawing?

2. On a model of Montu, the height of the loop is 13 inches. What is the scale?

3. In a scale drawing of Montu, the height of the roller coaster is 10 inches. What is the scale factor?

4. SheiKra has a hill that goes through a tunnel. On a model of the roller coaster, the hill is 23 inches tall and the scale is 1 inch = 6 feet. What is the actual height of the tunnel hill?

5. An engineer is building a model of SheiKra. She wants the model to be about 32 inches high. Choose an appropriate scale for the model. Then use it to find the loop height of the model.

Busch Gardens Africa

Roller Coaster	Coaster Height (ft)	Loop Height (ft)
SheiKra	200	145
Montu	150	104

FOLDABLES®
Study Organizer

Be sure the following Key Concepts are noted in your Foldable.

Rates

Proportional and Nonportional Relationships

Solve Proportions

Scale Dreawings

Similar Figures

Perimeter and Area of Similar Figures

Key Concepts

Proportions (Lesson 1)
- If two related quantities are proportional, then they have a constant ratio.
- The cross products of a proportion are equal.

Scale Drawings (Lesson 2)
- Scale drawings are used to represent objects that cannot be drawn at actual size.
- The scale is a ratio of a drawing's measurement to an object's measurement.

Similar Figures (Lesson 3)
- If figure *B* is similar to figure *A* by a scale factor, then

$$\frac{\text{perimeter of}}{\text{figure } B} = \frac{\text{perimeter of}}{\text{figure } A} \cdot \text{scale factor}$$

$$\frac{\text{area of}}{\text{figure } B} = \frac{\text{area of}}{\text{figure } A} \cdot (\text{scale factor})^2$$

a

b

Figure A   **Figure B**

Key Vocabulary

corresponding angles	rate
corresponding sides	scale
cross products	scale drawing
equivalent ratios	scale factor
indirect measurement	scale model
nonproportional	similar figures
proportion	unit rate
proportional	

Vocabulary Check

Choose the term from the list above that best matches each phrase.

1. two quantities that have a constant rate or ratio

2. a scale written as a ratio in simplest form without units of measurement

3. an equation that shows that two ratios or rates are equivalent

4. a ratio of two measurements with different units

5. a rate that is simplified so that it has a denominator of 1

6. two ratios that have the same value

7. the ratio of the distance on a map to the actual distance

8. figures with the same shape

9. a relationship in which the ratio is not constant

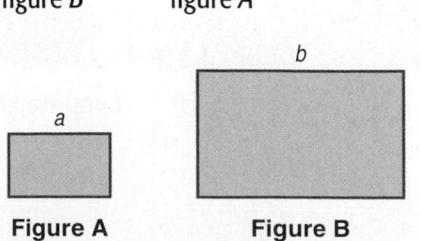

Multi-Part Lesson Review

Lesson 1 Proportions

Rates (Lesson 1B)

Find each unit rate.

10. $23.75 for 5 pounds

11. 810 miles in 9 days

12. **SHAMPOO** Which bottle of shampoo shown at the right costs the least per ounce?

Bottle	Price
8 oz	$1.99
12 oz	$2.59
16 oz	$3.19

EXAMPLE 1 Find the unit price of a 16-ounce box of pasta that is on sale for 96 cents.

$$16\text{-oz box for }96\text{ cents} = \frac{96 \text{ cents} \div 16}{16 \text{ ounces} \div 16}$$
$$= \frac{6 \text{ cents}}{1 \text{ ounce}}$$

The unit price is 6 cents per ounce.

Proportional and Nonproportional Relationships (Lesson 1C)

13. **INTERNET** An Internet company charges $30 a month. There is also a $30 installation fee. Is the number of months you can have Internet proportional to the total cost? Explain.

14. **WORK** On Friday, Jade washed 10 vehicles in 4 hours. The next day she washed 15 vehicles in 6 hours. Is the number of vehicles she washed proportional to the time it took her to wash them? Explain.

EXAMPLE 2 Is the amount of money earned proportional to the number of haircuts? Explain.

Earnings ($)	28	56	84	112
Haircuts	1	2	3	4

$$\frac{earnings}{haircuts} \longrightarrow \frac{28}{1}, \frac{56}{2}, \frac{84}{3}, \frac{112}{4}$$

Since these ratios are all equal to 28, the amount of money earned is proportional to the number of haircuts.

Solve Proportions (Lesson 1D)

Solve each proportion.

15. $\dfrac{3}{r} = \dfrac{6}{8}$

16. $\dfrac{30}{0.5} = \dfrac{y}{0.25}$

17. $\dfrac{7}{4} = \dfrac{n}{2}$

18. $\dfrac{k}{5} = \dfrac{72}{8}$

19. $\dfrac{2}{t} = \dfrac{8}{50}$

20. $\dfrac{12}{8} = \dfrac{a}{6}$

21. **SPEED** A squirrel can run 1 mile in 5 minutes. How far can it travel in 16 minutes?

22. **WEIGHT** If 3 televisions weigh 240.6 pounds, how much do 9 of the same televisions weigh?

EXAMPLE 3 Solve $\dfrac{9}{x} = \dfrac{4}{18}$.

$$\frac{9}{x} = \frac{4}{18} \qquad \text{Write the proportion.}$$
$$9 \cdot 18 = x \cdot 4 \qquad \text{Find the cross products.}$$
$$162 = 4x \qquad \text{Multiply.}$$
$$\frac{162}{4} = \frac{4x}{4} \qquad \text{Divide each side by 4.}$$
$$40.5 = x \qquad \text{Simplify.}$$

Lesson 2   Scale Drawings and Models

PSI: Draw a Diagram (Lesson 2A)

Solve each problem by drawing a diagram.

23. **WATER** A 500-gallon hot tub is being filled with water. Eighty gallons of water are in the hot tub after 4 minutes. At this rate, how long will it take to fill the hot tub?

24. **GEOMETRY** A diagonal is a line segment that connects two non adjacent vertices. Find the number of diagonals in an octagon.

EXAMPLE 4 Alvin has filled $\frac{1}{3}$ or 50 gallons of his fish tank. Find the total capacity of the fish tank.

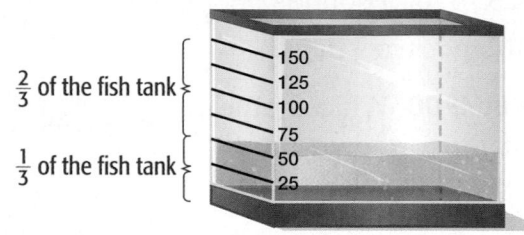

If $\frac{1}{3}$ of the fish tank is 50 gallons, then $\frac{2}{3}$ is 100 gallons. So, the total capacity of the fish tank is 50 + 100, or 150 gallons.

Scale Drawings (Lesson 2B)

The scale on a map is 3 centimeters = 7 kilometers. Find the actual distance for each map distance.

25. 9 cm   26. 21 cm

27. **ARCHITECTS** On an architect's blueprint, the dimensions of a room are 5 inches by 8 inches. If the actual dimensions of the room are 10 feet by 16 feet, what is the scale of the blueprint?

28. **MODELS** A Boeing 747 jet is 70.5 meters long and has a wingspan of 60 meters. A model of the 747 has a wingspan of 80 centimeters. What is the length of the model?

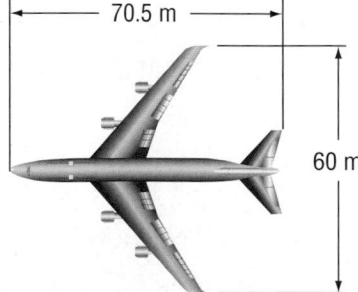

EXAMPLE 5 On a map, the distance between two cities is 10.9 centimeters. If the scale is 1 centimeter = 250 kilometers, what is the actual distance?

$$\text{map} \longrightarrow \frac{1 \text{ cm}}{250 \text{ km}} = \frac{10.9 \text{ cm}}{n \text{ km}} \longleftarrow \text{map}$$
$$\text{actual} \longrightarrow \qquad\qquad \longleftarrow \text{actual}$$
$$1 \cdot n = 250 \cdot 10.9$$
$$n = 2{,}725$$

The actual distance is 2,725 kilometers.

EXAMPLE 6 The scale on a model of a bullfrog is 2 centimeters = 25 millimeters. Find the actual length of the bullfrog if the model length is 11 centimeters.

$$\frac{2 \text{ cm}}{25 \text{ mm}} = \frac{11 \text{ cm}}{x \text{ mm}} \quad \begin{array}{l}\longleftarrow \text{model length}\\ \longleftarrow \text{actual length}\end{array}$$

$2 \cdot x = 25 \cdot 11$ Find the cross products.

$2x = 275$ Multiply.

$x = 137.5$ Divide each side by 2.

The actual length is 137.5 millimeters.

29. Which quadrilateral is similar to quadrilateral *ABCD*?

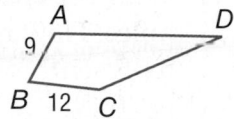

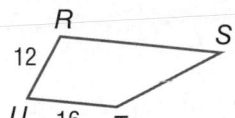

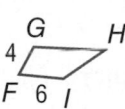

Find the value of *x* in each pair of similar figures.

30.

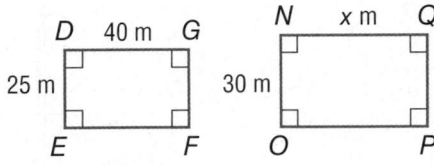

31.

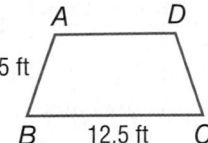

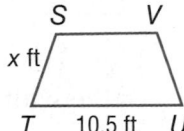

EXAMPLE 7 Find the value of *x* in the pair of similar figures.

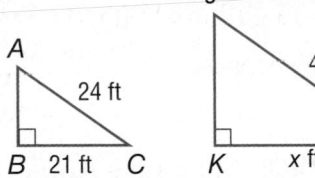

$$\frac{24 \text{ ft}}{42 \text{ ft}} = \frac{21 \text{ ft}}{x \text{ ft}} \quad \text{Write a proportion.}$$

$24 \cdot x = 42 \cdot 21$ Find the cross products.

$24x = 882$ Multiply.

$x = 36.75$ Divide each side by 24.

The length of $\overline{KL}$ is 36.75 feet.

Perimeter and Area of Similar Figures (Lesson 3B)

32. DOLLHOUSES Franklin is building a dollhouse modeled after his house for his niece. He would like the scale factor to be $\frac{1}{12}$. If the perimeter of his house is 1,800 inches, what will be the perimeter of the dollhouse?

33. MEASUREMENT Two similar triangles have a scale factor of $\frac{1}{5}$. Find the area of the larger triangle if the area of the smaller triangle is 6 square yards.

EXAMPLE 8 Cynthia is creating a poster of a photograph she took. The photograph is 5 inches by 7 inches. What will be the area of the poster if the poster has dimensions four times the photograph?

The scale factor relating the poster and the photograph is 4.

$$\begin{array}{ccc} \text{area of} & = & \text{area of} \\ \text{poster} & & \text{photo} \end{array} \cdot (\text{scale factor})^2$$

$$= \quad (5 \cdot 7) \quad \cdot \quad (4)^2$$

$$= \quad 560$$

The area of the poster is 560 square inches.

Practice Chapter Test

Find each unit rate. Round to the nearest hundredth if necessary.

1. 24 greeting cards for $4.80

2. 330 miles on 15 gallons of gas

3. 112 feet in 2.8 seconds

4. MULTIPLE CHOICE At Flynn's Apple Orchard, 16 acres of land produced 368 bushels of apples. Which rate represents the number of bushels per acre?

   A. $16:1$ **C.** $23:2$

   B. $23:1$ **D.** $46:1$

5. MEASUREMENT Nick rides his bike 20 miles every two days. Is the distance Nick rides proportional to the number of days? Explain.

Solve each proportion.

6. $\dfrac{3}{a} = \dfrac{9}{12}$ **7.** $\dfrac{5}{3} = \dfrac{20}{y}$

8. $\dfrac{2}{3} = \dfrac{x}{42}$ **9.** $\dfrac{t}{21} = \dfrac{15}{14}$

10. FOOD Of the 30 students in a life skills class, 19 like to cook main dishes, 15 like to bake desserts, and 7 like to do both. How many students like to cook main dishes but not bake desserts? Use the *draw a diagram* strategy.

Each pair of polygons is similar. Write a proportion to find each missing measure. Then solve.

11.

12.

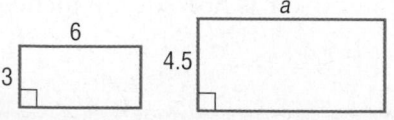

13. MULTIPLE CHOICE On a map, 1 inch = 7.5 miles. How many miles does 2.5 inches represent?

   F. $\dfrac{1}{3}$ **H.** 10

   G. 3   **I.** 18.75

14. BLUEPRINTS Use the following scale drawing of a room.

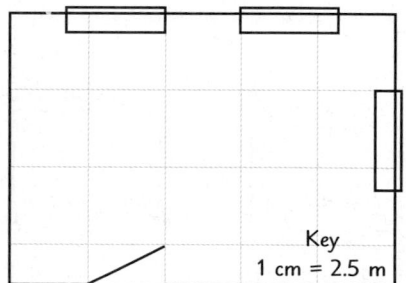

Key
1 cm = 2.5 m

   a. Use a centimeter ruler to find the length of the wall with two windows.

   b. How wide would a 1.4-meter-wide dresser appear on this drawing?

15. EXTENDED RESPONSE Layla is training for a marathon. Each week she runs about 40 miles. Her coach gave her a diagram of the route he would like her to run.

1 inch = 0.5 mile

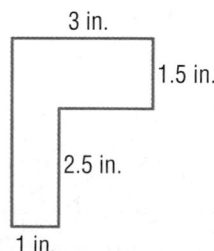

   Part A How many inches is the route on the diagram?

   Part B How many miles will Layla run in one loop of the route?

   Part C How many times per week will she need to complete the loop to run at least 40 miles?

310 Proportions and Similarity

Preparing for Standardized Tests

Multiple Choice: Unreasonable Answer Choices

Sometimes answer choices are not reasonable for the information given in a problem. These choices can be eliminated.

TEST EXAMPLE

A scenic trail circles Loon Lake. Hikers traveled on the trail from East Marina to Perry's Landing. This distance on a map is about 1 inch. Based on the scale below, how far did the hikers travel in miles?

Scale
$\frac{1}{2}$ in. = $6\frac{1}{4}$ mi

A. $2\frac{1}{2}$ miles   **C.** $6\frac{1}{4}$ miles

B. $3\frac{1}{8}$ miles   **D.** $12\frac{1}{2}$ miles

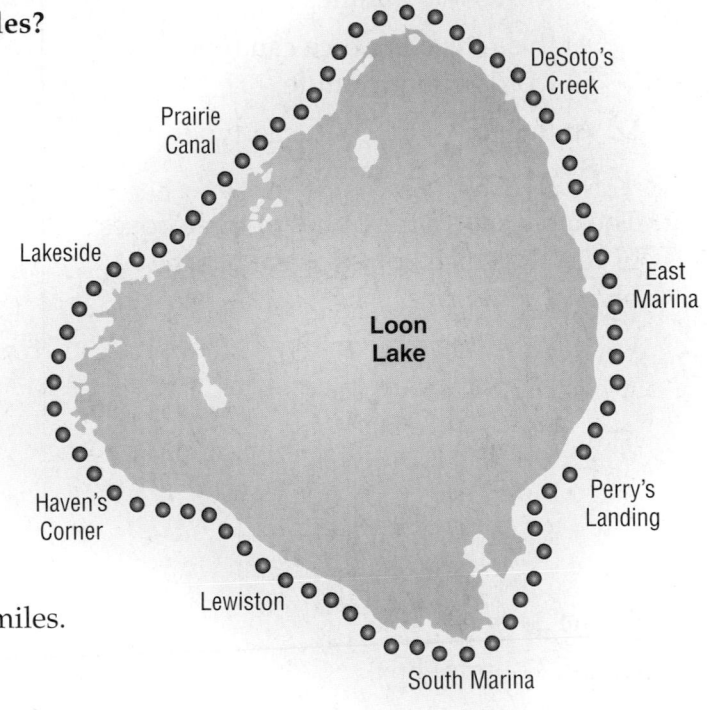

Since the distance on the map is greater than $\frac{1}{2}$ inch, the actual distance must be greater than 5 miles. So, choices A and B can be automatically eliminated.

Since 1 inch is twice as long as $\frac{1}{2}$ inch, the actual distance will be twice as long as $6\frac{1}{4}$ miles.

The correct answer is D.

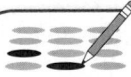 Work on It

On a model of a roller coaster, the first hill is 15 inches high. The actual height of the first hill is 240 feet. If the second hill on the roller coaster is 180 feet high, how high is the second hill on the model?

F. $11\frac{1}{4}$ inches   **H.** 16 inches

G. 12 inches   **I.** 20 inches

Test Hint

If your answer is not one of the choices given, then reread the question carefully to check that you understand what is being asked.

✓ Test Practice

Read each question. Then fill in the correct answer on the answer sheet provided by your teacher or on a sheet of paper.

1. Francesca typed 496 words in 8 minutes. Which of the following is a correct understanding of this rate?

   A. At this rate, it takes 62 minutes for Francesca to type one word.

   B. At this rate, Francesca can type 62 words in 8 minutes.

   C. At this rate, Francesca can type 62 words in one minute.

   D. At this rate, Francesca can type 8 words in one minute.

2. The table shows the prices of three boxes of cereal. Which box of cereal has the highest unit price?

Cereal Box Size (ounces)	Price ($)
48	5.45
32	3.95
20	3.10

   F. the 20-ounce box

   G. the 32-ounce box

   H. the 48-ounce box

   I. All three boxes have the same unit price.

3. ✎ GRIDDED RESPONSE A bakery sells 6 bagels for $2.99 and 4 muffins for $3.29. What is the total cost in dollars of 4 dozen bagels and 16 muffins, not including tax?

4. THINK SOLVE EXPLAIN SHORT RESPONSE A teacher plans to buy 5 pencils for each student in her class. Pencils come in packages of 18 and cost $1.99 per package. What other information is needed to find the cost of the pencils?

5. During a 3-hour period, 2,292 people rode the roller coaster at an amusement park. Which proportion can be used to find x, the number of people who rode the coaster during a 12-hour period, if the rate is the same?

   A. $\dfrac{3}{2{,}292} = \dfrac{x}{12}$ C. $\dfrac{3}{x} = \dfrac{12}{2{,}292}$

   B. $\dfrac{3}{2{,}292} = \dfrac{12}{x}$ D. $\dfrac{x}{3} = \dfrac{12}{2{,}292}$

6. An architect created the scale drawing below showing a wall of a child's playhouse.

   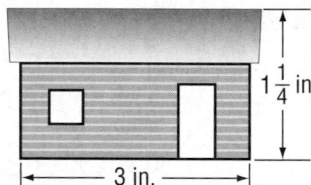

   Which of these was the scale used for the drawing if the actual height of the wall is $7\frac{1}{2}$ feet?

   F. 1 in. = 1 ft   H. 2 in. = 12 ft

   G. $\frac{1}{2}$ in. = 1 ft   I. $\frac{1}{4}$ in. = 1 ft

7. THINK SOLVE EXPLAIN SHORT RESPONSE You can drive your car 21.7 miles with one gallon of gasoline. At that rate, how many miles can you drive with 13.2 gallons of gasoline?

8. At 8 A.M., the temperature was 13°F below zero. The temperature rose 22°F by 1 P.M. and dropped 14°F by 6 P.M. What was the temperature at 6 P.M.?

   A. 5°F above zero

   B. 5°F below zero

   C. 21°F above zero

   D. 21°F below zero

9. Trapezoid *ABCD* is similar to trapezoid *WXYZ*. What is the length of $\overline{XY}$?

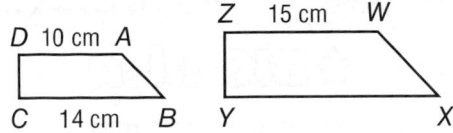

   F. 20 cm   **H.** 24 cm

   G. 21 cm   **I.** 27 cm

10. **GRIDDED RESPONSE** Find $15 \div (-5)$.

11. The bridge structure is supported by the triangular braces as shown.

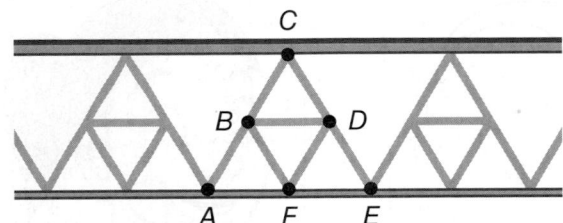

Triangles *ACE* and *ABF* are similar triangles. The scale factor is 0.5. If *CE* = 10 feet, what is the length of *BF*?

   A. 2.5 ft   **C.** 6 ft

   B. 5 ft   **D.** 12 ft

12. **SHORT RESPONSE** Thom has a scale model of his car. The scale is 1 : 12. If the actual car has 16-inch wheels, what size are the wheels on the scale model?

16 in.

13. A recipe calls for $2\frac{1}{3}$ packages of pudding. How many batches can be made if 20 packages of pudding are available?

   F. 8 batches

   G. 9 batches

   H. 10 batches

   I. 11 batches

14. Which line contains the ordered pair $(-1, 2)$?

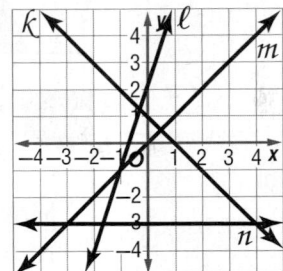

   A. line *k*

   B. line ℓ

   C. line *m*

   D. line *n*

15. **EXTENDED RESPONSE** Heath and Jerome are brothers. At the beginning of the school year, Heath wore a size 3 shoe and Jerome wore a size 5. At the end of the school year, Heath wore a size 4 and Jerome wore a size 6.

   Part A Jerome says they grew the same amount. Explain how he is right.

   Part B Heath thinks that he grew more. Explain how he is also correct.

NEED EXTRA HELP?															
If You Missed Question...	1	2	3	4	5	6	7	8	9	10	11	12	13	14	15
Go to Chapter-Lesson...	5-1B	5-1B	5-1D	5-1D	5-1D	5-2B	5-1D	2-2B	5-3A	2-3D	5-3A	5-2B	3-3D	2-1C	5-1C

Are You Ready for the Chapter?

You have two options for checking prerequisite skills for this chapter.

Text Option Take the Quick Check below. Refer to the Quick Review for help.

QUICK Check

Multiply.

1. $300 \times 0.02 \times 8$ **2.** $85 \times 0.25 \times 3$

3. $560 \times 0.6 \times 4.5$ **4.** $154 \times 0.12 \times 5$

5. MONEY If Nicole saves $0.05 every day, how much money will she have in 3 years?

Write each percent as a decimal.

6. 40%   **7.** 17%   **8.** 110%

9. 157%   **10.** 3.25%   **11.** 8.5%

12. FOOD Approximately 92% of a watermelon is water. What decimal represents this amount?

13. TAX A county sales tax is 7.5%. Write this percent as a decimal.

Write each decimal as a percent.

14. 0.7   **15.** 0.08   **16.** 0.95

17. 5.8   **18.** 0.675   **19.** 0.725

20. SPORTS A tennis player won 0.805 of the matches she played. What percent of the matches did she lose?

QUICK Review

EXAMPLE 1

Evaluate $240 \times 0.03 \times 5$.

$240 \times 0.03 \times 5$
$= 7.2 \times 5$ Multiply 240 by 0.03.
$= 36$ Simplify.

EXAMPLE 2

Write 9.8% as a decimal.

$9.8\% = 0.098$ Move the decimal point two places to the left and remove the percent symbol.

EXAMPLE 3

Write 0.35 as a percent.

$0.35 = 35\%$ Move the decimal point two places to the right and add the percent symbol.

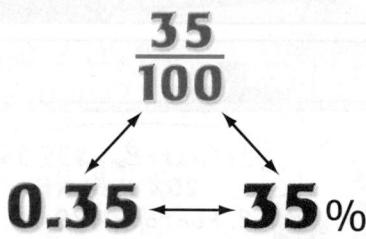

EXAMPLE 4

Write 0.2 as a percent.

$0.20 = 20\%$ Move the decimal point two places to the right and add the percent symbol.

 **Online Option** Take the Online Readiness Quiz.

Studying Math

Draw a Picture

Drawing a picture can help you better understand numbers in a word problem. For example, a *number map* can show how numbers are related to each other. Start by placing a number in the center of the map.

Have you heard the expression *a picture is worth a thousand words?*

Below is a number map that shows various meanings of the decimal 0.5. Notice that you can add both mathematical meanings and everyday meanings to the number map.

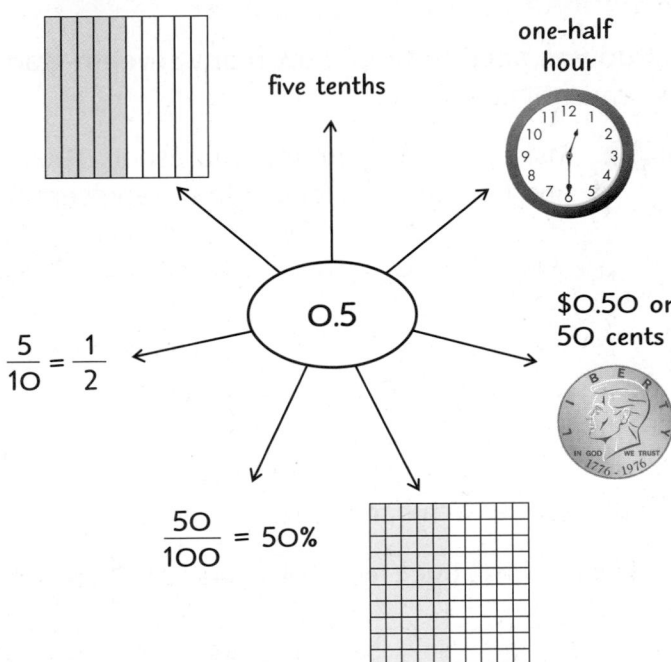

five tenths

one-half hour

0.5

$\frac{5}{10} = \frac{1}{2}$

$\$0.50$ or 50 cents

$\frac{50}{100} = 50\%$

Practice

Make a number map for each number. (*Hint*: For whole numbers, think of factors, prime numbers, divisibility, place value, and so on.)

1. 0.75   **2.** 0.1   **3.** 0.01

4. 1.25   **5.** 2.5   **6.** 25

7. 45   **8.** 60   **9.** 100

10. Refer to Exercise 1. Explain how each mathematical or everyday meaning on the number map relates to the decimal 0.75.

Explore Percent Diagrams

Main Idea

Use percent diagrams to solve problems.

Get ConnectED

MUSIC There are 500 seventh-grade students at Heritage Middle School. Sixty percent of them play a musical instrument. How many seventh-grade students play a musical instrument?

ACTIVITY

What do you need to find? how many seventh-grade students play a musical instrument

STEP 1 Make a bar diagram that represents 100%. Make another bar of equal length to represent 500 students.

percent		100%
students		500

STEP 2 Divide each bar into ten equal parts.

percent	10%	10%	10%	10%	10%	10%	10%	10%	10%	10%	100%
students	50	50	50	50	50	50	50	50	50	50	500

STEP 3 Shade 60% on the percent bar and an equal amount on the student bar.

|----------- 60% -----------|

percent	10%	10%	10%	10%	10%	10%	10%	10%	10%	10%	100%
students	50	50	50	50	50	50	50	50	50	50	500

|----------- 300 -----------|

Since 60% corresponds to the number 300, there are 300 seventh-grade students who play a musical instrument.

Analyze the Results

1. How many seventh-grade students do *not* play a musical instrument?

2. Suppose there had been 750 seventh-grade students.

   a. What would you do differently in Step 2?

   b. What is 60% of 750?

Practice and Apply

3. **SCHOOL SPORTS** The seventh-grade class at Fort Couch Middle School had a goal of selling 300 tickets to the annual student-teacher basketball game. The eighth-grade class had a goal of selling 400 tickets.

   a. By the end of the first week, the eighth-grade students sold 30% of their goal. Draw a bar diagram to represent this situation.

   b. How many tickets had the eighth grade sold?

   c. The seventh grade sold 40% of their goal. How many tickets do the students still need to sell? Use a bar diagram to justify your solution.

4. **ALLOWANCES** The graph shows the results of a survey asking 500 teens about their allowances.

   a. Draw a bar diagram to represent how many of the teens received between $10 and $20 as a weekly allowance. (*Hint*: Since $75\% = \frac{3}{4}$, divide the bar diagram into fourths.)

   b. How many teens did *not* receive between $10 and $20? Justify your solution.

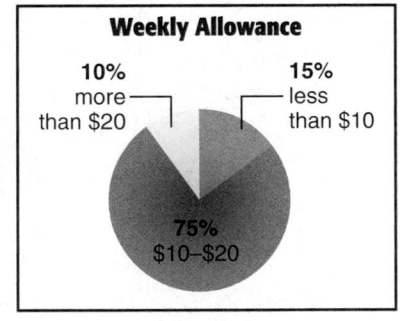

5. **SOCIAL STUDIES** Reggie has memorized 60% of the 50 state capitals for a social studies test.

   a. Draw a bar diagram to represent how many state capitals Reggie has memorized.

   b. How many more capitals does he need to memorize before the test?

WRITE MATH Write a real-world problem for each bar diagram. Then solve your problem.

6.

10%	10%	10%	10%	10%	10%	10%	10%	10%	10%	100%
25	25	25	25	25	25	25	25	25	25	250

7.

25%	25%	25%	25%	100%
15	15	15	15	60

Main Idea
Find the percent of a number.

 7.RP.3, 7.EE.3

Percent of a Number

 PETS Some students are collecting money for a local pet shelter. The model shows that they have raised 60% of their $2,000 goal or $1,200.

1. Sketch the model and label using decimals instead of percents.

2. Sketch the model using fractions instead of percents.

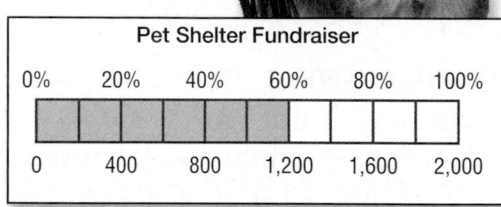

Pet Shelter Fundraiser

| 0% | 20% | 40% | 60% | 80% | 100% |

| 0 | 400 | 800 | 1,200 | 1,600 | 2,000 |

3. Use these models to write two multiplication sentences that are equivalent to 60% of 2,000 = 1,200.

To find the percent of a number such as 60% of 2,000, you can use either of the following methods.

• Write the percent as a fraction and then multiply.

• Write the percent as a decimal and then multiply.

EXAMPLE **Find the Percent of a Number**

1 Find 5% of 300.

Method 1   Write the percent as a fraction.

$5\% = \dfrac{5}{100}$ or $\dfrac{1}{20}$

$\dfrac{1}{20}$ of $300 = \dfrac{1}{20} \times 300$ or 15

Method 2   Write the percent as a decimal.

$5\% = \dfrac{5}{100}$ or 0.05

0.05 of $300 = 0.05 \times 300$ or 15

So, 5% of 300 is 15.

 CHOOSE Your Method

Find the percent of each number.

a. 40% of 70   **b.** 15% of 100   **c.** 55% of 160

 **EXAMPLE** **Use Percents Greater Than 100%**

Study Tip

Check for Reasonableness 120% is a little more than 100%. So, the answer should be a little more than 100% of 75 or a little more than 75.

(2) Find 120% of 75.

Method 1 Write the percent as a fraction.

$120\% = \frac{120}{100}$ or $\frac{6}{5}$

$\frac{6}{5}$ of $75 = \frac{6}{5} \times 75$

$= \frac{6}{5} \times \frac{75}{1}$ or 90

Method 2 Write the percent as a decimal.

$120\% = \frac{120}{100}$ or 1.2

1.2 of $75 = 1.2 \times 75$ or 90

So, 120% of 75 is 90. Use a model to check the answer.

 CHOOSE Your Method

Find each number.

d. 150% of 20   **e.** 160% of 35

 REAL-WORLD EXAMPLE

(3) **SURVEYS** Refer to the graph. If 275 students took the survey, how many can be expected to have 3 televisions each in their houses?

To find 23% of 275, write the percent as a decimal. Then multiply.

23% of $275 = 23\% \times 275$

$= 0.23 \times 275$

$= 63.25$

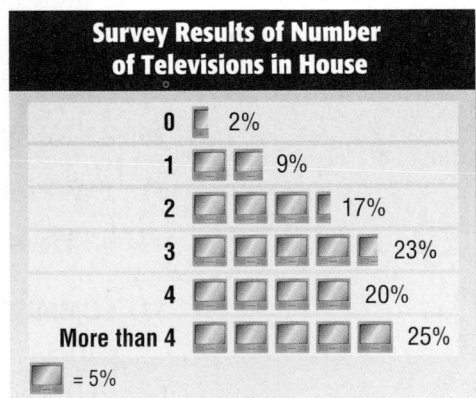

Survey Results of Number of Televisions in House	
0	2%
1	9%
2	17%
3	23%
4	20%
More than 4	25%
= 5%	

So, about 63 students can be expected to have 3 televisions each in their houses.

QUICK Review

Multiply Decimals

275 ← 0 decimal places
$\times\ 0.23$ ← 2 decimal places
825
$+ 5500$
63.25 ← 2 decimal places

 CHECK Your Progress

f. SURVEYS Refer to the graph above. Suppose 455 students took the survey. How many can be expected to have more than 4 televisions each in their houses?

✓ CHECK Your Understanding

Examples 1 and 2 Find each number. Round to the nearest tenth if necessary.

1. 8% of 50   **2.** 95% of 40   **3.** 42% of 263

4. 110% of 70   **5.** 115% of 20   **6.** 130% of 78

Example 3 **7. TAXES** Mackenzie wants to buy a new backpack that costs $50. If the tax rate is 6.5%, how much tax will she pay when she buys the backpack?

Practice and Problem Solving

 = **Step-by-Step Solutions** begin on page R1.
Extra Practice begins on page EP2.

Examples 1 and 2 Find each number. Round to the nearest tenth if necessary.

8. 65% of 186   **9.** 45% of $432   **10.** 23% of $640

11. 54% of 85   **12.** 12% of $230   **13.** 98% of 15

14. 130% of 20   **15** 175% of 10   **16.** 150% of 128

17. 250% of 25   **18.** 108% of $50   **19.** 116% of $250

20. 32% of 4   **21.** 5.4% of 65   **22.** 23.5% of 128

23. 75.2% of 130   **24.** 67.5% of 76   **25.** 18.5% of 500

Example 3 **26. BASEBALL** Tomás got on base 60% of the times he was up to bat. If he was up to bat 5 times, how many times did he get on base?

27. TELEVISION In a recent year, 17.7% of households watched the finals of a popular reality series. There are 110.2 million households in the United States. How many households watched the finals?

Find each number. Round to the nearest hundredth if necessary.

28. $\frac{4}{5}$% of 500   **29.** $5\frac{1}{2}$% of 60   **30.** $20\frac{1}{4}$% of 3

31. 1,000% of 99   **32.** 100% of 79   **33.** 520% of 100

34. 0.15% of 250   **35.** 0.3% of 80   **36.** 0.28% of 50

37. COMMISSION In addition to her salary, Ms. Lopez earns a 3% *commission*, or fee paid based on a percent of her sales, on every vacation package that she sells. One day, she sold the three vacation packages shown. What was her total commission?

Package #1	$2,375
Package #2	$3,950
Package #3	$1,725

38. INTERNET A family pays $19 each month for Internet access. Next month, the cost will increase by 5%. After this increase, what will be the cost for the Internet access?

39. BUSINESS A store sells a certain brand of lawn mower for $275. Next year, the cost of the lawn mower will increase by 8%. What will be the cost of the lawn mower next year?

40. RADIO The graph below shows the results of a poll of 2,632 listeners. Round answers to the nearest whole number.

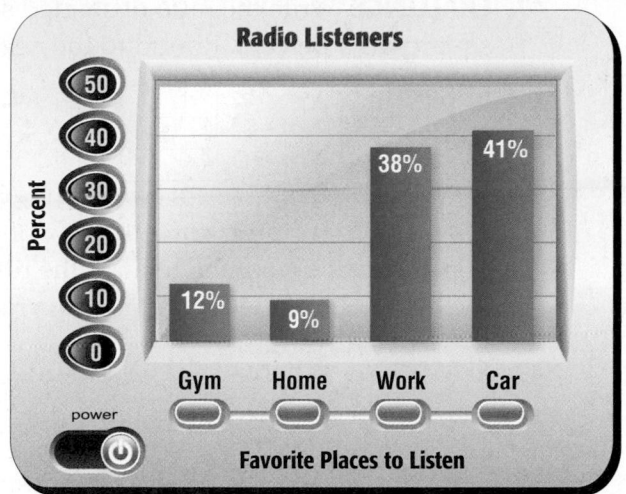

Radio Listeners

Favorite Places to Listen

a. How many people listen to the radio during work?

b. How many people like to listen to the radio while they are at the gym?

c. How many more people listen to the radio in the car than at home?

Use mental math to estimate each percent.

41. 53% of 60

42. 24% of 48

43. 75% of 19

44. FRUIT The table shows the results of a survey of 250 people about their favorite fruit.

a. Of those surveyed, how many people prefer peaches?

b. Which type of fruit did more than 100 people prefer?

c. Of those surveyed, how many people did *not* prefer cherries? Explain your answer.

Favorite Fruit	
Berries	44%
Peaches	32%
Cherries	24%

45 SCHOOL Suppose there are 20 questions on a multiple-choice test. If 25% of the answers are choice B, how many of the answers are *not* choice B?

46. GRAPHIC NOVEL Refer to the graphic novel frame below. Find the dollar amount of the group discount each student would receive at each park.

Marisol, Blake, and I want to get the best deal for our class.

1. Pirate Bay $35.95
   20% discount
2. Funtopia $29.75
   15% discount
3. Zoomland $38.49
   25% discount

First, let's figure out what each student would be able to save on their ticket at each park.

47. CHALLENGE Write the ratio of shaded squares to the total number of squares as a percent. Round to the nearest tenth if necessary.

a. b. c. d.

48. CHALLENGE Suppose you add 10% of a number to the number, and then you subtract 10% of the total. Is the result *greater than*, *less than*, or *equal to* the original number? Explain your reasoning.

49. REASONING When is it easiest to find the percent of a number using a fraction? using a decimal?

50. WRITE MATH Give two examples of real-world situations in which you would find the percent of a number.

Test Practice

51. Marcos earned $300 mowing lawns this month. Of his earnings, he plans to spend 18% repairing lawn equipment, put 20% in his savings, and use 35% for camp fees. He will spend the rest. How much will Marcos have left to spend?

A. $27.00 C. $81.00

B. $55.00 D. $100.00

52. GRIDDED RESPONSE Tanner has 200 baseball cards. Of those, 42% are in mint condition. How many of the cards are in mint condition?

53. The table shows the results of a survey of 200 movie rental customers.

Favorite Type of Movie	Percent of Customers
Comedy	15
Mystery	10
Horror	46
Science Fiction	29

How many customers prefer horror movies?

F. 20 H. 46

G. 30 I. 92

54. The graph shows the Ramirez family budget. Their budget is based on a monthly income of $3,000.

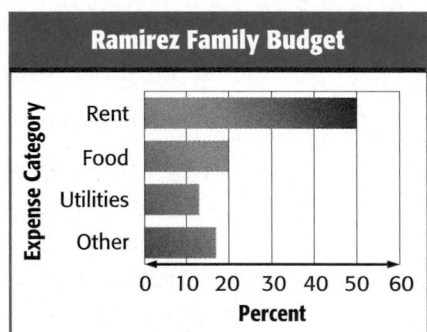

Which of the following statements is *true*?

A. The family budgeted $1,000 for rent.

B. The family budgeted $600 for food.

C. The family budgeted $100 less for utilities than for other expenses.

D. The family budgeted $900 more for food than for rent.

55. GRIDDED RESPONSE Jin Li has 310 cards in his card collection. Of his cards, 40% are football cards. How many football cards does Jin Li have?

Main Idea

Estimate percents by using fractions and decimals.

 7.RP.3, 7.EE.3

Percent and Estimation

MUSIC Refer to the graph below.

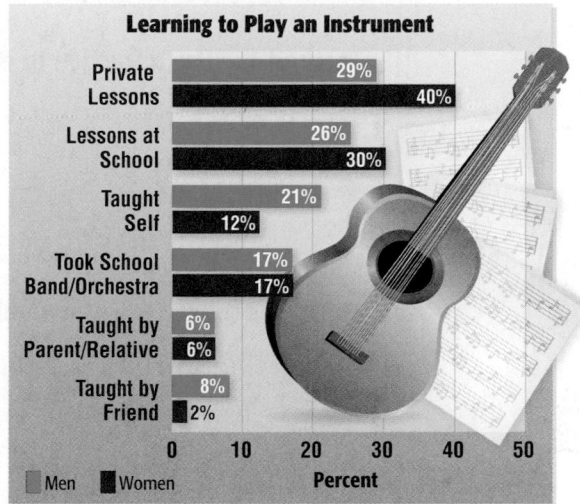

Learning to Play an Instrument

Private Lessons 29% / 40%
Lessons at School 26% / 30%
Taught Self 21% / 12%
Took School Band/Orchestra 17% / 17%
Taught by Parent/Relative 6% / 6%
Taught by Friend 8% / 2%

0 10 20 30 40 50 **Percent**

■ Men ■ Women

1. What fraction of women took lessons at school? If 200 women were surveyed, how many of them took lessons at school?

2. Use a fraction to estimate the number of men who took lessons at school. Assume 200 men were surveyed.

Sometimes an exact answer is not needed when using percents. One way to estimate the percent of a number is to use a fraction.

REAL-WORLD EXAMPLE

① **SPORTS** In a recent year, quarterback Carson Palmer completed 62% of his passes. He threw 520 passes. About how many did he complete?

62% of $520 \approx 60\%$ of 520 $62\% \approx 60\%$

$\approx \dfrac{3}{5} \cdot 520$ $60\% = \dfrac{6}{10}$ or $\dfrac{3}{5}$

≈ 312 Multiply.

So, Carson Palmer completed about 312 passes.

QUICK Review

Multiplying Fractions

Remember to divide out common factors.

$\dfrac{3}{5} \times 520 = \dfrac{3}{\cancel{5}} \times \dfrac{\cancel{520}^{104}}{1}$

$= \dfrac{3}{1} \times \dfrac{104}{1}$

$= 312$

CHECK Your Progress

a. **REPTILES** Box turtles have been known to live for 120 years. American alligators have been known to live 42% as long as box turtles. About how long can an American alligator live?

Another method for estimating the percent of a number is first to find 10% of the number and then multiply. For example, $70\% = 7 \cdot 10\%$. So, 70% equals 7 times 10% of the number.

REAL-WORLD EXAMPLE

2 **MONEY** Marita and four of her friends ordered a pizza that cost $14.72. She is responsible for 20% of the bill. About how much money will she need to pay?

Method 1   Use a fraction to estimate.

20% is $\frac{2}{10}$ or $\frac{1}{5}$.

20% of $\$14.72 \approx \frac{1}{5} \cdot \$15.00$ $20\% = \frac{1}{5}$ and round $14.72 to $15.00.

$\approx \$3.00$ Multiply.

Method 2   Use 10% of a number to estimate.

Step 1   Find 10% of $15.00.

10% of $\$15.00 = 0.1 \cdot \$15.00$ To multiply by 10%, move the
$= \$1.50$ decimal point one place to the left.

Step 2   Multiply.

20% of $15.00 is 2 times 10% of $15.00.

$2 \cdot \$1.50 = \$3.00$

So, Marita should pay about $3.00.

CHOOSE Your Method

b. MONEY Dante plans to put 80% of his paycheck into a savings account. His paycheck this week was $295. About how much money will he put into his savings account?

You can also estimate percents of numbers when the percent is greater than 100 or less than 1.

EXAMPLES Percents Greater Than 100 or Less Than 1

3 **Estimate 122% of 50.**

122% is about 120%.

120% of $50 = (100\% \text{ of } 50) + (20\% \text{ of } 50)$ $120\% = 100\% + 20\%$

$= (1 \cdot 50) + \left(\frac{1}{5} \cdot 50\right)$ $100\% = 1$ and $20\% = \frac{1}{5}$

$= 50 + 10$ or 60 Simplify.

So, 122% of 50 is about 60.

Estimate $\frac{1}{4}$% of 589.

$\frac{1}{4}$% is one fourth of 1%. 589 is about 600.

1% of $600 = 0.01 \cdot 600$ Write 1% as 0.01.

$\quad\quad\quad\quad\quad = 6$ To multiply by 1%, move the decimal point two places to the left.

One fourth of 6 is $\frac{1}{4} \cdot 6$ or 1.5. So, $\frac{1}{4}$% of 589 is about 1.5.

 CHECK Your Progress

Estimate.

c. 174% of 200   **d.** 298% of 45   **e.** 0.25% of 789

Real-World Link · · · ·
The American Camp Association (ACA) reports that 10 million children attend camp annually.

REAL-WORLD EXAMPLE

5 **CELL PHONES** In a recent year, there were about 200 million people in the U.S. with cell phones. Of those, about 0.5% used their phone as an MP3 player. Estimate the number of people who used their phone as an MP3 player.

0.5% is half of 1%.

1% of 200 million $= 0.01 \cdot 200{,}000{,}000$

$\quad\quad\quad\quad\quad\quad\quad\quad = 2{,}000{,}000$

So, 0.5% of 200,000,000 is about $\frac{1}{2}$ of 2,000,000 or 1,000,000.

About 1,000,000 people used their phone as an MP3 player.

 CHECK Your Progress

f. ATTENDANCE Last year, 639 students attended a summer camp. Of those who attended this year, 0.9% also attended last year. About how many students attended the camp two years in a row?

 CHECK Your Understanding

Examples 1–4   **Estimate.**

1. 52% of 10   **2.** 7% of 20   **3** 38% of 62

4. 79% of 489   **5.** 151% of 70   **6.** $\frac{1}{2}$% of 82

7. BIRTHDAYS Of the 78 teenagers at a youth camp, 63% have birthdays in the spring. About how many have birthdays in the spring?

Example 5   **8. GEOGRAPHY** About 0.8% of the land in Maine is federally owned. If Maine has 19,847,680 acres, about how many acres are federally owned?

Practice and Problem Solving

● = **Step-by-Step Solutions** begin on page R1.
Extra Practice begins on page EP2.

Example 1 **Estimate.**

9. 47% of 70   **10.** 39% of 120   ⑪ 21% of 90

12. 76% of 180   **13.** 57% of 29   **14.** 92% of 104

15. 24% of 48   **16.** 28% of 121   **17.** 88% of 207

18. 62% of 152   **19.** 65% of 152   **20.** 72% of 238

Example 2 **21. MONEY** Carlie spent $42 at the hair salon. Her mother loaned her the money. Carlie will pay her mother 15% each week until she has repaid the loan. About how much will Carlie pay each week?

22. HEALTH You use 43 muscles to frown. When you smile, you use 32% of these same muscles. About how many muscles do you use when you smile?

Examples 3 and 4 **Estimate.**

23. 132% of 54   **24.** 224% of 320   **25.** $\frac{1}{2}$% of 412

26. $\frac{3}{4}$% of 168   **27.** 0.4% of 510   **28.** 0.9% of 74

Example 5 **29. GEOGRAPHY** The United States has 12,383 miles of coastline. If 0.8% of the coastline is located in Georgia, about how many miles of coastline are in Georgia?

30. BIRDS During migration, 450,000 sandhill cranes stop to rest in Nebraska. About 0.6% of these cranes also stop to rest in Oregon. About how many sandhill cranes stop in Oregon during migration?

Estimate.

31. 67% of 8.7   **32.** 54% of 76.8   **33.** 32% of 89.9

34. 10.5% of 238   **35.** 22.2% of 114   **36.** 98.5% of 45

㊲ **ANALYZE GRAPHS** Use the graph shown.

a. About how many hours does Avery spend doing her homework each day?

b. About how many more hours does Avery spend sleeping than doing the activities in the "other" category? Justify your answer.

c. What is the approximate number of minutes Avery spends each day on extracurricular activities?

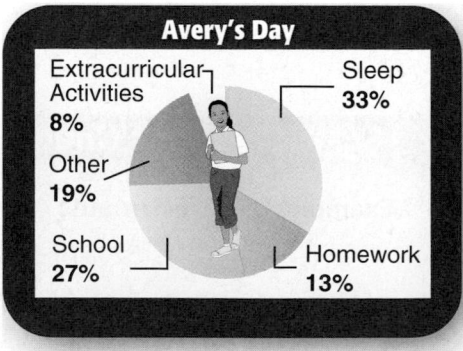

38. ANIMALS The average white rhinoceros gives birth to a single calf that weighs about 3.8% as much as its mother. If the mother rhinoceros weighs 3.75 tons, about how many pounds does its calf weigh?

🔊 **328** Percent

Real-World Link · · · ·

The NFL calculates a quarterback's pass rating using a formula based on completions, passing yards, interceptions, and passing touchdowns. Terry Bradshaw had a perfect passer rating in 1976.

··**39. FOOTBALL** The table shows the number of passes attempted and the percent completed by the top quarterbacks in the NFL for a recent season.

NFL Quarterbacks		
Player	Passes Attempted	Percent Completed
T. Brady	578	69
P. Manning	515	65
B. Roethlisberger	404	65
T. Romo	520	64
D. Garrard	325	64

  a. You can estimate the number of passes that Tom Brady completed by rounding 578 to 600 and 69% to 70%. Draw a bar diagram to represent this situation.

  b. Estimate the number of passes that Tom Brady completed.

  c. Is your estimate greater or less than the actual number of passes he completed? Explain.

  d. Without calculating, determine whether Ben Roethlisberger or Peyton Manning completed more passes. Justify your reasoning.

40. CLEANING A cleaning solution is made up of 0.9% chlorine bleach.

  a. About how many ounces of chlorine bleach are in 189 ounces of cleaning solution?

  b. About how many ounces of chlorine bleach would be found in 412 ounces of cleaning solution?

41. FOOD DRIVE The students at Monroe Junior High sponsored a canned food drive. The seventh-grade class collected 129% of its canned food goal.

  a. About how many canned foods did the seventh-graders collect if their goal was 200 cans?

  b. About how many canned foods did the seventh-graders collect if their goal was 595 cans?

42. COASTLINE The coastline of the Atlantic Coast is 2,069 miles long. Approximately $\frac{6}{10}$% of the coastline lies in New Hampshire. About how many miles of the coastline lie in New Hampshire? Explain how you estimated.

43. MP3 PLAYER The bar diagram shows the amount of money Marquis saved during a 4-month period to buy an MP3 player.

$60 in all			
Jan.	Feb.	Mar.	Apr.

The cost of the MP3 player is 130% of the amount shown in the bar diagram.

  a. Estimate the cost of the MP3 player.

  b. Is your estimate greater or lesser than the actual cost of the MP3 player? Explain.

H.O.T. Problems

44. **CHALLENGE** Explain how you could find $\frac{3}{8}$% of $800.

45. **FIND THE ERROR** Jamar is estimating 1.5% of 210. Find his mistake and correct it.

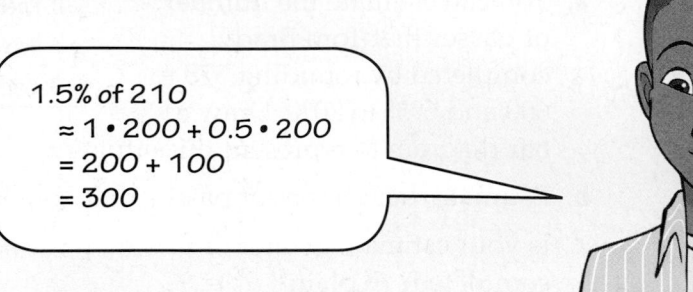

> 1.5% of 210
> ≈ 1 • 200 + 0.5 • 200
> = 200 + 100
> = 300

46. **NUMBER SENSE** Is an estimate for the percent of a number *always*, *sometimes*, or *never* greater than the actual percent of the number? Give an example or a counterexample to support your answer.

47. **WRITE MATH** Estimate 22% of 136 using two different methods. Justify the steps used in each method.

Test Practice

48. The graph shows the results of a survey of 510 students.

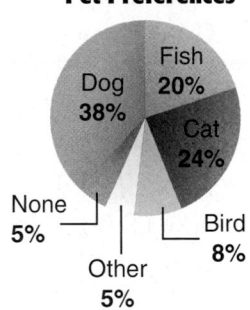

Pet Preferences

Fish 20%
Dog 38%
Cat 24%
None 5%
Other 5%
Bird 8%

Which is the best estimate for the percent of students who prefer cats?

A. 75

B. 125

C. 225

D. 450

49. Mallory is buying bedroom furniture for $1,789.43. The dresser is 39.7% of the total cost. Which is the best estimate for the cost of the dresser?

F. $540   H. $720

G. $630   I. $810

50. Abey asked 50 students to vote for the school issue that was most important to them. The results are shown below.

Issue	Votes
Library use	10%
Time to change classes	12%
Use of electronics	18%
Lunch room rules	20%
Dress code	40%

About how many students chose "Time to change classes" as the most important issue?

A. 3   C. 9

B. 5   D. 12

 **Find Percents**

Main Idea

Solve problems involving percents.

Get ConnectED

TICKETS The eighth grade had 300 tickets to sell to the school play and the seventh grade had 250 tickets to sell. One hour before the show, the eighth grade had sold 225 tickets and the seventh grade had sold 200 tickets. Which grade sold the greater percent of tickets?

ACTIVITY

What do you need to find? which grade sold the greater percent of tickets

STEP 1 Make a bar diagram for each grade to show 100%. Divide each bar into 10 equal parts to show 10%.

STEP 2 Find the value of each part.

8th grade: $300 \div 10 = 30$ 7th grade: $250 \div 10 = 25$

STEP 3 Shade each bar to represent the number of tickets sold.

Eighth grade: $225 \div 30 = 7.5$ Shade 7.5 parts.
Seventh grade: $200 \div 25 = 8$ Shade 8 parts.

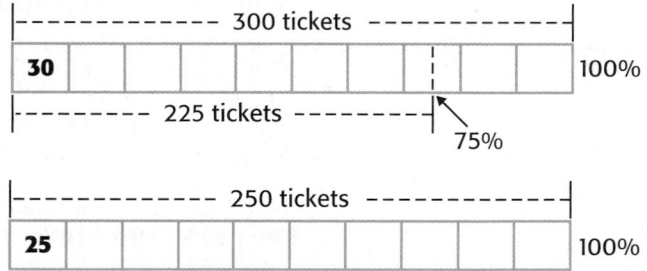

Since 80% > 75%, the seventh grade sold the greater percent.

Analyze the Results

1. Why is the percent for the seventh-grade sales greater even though they sold fewer tickets?

2. **SCHOOL** Vanlue Middle School has 600 students and Memorial Middle School has 450 students. Vanlue has 270 girls and Memorial has 243 girls. Draw a bar diagram to determine which school has the greater percent of girls. Explain.

Main Idea
Solve problems using the percent proportion.

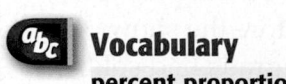
Vocabulary
percent proportion

7.RP.3, 7.EE.3

The Percent Proportion

🏃 **MONSTER TRUCKS** The tires on a monster truck weigh approximately 3,600 pounds. The entire truck weighs about 11,000 pounds.

1. Write the ratio of tire weight to total weight as a fraction.

2. Use a calculator to write the fraction as a decimal to the nearest hundredth.

3. What percent of the monster truck's weight is the tires?

In a **percent proportion**, one ratio or fraction compares part of a quantity to the whole quantity. The other ratio is the equivalent percent written as a fraction with a denominator of 100.

4 out of **5** is **80**%.

$$\frac{\text{part}}{\text{whole}} \longrightarrow \frac{4}{5} = \frac{80}{100} \Big\} \text{percent}$$

| 10% | 10% | 10% | 10% | 10% | 10% | 10% | 10% | 10% | 10% | 100% |

| | | | | | 5 |

Key Concept	**Types of Percent Problems**	
Type	**Example**	**Proportion**
Find the Percent	What percent of 5 is 4?	$\frac{4}{5} = \frac{n}{100}$
Find the Part	What number is 80% of 5?	$\frac{p}{5} = \frac{80}{100}$
Find the Whole	4 is 80% of what number?	$\frac{4}{w} = \frac{80}{100}$

 **EXAMPLE** Find the Percent

1 What percent of $15 is $9?

Words	What percent of $15 is $9?
Variable	Let n represent the percent.
Proportion	$\frac{part \rightarrow}{whole \rightarrow} \frac{9}{15} = \frac{n}{100}$ } percent

$\dfrac{9}{15} = \dfrac{n}{100}$ Write the proportion.

$9 \cdot 100 = 15 \cdot n$ Find the cross products.

$900 = 15n$ Simplify.

$\dfrac{900}{15} = \dfrac{15n}{15}$ Divide each side by 15.

$60 = n$

So, $9 is 60% of $15.

CHECK Your Progress

   a. What percent of 25 is 20?   **b.** $12.75 is what percent of $50?

Study Tip

The Percent Proportion The whole usually comes after the word *of*.

 **EXAMPLE** Find the Part

2 What number is 40% of 120?

Words	What number is 40% of 120?
Variable	Let p represent the part.
Proportion	$\frac{part \rightarrow}{whole \rightarrow} \frac{p}{120} = \frac{40}{100}$ } percent

$\dfrac{p}{120} = \dfrac{40}{100}$ Write the proportion.

$p \cdot 100 = 120 \cdot 40$ Find the cross products.

$100p = 4,800$ Simplify.

$\dfrac{100p}{100} = \dfrac{4,800}{100}$ Divide each side by 100.

$p = 48$

So, 48 is 40% of 120.

CHECK Your Progress

   c. What number is 5% of 60?   **d.** 12% of 85 is what number?

 **EXAMPLE** Find the Whole

3 18 is 25% of what number?

Words	18 is 25% of what number?
Variable	Let w represent the whole.
Proportion	$\dfrac{part}{whole} \rightarrow \dfrac{18}{w} = \dfrac{25}{100}$ } percent

$\dfrac{18}{w} = \dfrac{25}{100}$ Write the proportion.

$18 \cdot 100 = w \cdot 25$ Find the cross products.

$1{,}800 = 25w$ Simplify.

$\dfrac{1{,}800}{25} = \dfrac{25w}{25}$ Divide each side by 25.

$72 = w$

So, 18 is 25% of 72.

CHECK Your Progress

e. 40% of what number is 26?   **f.** 84 is 75% of what number?

 REAL-WORLD EXAMPLE

4 **ANIMALS** The average adult male Western Lowland gorilla eats about 33.5 pounds of fruit each day. How much food does the average adult male gorilla eat each day?

You know that 33.5 pounds is the part. You need to find the whole.

$\dfrac{33.5}{w} = \dfrac{67}{100}$ Write the proportion.

$33.5 \cdot 100 = w \cdot 67$ Find the cross products.

$3{,}350 = 67w$ Simplify.

$\dfrac{3{,}350}{67} = \dfrac{67w}{67}$ Divide each side by 67.

$50 = w$

The average adult male gorilla eats 50 pounds of food each day.

Western Lowland Gorilla's Diet	
Food	**Percent**
Fruit	67%
Seeds, leaves, stems, and pith	17%
Insects/ insect larvae	16%

Real-World Link · · · ·
Male Western Lowland gorillas weigh about 350–400 pounds. Females weigh about 160–200 pounds.

CHECK Your Progress

g. **ZOO** If 200 of the 550 reptiles in a zoo are on display, what percent of the reptiles are on display? Round to the nearest whole number.

Examples 1–3   **Find each number. Round to the nearest tenth if necessary.**

1. What percent of 50 is 18?   **2.** What percent of $90 is $9?

3. What number is 2% of 35?   **4.** What number is 25% of 180?

5 9 is 12% of which number?   **6.** 62 is 90.5% of what number?

Example 4   **7. MEASUREMENT** If a box of Brand A cereal contains 10 cups of cereal, how many more cups of cereal are in a box of Brand B cereal?

Practice and Problem Solving

● = **Step-by-Step Solutions** begin on page R1.
Extra Practice begins on page EP2.

Examples 1–3   **Find each number. Round to the nearest tenth if necessary.**

8. What percent of 60 is 15?   **9.** $3 is what percent of $40?

10. What number is 15% of 60?   **11.** 12% of 72 is what number?

12. 9 is 45% of what number?   **13.** 75 is 20% of what number?

Example 4   **14. SCHOOL** Roman has 2 red pencils in his backpack. If this is 25% of the total number of pencils, how many pencils are in his backpack?

15. BASKETBALL Eileen and Michelle scored 48% of their team's points. If their team had a total of 50 points, how many points did they score?

16. SHOES A pair of sneakers is on sale as shown. This is 75% of the original price. What was the original price of the shoes?

17. BOOKS Of the 60 books on a bookshelf, 24 are nonfiction. What percent of the books are nonfiction?

Sale Price $51

Find each number. Round to the nearest hundredth if necessary.

18. What percent of 25 is 30?   **19.** What number is 8.2% of 50?

20. 40 is 50% of what number?   **21.** 12.5% of what number is 24?

22. What number is 0.5% of 8?   **23** What percent of 300 is 0.6?

24. ASTRONOMY Use the table shown.

a. Mercury's radius is what percent of Jupiter's radius?

b. If the radius of Mars is about 13.7% of Neptune's radius, what is the radius of Neptune?

c. Earth's radius is about 261.4% of Mercury's radius. What is the radius of Earth?

Planet	Radius (km)
Mercury	2,440
Mars	3,397
Jupiter	71,492

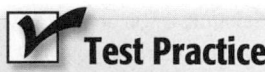
H.O.T. Problems

25. **REASONING** Seventy percent of the 100 students in a middle school cafeteria bought their lunch. Some of these students leave the cafeteria to attend an assembly. Now only 60% of the remaining students bought their lunch. How many students are remaining in the cafeteria? Explain.

26. **CHALLENGE** Without calculating, arrange the following from greatest to least value. Justify your reasoning.

20% of 100, 20% of 500, 5% of 100

27. **WRITE MATH** Create a problem involving a percent that can be solved by using the proportion $\frac{3}{b} = \frac{15}{100}$.

✓ Test Practice

28. Of the 273 students in a school, 95 volunteered to work the book sale. About what percent of the students volunteered?

   A. 35%

   B. 65%

   C. 70%

   D. 75%

29. One hundred ninety-two students were surveyed about their favorite kind of TV programs. The results are shown in the table. Which kind of program did 25% of the students report as their favorite?

Favorite TV Programs	
Kind	**Number**
Music	48
Reality	44
Comedy	41
Sports	36
Drama	23

   F. Music   H. Comedy

   G. Reality   I. Sports

Spiral Review

30. **FOOD** Out of 823 students, 47.2% chose pizza as their favorite food. What is a reasonable estimate for the number of students who chose pizza as their favorite food? Explain. (Lesson 1C)

Find each number. Round to the nearest tenth if necessary. (Lesson 1B)

31. What is 25% of 120?   32. Find 45% of 70.

33. **FLOWERS** Use the information in the table to write the percent of each type of flower in the arrangement. (Lesson 1B)

   a. lilies

   b. snapdragons

   c. roses

Flower Arrangement	
Lilies	4
Roses	15
Snapdragons	6

Main Idea
Solve problems by using the percent equation.

 Vocabulary
percent equation

 Get Connect**ED**

 CCSS 7.RP.2c, 7.EE.3

The Percent Equation

ARTHROPODS Suppose there are 854,000 different species of spiders, insects, crustaceans, millipedes, and centipedes on Earth. The graph shows that 88% of the total number of species of arthropods are insects.

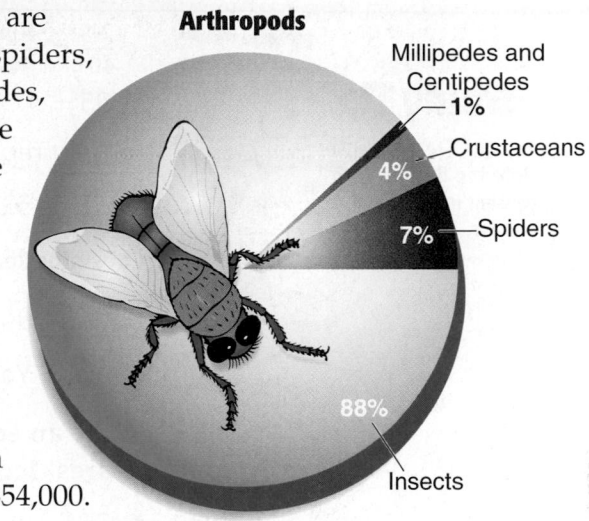

Arthropods

Millipedes and Centipedes — 1%
Crustaceans — 4%
Spiders — 7%
88% Insects

1. Use the percent proportion to find how many species are insects.

2. Express the percent of insects as a decimal. Then multiply the decimal by 854,000. What do you notice?

You have used a percent proportion to find the missing part, percent, or whole. You can also use a percent equation.

$$\frac{\text{part}}{\text{whole}} = \text{percent}$$ The percent must be written as a decimal or fraction.

$$\frac{\text{part}}{\text{whole}} \cdot \textbf{whole} = \text{percent} \cdot \textbf{whole}$$ Multiply each side by the whole.

$$\text{part} = \text{percent} \cdot \text{whole}$$ ← This form is called the **percent equation.**

 EXAMPLE Find the Part

 What number is 12% of 150?

Estimate 12% of 150 ≈ 0.1 • 150 or 15

$\underbrace{\text{part}}_{} = \underbrace{\text{percent}}_{} \cdot \underbrace{\text{whole}}_{}$

$p = 0.12 \cdot 150$ Write the percent equation. 12% = 0.12

$p = 18$ Multiply.

So, 18 is 12% of 150.

Study Tip
Percent Equation
A percent must always be converted to a decimal or a fraction when it is used in an equation.

 CHECK Your Progress

Write an equation for each problem. Then solve.

a. What is 6% of 200?

b. Find 72% of 50.

 **EXAMPLE** Find the Percent

2 **21 is what percent of 40?**

Estimate $\frac{21}{40} \approx \frac{1}{2}$ or 50%

$\underbrace{\text{part}}_{} = \underbrace{\text{percent}}_{} \cdot \underbrace{\text{whole}}_{}$

$21 = n \cdot 40$ Write the percent equation.

$\frac{21}{40} = \frac{40n}{40}$ Divide each side by 40.

$0.525 = n$ Simplify.

Since n represents the decimal form, the percent is 52.5%.

So, 21 is 52.5% of 40.

Check for Reasonableness 52.5% ≈ 50% ✓

 CHECK Your Progress

Write an equation for each problem. Then solve. Round to the nearest tenth if necessary.

c. 35 is what percent of 70?   **d.** What percent of 125 is 75?

e. What percent of 40 is 9?   **f.** 27 is what percent of 150?

 **EXAMPLE** Find the Whole

3 **13 is 26% of what number?**

Estimate $\frac{1}{4}$ of 48 = 12

$\underbrace{\text{part}}_{} = \underbrace{\text{percent}}_{} \cdot \underbrace{\text{whole}}_{}$

$13 = 0.26 \cdot w$ Write the percent equation. 26% = 0.26

$\frac{13}{0.26} = \frac{0.26w}{0.26}$ Divide each side by 0.26.

$50 = w$ Simplify.

So, 13 is 26% of 50.

Check for Reasonableness 50 is close to 48. ✓

 CHECK Your Progress

Write an equation for each problem. Then solve. Round to the nearest tenth if necessary.

g. 39 is 84% of what number?   **h.** 26% of what number is 45?

i. 14% of what number is 7?   **j.** 24 is 32% of what number?

 REAL-WORLD EXAMPLE

4 CELL PHONES A survey found that 25% of people aged 18–24 gave up their home phone and only use a cell phone. If 3,264 people only use a cell phone, how many people were surveyed?

Words	3,264 people is 25% of what number of people?
Variable	Let n represent the number of people.
Equation	$3{,}264 \quad = \quad 0.25 \quad \cdot \quad n$

$3{,}264 = 0.25 \cdot n$ Write the percent equation. 25% = 0.25

$\dfrac{3{,}264}{0.25} = \dfrac{0.25n}{0.25}$ Divide each side by 0.25. Use a calculator.

$13{,}056 = n$ Simplify.

About 13,056 people were surveyed.

CHECK Your Progress

k. POPULATION The Miami-Dade County metropolitan area contains 13.3% of the population of Florida. If the population of Florida is about 18,089,888 people, what is the population of the Miami-Dade County metropolitan area?

Key Concept	Types of Percent Problems	
Type	**Example**	**Equation**
Find the Percent	3 is what percent of 6?	$3 = n \cdot 6$
Find the Part	What number is 50% of 6?	$p = 0.5 \cdot 6$
Find the Whole	3 is 50% of what number?	$3 = 0.5 \cdot w$

CHECK Your Understanding

Examples 1–3 Write an equation for each problem. Then solve. Round to the nearest tenth if necessary.

1. What number is 88% of 300?
2. What number is 12% of 250?
3. 75 is what percent of 150?
4. 24 is what percent of 120?
5. 3 is 12% of what number?
6. 84 is 60% of what number?

Example 4 7. **BUSINESS** A local bakery sold 60 loaves of bread in one day. If 65% of these were sold in the afternoon, how many loaves were sold in the afternoon?

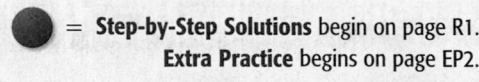

● = **Step-by-Step Solutions** begin on page R1.
Extra Practice begins on page EP2.

Examples 1–3   **Write an equation for each problem. Then solve. Round to the nearest tenth if necessary.**

8. What number is 65% of 98?   **9.** Find 39% of 65.

10. Find 24% of 25.   **11.** What number is 53% of 470?

12. 9 is what percent of 45?   **13.** What percent of 96 is 26?

14. What percent of 392 is 98?   **15.** 30 is what percent of 64?

16. 33% of what number is 1.45?   **17.** 84 is 75% of what number?

18. 17 is 40% of what number?   **19.** 80% of what number is 64?

Example 4   **20.** **BOOKS** Ruben bought 6 new books for his collection. This increased his collection by 12%. How many books did he have before his purchases?

21. **VIDEO GAMES** A store sold 550 video games during the month of December. If this made up 12.5% of their yearly video game sales, about how many video games did the store sell all year?

22. **MEASUREMENT** The length of Giselle's arm is 27 inches. The length of her lower arm is 17 inches. About what percent of Giselle's arm is her lower arm?

23. **LOBSTERS** Approximately 0.02% of North Atlantic lobsters are born bright blue in color. Out of 5,000 North Atlantic lobsters, how many would you expect to be blue in color?

Write an equation for each problem. Then solve. Round to the nearest tenth if necessary.

24. Find 135% of 64.   **25.** What number is 0.4% of 82.1?

26. 450 is 75.2% of what number?   **27.** What percent of 200 is 230?

28. **SALARY** Suppose you earn $6 per hour at your part-time job. What will your new hourly rate be after a 2.5% raise?

㉙ **ONLINE VIDEOS** About 142 million people in the United States watch online videos. Use the graph that shows what type of videos they watch.

   a. About what percent of viewers watch comedy, jokes, and bloopers?

   b. About what percent watch news stories?

   c. What percent of the viewers do *not* watch movie previews?

30. OPEN ENDED Write a percent problem for which the percent is greater than 100 and the part is known. Use the percent equation to solve your problem to find the whole.

31. CHALLENGE If you need to find the percent of a number, explain how you can predict whether the part will be less than, greater than, or equal to the number.

32.  **WRITE MATH** Compare the percent equation and the percent proportion. Then explain when it might be easier to use the percent equation rather than the percent proportion.

Test Practice

33. In a survey, 100 students were asked to choose their favorite take-out food. The table shows the results.

Favorite Take-Out Food	
Type of Food	**Percent**
Pizza	40
Sandwiches	32
Chicken	28

Based on these data, predict how many out of 1,800 students would choose sandwiches.

- **A.** 504
- **B.** 576
- **C.** 680
- **D.** 720

34. If 60% of a number is 18, what is 90% of the number?

- **F.** 3
- **G.** 16
- **H.** 27
- **I.** 30

35. Taryn's grandmother took her family out to dinner. If the dinner was $74 and Taryn's dinner was 20% of the bill, how much was Taryn's dinner?

- **A.** $6.80
- **B.** $7.20
- **C.** $9.50
- **D.** $14.80

Spiral Review

Find each number. Round to the nearest hundredth if necessary. (Lesson 2B)

36. What percent of 15 is 20?

37. 20.5% of what number is 35?

38. What number is 0.5% of 10?

39. SPORTS The graph shows the number of student athletes participating in fall sports. In which sport do 15% of the student athletes compete? Justify your answer. (Lesson 2B)

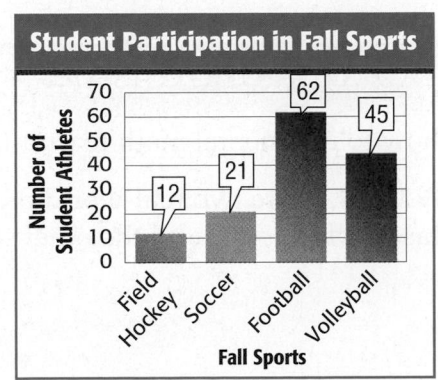

Student Participation in Fall Sports

40. TRAVEL On a 511-mile trip, Mya drove about 68% on Monday. Determine a reasonable estimate for the number of miles she drove on Monday. (Lesson 1C)

Problem-Solving Investigation

Main Idea Solve problems by determining reasonable answers.

P.S.I. TEAM +

Determine Reasonable Answers

DOUG: My dad and I are at a restaurant. Our bill is $28. We want to leave a tip that is 15% of the bill. I think he should leave $5 for the tip.

YOUR MISSION: Determine whether it is reasonable for Doug's dad to leave $5 as the tip.

Understand	Doug and his dad spent $28. Doug thinks that $5 is about 15% of the bill.
Plan	Make a bar diagram to represent 100%.
Solve	Divide it into ten parts, each representing 10%.

|------------------------ $28 ------------------------|

| $2.80 | $2.80 | $2.80 | $2.80 | $2.80 | $2.80 | $2.80 | $2.80 | $2.80 | $2.80 |

So, a 10% tip would be $2.80.

If 10% = $2.80, 5% is $2.80 ÷ 2 or $1.40.

So, a 15% tip would be $2.80 + $1.40 or $4.20. A $5 tip is reasonable.

Check	$28 × 0.15 = $4.20. So, $5 is a reasonable estimate.

Analyze the Strategy

1. Explain how to use mental math to find 15% of a number.

2. **WRITE MATH** Write two real-world problems involving tips. One should have a reasonable answer and the other should not have a reasonable answer.

= **Step-by-Step Solutions** begin on page R1.
Extra Practice begins on page EP2.

- Determine reasonable answers.
- Guess, check, and revise.
- Look for a pattern.
- Choose an operation.

Use the *determine reasonable answers* strategy to solve Exercises 3–6.

3. **TIPS** Brett decides to leave a 20% tip on a restaurant bill of $17.50. How much should he tip the restaurant server?

`|------------- $17.50 -------------|`
`|-- tip --|`

4. **SCHOOL** Of 423 students, 57.6% live within 5 miles of the school. What is a reasonable estimate for the number of students living within 5 miles of the school? Explain.

5. **EXERCISE** A survey showed that 61% of middle school students do some kind of physical activity every day. If there are 828 middle school students in your school, would the number of students who exercise be about 300, 400, or 500? Explain.

6. **TRAVEL** A travel agency surveyed 140 families about their favorite vacation spots. Is 60, 70, or 80 families a reasonable estimate for the number of families that did *not* choose Hawaii?

Favorite Vacation Spots

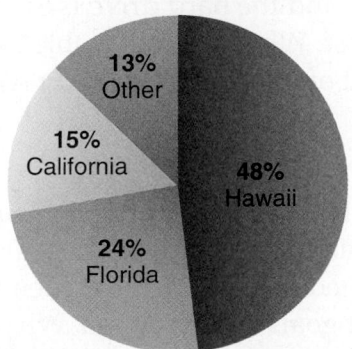

13% Other
15% California
48% Hawaii
24% Florida

Use any strategy to solve Exercises 7–13.

7. **COINS** Gavin has 10 coins that total $0.83. What are the coins?

8. **BAKING** Refer to the graph. A pie is set out to cool. Is it reasonable to estimate that the pie will be 90°F after ten minutes of cooling? Explain.

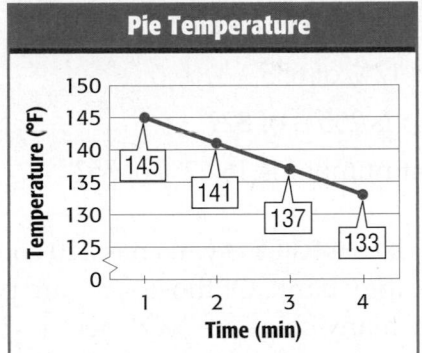

Pie Temperature

145, 141, 137, 133

9. **SHOPPING** Deshawn wants to buy a shirt that has a regular price of $41 but is now on sale for 25% off the original price. Is $25, $30, or $35 the best estimate for the cost of the shirt?

10. **BOWLING** In bowling, you get a spare when you knock down the ten pins in two throws. How many possible ways are there to get a spare?

11. **SAVING** Aliayah saves $11 each month for her class trip. What is a reasonable estimate for the amount of money she will have saved after a year: about $100, $120, or $160? Explain.

12. **FUNDRAISER** During a popcorn sale for a fundraiser, the soccer team gets to keep 25% of the sales. One box of popcorn sells for $1.50 and the team has sold 510 boxes so far. Has the team raised a total of $175? Explain.

13. **MEASUREMENT** How many square yards of carpet are needed to carpet the two rooms described below? Explain.

Room	Dimensions
Living room	15 ft by 18 ft
TV room	18 ft by 20 ft

Mid-Chapter Check

Find each number. Round to the nearest tenth if necessary. (Lesson 1B)

1. Find 17% of 655.

2. What is 235% of 82?

3. What number is 162.2% of 55?

4. **MULTIPLE CHOICE** Ayana has 220 coins in her piggy bank. Of those, 45% are pennies. How many coins are NOT pennies? (Lesson 1B)

   A. 121   C. 109

   B. 116   D. 85

Estimate. (Lesson 1C)

5. 20% of 392   6. 78% of 112   7. 30% of 42

Find each number. Round to the nearest tenth if necessary. (Lesson 2B)

8. What percent of 84 is 12?

9. 15 is 25% of what number?

10. 85% of 252 is what number?

11. **ANALYZE GRAPHS** Refer to the graph that shows the results of a survey of 200 students' favorite DVDs. (Lesson 2C)

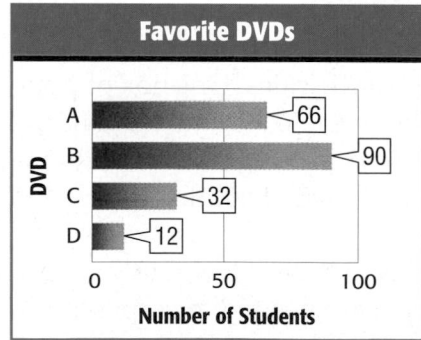

Favorite DVDs

A — 66
B — 90
C — 32
D — 12

Number of Students

a. What percent of students preferred DVD A?

b. Which DVD did about 15% of students prefer?

Write an equation for each problem. Then solve. Round to the nearest tenth if necessary. (Lesson 2C)

12. What number is 35% of 72?

13. 16.1 is what percent of 70?

14. 27.2 is 68% of what number?

15. 16% of 32 is what number?

16. 55% of what number is 1.265?

17. 15 is 35% of what number?

18. 12 is what percent of 20?

19. **ANALYZE TABLES** The table shows the costs of owning a dog over an average 11-year lifespan. What percent of the total cost is veterinary bills? (Lesson 2C)

Dog Ownership Costs	
Item	**Cost ($)**
Food	4,020
Veterinary bills	3,930
Grooming equipment	2,960
Training	1,220
Other	2,470

20. **SHOPPING** A desktop computer costs $849.75 and the hard drive is 61.3% of the total cost. What is a reasonable estimate for the cost of the hard drive? (Lesson 2D)

21. **MULTIPLE CHOICE** A football player has made about 75% of the field goals he has attempted in his career. He attempts 41 field goals in one season. What is a reasonable estimate for the number of field goals he is expected to make? (Lesson 2D)

   F. 35   H. 25

   G. 30   I. 20

 **Percent of Change**

Main Idea

Use bar diagrams to solve problems involving percent of change.

CCSS 7.RP.3

ADMISSION The admission price for the state fair has increased by 50% in the last five years. If the admission price was $6 five years ago, what is the current admission price?

ACTIVITY

What do you need to find? current admission price for the state fair after a 50% increase

STEP 1 Draw a bar diagram that will represent 100%. Cut it out and label it.

price 5 years ago = $6	100%

STEP 2 Draw a second bar diagram that is half the length of the first. It represents 50%. Cut it out and label it.

increase = $3	50%

STEP 3 Tape the two bars together, end to end. This new bar represents a 50% increase or 150% of the first bar.

price 5 years ago = $6	increase = $3	150%

The current admission price for the state fair is $6 + $3 or $9.

Analyze the Results

1. Suppose the price had increased 10%. What would you do differently in Step 2? What is the new admission price?
2. Describe how this process would change to show percent of decrease.

Practice and Apply

3. Model each percent of change with bar diagrams.

   a. 25% increase   **b.** 75% increase
   c. 30% decrease   **d.** 40% decrease

Main Idea
Find the percent of increase or decrease.

Vocabulary
percent of change
percent of increase
percent of decrease

7.RP.3, 7.EE.3

Percent of Change

FOOTBALL The table shows about how many people attended the home games of a high school football team each year.

Attendance of Home Games	
Year	Total Attendance (thousands)
2007	16.6
2008	16.4
2009	16.9
2010	17.4
2011	17.6

1. How much did the attendance increase from 2009 to 2010?

2. Write the ratio $\frac{\text{amount of increase}}{\text{attendance in 2009}}$. Then write the ratio as a percent. Round to the nearest hundredth.

3. How much did the attendance increase from 2008 to 2009?

4. Write the ratio $\frac{\text{amount of increase}}{\text{attendance in 2008}}$. Then write the ratio as a percent. Round to the nearest hundredth.

5. **MAKE A CONJECTURE** Why are the amounts of increase the same but the percents different?

When you subtracted the original amount from the final amount, you found the *amount* of change. When you compared the change to the original amount in a ratio, you found the *percent* of change.

> ## Key Concept   Percent of Change
>
> **Words**   A **percent of change** is a ratio that compares the change in quantity to the original amount.
>
> **Equation** $\text{percent of change} = \dfrac{\text{amount of change}}{\text{orginal amount}}$
>
> $= \dfrac{\text{final amount} - \text{original amount}}{\text{original amount}}$

When the percent of change is positive, then it is called a **percent of increase**. When the percent of change is negative, then it is called a **percent of decrease**.

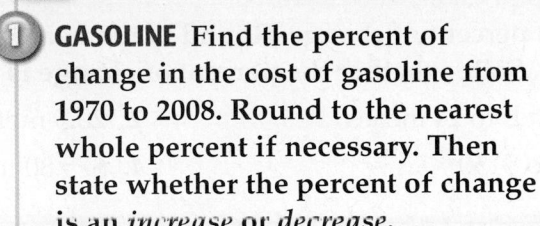 **REAL-WORLD EXAMPLES** **Find Percent of Change**

1 **GASOLINE** Find the percent of change in the cost of gasoline from 1970 to 2008. Round to the nearest whole percent if necessary. Then state whether the percent of change is an *increase* or *decrease*.

Step 1 Find the amount of change.
$3.95 − $1.30 = $2.65

Step 2 Find the percent of change.

$$\text{percent of change} = \frac{\text{amount of change}}{\text{original amount}}$$

$$= \frac{\$2.65}{\$1.30} \qquad \text{Substitution}$$

$$\approx 2.04 \qquad \text{Simplify.}$$

$$\approx 204\% \qquad \text{Write 2.04 as a percent.}$$

The percent of change is 204%. Since the percent of change is positive, this is a percent of *increase*.

Study Tip

Percents
In the percent of change formula, the decimal repesenting the percent of change must be written as a percent.

2 **DVD RECORDER** Yusuf bought a DVD recorder for $280. Now, it is on sale for $220. Find the percent of change in the price. Round to the nearest whole percent if necessary. Then state whether the percent of change is an *increase* or *decrease*.

Step 1 Find the amount of change.
$220 − $280 = −$60

Step 2 Find the percent of change.

$$\text{percent of change} = \frac{\text{amount of change}}{\text{original amount}}$$

$$= \frac{-\$60}{\$280} \qquad \text{Substitution}$$

$$\approx -0.21 \qquad \text{Simplify.}$$

$$\approx -21\% \qquad \text{Write } -0.21 \text{ as a percent.}$$

The percent of change is −21%. Since the percent of change is negative, this is a percent of *decrease*.

CHECK Your Progress

a. **MEASUREMENT** Find the percent of change from 10 yards to 13 yards. Then state whether the percent of change is an *increase* or *decrease*.

b. **MONEY** Find the percent of change from $20 to $15. Then state whether the percent of change is an *increase* or *decrease*.

Examples 1 and 2 Find each percent of change. Round to the nearest whole percent if necessary. State whether the percent of change is an *increase* or a *decrease*.

1. 30 inches to 24 inches

2. 20.5 meters to 35.5 meters

3. $126 to $150

4. $75.80 to $94.75

5. SOCCER The table shows the number of youth 7 years and older who played soccer from 2000 to 2007.

Playing Soccer	
Year	Number (millions)
2000	12.9
2002	13.7
2004	13.3
2006	14.0
2007	13.8

  a. Find the percent of change from 2004 to 2007. Round to the nearest tenth of a percent. Is it a change of increase or decrease?

  b. Find the percent of change from 2002 to 2004. Round to the nearest tenth of a percent. Is it a change of increase or decrease?

Practice and Problem Solving

● = **Step-by-Step Solutions** begin on page R1.
Extra Practice begins on page EP2.

Examples 1 and 2 Find each percent of change. Round to the nearest whole percent if necessary. State whether the percent of change is an *increase* or a *decrease*.

6. 15 yards to 18 yards

7. 100 acres to 140 acres

8. $12 to $6

9. 48 notebooks to 14 notebooks

10. 125 centimeters to 87.5 centimeters

11 $15.60 to $11.70

12. 1.6 hours to 0.95 hour

13. 132 days to 125.4 days

14. $240 to $320

15. 624 feet to 702 feet

Find each percent of change. Round to the nearest whole percent if necessary. State whether the percent of change is an *increase* or a *decrease*.

16. BOOKS On Monday, Kenya spent 60 minutes reading her favorite book. Today, she spent 45 minutes reading this book.

17. EXERCISE Three months ago, Santos could walk 2 miles in 40 minutes. Today he can walk 2 miles in 25 minutes.

18. SCHOOL Last school year the enrollment of Genoa Middle School was 465 students. This year the enrollment is 525.

19. MONEY Jake had $782 in his checking account. He now has $798.

20. MEASUREMENT Refer to the rectangle at the right. Suppose the side lengths are doubled.

4 in.

2 in.

  a. Find the percent of change in the perimeter.

  b. Find the percent of change in the area.

21. **MATH IN THE MEDIA** Find examples of data reflecting change over a period of time in a newspaper or magazine, on television, or on the Internet. Determine the percent of change. Explain whether the data show a percent of increase or decrease.

Study Tip

Percent of Change
Always use the original amount as the whole when finding percent of change.

22. **SALES** Use the graph shown to find the percent of change in CD sales from 2011 to 2012.

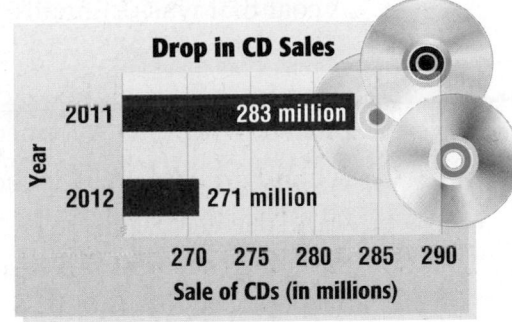

Drop in CD Sales

23 **SHOES** Shoe sales for a certain company were $25.9 billion. Sales are expected to increase by about 20% in the next year. Find the projected amount of shoe sales next year.

24. **BABYSITTING** The table shows how many hours Catalina spent babysitting during the months of April and May.

Month	Hours Worked
April	40
May	32
June	42

a. If Catalina charges $6.50 per hour, what is the percent of change in the amount of money earned from April to May? Is it a change of increase or decrease?

b. What is the percent of change in the amount of money earned from May to June? Round to the nearest percent if needed. Is it a change of increase or decrease?

c. Compare the percent of change from April to May and then May to June. Which is a greater percent of change?

H.O.T. Problems

25. **NUMBER SENSE** The costs of two different sound systems were decreased by $10. The original costs of the systems were $90 and $60, respectively. Without calculating, which had a greater percent of decrease? Explain.

26. **FIND THE ERROR** Dario is finding the percent of change from $52 to $125. Find his mistake and correct it.

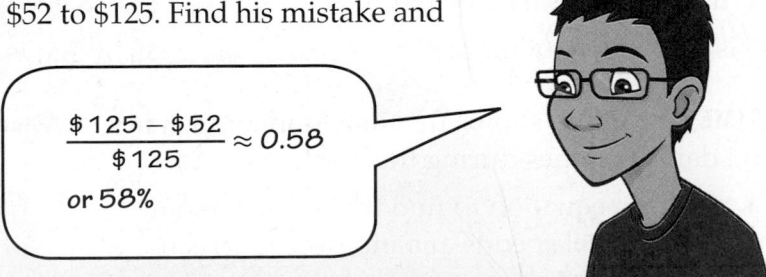

$$\frac{\$125 - \$52}{\$125} \approx 0.58$$

or 58%

27. **WRITE MATH** Explain how, when comparing data, two amounts of change can be the same but the percents of change can be different.

28. Which of the following represents the least percent of change?

 A. A coat that was originally priced at $90 is now $72.

 B. A puppy who weighed 6 ounces at birth now weighs 96 ounces.

 C. A child grew from 54 inches to 60 inches in 1 year.

 D. A savings account increased from $500 to $550 in 6 months.

29. THINK SOLVE EXPLAIN **SHORT RESPONSE** A music video Web site received 5,000 comments on a new song they released. After the artist performed the song on television, the number of comments increased by 30% the next day. How many new comments were on the Web site at the end of the next day?

30. Students in a reading program gradually increased the amount of time they read. The first week, they read 20 minutes per day. Each week thereafter, they increased their reading time by 50% until they read an hour per day. In what week of the program did the students begin reading an hour per day?

 F. Week 2   **H.** Week 4

 G. Week 3   **I.** Week 5

31. ✏️ **GRIDDED RESPONSE** Find the percent of change in the perimeter of the square below if its side length is tripled.

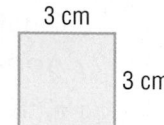

3 cm

3 cm

32. SURVEY There are 622 students at Jackson Middle School. About 52.3% of them selected art class as their favorite class. What is a reasonable estimate for the number of students who chose art as their favorite? (Lesson 2D)

ALGEBRA Write an equation for each problem. Then solve. Round to the nearest tenth if necessary. (Lesson 2C)

33. 30% of what number is 17?

34. What is 21% of 62?

35. What number is 64% of 150?

36. 18 is what percent of 200?

37. 5 is 20% of what number?

38. What is 60% of 35?

39. TIME The graph shows the time Manuel spends on daily activities during the week. (Lesson 2C)

 a. Write an equation to find the percent of time that Manuel spends at band practice. Then solve the equation.

 b. About what percent of time does Manuel spend sleeping?

Manuel's Daily Activities

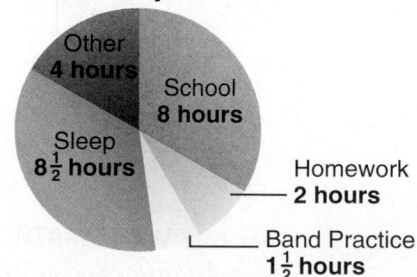

Other
4 hours

School
8 hours

Sleep
8½ hours

Homework
2 hours

Band Practice
1½ hours

Main Idea

Solve problems involving sales tax and tips.

 Vocabulary

sales tax
tip
gratuity

 Get ConnectED

 CCSS 7.RP.3, 7.EE.3

Sales Tax and Tips

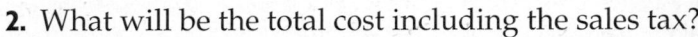

KAYAKS Alonso plans to buy a new kayak that costs $1,849. He lives in a county where there is a 7.5% sales tax.

1. Calculate the sales tax by finding 7.5% of $1,849. Round to the nearest cent.

2. What will be the total cost including the sales tax?

3. Multiply 1.075 and 1,849. How does the result compare to your answer in Exercise 2?

Sales tax is an additional amount of money charged on items that people buy. The total cost of an item is the regular price plus the sales tax.

 EXAMPLE **Find the Total Cost**

 ELECTRONICS A DVD player costs $140 and the sales tax is 5.75%. What is the total cost of the DVD player?

Method 1   **Add sales tax to the regular price.**

First, find the sales tax.

5.75% of $\$140 = 0.0575 \times 140$ Write 5.75% as a decimal.
$= 8.05$ The sales tax is $8.05.

Next, add the sales tax to the regular price.
$\$8.05 + \$140 = \$148.05$

Method 2   **Add the percent of tax to 100%.**

$100\% + 5.75\% = 105.75\%$ Add the percent of tax to 100%.

The total cost is 105.75% of the regular price.

105.75% of $\$140 = 1.0575 \times \$140$ Write 105.75% as a decimal.
$= \$148.05$ Multiply.

The total cost of the DVD player is $148.05.

 CHOOSE Your Method

a. **CLOTHES** What is the total cost of a sweatshirt if the regular price is $42 and the sales tax is $5\frac{1}{2}\%$?

Real-World Link · · · ·
In a recent year, the Internal Revenue Service estimated that Americans paid $15.37 billion in tips.

A **tip** or **gratuity** is a small amount of money in return for a service. The total price is the regular price of the service plus the tip.

 EXAMPLE **Find the Tip**

2 **TIPPING** A customer wants to tip 15% of the restaurant bill. What will be the total bill with tip?

Sal's Bistro
Check 004322

Herbed Salmon	16.25
Chicken Pasta	15.25
Iced Tea	1.75
Iced Tea	1.75

Total 35.00

THANK YOU

Method 1 **Add the tip to the regular price.**

First, find the tip.

15% of $\$35 = 0.15 \times 35$ Write 15% as a decimal.
$\qquad\qquad = 5.25$ The tip is $5.25.

$\$5.25 + \$35 = \$40.25$ Add the tip to the bill.

Method 2 **Add the percent of tip to 100%.**

$100\% + 15\% = 115\%$ Add the percent of tip to 100%.

The total cost is 115% of the bill.

115% of $\$35 = 1.15 \times \$35$ Write 115% as a decimal.
$\qquad\qquad = \$40.25$ Multiply.

The total cost of the bill with tip is $40.25.

CHOOSE Your Method

b. **TAXICAB** Scott wants to tip his taxicab driver. If his commute costs $15 and he wants to give the driver a 20% tip, what is the total cost?

 REAL-WORLD EXAMPLE

3 **HAIRCUTS** A haircut costs $20. Sales tax is 4.75%. Is $25 sufficient to cover the haircut with tax and a 15% tip?

Sales tax is 4.75% and the tip is 15%, so together they will be 19.75%.

19.75% of $\$20 = 0.1975 \times 20$ Write 19.75% as a decimal.
$\qquad\qquad\quad = 3.95$ Multiply.

$\$20 + \$3.95 = \$23.95$ Add.

Since $23.95 < $25, $25 is sufficient to cover the total cost.

CHECK Your Progress

c. **SPA** Find the total cost of a spa treatment of $42 including 6% tax and 20% tip.

Study Tip

Mental Math 10% of a number can be found by moving the decimal one place to the left. 10% of $20 is $2. So, 20% of $20 is $4.

Examples 1 and 2 Find the total cost to the nearest cent.

1. $2.95 notebook; 5% tax
2. $46 shoes; 2.9% tax
3. $28 lunch; 15% tip
4. $98 catered dinner; 18% gratuity

Example 3 5. **MANICURE** Jaimi went to have a manicure that cost $30. She wanted to tip the technician 20% and tax is 5.75%. How much did she spend total for the manicure?

Practice and Problem Solving

● = **Step-by-Step Solutions** begin on page R1.
Extra Practice begins on page EP2.

Examples 1 and 2 Find the total cost to the nearest cent.

6. $58 bill; 20% tip
7. $1,500 computer; 7% tax
8. $99 CD player; 5% tax
9. $13 haircut; 15% tip
10. $43 dinner; 18% gratuity
11. $7.50 meal; 6.5% tax
12. $39 pizza order; 15% tip
13. $89.75 scooter; $7\frac{1}{4}$% tax

Example 3 14. **PET GROOMING** Toru takes his dog to be groomed. The fee to groom the dog is $75 plus 6.75% tax. Is $80 enough to pay for the service? Explain.

15. **CLEANING** Diana and Sujit clean homes for a summer job. They charge $70 for the job plus 5% for supplies. A homeowner gave them a 15% tip. Did they receive more than $82 for their job? Explain.

16. **VIDEO GAMES** What is the sales tax of a $178.90 video game system if the tax rate is 5.75%?

17. **RESTAURANTS** A restaurant bill comes to $28.35. Find the total cost if the tax is 6.25% and a 20% tip is left on the amount before tax.

18. **GRAPHIC NOVEL** Refer to the graphic novel frame below.

We want the cheapest admission for our trip. Refer to the calculations you made at the end of Lesson 1B.

I think we have it! Guess where we are going?

a. Find the price that a student would pay including the group discount for each amusement park.

b. Which is the best deal?

19. CHALLENGE The Leather Depot buys a coat from a supplier for $90 wholesale and marks up the price by 40%. What is the retail price including 7% tax?

20. OPEN ENDED Give an example of the regular price of an item and the total cost including sales tax if the tax rate is 5.75%.

21. Which One Doesn't Belong? In each pair, the first value is the regular price of an item and the second value is the price with gratuity. Identify the pair that does not belong with the other three. Explain.

| $30, $34.50 | $54, $64.80 | $16, $18.40 | $90, $103.50 |

22. **WRITE MATH** Describe two methods for finding the total price of a bill that includes a 20% tip. Which method do you prefer? Explain.

Test Practice

23. Ms. Taylor bought a water tube to pull behind her boat. The tube cost $87.00 and 9% sales tax was added at the register. Ms. Taylor gave the cashier five $20 bills. How much change should she have received?

  A. $4.83

  B. $5.17

  C. $94.83

  D. $117.00

24. Prices for several cell phones are listed in the table below. It shows the regular price p and the price with tax t.

Phone	Regular Price (p)	Price with Tax (t)
Flip phone	$80	$86.40
Slide phone	$110	$118.80
Picture phone	$120	$129.60

Which formula can be used to calculate the price with tax?

  F. $t = p \times 0.8$ **H.** $t = p \times 0.08$

  G. $t = p - 0.8$ **I.** $t = p \times 1.08$

Find each percent of change. Round to the nearest whole percent if necessary. State whether the percent of change is an *increase* or *decrease*. (Lesson 3B)

25. 4 hours to 6 hours

26. $500 to $456

27. 20.5 meters to 35.5 meters

28. 80 books to 110 books

29. FINANCIAL LITERACY Bethany has to pay a 20% handling fee on a book she ordered online. Explain how to use mental math to find a reasonable handling fee on Bethany's book. (Lesson 2D)

Main Idea

Solve problems involving discount.

 Vocabulary

discount

 **Get Connect**ED

 **7.RP.3, 7.EE.3**

Discount

WATER PARKS A pass at a water park is $58. Halfway through the season, the pass is discounted by 20%.

1. Calculate the discount by finding 20% of $58. Round to the nearest cent.

2. What will be the discounted price?

3. Multiply 0.8 and $58. How does the result compare to your answer in Exercise 2?

Discount is the amount by which the regular price of an item is reduced. The sale price is the regular price minus the discount.

 REAL-WORLD EXAMPLE

1 **DVDs** A DVD normally costs $22. This week it is on sale for 25% off the original price. What is the sale price of the DVD?

Method 1   Subtract the discount from the regular price.

First, find the amount of the discount.

25% of $22 = 0.25 × $22   Write 25% as a decimal.
            = $5.50   The discount is $5.50.

Next, subtract the discount from the regular price.
$22 − $5.50 = $16.50

Method 2   Subtract the percent of discount from 100%.

100% − 25% = 75%   Subtract the discount from 100%.

The sale price is 75% of the regular price.

75% of $22 = 0.75 × $22   Write 75% as a decimal.
            = $16.50   Multiply.

The sale price of the DVD is $16.50.

 CHOOSE Your Method

a. **CLOTHES** A shirt is regularly priced at $42. It is on sale for 15% off. What is the sale price of the shirt?

EXAMPLE Find the Sale Price

Study Tip

Sales Tax and Discount
If both are represented as percents, sales tax is a percent of increase and discount is a percent of decrease.

② **BOOGIE BOARDS** A boogie board that has a regular price of $69 is on sale at a 35% discount. What is the sale price with 7% tax?

Step 1 Find the amount of the discount.

35% of $69 = 0.35 · $69   Write 35% as a decimal.

= $24.15   The discount is $24.15.

Step 2 Subtract the discount from the regular price.

$69 − $24.15 = $44.85

Step 3 The percent of tax is applied after the discount is taken.

7% of $44.85 = 0.07 · 44.85   Write 7% as a decimal.

= 3.14   The tax is $3.14.

Add the tax to the sale price of the boogie board.

$44.85 + $3.14 = $47.99

The sale price of the boogie board including tax is $47.99.

CHECK Your Progress

b. MUSIC A CD that has a regular price of $15.50 is on sale at a 25% discount. What is the sale price with 6.5% tax?

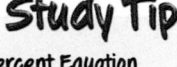 **EXAMPLE** Find the Original Price

Study Tip

Percent Equation
Remember that in the percent equation, the percent must be written as a decimal. Since the sale price is 70% of the original price, use 0.7 to represent 70% in the percent equation.

③ **CELL PHONES** A cell phone is on sale for 30% off. If the sale price is $239.89, what is the original price?

The sale price is 100% − 30% or 70% of the original price.

Words	$239.89 is 70% of what price?
Variable	Let p represent the original price.
Equation	$239.89 = 0.7 \times p$

$239.89 = 0.7p$ Write the equation.

$\dfrac{239.89}{0.7} = \dfrac{0.7p}{0.7}$ Divide each side by 0.7.

$342.70 = p$ Simplify.

The original price is $342.70.

CHECK Your Progress

c. Find the original price if the sale price of the cell phone is $205.50.

Examples 1 and 2   **Find the sale price to the nearest cent.**

    1. $210 bicycle; 25% discount   **2.** $40 sweater; 33% discount

    3. $1,575 computer; 15% discount; 4.25% tax

    4. $119.50 skateboard; 20% off; 7% tax

Example 3   **5. IN-LINE SKATES** A pair of in-line skates is on sale for $90. If this price represents a 9% discount from the original price, what is the original price to the nearest cent?

Practice and Problem Solving

 = **Step-by-Step Solutions** begin on page R1.
Extra Practice begins on page EP2.

Examples 1 and 2   **Find the sale price to the nearest cent.**

    6. $64 jacket; 20% discount   **7.** $1,200 TV; 10% discount

    8. $199 MP3 player; 15% discount   **9.** $12.25 pen set; 60% discount

    10. $4.30 makeup; 40% discount; 6% tax

    11 $7.50 admission; 20% off; 5.75% tax

    12. $39.60 sweater; 33% discount; 4.5% tax

    13. $90.00 skateboard; $33\frac{1}{3}$% off; 8% tax

Example 3   **14. COSMETICS** A bottle of hand lotion is on sale for $2.25. If this price represents a 50% discount from the original price, what is the original price to the nearest cent?

    15. TICKETS At a movie theater, the cost of admission to a matinee is $5.25. If this price represents a 30% discount from the evening price, find the evening price to the nearest cent.

Find the original price to the nearest cent.

    16. calendar: discount, 75%   **17.** telescope: discount, 30%
              sale price, $2.25   sale price, $126

    18. COMPUTERS The Wares want to buy a new computer. The regular price is $1,049. The store is offering a 20% discount and a sales tax of 5.25% is added after the discount. What is the total cost?

    19 **MULTIPLE REPRESENTATIONS** An online store is having a sale on digital cameras. The table shows the regular price and the sale price for the cameras.

Camera Model	Regular Price	Sale Price
A	$97.99	$83.30
B	$102.50	$82.00
C	$75.99	$65.35
D	$150.50	$135.45

    a. TABLE Copy the table including a column for the discount.

    b. WORDS Write a rule that can be used to find the percent of decrease for any of the cameras.

    c. NUMBERS Which model has the best discount?

20. CHALLENGE A gift store is having a sale in which all items are discounted 20%. Including tax, Colin paid $21 for a picture frame. If the sales tax rate is 5%, what was the original price of the picture frame?

21. OPEN ENDED Give an example of the sale price of an item and the total cost including sales tax if the tax rate is 5.75% and the item is 25% off.

22. REASONING Two department stores, The James Store and Ratcliffe's, are having sales. The stores sell the same brand of sneakers. The James Store usually sells them for $50, but has marked them at 40% off. At Ratcliffe's, the sneakers are marked down to 30% off of the usual price of $30. Which store has the better sale price? Explain.

23. WRITE MATH Describe two methods for finding the sale price of an item that is discounted 30%. Which method do you prefer? Explain.

Test Practice

24. A computer software store is having a sale. The table shows the regular price *r* and the sale price *s* of various items.

Item	Regular Price (*r*)	Sale Price (*s*)
A	$5.00	$4.00
B	$8.00	$6.40
C	$10.00	$8.00
D	$15.00	$12.00

Which formula can be used to calculate the sale price?

A. $s = r \times 0.2$ **C.** $s = r \times 0.8$

B. $s = r - 0.2$ **D.** $s = r - 0.8$

25. A chair that costs $210 was reduced by 40% for a one-day sale. After the sale, the sale price was increased by 40%. What is the price of the chair?

F. $176.40   **H.** $205.50

G. $185.30   **I.** $210.00

26. Carmen paid $10.50 for a T-shirt at the mall. It was on sale for 30% off. What was the original price before the discount?

A. $3.15   **C.** $15.00

B. $7.35   **D.** $35.00

Spiral Review

27. RESTAURANTS Mitchell spent $13 on dinner. He wants to tip the server 15%. About how much money should he leave as the tip? (Lesson 3C)

Find the percent of change. Round to the nearest whole percent if necessary. State whether the percent of change is an *increase* or *decrease*. (Lesson 3B)

28. 35 birds to 45 birds   **29.** 60 inches to 38 inches   **30.** $2.75 to $1.80

Main Idea
Solve problems involving simple interest.

 Vocabulary
principal
simple interest

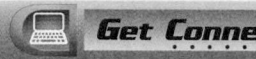

 7.RP.3, 7.EE.3

Financial Literacy: Simple Interest

INVESTING Suni plans to save the $200 she received for her birthday. The table shows the average yearly rates at three different banks.

Bank	Interest Rate
Nation Bank	3%
Federal Credit Union	2.50%
First Bank	2.75%

1. Calculate 2.50% of $200 to find the amount of money Suni can earn in one year at Federal Credit Union.

2. Calculate 2.75% of $200 to find the amount of money Suni can earn in one year at First Bank.

Principal is the amount of money deposited or borrowed. **Simple interest** is the amount paid or earned for the use of money. To find simple interest I, use the following formula.

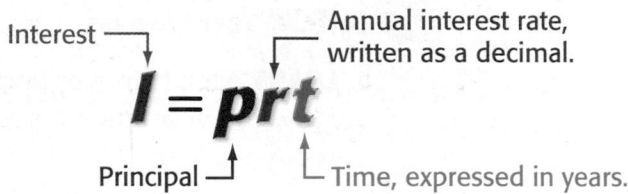

Interest ———

Annual interest rate, written as a decimal.

$$I = prt$$

Principal ——

Time, expressed in years.

EXAMPLES **Find Interest Earned**

CHECKING Arnold has $580 in a savings account that pays 3% interest. How much interest will he earn in each amount of time?

1 **5 years**

$I = prt$	Formula for simple interest
$I = 580 \cdot 0.03 \cdot 5$	Replace p with $580, r$ with 0.03, and t with 5.
$I = 87$	Simplify.

Arnold will earn $87 in interest in 5 years.

2 **6 months**

6 months $= \frac{6}{12}$ or 0.5 year	Write the time as years.
$I = prt$	Formula for simple interest
$I = 580 \cdot 0.03 \cdot 0.5$	$p = \$580, r = 0.03, t = 0.5$
$I = 8.7$	Simplify.

Arnold will earn $8.70 in interest in 6 months.

CHECK Your Progress

    a. SAVINGS Jenny has $1,560 in a savings account that pays 2.5% simple interest. How much interest will she earn in 3 years?

The formula $I = prt$ can also be used to find the interest owed when you borrow money. In this case, p is the amount of money borrowed and t is the amount of time the money is borrowed.

 **EXAMPLE** Find Interest Paid on a Loan

Real-World Link · · · ·
There are over 250 million registered passenger vehicles in the United States.

③ LOANS Rondell's parents borrow $6,300 from the bank for a new car. The interest rate is 6% per year. How much simple interest will they pay if they take 2 years to repay the loan?

$I = prt$	Formula for simple interest
$I = 6{,}300 \cdot 0.06 \cdot 2$	Replace p with $6,300, r$ with 0.06, and t with 2.
$I = 756$	Simplify.

Rondell's parents will pay $756 in interest in 2 years.

 CHECK Your Progress

    b. LOANS Mrs. Hanover borrows $1,400 at a rate of 5.5% per year. How much simple interest will she pay if it takes 8 months to repay the loan?

 EXAMPLE Find Total Paid on a Credit Card

④ CREDIT CARDS Derrick's dad bought new tires for $900 using a credit card. His card has an interest rate of 19%. If he has no other charges on his card and does not make a payment, how much money will he owe after one month?

 **Study Tip**

Fractions of Years
Remember to express 1 month as $\frac{1}{12}$ year in the formula.

$I = prt$	Formula for simple interest
$I = 900 \cdot 0.19 \cdot \dfrac{1}{12}$	Replace p with $900, r$ with 0.19, and t with $\frac{1}{12}$.
$I = 14.25$	Simplify.

The interest owed after one month is $14.25. So, the total amount owed would be $900 + $14.25 or $914.25.

 CHECK Your Progress

    c. CREDIT CARDS An office manager charged $425 worth of office supplies on a credit card with an interest rate of 9.9%. How much money will he owe at the end of one month if he makes no other charges on the card and does not make a payment?

Examples 1 and 2 Find the simple interest earned to the nearest cent for each principal, interest rate, and time.

1. $640, 3%, 2 years
2. $1,500, 4.25%, 4 years
3. $580, 2%, 6 months
4. $1,200, 3.9%, 8 months

Example 3 Find the simple interest paid to the nearest cent for each loan, interest rate, and time.

5 $4,500, 9%, 3.5 years
6. $290, 12.5%, 6 months

Example 4 7. **FINANCES** The Masters family financed a computer that cost $1,200. If the interest rate is 19%, how much will the family owe for the computer after one month if no payments are made?

Practice and Problem Solving

 = **Step-by-Step Solutions** begin on page R1.
Extra Practice begins on page EP2.

Examples 1 and 2 Find the simple interest earned to the nearest cent for each principal, interest rate, and time.

8. $1,050, 4.6%, 2 years
9. $250, 2.85%, 3 years
10. $500, 3.75%, 4 months
11. $3,000, 5.5%, 9 months

Example 3 Find the simple interest paid to the nearest cent for each loan, interest rate, and time.

12. $1,000, 7%, 2 years
13. $725, 6.25%, 1 year
14. $2,700, 8.2%, 3 months
15. $175.80, 12%, 8 months

Example 4 16. **CREDIT CARDS** Leon charged $75 at an interest rate of 12.5%. How much will Leon have to pay after one month if he makes no payments?

17. **TRAVEL** A family charged $1,345 in travel expenses to a credit card with a 7.25% interest rate. If no payments are made, how much will they owe after one month for their travel expenses?

18. **BANKING** The table shows interest rates based on time invested.

    a. What is the simple interest earned on $900 for 9 months?

    b. Find the simple interest earned on $2,500 for 18 months.

Home Savings and Loan

Time	Rate
6 months	2.4%
9 months	2.9%
12 months	3.0%
18 months	3.1%

19 **INVESTING** Ramon has $4,200 to invest for college.

    a. If Ramon invests $4,200 for 3 years and earns $630, what is the simple interest rate?

    b. Ramon's goal is to have $5,000 after 4 years. Is this possible if he invests with a rate of return of 6%? Explain.

H.O.T. Problems

20. OPEN ENDED Suppose you earn 3% on a $1,200 deposit for 5 years. Explain how the simple interest is affected if the rate is increased by 1%. What happens if the time is increased by 1 year?

21. CHALLENGE Mrs. Antil deposits $800 in a savings account that earns 3.2% interest annually. At the end of the year, the interest is added to the principal or original amount. She keeps her money in this account for three years without withdrawing any money. Find the total in her account after each year for three years.

22.  **WRITE MATH** List the steps you would use to find the simple interest on a $500 loan at a 6% interest rate for 18 months. Then find the simple interest.

Test Practice

23. Jada invests $590 in a money market account. Her account pays 7.2% simple interest. If she does not add or withdraw any money, how much interest will Jada's account earn after 4 years of simple interest?

    A. $75.80

    B. $158.67

    C. $169.92

    D. $220.67

24. Mr. Sprockett borrows $3,500 from his bank to buy a used car. The loan has a 7.4% annual simple interest rate. If it takes Mr. Sprockett two years to pay back the loan, what is the total amount he will be paying?

    F. $3,012

    G. $3,598

    H. $4,018

    I. $4,550

25. ENTERTAINMENT The flat screen television shown is on sale for a discount of 20% off its original price. Find the total cost of the TV with a 5.5% sales tax. (Lesson 3D)

Find the total cost of each of the following. (Lesson 3C)

26. backpack, $25 with 7% tax

27. car, $8,000 with $5\frac{1}{2}$% tax

28. dinner, $50 with 18% tip

29. car wash, $25 with 15% tip

$850

30. TIME Tim spent 90 minutes completing his chores. If 40% of this time he was cleaning, how many minutes did he spend *not* cleaning? (Lesson 2C)

Main Idea

Use a spreadsheet to calculate simple interest.

 7.RP.3

Spreadsheet: Simple Interest

A computer spreadsheet is a useful tool for quickly calculating simple interest for different values of principal, rate, and time.

ACTIVITY

SAVINGS Joel plans on opening a "Young Savers" account at his bank. The current rate on the account is 4%. To find the balance at the end of 2 years for different principal amounts, he enters the values B2 = 4 and C2 = 2 into the spreadsheet below.

Simple Interest

◇	A	B	C	D	E
1	Principal (p)	Rate (r)	Time (t)	Interest (I)	New Balance
2					
3	500	=B2/100	=C2	=A3*B3*C3	=A3+D3
4	1000	=B2/100	=C2	=A4*B4*C4	=A4+D4
5	1500	=B2/100	=C2	=A5*B5*C5	=A5+D5
6	2000	=B2/100	=C2	=A6*B6*C6	=A6+D6
7	2500	=B2/100	=C2	=A7*B7*C7	=A7+D7

Sheet 1 / Sheet 2 / Sheet 3

For each principal given in column A, simple interest is calculated for any values of rate and time entered in B2 and C2, respectively.

The spreadsheet adds simple interest to the principal.

Analyze the Results

1. Why is the rate in column B divided by 100?

2. What is the balance in Joel's account after 2 years if the principal is $1,500 and the simple interest rate is 4%?

3. How much interest does Joel earn in 2 years if his account has a principal of $2,000 and a simple interest rate of 4%?

4. Is the amount of principal proportional to the interest Joel earns if his account earns 4% simple interest over 2 years? Explain.

5. Is the amount of principal proportional to the balance in Joel's account if it earns 4% simple interest over 2 years? Explain.

Problem Solving in Video Game Design

ALL FUN AND GAMES

Are you passionate about computer gaming? You might want to explore a career in video game design. A video game designer is responsible for a game's concept, layout, character development, and game-play. Game designers use math and logic to compute how different parts of a game will work.

21st Century Careers

Are you interested in a career as a video game designer? Take some of the following courses in high school.

- 3-D Digital Animation
- Introduction to Computer Literacy
- Introduction to Game Development

Video Game Sales History

Week	Japan Sales ($)	U.S. Sales ($)
1	580,510	1,213,264
2	185,528	415,320
3	149,045	263,825

Real-World Math

Use the information in the circle graph and the table to solve each problem.

1. How many of the top 20 video games sold were sports games?

2. Out of the top 20 video games sold, how many more music games were there than racer games?

3. In Week 1, the total sales for a video game were $2,374,136. What percent of the total sales was from the United States? Round to the nearest whole percent.

4. Find the percent of change in sales of the video game from Week 1 to Week 3 in Japan. Round to the nearest whole percent. State whether the percent of change is an *increase* or a *decrease*.

5. Which country had a greater percent decrease in sales from Week 1 to Week 2: Japan or the United States? Explain.

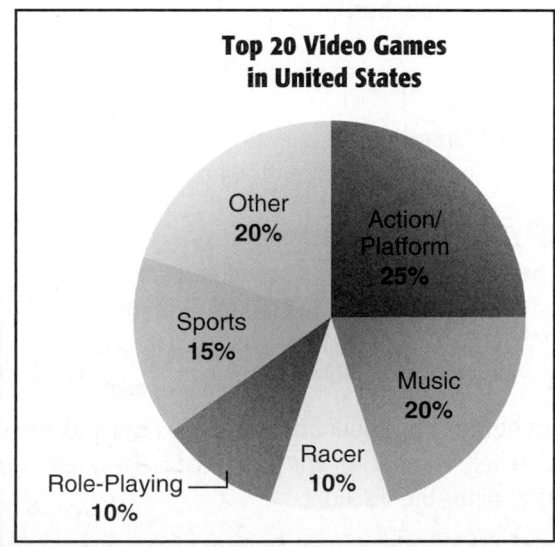

Top 20 Video Games in United States

Other 20%
Action/Platform 25%
Sports 15%
Music 20%
Racer 10%
Role-Playing 10%

Chapter Study Guide and Review

FOLDABLES® Study Organizer

Be sure the following Key Concepts are noted in your Foldable.

Key Concepts

Percents (Lesson 1)
- To find the percent of a number, first write the percent as either a fraction or decimal and then multiply.

Proportions and Equations (Lesson 2)
- **Percent Proportion**

$$\frac{\text{part}}{\text{whole}} = \frac{n}{100} \Big\} \text{percent}$$

- **Percent Equation**

$$\text{part} = \text{percent} \cdot \text{whole}$$

Applying Percents (Lesson 3)
- A **percent of change** is a ratio that compares the change in quantity to the original amount.

$$\text{percent of change} = \frac{\text{amount of change}}{\text{original amount}}$$

- **Discount** is the amount by which the regular price of an item is reduced. The sale price is the regular price minus the discount.

- **Sales tax** is an additional amount of money charged on items. The total cost of an item is the regular price plus the sales tax.

Key Vocabulary

discount	percent proportion
gratuity	principal
percent equation	sales tax
percent of change	simple interest
percent of decrease	tip
percent of increase	

Vocabulary Check

State whether each sentence is *true* or *false*. If *false*, replace the underlined word or number to make a true sentence.

1. The sale price of a discounted item is the regular price <u>minus</u> the discount.

2. A ratio that compares the change in quantity to the original amount is called the <u>percent of change</u>.

3. A <u>percent proportion</u> compares part of a quantity to the whole quantity using a percent.

4. The formula for simple interest is <u>$I = prt$</u>.

5. One way to find the total bill with tip is to add the percent of tip to <u>10%</u>.

6. The equation part = percent · whole is known as the <u>principal</u> equation.

7. The <u>principal</u> is the amount of money deposited or borrowed.

8. A <u>tax</u> is the amount by which the regular price of an item is reduced.

9. To find a percent of increase, compare the amount of the increase to the <u>new</u> amount.

10. If the new amount is greater than the original amount, then the percent of change is percent of <u>decrease</u>.

Multi-Part Lesson Review

Lesson 1   Percents

Percent of a Number (Lesson 1B)

Find each number. Round to the nearest tenth if necessary.

11. Find 78% of 50.

12. 45.5% of 75 is what number?

13. What is 225% of 60?

14. 0.75% of 80 is what number?

EXAMPLE 1   Find 24% of 200.

24% of 200

$= 24\% \times 200$ Write the expression.

$= 0.24 \times 200$ Write 24% as a decimal.

$= 48$ Multiply.

So, 24% of 200 is 48.

Percent and Estimation (Lesson 1C)

Estimate.

15. 25% of 81   **16.** 33% of 122

17. 77% of 38   **18.** 19.5% of 96

Estimate by using 10%.

19. 12% of 77   **20.** 88% of 400

EXAMPLE 2   Estimate 52% of 495.

$52\% \approx 50\%$ or $\frac{1}{2}$, and $495 \approx 500$.

52% of $495 \approx \frac{1}{2} \cdot 500$ or 250.

So, 52% of 495 is about 250.

Lesson 2   Proportions and Equations

The Percent Proportion (Lesson 2B)

Find each number. Round to the nearest tenth if necessary.

21. 6 is what percent of 120?

22. Find 0.8% of 35.

23. What percent of 375 is 40?

24. SOCCER A soccer team lost 30% of their games. If they played 20 games, how many did they win?

25. PHONE SERVICE A family pays $21.99 each month for its long distance phone service. This is 80% of the original price of the phone service. What is the original price of the phone service? Round to the nearest cent if necessary.

EXAMPLE 3   What percent of 90 is 18?

$\frac{18}{90} = \frac{n}{100}$ Write the proportion.

$18 \cdot 100 = 90 \cdot n$ Find the cross products.

$\frac{1,800}{90} = \frac{90n}{90}$ Divide each side by 90.

$20 = n$

So, 18 is 20% of 90.

EXAMPLE 4   52 is 65% of what number?

$\frac{52}{w} = \frac{65}{100}$ Write the proportion.

$52 \cdot 100 = w \cdot 65$ Find the cross products.

$\frac{5,200}{65} = \frac{65w}{65}$ Divide each side by 65.

$80 = w$

So, 52 is 65% of 80.

Lesson 2 ## Proportions and Equations (continued)

The Percent Equation (Lesson 2C)

Write an equation for each problem. Then solve. Round to the nearest tenth if necessary.

26. 32 is what percent of 50?

27. 65% of what number is 39?

28. SALONS A salon increased the sales of hair products by about 12.5% this week. If they sold 43 hair products last week, how many did they sell this week?

EXAMPLE 5 **27 is what percent of 90?**

27 is the part and 90 is the base. Let n represent the percent.

$$\underbrace{\text{part}}_{} = \underbrace{\text{percent}}_{} \cdot \underbrace{\text{base}}_{}$$

| $27 =$ | n | $\cdot$ | 90 | Write an equation. |

$$\frac{27}{90} = \frac{90n}{90}$$ Divide each side by 90.

$$0.3 = n$$ The percent is 30%.

So, 27 is 30% of 90.

PSI: Determine Reasonable Answers (Lesson 2D)

Determine a reasonable answer for the problem.

29. CABLE TV In a survey of 1,813 consumers, 18% said that they would be willing to pay more for cable if they got more channels. Is 3.3, 33, or 333 a reasonable estimate for the number of consumers willing to pay more for cable?

EXAMPLE 6 **Out of 394 students, 24.8% participate in music programs. What is a reasonable estimate for the number of students who participate in music programs?**

$394 \times 0.248 \rightarrow 400 \times 0.25$ or 100 students

So, about 100 students.

Lesson 3 ## Applying Percents

Percent of Change (Lesson 3B)

Find each percent of change. Round to the nearest whole percent if necessary. State whether the percent of change is an *increase* or *decrease*.

30. original: 172
new: 254

31. original: $200
new: $386

32. original: 75
new: 60

33. original: $49.95
new: $54.95

34. COMICS Tyree bought a collectible comic book for $49.62 last year. This year, he sold it for $52.10. Find the percent of change of the price of the comic book. Round to the nearest percent.

EXAMPLE 7 **A magazine that originally cost $2.75 is now $3.55. Find the percent of change. Round to the nearest whole percent.**

Find the amount of change.
$3.55 - 2.75 = 0.80$

$$\text{percent of change} = \frac{\text{amount of change}}{\text{original amount}}$$

$$= \frac{0.80}{2.75}$$ Substitution

$$\approx 0.29$$ Simplify.

The percent of change is about 29%. Since the percent of change is positive, this is a percent of *increase*.

Applying Percents (continued)

Sales Tax and Tips (Lesson 3C)

Find the total cost to the nearest cent.

35. $12 haircut; 15% tip

36. $48 dinner; 18% gratuity

37. **RESTAURANTS** A restaurant bill comes to $42.75. Find the total cost if the tax is 6% and a 15% tip is left on the amount after the tax is added.

EXAMPLE 8 Raymond wants to tip the server for his dinner. If the dinner costs $24 and he wants to give the server a 15% tip, what is the total cost?

15% of 24 = 0.15 × 24	Write 15% as a decimal.
= 3.60	The tip is $3.60.
$24.00 + 3.60 = $27.60	Add the tip and cost of the dinner.

The total cost is $27.60.

Discount (Lesson 3D)

Find the sale price to the nearest cent.

38. $45 roller blades; 20% discount

39. $15 T-shirt; 25% discount

40. **ELECTRONICS** A new radio is priced at $30. The store has an end-of-year sale and all their items are 40% off. What is the sale price of the radio?

EXAMPLE 9 A pair of shoes normally costs $40. This week they are on sale for 20% off. What is the sale price of the shoes?

20% of 40 → 0.20 × 40	Write 20% as a decimal.
= 8.00	The discount is $8.00.
$40 − $8 = $32	Subtract the discount from the cost.

The sale price is $32.

Financial Literacy: Simple Interest (Lesson 3E)

Find the interest earned to the nearest cent for the principal, interest rate, and time.

41. $475, 5%, 2 years

Find the interest paid to the nearest cent for the loan balance, interest rate, and time.

42. $3,200, 8%, 4 years

43. **SAVINGS** Aleta deposited $450 into a savings account earning 3.75% annual simple interest. How much interest will she earn in 6 years?

EXAMPLE 10 Find the interest earned on $400 at 9% for 3 years.

$I = prt$	Simple interest formula
$I = 400 \cdot 0.09 \cdot 3$	$p = \$400, r = 0.09, t = 3$
$I = 108$	Simplify.

The interest earned is $108.

Find each number. Round to the nearest tenth if necessary.

1. Find 55% of 164.

2. What is 355% of 15?

3. Find 25% of 80.

4. **MULTIPLE CHOICE** Out of 365 students, 210 bought a hot lunch. About what percent of the students did NOT buy a hot lunch?

   A. 35%   C. 56%

   B. 42%   D. 78%

Estimate.

5. 18% of 246   6. 145% of 81

7. 71% of 324   8. 56% of 65.4

9. **COMMUNICATION** Carla makes a long distance phone call and talks for 50 minutes. Of these minutes, 25% were spent talking to her brother. Would the time spent talking with her brother be about 8, 12, or 15 minutes? Explain your reasoning.

Write an equation for each problem. Then solve. Round to the nearest tenth if necessary.

10. Find 14% of 65.

11. What number is 36% of 294?

12. 82% of what number is 73.8?

13. 75 is what percent of 50?

Find each percent of change. Round to the nearest whole percent if necessary. State whether the percent of change is an *increase* or a *decrease*.

14. $60 to $75

15. 145 meters to 216 meters

16. 48 minutes to 40 minutes

Find the total cost or sale price to the nearest cent.

17. $2,200 computer, $6\frac{1}{2}$% sales tax

18. $16 hat, 55% discount

19. $35.49 jeans, 33% discount

Find the simple interest earned to the nearest cent for each principal, interest rate, and time.

20. $750, 3%, 4 years

21. $1,050, 4.6%, 2 years

22. $2,600, 4%, 3 months

23. **MULTIPLE CHOICE** Mr. Glover borrows $3,500 to renovate his home. His loan has an annual simple interest rate of 15%. If he pays off the loan after 6 months, about how much will he pay in all?

   F. $3,500

   G. $3,720

   H. $3,763

   I. $4,025

24. THINK SOLVE EXPLAIN **EXTENDED RESPONSE** Use the table below. It shows the results of a survey in which 175 students were asked what type of food they wanted for their class party.

Type of Food	Percent
Subs	32%
Tex-Mex	56%
Italian	12%

Part A How many of the 175 students chose Italian food? Explain.

Part B How many students chose Tex-Mex food?

 ## Gridded Response: Percents

When the answer to a gridded-response question is a percent, do not convert the percent to a decimal or fraction. Grid in the percent value without the % symbol.

TEST EXAMPLE

Noah took 72 pictures with his new camera. Of those, 54 were taken while he was at summer camp. What percent of Noah's pictures were taken at summer camp?

$$\frac{54}{72} = \frac{n}{100}$$ Write the proportion.

$54 \cdot 100 = 72 \cdot n$ Find the cross products.

$$\frac{5400}{72} = \frac{72n}{72}$$ Simplify and divide each side by 72.

$$75 = n$$

So, 75% of Noah's pictures were taken at summer camp. Grid in 75.

Correct

NOT Correct

 ## Work on It

Out of the 215 students who play a sport, 86 play more than one sport. What percent of the student athletes play more than one sport? Fill in your answer on a grid like the one shown above.

Test Hint

If your answer will not fit in the response grid, then solve the problem again.

✓ Test Practice

Read each question. Then fill in the correct answer on the answer sheet provided by your teacher or on a sheet of paper.

1. Sarah wants to buy new pillows for her room. Which store offers the best buy on pillows?

Store	Sale Price
A	3 pillows for $40
B	4 pillows for $50
C	2 pillows for $19
D	1 pillow for $11

- **A.** Store A
- **B.** Store B
- **C.** Store C
- **D.** Store D

2. The graph shows the attendance at a summer art festival from 2002 to 2007. If the trend in attendance continues, which of the following is a reasonable prediction for the attendance in 2011?

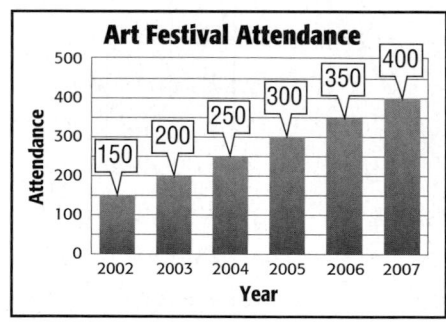

- **F.** Fewer than 200
- **G.** Between 500 and 600
- **H.** Between 700 and 800
- **I.** More than 800

3. At their annual car wash, the science club washes 30 cars in 45 minutes. At this rate, how many cars will they wash in 1 hour?

- **A.** 40
- **B.** 45
- **C.** 50
- **D.** 60

4. ✎ **GRIDDED RESPONSE** A necklace regularly sells for $18.00. The store advertises a 15% discount. What is the sale price of the necklace in dollars?

5. ✎ **GRIDDED RESPONSE** At a middle school, 38% of all seventh graders have taken swimming lessons. There are 250 students in the seventh grade. How many of them have taken swimming lessons?

6. The cost of Ken's car wash was $23.95. If he wants to give his detailer a 15% tip, about how much of a tip should he leave?

- **F.** $2.40
- **G.** $3.60
- **H.** $4.60
- **I.** $4.80

7. An architect made a model of an office building using a scale of 1 inch equals 3 meters. If the height of the model is 12.5 inches, which of the following represents the actual height of the building?

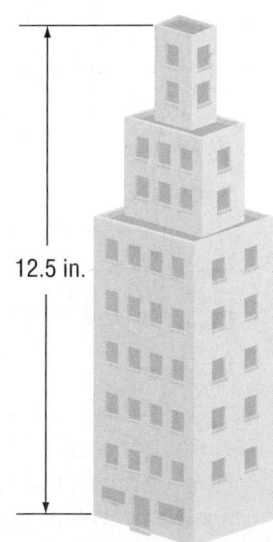

12.5 in.

- **A.** 40.0 m
- **B.** 37.5 m
- **C.** 36.0 m
- **D.** 28.4 m

8. At a pet store, 38% of the animals are dogs. If there are a total of 88 animals at the pet store, which proportion can be used to find x, the number of dogs at the pet store?

F. $\dfrac{x}{88} = \dfrac{100}{38}$

G. $\dfrac{38}{88} = \dfrac{100}{x}$

H. $\dfrac{x}{88} = \dfrac{38}{100}$

I. $\dfrac{100}{88} = \dfrac{x}{38}$

9. ✏️ **GRIDDED RESPONSE** A wrestler competes in 25 matches. Of those matches, he wins 17. What percent of the matches did the wrestler win?

10. **THINK SOLVE EXPLAIN** **SHORT RESPONSE** The average cost per month of a 2-bedroom apartment in Grayson was $625 last year. This year, the average cost is $650. What is the percent of increase from last year to this year?

11. Mr. Cooper asked his students whether they prefer to go to the aquarium or the planetarium for a field trip. The table shows the results.

Response	Percent
Aquarium	50
Planetarium	25

Suppose the rest of the class had no preference. What is the ratio of students who have no preference to the students who prefer to go to the aquarium?

A. 1:5   C. 1:3

B. 1:4   D. 1:2

12. In Nadia's DVD collection, she has 8 action DVDs, 12 comedy DVDs, 7 romance DVDs, and 3 science fiction DVDs. What percent of Nadia's DVD collection is comedies?

F. 25%

G. 30%

H. 35%

I. 40%

13. A salesman needed to sell a four-wheeler. He priced it at $3,500 the first day it was on the market. The second day he reduced the price by 10%. What was the price of the four-wheeler after this reduction?

A. $3,850

B. $3,465

C. $3,150

D. $3,000

14. **THINK SOLVE EXPLAIN** **EXTENDED RESPONSE** Cable Company A increases their rates from $98 a month to $101.92 a month.

Part A What is the percent of increase?

Part B Cable Company B offers their cable for $110 dollars a month but gives a 10% discount for new customers. Describe two ways to find the cost for new customers.

Part C If you currently use Cable Company A, would it make sense to change to Cable Company B? Explain.

NEED EXTRA HELP?

If You Missed Question...	1	2	3	4	5	6	7	8	9	10	11	12	13	14
Go to Chapter-Lesson...	5-1B	6-2D	5-1B	6-3D	6-2B	6-1B	5-2B	6-2B	6-2C	6-3B	5-1B	6-2B	6-3D	6-3B

Linear Functions

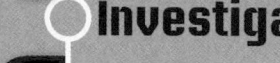
connectED.mcgraw-hill.com

Investigate

Animations

Vocabulary

Multilingual eGlossary

Learn

Personal Tutor

Virtual Manipulatives

Graphing Calculator

Audio

Foldables

Practice

Self-Check Practice

Worksheets

Assessment

The ☆BIG Idea

What are the different ways to represent the relationship between two sets of numbers?

FOLDABLES Study Organizer

Make this Foldable to help you organize your notes.

Review Vocabulary

expression **expresión** a combination of numbers, variables, and at least one operation

$$2x + 5$$

equation **equación** a mathematical sentence that contains an equal sign, =

$$2x + 5 = 13$$

Key Vocabulary

English	Español
function	función
linear function	función lineal
rate of change	tasa de cambio
slope	pendiente

When Will I Use This?

Are You Ready for the Chapter?

You have two options for checking prerequisite skills for this chapter.

Text Option Take the Quick Check below. Refer to the Quick Review for help.

QUICK Check

Evaluate each expression if $a = 4$, $b = 10$, and $c = 8$.

1. $a + b + c$ **2.** $bc - ac$

3. $b + ac$ **4.** $4c + 3b$

5. $2b - (a + c)$ **6.** $2c - b + a$

Evaluate each expression if $x = 2$, $y = 5$, and $z = -1$.

7. $x + 12$ **8.** $z + (-5)$

9. $4y + 8$ **10.** $10 + 3z$

11. $(2 + y)9$ **12.** $6(x - 4)$

13. $3xy$ **14.** $2z + y$

Use the coordinate grid to name the point for each ordered pair.

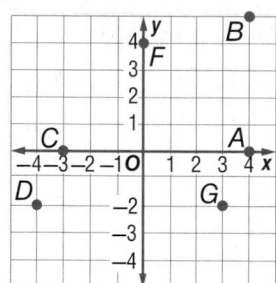

15. $(-3, 0)$

16. $(3, -2)$

17. $(-4, -2)$

18. $(0, 4)$

19. $(4, 5)$

20. $(4, 0)$

QUICK Review

EXAMPLE 1

Evaluate $3rt - s$ if $r = 3$, $s = -4$, and $t = 4$.

$3rt - s$ Write the expression.

$= 3(3)(4) - (-4)$ Substitute.

$= 36 - (-4)$ Multiply.

$= 36 + 4$ To subtract -4, add 4.

$= 40$ Add.

EXAMPLE 2

Graph and label the point $P(-3, 4)$.

Start at the origin. The first number in the ordered pair is the distance to the left or right on the x-axis. Since the number is -3, move three units to the *left*.

The second number in the ordered pair is the distance up or down on the y-axis. Since the number is 4, move *up* four units.

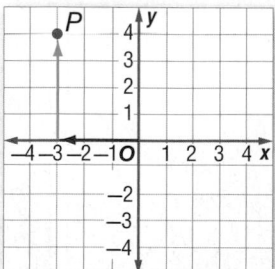

 **Online Option** Take the Online Readiness Quiz.

Explore Relations and Functions

Main Idea

Determine whether a relation is a function.

A *relation* expresses how objects in one group, *inputs,* are assigned or related to objects in another group, *outputs.*

The mapping diagram below shows three possible relations. If each item, or input, has only one price, or output, then the relation is also a *function.*

Function		Function		Not a Function	
Input	Output	Input	Output	Input	Output
Binder → $3.00		Binder → $3.00		Binder → $3.00	
Gel Pens → $1.50		Gel Pens → $3.00		Gel Pens → $1.50	
Notebook → $0.75		Notebook → $0.75		Notebook → $0.75	

The first two relations are functions because each item has only one price. The third relation is not a function because the binder has two prices, $3 and $1.50.

ACTIVITY

STEP 1   Copy the relation diagram. Draw lines from the input values to the output values so that the relation is a function.

STEP 2   Copy the relation diagram from Step 1. Draw lines from the input values to the output values so that the relation is *not* a function.

Input	Output
1	2
3	5
6	7
8	10

Analyze the Results

1. A relation can be written as a set of ordered pairs, with the input as the *x*-coordinate and the output as the *y*-coordinate. For each relation diagram you drew in the Activity above, write the relation as a set of ordered pairs.

2. Describe why each relation is or is not a function. Explain your reasoning in terms of the ordered pairs.

Main Idea
Make function tables and write equations.

Vocabulary
relation
function
function rule
function table
domain
range
independent variable
dependent variable
function notation

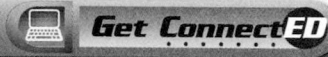

Equations and Functions

MAGAZINES Suppose you can buy magazines for $4 each.

1. Copy and complete the table to find the cost of 2, 3, and 4 magazines.

2. Describe the pattern in the table between the cost and the number of magazines.

Number	Multiply by 4	Cost ($)
1	4 × 1	4
2		
3		
4		

A **relation** is a set of ordered pairs. A **function** is a relation in which each member of the input is paired with *exactly* one member of the output. A **function rule** is the operation(s) performed on the input value to get the output value.

You can organize the input numbers, output numbers, and the function rule in a **function table**. The set of input values is called the **domain**, and the set of output values is called the **range**.

 EXAMPLE **Make a Function Table**

 MONEY Javier saves $20 each month. Make a function table to show his savings after 1, 2, 3, and 4 months. Then identify the domain and range.

The domain is {1, 2, 3, 4}, and the range is {20, 40, 60, 80}.

Input	Function Rule	Output
Number of Months	Multiply by 20	Total Savings ($)
1	20 × 1	20
2	20 × 2	40
3	20 × 3	60
4	20 × 4	80

 CHECK Your Progress

a. **MOVIES** A student movie ticket costs $3. Make a function table that shows the total cost for 1, 2, 3, and 4 tickets. Then identify the domain and range.

Functions are often written as equations with two variables—one to represent the input and one to represent the output. Here's an equation for the situation in Example 1.

Function rule: multiply by 20

$$20x = y$$

Input: number of months ⟶   ⟵ Output: total savings

REAL-WORLD EXAMPLES

2 **ANIMALS** An armadillo sleeps 19 hours each day. Write an equation using two variables to show the relationship between the number of hours h an armadillo sleeps in d days.

Input	Function Rule	Output
Number of Days (d)	Multiply by 19	Number of Hours Slept (h)
1	1×19	19
2	2×19	38
3	3×19	57
d	$d \times 19$	$19d$

Real-World Link · · · ·
Armadillos have the ability to remain underwater for as long as six minutes. Because of the density of its armor, an armadillo will sink in water unless it inflates its stomach with air, which often doubles its size.

Words	Number of hours slept equals number of days times 19 hours each day.
Variable	Let d represent the number of days. Let h represent the number of hours.
Equation	$h = 19d$

3 **How many hours does an armadillo sleep in 4 days?**

$h = 19d$ Write the equation.

$h = 19(4)$ Replace d with 4.

$h = 76$ Multiply.

An armadillo sleeps 76 hours in 4 days.

 CHECK Your Progress

BOTANIST A botanist discovers that a certain species of bamboo grows 4 inches each hour.

b. Write an equation using two variables to show the relationship between the growth g in inches of this bamboo plant in h hours.

c. Use your equation to explain how to find the growth in inches of this species of bamboo after 6 hours. Then solve.

Example 1   Copy and complete each function table. Then identify the domain and range.

1. $y = 3x$

x	3x	y
1	3 • 1	3
2	3 • 2	
3	3 • 3	
4		

2. $y = 4x$

x	4x	y
0	4 • 0	
1	4 • 1	
2		
3		

3. MUSIC Jonas downloads 8 songs each month onto his digital music player. Make a function table that shows the total number of songs downloaded after 1, 2, 3, and 4 months. Then identify the domain and range.

Examples 2 and 3   **4. SPORTS** The top speed reached by a race car is 231 miles per hour.

a. Write an equation using two variables to show the relationship between the number of miles *m* that a race car can travel in *h* hours.

b. Use your equation to explain how to find the distance in miles the race car will travel in 3 hours. Then solve.

Practice and Problem Solving

● = **Step-by-Step Solutions** begin on page R1.
Extra Practice begins on page EP2.

Example 1   Copy and complete each function table. Then identify the domain and range.

5. $y = 2x$

x	2x	y
0	2 • 0	0
1	2 • 1	
2		
3		

6. $y = 6x$

x	6x	y
1		
2		
3		
4		

 $y = 9x$

x	9x	y
1		
2		
3		
4		

Make a function table for each situation. Then identify the domain and range.

8. PIZZA A pizza shop sells 25 pizzas each hour. At the same rate, find the number of pizzas sold after 1, 2, 3, and 4 hours.

9. TYPING Suppose you can type 60 words per minute. What is the total number of words typed after 5, 10, 15, and 20 minutes?

Examples 2 and 3   **10. CELL PHONES** A cell phone provider charges a customer $40 for each month of service.

a. Write an equation using two variables to show the relationship between the total amount charged *c* after *m* months of cell phone service.

b. Use your equation to explain how to find the total cost for 6 months of cell phone service. Then solve.

11. INSECTS A cricket will chirp 35 times per minute when the outside temperature is 72°F.

  a. Write an equation using two variables to show the relationship between the total number of times a cricket will chirp t after m minutes at this temperature.

  b. Use your equation to explain how to find the number of times a cricket will have chirped after 15 minutes. Then solve.

Real-World Link · · · ·

Crickets are among the 800,000 different types of insects in the world.

Copy and complete each function table. Then identify the domain and range.

12. $y = 0.5x$

x	0.5x	y
1		
2		
3		
4		

13. $y = 1.3x$

x	1.3x	y
1		
2		
3		
4		

14. $y = -0.2x$

x	−0.2x	y
0		
1		
2		
3		

15. $y = \frac{2}{3}x$

x	$\frac{2}{3}x$	y
2		
3		
4		
5		

16. MEASUREMENT The formula for the area of a rectangle with a length of 6 units and width w is $A = 6w$.

  a. Make a function table that shows the area in square units of a rectangle with a width of 2, 3, 4, and 5 units.

  b. Study the pattern in your table. Explain how the area of a rectangle with a length of 6 units changes when the width is increased by 1 unit.

17 SCIENCE The table shows the approximate velocity of certain planets as they orbit the Sun.

  a. Write an equation to show the relationship between the total number of miles m Jupiter travels in s seconds as it orbits the Sun.

  b. What equation can be used to show the total number of miles m Earth travels in s seconds?

  c. Use your equation to explain how to find the number of miles Jupiter and Earth each travel in 1 minute. Then solve.

Orbital Velocity Around Sun	
Planet	**Velocity (mi/s)**
Mercury	30
Earth	19
Jupiter	8
Saturn	6
Neptune	5

18. RUNNING The average marathon running speed is 5.5 miles per hour. Find the number of miles run after 1, 2, 3, and 4 hours at this rate.

19. FIND THE ERROR Divya is finding the range for the function $y = 3x$ given the domain $\{2, 4, 6, 8\}$. Find her mistake and correct it.

> If $y = 3x$ and I have the inputs of 2, 4, 6, and 8, then the outputs should be 5, 7, 9, and 11.

20. OPEN ENDED Write about a real-world situation that could be represented by the function $12n$.

21. CHALLENGE A function is shown in the table below.

x	2	4	6	8
y	0.5	1.0	1.5	2

   a. Write a function rule for the function.

   b. Use the rule from part **a.** Find the value of y, if $x = 10, 12$, and 14.

22. CHALLENGE Write a rule for the function represented by $\{(1, 2), (2, 5), (3, 10), (4, 17)\}$.

23. **WRITE MATH** Explain the relationship among an *input*, an *output*, and a *function rule*.

Test Practice

24. The table shows the number of hand painted T-shirts Mi-Ling can make after a given number of days.

Number of Days, x	Total Number of T-Shirts, y
1	6
2	12
3	18
4	24

Which function rule represents the data?

A. $y = 4x$

B. $y = 5x$

C. $y = 6x$

D. $y = 12x$

25. Rosa needs to have 50 posters printed to advertise a community book fair. The printing company charges $3 to print each poster. Which table represents this situation?

F.

Posters	Cost ($)
3	3
6	6
9	9
p	p

G.

Posters	Cost ($)
1	3
2	6
3	9
p	3p

H.

Posters	Cost ($)
1	3
2	6
3	9
p	3 + p

I.

Posters	Cost ($)
3	1
6	2
9	3
p	p ÷ 3

More About Relations and Functions

Refer to the beginning of the lesson. The total cost (output) of the magazines depends on, or is a function of, the number of magazines bought (input). The number of magazines bought is called the **independent variable** because the values are chosen and do not depend upon the other variable. The total cost is called the **dependent variable** because it depends on the input value.

In a set of ordered pairs, the *x*-value is the independent variable, and the *y*-value is the dependent variable.

EXAMPLE

1 **Determine whether each relation is a function. Explain.**

a. {(2, 1), (4, 3), (5, 4), (9, 7)}

b.

x	2	5	9	2
y	1	4	7	3

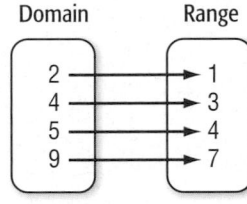

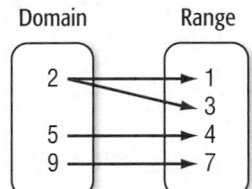

The set of ordered pairs is a function because each *x*-value has only one corresponding *y*-value.

This is not a function because the *x*-value has two different *y*-values.

Another way to determine whether a relation is a function is to apply the *vertical line test* to the graph of the relation. Use a pencil or straightedge to represent a vertical line.

Place the pencil at the left of the graph. Move it to the right across the graph. If, for each value of *x* in the domain, the pencil passes through only one point on the graph, then the graph represents a function.

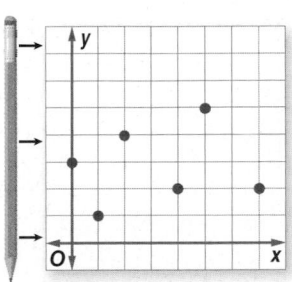

EXAMPLE

2 **Determine whether the graph is a function. Explain your answer.**

The graph represents a relation that is not a function because it does not pass the vertical line test.

By looking at the graph, you can see that when *x* = 3, there are two different *y*-values, 2 and 6.

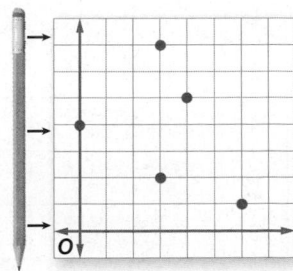

A function that is written as an equation can also be written in a form called **function notation**. Consider the equation $y = 4x$.

Equation	Function notation
$y = 4x$	$f(x) = 4x$

The function notation, $f(x)$ is read *f of x*.

The variable y and $f(x)$ both represent the dependent variable. To find the value of the function for a certain input, substitute the input for the variable x.

 EXAMPLE

3 Find $f(3)$ if $f(x) = 5x$.

$f(x) = 5x$ Write the function.

$f(3) = 5(3)$ Substitute 3 for *x* into the function rule.

So, $f(3) = 15$.

Determine whether each relation is a function. Explain.

26. $\{(4, 12), (5, 16), (5, 17), (6, 10)\}$

27. $\{(8, 8), (9, 9), (10, 10), (11, 11)\}$

28.

x	2	5	8	11
y	3	3	5	7

29.

x	12	16	20	12
y	4	8	12	16

30.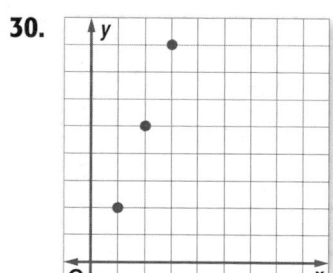

31.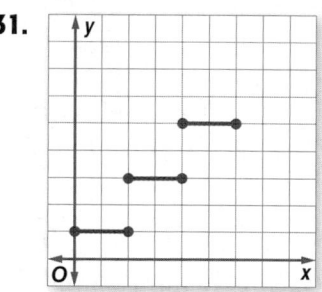

Find the value of each function.

32. $f(4)$ if $f(x) = 2x$

33. $f(-5)$ if $f(x) = 20x$

34. $f(-8)$ if $f(x) = -5x + 10$

Determine a rule for each set of ordered pairs. Then copy and complete each function table.

35.

x	?	f(x)
2		−1
4		1
6		3
8		

36.

x	?	f(x)
1		3
2		7
3		11
4		

37.

x	?	f(x)
0		6
1		8
3		12
5		

Main Idea

Graph data to demonstrate relationships.

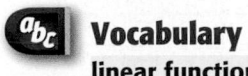

Vocabulary

linear function

Functions and Graphs

MONEY The Westerville Marching Band is going on a year-end trip to an amusement park. Each band member must pay an admission price of $15. In the table, this is represented by 15m.

Total Cost of Admission		
Number of Members	15m	Total Cost ($)
1	15(1)	15
2	15(2)	30
3	15(3)	
4		
5		
6		

1. Copy and complete the function table for the total cost of admission.

2. Graph the ordered pairs (number of members, total cost).

3. Describe how the points appear on the graph.

The total cost is a *function* of the number of band members. In general, the output y is a function of the input x.

The graph of the function consists of the points in the coordinate plane that correspond to *all* the ordered pairs of the form (input, output) or (x, y).

REAL-WORLD EXAMPLE

1 **TEMPERATURE** The table shows temperatures in Celsius and the corresponding temperatures in Fahrenheit. Make a graph of the data to show the relationship between Celsius and Fahrenheit.

The ordered pairs (5, 41), (10, 50), (15, 59), (20, 68), (25, 77), and (30, 86) represent this function. Graph the ordered pairs.

Celsius (input)	Fahrenheit (output)
5	41
10	50
15	59
20	68
25	77
30	86

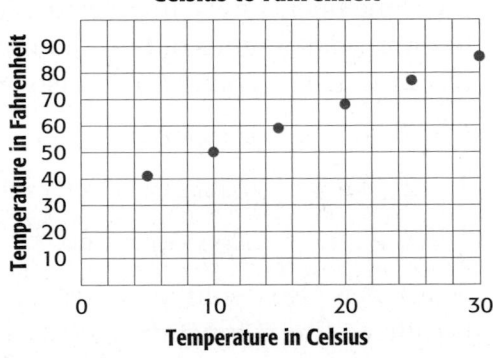

Celsius to Fahrenheit

 **CHECK** Your Progress

a. MUSIC The table shows the money remaining on a $75 gift certificate after a certain number of CDs are bought. Make a graph to show how the number of CDs bought and the remaining balance are related.

$75 Music Gift Certificate	
Number of CDs	**Balance ($)**
1	63
2	51
3	39
4	27
5	15

Vocabulary Link

Everyday Use

Function The purpose for which something is designed or exists, or its role.

Math Use

Function A relationship that assigns one output value for each input value.

The solution of an equation with two variables consists of two numbers, one for each variable, that make the equation true. The solution is usually written as an ordered pair (x, y).

 **EXAMPLE** **Graph Solutions of Linear Equations**

 Graph $y = 2x + 1$.

Select any four values for the input x. We chose 2, 1, 0, and -1. Substitute these values for x to find the output y.

x	$2x + 1$	y	(x, y)
2	$2(2) + 1$	5	$(2, 5)$
1	$2(1) + 1$	3	$(1, 3)$
0	$2(0) + 1$	1	$(0, 1)$
-1	$2(-1) + 1$	-1	$(-1, -1)$

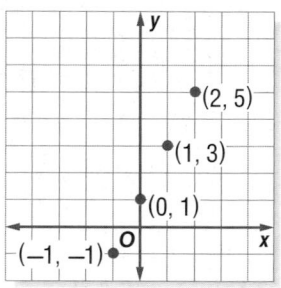

The four inputs correspond to the solutions $(2, 5)$, $(1, 3)$, $(0, 1)$, and $(-1, -1)$. Graph these ordered pairs to graph $y = 2x + 1$.

 CHECK Your Progress

Graph each equation.

b. $y = x - 3$ **c.** $y = -3x$ **d.** $y = -3x + 2$

Study Tip

Graphing Equations Only two points are needed to graph the line. However, you can graph more points to check accuracy.

Notice that all four points in the graph lie on the same line. Draw a line through the points to graph *all* solutions of the equation $y = 2x + 1$. The point $(3, 7)$ is also on this line.

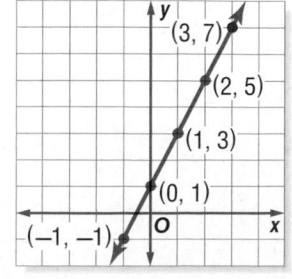

$y = 2x + 1$ Write the equation.

$7 \stackrel{?}{=} 2(3) + 1$ Replace x with 3 and y with 7.

$7 = 7$ ✓   This sentence is true.

So, $(3, 7)$ is also a solution of $y = 2x + 1$. A function like $y = 2x + 1$ is called a **linear function** because its graph is a line.

3 **SWIMMING** Michael Phelps swims the 400-meter individual medley at an average speed of 100 meters per minute. The equation $d = 100t$ describes the distance d that he can swim in t minutes at this speed. Represent the function by a graph.

Step 1 Select any four values for t. Select only positive numbers because t represents time. Make a function table.

t	$100t$	d	(t, d)
1	100(1)	100	(1, 100)
2	100(2)	200	(2, 200)
3	100(3)	300	(3, 300)
4	100(4)	400	(4, 400)

Step 2 Graph the ordered pairs and draw a line through the points.

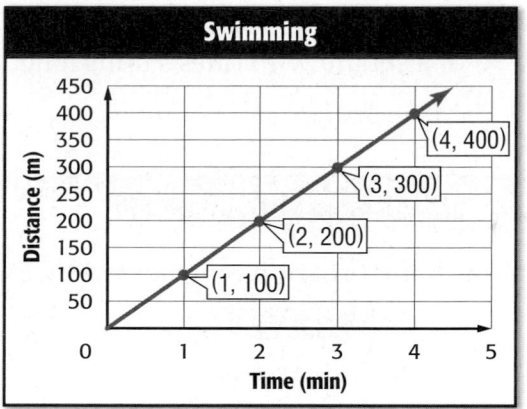

QUICK Review

When drawing a graph, include a title and labels for the horizontal and vertical axes.

CHECK Your Progress

e. JOBS Sandi makes $6 an hour babysitting. The equation $m = 6h$ describes how much money m she earns babysitting for h hours. Represent this function by a graph.

Key Concept **Representing Functions**

Words There are 12 inches in one foot.

Table

Feet (f)	Inches (n)
1	12
2	24
3	36
4	48

Graph

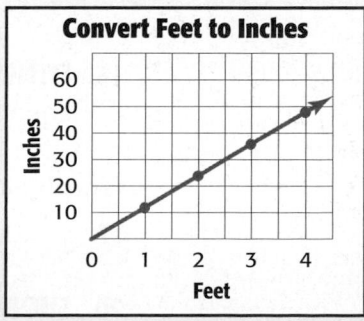

Equation $n = 12f$, where f represents the number of feet and n represents the number of inches.

Example 1 Graph the function represented by each table.

1.

Total Cost of Baseballs	
Baseball	Total Cost ($)
1	4
2	8
3	12
4	16

2.

Savings Account	
Week	Amount ($)
0	300
1	350
2	400
3	450

Example 2 Graph each equation.

3 $y = x - 1$ **4.** $y = -1x$ **5.** $y = -2x + 3$

Example 3 **6. MEASUREMENT** The perimeter of a square is 4 times as great as the length of any of its sides. The equation $p = 4s$ describes the perimeter p of a square with sides s units long. Represent this function by a graph.

Practice and Problem Solving

 = **Step-by-Step Solutions** begin on page R1.
Extra Practice begins on page EP2.

Example 1 Graph the function represented by each table.

7.

Total Phone Bill	
Time (min)	Total ($)
0	10.00
1	10.08
2	10.16
3	10.24

8.

Calories in Fruit Cups	
Servings	Total Calories
1	70
3	210
5	350
7	490

Example 2 Graph each equation.

9. $y = x + 1$ **10.** $y = x + 3$ **11.** $y = x$

12. $y = -2x$ **13.** $y = 2x + 3$ **14.** $y = 3x - 1$

Example 3 Represent each function by a graph.

15. CARS A car averages 36 miles per gallon of gasoline. The function $m = 36g$ represents the miles m driven using g gallons of gasoline.

16. FITNESS A health club charges $35 a month for membership fees. The equation $c = 35m$ describes the total charge c for m months of membership.

Graph each equation.

17. $y = 0.25x$ **18.** $y = x + 0.5$ **19** $y = 0.5x - 1$

20. SHOPPING You buy a DVD for $14 and CDs for $9 each. The equation $t = 14 + 9c$ represents the total amount t that you spend if you buy 1 DVD and c CDs. Represent this function by a graph.

21. **OPEN ENDED** Draw the graph of a linear function. Name three ordered pairs that satisfy the function.

22. **CHALLENGE** Darrell has recorded the greenhouse temperature at certain times of the day in the table. The greenhouse was advertised to maintain temperatures between 65°F and 85°F.

    a. What is the rate of increase as a unit rate?

    b. Create a new table that shows the temperatures recorded hourly from 1:00 A.M. to 8:00 A.M. using the given data in the table.

    c. Create a graph of the time and temperatures from midnight to 9:00 A.M. Is the relationship linear? Explain.

Time	Temperature (°F)
1:00 A.M.	66
3:00 A.M.	71
6:00 A.M.	78.5
8:00 A.M.	83.5

23. **WRITE MATH** Name a set of ordered pairs that does *not* represent a function. Explain your reasoning.

Test Practice

24. The graph shows the relationship between the number of hours Serefina spent jogging and the total number of miles she jogged. Which table **best** represents the data in the graph?

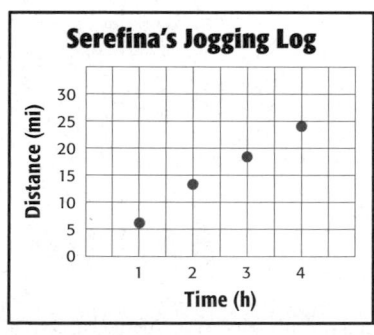

A.

Time (h)	Distance (mi)
6	4
12	3
18	2
24	1

B.

Time (h)	Distance (mi)
2	6
3	12
4	18
5	24

C.

Time (h)	Distance (mi)
1	6
2	14
3	18
4	24

D.

Time (h)	Distance (mi)
4	6
3	6
2	6
1	6

25. Which ordered pair is NOT a solution of $y = 5x + 2$?

    F. $(0, 5)$ **H.** $(2, 12)$

    G. $(1, 7)$ **I.** $(3, 17)$

Extend

Main Idea

Use technology to graph relationships involving conversions of measurement.

Get ConnectED

Graphing Technology:
Graphing Relationships

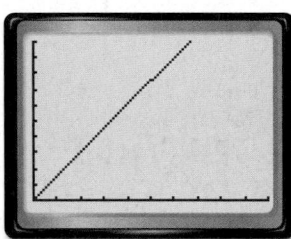

ACTIVITY

MEASUREMENT Use the table at the right to write a function that relates the number of yards x to the number of feet y. Then graph your function.

Yards (x)	Feet (y)
1	3
2	6
3	9
4	12

STEP 1 By examining the table, you can see that the number of feet is 3 times the number of yards. Write a function.

The number of feet	is	3 times	the number of yards.
y	$=$	3	x

STEP 2 Press $\boxed{Y=}$ and enter the function $y = 3x$ into Y1.

STEP 3 Press $\boxed{\text{WINDOW}}$ and enter the following values: Xmin $= 0$, Xmax $= 10$, Xscl: 1, Ymin: 0, Ymax: 20, Yscl: 2.

STEP 4 Graph the function by pressing $\boxed{\text{GRAPH}}$.

Analyze the Results

1. Test the function above using one of the values from the table and the CALC feature on your calculator. Press $\boxed{\text{2nd}}$ [CALC] 1 and then enter an x-value of 3. What y-value is displayed? What do each of these values represent and how are they represented on the graph?

2. Use your graph to convert 7 yards into feet. Explain your method.

3. **MAKE A CONJECTURE** Write a function that could be used to convert feet into yards. What is an appropriate window for a graph of this function? Graph and test your function.

4. Use your function from Exercise 3 to convert 16 feet into yards.

5. Write a function that could be used to convert 36 ounces to pounds. Indicate an appropriate window, then use a graph of the function to convert 36 ounces to pounds. (*Hint*: 1 pound = 16 ounces)

Explore Rate of Change

Main Idea

Understand slope as it relates to rate of change.

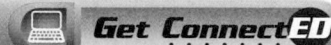 **Get Connect**ED

PET CARE The Happy Hound is a doggie daycare where people can drop off their dogs while they are at work. Farah takes her dog to the Happy Hound several days a week. The table shows their prices. Use a graph to determine how the number of hours is related to the cost.

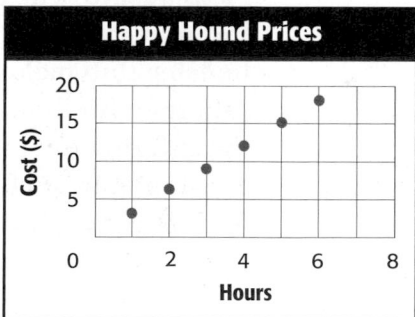

Pet Care 🐾	
Number of Hours	Cost ($)
1	3.00
2	6.00
3	9.00
4	12.00
5	15.00
6	18.00

ACTIVITY

What do you need to find? the relationship between number of hours and cost for pet care at Happy Hound

The cost is a function of the number of hours. So, the cost is the output *y*, and the number of hours is the input *x*. Create a graph of the data.

Happy Hound Prices

Graph with y-axis labeled Cost ($) marked 5, 10, 15, 20 and x-axis labeled Hours marked 0, 2, 4, 6, 8, showing points at (1,3), (2,6), (3,9), (4,12), (5,15), (6,18).

Analyze the Results

1. Is the graph linear? Explain.

2. What is the cost per hour, or unit rate, charged by the Happy Hound?

3. Examine any two consecutive ordered pairs from the table. How do the values change?

4. Is this relationship true for any two consecutive values in the table?

5. Use the graph to examine any two consecutive points. By how much does *y* change? By how much does *x* change?

6. How does this change relate to the unit rate?

Main Idea
Identify constant rate of change using tables and graphs.

 Vocabulary
rate of change
constant rate of change
nonlinear function

Constant Rate of Change

HEIGHTS The table shows Horacio's height at ages 9 and 12.

Age (yr)	9	12
Height (in.)	53	59

1. What is the change in Horacio's height from ages 9 to 12?

2. Over what number of years did this change take place?

3. Write a rate that compares the change in Horacio's height to the change in age. Express your answer as a unit rate and explain its meaning.

A **rate of change** is a rate that describes how one quantity changes in relation to another. A rate of change is usually expressed as a unit rate. A **constant rate of change** is the rate of change in a linear relationship.

 EXAMPLE Use a Table

1. **FUNDRAISING** The table shows the amount of money a booster club makes washing cars for a fundraiser. Use the information to find the constant rate of change in dollars per car.

Cars Washed	
Number	**Money ($)**
5	40
10	80
15	120
20	160

+5 (...) +40
+5 (...) +40
+5 (...) +40

Find the unit rate to determine the constant rate of change.

$$\frac{\text{change in money}}{\text{change in cars}} = \frac{40 \text{ dollars}}{5 \text{ cars}}$$ The money earned increases by $40 for every 5 cars.

$$= \frac{8 \text{ dollars}}{1 \text{ car}}$$ Write as a unit rate.

So, the number of dollars earned increases by $8 for every car washed.

CHECK Your Progress

a. **PLANES** The table shows the number of miles a plane traveled while in flight. Use the information to find the constant rate of change in miles per minute.

Time (min)	30	60	90	120
Distance (mi)	290	580	870	1,160

 **EXAMPLE** **Use a Graph**

 **Read Math**

Ordered Pairs The ordered pair (2, 120) represents traveling 120 miles in 2 hours.

2 **DRIVING** The graph represents the distance traveled while driving on a highway. Use the graph to find the constant rate of change in miles per hour.

To find the rate of change, pick any two points on the line, such as (1, 60) and (2, 120).

$$\frac{\text{change in miles}}{\text{change in hours}} = \frac{(120 - 60) \text{ miles}}{(2 - 1) \text{ hours}}$$

$$= \frac{60 \text{ miles}}{1 \text{ hour}}$$

The distance increases by 60 miles in 1 hour. So, the rate of traveling on a highway is 60 miles per hour.

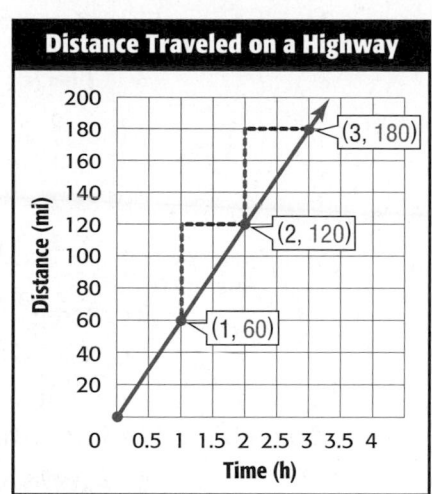

Distance Traveled on a Highway

CHECK Your Progress

b. DRIVING Use the graph to find the constant rate of change in miles per hour while driving in the city.

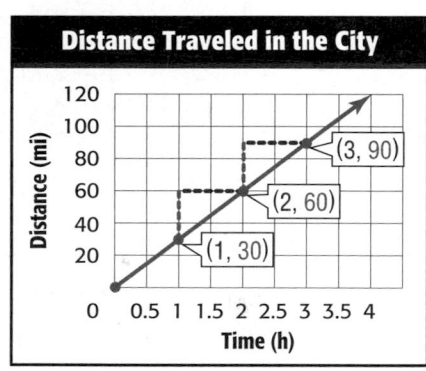

Distance Traveled in the City

Notice that the graph in Example 2 about driving on a highway represents a rate of change of 60 mph. The graph in Check Your Progress about driving in the city is not as steep. It represents a rate of change of 30 mph.

✓ CHECK Your Understanding

Example 1 **1** Use the information in the table to find the constant rate of change in degrees per hour.

Temperature (°F)	54	57	60	63
Time	6 A.M.	8 A.M.	10 A.M.	12 A.M.

Example 2 **2. DISTANCE** The graph shows Benito's distance from the starting line. Use the graph to find the constant rate of change.

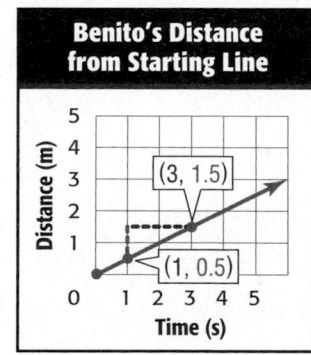

Benito's Distance from Starting Line

Practice and Problem Solving

= **Step-by-Step Solutions** begin on page R1.
Extra Practice begins on page EP2.

Example 1 Find the constant rate of change for each table.

3.

Time (s)	Distance (m)
0	6
1	12
2	18
3	24

4.

Time (h)	Wage ($)
0	0
1	9
2	18
3	27

5.

Cost ($)	38	50	62	74	86
Minutes	1,000	1,500	2,000	2,500	3,000

Example 2 Find the constant rate of change for each graph.

6.

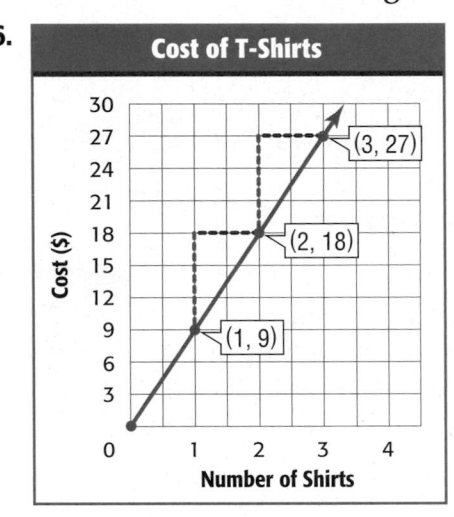

Cost of T-Shirts

7.

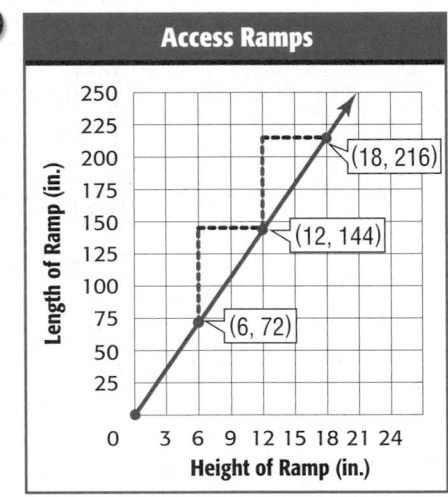

Access Ramps

8. WATER At 1:00 P.M., the water level in a pool is 13 inches. At 2:30 P.M., the water level is 28 inches. What is the constant rate of change?

9 **GRAPHIC NOVEL** Refer to the lap times for Exercises **a** and **b**.

The race is 20 laps, which is 5 miles. Assuming your speed is constant...

Let's calculate Seth's times.

Lap	4	8	12	16	20
Distance	1 mile	2 miles	3 miles	4 miles	5 miles
Time	57.1s	114.2 s	171.3s		

a. How long does it take Seth to complete 1 mile if 20 laps equal 5 miles? Write the constant rate of change in miles per second.

b. Graph the distance and time on a coordinate plane with the distance on the y-axis and the time on the x-axis.

H.O.T. Problems

10. **OPEN ENDED** Make a table where the constant rate of change is 6 inches for every foot.

11. **WRITE MATH** Write a problem to represent a constant rate of change of $15 per item.

☑ Test Practice

12. Use the information in the table to find the constant rate of change.

Number of Apples	3	7	10
Number of Seeds	30	70	110

A. $\frac{10}{1}$ B. $\frac{1}{10}$ C. $\frac{40}{4}$ D. $\frac{4}{40}$

13. ▤✎ **GRIDDED RESPONSE** Reggie started a running program to prepare for track season. Every day for 60 days he ran a half hour in the morning and a half hour in the evening. He averaged 6.5 miles per hour. At this rate, what is the total number of miles Reggie ran over the 60-day period?

More About Rates of Change

Recall that linear functions have graphs that are straight lines. **Nonlinear functions** have graphs that are not straight lines.

📝 REAL-WORLD EXAMPLE

GARDEN Determine whether the table represents a *linear* or *nonlinear function*. Explain.

Graph the points on a coordinate plane.

Garden Length (ft)	Garden Area (sq ft)
1	1
2	4
3	9
4	16

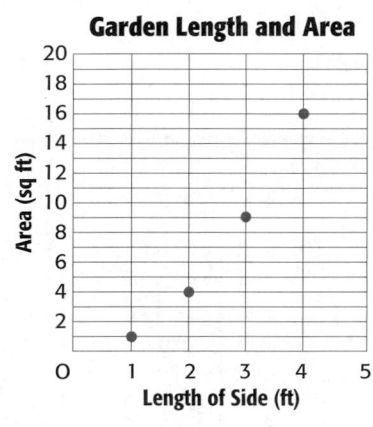

Garden Length and Area

The change in area from a side length of 1 foot to 2 feet and from 2 feet to 3 feet is *not* constant. Therefore, the graph is not a straight line. So, the table represents a nonlinear function.

Determine whether the table represents a *linear* or *nonlinear* function. Explain.

14.

x	1	2	3	4
y	1	8	27	64

15.

x	1	2	3	4
y	0	2	5	9

Chapter 7 Lesson 2B Slope **395**

Main Idea

Identify slope using tables and graphs.

Vocabulary

slope

Slope

BOOKS Hero Comics is a company that publishes comic books. They typically recycle the original copies after the binding process. The table below shows the total number of pounds of paper that has been recycled each day during the month.

Day of Month	Total Recycled (lbs)
3	36
5	60
6	72
7	84
12	144
15	180

1. Is the relationship linear? Explain.

2. Create a graph of the function. Use two points from the graph to find the constant rate of change.

In a linear relation, the vertical change (change in y-value) per unit of horizontal change (change in x-value) is always the same. This ratio is called the **slope** of the function. The constant rate of change, or unit rate of a function, is the same as the slope of the related linear function. The slope tells how steep the line is.

Key Concept   Slope

Slope is the rate of change between any two points on a line.

$$\text{slope} = \frac{\text{change in } y}{\text{change in } x} \quad \longleftarrow \text{ vertical change}$$
$$\longleftarrow \text{ horizontal change}$$

$$= \frac{2}{1} \text{ or } 2$$

The vertical change is sometimes called "rise" while the horizontal change is called "run." You can say that slope $= \dfrac{\text{rise}}{\text{run}}$.

🏃 ✏️ **REAL-WORLD EXAMPLE** **Find Slope**

1 **PHYSICAL SCIENCE** The table below shows the relationship between the number of seconds y it takes to hear thunder after a lightning strike and the miles x you are from the lightning.

Miles (x)	0	1	2	3	4	5
Seconds (y)	0	5	10	15	20	25

Graph the data and find the slope. Explain what the slope represents.

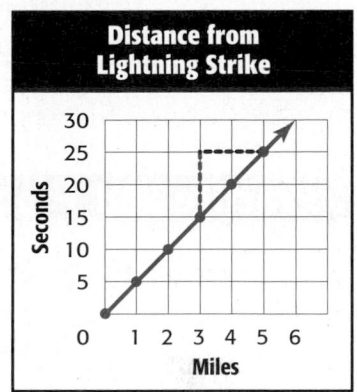

Distance from Lightning Strike

$$\text{slope} = \frac{\text{change in } y}{\text{change in } x} \quad \text{Definition of slope}$$

$$= \frac{25 - 15}{5 - 3} \quad \text{Use (3, 15) and (5, 25).}$$

$$= \frac{10}{2} \quad \leftarrow \text{seconds} \\ \quad \quad \leftarrow \text{miles}$$

$$= \frac{5}{1} \quad \text{Simplify.}$$

So, for every 5 seconds between a lightning flash and the sound of thunder, there is 1 mile between you and the lightning strike.

 CHECK Your Progress

a. **PLANTS** Graph the data about plant height for a science fair project. Then find the slope of the line. Explain what the slope represents.

Week	Plant Height (cm)
1	1.5
2	3
3	4.5
4	6

✏️ **EXAMPLE** **Interpret Slope**

2 **BANKING** Renaldo opened a savings account with the $300 he earned mowing yards over the summer. Each week he withdraws $20 for expenses. Draw a graph of the account balance versus time. Find the numerical value of the slope and interpret it in words.

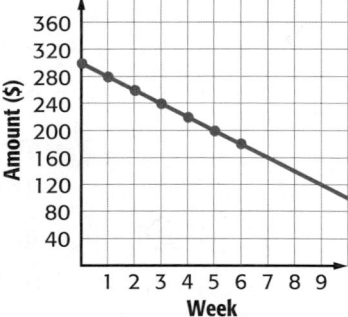

Renaldo's Savings Account

The slope of the line is the rate at which the account balance falls, or $-\frac{\$20}{1 \text{ week}}$. This is a negative slope and therefore a negative rate of change.

 CHECK Your Progress

b. Jessica buys a $10 pre-paid card for her cell phone. She pays a rate of $0.05 per text. Draw a graph of the card balance versus the number of text messages. Find the numerical value of the slope and interpret it in words.

Example 1 **①** **SNACKS** The table below shows the number of small packs of fruit snacks y per box x. Graph the data. Then find the slope of the line. Explain what the slope represents.

Boxes, x	3	5	7	9
Packs, y	24	40	56	72

Example 2 **2. TEMPERATURE** At 2:00 P.M., the temperature is 81°F. At 3:00 P.M., the temperature is 78°F. Draw a graph of temperature versus time if the temperature continues to decrease at this rate from 2:00 P.M. to 6:00 P.M. Find the value of the slope and interpret it in words.

Practice and Problem Solving

● = **Step-by-Step Solutions** begin on page R1.
Extra Practice begins on page EP2.

Example 1 **3. CYCLING** The table shows the distance y Adriano traveled in x minutes while competing in the cycling portion of a triathlon. Graph the data. Then find the slope of the line. Explain what the slope represents.

Time (min)	45	90	135	180
Distance (km)	5	10	15	20

4. MARKERS The table below shows the number of markers per box. Graph the data. Then find the slope of the line. Explain what the slope represents.

Boxes	1	2	3	4
Markers	8	16	24	32

Example 2 **5. MEASUREMENT** There are 3 feet for every yard. Draw a graph of feet versus yards. Find the numerical value of the slope and interpret it in words.

6. SWIMMING Joshua swims 25 meters in 1 minute. Draw a graph of meters swam versus time. Find the value of the slope and interpret it in words.

⑦ **SPEED** The graph shows the average speed of two cars on the highway.

   a. What does (2, 120) represent?

   b. What does (1.5, 67.5) represent?

   c. What does the ratio of the y-coordinate to the x-coordinate for each pair of points on the graph represent?

   d. What does the slope of each line represent?

   e. Which car is traveling faster? How can you tell from the graph?

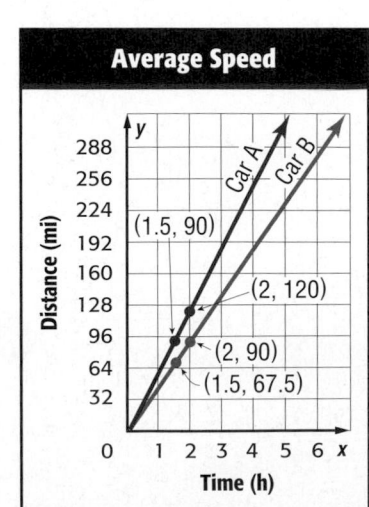

H.O.T. Problems

8. **CHALLENGE** Flavio is saving money at a rate of $30 per month. Edgardo is saving money at a rate of $35 per month. They both started saving at the same time. If you were to create a table of values and graph each function, what would be the slope of each graph?

9. **FIND THE ERROR** Marisol is finding the slope of the line containing the points (3, 7) and (5, 10). Find her mistake and correct it.

> The slope between the two points (3, 7) and (5, 10) is found like this:
> $$slope = \frac{rise}{run} = \frac{5-3}{10-7} = \frac{2}{3}$$

10. **WRITE MATH** Describe the relationship between the rate of change and the slope. How do you find each?

Test Practice

11. **SHORT RESPONSE** Find the slope of the line below that shows the distance Jairo traveled while jogging.

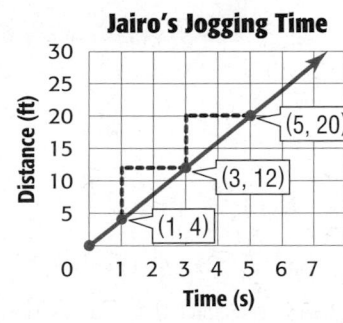

Jairo's Jogging Time

(5, 20)
(3, 12)
(1, 4)

12. Line *RS* represents a bike ramp.

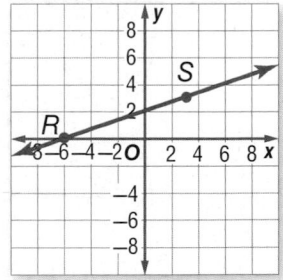

What is the slope of the ramp?

A. $-\frac{3}{1}$ C. $\frac{1}{3}$

B. $-\frac{1}{3}$ D. $\frac{3}{1}$

13. **WAGES** Use the information in the table to find the constant rate of change in dollars per hour. (Lesson 2B)

Wage ($)	0	9	18	27
Time (h)	0	1	2	3

14. Graph $y = 4x$. (Lesson 1C)

Extend

Main Idea

Use technology to compare and contrast graph relationships.

Graphing Technology: Compare Graphs

You can use a graphing calculator to compare and contrast relationships of graphs.

ACTIVITY

Compare and contrast the graphs of $y = 5x$ and $y = 7x$.

STEP 1 Press Y= and enter the function $y = 5x$ into Y1. Then enter $y = 7x$ into Y2.

STEP 2 Adjust your viewing window. Press WINDOW and change the values to allow you to compare the different graphs.

STEP 3 Finally, graph the functions by pressing GRAPH. Compare and contrast the graphs.

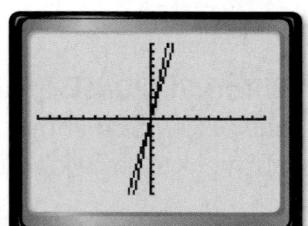

Analyze the Results

1. What happens to the graph of $y = 5x$ and $y = 7x$ as the coefficient changes in each function? Explain your reasoning.

2. Graph $y = 2x$ and $y = 2x + 1$. What happens to the graph of $y = 2x$ when you add 1 to the function? What would you expect to happen if you add 2? subtract 1?

3. What similarities do you notice among all of the functions? differences?

4. Graph $y = 3x$ and $y = -3x$. Compare and contrast the two graphs.

5. **MAKE A CONJECTURE** Explain in two or more sentences how positive and negative coefficients affect the graphs of the functions.

Mid-Chapter Check

1. **TRAVEL** Beth drove at an average rate of 65 miles per hour for several hours. Make a function table that shows her distance traveled after 2, 3, 4, and 5 hours. Then identify the domain and range. (Lesson 1B)

2. **MONEY** Anthony earns extra money after school doing yard work for his neighbors. He charges $12 for each lawn he mows. (Lesson 1B)

   a. Write an equation to show the relationship between the number of lawns mowed m and number of dollars earned d.

   b. Find the number of dollars earned if he mows 14 lawns.

Graph each equation. (Lesson 1C)

3. $y = x + 1$

4. $y = 2x$

5. $y = 2x - 3$

6. $y = -x + 1$

7. **MOVIES** A student ticket to the movies costs $6. The equation $c = 6t$ describes the total cost c for t tickets. Make a function table that shows the total cost for 1, 2, 3, and 4 tickets, and then graph the equation. (Lesson 1C)

8. **MAGAZINES** Use the graph to find the constant rate of change in cost per magazine. (Lesson 2B)

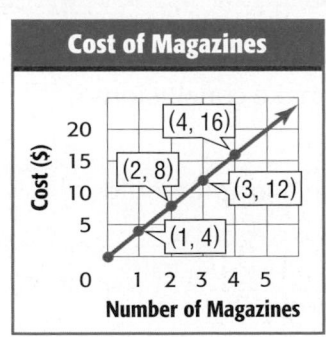

9. **MULTIPLE CHOICE** The graph shows the relationship between time and water level of a pool. Which represents the rate of change? (Lesson 2B)

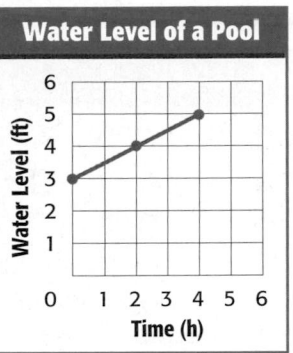

   A. increases 2 feet every 1 hour

   B. increases 1 foot every 2 hours

   C. increases 1 foot every 1 hour

   D. increases 2 feet every 2 hours

10. **MAPS** The table shows the key for a map. Graph the data. Then find the slope of the line. (Lesson 2C)

Distance on Map (cm)	1	2	3	4
Actual Distance (km)	20	40	60	80

11. **MULTIPLE CHOICE** Line AB represents a ramp for loading a truck. What is the slope of the ramp? (Lesson 2C)

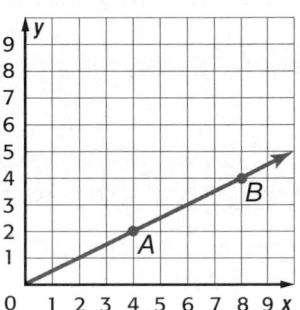

   F. $\frac{2}{1}$ H. $-\frac{2}{1}$

   G. $\frac{1}{2}$ I. $-\frac{1}{2}$

Problem-Solving Investigation

Main Idea Solve problems by using a graph.

P.S.I. TEAM +

Use a Graph

RICK: The table shows the study times and test scores of 13 students in Mrs. Collins's English class.

YOUR MISSION: Use a graph to predict the test score of a student who studied for 80 minutes.

Study Time and Test Scores											
Study Time (min)	120	30	60	95	70	55	90	45	75	60	10
Test Score (%)	98	75	80	93	82	78	95	74	87	83	65

Understand	You know the number of minutes studied. You need to predict the test score.
Plan	Organize the data in a graph so you can easily see any trends.
Solve	As the study times progress, the test scores increase. Draw a line that is close to as many of the points as possible. You can predict that the test score of a student who studied for 80 minutes is about 88%.
Check	The estimate is close to the line, so the prediction is reasonable.

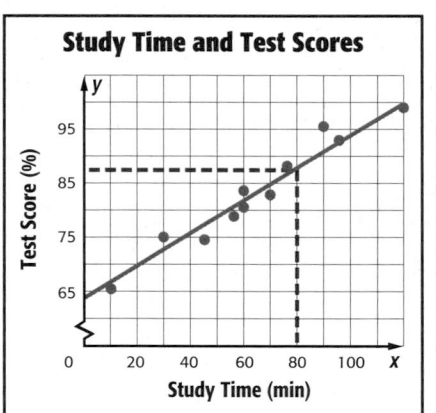

Study Time and Test Scores

Analyze the Strategy

1. Explain why analyzing a graph is a useful way to quickly make conclusions about a set of data.

2. ✏ **WRITE MATH** Write a real-world problem in which using a graph would be a useful way to check a solution.

• Use a graph.
• Guess, check, and revise.
• Look for a pattern.
• Choose an operation.

= **Step-by-Step Solutions** begin on page R1.
Extra Practice begins on page EP2.

Use a graph to solve Exercises 3 and 4.

3. The table shows the relationship between Celsius and Fahrenheit temperatures.

Temperature	
Celsius	**Fahrenheit**
0	32
10	50
20	68
30	86
40	104

a. Make a graph of the data.

b. Suppose the temperature is 25° Celsius. Estimate the temperature in Fahrenheit.

4. SCHOOLS The graph shows the number of students enrolled at Yorktown Middle School over a five-year period. Predict the school's enrollment in 2015.

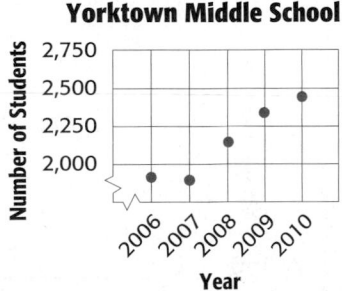

Use any strategy to solve Exercises 5–11.

5. ALGEBRA What are the next two numbers in the pattern 8, 18, 38, 78, … ?

6. FARMING A farmer sells a bushel of corn for $3. Suppose the farmer wants to earn $165 from bushels of corn sales on Saturday. How many bushels does he need to sell to make his goal?

7. EXERCISE Flora walked 8 minutes on Sunday and, each day, plans to walk twice as long as she did the previous day. On what day will she walk over 1 hour?

8. OIL SPILLS The graph shows the number of oil spills each year.

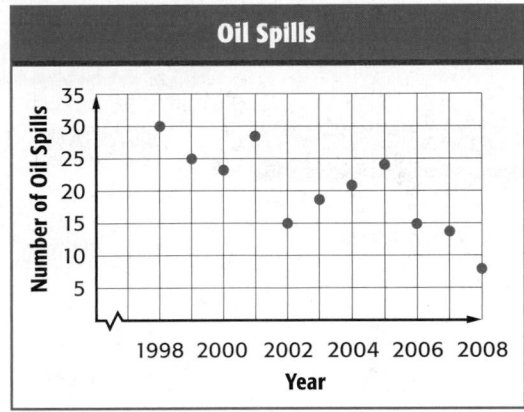

a. What does the graph indicate happened from 2002 to 2004?

b. What is the overall trend in the data?

c. Is the relationship linear?

d. Explain the difference between a linear relationship and a trend.

9 HELICOPTERS A helicopter has a maximum freight capacity of 2,400 pounds. How many crates, each weighing about 75 pounds, can the helicopter hold?

10. MOVIES Moses and some of his friends are going to the movies. Suppose they each buy nachos and a beverage. They spend $36. How many friends are going to the movies with Moses?

Movie Costs	
Item	**Price**
Ticket	$6.00
Beverage	$2.25
Nachos	$3.75

11. NUMBER THEORY A whole number is squared and the result is 324. Find the number.

Main Idea

Compare and contrast proportional and nonproportional linear functions.

CCSS 7.RP.2, 7.RP.2a

Proportional and Nonproportional Relationships

In this activity, you will use models to explore two different functions.

 **ACTIVITY**

STEP 1   Using centimeter cubes, arrange the cubes in towers as shown in the diagrams below.

Pattern	A				B			
Figures								
Figure Number	0	1	2	3	0	1	2	3

STEP 2   Let x represent the figure number and y represent the number of centimeter cubes in each tower. Copy and complete the table below for each pattern. Then graph the data on separate coordinate planes.

x	Process	y
0		
1		
2		
3		
4		
x		

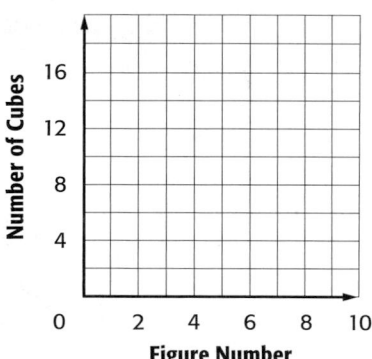

Analyze the Results

1. Compare and contrast the models, processes, and graphs of Patterns A and B.

2. Which pattern represents a proportional relationship? Which represents a nonproportional relationship? How can you tell this from the data shown in the table? from the graph?

Main Idea
Use direct variation to solve problems.

 Vocabulary
direct proportion
direct variation
constant of variation
slope-intercept form
y-intercept

 *Get Connect*ED

 CCSS 7.RP.2, 7.RP.2a, 7.RP.2b, 7.RP.2c

Direct Variation

SPEED A car travels 130 miles in 2 hours, 195 miles in 3 hours, and 260 miles in 4 hours, as shown.

1. What is the constant rate of change, or slope, of the line?

2. Is the distance traveled always proportional to the driving time? What is the constant ratio?

3. Compare the constant rate of change to the constant ratio.

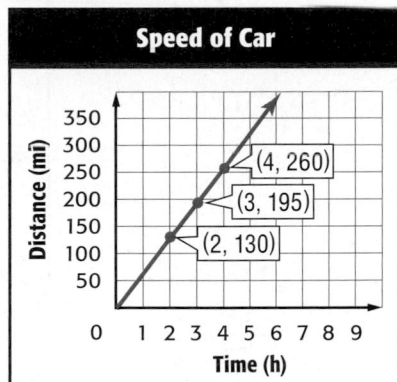

When two variable quantities have a constant ratio, their relationship is called a **direct variation** or a **direct proportion**. The constant ratio is called the **constant of variation**.

 REAL-WORLD EXAMPLE **Find a Constant Ratio**

1. **POOLS** The height of the water as a pool is being filled is shown in the graph. Determine the rate in inches per minute.

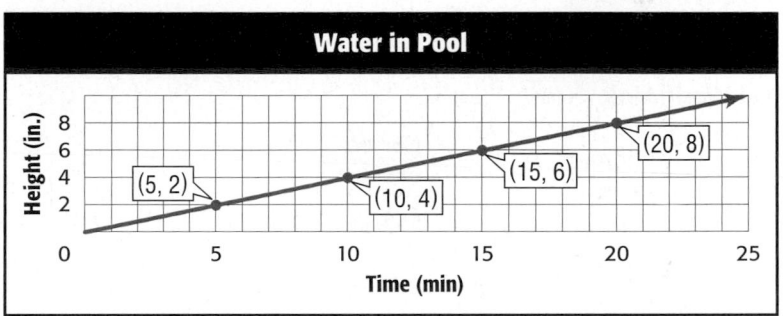

Since the graph of the data forms a line, the rate of change is constant. Use the graph to find the constant ratio.

$\frac{\text{height}}{\text{time}} \rightarrow$ $\frac{2}{5}$ or $\frac{0.4}{1}$ $\frac{4}{10}$ or $\frac{0.4}{1}$ $\frac{6}{15}$ or $\frac{0.4}{1}$ $\frac{8}{20}$ or $\frac{0.4}{1}$

The pool fills at a rate of 0.4 inch every minute.

 CHECK Your Progress

a. **SCUBA DIVING** Two minutes after a diver enters the water, he has descended 52 feet. After 5 minutes, he has descended 130 feet. At what rate is the scuba diver descending?

In a direct variation equation, the constant rate of change, or slope, is assigned a special variable, k.

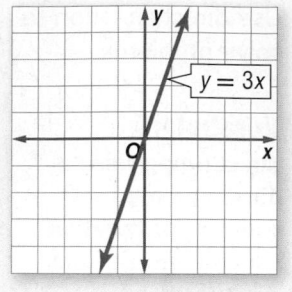
As you have already learned, not all situations with a constant rate of change are proportional relationships. Likewise, not all linear functions are direct variations.

EXAMPLE **Determine Direct Variation**

(2) **PIZZA** Pizzas cost \$8 each plus a \$3 delivery charge. Make a table and graph to show the cost of 1, 2, 3, and 4 pizzas. Is there a constant rate? a direct variation?

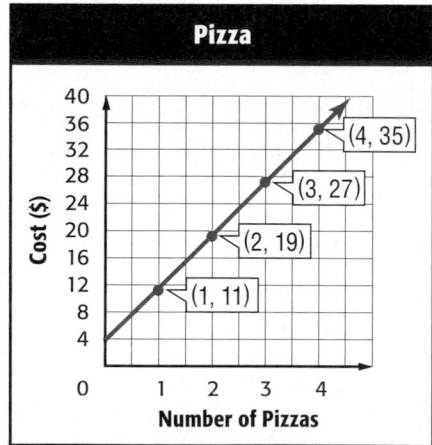

Number of Pizzas	1	2	3	4
Cost (\$)	\$11	\$19	\$27	\$35

$\dfrac{\text{number of pizzas}}{\text{cost}}$ → $\dfrac{11}{1}, \dfrac{19}{2}$ or 9.5, $\dfrac{27}{3}$ or 9, $\dfrac{35}{4}$ or 8.75

Because there is no constant rate, there is no direct variation.

CHECK Your Progress

b. FOOD COSTS Two pounds of cheese cost \$8.40. Make a table and graph to show the cost of 1, 2, 3, and 4 pounds of cheese. Is there a constant rate? a direct variation?

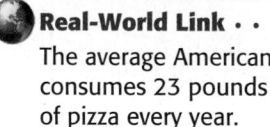

Real-World Link • • •

The average American consumes 23 pounds of pizza every year.

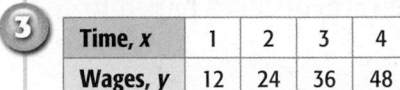 **EXAMPLES** Identify Direct Variation

Determine whether each linear function is a direct variation. If so, state the constant of variation.

3

Time, x	1	2	3	4
Wages, y	12	24	36	48

Compare the ratios to check for a common ratio.

$\frac{wages}{time}$ ⟶ $\frac{12}{1}$ $\frac{24}{2}$ or $\frac{12}{1}$ $\frac{36}{3}$ or $\frac{12}{1}$ $\frac{48}{4}$ or $\frac{12}{1}$

Since the ratios are the same, the function is a direct variation. The constant of variation is $\frac{12}{1}$.

4

Time (h), x	2	3	4	5
Temperature Change (°), y	4	5	7	11

$\frac{temperature}{time}$ ⟶ $\frac{4}{2}$ or $\frac{2}{1}$ $\frac{5}{3}$ or $\frac{1.67}{1}$ $\frac{7}{4}$ or $\frac{1.75}{1}$ $\frac{11}{5}$ or $\frac{2.2}{1}$

The ratios are not the same, so the function is not a direct variation.

CHECK Your Progress

c.

Year, x	5	10	15	20
Height, y	12.5	25	37.5	50

Study Tip

Direct Variation When a relationship varies directly, the graph of the function will always go through the origin. So, (0, 0) is a solution of $y = 3x$.

Concept Summary Linear Functions

	Table			Graph	Equation

Direct Variation Functions

x	y	$\frac{y}{x}$
−2	−6	3
−1	−3	3
1	3	3
2	6	3

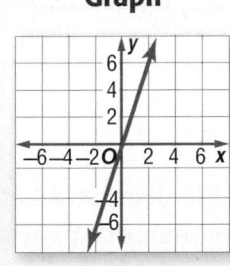

$y = 3x$

Not Direct Variation Functions

x	y	$\frac{y}{x}$
−2	−7	$\frac{-7}{2}$
−1	−4	4
1	2	2
2	5	$\frac{5}{2}$

$y = 3x - 1$

CHECK Your Understanding

Example 1

1. **BAKING** The number of cakes baked varies directly with the number of hours the caterers work. What is the ratio of cakes baked to hours worked?

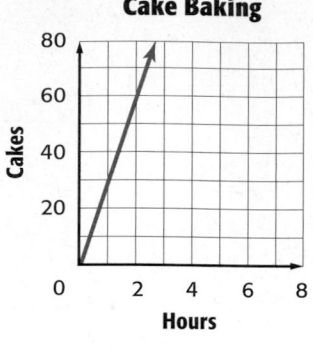

Cake Baking

Example 2

2. **TRANSPORTATION** An airplane travels 780 miles in 4 hours. Make a table and graph to show the mileage for 2, 8, and 12 hours. Is there a direct variation?

Examples 3 and 4

3. Determine whether the linear function is a direct variation. If so, state the constant of variation.

Hours, x	3	5	7	9
Miles, y	108	180	252	324

Practice and Problem Solving

= **Step-by-Step Solutions** begin on page R1.
Extra Practice begins on page EP2.

Example 1

4. **GARDENING** Veronica is mulching her front yard. The total weight of mulch varies directly with the number of bags of mulch. What is the rate of change?

⑤ **DOG WALKING** The money Shelley earns varies directly with the number of dogs she walks. How much does Shelley earn for each dog she walks?

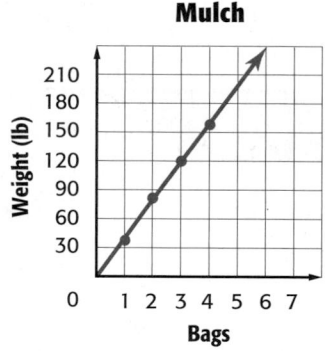

Mulch

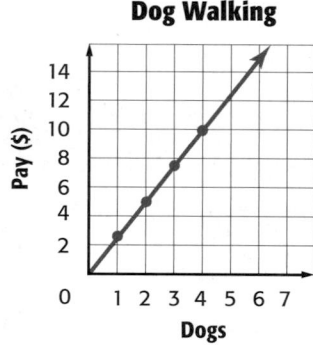

Dog Walking

Examples 2–4 Determine whether each linear function is a direct variation. If so, state the constant of variation.

6.

Pictures, x	3	4	5	6
Profit, y	24	32	40	48

7.

Minutes, x	185	235	275	325
Cost, y	60	115	140	180

8.

Age, x	11	13	15	19
Grade, y	5	7	9	11

9.

Price, x	20	25	30	35
Tax, y	4	5	6	7

10. **PRESSURE** At a 33-foot depth underwater, the pressure is 29.55 pounds per square inch (psi). At a depth of 66 feet, the pressure reaches 44.4 psi. At what rate is the pressure increasing?

408 Linear Functions

11. **BOOKS** One month, the Allen family had a total of $3.20 in fines for 4 books. The next month, they had 7 books overdue for $5.60. What was the overdue charge per book?

ALGEBRA If y varies directly with x, write an equation for the direct variation. Then find each value.

12. If $y = -14$ when $x = 8$, find y when $x = -12$.

13. Find y when $x = 15$ if $y = 6$ when $x = 30$.

14. If $y = -6$ when $x = -24$, what is the value of x when $y = -7$?

15. Find x when $y = 14$, if $y = 7$ when $x = 8$.

16. Find y when $x = 9$, if $y = 16.4$ when $x = 8.2$.

17. **MEASUREMENT** The number of yards in a measure varies directly with the number of feet. Using the graph at the right, find the measure of an object in yards if it is 78 feet long.

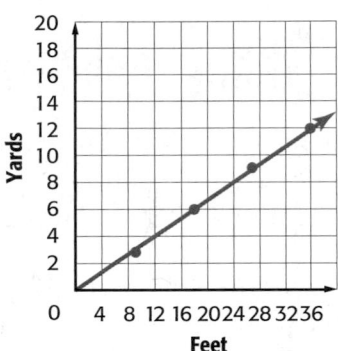

18. **BAKING** A cake recipe requires $3\frac{1}{4}$ cups of flour for 13 servings and $4\frac{1}{2}$ cups of flour for 18 servings. How much flour is required to make a cake that serves 28?

Real-World Link
The typical American consumes four times as many Calories from sugar, flour, and seed oils as from meat.

19. **MULTIPLE REPRESENTATIONS** Robert is in charge of the community swimming pool. Each spring he drains it in order to clean it. He then must refill the pool, which holds 120,000 gallons of water. Robert fills the pool at a rate of 10 gallons each minute.

    a. **WORDS** What is the rate at which Robert will fill the pool? Is it constant?

    b. **GRAPH** Create a graph of the relationship.

    c. **ALGEBRA** Write an equation for the direct variation.

20. **MATH IN THE MEDIA** Find an example of a graph of a linear function in a newspaper or magazine, or on the Internet. Determine if the graph represents a direct variation. Explain your reasoning.

H.O.T. Problems

21. **OPEN ENDED** Identify two additional values for x and y in a direct variation relationship where $y = 11$ when $x = 18$.

22. **CHALLENGE** Find y when $x = 14$ if y varies directly as the square of x, and $y = 72$ when $x = 6$.

23. **WRITE MATH** Write a real-world problem involving a direct variation. Then solve your problem.

24. To make lemonade, Andy adds 8 tablespoons of sugar for every 12 ounces of water. If he uses 32 ounces of water, which proportion can he use to find the number of tablespoons of sugar x he should add to make the lemonade?

    A. $\dfrac{8}{12} = \dfrac{32}{x}$ C. $\dfrac{8}{12} = \dfrac{x}{32}$

    B. $\dfrac{8}{x} = \dfrac{32}{12}$ D. $\dfrac{x}{12} = \dfrac{8}{32}$

25. Anjuli read 22 pages during a 30-minute study hall. At this rate, how many pages would she read in 45 minutes?

    F. 30

    G. 33

    H. 45

    I. 48

More About Graphing Linear Equations

Some situations start with a given amount and increase at a certain rate. For example, in Example 2, pizzas cost $8 each with a $3 delivery charge. The equation $y = 8x + 3$ represents the total cost y of x pizzas.

Equations such as $y = 8x + 3$ are written in the form $y = mx + b$, where m is the slope and b is the y-intercept. It is called **slope-intercept form**.

slope $y = mx + b$ ← y-intercept, the y-coordinate of the point where the graph crosses the x-axis

$y = 8x + 3$

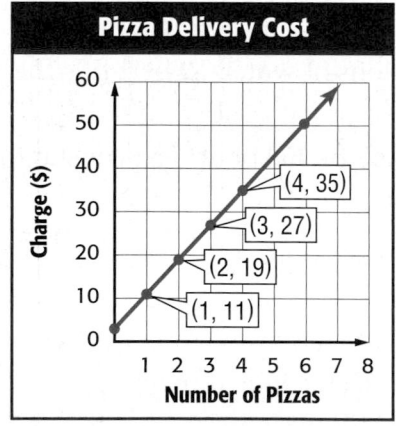

Pizza Delivery Cost

The slope is 8 or $\dfrac{8}{1}$. This means the cost is $8 per pizza.

The y-intercept is at $(0, 3)$. This means the delivery charge is $3.

To find the total cost of 6 pizzas, replace x with 6. Since $8(6) + 3 = 51$, six pizzas cost $51.

ROAD TRIP On Day 1, the Lopez family drove 100 miles. On Day 2, they drove x hours at 45 miles per hour.

26. Write and graph an equation for the total distance the Lopez family drove by the end of Day 2.

27. Interpret the meaning of the slope and y-intercept for your equation.

28. If the Lopez family drove 6 hours on Day 2, how many miles did they drive altogether? Show how you solved.

29. Does your equation show a proportional relationship? Explain.

 **Inverse Variation**

Main Idea

Graph inverse variations.

Get Connect**ED**

Jackie rides her bike at an average rate of 4 miles per hour. This situation can be represented by the direct variation equation $d = 4t$, where d is the distance in miles and t is the time in hours. The rate is a constant.

Suppose Jackie wants to bike 12 miles each day. Some days she rides faster than others. This situation can be represented by the equation $12 = rt$, where r is the rate.

ACTIVITY

STEP 1 Copy and complete the table for the equation $12 = rt$.

t (hours)	12	8	6	■	■	■	■
r (miles per hour)	1	1.5	2	2.5	3	3.5	4

STEP 2 Copy and complete the graph of the ordered pairs from Step 1. Connect the points with a smooth curve. The first three points are done for you.

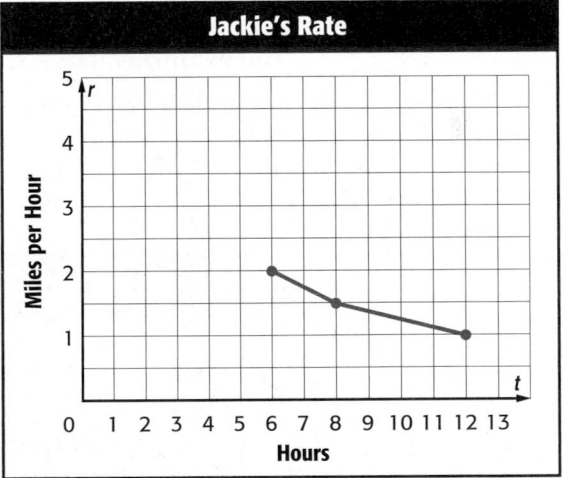

Jackie's Rate

Analyze the Results

1. Is the time proportional to the rate? Explain why or why not.

2. When the product of two variables is a constant, the relationship is an *inverse variation*. Which situation is an inverse variation: Jackie biking at 4 miles per hour or biking 12 miles at varying rates? Identify the constants in each situation.

Main Idea
Use inverse variation to solve problems.

Vocabulary
inverse variation
inverse proportion

*Get Connect*ED

Inverse Variation

Explore Refer to the rectangles below.

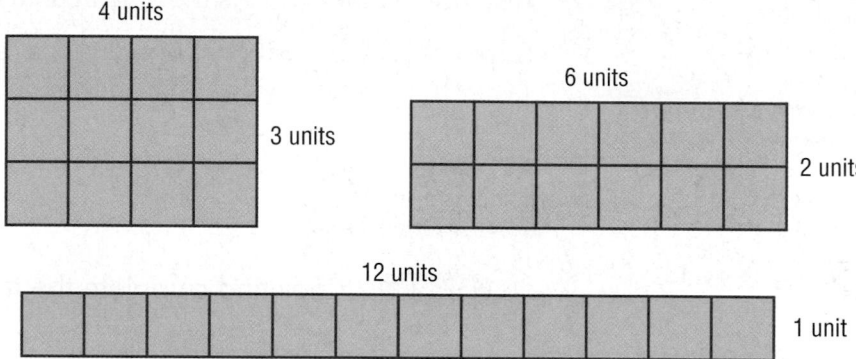

1. What is the constant in each rectangle?
2. What happens to the width as the length increases?
3. What happens to the length as the width increases?

The example above shows an inverse variation. In an **inverse variation**, the product of x and y is a constant. We say that y is inversely proportional to x. An inverse variation is also known as an **inverse proportion**.

Key Concept — Inverse Variation

Words
An inverse variation is a relationship where the product of x and y is a constant k.

Symbols
$xy = k$ or $y = \dfrac{k}{x}$, where $k \neq 0$

Example
$xy = 3$ or $y = \dfrac{3}{x}$

Model

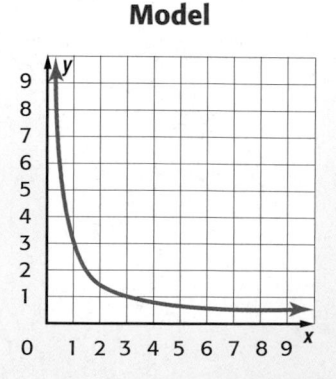

As x increases, the value of y decreases, but not at a constant rate. So, the graph of an inverse variation is not a straight line as in a direct variation.

1 **MUSIC** The table shows the relationship between the frequency and wavelength of a musical tone. Determine if the relationship is an inverse variation. Justify your response.

Graph the data in the table.

Frequency (vibrations per second)	220	440	660	880
Wavelength (feet)	4	2	$1\frac{1}{3}$	1

Real-World Link · · · ·

Singers and musicians use the word *pitch* when referring to frequency. They often refer to frequencies by using note names, such as middle C.

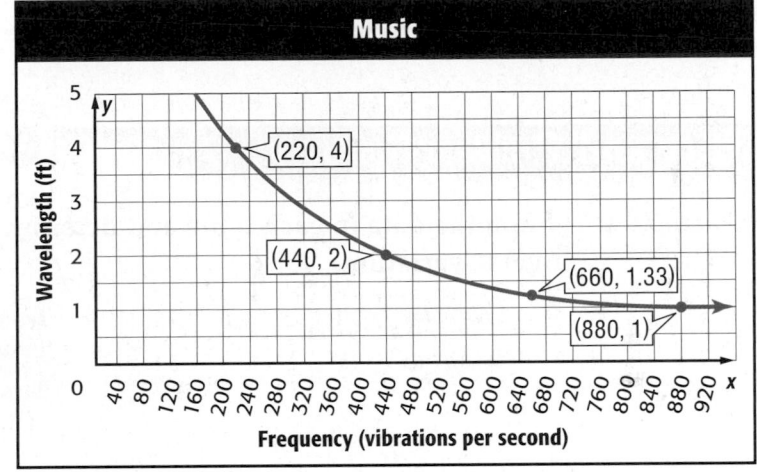

The graph shows that this is an inverse variation. The *x*- and *y*-coordinates each have a product of 880.

2 **CONSTRUCTION** The number of carpenters needed to frame a house varies inversely as the number of days needed to complete the project. Suppose 5 carpenters can frame a certain house in 16 days. How many days will it take 8 carpenters to frame the house? Assume that they all work at the same rate.

Solve the problem by using inverse variation. Let *x* be the number of carpenters. Let *y* be the number of days.

Step 1 Find the value of *k*.

$xy = k$ Inverse variation

$(5)(16) = k$ Replace *x* with 5 and *y* with 16.

$80 = k$ Simplify to determine *k*.

Step 2 Find the number of days.

$y = \dfrac{k}{x}$ Definition of inverse variation

$y = \dfrac{80}{8}$ Replace *k* with 80 and *x* with 8.

$y = 10$ Simplify.

A crew of 8 carpenters can frame the house in 10 days.

 CHECK Your Progress

a. **CANDLES** Graph the data in the table. Then determine if the relationship is an inverse variation.

Candle's Volume			
Burn Time (h)	1	2	4	8
Volume (in³)	128	64	32	16

Example 1 **1 MUSIC** The table shows the relationship between the length of a piano string and the frequency of its vibrations. Graph the data in the table and determine if the relationship is an inverse variation.

Length (in.)	48	36	24	12
Frequency (cycles/s)	360	480	720	1440

Example 2 **2. TRAVEL** The time it takes to travel a certain distance varies inversely with the speed at which you are traveling. Suppose it takes 3 hours to drive from one city to another at a rate of 65 miles per hour. How long will the return trip take traveling at 55 miles per hour?

Practice and Problem Solving

● = **Step-by-Step Solutions** begin on page 614.
Extra Practice begins on page 520.

Example 1 Graph the data in each table and determine if the relationship is an inverse variation.

3.
Length (m)	2.4	3	6	10
Width (m)	15	12	6	3.6

4.
Gift Card Balance ($)	50.00	42.50	27.50	5.00
Number of Movies	0	1	3	6

Example 2 **5. BRICKS** The number of bricklayers needed to build a brick wall varies inversely as the number of hours needed. Four bricklayers can build a brick wall in 30 hours. How long would it take 5 bricklayers to build a wall?

6. RUNNING In the formula $d = rt$, the time t varies inversely with the rate r. A student running at 5 miles per hour runs one lap around the school campus in 8 minutes. If a second student takes 10 minutes to run one lap around the school campus, how fast is she running?

7 GRAPHIC NOVEL Refer to the graphic novel frame below. Seth applies his brakes and begins slowing. Suppose the car travels 88 feet after 2 seconds. After 4 seconds the car travels another 44 feet. It takes several more seconds for the car to come to a complete stop. Is this an example of direct or inverse variation? Explain.

Slowing down after racing is mathematical too.

SKREEECH!

8. **CHALLENGE** When does a graph representing an inverse variation cross the x-axis? Explain.

9. **OPEN ENDED** Identify three sets of values of x and y for the inverse variation $xy = 24$.

10. **WRITE MATH** Write a few sentences that compare and contrast inverse variation and direct variation.

Test Practice

11. Several people share twenty-five pieces of candy. If the situation is an inverse variation, which equation represents the situation?

    A. $xy = 125$

    B. $y = \dfrac{25}{x}$

    C. $y = 125x$

    D. $25x = y$

12. Determine k if y varies inversely as x and $y = 4.2$ when $x = -1.3$.

    F. -3.27

    G. -5.46

    H. -0.31

    I. -2.95

13. Identify the graph of $xy = k$ if $x = 2$ when $y = 4$.

    A.
    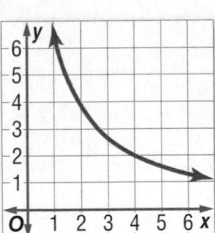

    C.
    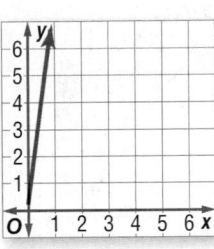

    B.
    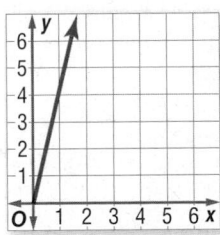

    D.
    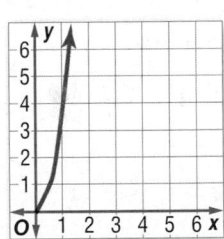

Spiral Review

14. **TRAVEL** A train traveled 203 miles in $1\frac{1}{2}$ hours. At this rate, how far will the train travel after 5 hours? Assuming that the distance traveled varies directly with the time traveled, write and solve an equation to represent the situation. (Lesson 3C)

15. **STAMP COLLECTING** Sun Li collects postage stamps from European countries. Use the graph to predict the number of postage stamps Sun Li will have in Year 10. (Lesson 3A)

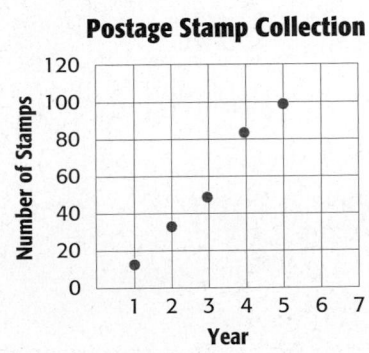

Problem Solving in Engineering

START OFF ON THE RIGHT FOOT

Did you know that more than 700 pounds of force are exerted on a 140-pound long-jumper during the landing? Biomechanical engineers understand how forces travel through the shoe to an athlete's foot and how the shoes can help reduce the impact of those forces on the legs. If you are curious about how engineering can be applied to the human body, a career in biomechanical engineering might be a great fit for you.

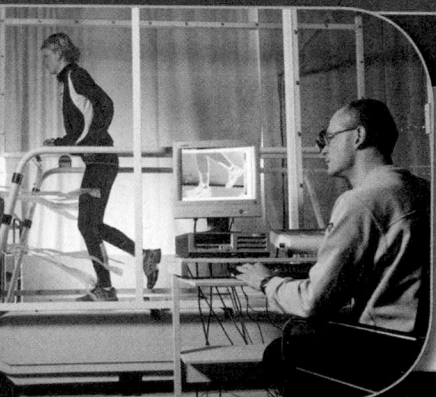

21st Century Careers

Are you interested in a career as a biomechanical engineer? Take some of the following courses in high school.

- Biology
- Calculus
- Physics
- Trigonometry

 Get ConnectED

Real-World Math

Use the information in the graph to solve each problem.

1. Find the constant rate of change for the data shown in the graph. Interpret its meaning.

2. Is there a proportional relationship between the weight of an athlete and the forces that are generated from running? Explain your reasoning.

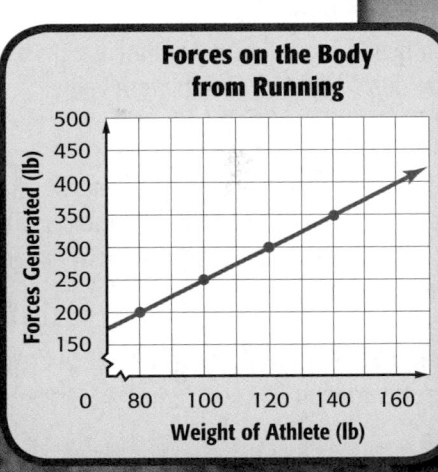

Forces on the Body from Running

FOLDABLES®
Study Organizer

Be sure the following Key Concepts are noted in your Foldable.

Key Concepts

Rates and Functions (Lesson 1)
• A function is a relationship that assigns exactly one *output* value for each *input* value.

Slope (Lesson 2)
• A rate of change is a rate that describes how one quantity changes in relation to another. The slope of a graph is the rate of change between any two points on a line.

Variation (Lesson 3)
• Direct variation
$\frac{y}{x}$ is a constant.

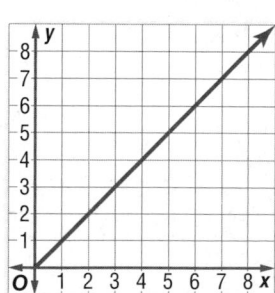

• Inverse variation xy is a constant.

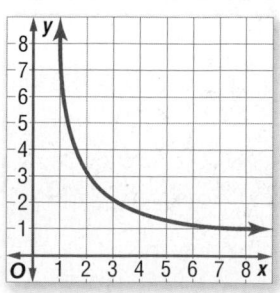

Key Vocabulary

constant of variation

constant rate of change

direct variation

domain

function

function rule

function table

inverse variation

linear function

range

rate of change

slope

Vocabulary Check

Choose the term from the list above that best matches each phrase.

1. the rate of change between any two points on a line

2. the set of input values for a function

3. a relationship when corresponding x and y values have a constant product

4. the constant ratio

5. a relation in which each element of the input is paired with exactly one element of the output according to a specified rule

6. the set of output values for a function

7. a table used to organize the input numbers, output numbers, and the function rule

8. two variable quantities have a constant ratio

9. the rate of change in a linear relationship

10. the operation performed on the input of a function

11. rate that describes how one quantity changes in relation to another

12. graph of a straight line

Multi-Part Lesson Review

Lesson 1 — Rates and Functions

Equations and Functions (Lesson 1B)

Copy and complete each function table. Then identify the domain and range.

13. $y = 2x$

x	2x	y
1	2 × 1	2
2	2 × 2	
3	2 × 3	
4		
5		

14. $y = 11x$

x	11x	y
0	11 × 0	0
1	11 × 1	
2		
3		
4		
5		

EXAMPLE 1 Curtis can drive his car 28 miles for every gallon of gasoline. Make a function table to show how many miles he can drive on 1, 2, 3, and 4 gallons. Then identify the domain and range.

Input	Function Rule	Output
Gallons of Gasoline	Multiply by 28	Total Miles
1	28 × 1	28
2	28 × 2	56
3	28 × 3	84
4	28 × 4	112

The domain is {1, 2, 3, 4} and the range is {28, 56, 84, 112}.

Functions and Graphs (Lesson 1C)

Graph the function represented by the table.

15.

Texting	
Time (min)	Total Words
0	0
1	20
2	40
3	60

Graph each equation.

16. $y = 4x - 2$ **17.** $y = x + 6$

18. RUNNING Damon runs the 100-meter dash at an average speed of 8 meters per second. The equation $d = 8t$ describes the distance d that Damon can run in t seconds at this speed. Represent the function by a graph.

EXAMPLE 2 Graph $y = 3x + 2$.

Select any four values for the input x. We chose 0, 1, 2, and 3. Substitute these values for x to find the output y.

x	3x + 2	(x, y)
0	3(0) + 2	(0, 2)
1	3(1) + 2	(1, 5)
2	3(2) + 2	(2, 8)
3	3(3) + 2	(3, 11)

Graph the ordered pairs.

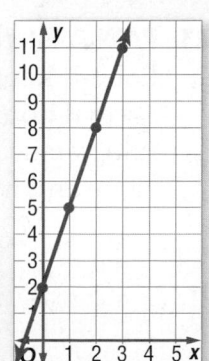

Constant Rate of Change (Lesson 2B)

Find the constant rate of change for each table.

19.

Time (min)	Laps Swam
0	0
1	2
2	4
3	6

20.

Package	Cookies
5	60
10	120
15	180
20	240

21.

Yards	1	2	3	4
Feet	3	6	9	12

EXAMPLE 3   The table shows the amount of money Hallie made doing yard work. Find the constant rate of change in dollars per hour.

Work

Hours	Money ($)
2	12
4	24
6	36
8	48

+2 (+12
+2 (+12
+2 (+12

Find the unit rate.

$$\frac{\text{change in money}}{\text{change in hours}} = \frac{12 \text{ dollars}}{2 \text{ hours}}$$

$$= \frac{6 \text{ dollars}}{1 \text{ hour}}$$

So, the number of dollars earned increases by $6 for every hour worked.

Slope (Lesson 2C)

22. CONCRETE The table shows the amount y of concrete that a construction company can pour per x minutes. Graph the data. Then find the slope of the line. Explain what the slope means.

Time (min)	15	30	45	60
Yards of Concrete	3	6	9	12

23. RAIN Yesterday, 2 inches of rain fell in 1 hour. It kept raining at the same rate for several hours. Draw a graph of inches of rain versus time. Find the value of the slope and interpret it in words.

EXAMPLE 4   The graph shows the relationship between the minutes spent reading x and the average amount of words read y. Find the slope of the line. Explain what the slope represents.

$$\text{slope} = \frac{\text{rise}}{\text{run}}$$

$$= \frac{65}{1}$$

So, for every minute of reading, there are 65 words read.

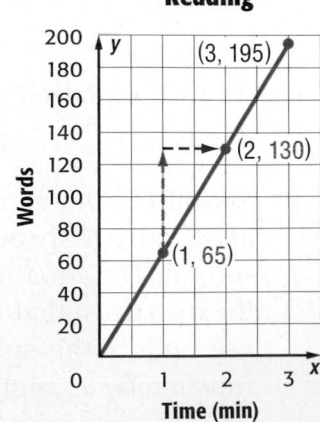

Reading

Lesson 3 Variation

PSI: Use a Graph (Lesson 3B)

STATUES For Exercises 24 and 25, use the graph that shows the heights of four freestanding statues in the world.

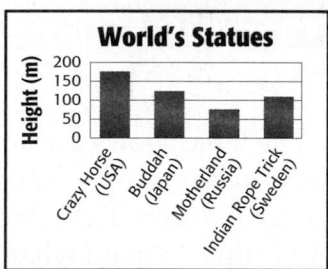

24. Which statue is the tallest?

25. Compare the heights of the Motherland statue and the Crazy Horse statue.

EXAMPLE 5 In the graph below, which place was favored by most students?

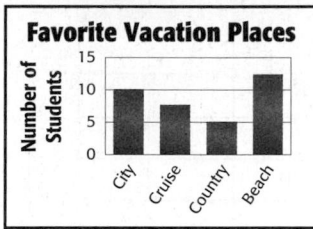

The beach was favored by 12 students, which was the greatest number.

Direct Variation (Lesson 3C)

Determine whether the linear function is a direct variation. If so, state the constant of variation.

26.

Tickets Available	80	60	40	20
Tickets Sold	40	30	20	10

27.

Distance (mi)	1,000	2,000	3,000	4,000
Time (h)	2	4	6	8

EXAMPLE 6 Determine whether the linear function is a direct variation. If so, state the constant of variation.

Number of Iron Weights	1	2	3	4	5
Total Weight (lb)	5	10	15	20	25

Since the ratios $\dfrac{\text{total weight}}{\text{number of weights}}$ are all $\dfrac{5}{1}$, the function is a direct variation.

Inverse Variation (Lesson 3E)

28. DRIVING The time it takes to complete a driving course varies inversely with the speed at which you drive. A driver can run the course in 2 hours at an average rate of 55 miles per hour. How long will it take to drive the same course at 50 miles per hour?

EXAMPLE 7 The time it takes to read a book varies inversely with the speed at which you are reading. Suppose it took 6 days to read a book at a rate of 35 pages per day. How long will it take to read the same book at 40 pages per day?

$$xy = k \qquad\qquad y = \frac{k}{x}$$

$$(6) \cdot (35) = k \qquad\qquad y = \frac{210}{40}$$

$$210 = k \qquad\qquad y = 5\frac{1}{4} \text{ days}$$

1. Complete the function table for $y = 4x$ and identify the domain and range.

x	4x	y
5		
6		
7		
8		

2. NAME TAGS Charmaine can make 32 name tags per hour. Make a function table that shows the number of name tags she can make in 3, 4, 5, and 6 hours.

Graph each equation.

3. $y = x + 5$

4. $y = 3x + 2$

5. $y = -2x + 3$

6. FINANCIAL LITERACY Clara earns $9 per hour mowing lawns. Make a function table and graph that show her total earnings for 2, 4, 6, and 8 hours.

7. MULTIPLE CHOICE During a bike-a-thon, Shalonda cycled at a constant rate. The table shows the distance she covered in half-hour intervals.

Time (h)	Distance (mi)
$\frac{1}{2}$	6
1	12
$1\frac{1}{2}$	18
2	24

Which of the following equations represents the distance d Shalonda covered after h hours?

A. $d = 6 + h$ **C.** $d = 12 + h$

B. $d = 6h$ **D.** $d = 12h$

8. TEMPERATURE Find the rate of change in degrees per hour.

Time (h)	0	1	2
Temperature (°C)	50	52	54

9. MONEY The table shows the amount of money José saved over a period of time. Graph the data. Then find the slope of the line. Explain what the slope represents.

Amount ($)	30	60	90
Weeks	1	2	3

10. TECHNOLOGY The time it takes to burn information onto a CD varies directly with the amount of information. If 2.5 megabytes of information take 10 seconds to burn, how long does it take to burn 10 megabytes of information?

11. RACING In the formula $d = rt$, the time t varies inversely with the rate r. A race car traveling 125 miles per hour completed one lap around a racetrack in 1.2 minutes. How fast was the car traveling if it completed the next lap in 0.8 minute?

12. THINK SOLVE EXPLAIN **EXTENDED RESPONSE** The function $m = 6h$ represents how much money Martha makes for every hour she works.

Part A What is her hourly rate, or her unit rate, of pay?

Part B Create an input/output table of her earnings from 5 to 10 hours.

Part C Graph the function.

Part D How is the rate of change related to the graph?

 ## Gridded Response: Decimals

When a gridded-response answer is a decimal, write the decimal point in an answer box at the top of the grid. Then fill in the decimal point bubble.

TEST EXAMPLE

The graph represents a cyclist's speed in miles per hour. What is the slope of the line?

$$\text{slope} = \frac{57 - 28.5}{4 - 2} \quad \longleftarrow \quad \frac{\text{change in } y}{\text{change in } x}$$

$$= \frac{28.5}{2} \text{ or } 14.25 \quad \text{The slope is } 14.25.$$

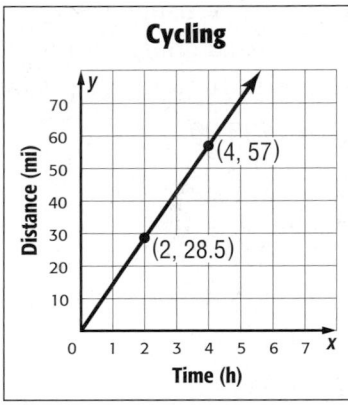

Cycling

Correct

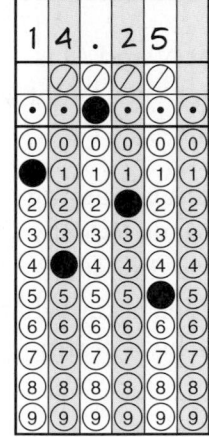

NOT Correct

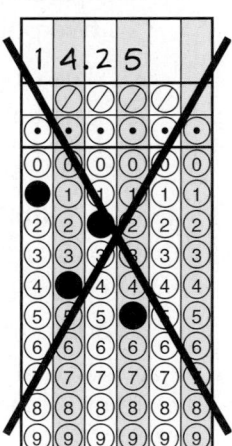

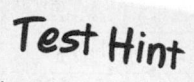 ## Work on It

The table below shows the costs of dog treats.

Number of Treats	3	6	9
Cost ($)	2.64	5.28	7.92

What is the slope of the line that represents the cost per treat? Fill in your answer on a grid like the one shown above.

Test Hint

A decimal like 0.75 could be correctly gridded as 0.75 or as .75.

Read each question. Then fill in the correct answer on the answer document provided by your teacher or on a sheet of paper.

1. Lisa's monthly charge for phone calls c can be found using the following equation.

$$c = 15 + 1.5h$$

In the equation, h represents the number of hours of usage during a month. What is the total charge for a month in which Lisa made 10 hours of phone calls?

A. $25.00

B. $27.50

C. $30.00

D. $32.00

2. The graph of the line $y = 2x + 2$ is shown below. Which table of ordered pairs contains only points on this line?

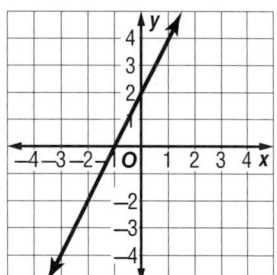

F.

x	−2	1	2
y	2	4	6

G.

x	−2	1	2
y	−2	4	6

H.

x	−2	1	2
y	−2	−4	−6

I.

x	−2	1	2
y	−2	−4	6

3. Which statement is true about the slope of line *LN*?

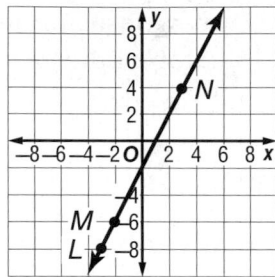

A. The slope is negative.

B. The slope between point *L* and point *M* is greater than the slope between point *M* and point *N*.

C. The slope is the same between any two points.

D. The slope between point *N* and *M* is greater than the slope between point *M* and *L*.

4. ▤ **GRIDDED RESPONSE** John is training for a marathon and has 50 days to train. He plans to run for 3 hours a day at an average speed of 4.5 miles per hour. What is the total number of miles he will run over the next 50 days?

5. ▤ **GRIDDED RESPONSE** Elia is in an airplane on her way to the Caribbean. She is currently at 7,000 feet. She looks down 150 feet and sees a cloud. How high in feet is the cloud above the ground?

6. THINK SOLVE EXPLAIN **SHORT RESPONSE** A dinner is served at an athletic booster fundraiser. The constant relationship between the number of people served at dinner n and the number of ounces of beef used b is shown in the table below. How many people were served if 760 ounces of beef were used?

n	5	20	150	?
b	20	80	600	760

7. Line *AB* represents a walking ramp.

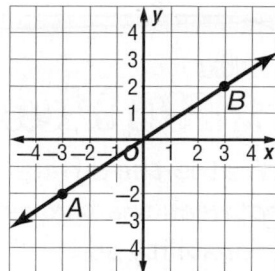

What is the slope of the ramp?

F. 1

G. $\frac{3}{4}$

H. $\frac{2}{3}$

I. $\frac{1}{3}$

8. Which expression represents the least value?

A. $430 \div \frac{1}{4}$

B. $430 + \frac{1}{4}$

C. $430 \times \frac{1}{4}$

D. $430 - \frac{1}{4}$

9. Which rule represents the function shown in the table?

x	y
1	5
2	8
3	11

F. $y = 5x$

G. $y = x \div 5$

H. $y = 3x + 2$

I. $y = 2x + 3$

10. A family went on a vacation and used 5.4 gallons of gasoline to travel 150 miles. How many total gallons of gasoline will they need to travel 200 more miles?

A. 12.6 gallons

B. 13.1 gallons

C. 14.3 gallons

D. 16.2 gallons

11. **GRIDDED RESPONSE** Heidi went to a parade and came back with 250 pieces of candy. Heidi then gave 150 pieces of candy to her brothers who could not go to the parade. Heidi's friend then gave her 25 pieces of candy. How many pieces of candy does Heidi now have?

12. **SHORT RESPONSE** Mrs. Smith has two pies for her fall feast. She has one apple pie and one pumpkin pie. After everyone got dessert, $\frac{2}{3}$ of the apple pie was eaten and $\frac{4}{5}$ of the pumpkin pie was eaten. How much total pie does Mrs. Smith have left?

13. **EXTENDED RESPONSE** Use the graph below to answer the following questions.

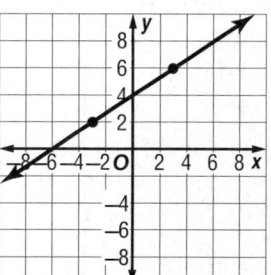

Part A What is the slope of the line?

Part B Describe the line.

NEED EXTRA HELP?													
If You Missed Question...	1	2	3	4	5	6	7	8	9	10	11	12	13
Go to Chapter-Lesson...	7-1B	7-1C	7-2C	7-2B	7-1B	2-2B	7-2C	3-3B	2-1C	5-1D	2-2B	3-2C	7-2C

connectED.mcgraw-hill.com

Investigate

 Animations

 Vocabulary

 Multilingual eGlossary

Learn

 Personal Tutor

 Virtual Manipulatives

 Graphing Calculator

 Audio

 Foldables

Practice

 Self-Check Practice

 Worksheets

 Assessment

Probability and Predictions

The ☆BIG Idea

What is the difference between the theoretical probability and experimental probability of an event?

FOLDABLES®
Study Organizer

Make this Foldable to help you organize your notes.

1
Probability

Review Vocabulary

ratio   razón a comparison of two quantities by division

The ratio of suns to stars is 2 to 3, 2:3, or $\frac{2}{3}$.

Key Vocabulary

English	Español
outcome	resultado
population	pablación
probability	probabilidad
sample space	espacio muestral

When Will I Use This?

Are You Ready for the Chapter?

You have two options for checking prerequisite skills for this chapter.

Text Option Take the Quick Check below. Refer to the Quick Review for help.

QUICK Check

Find each value.

1. 7×15
2. 24×6
3. 13×4
4. 8×21
5. $6 \times 5 \times 4 \times 3$
6. $4 \times 3 \times 2 \times 1$
7. $10 \times 9 \times 8 \times 7$
8. $11 \times 10 \times 9$
9. $\dfrac{6 \times 5}{3 \times 2}$
10. $\dfrac{9 \times 8 \times 7}{5 \times 4 \times 3}$
11. $\dfrac{4 \times 3 \times 2}{3 \times 2 \times 1}$
12. $\dfrac{7 \times 6 \times 5 \times 4}{4 \times 3 \times 2 \times 1}$

13. **MUSIC** If you download 9 songs an hour for 5 hours every day this week, how many songs will you have downloaded this week?

Write each fraction in simplest form.

14. $\dfrac{8}{12}$
15. $\dfrac{3}{18}$
16. $\dfrac{4}{16}$
17. $\dfrac{5}{15}$

18. **SLEEP** Use the table shown. What fraction of the day, in simplest form, does a guinea pig spend sleeping?

Average Sleep Time Per Day	
Animal	**Number of Hours**
Baboon	10
Guinea Pig	9
Gray Seal	6

QUICK Review

EXAMPLE 1

Find $7 \times 6 \times 5 \times 4$.

$7 \times 6 \times 5 \times 4 = 42 \times 5 \times 4$ Multiply from left to right.
$= 210 \times 4$
$= 840$

EXAMPLE 2

Find the value of $\dfrac{6 \times 5 \times 4}{3 \times 2 \times 1}$.

$\dfrac{6 \times 5 \times 4}{3 \times 2 \times 1} = \dfrac{\overset{2}{\cancel{6}} \times 5 \times \overset{2}{\cancel{4}}}{\underset{1}{\cancel{3}} \times \underset{1}{\cancel{2}} \times 1}$ Simplify. Divide out common factors.

$= \dfrac{20}{1}$ ← Multiply the numerator.
        ← Multiply the denominator.

$= 20$

EXAMPLE 3

Write $\dfrac{21}{28}$ in simplest form.

$\overset{\div 7}{\frown}$
$\dfrac{21}{28} = \dfrac{3}{4}$ Divide the numerator and denominator by the GCF, 7.
$\underset{\div 7}{\smile}$

Online Option Take the Online Readiness Quiz.

Main Idea

Find the probability of a simple event.

Vocabulary

outcome
simple event
probability
random
complementary event
geometric probability

Get ConnectED

CCSS 7.SP.5, 7.SP.7a

Probability and Simple Events

FOOD A cheesecake has four equal slices of each type as shown.

1. What fraction of the cheesecake is chocolate? Write in simplest form.

2. Suppose your friend gives you the first piece of cheesecake without asking which type you prefer. Are your chances of getting original the same as getting raspberry? Explain.

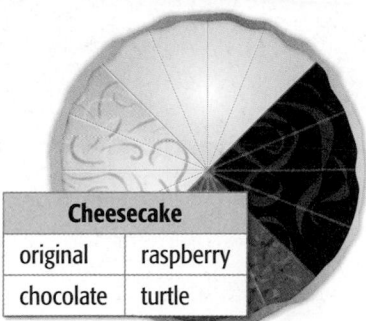

Cheesecake	
original	raspberry
chocolate	turtle

An **outcome** is any one of the possible results of an action. A **simple event** has one outcome or a collection of outcomes. For example, getting a piece of chocolate cheesecake is a simple event. The chance of that event happening is called its **probability**.

Key Concept **Probability**

Words If all outcomes are equally likely, the probability of a simple event is a ratio that compares the number of favorable outcomes to the number of possible outcomes.

Symbols $P(\text{event}) = \dfrac{\text{number of favorable outcomes}}{\text{number of possible outcomes}}$

 EXAMPLE **Find Probability**

1 **What is the probability of rolling an even number on a number cube?**

$$P(\text{even number}) = \frac{\text{even numbers possible}}{\text{total numbers possible}}$$
$$= \frac{3}{6} \text{ or } \frac{1}{2}$$

The probability of rolling an even number is $\frac{1}{2}$, 0.5, or 50%.

 CHECK Your Progress

Use a number cube to find the probability of each event. Write as a fraction in simplest form, a decimal, and a percent.

a. $P(\text{odd number})$ **b.** $P(5 \text{ or } 6)$ **c.** $P(\text{prime number})$

Outcomes occur at **random** if each outcome occurs by chance. For example, the number that results when rolling a number cube is a *random* outcome.

 REAL-WORLD EXAMPLE

2 **TALENT COMPETITION** Simone and her three friends were deciding how to pick the song they will sing for their school's talent show. They decide to roll a number cube. The person with the lowest number chooses the song. If her friends rolled a 6, 5, and 2, what is the probability that Simone will get to choose the song?

The possible outcomes of rolling a number cube are 1, 2, 3, 4, 5, and 6.

In order for Simone to be able to choose the song, she will need to roll a 1.

Let $P(A)$ be the probability that Simone chooses the song.

$$P(A) = \frac{\text{number of favorable outcomes}}{\text{number of possible outcomes}}$$

$$= \frac{1}{6} \quad \text{There are 6 possible outcomes, and 1 of them is favorable.}$$

The probability that Simone will choose the song is $\frac{1}{6}$, $0.1\overline{6}$, or about 17%.

Real-World Link · · · ·
Founded in 1842, the New York Philharmonic is the oldest symphony orchestra in the United States. They have performed over 14,000 concerts.

✓ **CHECK Your Progress**

MUSIC The table shows the numbers of brass instrument players in the New York Philharmonic. Suppose one brass instrument player is randomly selected to be a featured performer. Find the probability of each event. Write as a fraction in simplest form.

New York Philharmonic Brass Instrument Players	
Horn	6
Trombone	4
Trumpet	3
Tuba	1

d. P(trumpet)   **e.** P(brass)

f. P(flute)   **g.** P(horn or tuba)

The probability that an event will happen can be any number from 0 to 1, including 0 and 1, as shown on the number line below. Notice that probabilities can be written as fractions, decimals, or percents.

Impossible	Unlikely	As likely to happen as not	Likely	Certain
0	$\frac{1}{4}$	$\frac{1}{2}$	$\frac{3}{4}$	1
0	0.25	0.5	0.75	1
0%	25%	50%	75%	100%

In Example 2, either Simone will go first or she will *not* go first. These two events are complementary events. Two events are **complementary events** if they are the only two possible outcomes. The sum of the probabilities of an event and its complement is 1 or 100%. In symbols, $P(A) + P(not\ A) = 1$.

Study Tip

Complement of an Event
The probability that event A will not occur is noted as P(not A). Since P(A) + P(not A) = 1, P(not A) = 1 - P(A). P(not A) is read as the probability of the complement of A.

 EXAMPLE **Complementary Events**

③ **TALENT COMPETITION** Refer to Example 2. Find the probability that Simone will *not* choose the song.

$P(A) + P(not\ A) =$	1	Definition of complementary events
$\frac{1}{6} + P(not\ A) =$	1	Replace $P(A)$ with $\frac{1}{6}$.
$-\frac{1}{6}\qquad\quad = -\frac{1}{6}$		Subtract $\frac{1}{6}$ from each side.
$P(not\ A) =$	$\frac{5}{6}$	$1 - \frac{1}{6}$ is $\frac{6}{6} - \frac{1}{6}$ or $\frac{5}{6}$

CHECK Your Progress

SCHOOL Ricardo's teacher uses a spinner similar to the one shown at the right to determine the order in which each group will make its presentation. Use the spinner to find each probability. Write as a fraction in simplest form.

h. $P(not$ group 4) **i.** $P(not$ group 1 or group 3)

 **Your Understanding**

Example 1 Use the spinner to find each probability. Write as a fraction in simplest form, a decimal, and a percent.

1. $P(M)$ **2.** $P(Q$ or $R)$ **3.** $P(vowel)$

Examples 2 and 3 **MARBLES** Roberto has a bag that contains 7 blue, 5 purple, 12 red, and 6 orange marbles. Find each probability if he draws one marble at random from the bag. Write as a fraction in simplest form.

4. $P(purple)$ **⑤** $P(red$ or orange) **6.** $P(green)$

7. $P(not$ blue) **8.** $P(not$ red or orange) **9.** $P(not$ yellow)

10. **SURVEYS** Shanté asked her classmates how many pets they own. The responses are in the table. If a student in her class is selected at random, what is the probability that the student does *not* own 3 or more pets? Write as a fraction in simplest form.

Number of Pets	Response
None	6
1–2	15
3 or more	4

● = **Step-by-Step Solutions** begin on page R1.
Extra Practice begins on page EP2.

Example 1 A set of 20 cards is numbered 1, 2, 3, . . . 20. Suppose you pick a card at random without looking. Find the probability of each event. Write as a fraction in simplest form, a decimal, and a percent.

11. $P(1)$ **12.** $P(3 \text{ or } 13)$ **13.** $P(\text{multiple of 3})$

14. $P(\text{even number})$ **15.** $P(not\ 20)$ **16.** $P(not\ \text{a factor of 10})$

Examples 2 and 3 **RAFFLE** The table shows the number of students in seventh grade who entered the school drawing to win lunch with the principal. Suppose that only one student is randomly selected to win. Find the probability of each event. Write as a fraction in simplest form.

Lunch Raffle	
Room 8	24
Room 9	20
Room 10	10
Room 11	16
Room 12	14

17. $P(\text{Room 8})$ **18.** $P(\text{Room 9})$

19. $P(\text{Room 12})$ **20.** $P(\text{Room 10})$

21. $P(not\ \text{Room 8})$ **22.** $P(\text{Room 11})$

23 $P(not\ \text{Room 10})$ **24.** $P(\text{Room 10 or 11})$

25. SOUP A cupboard contains 20 soup cans. Seven are tomato, 4 are cream of mushroom, 5 are chicken, and 4 are vegetable. If one can is chosen at random from the cupboard, what is the probability that it is *neither* cream of mushroom *nor* vegetable soup? Write as a percent.

26. VIDEOS In a drawing, one name is randomly chosen from a jar of 75 names to receive free video rentals for a month. If Enola entered her name 8 times, what is the probability that she is *not* chosen to receive the free rentals? Write as a fraction in simplest form.

27. PETS The graph shows the last 33 types of pets that were purchased at a local pet store. What is the probability that a receipt chosen at random from these sales would be for a cat or dog? Express as a percent.

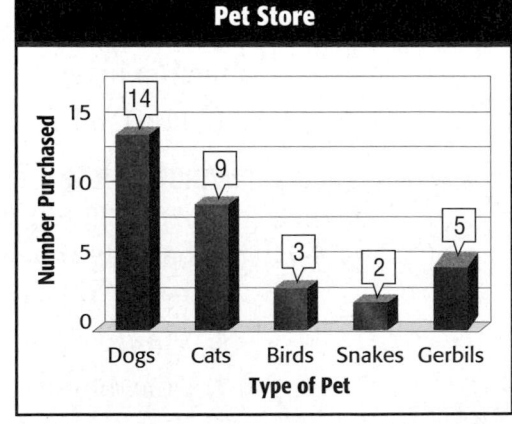

28. GAMES For a certain game, the probability of choosing a card with the number 13 is 0.008. What is the probability of *not* choosing card 13? Then describe the likelihood of the event occurring.

29. WEATHER The forecast for tomorrow says that there is a 40% chance of rain. Describe the complementary event and predict its probability.

30. **FLOWERS** Tatiana and her brother are each looking for flowers to give their aunt for her birthday. Tatiana likes either red roses or yellow tulips. Her brother likes blue irises, yellow daisies, red tulips, or white gardenias. Suppose Tatiana and her brother each choose one of their favorite flowers at random.

    a. What is the probability that Tatiana will choose a yellow flower? a red flower? Then describe the likelihood of each event.

    b. What is the probability that her brother will choose a yellow flower? a red flower? Describe the likelihood of each choice.

31. **GRAPHIC NOVEL** Refer to the graphic novel frame below. If Jamar and Theresa decide to create a music mix for a school dance and include an equal number of each genre, what would be the probability that any given song would be from the hip-hop genre?

Refer to the start of the chapter.

Other surveys in cities like ours say there are five kinds of music that the kids like—country, classical, hip-hop, oldies, and alternative.

CD-R WRITING...

We want to make sure there are all five types on this mix CD for the school dance.

H.O.T. Problems

32. **REASONING** A *leap year* has 366 days and occurs in non-century years that are divisible by 4. The extra day is added as February 29th. Determine whether each probability is 0 or 1. Explain your reasoning.
    a. P(there will be 29 days in February in 2032)
    b. P(there will be 29 days in February in 2058)

33. **CHALLENGE** A bag contains 6 red marbles, 4 blue marbles, and 8 green marbles. How many marbles of each color should be added so that the total number of marbles is 27, but the probability of randomly selecting one marble of each color remains unchanged? Explain your reasoning.

34. **WRITE MATH** Marissa has 5 black T-shirts, 2 purple T-shirts, and 1 orange T-shirt. Without calculating, determine whether each of the following probabilities is reasonable if she randomly selects one T-shirt. Explain your reasoning.
    a. P(black T-shirt) $= \frac{1}{3}$ **b.** P(orange T-shirt) $= \frac{4}{5}$ **c.** P(purple T-shirt) $= \frac{1}{4}$

35. A bag contains 8 blue marbles, 15 red marbles, 10 yellow marbles, and 3 brown marbles. If a marble is randomly selected, what is the probability that it will be brown?

A. 0.27   **C.** $0.08\overline{3}$

B. 22%   **D.** $\frac{3}{8}$

36. The records of a sporting goods store show that 1 out of every 4 balls purchased is a football. What is the probability that a football will NOT be purchased when a ball is purchased?

F. $\frac{1}{1}$ **H.** $\frac{3}{4}$

G. $\frac{1}{4}$ **I.** $\frac{1}{25}$

37. What is the probability of the spinner landing on a number less than 3?

A. 25%   **C.** 50%

B. 37.5%   **D.** 75%

38. The desserts at a picnic include 8 ice cream sandwiches, 9 chocolate frozen bananas, 12 ice cream cones, 11 frozen pops, 16 push-up sticks, and 14 shaved ice cones. What is the probability of choosing an ice cream cone at random?

F. $\frac{6}{29}$ **H.** $\frac{1}{6}$

G. $\frac{6}{35}$ **I.** $\frac{1}{12}$

 **Probability** ....

Geometric probability deals with the areas of figures instead of the number of outcomes. For example, the probability of landing in a specific region of a target is the ratio of the area of the specific region to the area of the target.

$$P(\text{specific region}) = \frac{\text{area of a specific region}}{\text{area of the target}}$$

EXAMPLE

Find the probability that the point of a pin dropped on the board will land within the shaded region. Write as a fraction in simplest form.

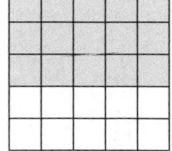

$P(\text{shaded region}) = \dfrac{\text{area of shaded region}}{\text{area of the target}}$

$= \dfrac{15}{25}$ or $\dfrac{3}{5}$

39. Find the probability that a randomly dropped pin will land with the point in the shaded region of each board. Write as a fraction in simplest form.

a. 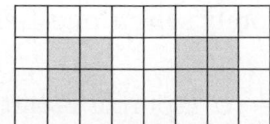   **b.** 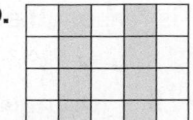   **c.**

Main Idea

Find sample spaces and probabilities.

Vocabulary

sample space
tree diagram
odds in favor

Get Connect ED

CCSS 7.SP.8, 7.SP.8a, 7.SP.8b

Sample Spaces

Explore Here is a probability game for two players.

• Place two green marbles into Bag A. Place one green and one red marble into Bag B.

• Without looking, Player 1 chooses a marble from each bag. If both marbles are the same color, Player 1 wins a point. If the marbles are different colors, Player 2 wins a point. Record your results and place the marbles back in the bag.

• Player 2 then pulls a marble from each bag and records the results. Continue alternating turns until each player has pulled from the bag 10 times. The player with the most points wins.

Now, play the game. Who won? What was the final score?

The set of all of the possible outcomes in a probability experiment is called the **sample space**. A **tree diagram** is a display that represents the sample space.

🏃 📝 **EXAMPLE** Find a Sample Space

① **ICE CREAM** A vendor sells vanilla and chocolate ice cream. Customers can choose from a waffle or sugar cone. Find the sample space for all possible orders of one scoop of ice cream in a cone.

Make a tree diagram that shows all of the possible outcomes.

Ice Cream	Cone	Sample Space
vanilla	waffle	vanilla, waffle
	sugar	vanilla, sugar
chocolate	waffle	chocolate, waffle
	sugar	chocolate, sugar

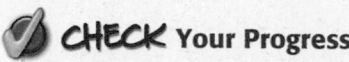

CHECK Your Progress

a. **PETS** The animal shelter has both male and female Labradors in yellow, brown, or black. Find the sample space for all possible Labradors available at the shelter.

 EXAMPLE Find Probability

2 **GAMES** Refer to the Explore at the start of this lesson. Find the sample space. Then find the probability that player 2 wins.

Make a tree diagram to show the sample space.

	Bag A	Bag B	Sample Space

```
Bag A      Bag B      Sample
                       Space
              G ——————— GG
        G <
              R ——————— GR
              G ——————— GG
        G <
              R ——————— GR
```

There are 4 equally likely outcomes with 2 favoring each player. So, the probability that player 2 wins is $\frac{2}{4}$, or $\frac{1}{2}$, or 0.5, or 50%.

Study Tip

Fair Game A fair game is one in which each player has an equal chance of winning. This game is a fair game.

CHECK Your Progress

b. **GAMES** Delmar tosses three coins. If all three coins show up heads, Delmar wins. Otherwise, Kara wins. Find the sample space. Then find the probability that Delmar wins.

 CHECK Your Understanding

Example 1 For each situation, find the sample space using a tree diagram.

1 A coin is tossed twice.

2. A pair of brown or black sandals are available in sizes 7, 8, or 9.

3. **FOOD** The table shows the sandwich choices for the school picnic. Find the sample space for a sandwich consisting of one type of meat and one type of bread.

Meat	Bread
ham	rye
turkey	sourdough
	white

Example 2 4. **GAMES** Gerardo spins a spinner with four sections of equal size, labeled A, B, C, and D, twice. If letter A is spun at least once, Gerardo wins. Otherwise, Odell wins. Find the sample space. Then find the probability that Odell wins.

= **Step-by-Step Solutions** begin on page R1.
Extra Practice begins on page EP2.

Example 1 **For each situation, find the sample space using a tree diagram.**

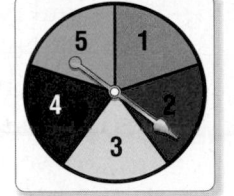

5. tossing a coin and spinning the spinner from the choices at the right

6. picking a number from 1 to 5 and choosing the color red, white, or blue

7. choosing a purple, green, black, or silver bike having 10, 18, 21, or 24 speeds

8. choosing a letter from the word SPACE and choosing a consonant from the word MATH

9. **CLOTHES** Jerry can buy a school T-shirt with either short sleeves or long sleeves in either gray or white and in small, medium, or large. Find the sample space for all possible T-shirts he can buy.

10. **FOOD** Three-course dinners can be made from the menu shown. Find the sample space for a dinner consisting of an appetizer, entrée, and dessert.

Appetizers	Entrees	Desserts
Soup	Steak	Carrot cake
Salad	Chicken	Apple pie
	Fish	

Example 2 **For each game, find the sample space. Then find the indicated probability.**

11. Elba tosses a quarter, a dime, and a nickel. If tails comes up at least twice, Steve wins. Otherwise Elba wins. Find P(Elba wins).

12. Ming rolls a number cube, tosses a coin, and chooses a card from two cards marked A and B. If an even number and heads appears, Ming wins, no matter which card is chosen. Otherwise Lashonda wins. Find P(Ming wins).

13. **FAMILIES** Mr. and Mrs. Romero are expecting triplets. Suppose the chance of each child being a boy is 50% and of being a girl is 50%. Find the probability of each event.

   a. P(all three children will be boys)

   b. P(at least one boy and one girl)

   c. P(two boys and one girl)

   d. P(at least two girls)

   e. P(the first two born are boys and the last born is a girl)

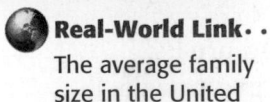

Real-World Link·····
The average family size in the United States is 2.59 people.

14. **GAMES** The following is a game for two players.

   • Three counters are labeled according to the table at the right.

   • Toss the three counters.

   • If exactly 2 counters match, Player 1 scores a point. Otherwise, Player 2 scores a point.

Counters	Side 1	Side 2
Counter 1	red	blue
Counter 2	red	yellow
Counter 3	blue	yellow

   Find the probability that each player scores a point.

15. UNIFORMS The University of Oregon's football team has many different uniforms. The coach can choose from four colors of jerseys and pants: green, yellow, white, and black. There are three helmet options: green, white, and yellow. Also, there are the same four colors of socks and two colors of shoes, black and yellow.

a. How many jersey/pant combinations are there?

b. If the coach picks a jersey/pant combination at random, what is the probability he will pick a yellow jersey with green pants?

c. Use a tree diagram to find all of the possible shoe and sock combinations.

d. MATH IN THE MEDIA Use a newspaper or magazine, the television, or the Internet to find the number of jerseys and pants your favorite college or professional team has as part of its uniform. Determine the number of jersey/pant combinations there are for your team.

Study Tip

Tree Diagrams To save time and space, you can use a single letter to represent each outcome in a tree diagram.

H.O.T. Problems

16. CHALLENGE Refer to Exercise 14. Do the two players both have an equal chance of winning? Explain.

17. FIND THE ERROR Caitlyn wants to determine the probability of guessing correctly on two true-false questions on her history test. She draws the tree diagram below using C for correct and I for incorrect. Find her mistake and correct it.

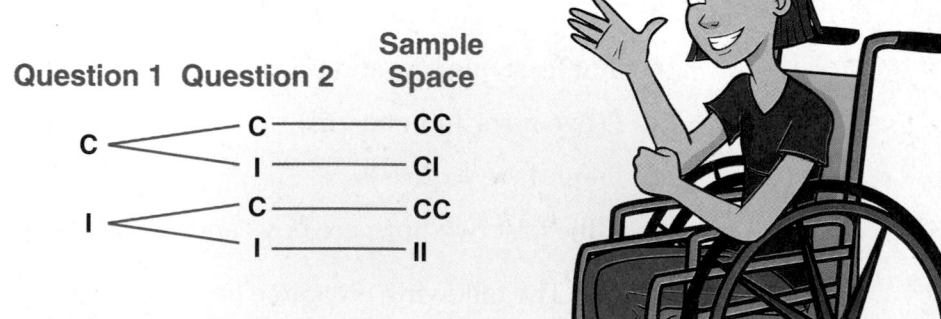

18. REASONING In the English language, 13% of the letters used are Es. Suppose you are guessing the letters of a two-letter word in a puzzle. Would you guess an E? Explain.

19. WRITE MATH Describe a game between two players using one coin in which each player has an equal chance of winning.

20. Mr. Zajac will choose one student from each of the two groups below to present their history report to the class.

Group 1	Group 2
Ava	Mario
Antoine	Brooke
Greg	

Which set shows all the possible choices?

A. {(Ava, Mario), (Antoine, Mario), (Greg, Mario)}

B. {(Ava, Antoine), (Antoine, Greg), (Brooke, Mario)}

C. {(Ava, Mario), (Antoine, Mario), (Greg, Mario), (Ava, Brooke), (Antoine, Brooke), (Greg, Brooke)}

D. {(Brooke, Antoine), (Mario, Greg), (Ava, Brooke), (Mario, Antoine)}

21. 📝 **GRIDDED RESPONSE** A coffee shop offers 2 types of coffee: regular and decaffeinated; 3 types of flavoring: vanilla, hazelnut, and caramel; and 2 choices of topping: with or without whipped cream. How many possible choices are there?

22. **THINK SOLVE EXPLAIN** **SHORT RESPONSE** Miranda needs to get dressed (D), brush her teeth (T), pack her lunch (L), and make her bed (B) before she leaves for school.

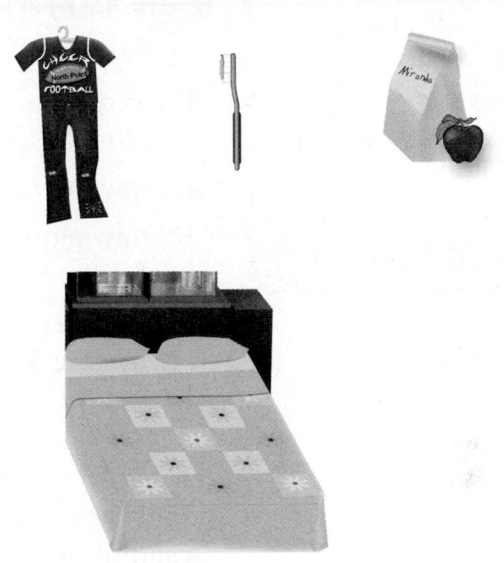

She always makes her bed right after she packs her lunch. List all the different combinations of tasks Miranda could do. Use the given letters of each task in your list (D, T, L, B).

More About **Sample Spaces**

Another way to describe the chance of an event occurring is with odds. The **odds in favor** of an event is a ratio that compares the number of ways the event *can* occur to the number of ways that the event *cannot* occur.

Suppose you want to find the odds in favor of rolling a 5 or 6 on a number cube.

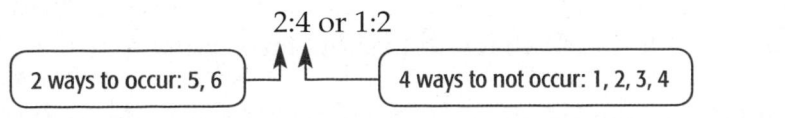

2:4 or 1:2

| 2 ways to occur: 5, 6 | 4 ways to not occur: 1, 2, 3, 4 |

So, the odds in favor of rolling a 5 or 6 is 1:2.

23. A letter from the alphabet is chosen at random. Find the odds in favor of picking an A, E, U, or O.

Main Idea

Use multiplication to count outcomes and find probabilities.

 Vocabulary

Fundamental Counting Principle

 Get Connect ED

 CCSS 7.SP.8a

Count Outcomes

 SALES The Shoe Warehouse sells sandals in different colors and styles.

Color	Style
Black	Platform
Brown	Slide
Tan	Wedge
White	
Red	

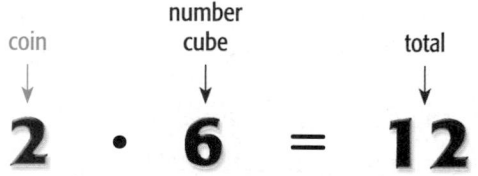

1. According to the table, how many colors of sandals are available?

2. How many styles are available?

3. Draw a tree diagram to find the number of different color and style combinations.

4. Find the product of the two numbers you found in Exercises 1 and 2. How does the number of outcomes compare to the product?

In the activity above, you discovered that multiplication, instead of a tree diagram, can be used to find the number of possible outcomes in a sample space. This is called the **Fundamental Counting Principle**.

Key Concept **Fundamental Counting Principle**

Words If event M has m possible outcomes and event N has n possible outcomes, then event M followed by event N has $m \times n$ possible outcomes.

EXAMPLE **Find the Number of Outcomes**

① Find the total number of outcomes when a coin is tossed and a number cube is rolled.

coin   number cube   total

$$2 \cdot 6 = 12$$ Fundamental Counting Principle

There are 12 different outcomes.

Check Draw a tree diagram to show the sample space.

 CHECK Your Progress

a. Find the total number of outcomes when choosing from bike helmets that come in three colors and two styles.

Study Tip

Jean Size In men's jeans, the size is labeled waist x length. So, a 32 x 34 is a 32-inch waist with a 34-inch length.

② **JEANS** The Jeans Shop sells young men's jeans in different sizes, styles, and lengths. Find the number of jeans available. Then find the probability of randomly selecting a size 32 × 34 slim fit. Is it likely or unlikely that the size would be chosen?

The Jeans Shop		
Waist Size	Length (in.)	Style
30	30	slim fit
32	32	bootcut
34	34	loose fit
36		
38		

size   length   style   total

$$5 \cdot 3 \cdot 3 = 45 \quad \text{Fundamental Counting Principle}$$

There are 45 different types of jeans to choose from. Out of the 45 possible outcomes, only one is favorable. So, the probability of randomly selecting a 32 × 34 slim fit is $\frac{1}{45}$ or about 2%.

It is very unlikely that the size would be chosen at random.

CHECK Your Progress

b. JEANS If the Jeans Shop adds relaxed fit jeans to its selection, find the number of available jeans. Then find the probability of randomly selecting a 36 × 30 relaxed fit pair of jeans. Is it likely or unlikely that the size would be chosen?

✓ **CHECK Your Understanding**

Example 1 Use the Fundamental Counting Principle to find the total number of outcomes for each situation.

① tossing a quarter, a dime, and a nickel

2. choosing a number on a number cube and picking a marble from the bag at the right

Example 2 **3. CLOTHES** Mira has 3 sweaters, 4 blouses, and 6 skirts that coordinate. Find the number of different outfits consisting of a sweater, blouse, and skirt that are possible. Then find the probability of randomly selecting a particular sweater-blouse-skirt outfit. State the probability as a fraction and percent. Is the probability of this event likely or unlikely?

= **Step-by-Step Solutions** begin on page R1.
Extra Practice begins on page EP2.

Example 1 Use the Fundamental Counting Principle to find the total number of outcomes for each situation.

4. choosing a bagel with one type of cream cheese from the list shown in the table

5. choosing a number from 1 to 20 and a color from 7 colors

6. picking a month of the year and a day of the week

7. choosing from a comedy, horror, or action movie each shown in four different theaters

8. rolling a number cube and tossing two coins

9. choosing iced tea in regular, raspberry, lemon, or peach flavors; sweetened or unsweetened; and in a glass or a plastic container

Bagels	Cream Cheese
Plain	Plain
Blueberry	Chive
Cinnamon raisin	Sun-dried tomato
Garlic	

Example 2

10. **ROADS** Two roads, Broadway and State, connect the towns of Eastland and Harping. Three roads, Park, Fairview, and Main, connect the towns of Harping and Johnstown. Find the number of possible routes from Eastland to Johnstown that pass through Harping. Then find the probability that State and Fairview will be used if a route is selected at random. State the probability as a fraction and percent and its likelihood.

11. **APPLES** An orchard makes apple nut bread, apple pumpkin nut bread, and apple buttermilk bread using 6 different varieties of apples, including Fuji. Find the number of possible bread choices. Then find the probability of selecting a Fuji apple buttermilk bread if a customer buys a loaf of bread at random. How likely is the probability of buying this bread at random?

12. **PASSWORDS** Find the number of possible choices for a 2-digit password that is greater than 19. Then find the number of possible choices for a 4-digit Personal Identification Number (PIN) if the digits cannot be repeated.

13. **T-SHIRTS** A store advertises that they have a different T-shirt for each day of the year. The store offers 32 different T-shirt designs and 11 choices of color. Is the advertisement true? Explain.

14. **CELL PHONES** The table shows cell phone options offered by a wireless phone company. If a phone with one payment plan and one accessory is given away at random, predict the probability that it will be Brand B and have a headset. Explain your reasoning.

Phone Brands	Payment Plans	Accessories
Brand A	Individual	Leather case
Brand B	Family	Car mount
Brand C	Business	Headset
	Government	Travel charger

Real-World Link
A popular orchard in Wapato, Washington, grows 7 varieties of apples including Granny Smith and Fuji.

H.O.T. Problems

15. **CHALLENGE** Determine the number of possible outcomes when tossing one coin, two coins, and three coins. Then determine the number of possible outcomes for tossing n coins. Describe the strategy you used.

16. **Which One Doesn't Belong?** Identify the choices for events M and N that do not result in the same number of outcomes as the other two. Explain your reasoning.

| 9 drinks, 8 desserts | 18 shirts, 4 pants | 10 groups, 8 activities |

17. ✍ **WRITE MATH** Compare and contrast the Fundamental Counting Principle and tree diagrams.

Test Practice

18. A bakery offers white, chocolate, or yellow cakes with white or chocolate icing. There are also 24 designs that can be applied to a cake. If all orders are equally likely, what is the probability that a customer will order a white cake with white icing in a specific design?

    A. $\dfrac{1}{30}$

    B. $\dfrac{1}{64}$

    C. $\dfrac{1}{120}$

    D. $\dfrac{1}{144}$

19. 📋 **SHORT RESPONSE** Hat Shack sells 9 different styles of hats in several different colors for 2 different sports teams. If the company makes 108 kinds of hats, how many different colors do they make?

Hat Shack		
Styles	**Colors**	**Teams**
9	?	2

20. **SCHOOL** Corinna can choose from 2 geography, 3 history, and 2 statistics classes. Find the sample space for all possible schedules. (Lesson 1B)

PROBABILITY Find the probability that the spinner shown at the right will stop on each of the following. Write as a fraction in simplest form. (Lesson 1A)

21. P(a vowel)

22. P(red)

23. P(R)

24. P(S)

Main Idea
Find the number of permutations of a set of objects and find probabilities.

 Vocabulary
permutation
combination

 Get ConnectED

 CCSS 7.SP.8a

Permutations

 Explore How many different ways are there to arrange your first three classes if they are math, science, and language arts?

STEP 1 Write math, science, and language arts on the index cards.

MATH	SCIENCE	LANGUAGE ARTS

STEP 2 Find and record all arrangements of classes by changing the order of the index cards.

1. When you first started to make your list, how many choices did you have for your first class?

2. Once your first class was selected, how many choices did you have for the second class? then, the third class?

A **permutation** is an arrangement, or listing, of objects in which order is important. The arrangement of science, math, language arts is a permutation of math, science, language arts because the order of the classes is different.

You can use the Fundamental Counting Principle to find the number of permutations.

 EXAMPLE **Find a Permutation**

1 **SCHEDULES** Find the number of possible arrangements of classes in the Explore Activity. Use the Fundamental Counting Principle.

There are **3** choices for the first class.

There are **2** choices that remain for the second class.

There is **1** choice that remains for the third class.

$$3 \cdot 2 \cdot 1 = 6 \longleftarrow \text{the number of permutations of 3 classes}$$

There are 6 possible arrangements, or permutations, of the 3 classes.

 CHECK Your Progress

a. **VOLLEYBALL** In how many ways can the starting six players of a volleyball team stand in a row for a picture?

You can use the Fundamental Counting Principle to find the probability of an event.

REAL-WORLD EXAMPLE Find Probability

2 **SWIMMING** The finals of the Northwest Swimming League features 8 swimmers. If each swimmer has an equally likely chance of finishing in the top two, what is the probability that Yumii will be in first place and Paquita in second place?

Northwest League Finalists	
Octavia	Eden
Natasha	Paquita
Calista	Samantha
Yumii	Lorena

There are **8** choices for first place.

There are **7** choices that remain for second place.

$$8 \cdot 7 = 56 \leftarrow \text{the number of permutations of the two places}$$

There are 56 possible arrangements, or permutations, of the two places.

Since there is only one way of having Yumii come in first and Paquita second, the probability of this event is $\frac{1}{56}$.

Study Tip

Reasonable Answers A possible probability of $\frac{1}{56}$ indicates that it is *unlikely* that Yumii will finish first and Paquita will finish second.

CHECK Your Progress

b. **LETTERS** Two different letters are randomly selected from the letters in the word *math*. What is the probability that the first letter selected is *m* and the second letter is *h*?

CHECK Your Understanding

Example 1   1. **AMUSEMENT PARKS** Seven friends are waiting to ride the new roller coaster. In how many ways can they board the ride, once it is their turn?

2. **COMMITTEES** In how many ways can a president, vice president, and secretary be randomly selected from a class of 25 students?

Example 2   3 **DVDs** You have five seasons of your favorite TV show on DVD. If you randomly select two of them from a shelf, what is the probability that you will select season one first and season two second?

4. **PASSWORDS** A password consists of four letters, none of which are repeated. What is the probability that a person could guess the entire password by randomly selecting the four letters?

Practice and Problem Solving

● = **Step-by-Step Solutions** begin on page R1.
Extra Practice begins on page EP2.

Example 1 **5** **CONTESTS** In the Battle of the Bands contest, in how many ways can the four participating bands be ordered?

6. **CODES** A garage door code has 5 digits. If no digit is repeated, how many codes are possible?

7. **LETTERS** How many permutations are possible of the letters in the word *friend?*

8. **NUMBERS** How many different 3-digit numbers can be formed using the digits 9, 3, 4, 7, and 6? Assume no number can be used more than once.

Example 2 9. **CAPTAINS** The members of the Evergreen Junior High Quiz Bowl team are listed in the table. If a captain and an assistant captain are chosen at random, what is the probability that Walter is selected as captain and Mi-Ling as co-captain?

Evergreen Junior High Quiz Bowl Team	
Jamil	Luanda
Savannah	Mi-Ling
Tucker	Booker
Ferdinand	Nina
Walter	Meghan

10. **BASEBALL** Adriano, Julián, and three of their friends will sit in a row of five seats at a baseball game. If each friend is equally likely to sit in any seat, what is the probability that Adriano will sit in the first seat and Julián will sit in the second seat?

11. **GAMES** Alex, Aiden, Dexter, and Dion are playing a video game. If they each have an equally likely chance of getting the highest score, what is the probability that Dion will get the highest score and Alex the second highest?

12. **BLOCKS** A child has wooden blocks with the letters shown below. Find the probability that the child randomly arranges the letters in the order TIGER.

13 **PHOTOGRAPHY** The Coughlin family discovered they can stand in a row for their family portrait in 720 different ways. How many members are in the Coughlin family?

14. **SCHOOL** Hamilton Middle School assigns a four-digit identification number to each student. The number is made from the digits 1, 2, 3, and 4, and no digit is repeated. If assigned randomly, what is the probability that an ID number will end with a 3?

15. **OPEN ENDED** Describe a real-world situation that has 6 permutations.

16. **CHALLENGE** There are 1,320 ways for three students to win first, second, and third place during a debate match. How many students are there on the debate team? Explain your reasoning.

17. **WRITE MATH** The symbol $P(10, 3)$ represents the number of permutations of 10 things taken 3 at a time. What does the symbol $P(20, 4)$ represent? Explain how you would find $P(20, 4)$.

Test Practice

18. **THINK SOLVE EXPLAIN** **SHORT RESPONSE** The five finalists in a random drawing are shown. Find the probability that Sean is awarded first prize and Teresa is awarded second prize.

Finalists
Cesar
Teresa
Sean
Nikita
Alfonso

19. A baseball coach is deciding on the batting order for his nine starting players with the pitcher batting last. How many batting orders are possible?

A. 8
B. 72
C. 40,320
D. 362,880

More About Outcomes

An arrangement, or listing, of objects in which order is *not* important is called a **combination**. You can find the number of combinations of objects by dividing the number of permutations of the entire set by the number of ways the smaller set can be arranged.

REAL-WORLD EXAMPLE

BASKETBALL A one-on-one basketball tournament has 6 players. **How many different pairings are there in the first round?**

Find the number of ways 2 players can be chosen from a group of 6.

There are 6 • 5 ways to choose 2 people.
There are 2 • 1 ways to arrange 2 people. $\dfrac{6 \times 5}{2 \times 1} = \dfrac{30}{2}$ or 15

So, there 15 different pairings in the first round.

20. **INTERNET** Of 12 Web sites, in how many ways can you choose to visit 6?

21. **TEAMS** There are 8 members on the debate team. How many different 5-player teams are possible?

Mid-Chapter Check

A number cube is rolled. Find each probability. Write as a fraction in simplest form. (Lesson 1A)

1. P(an odd number)

2. P(a number not greater than 4)

3. P(a number less than 6)

4. P(a multiple of 2)

5. **BOOKS** Brett owns 5 science-fiction books, 3 biographies, and 12 mysteries. If he randomly picks a book to read, what is the probability that he will not pick a science-fiction book? Write as a percent. (Lesson 1A)

For each situation, find the sample space using a tree diagram. (Lesson 1B)

6. The spinner shown is spun, and a digit is randomly selected from the number 803.

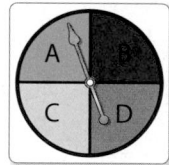

7. Two coins are tossed.

8. **MULTIPLE CHOICE** At a diner, a customer can choose from eggs or pancakes as an entrée and from ham or sausage as a side. Which set shows all the possible choices of one entrée and one side? (Lesson 1B)

   A. {(eggs, pancakes), (ham, sausage)}

   B. {(eggs, ham), (eggs, sausage), (pancakes, ham), (pancakes, sausage)}

   C. {(eggs, ham), (eggs, pancakes), (sausage, pancakes)}

   D. {(eggs, ham), (pancakes, sausage)}

9. **GAMES** Abbey rolls a number cube and chooses a card from among cards marked A, B, and C. If an odd number and a vowel turn up, Abbey wins. Otherwise, Benny wins. Find the sample space. Then find the probability that Benny wins. (Lesson 1B)

Use the Fundamental Counting Principle to find the total number of possible outcomes for each situation. (Lesson 1C)

10. A customer chooses a paper color, size, and binding style for some copies.

Color	Size	Binding
white	8.5″ x 11″	paper clip
yellow	8.5″ x 14″	binder clip
green	8.5″ x 17″	staple

11. A number cube is rolled and three coins are tossed.

12. **CARS** A certain car model comes in the colors in the table and with an automatic or manual transmission. Find the probability that a randomly selected car will have a black exterior, tan interior, and manual transmission if all outcomes are equally likely. (Lesson 1C)

Exterior	Interior
Black	Gray
White	Tan
Red	
Silver	

13. **COMPETITION** How many ways can 6 swimmers come in first, second, or third place? (Lesson 1D)

14. **MULTIPLE CHOICE** Noriko packed five different sweaters for her three-day trip. If she randomly selects one sweater to wear each day, what is the probability that she will select the brown sweater on Day 1, the orange sweater on Day 2, and the pink sweater on Day 3? (Lesson 1D)

   F. $\frac{1}{60}$

   G. $\frac{1}{120}$

   H. $\frac{3}{5}$

   I. $\frac{1}{5}$

Explore

Main Idea

Explore the probability of independent and dependent events.

 Get ConnectED

 CCSS 7.SP.7b, 7.SP.8b

Independent and Dependent Events

Place two red counters and two white counters in a paper bag. Then complete the following activities.

 ACTIVITIES

① **STEP 1** Without looking, remove a counter from the bag and record its color. Place the counter back in the bag.

STEP 2 Without looking, remove a second counter and record its color. The two colors are one trial. Place the counter back in the bag.

STEP 3 Repeat until you have 50 trials. Count and record the number of times you chose a red counter followed by a white counter.

② **STEP 1** Without looking, remove a counter from the bag and record its color. This time do not replace the counter.

STEP 2 Without looking, remove a second counter and record its color. The two colors are one trial. Place both counters back in the bag.

STEP 3 Repeat until you have 50 trials. Count and record the number of times you chose a red counter followed by a white counter.

Analyze the Results

1. In Activity 1, how many times did you choose a red counter followed by a white counter?

2. Were the results of Activity 2 the same or different than Activity 1?

3. Explain what caused any differences between Activity 2 and Activity 1.

Main Idea
Find the probability of independent and dependent events.

 Vocabulary
compound event
independent events
dependent events
disjoint events

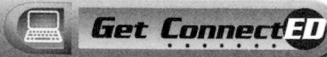

CCSS 7.SP.8, 7.SP.8a, 7.SP.8b

Independent and Dependent Events

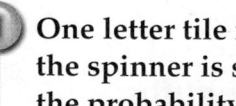 **SALES** A sale advertises that if you buy an item from the column on the left, you get a tote bag free. Suppose you choose items at random.

Type of Item	Tote Bag Colors
T-shirt	green
jacket	red
hat	white
beach towel	
visor	
polo shirt	

1. What is the probability of buying a beach towel? receiving a red tote bag?

2. What is the product of the probabilities in Exercise 1?

3. Draw a tree diagram to determine the probability that someone buys a beach towel and receives a red tote bag.

The combined action of buying an item and receiving a free tote bag is a compound event. A **compound event** consists of two or more simple events. These events are **independent events** because the outcome of one event does not affect the other event.

Key Concept Probability of Independent Events

Words The probability of two independent events can be found by multiplying the probability of the first event by the probability of the second event.

Symbols $P(A \text{ and } B) = P(A) \cdot P(B)$

EXAMPLES Independent Events

1. One letter tile is selected and the spinner is spun. What is the probability that both will be a vowel?

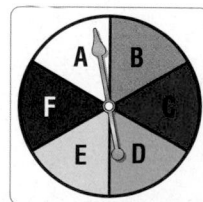

$P(\text{selecting a vowel}) = \frac{2}{7}$

$P(\text{spinning a vowel}) = \frac{2}{6} \text{ or } \frac{1}{3}$

$P(\text{both letters are vowels}) = \frac{2}{7} \cdot \frac{1}{3} \text{ or } \frac{2}{21}$

A spinner and a number cube are are used in a game. The spinner has an equal chance of landing on one of five colors: red, yellow, blue, green, or purple. The faces of the cube are labeled 1 through 6. What is the probability of a player spinning blue and then rolling a 3 or 4?

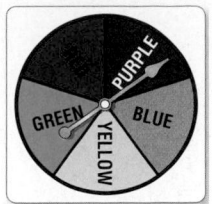

Vocabulary Link

Everyday Use

Independent not under control by others

Math Use

Independent outcome does not rely on another event

You are asked to find the probability of the spinner landing on blue and rolling a 3 or 4 on a number cube. The events are independent because spinning the spinner does not affect the outcome of rolling a number cube.

First, find the probability of each event.

$P(\text{blue}) = \dfrac{1}{5}$ ← number of ways to spin blue / number of possible outcomes

$P(3 \text{ or } 4) = \dfrac{2}{6} \text{ or } \dfrac{1}{3}$ ← number of ways to roll 3 or 4 / number of possible outcomes

Then, find the probability of both events occurring.

$P(\text{blue and 3 or 4}) = \dfrac{1}{5} \cdot \dfrac{1}{3}$ $P(A \text{ and } B) = P(A) \cdot P(B)$

$\qquad\qquad\qquad\qquad = \dfrac{1}{15}$ Multiply.

The probability is $\dfrac{1}{15}$.

 CHECK Your Progress

a. A game requires players to roll two number cubes to move the game pieces. The faces of the cubes are labeled 1 through 6. What is the probability of rolling a 2 or 4 on the first number cube and then rolling a 5 on the second?

Vocabulary Link

Everyday Use

Dependent under the control of others

Math Use

Dependent relying on another quantity or action

If the outcome of one event affects the outcome of another event, the events are called **dependent events**.

Key Concept   **Probability of Dependent Events**

Words   If two events A and B are dependent, then the probability of both events occurring is the product of the probability of A and the probability of B after A occurs.

Symbols $P(A \text{ and } B) = P(A) \cdot P(B \text{ following } A)$

3 FRUIT There are 4 oranges, 7 bananas, and 5 apples in a fruit basket. Ignacio selects a piece of fruit at random and then Terrance selects a piece of fruit at random. Find the probability that two apples are chosen.

Since the first piece of fruit is not replaced, the first event affects the second event. These are dependent events.

$P(\text{first piece is an apple}) = \dfrac{5}{16}$ ← number of apples
← total pieces of fruit

$P(\text{second piece is an apple}) = \dfrac{4}{15}$ ← number of apples left
← total pieces of fruit left

$P(\text{two apples}) = \dfrac{\overset{1}{\cancel{5}}}{\underset{4}{\cancel{16}}} \cdot \dfrac{\overset{1}{\cancel{4}}}{\underset{3}{\cancel{15}}}$ or $\dfrac{1}{12}$

The probability that two apples are chosen is $\dfrac{1}{12}$.

> **Study Tip**
>
> **Replacement**
> When finding the probability of an event, "with replacement" means to put the first piece of fruit chosen back into the basket before choosing the second piece. "Without replacement" means to choose the second piece of fruit without putting the first piece back in the basket.

 CHECK Your Progress

Refer to the situation above. Find each probability.

b. $P(\text{two bananas})$ **c.** $P(\text{orange then apple})$

d. $P(\text{apple then banana})$ **e.** $P(\text{two oranges})$

CHECK Your Understanding

Example 1 A penny is tossed and a number cube is rolled. Find each probability.

  1 $P(\text{tails and 3})$ **2.** $P(\text{heads and odd})$

Example 2 **3.** A spinner and a number cube are used in a game. The spinner has an equal chance of landing on 1 of 3 colors: red, yellow, and blue. The faces of the cube are labeled 1 through 6. What is the probability of a player spinning red and then rolling an even number?

Example 3 A card is drawn at random from the cards shown and not replaced. Then, a second card is drawn at random. Find each probability.

  4. $P(\text{two even numbers})$

  5. $P(\text{a number less than 4 and then a number greater than 4})$

Practice and Problem Solving

● = **Step-by-Step Solutions** begin on page R1.
  Extra Practice begins on page EP2.

Example 1 A number cube is rolled and a marble is selected at random from the bag at the right. Find each probability.

6. P(1 and red)

7. P(3 and purple)

8. P(even and yellow)

9. P(odd and *not* green)

10. P(less than 4 and blue)

11. P(greater than 1 and red)

Example 2 **12. GAMES** Corbin is playing a board game that requires rolling two number cubes to move a game piece. He needs to roll a sum of 6 on his next turn and then a sum of 10 to land on the next two bonus spaces. What is the probability that Corbin will roll a sum of 6 and then a sum of 10 on his next two turns?

Example 3 **13. LAUNDRY** A laundry basket contains 18 blue socks and 24 black socks. What is the probability of randomly picking 2 black socks from the basket?

Mrs. Ameldo's class has 5 students with blue eyes, 7 with brown eyes, 4 with hazel eyes, and 4 with green eyes. Two students are selected at random. Find each probability.

14. P(two blue)

15 P(green then brown)

16. P(hazel then blue)

17. P(brown then blue)

18. P(two green)

19. P(two *not* hazel)

20. MARKETING A discount supermarket has found that 60% of their customers spend more than $75 each visit. What is the probability that the next two customers will each spend more than $75?

21. SCHOOL Use the information below and the table at the right.

At Clearview Middle School, 56% of the students are girls and 44% are boys.

a. If two students are chosen at random, what is the probability that the first student is a girl and that the second student's favorite subject is science?

b. What is the probability that of two randomly selected students, one is a boy and the other is a student whose favorite subject is *not* art or math?

Clearview Middle School	
Favorite Subject	
Art	16%
Language Arts	13%
Math	28%
Music	7%
Science	21%
Social Studies	15%

22. MOVIES You and a friend plan to see 2 movies over the weekend. You can choose from 6 comedy, 2 drama, 4 romance, 1 science fiction, or 3 action movies. You write the movie titles on pieces of paper, place them in a bag, and each randomly select a movie. What is the probability that neither of you selects a comedy? Is this a dependent or independent event? Explain.

Chapter 8 Lesson 2B Compound Events **453**

23. **MONEY** Donoma had 8 dimes and 6 pennies in her pocket. If she randomly selected one coin and then a second coin without replacing the first, what is the probability that both coins were dimes? Is this a dependent or independent event? Explain.

24. **POPULATION** Use the information in the table.

Assume that age is *not* dependent on the region.

a. A resident of Lewburg County is picked at random. What is the probability that the person is under 64 years old and from an urban area?

b. What is the probability that the person is less than 18 years old or 65 years or older and from a rural area?

Lewburg County Population	
Demographic Group	**Fraction of the Population**
Under age 18	$\frac{3}{10}$
18 to 64 years old	$\frac{3}{5}$
65 years or older	$\frac{1}{10}$
Rural Area	$\frac{4}{5}$
Urban Area	$\frac{1}{5}$

25. **DOMINOES** A standard set of dominoes contains 28 tiles, with each tile having two sides of dots from 0 to 6. Of these tiles, 7 have the same number of dots on each side. If four players each randomly choose a tile, without replacement, what is the probability that each chooses a tile with the same number of dots on each side?

Real-World Link
The game of dominoes is believed to have originated in 12th century China.

H.O.T. Problems

26. **OPEN ENDED** There are 9 marbles representing 3 different colors. Write a problem where 2 marbles are selected at random without replacement and the probability is $\frac{1}{6}$.

27. **FIND THE ERROR** The spinner below is spun twice. Raul is finding the probability that both spins will result in an even number. Find his mistake and correct it.

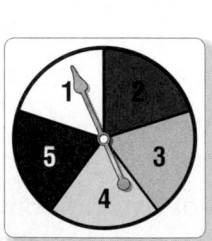

$$\frac{2}{5} \cdot \frac{1}{4} = \frac{2}{20}$$

28. **CHALLENGE** Determine whether the following statement is *true* or *false*. If false, provide a counterexample.

If two events are independent, then the probability of both events is less than 1.

29. **WRITE MATH** Explain the difference between independent events and dependent events.

30. Juan is holding four straws of different lengths. He has asked four friends to each randomly pick a straw. Milo picks first, and keeps the second longest straw. What is the probability that Felipe will get the longest straw if he picks second?

A. $\frac{1}{4}$ C. $\frac{1}{3}$

B. $\frac{1}{2}$ D. $\frac{1}{5}$

31. The spinner is spun twice. What is the probability that the spinner will land on a section for $10 off and then $5 off?

F. $\frac{3}{8}$ H. $\frac{5}{64}$

G. $\frac{1}{4}$ I. $\frac{3}{32}$

32. The spinners are each spun once.

 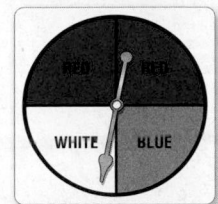

What is the probability of spinning 2 and white?

A. $\frac{1}{16}$ C. $\frac{2}{5}$

B. $\frac{1}{4}$ D. $\frac{3}{5}$

33. A jar of beads contains 6 aqua beads and 4 black beads. If two beads are selected at random, with replacement, what is the probability that both beads will be aqua?

F. $\frac{1}{3}$ H. $\frac{3}{5}$

G. $\frac{9}{25}$ I. $\frac{9}{10}$

More About Compound Events

Sometimes two events cannot happen at the same time. For example, when a coin is tossed, either heads *or* tails will turn up. Tossing heads and tossing tails are examples of **disjoint events**, or events that cannot happen at the same time. Disjoint events are also called *mutually exclusive events*.

EXAMPLE

A number cube is rolled. What is the probability of rolling a multiple of 3 or a 1?

$P(\text{multiple of 3 or 1}) = \dfrac{3}{6}$ ←— There are three favorable outcomes: 1, 3, or 6.
                           ←— There are 6 total outcomes.

So, the probability of rolling a multiple of 3 or a 1 is $\dfrac{3}{6}$ or $\dfrac{1}{2}$.

A number cube is rolled. Find each probability. Write as a fraction in simplest form.

34. $P(4 \text{ or } 5)$ **35.** $P(3 \text{ or even number})$ **36.** $P(1 \text{ or multiple of } 2)$

Problem Solving in Market Research

Keeping Your Eye on the Target Market

Do you think that gathering and analyzing information about people's opinions, tastes, likes, and dislikes sounds interesting? If so, then you should consider a career in market research. Market research analysts help companies understand what types of products and services consumers want. They design Internet, telephone, or mail response surveys and then analyze the data, identify trends, and present their conclusions and recommendations. Market research analysts must be analytical, creative problem-solvers, have strong backgrounds in mathematics, and have good written and verbal communication skills.

21st Century Careers

Are you interested in a career as a market research analyst? Take some of the following courses in high school.

- Algebra
- Calculus
- Computer Science
- English
- Statistics

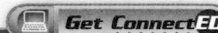
Get ConnectED

Survey Results: Teens and Social Networking

Reason to Use Social Networks	Percent of Respondents
Stay in touch with friends	91%
Make plans with friends	72%
Make new friends	49%

Friends on Social Networking Sites	Average Number
People who are regularly seen	43
People who are occasionally seen	23
People who are never seen in person	33
Total	99

Real-World Math

Use the results of the survey to solve each problem.

1. At Hastings Middle School, 560 of the students use social networking sites. Predict how many of them use the sites to make plans with friends.

2. Suppose 17.9 million teens use online social networks. Predict how many will be using the sites to make new friends.

3. According to the survey, what percent of a teen's networking site friends are people they regularly see?

4. Landon randomly selects a friend from his social networking site. What is the probability that it is someone he never sees in person? Write as a percent.

5. Paris wants to leave a message on 8 of her friends' social networking sites. In how many ways can she leave a message on her friends' sites?

Main Idea
Find and compare experimental and theoretical probabilities.

 Vocabulary
theoretical probability
experimental probability

 **Get ConnectED**

CCSS 7.SP.7, 7.SP.7a, 7.SP.7b

Theoretical and Experimental Probability

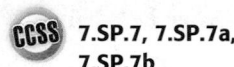 **Explore** Follow the steps to determine how many times doubles are expected to turn up when two number cubes are rolled.

Step 1 Use the table to help you find the expected number of times doubles should turn up when rolling two number cubes 36 times. The top row represents one number cube, and the left column represents the other number cube.

	1	2	3	4	5	6
1	1, 1	1, 2	1, 3	1, 4	1, 5	1, 6
2	2, 1	2, 2	2, 3	2, 4	2, 5	2, 6
3	3, 1	3, 2	3, 3	3, 4	3, 5	3, 6
4	4, 1	4, 2	4, 3	4, 4	4, 5	4, 6
5	5, 1	5, 2	5, 3	5, 4	5, 5	5, 6
6	6, 1	6, 2	6, 3	6, 4	6, 5	6, 6

Step 2 Roll two number cubes 36 times. Record the number of times doubles turn up.

1. Compare the number of times you *expected* to roll doubles with the number of times you *actually* rolled doubles.

2. Write the probability of rolling doubles out of 36 rolls using the number of times you *expected* to roll doubles from Step 1. Then write the probability of rolling doubles out of 36 rolls using the number of times you *actually* rolled doubles from Step 2.

Theoretical probability is based on what *should* happen when conducting a probability experiment. This is the probability you have been using since Lesson 1A. **Experimental probability** is based on what *actually* occurred during such an experiment.

Theoretical Probability	**Experimental Probability**
$\dfrac{6}{36}$ ← 6 rolls *should* occur	$\dfrac{n}{36}$ ← *n* rolls *actually* occurred

The theoretical probability and the experimental probability of an event may or may not be the same. As the number of times an experiment is conducted increases, the theoretical probability and the experimental probability should become closer in value.

EXAMPLE **Experimental Probability**

1 When two number cubes are rolled together 75 times, a sum of 9 is rolled 10 times. What is the experimental probability of rolling a sum of 9?

$$P(9) = \frac{\text{number of times a sum of 9 occurs}}{\text{total number of rolls}}$$

$$= \frac{10}{75} \text{ or } \frac{2}{15}$$

The experimental probability of rolling a sum of 9 is $\frac{2}{15}$.

CHECK Your Progress

a. In the above experiment, what is the experimental probability of rolling a sum that is *not* 9?

Study Tip

Trials A trial is one experiment in a series of successive experiments.

 **EXAMPLES** **Experimental and Theoretical Probability**

2 The graph shows the results of an experiment in which a spinner with 3 equal sections is spun sixty times. Find the experimental probability of spinning red for this experiment.

The graph indicates that the spinner landed on red 24 times, blue 15 times, and green 21 times.

$$P(\text{red}) = \frac{\text{number of times red occurs}}{\text{total number of spins}}$$

$$= \frac{24}{60} \text{ or } \frac{2}{5}$$

The experimental probability of spinning red is $\frac{2}{5}$.

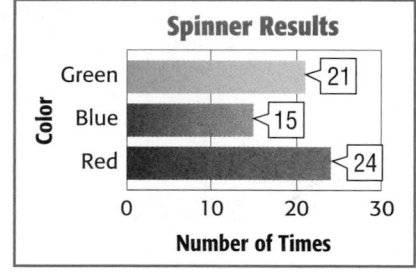

Spinner Results

(Color vs. Number of Times)
Green — 21
Blue — 15
Red — 24

3 Compare the experimental probability you found in Example 2 to its theoretical probability.

The spinner has three equal sections: red, blue, and green. So, the theoretical probability of spinning red is $\frac{1}{3}$. Since $\frac{2}{5} \approx \frac{1}{3}$, the experimental probability is close to the theoretical probability.

 CHECK Your Progress

b. Refer to Example 2. If the spinner was spun 3 more times and landed on green each time, find the experimental probability of spinning green for this experiment.

c. Compare the experimental probability you found in Exercise **b** to its theoretical probability.

Theoretical and experimental probability can be used to make predictions about future events.

REAL-WORLD EXAMPLES **Predict Future Events**

④ **MEDIA** Last year's sales at the Buy More DVD Store are shown. What is the probability that a customer bought a comedy DVD last year?

DVDs Sold	
Type	**Number**
Action	670
Comedy	580
Drama	450
Horror	300

2,000 DVDs were sold and 580 were comedy. So, the probability is $\frac{580}{2,000}$ or $\frac{29}{100}$.

⑤ Suppose a media buyer expects to sell 5,000 DVDs this year. How many comedy DVDs should she buy?

$\frac{580}{2,000} = \frac{x}{5,000}$ Write a proportion.

$580 \cdot 5,000 = 2,000 \cdot x$ Find the cross products.

$2,900,000 = 2,000x$ Multiply.

$1,450 = x$ Divide each side by 2,000.

She should buy about 1,450 comedy DVDs.

QUICK Review

Solving Proportions
The cross products of any proportion are equal.

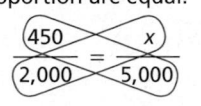

$450 \cdot 5,000 = 2,000 \cdot x$

CHECK Your Progress

d. What is the probability that a person buys a horror DVD?

e. If the media buyer expects to sell 3,000 DVDs this year, about how many action movies should she buy?

✓ CHECK Your Understanding

Examples 1–3   ① A coin is tossed 50 times, and it lands on heads 28 times.

    a. Find the experimental probability of the coin landing on heads.

    b. Find the theoretical probability of the coin landing on heads.

    c. Compare the probabilities in Exercises **a** and **b**.

Examples 4 and 5   **2. FOOD** The table shows the types of muffins that customers bought one morning from a bakery.

Muffin	Number of People
Blueberry	22
Poppyseed	17
Banana	11

    a. What is the probability that a customer bought a blueberry muffin?

    b. If 100 customers buy muffins tomorrow, about how many would you expect to buy a banana muffin?

Practice and Problem Solving

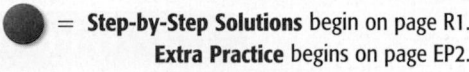
● = **Step-by-Step Solutions** begin on page R1.
Extra Practice begins on page EP2.

Examples 1–3　**3.** A number cube is rolled 20 times and lands on 1 two times and on 5 four times.

　　　a. Find the experimental probability of landing on 5. Compare the experimental probability to the theoretical probability.

　　　b. Find the experimental probability of *not* landing on 1. Compare the experimental probability to the theoretical probability.

Examples 4 and 5　**4. ZOO** Use the graph of a survey of 70 zoo visitors who were asked to name their favorite animal exhibit.

　　　a. What is the probability that the elephant exhibit is someone's favorite?

　　　b. What is the probability that the bear exhibit is someone's favorite?

　　　c. Suppose 540 people visit the zoo. Predict how many people will choose the monkey exhibit as their favorite.

What is your Favorite Animal Exhibit?		
Exhibit	**Tally**	**Frequency**
Bears	𝍠𝍠 I	6
Elephants	𝍠𝍠 𝍠𝍠 𝍠𝍠 II	17
Monkeys	𝍠𝍠 𝍠𝍠 𝍠𝍠 𝍠𝍠 I	21
Penguins	𝍠𝍠 𝍠𝍠 III	13
Snakes	𝍠𝍠 𝍠𝍠 III	13

　　　d. Suppose 720 people visit the zoo. Predict how many people will choose the penguin exhibit as their favorite.

5 ⟳ **MULTIPLE REPRESENTATIONS** A spinner with three equal-sized sections marked A, B, and C is spun 100 times.

Section	Frequency
A	24
B	50
C	26

　　　a. NUMBERS What is the theoretical probability of landing on A?

　　　b. NUMBERS The results of the experiment are shown in the table. What is the experimental probability of landing on A? on C?

　　　c. MODELS Make a drawing of what the spinner might look like based on its experimental probabilities. Explain.

6. GIFTS Use the graph at the right.

　　　a. What is the probability that a mother received a gift of flowers or plants? Write the probability as a fraction in simplest form.

　　　b. Out of 400 mothers that receive gifts, predict how many will receive flowers or plants.

Most Popular Mother's Day Gifts

card — 40%
flowers/plants — 28%
dinner/brunch — 8%
gardening items — 8%
apparel — 7%
jewelry — 6%
home décor — 3%

7. **CHALLENGE** The experimental probability of a coin landing on heads is $\frac{7}{12}$. If the coin landed on tails 30 times, find the number of tosses.

8. **REASONING** Twenty sharpened pencils are placed in a box containing an unknown number of unsharpened pencils. Suppose 15 pencils are taken out at random, of which five are sharpened. Based on this, is it reasonable to assume that the number of unsharpened pencils was 40? Explain your reasoning.

9. **WRITE MATH** Compare and contrast experimental probability and theoretical probability.

Test Practice

10. The table shows Mitch's record for the last thirty par-3 holes he has played.

Mitch's Golf Results	
Score	Number of Holes
2	4
3	14
4	9
5	3

Based on this record, what is the probability that Mitch will score a 2 or 3 on the next par-3 hole?

A. $\frac{7}{9}$ C. $\frac{3}{10}$

B. $\frac{3}{5}$ D. $\frac{9}{50}$

11. J.R. tossed a coin 100 times.

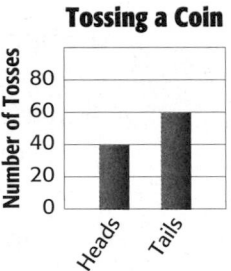

Tossing a Coin

Based on his results, what is the experimental probability of tossing tails on the next toss?

F. $\frac{1}{5}$ H. $\frac{3}{5}$

G. $\frac{2}{3}$ I. $\frac{4}{5}$

Eight cards numbered 1–8 are shuffled together. A card is drawn at random. It is not replaced. Find each probability. (Lesson 2B)

12. P(8 then 4)

13. P(even then odd)

14. **PUMPKINS** The table shows the number of participants at the pumpkin festival. A ribbon will be awarded to the top three participants in the perfect pumpkin contest. Find the number of ways the ribbons can be awarded in the perfect pumpkin contest. (Lesson 1D)

15. What is the probability of rolling a number greater than 4 on a number cube? (Lesson 1A)

Pumpkin Festival	
Competition	Number of Participants
Perfect Pumpkin	18
Biggest Pumpkin	12

Simulations

Main Idea

Investigate experimental probability by conducting a simulation.

CCSS 7.SP.6, 7.SP.7a, 7.SP.7b, 7.SP.8

A *simulation* is a way of modeling a problem situation. Simulations often mimic events that would be difficult or impractical to perform. In this lab, you will simulate purchasing a box of cereal and getting one of four possible prizes inside.

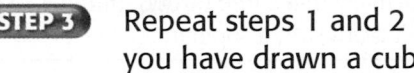

STEP 1 Place four different colored cubes into a paper bag.

STEP 2 Without looking, draw a cube from the bag, record its color, and then place the cube back in the bag.

STEP 3 Repeat steps 1 and 2 until you have drawn a cube from the bag a total of four times.

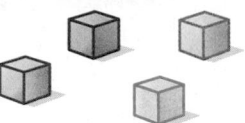

Analyze the Results

1. Based on your results, what is the experimental probability of drawing each cube?

2. **MAKE A PREDICTION** Predict the probability of selecting all four cubes in four boxes of cereal.

3. Repeat the simulation above 20 times. Use this data to predict the probability of selecting all four cubes in four choices.

4. Calculate the probability you found in Exercise 3 using the combined data of five different groups. How does this probability compare with your prediction?

5. Describe a simulation that could be used to predict the probability of taking a five-question true/false test and getting all five questions correct by guessing. Choose from two-sided counters, number cubes, coins, or spinners for your model.

6. **COLLECT THE DATA** Conduct 50 trials of the experiment you described in Exercise 5. Then use your results to predict the number of tests out of 20 in which you would get all 5 questions correct. Use a proportion.

Problem-Solving Investigation

CCSS 7.SP.6, 7.SP.7b

Main Idea Solve problems by acting them out.

P.S.I. TEAM ✛

Act It Out

EDDIE: I've been practicing free throws every day after school. Now I can make an average of 3 out of 4 free throws I try. I wonder how many times I usually make two free throws in a row.

YOUR MISSION: Act it out to determine the probability that Eddie makes two free throws in a row.

Understand	You know that Eddie makes an average of 3 out of 4 free throws. You could have Eddie actually make free throws, but that requires a basketball hoop. You could also act it out with a spinner.
Plan	Spin a spinner, numbered 1 to 4, two times. If the spinner lands on 1, 2, or 3, he makes the free throw. If the spinner lands on 4, he doesn't make it. Repeat the experiment 10 times.
Solve	Spin the spinner and make a table of the results.

Trials	1	2	3	4	5	6	7	8	9	10
First Spin	4	1	4	3	1	2	2	1	3	2
Second Spin	2	3	3	2	1	4	1	4	3	3

The circled columns show that six out of the 10 trials resulted in two free throws in a row. So, the probability is 60%.

Check	Repeat the experiment several times to see whether the results agree.

Analyze the Strategy

1. Would the results of the experiment be the same if it were repeated?

2. **WRITE MATH** Write a real-world problem that can be solved by acting it out. Then solve the problem by acting it out.

• Act it out.
• Draw a diagram.
• Use reasonable answers.
• Choose an operation.

● = **Step-by-Step Solutions** begin on page R1.
Extra Practice begins on page EP2.

Use the *act it out* strategy to solve Exercises 3–6.

3. TESTS Determine whether using a spinner with four equal sections is a good way to answer a five-question multiple-choice quiz if each question has choices A, B, C, and D. Justify your answer.

4. BOOKS There are 6 students in a book club. Two of them order books, and the delivery comes to the classroom teacher. However, the teacher cannot remember which 2 students ordered the books. Is it a good idea for the teacher to randomly pass out the books to any two students? Explain. What is the probability that the teacher will give the books to the correct students?

5. RUNNING Six runners are entered in a race. Assuming there are no ties, in how many different ways can first and second places be awarded?

6. MOVIES In how many different ways can four friends sit in a row of four seats at the movies if two of the friends insist on sitting next to each other?

Use any strategy to solve Exercises 7–12.

7 FESTIVALS The Student Council surveyed 160 students about their booth preference. The results are shown below. Is 35, 65, or 95 a reasonable answer for the number of students who would prefer a dunking booth? Explain.

Town Festival Survey Results

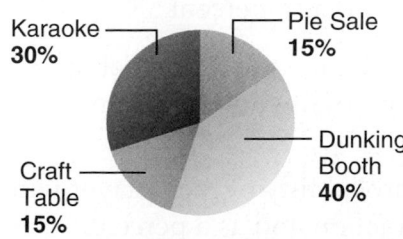

Karaoke 30%
Pie Sale 15%
Dunking Booth 40%
Craft Table 15%

8. ALGEBRA The pattern below is known as Pascal's Triangle. Would 1, 6, 10, 10, 6, and 1 be a reasonable conjecture for the numbers in the 6th row? Justify your answer.

9. CHESS A chess tournament will be held and 32 students will participate. The tournament will be single-elimination, which means if a player loses one match, he or she will be eliminated. How many total games will be played in the tournament?

10. SCHOOL Suppose rolling an even number on a number cube corresponds to an answer of *true* and rolling an odd number corresponds to an answer of *false*. Determine whether rolling this number cube is a good way to answer a five-question true-false quiz. Justify your answer.

11. CABLE A cable company is running a special for new customers. For the first 3 months they will get a discount of 20% off their regular bill. Would $50, $60, or $70 be a reasonable estimate for the first bill if their regular bill is $95? Explain.

12. MUSIC Liseta has an equal number of jazz, country, rap, pop, and R&B songs on her MP3 player. She listens to her MP3 player on random mode on both Wednesday and Thursday nights. What is the probability Liseta will hear a rap song first on Wednesday night?

Explore   Fair and Unfair Games

Main Idea

Use experimental and theoretical probabilities to decide whether a game is fair or unfair.

 **Vocabulary**

fair game
unfair game

 Get ConnectED

CCSS 7.SP.7b

Mathematically speaking, a two-player game is **fair** if each player has an equal chance of winning. A game is **unfair** if there is not such a chance. In this activity, you will analyze two simple games and determine whether each game is fair or unfair.

ACTIVITY

① **In a counter-toss game, players toss three two-color counters. The winner of each game is determined by how many counters land with either the red or yellow side facing up. Play this game with a partner.**

STEP 1   Player 1 tosses the counters. If 2 or 3 counters land red-side up, Player 1 wins. If 2 or 3 counters land yellow-side up, Player 2 wins. Record the results in a table like the one shown below. Place a check in the winner's column for each game.

Game	Player 1	Player 2
1		
2		

STEP 2   Player 2 then tosses the counters and the results are recorded.

STEP 3   Continue alternating the tosses until each player has tossed the counters 10 times.

Analyze the Results

1. Make an organized list of all the possible outcomes resulting from one toss of the 3 counters.

2. Calculate the theoretical probability of each player winning. Write each probability as a fraction and as a percent.

3. **MAKE A CONJECTURE** Based on the theoretical probabilities of each player winning, is this a fair or unfair game? Explain your reasoning.

4. Calculate the experimental probability of each player winning. Write each probability as a fraction and as a percent.

5. Compare the probabilities in Exercises 2 and 4.

6. **GRAPH THE DATA** Make a coordinate graph of the experimental probabilities of Player 1 winning for 5, 10, 15, and 20 games. Graph the ordered pairs (games played, Player 1 wins) using a blue pencil, pen, or marker. Describe how the points appear on your graph.

7. Add to the graph you created in Exercise 6 the theoretical probabilities of Player 1 winning for 5, 10, 15, and 20 games. Graph the ordered pairs (games played, Player 1 wins) using a red pencil, pen, or marker. Connect these red points and describe how they appear on your graph.

8. **MAKE A PREDICTION** Predict the number of times Player 1 would win if the game were played 100 times. Is this a fair or unfair game? Explain.

ACTIVITY

2 **In a number-cube game, players roll two number cubes. Play this game with a partner.**

STEP 1 Player 1 rolls the number cubes. Player 1 wins if the total of the numbers rolled is 5 or if a 5 is shown on one or both number cubes. Otherwise, Player 2 wins. Record the results in a table like the one shown below.

Game	Player 1	Player 2
1		
2		

STEP 2 Player 2 then rolls the number cubes and the results are recorded.

STEP 3 Continue alternating the rolls until each player has rolled the number cubes 10 times.

Analyze the Results

9. Make an organized list of all the possible outcomes resulting from one roll.

10. Calculate the theoretical probability of each player winning and the experimental probability of each player winning. Write each probability as a fraction and as a percent. Then compare these probabilities.

11. **MAKE A CONJECTURE** Based on the theoretical and experimental probabilities of each player winning, is this a fair or unfair game? Explain your reasoning.

12. **✍ WRITE MATH** If the game is fair, explain how you could change the game so that it is unfair. If the game is unfair, explain how you could change the game to make it fair. Explain.

Main Idea

Predict actions of a larger group by using a sample.

 Vocabulary

survey
population
sample

 Get Connect**ED**

CCSS 7.SP.1, 7.SP.2

Use Data to Predict

 TELEVISION The circle graph shows the results of a survey in which children ages 8 to 12 were asked whether they have a television in their bedroom.

1. Can you tell how many were surveyed? Explain.

2. Describe how you could use the graph to predict how many students in your school have a television in their bedroom.

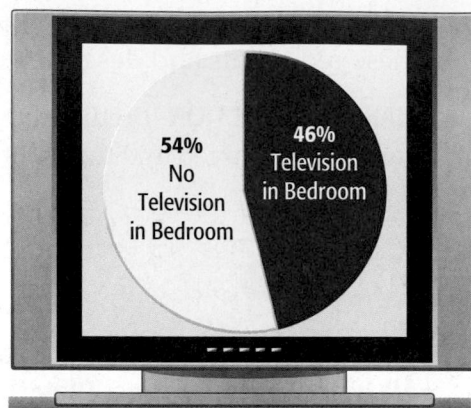

54% No Television in Bedroom

46% Television in Bedroom

A **survey** is designed to collect data about a specific group of people, called the **population**. A smaller group called a **sample** must be chosen. A sample is used to represent a population. If a survey is conducted at random, or without preference, you can assume that the survey represents the population. In this lesson, you will use the results of randomly conducted surveys to make predictions about the population.

REAL-WORLD EXAMPLE

① **TELEVISION Refer to the graphic above. Predict how many out of 1,725 students would not have a television in their bedroom.**

You can use the percent equation and the survey results to predict what part p of the 1,725 students have no TV in their bedroom.

$\underbrace{\text{part}}_{} = \underbrace{\text{percent}}_{} \cdot \underbrace{\text{whole}}_{}$

$p = 0.54 \cdot 1{,}725$ Survey results: 54%

$p = 931.5$ Multiply.

About 932 students do not have a television in their bedroom.

 CHECK Your Progress

a. TELEVISION Refer to the same graphic. Predict how many out of 1,370 students have a television in their bedroom.

Real-World Link · · · · ·
A survey found that 85% of people use emoticons on their instant messengers.

2 **INSTANT MESSAGING** Use the information at the left to predict how many of the 2,450 students at Washington Middle School use emoticons on their instant messengers.

You need to predict how many of the 2,450 students use emoticons.

Words	What number of students is 85% of 2,450 students?
Variable	Let n represent the number of students.
Equation	$n \quad = \quad 0.85 \quad \cdot \quad 2,450$

$n = 0.85 \cdot 2,450$ Write the percent equation.

$n = 2,082.5$ Multiply.

About 2,083 of the students use emoticons.

CHECK Your Progress

b. INSTANT MESSAGING This same survey found that 59% of people use sound on their instant messengers. Predict how many of the 2,450 students use sound on their instant messengers.

CHECK Your Understanding

Example 1 **1** **SPENDING** Use the circle graph that shows the results of a poll to which 60,000 teens responded.

    a. How many of the teens surveyed said that they would save their money?

    b. Predict how many of the approximately 28 million teens in the United States would buy a music CD if they were given $20.

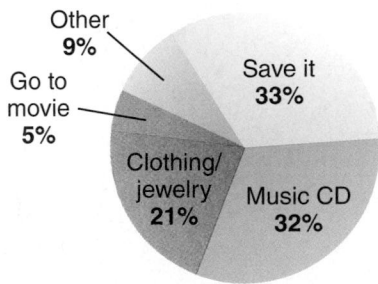

How Would You Spend a Gift of $20?

Other 9%
Go to movie 5%
Save it 33%
Clothing/jewelry 21%
Music CD 32%

Example 2 **2. FOOD** Use the bar graph that shows the results of a survey in which 538 students responded about their favorite ice cream flavor.

    a. Predict how many students prefer strawberry ice cream.

    b. Predict how many students prefer chocolate ice cream.

Favorite Ice Cream Flavor

Chocolate 19%
Vanilla 61%
Strawberry 11%
Butter Pecan 9%

Flavor

0 10 20 30 40 50 60 70
Percent

Practice and Problem Solving

⬤ = **Step-by-Step Solutions** begin on page R1.
Extra Practice begins on page EP2.

Examples 1 and 2

3. RECREATION In a survey, 250 people from a town were asked if they thought the town needed a recreation center. The results are shown in the table.

Recreation Center Needed	
Response	**Percent**
Yes	44%
No	38%
Undecided	18%

  a. Predict how many of the 3,225 people in the town think a recreation center is needed.

  b. About how many of the 3,225 people would be undecided?

4. VOLUNTEERING A survey showed that 90% of teens donate money to a charity during the holidays. Based on that survey, how many teens in a class of 400 will donate money the next holiday season?

Match each situation with the appropriate equation or proportion.

5 27 MP3s is what percent of 238 MP3s?

6. 238% of 27 is what number?

7. 27% of MP3 owners download music weekly. Predict how many MP3 owners out of 238 owners download music weekly.

  a. $n = 27 \cdot 2.38$

  b. $\dfrac{27}{100} = \dfrac{p}{238}$

  c. $\dfrac{27}{238} = \dfrac{n}{100}$

8. CATS Use the graph that shows the percent of cat owners who train their cats in each category.

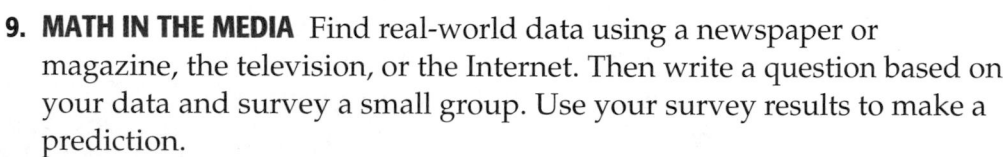

Training Cats

  a. Out of 255 cat owners, predict how many owners trained their cat not to climb on furniture.

  b. Out of 316 cat owners, predict how many more cat owners have trained their cat not to claw on furniture than have trained their cat not to fight with other animals.

9. MATH IN THE MEDIA Find real-world data using a newspaper or magazine, the television, or the Internet. Then write a question based on your data and survey a small group. Use your survey results to make a prediction.

H.O.T. Problems

10. CHALLENGE A survey found that 80% of teens enjoy going to the movies in their free time. Out of 5,200 teens, predict how many said that they do not enjoy going to the movies in their free time.

11. ✍ **WRITE MATH** Explain how to use a sample to predict what a group of people prefer. Then give an example of a situation in which it makes sense to use a sample.

12. The table shows how students spend time with their family.

How Students Spend Time with Family	
Dinner	34%
TV	20%
Talking	14%
Sports	14%
Taking Walks	4%
Other	14%

Of the 515 students surveyed, about how many spend time with their family at dinner?

A. 17   C. 119

B. 34   D. 175

13. Yesterday, a bakery baked 54 loaves of bread in 20 minutes. Today, the bakery needs to bake 375 loaves of bread. At this rate, predict how long it will take to bake the bread.

F. 1.5 hours   H. 3.0 hours

G. 2.3 hours   I. 3.75 hours

14. Of the 357 students in a freshman class, about 82% plan to go to college. How many students plan on going to college?

A. 224   C. 314

B. 293   D. 325

15. **THINK SOLVE EXPLAIN** **EXTENDED RESPONSE** Mr. Freisen surveyed his middle school students about how much time they spend playing video games each week. The results are shown below.

Video Game Time	Percent
Do not play	14
Around 1–5 hours per week	42
5–10 hours per week	32
11 or more hours per week	12

Part A What percent of students play at least 5 hours or more per week?

Part B What percent of students play video games during the week?

Part C Explain how you solved part b.

16. **SPINNERS** In how many ways could the colors in the spinner shown be arranged so that red and blue remain in the same place? (Lesson 3C)

17. **TOOTHBRUSHES** A dental hygienist randomly chooses a toothbrush in a drawer containing 12 purple toothbrushes, 9 aqua toothbrushes, and 6 teal toothbrushes. What is the theoretical probability that an aqua toothbrush is chosen? Write as a fraction in simplest form. (Lesson 3B)

18. A coin is tossed and a number cube is rolled. Find the probability of tossing tails and rolling an even number. (Lesson 2B)

19. **READING** Mr. Steadman plans to read eight children's novels to his second graders during the school year. In how many ways can he arrange the books to be read? (Lesson 1D)

Main Idea

Predict the actions of a larger group by using a sample.

 Vocabulary

unbiased sample
simple random sample
systematic random sample
biased sample
convenience sample
voluntary response sample

 Get ConnectED

 7.SP.1

Unbiased and Biased Samples

ENTERTAINMENT The manager of a television station wants to conduct a survey to determine which sport people consider their favorite to watch.

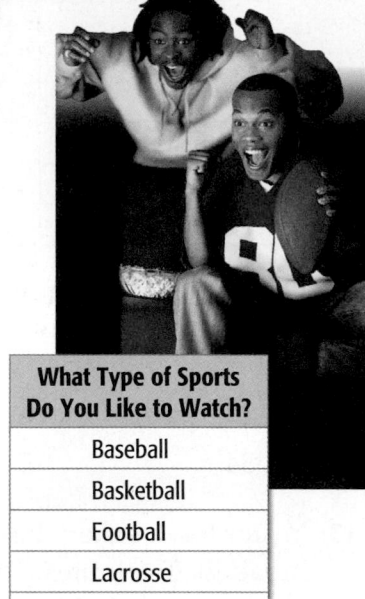

1. Suppose she surveys a group of 100 people at a basketball game. Do the results represent all of the people in the viewing area? Explain.

2. Suppose she surveys 100 students at your middle school. Do the results represent all of the people in the viewing area? Explain.

What Type of Sports Do You Like to Watch?
Baseball
Basketball
Football
Lacrosse
Soccer

3. Suppose she calls every 100th household in the telephone book. Do the results represent all of the people in the viewing area? Explain.

The manager of the television station cannot survey everyone in the viewing area. To get valid results, a sample must be chosen very carefully. An **unbiased sample** is selected so that it accurately represents the entire population. Two ways to pick an unbiased sample are listed below.

Concept Summary · Unbiased Samples

Type	Description	Example
Simple Random Sample	Each item or person in the population is as likely to be chosen as any other.	Each student's name is written on a piece of paper. The names are placed in a bowl, and names are picked without looking.
Systematic Random Sample	The items or people are selected according to a specific time or item interval.	Every 20th person is chosen from an alphabetical list of all students attending a school.

Vocabulary Link

Everyday Use

Bias a tendency or prejudice

Math Use

Bias error introduced by selecting or encouraging a specific outcome

In a **biased sample**, one or more parts of the population are favored over others. Two ways to pick a biased sample are listed below.

Concept Summary — Biased Samples

Type	Description	Example
Convenience Sample	A convenience sample consists of members of a population that are easily accessed.	To represent all the students attending a school, the principal surveys the students in one math class.
Voluntary Response Sample	A voluntary response sample involves only those who want to participate in the sampling.	Students at a school who wish to express their opinions complete an online survey.

 EXAMPLES Determine Validity of Conclusions

Determine whether each conclusion is valid. Justify your answer.

1) Every tenth person who walks into a department store is surveyed to determine his or her music preference. Out of 150 customers, 70 stated that they prefer rock music. The manager concludes that about half of all customers prefer rock music.

Since the population is every tenth customer of a department store, the sample is an unbiased, systematic random sample. The conclusion is valid.

2) The customers of a music store are surveyed to determine their favorite leisure time activity. Of these, 85% said that they like to listen to music, so the store manager concludes that most people prefer to listen to music in their leisure time.

The customers of a music store probably like to listen to music in their leisure time. The sample is a biased, convenience sample since all of the people surveyed are in one specific location. The conclusion is not valid.

 CHECK Your Progress

a. A radio station asks its listeners to indicate their preference for one of two candidates in an upcoming election. Seventy-two percent of the listeners who responded preferred candidate A, so the radio station announced that candidate A would win the election. Is the conclusion valid? Justify your answer.

A valid sampling method uses unbiased samples. If a sampling method is valid, you can make generalizations about the population.

 REAL-WORLD EXAMPLE **Use Sampling to Predict**

3 **STORES** A store sells 4 styles of pants: jeans, capris, cargos, and khakis. The store workers survey 50 customers at random. The survey responses are indicated at the right. If 450 pairs of pants are to be ordered, how many should be jeans?

Type	Number
Jeans	25
Capris	10
Cargos	8
Khakis	7

First, determine whether the sample method is valid. The sample is a simple random sample since customers were randomly selected. Thus, the sample method is valid.

$\frac{25}{50}$ or 50% of the customers prefer jeans. So, find 50% of 450.

$0.5 \times 450 = 225$, so about 225 pairs of jeans should be ordered.

CHECK Your Progress

b. RECREATION An instructor at a swimming pool asked her students if they would be interested in an advanced swimming course, and 60% stated that they would. Is the sample method valid? If so, suppose there are 870 pool members. How many people can the instructor expect to take the course?

CHECK Your Understanding

Examples 1 and 2   **Determine whether each conclusion is valid. Justify your answer.**

1. To determine how much money the average American family spends to cool their home, 100 Alaskan households are surveyed at random. Of the households, 85 said that they spend less than $75 per month on cooling. The researcher concluded that the average American household spends less than $75 on cooling per month.

2. To determine the most important company benefit, one out of every five employees is chosen at random. Medical insurance was listed as the most important benefit by 67% of the employees. The company managers conclude that medical insurance should be provided to all employees.

Example 3   **3** **GOLF** Zach is trying to decide which of three different golf courses is the best. He randomly surveyed people at a golf store and recorded the results in the table. Is the sample method valid? If so, suppose Zach surveyed 150 more people. How many people would be expected to vote for Rolling Meadows?

Course	Number
Whispering Trail	10
Tall Pines	8
Rolling Meadows	7

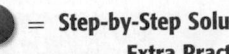 = **Step-by-Step Solutions** begin on page R1.
Extra Practice begins on page EP2.

Examples 1 and 2 **Determine whether each conclusion is valid. Justify your answer.**

4. To evaluate the quality of their product, a manufacturer of cell phones checks every 50th phone off the assembly line. Out of 200 phones tested, 4 are defective. The manager concludes that about 2% of the cell phones produced will be defective.

5. To determine whether the students will attend an arts festival at the school, Oliver surveys his friends in the art club. All of his friends plan to attend, so Oliver assumes that all the students at his school will also attend.

6. A magazine asks its readers to complete and return a questionnaire about popular television actors. The majority of those who replied liked one actor the most, so the magazine decides to write more articles about that actor.

7. To determine what people in California think about a proposed law, 5,000 people from the state are randomly surveyed. Of the people surveyed, 58% are against the law. The legislature concludes that the law should not be passed.

Example 3 **Determine if the sample method is valid. If so, solve the problem.**

8. **COMMUNICATION** The Student Council advisor asked every tenth student in the lunch line how they preferred to be contacted with school news. The results are shown in the table. If there are 680 students at the school, how many can be expected to prefer E-mail?

Method	Number
E-mail	16
Newsletter	12
Announcement	5
Telephone	3

9. **TRAVEL** A random sample of people at a mall shows that 22 prefer to take a family trip by car, 18 prefer to travel by plane, and 4 prefer to travel by bus. Out of 500 people, how many would you expect to say they prefer to travel by plane?

10. **GRAPHIC NOVEL** Refer to the graphic novel frame to answer Exercises a–b.

a. Which strategy gives Jamar and Marisol an appropriate sample for valid data?

b. What makes the sample appropriate?

Are You Ready for the Chapter?

You have two options for checking prerequisite skills for this chapter.

Text Option Take the Quick Check below. Refer to the Quick Review for help.

QUICK Check

SCAVENGER HUNT The bar graph below shows the number of items each student obtained from a scavenger hunt.

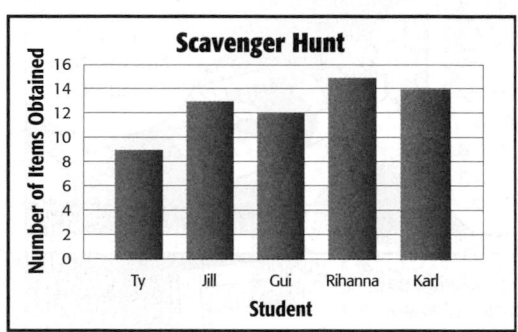

1. Who obtained the most items?

2. Who obtained the least items?

3. **TESTS** The table shows the results of Mr. Horowitz's first period science class test scores.

Science Test Scores				
91	86	72	90	81
88	79	93	66	83
77	92	70	80	83
75	82	73	98	84

What percent of students received a score of at least 90 on the science test?

QUICK Review

EXAMPLE 1

SPORTS Which player(s) average more than 10 points per game?

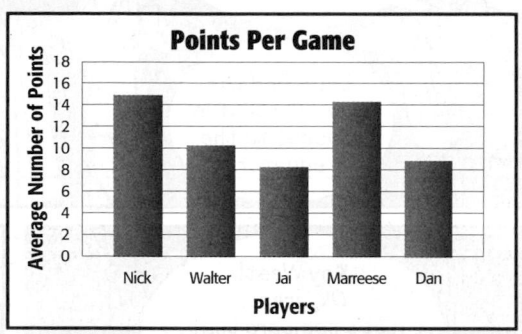

Nick, Walter, and Marreese averaged more than 10 points per game.

EXAMPLE 2

SOCIAL NETWORKING Use the circle graph. If 300 people were surveyed, how many people have two accounts?

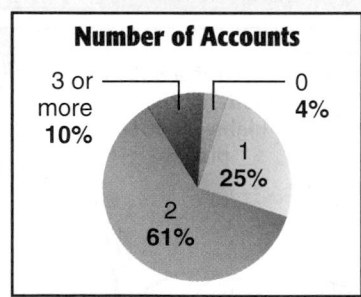

Find 61% of 300.
61% of 300 = 61% × 300
= 0.61 × 300 or 183
So, 183 people have two accounts.

 **Online Option** Take the Online Readiness Quiz.

Writing Math

Describe Data

When you *describe* something, you represent it in words.

What can you write about the data displayed in the table below?

TAKEOUT The table shows the menu for Lombardo's Restaurant.

Takeout	Price ($)
Main Dish	8.00
Side Dish	2.50
Dessert	4.00

- The price of a dessert is $4.00.

- The price of a main dish is twice as much as the price of a dessert.

- A side dish is the least expensive item.

- If you buy one of each item, the cost is more than $10.

All of these statements describe the data. In what other ways can you describe the data?

Practice

1. **ADVERTISING** The table shows the results of a survey in which teens were asked to which types of advertising they pay attention. Describe the data.

Type of Advertising	Percent of Teens
Television	80
Magazine	62
Product in a movie	48
E-mail	24

2. **SPORTS** The table shows the number of participants ages 7–17 for each sport. Describe the data.

Sport	Number (millions)
Karate	8.1
Roller hockey	2.7
Snowboarding	9.3
Golf	11.4

3. **WRITE MATH** Use the Internet or other resources to find real-world data. Display your data in a table. Then have a classmate describe your data.

Explore

Changes in Data Values

Main Idea

Explore how changes in data values affect mean, median, and mode.

Get Connect ED

FISHING Quin and four of his friends went fishing on a Saturday morning. Each bucket shows the number of fish that each person caught.

There are many ways to describe this data.

ACTIVITY

STEP 1 Place counters in five cups to represent the five buckets of fish. Find the mode.

STEP 2 Rearrange the cups from least to greatest to find the amount of counters in the middle cup.

STEP 3 Move the counters among the cups so that each cup has the same number of counters to find the average.

Analyze the Results

1. Was there a number of fish that occurred the most times for the data set? Explain your reasoning.

2. Would your answer to Exercise 1 change if you included a sixth friend who caught 9 fish? Explain.

3. How many counters are in the middle cup after you rearrange the cups from least to greatest? Compare this to your answer if you were to include the sixth friend who caught 9 fish.

4. How many counters are in each cup after moving the counters? How many counters would be in each cup if a sixth friend caught 9 fish?

Main Idea

Determine and describe how changes in data values impact measures of central tendency.

 Vocabulary

measures of central tendency
mean
median
mode

 Get Connect**ED**

Measures of Central Tendency

BICYCLES The sizes of students' bicycles are listed in the table.

1. Which number appears most often in the table?

2. Order the numbers from least to greatest. Which number(s) is in the middle?

Students' Bicycle Sizes (in.)			
20	24	20	26
24	24	24	26
24	29	26	24

Measures of central tendency are numbers that describe the center of a data set. They are the mean, median, and mode.

Key Concept	Measures of Central Tendency
mean	sum of the data divided by the number of items in the set; commonly called the *average*
median	middle number of the data ordered from least to greatest, or the mean of the middle two numbers
mode	number or numbers that occur most often

 REAL-WORLD EXAMPLE

 DVDS The numbers of DVDs rented during a week are listed in the table. Find the mean, median, and mode.

Daily DVD Rentals						
S	M	T	W	TH	F	S
55	34	35	34	57	78	106

Mean $\dfrac{55 + 34 + 35 + 34 + 57 + 78 + 106}{7} = \dfrac{399}{7}$ or 57

Median 34, 34, 35, (55), 57, 78, 106　First, write the data in order.

Mode 34　It is the only value that occurs more than once.

The mean is 57 DVDs, the median is 55 DVDs, and the mode is 34 DVDs.

 CHECK Your Progress

a. **FOOTBALL** The points scored in each game by Darby Middle School's football team are 21, 35, 14, 17, 28, 14, 7, 21, and 14. Find the mean, median, and mode.

   **Effect of Extreme Values**

2 **MUSIC DOWNLOADS**
The number of songs
downloaded during
one week are shown
in the table. Which measure is most affected by Saturday's
downloads?

Music Downloads						
S	M	T	W	TH	F	S
5	2	3	0	6	4	36

With Saturday's Downloads

Mean $\dfrac{5 + 2 + 3 + 0 + 6 + 4 + 36}{7} = \dfrac{56}{7}$ or 8

Median 0, 2, 3, ④, 5, 6, 36
Mode no mode

Without Saturday's Downloads

Mean $\dfrac{5 + 2 + 3 + 0 + 6 + 4}{6} = \dfrac{20}{6}$ or $3\dfrac{1}{3}$

Median 0, 2, ③, ④, 5, 6
          ↓
        3.5
Mode no mode

> **Study Tip**
> **Mode** If every value occurs once, then there is no mode, not a mode of 0.

So, Saturday's downloads affect the mean and the median. The
mean is affected the most in this example.

 CHECK Your Progress

b. LIBRARIES The number of library
books returned at Edison Middle
School is shown in the table.
Which measure is most affected by
Thursday's returned books?

Library Books Returned				
M	T	W	TH	F
35	23	18	5	29

The table lists some guidelines for using each measure of central
tendency.

Key Concept   **Mean, Median, and Mode**

Measure	Most Useful When...
Mean	the data set has no extreme values
Median	the data set has extreme values there are no big gaps in the middle of the data
Mode	the data set has many identical numbers

 **REAL-WORLD EXAMPLE** **Choose an Appropriate Measure**

(3) **EXERCISE** The following set of data shows the number of push-ups Rufio did in one minute for the past 5 days: 24, 21, 28, 27, and 26. Which measure of central tendency best represents the data? Justify your selection and then find the measure.

Since the data set has no extreme values or numbers that are identical, the mean would best represent the data.

Mean $\dfrac{24 + 21 + 28 + 27 + 26}{5} = \dfrac{126}{5}$ or 25.2

The measure 25.2 push-ups best represents the data. The median, 26, is also appropriate.

 CHECK Your Progress

c. CLOTHING The following set of data shows the costs of pairs of jeans for various stores: $25.99, $29.99, $34.99, $19.99, $45.99. Which measure of central tendency best represents the data? Justify your selection and then find the measure.

✓ CHECK Your Understanding

Example 1 Find the mean, median, and mode for each data set. Round to the nearest tenth if necessary.

1. number of states visited: 4, 0, 3, 19, 2, 0, 5, 7

2. inches of rainfall: 3.5, 2.8, 1.4, 1.2, 12.5, 2.3, 0.4

3.

Number of Calories in Vegetables
(1 serving)

12	6	15	14	5	18	14
20	24	10	19	55	14	19

Example 2 **4. RETAIL** A department store recorded the time, in minutes, it took to help customers: 4, 3.5, 18, 1.5, 2.5, 5, 4.5, 6, 2.5, 3.5. Which measure is most affected by the extreme value?

Example 3 **5. VIDEO GAMES** The table shows the cost of various video games. Which measure of central tendency best represents the data? Justify your selection and then find the measure.

Video Game Prices			
$28	$19	$14	$31
$15	$29	$37	$22

Practice and Problem Solving

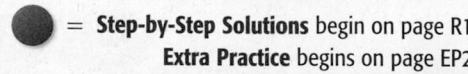

= **Step-by-Step Solutions** begin on page R1.
Extra Practice begins on page EP2.

Example 1 **Find the mean, median, and mode for each data set. Round to the nearest tenth if necessary.**

6. scores earned on a math test: 87, 71, 95, 98, 48, 74, 83, 92, 87, 79

7 ages of professional football players: 25, 23, 23, 27, 39, 27, 23, 31, 26, 28

8.

Total Points Scored During Recent Basketball Season

Team	Points
Duke	2,830
Kansas	2,771
New Mexico State	2,793
North Carolina	3,454
Memphis	2,709
Tennessee	2,738

9.

Gas Mileage of Various Vehicles

Vehicles	
compact	28
hybrid	47
sedan	24
sports	21
SUV	17

Gas Mileage (miles per gallon)

Example 2 10. **HEIGHT** The heights of plants, in inches, are 13, 4, 6, 9, 11, 23, and 7. Which measure is most affected by the extreme value?

11. **TEST SCORES** Patty had scores of 90, 95, 65, 90, and 85 on five math tests. Which measure is most affected by the extreme value?

Example 3 12. **STUDENTS** The ages of the students in Ms. Watson's seventh-grade class are given in the table. Which measure of central tendency best represents the ages? Justify your selection and then find the measure.

Ages of Students

12	13	12	12	12
12	14	12	12	12
13	12	12	13	12
12	12	13	11	12

13. **SONGS** The number of weeks that songs have been on the Top 20 Country Songs list is shown in the table. Would the mean, median, or mode best represent the data? Explain.

Top 20 Country Songs

6	14	17	24	28
8	15	20	25	31
9	15	20	25	36
13	16	22	26	43

14. **INCOME** The five employees at Burger Bun each earn $8.50 per hour. The assistant manager earns $14.00 per hour. How would each measure of central tendency change if the manager's pay of $18.55 per hour was included?

15. The mean of a set of fifteen numbers is 5.2. What is the sum of the fifteen numbers?

16. **GRAPHIC NOVEL** Refer to the graphic novel frame below. Compare and contrast the mean, median, and mode of the data from the two cities. Round to the nearest tenth if necessary.

17. **MATH IN THE MEDIA** Find a data set in a newspaper or magazine, on television, or on the Internet. Which measure of central tendency best represents the data? Justify your selection and then find the measure.

18. **GYMNASTICS** Rama needs an average score of 9.5 from 5 judges to win a gymnastics meet. The mean score from 4 judges was 9.48. What is the lowest score Rama can get from the 5th judge and still win?

19. **MINERALS** The Mohs Hardness Scale is used to identify the hardness of a mineral. A mineral with a hardness of 1 on the scale, such as talc, is the softest, while a mineral with a hardness of 10, such as a diamond, is the hardest. What is the mean hardness of a group of minerals with the following hardness levels: 3, 5, 8, 2, 1, 1, 7, 3, 9, 2, 3, 10?

20. **SCIENCE** Mr. Gallaher's science class was observing the effects of fertilizer on plants. The results are shown in the table. Find the mean for each data set to the nearest tenth. Compare your results with the mean of all the data. How does combining the data affect the mean?

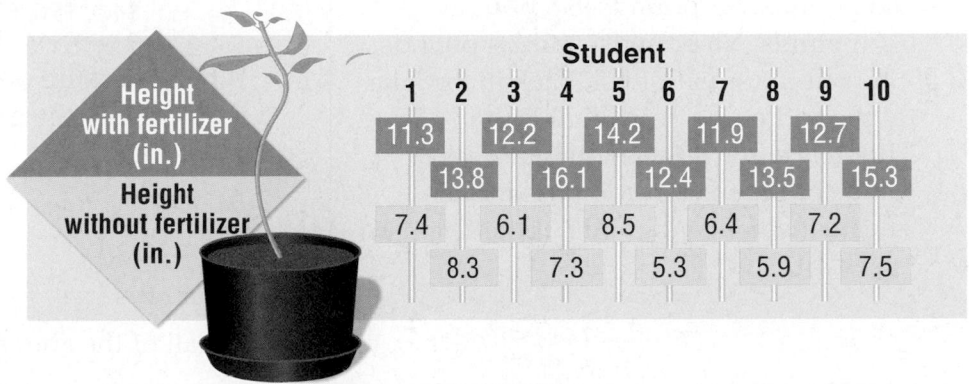

	Student									
	1	2	3	4	5	6	7	8	9	10
Height with fertilizer (in.)	11.3	12.2	14.2	11.9	12.7					
		13.8	16.1	12.4	13.5	15.3				
Height without fertilizer (in.)	7.4	6.1	8.5	6.4	7.2					
		8.3	7.3	5.3	5.9	7.5				

21. **OPEN ENDED** Write a data set for each of the following.

    a. at least four numbers that have a mean of 5 and a median that is not 5

    b. at least six numbers that have a mean of 10 and a mode of 10

22. **REASONING** Determine whether the following statement is *sometimes*, *always*, or *never* true. Explain your reasoning.

    The median must be a member of the data set.

23. **CHALLENGE** The range of a data set is the difference between the greatest and least values. A set of three numbers has a mean of 40, a median of 41, and a range of 9. What are the three numbers?

24. **WRITE MATH** Explain why the mean is the measure of central tendency that is most affected by extreme values.

Test Practice

25. A group of 99 students were asked how many siblings they each have. The results are shown in the table.

Siblings	
zero	5
one	37
two	28
three	29

If the 100th student answered 5 siblings, which measure below would be most affected?

    A. mean   **C.** mode

    B. median   **D.** none of them

26. **THINK SOLVE EXPLAIN SHORT RESPONSE** Vicki went fishing and caught 6 fish that weighed 7.3 pounds, 5.1 pounds, 8.8 pounds, 4.5 pounds, 5.6 pounds, and 2.4 pounds. She put the smallest fish back in the lake at the end of the day. What is the difference in the mean, median, and mode of the six fish compared to the mean, median, and mode of the five fish she kept?

27. Hiroshi's bowling scores are shown.

179, 178, 210, 180, 182, 190

If he bowled a 150 in his next game, which of the following statements would be true?

    F. The median would increase.

    G. The mean would increase.

    H. The median would decrease

    I. The mode would decrease.

28. The amount of money Mario earned each week mowing grass is shown.

$40, $56, $36, $44, $36

Which measure of central tendency would show the greatest amount of money earned?

    A. mean

    B. median

    C. mode

    D. all of the above

Extend

Main Idea

Use technology to calculate the mean, median, and mode of a set of data.

 Get Connect**ED**

Spreadsheet Lab: Mean, Median, Mode

ALLOWANCE Mrs. Jenson's seventh-grade class was surveyed about how much allowance each student receives each week. The results are shown in the table. Make a spreadsheet for the data and find the mean, median, and mode.

Allowance Per Week ($)				
15	10	11	9	12.50
28	12	10	10	15

ACTIVITY

STEP 1 Open a new spreadsheet. Create four columns labeled DATA, MEAN, MEDIAN, and MODE.

Use =AVERAGE (A2:A11) to find the mean.

Use =MEDIAN (A2:A11) to find the median.

Use =MODE (A2:A11) to find the mode.

Spreadsheet sample ⬚ ⬚ ☒

◇	A	B	C	D
1	DATA	MEAN	MEDIAN	MODE
2	28	13.25	11.5	10
3	15			
4	15			
5	12.5			
6	12			
7	11			
8	10			
9	10			
10	10			
11	9			

|◀ ◀ ▶ ▶| Sheet 1 ╱ Sheet 2 ╱ Sheet 3 ╱

STEP 2 Enter each allowance amount in the DATA column.

STEP 3 In cell B2, enter =AVERAGE(A2:A11). In cell C2, enter =MEDIAN(A2:A11). In cell D2, enter =MODE(A2:A11). Each of these will find the mean, median, and mode of the data set.

Analyze the Results

1. What data value is an extreme for the set? Explain your reasoning.

2. Describe how the measures of central tendency would change if the extreme value was not included in the data set.

Main Idea
Find the measures of variation of a set of data.

 Vocabulary
measures of variation
range
quartile
lower quartile
upper quartile
interquartile range
outlier

Measures of Variation

SURVEYS Jamie asked her classmates how many glasses of water they drink on a typical day.

1. What is the median of the data set?

2. Organize the data into two groups: the top half and the bottom half. How many data values are in each group?

3. What is the median of each group?

4. Find the difference between the two numbers from Exercise 3.

Glasses of Water Consumed

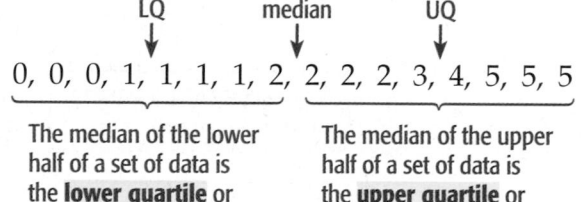

Measures of variation are used to describe the distribution of the data. The **range** is the difference between the greatest and least data values. **Quartiles** are values that divide the data set into four equal parts.

$$\underbrace{0,\ 0,\ 0,\ 1,\ 1,\ 1,\ 1,\ 2,}\ \overset{\text{LQ}}{\downarrow}\quad \overset{\text{median}}{\downarrow}\quad \overset{\text{UQ}}{\downarrow}\ \underbrace{2,\ 2,\ 2,\ 3,\ 4,\ 5,\ 5,\ 5}$$

The median of the lower half of a set of data is the **lower quartile** or LQ; in this case, 1.

The median of the upper half of a set of data is the **upper quartile** or UQ; in this case, 3.5.

So, one half of the data lie between the lower quartile and upper quartile. This is called the **interquartile range**.

> ### Key Concept — Measures of Variation
>
> **Upper and Lower Quartiles**
> The upper and lower quartiles are the medians of the upper half and lower half of a set of data, respectively.
>
> 14,   18,   19,   20,   24,   29,   31
>
>   lower quartile   median   upper quartile
>
> **Interquartile Range**
> The range of the middle half of the data. It is the difference between the upper quartile and the lower quartile; in this case, 29 − 18 or 11. The interquartile range is 29 − 18 or 11.
>
> **Range**
> The difference between the greatest and least data values; in this case, 31 − 14 or 17.

 **EXAMPLE** **Find Measures of Variation**

1 **SPEED** Find the measures of variation for the data.

Range 70 − 1 or 69 mph

Quartiles

Order the numbers from least to greatest.

Animal Speeds	
Animal	**Speed (mph)**
cheetah	70
lion	50
cat	30
elephant	25
mouse	8
spider	1

lower half	median	upper half

1 8 25 30 50 70

↑ LQ $\frac{25 + 30}{2} = 27.5$ ↑ UQ

Interquartile Range 50 − 8 or 42 UQ − LQ

The range is 69, the median is 27.5, the lower quartile is 8, the upper quartile is 50, and the interquartile range is 42.

Study Tip

Interquartile Range If the interquartile range is low, the middle data are grouped closely together.

 **CHECK** Your Progress

a. SPORTS Determine the measures of variation for the data in the table.

Basketball Scores				
64	61	67	59	60
58	57	71	56	62

An **outlier** is a data value that is either much *greater* or much *less* than the median. If a data value is more than 1.5 times the value of the interquartile range beyond the quartiles, it is an outlier.

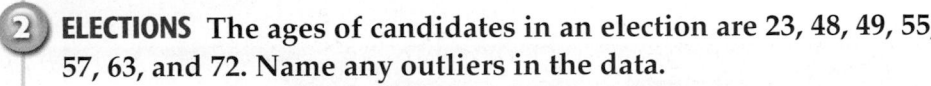 **EXAMPLE** **Find Outliers**

2 **ELECTIONS** The ages of candidates in an election are 23, 48, 49, 55, 57, 63, and 72. Name any outliers in the data.

Find the interquartile range.
   63 − 48 = 15

Multiply the interquartile range by 1.5.
   15 × 1.5 = 22.5

Subtract 22.5 from the lower quartile and add 22.5 to the upper quartile.
   48 − 22.5 = 25.5   63 + 22.5 = 85.5

The limits for the outliers are between 25.5 and 85.5. The only age beyond this is 23. So, it is the only outlier.

 Real-World Link · · · ·

The 26th Amendment to the United States Constitution lowered the voting age to 18.

 CHECK Your Progress

b. BRIDGES The lengths, in feet, of various bridges are 88, 251, 275, 354, and 1,121. Name any outliers in the data set.

3 **SCIENCE** The table shows a set of scores on a science test in two different classrooms. Compare and contrast their measures of variation.

Room A	Room B
72	63
100	93
67	79
84	83
65	98
78	87
92	73
87	81
80	65

Find the measures of variation for both rooms.

	Room A	**Room B**
Range	$100 - 65 = 35$	$98 - 63 = 35$
Median	80	81
UQ	$\dfrac{87 + 92}{2} = 89.5$	$\dfrac{87 + 93}{2} = 90$
LQ	$\dfrac{67 + 72}{2} = 69.5$	$\dfrac{65 + 73}{2} = 69$
Interquartile Range	$89.5 - 69.5 = 20$	$90 - 69 = 21$

Both classrooms have a range of 35, but Room B has an interquartile range of 21 while Room A's interquartile range is 20. There are slight differences in the medians as well as upper and lower quartiles.

CHECK Your Progress

c. WEATHER Temperatures for the first half of the year are given for Antelope, Montana, and Augusta, Maine. Compare and contrast the measures of variation of the two cities.

Month	Antelope, MT	Augusta, ME
January	21	28
February	30	32
March	42	41
April	58	53
May	70	66
June	79	75

✓ CHECK Your Understanding

Examples 1 and 2

1. WIND SPEED The average wind speeds for several cities in Pennsylvania are given in the table.

  a. Find the range of the data.

  b. Find the median and the upper and lower quartiles.

  c. Find the interquartile range.

  d. Identify any outliers in the data.

Wind Speed

Pennsylvania City	Speed (mph)
Allentown	8.9
Erie	11.0
Harrisburg	7.5
Middletown	7.7
Philadelphia	9.5
Pittsburgh	9.0
Williamsport	7.6

Example 3

2. TREES The heights of several types of palm trees, in feet, are 40, 25, 15, 22, 50, and 30. The heights of several types of pine trees, in feet, are 60, 75, 45, 80, 75, and 70. Compare and contrast the measures of variation of both kinds of trees.

Practice and Problem Solving

 = **Step-by-Step Solutions** begin on page R1.
Extra Practice begins on page EP2.

Examples 1 and 2 **3** **GOLF COURSES** The table shows the number of golf courses in various states.

    a. Find the range of the data.

    b. Find the median and the upper and lower quartiles.

    c. Find the interquartile range.

    d. Name any outliers in the data.

Number of Golf Courses	
California	1,117
Florida	1,465
Georgia	513
Iowa	437
Michigan	1,038
New York	954
North Carolina	650
Ohio	893
South Carolina	456
Texas	1,018

4. INTERNET The table shows the countries with the most Internet users.

    a. Find the range of the data.

    b. Find the median and the upper and lower quartiles.

    c. Find the interquartile range.

    d. Name any outliers in the data.

Millions of Internet Users	
China	99.8
Germany	41.88
India	36.97
Japan	78.05
South Korea	31.67
United Kingdom	33.11
United States	185.55

Example 3 **5. EXERCISE** The table shows the number of minutes of exercise for each person. Compare and contrast the measures of variation for both weeks.

Minutes of Exercise		
	Week 1	Week 2
Tanika	45	30
Tasha	40	55
Tyrone	45	35
Uniqua	55	60
Videl	60	45
Wesley	90	75

6. FOOTBALL The table shows the top teams in the National Football Conference (NFC) and the American Football Conference (AFC).

    a. Which conference had a greater range of penalties?

    b. Find the measures of variation for each conference.

    c. Compare and contrast the measures of variation for each conference.

Penalties By NFL Teams			
NFC		AFC	
Dallas Cowboys	104	New England Patriots	78
Arizona Cardinals	137	Indianapolis Colts	67
Green Bay Packers	113	Jacksonville Jaguars	76
New Orleans Saints	68	San Diego Chargers	94
New York Giants	77	Cleveland Browns	114
Seattle Seahawks	59	Pittsburgh Steelers	80
Minnesota Vikings	86	Houston Texans	82

For each data set, find the median, the upper and lower quartiles, and the interquartile range.

7. daily attendance at the water park: 346, 250, 433, 369, 422, 298

8. texts per day: 24, 53, 38, 12, 31, 19, 26

9. cost of admission: $13.95, $24.59, $19.99, $29.98, $23.95, $28.99

10. **SCIENCE** The table shows the number of known moons for each planet in our solar system. Use the measures of variation to describe the data.

Known Moons of Planets			
Mercury	0	Jupiter	63
Venus	0	Saturn	34
Earth	1	Uranus	27
Mars	2	Neptune	13

11. **EXERCISE** Lucy and Dena are training for a bike race and recorded their mileage for a week. Find the measures of variation of each person's mileage. Which measures of variation show the girls' similarities in their training? the differences? Explain.

	Monday	Tuesday	Wednesday	Thursday	Friday	Saturday	Sunday
Lucy	7 mi	3 mi	5 mi	8 mi	6 mi	10 mi	9 mi
Dena	6 mi	4 mi	6 mi	8 mi	11 mi	9 mi	7 mi

Real-World Link. The average twelve-year-old should spend about one hour a day doing moderate to intense physical activity.

H.O.T. Problems

12. **FIND THE ERROR** Hiroshi was finding the measures of variation of the following set of data: 89, 93, 99, 110, 128, 135, 144, 152, and 159. Find his mistake and correct it.

median = 128
lower quartile = 99
upper quartile = 144
interquartile range = 45
range = 70

13. **OPEN ENDED** Create a list of data with at least six numbers that has an interquartile range of 15 and two outliers.

14. **WRITE MATH** Explain why the median is not affected by very high or very low values in the data.

15. The number of games won by 10 chess players is given.

> 13, 15, 2, 7, 5, 9, 11, 10, 12, 11

Which of the following statements is NOT supported by these data?

A. Half of the players won more than 10.5 games and half won less than 10.5 games.

B. The range of the data is 13 games.

C. There are no outliers.

D. One fourth of the players won more than 7 games.

16. **SHORT RESPONSE** The ages in months of dogs enrolled in obedience class are: 8, 12, 20, 10, 6, 15, 12, 9, and 10. Find the range, median, upper and lower quartiles, and interquartile range of the dogs' ages.

17. The normal monthly rainfall in inches for a city are given in the table. What values, if any, are outliers?

Jan	Feb	Mar	Apr	May	June
0.65	1.39	0.63	2.16	2.82	4.21
July	Aug	Sept	Oct	Nov	Dec
3.22	1.20	9.31	11.25	0.70	0.80

F. 9.31

G. 11.25

H. 9.31 and 11.25

I. There are no outliers.

18. Which of the following sets of data has an interquartile range of 10?

A. 3, 4, 9, 16, 17, 24, 31

B. 41, 43, 49, 49, 50, 53, 55

C. 12, 14, 17, 19, 19, 20, 21

D. 55, 56, 56, 57, 58, 59, 62

More About Measures of Central Tendency

Data can be described by measures of central tendency. But sometimes using one particular measure might misrepresent the data.

REAL-WORLD EXAMPLE

BASKETBALL The table shows the number of points a basketball team scored in their first five games. The team says that the average number of points they score is 51 points. Explain how this might be misleading.

Game	Number of Points
1	38
2	40
3	95
4	39
5	43

Mean $\dfrac{38 + 39 + 40 + 43 + 95}{5} = \dfrac{255}{5} = 51$

Median 38, 39, ⃝40 43, 95

The measure of central tendency used by the team was the mean. This is greater than most number of points scored because it includes the outlier, 95. So, it is misleading to use this measure.

19. MARKETING An ice cream shop claims its average menu price is $3. Use the data in the table and explain how this might be misleading.

Kid's Cone	Single Cone	Double Cone	Triple Cone	Deluxe Sundae
$1.00	$2.75	$3.25	$3.75	$4.25

Main Idea

Display and interpret data in box-and-whisker plots.

Vocabulary

box-and-whisker plot

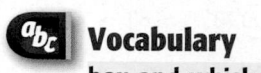

Box-and-Whisker Plots

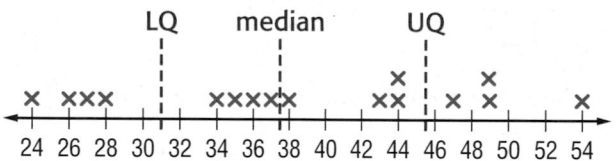 **FOOTBALL** The line plot shows the number of touchdowns scored by each of the 16 teams in the National Football Conference in a recent year.

```
 x  xxx        xxxxx      xx   x  x        x
+--+--+--+--+--+--+--+--+--+--+--+--+--+--+--+--+--
 24 26 28 30 32 34 36 38 40 42 44 46 48 50 52 54
```

1. Find the median, quartiles, and range of the data.

2. What percent of the teams scored less than 30 touchdowns?

3. What percent of the teams scored more than 37 touchdowns?

The medians and quartiles divide the data into four equal parts. The line plot shows that there are four pieces of data in each part.

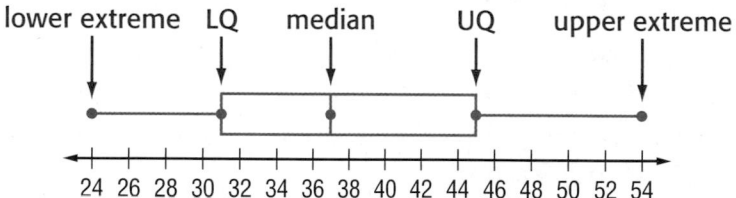

A **box-and-whisker plot** is a diagram that is constructed using the median, quartiles, and extreme values. A *box* is drawn around the quartile values, and the *whiskers* extend from each quartile to the extreme values. The median is marked with a vertical line. The figure below is a box-and-whisker plot of the football data.

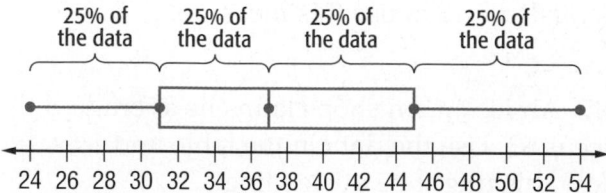

You can see all the pieces of data in a line plot, but in the box-and-whisker plot you can see only the median, quartiles, and extreme values. Box-and-whisker plots separate data into four parts. Even though the parts may differ in length, each contains 25% of the data.

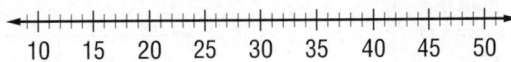 **REAL-WORLD EXAMPLE** **Construct a Box-and-Whisker Plot**

① **DRIVING** The list below shows the speeds of eleven cars. Draw a box-and-whisker plot of the data.

25   35   27   22   34   40   20   19   23   25   30

Step 1   Order the numbers from least to greatest. Then draw a number line that covers the range of the data.

```
 +++++++++++++++++++++++++++++++++++++++++++++++++
 10   15   20   25   30   35   40   45   50
```

Step 2   Find the median, the extremes, and the upper and lower quartiles. Mark these points above the number line.

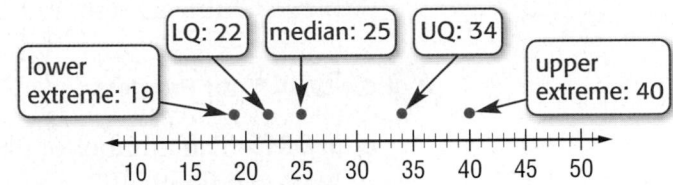

LQ: 22   median: 25   UQ: 34

lower extreme: 19   upper extreme: 40

```
 +++++++++++++++++++++++++++++++++++++++++++++++++
 10   15   20   25   30   35   40   45   50
```

Step 3   Draw the box so that it includes the quartile values. Draw a vertical line through the box at the median value. Extend the whiskers from each quartile to the extreme data points.

```
 +++++++++++++++++++++++++++++++++++++++++++++++++
 10   15   20   25   30   35   40   45   50
```

✓ **CHECK Your Progress**

a. Draw a box-and-whisker plot of the data set below.
{$20, $25, $22, $30, $15, $18, $20, $17, $30, $27, $15}

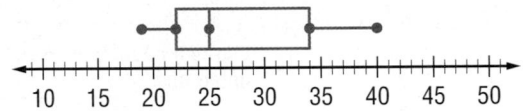 **REAL-WORLD EXAMPLES** **Interpret Data**

DRIVING Refer to the box-and-whisker plot in Example 1.

② **Half of the drivers were driving faster than what speed?**

Half of the drivers were driving faster than 25 miles per hour.

③ **What does the box-and-whisker plot's length tell about the data?**

The length of the left half of the box-and-whisker plot is short. This means that the speeds of the slowest half of the cars are concentrated. The speeds of the fastest half of the cars are spread out.

✓ **CHECK Your Progress**

b. What percent were driving faster than 34 miles per hour?

Double Box-and-Whisker Plots

④ The double box-and-whisker plot below shows the daily attendance of two fitness clubs. Compare and contrast the range and variance of the attendance at Super Fit versus Athletic Club.

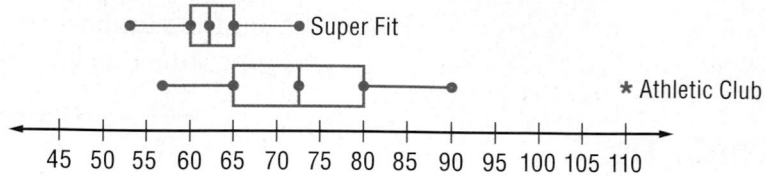

Study Tip

Outliers If the data set includes outliers, then the whiskers will not extend to the outliers, just to the previous data point. Outliers are represented with an asterisk (*) on the box-and-whisker plot.

Super Fit had an attendance between 53 and 72.5. The Athletic Club had an attendance between 57 and 110. The attendance at the Athletic Club varies more than the attendance at Super Fit.

CHECK Your Progress

c. **SPORTS** The number of games won in each conference of the National Football League is displayed below. Compare and contrast the range and variance of each conference.

National Football League Wins

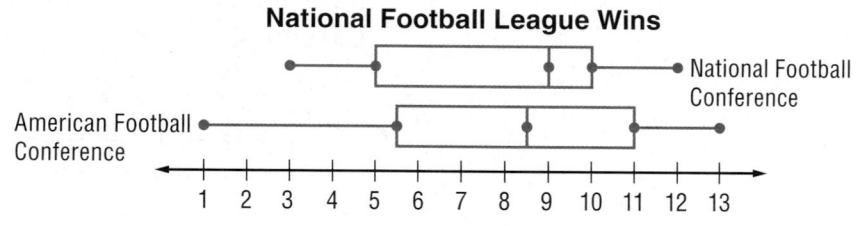

CHECK Your Understanding

Examples 1–3

1. **EARTH SCIENCE** Use the table.

   a. Make a box-and-whisker plot of the data.

Depth of Recent Earthquakes (km)						
5	15	1	11	2	7	3
9	5	4	9	10	5	7

   b. What percent of the earthquakes were between 4 and 9 kilometers deep?

   c. Write a sentence explaining what the length of the box-and-whisker plot means.

Example 4

2. **GAS MILEAGE** Use the box-and-whisker plots shown.

Average Gas Mileage for Various Sedans and SUVs

Sedans

SUVs

15 17 19 21 23 25 27 29 31 33 35 37 39 41 43

   a. Which types of vehicles tend to be less fuel-efficient?

   b. Compare the most fuel-efficient SUV to the least fuel-efficient sedan.

Practice and Problem Solving

= **Step-by-Step Solutions** begin on page R1.
Extra Practice begins on page EP2.

Example 1   Draw a box-and-whisker plot for each set of data.

3 {65, 92, 74, 61, 55, 35, 88, 99, 97, 100, 96}

4. {26, 22, 31, 36, 22, 27, 15, 36, 32, 29, 30}

5.

Height of Waves (in.)		
80	51	77
72	55	65
42	78	67
40	81	68
63	73	59

6.

Cost of MP3 Players ($)	
95	55
105	100
85	158
122	174
165	162

Examples 2 and 3

7. GEOGRAPHY The table shows the length of coastline for the 13 states along the Atlantic Coast.

a. Make a box-and-whisker plot of the data.

b. What percent of the coastline states have coastlines greater than 210 miles?

c. Half of the states have a coastline less than how many miles?

d. Write a sentence describing what the length of the box-and-whisker plot tells about the number of miles of coastline for states along the Atlantic coast.

Length of Coastline (mi)	
28	130
580	127
100	301
228	40
31	187
192	112
13	

Real-World Link

The total lengths of U.S. coastlines are shown below.

Atlantic Coast:
2,069 mi

Gulf Coast:
1,631 mi

Pacific Coast:
7,623 mi

Arctic Coast:
1,060 mi

8. TESTS Use the box-and-whisker plot. It summarizes math test scores.

Math Test Scores

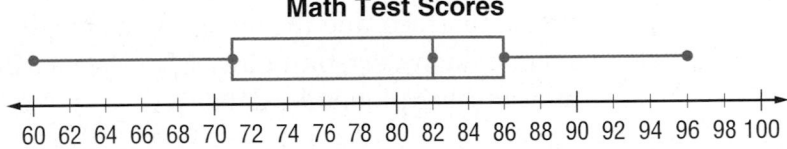

60 62 64 66 68 70 72 74 76 78 80 82 84 86 88 90 92 94 96 98 100

a. What was the greatest test score?

b. Explain why the median is not in the middle of the box.

c. What percent of the scores were between 71 and 96?

d. Half of the scores were higher than what score?

Example 4   **9. NUTRITION** The amount of Calories for certain fruits and vegetables is displayed. How do the Calories of fruits compare to vegetables?

Number of Calories

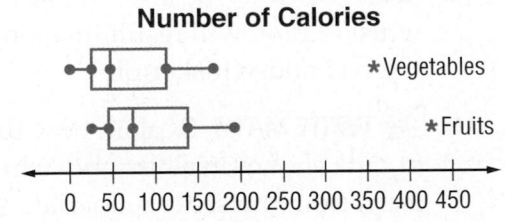

*Vegetables

*Fruits

0   50 100 150 200 250 300 350 400 450

10. ANIMALS Use the double box-and-whisker plot. It summarizes the weights of Asiatic black bears and pandas.

Weights of Animals (lb)

Asiatic Black Bears

Pandas

175 200 225 250 275 300 325 350 375 400

a. How do the weights of the two types of animals compare? Discuss how much the majority of the animals of each type weigh and the greatest weight of each.

b. Compare and contrast the range and variance of the weights of each animal.

11 TEMPERATURE Use the table. It shows the average monthly temperatures for two cities.

Average Monthly Temperatures (°F)

	J	F	M	A	M	J	J	A	S	O	N	D
Brownsville, TX	59	62	69	75	80	83	85	85	82	76	69	62
Caribou, ME	9	12	25	38	51	61	66	63	54	43	31	15

a. Find the low, high, and median temperatures and the upper and lower quartiles for each city.

b. On the same number line, draw a double box-and-whisker plot for each set of data. Place the Caribou data above the Brownsville data.

c. Write a few sentences comparing the average monthly temperatures displayed in the box-and-whisker plots.

12. WORD PROCESSING Find the median, upper and lower quartiles, and the interquartile range for the set of data in the table at the right. Create a box-and-whisker plot of the data.

Words Typed Per Minute				
80	72	67	51	77
42	63	73	68	55
65	81	40	59	78

H.O.T. Problems

13. CHALLENGE Write a set of data that contains 12 values for which the box-and-whisker plot has no whiskers. State the median, lower and upper quartiles, and lower and upper extremes.

14. REASONING Determine whether the statement is *true* or *false*. Explain.

The median divides the box of a box-and-whisker plot in half.

15. OPEN ENDED Write a set of data that, when displayed in a box-and-whisker plot, will result in a long box and short whiskers. Draw the box-and-whisker plot.

16. WRITE MATH Explain how the information you can learn from a set of data shown in a box-and-whisker plot is different from what you can learn from the same set of data shown in a line plot.

Real-World Link Asiatic black bears range from 4 to 6 feet in length.

17. Which box-and-whisker plot represents the data set 14, 18, 21, 24, and 29?

A.

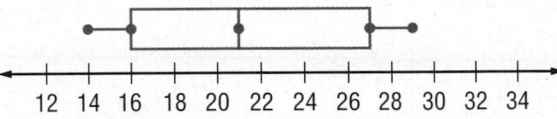

B.

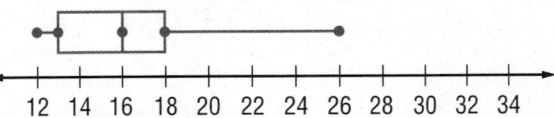

C.

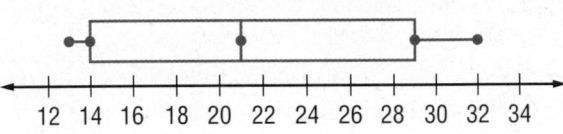

D.

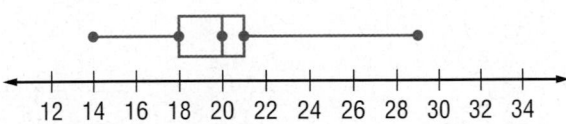

18. THINK SOLVE EXPLAIN **SHORT RESPONSE** Construct a box-and-whisker plot with the data set 35, 42, 44, 47, and 54.

19. Which of the following statements is NOT true concerning the box-and-whisker plot below?

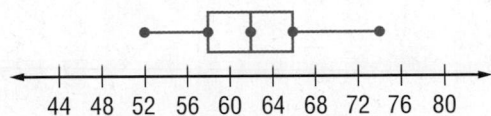

F. The value 74 is an extreme value.

G. Half of the data are above 62.

H. Half of the data are in the interval 62–74.

I. There are more data values in the interval 52–62 than there are in the interval 62–74.

20. The box-and-whisker plot shows distances traveled by vacationers. Half of the drivers traveled farther than what distance?

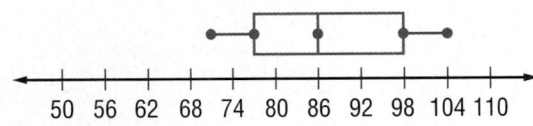

A. 68 miles

B. 75 miles

C. 86 miles

D. 99 miles

Spiral Review

For each set of data, find the range, median, upper and lower quartile, and interquartile range. (Lesson 2A)

21. number of instruments in a band: 3, 5, 11, 7, 8, 4, 15, 7

22. number of animals in a pet store: 25, 54, 32, 46, 65, 24, 39

Find the mean, median, and mode for each data set. Round to the nearest tenth if necessary. (Lesson 1B)

23. number of As on a test: 4, 7, 5, 2, 6, 10, 2, 9

24. miles run: 2, 5.1, 3, 4.8, 1, 3, 1.2, 3.4, 2.3

25. number of tracks on a CD: 10, 11, 13, 12, 10, 11, 15, 10, 12

26. daily number of boats in a harbor: 93, 84, 80, 91, 94, 90, 78, 93, 80

Problem Solving in Environmental Science

Thinking Green

Are you concerned about protecting the environment? If so, you should think about a career in environmental science. **Environmental engineers apply engineering principles along with biology and chemistry to develop solutions for improving the air, water, and land. They are involved in pollution control, recycling, and waste disposal. Environmental engineers also determine methods for conserving resources and for reducing environmental damage caused by construction and industry.**

21st Century Careers

Are you interested in a career as an environmental engineer? Take some of the following courses in high school.

- Algebra
- Biology
- Environmental Science
- Physics

Get ConnectED

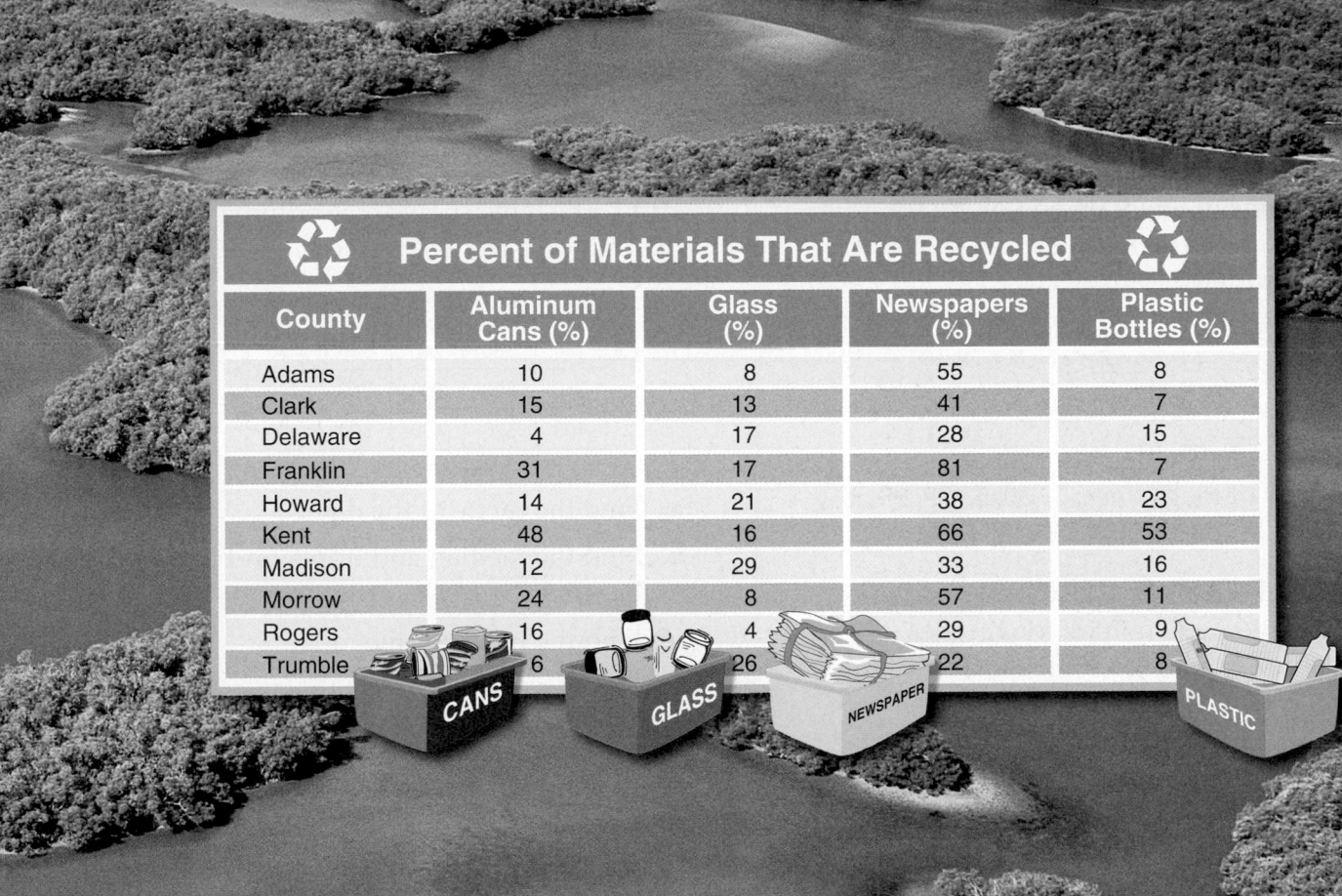

Percent of Materials That Are Recycled

County	Aluminum Cans (%)	Glass (%)	Newspapers (%)	Plastic Bottles (%)
Adams	10	8	55	8
Clark	15	13	41	7
Delaware	4	17	28	15
Franklin	31	17	81	7
Howard	14	21	38	23
Kent	48	16	66	53
Madison	12	29	33	16
Morrow	24	8	57	11
Rogers	16	4	29	9
Trumble	6	26	22	8

Real-World Math

Use the information in the table to solve each problem. Round to the nearest tenth if necessary.

1. Find the mean, median, and mode of the percent of recycled glass data.

2. If Kent County is removed from the recycled aluminum cans data, which changes the most: the mean, median, or mode? Does this make sense? Explain your reasoning.

3. Find the range, quartiles, and interquartile range of the percent of recycled newspapers data.

4. Find any outliers in the percent of recycled plastic bottles data.

5. Make a box-and-whisker plot of the percent of recycled glass data.

6. Refer to the box-and-whisker plot you made in Exercise 5. Compare the parts of the box and the lengths of the whiskers. What does this tell you about the data?

Mid-Chapter Check

Find the mean, median, and mode for each set of data. Round to the nearest tenth if necessary. (Lesson 1B)

1. number of suits dry cleaned each week: 22, 14, 37, 25, 21, 22, 20

2. number of cars washed each week: 65, 50, 57, 75, 76, 66, 64

3. prices of magazines: $3.50, $3.75, $3.50, $4.00, $3.00, $3.50, $3.25

4. **MULTIPLE CHOICE** The table shows the average April rainfall for 12 cities. If the value 4.2 is added to this list, which of the following would be true? (Lesson 1B)

Average Rainfall (in.)					
0.5	0.6	1.0	1.0	2.5	3.7
2.6	3.3	2.0	1.4	0.7	0.4

   A. The mode would increase.

   B. The mean would increase.

   C. The mean would decrease.

   D. The median would decrease.

5. **QUIZ** Cedro had quiz scores of 9, 10, 7, 7, 2, 6, 9, 10, 10, and 8. Which measure of central tendency would Cedro want his teacher to use to compute his final grade? Explain. (Lesson 1B)

6. **TEST PRACTICE** The table shows the pace for a half marathon. What measure of central tendency is most affected by the extreme value? (Lesson 1B)

Pace for Half Marathon (mph)				
6.7	7.1	7.5	7.6	3.3
7.4	7.2	7.4	7.1	7.4
6.8	6.9	7.3		

   F. mean

   G. median

   H. mode

   I. no effect

7. **FOOD** The table shows the sodium content of items on a restaurant menu. (Lesson 2A)

Sodium (mg)				
180	380	240	330	220
210	290	500	260	440
400	320	490	300	480

   a. Find the range of the data.

   b. Find the median and upper and lower quartiles.

   c. Find the interquartile range.

   d. Identify any outliers of the data.

8. **NUTRITION** The box-and-whisker plot shows the number of Calories in various low-fat granola bars. (Lesson 2B)

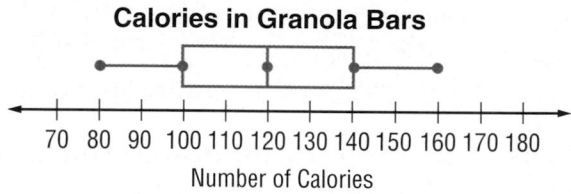

Calories in Granola Bars

70 80 90 100 110 120 130 140 150 160 170 180
Number of Calories

   a. What is the least number of Calories?

   b. What is the upper and lower quartile?

   c. What percent of the granola bars had Calories between 120 and 160?

   d. A quarter of the granola bars had more than what number of Calories?

9. **MIDDLE SCHOOL** The number of students in each seventh grade homeroom is shown below. (Lesson 2B)

   29, 15, 22, 30, 32, 46, 26, 22, 36, 31

   a. Draw a box-and-whisker plot for the data.

   b. Describe the data.

Explore

Main Idea

Use technology to create circle graphs.

Get ConnectED

Spreadsheet Lab:
Circle Graphs

A type of display used to compare categorical data is a *circle graph*. Circle graphs are useful when comparing parts of a whole.

ACTIVITY

MAGAZINES The spreadsheet below shows the results of a survey in which students were asked about their favorite type of magazine. Make a circle graph of the data.

STEP 1   Enter the data in a spreadsheet as shown.

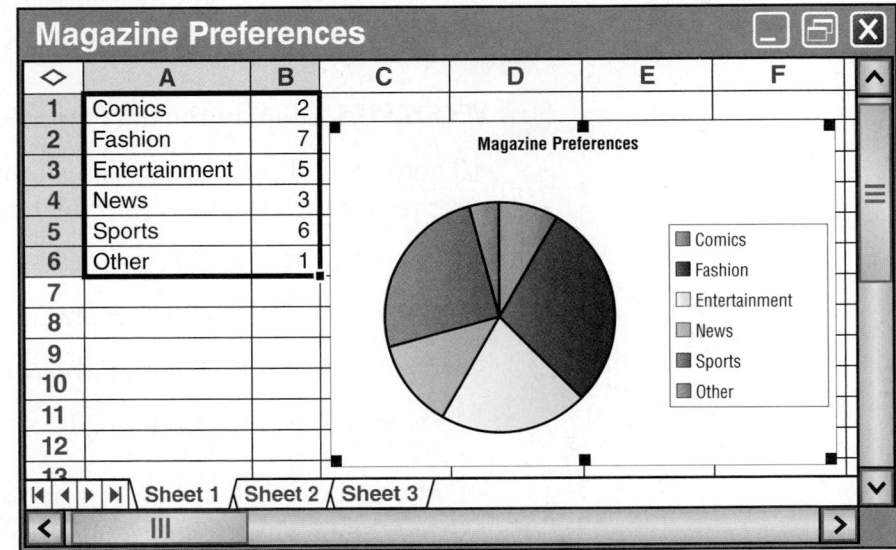

STEP 2   Select the information in cells A1 to B6. Click on the Chart Wizard icon. Choose the Pie chart type. Click Next twice. Enter the title. Then click Next and Finish.

1. **MAKE A CONJECTURE** Use the graph to determine which types of magazines were preferred by $\frac{1}{3}$ and 25% of the students. Explain your reasoning.

2. **COLLECT THE DATA** Collect some data that can be displayed in either a circle or bar graph. Then use a spreadsheet to make both types of displays. When would a bar graph be more useful? a circle graph? Justify your selection.

Main Idea

Construct and analyze circle graphs.

Vocabulary

circle graph

 Get ConnectED

Circle Graphs

VEGETABLES The students at Pine Ridge Middle School were asked to identify their favorite vegetable. The table shows the results of the survey.

1. Explain how you know that each student selected only one favorite vegetable.

2. If 400 students participated in the survey, how many students preferred carrots?

Favorite Vegetable	
Vegetable	Percent
Carrots	45%
Green Beans	23%
Peas	17%
Other	15%

One graph that shows data as parts of a whole is called a **circle graph**. In a circle graph, the percents add up to 100.

EXAMPLE **Display Data in a Circle Graph**

1. **VEGETABLES** Display the data above in a circle graph.

- There are 360° in a circle. Determine what part of the circle will represent each percent from the table above.

  45% of 360° = 0.45 · 360° or 162°

  23% of 360° = 0.23 · 360° or about 83°   Round to the nearest whole degree.

  17% of 360° = 0.17 · 360° or about 61°

  15% of 360° = 0.15 · 360° or 54°

- Draw a circle with a radius as shown. Then use a protractor to draw the first angle, in this case 162°. Repeat this step.

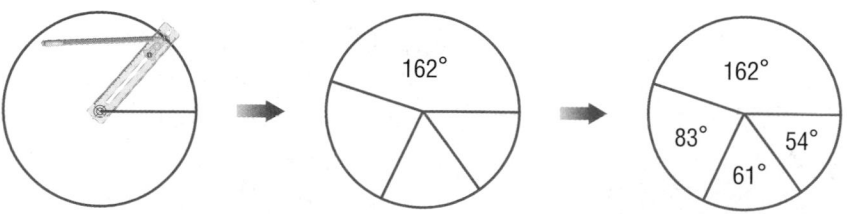

- Label each section of the graph with the category and percent. Label the graph with a title.

Check

162° + 83° + 61° + 54° = 360° ✔

Favorite Vegetable

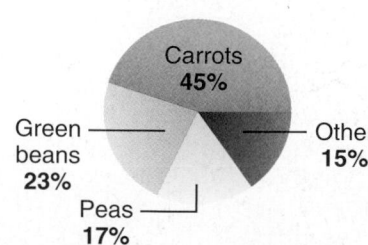

CHECK Your Progress

a. SCIENCE The table shows the present composition of Earth's atmosphere. Display the data in a circle graph.

Composition of Earth's Atmosphere	
Element	**Percent**
Nitrogen	78%
Oxygen	21%
Other gases	1%

When constructing a circle graph, you first may need to convert the data to ratios and decimals and then to degrees and percents.

EXAMPLE **Construct a Circle Graph**

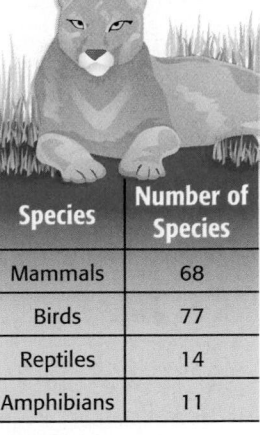

② **ANIMALS** The table shows endangered species in the United States. Make a circle graph of the data.

- Find the total number of species.
  $68 + 77 + 14 + 11 = 170$

- Find the ratio that compares each number with the total. Write the ratio as a decimal rounded to the nearest hundredth.

  mammals: $\frac{68}{170} = 0.40$ birds: $\frac{77}{170} \approx 0.45$

  reptiles: $\frac{14}{170} \approx 0.08$ amphibians: $\frac{11}{170} \approx 0.06$

Species	Number of Species
Mammals	68
Birds	77
Reptiles	14
Amphibians	11

- Find the number of degrees for each section of the graph.

  mammals: $0.40 \cdot 360° = 144°$

  birds: $0.45 \cdot 360° \approx 162°$

  reptiles: $0.08 \cdot 360° \approx 29°$

  amphibians: $0.06 \cdot 360° \approx 22°$

  Because of rounding, the sum of the degrees is 357°.

- Draw the circle graph. Label each piece with the title and percent. Label the graph with a title.

Check After drawing the first three sections, you can measure the last section of a circle graph to verify that the angle has the correct measure.

Endangered Species

Amphibians
Reptiles
6%
8%
Mammals
40%
Birds
45%

Real-World Link · · · ·
The Carolina Northern and Virginia Northern Flying Squirrel are both endangered. The northern flying squirrel is a small nocturnal gliding mammal that is about 10 to 12 inches in total length and weighs about 3–5 ounces.

 CHECK Your Progress

b. OLYMPICS The number of Winter Olympic medals won by the U.S. from 1924 to 2006 is shown in the table. Display the data in a circle graph.

U.S. Winter Olympic Medals	
Type	**Number**
Gold	78
Silver	81
Bronze	59

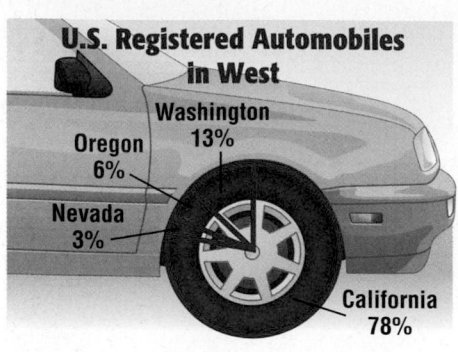

EXAMPLES Analyze a Circle Graph

AUTOMOBILES The graph shows the percent of automobiles registered in the western United States in a recent year.

U.S. Registered Automobiles in West

Washington 13%
Oregon 6%
Nevada 3%
California 78%

③ **Which state had the most registered automobiles?**

The largest section of the circle is the one representing California. So, California has the most registered automobiles.

Study Tip

Check for Reasonableness
To check Example 4, you can estimate and solve the problem another way.

78% - 6% ≈ 70%
70% of 24 is 17

Since 17.28 is about 17, the answer is reasonable.

④ **If 24 million automobiles were registered in these states, how many more automobiles were registered in California than in Oregon?**

California: 78% of 24 million → 0.78×24, or 18.72 million

Oregon: 6% of 24 million → 0.06×24, or 1.44 million

There were 18.72 million − 1.44 million, or 17.28 million more registered automobiles in California than in Oregon.

CHECK Your Progress

c. Which state had the least number of registered automobiles? Explain.

d. What was the total number of registered automobiles in Washington and Oregon?

CHECK Your Understanding

Examples 1 and 2 **Display each set of data in a circle graph.**

1.

Blood Types in the U.S.	
Blood Type	**Percent**
O	44
A	42
B	10
AB	4

2.

Favorite Musical Instrument	
Type	**Number of Students**
Piano	54
Guitar	27
Drum	15
Flute	24

Examples 3 and 4 **3. COLORS** Use the graph that shows the results of a survey.

a. What color is most favored?

b. If 400 people were surveyed, how many more people favored purple than red?

Favorite Color

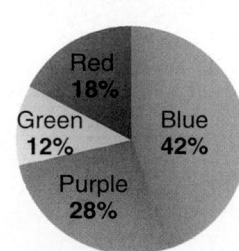

Red 18%
Green 12%
Blue 42%
Purple 28%

Practice and Problem Solving

= **Step-by-Step Solutions** begin on page R1.
Extra Practice begins on page EP2.

Examples 1 and 2 Display each set of data in a circle graph.

4.

U.S. Steel Roller Coasters	
Type	Percent
Sit down	86%
Inverted	8%
Other	6%

5.

U.S. Orange Production	
State	Orange Production
California	18%
Florida	81%
Texas	1%

6.

Animals in Pet Store	
Animal	Number of Pets
Birds	13
Cats	11
Dogs	9
Fish	56
Other	22

7.

Favorite Games	
Type of Game	Number of Students
Card	7
Board	9
Video	39
Sports	17
Drama	8

Examples 3 and 4 **8. LANDFILLS** Use the circle graph that shows what is in U.S. landfills.

  a. What takes up the most space in landfills?

  b. About how many times more paper is there than food and yard waste?

  c. If a landfill contains 200 million tons of trash, how much of it is plastic?

What is in U.S. Landfills?

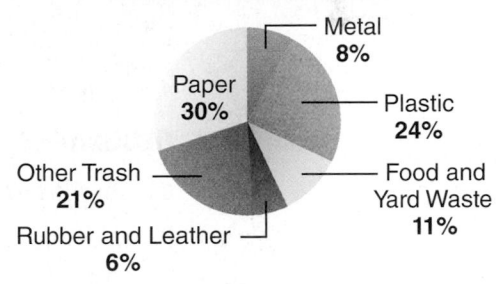

9 MONEY Use the graph that shows the results of a survey.

Do Americans Favor Common North American Currency?

  a. What percent of Americans favor a common North American currency?

  b. About how many of the approximately 298 million Americans would say "Don't Know" in response to this survey?

  c. About how many more Americans oppose a common currency than favor it?

DATA SENSE For each graph, find the missing values.

10. **Dog Expenses**

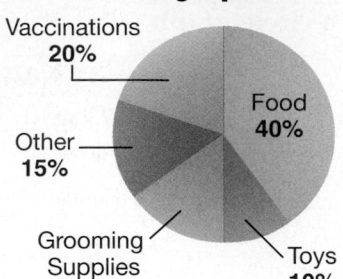

Vaccinations 20%
Food 40%
Other 15%
Grooming Supplies x%
Toys 10%

11. **Family Budget**

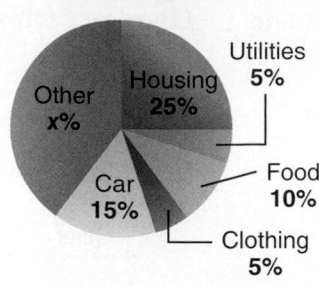

Utilities 5%
Other x%
Housing 25%
Food 10%
Car 15%
Clothing 5%

QUICK Review

Bar Graphs and Line Graphs

Bar graphs show the number of items in specific categories. Line graphs show change over a period of time.

Select an appropriate type of graph to display each set of data: line graph, bar graph, or circle graph. Then display the data using the graph.

12.

Top 5 Presidential Birth States	
Place	**Presidents**
Virginia	8
Ohio	7
Massachusetts	4
New York	4
Texas	3

13.

Tanya's Day	
Activity	**Percent**
School	25%
Sleep	33%
Homework	12%
Sports	8%
Other	22%

14. **GEOGRAPHY** Use the table.

   a. Display the data in a circle graph.

   b. Use your graph to find which two lakes equal the size of Lake Superior.

   c. Find the median size of the Great Lakes.

Sizes of U.S. Great Lakes	
Lake	**Size (sq mi)**
Erie	9,930
Huron	23,010
Michigan	22,400
Ontario	7,520
Superior	31,820

15. **POLITICS** A group of students was asked whether people their age could make a difference in the political decisions of elected officials. The results are shown in the graph.

   a. How many students participated in the survey?

   b. Write a convincing argument explaining whether or not it is reasonable to say that 50% more students said they could make a difference than those who said they could not make a difference.

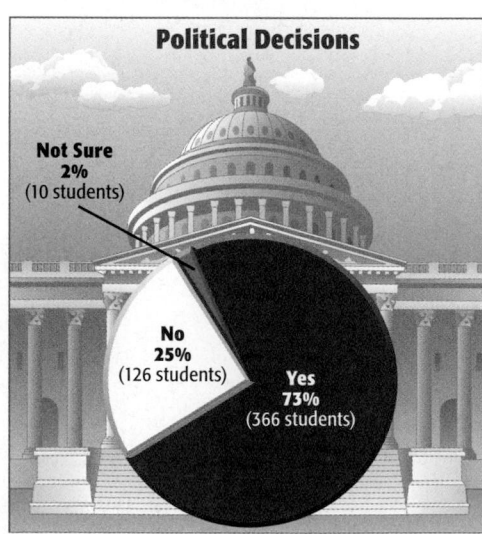

Political Decisions

Not Sure 2% (10 students)
No 25% (126 students)
Yes 73% (366 students)

16. **MATH IN THE MEDIA** Find some data in a newspaper or magazine, on television, or on the Internet that can be displayed in a circle graph. Then display the data in a circle graph and write a statement analyzing the data.

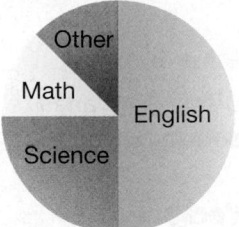

H.O.T. Problems

17. **CHALLENGE** The graph shows the results of a survey about students' favorite school subjects. About what percent of those surveyed said that math was their favorite subject? Explain your reasoning.

Favorite Subject

18. **COLLECT THE DATA** Collect some data from your classmates that can be represented in a circle graph. Then create the circle graph and write one statement analyzing the data.

19. **WRITE MATH** The table shows the percent of people who like each type of fruit juice. Can the data be represented in a circle graph? Justify your answer.

Fruit Juice	Percent
Apple	54%
Grape	48%
Orange	37%
Cranberry	15%

Test Practice

20. The graph shows the types of vehicles that crossed a bridge during one month.

Types of Vehicles

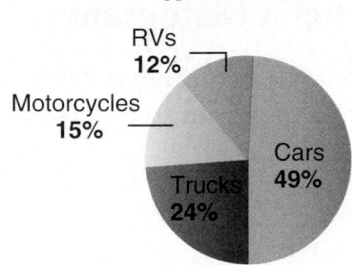

Which of the following shows the number of trucks if 480,000 vehicles cross the bridge each month?

A. 57,600   C. 115,200

B. 72,000   D. 235,200

21. **GRIDDED RESPONSE** The circle graph shows the results of a survey in which middle school students were asked to name one of their favorite pizza toppings. What percent of the people surveyed named mushrooms as their favorite topping?

Pizza Toppings

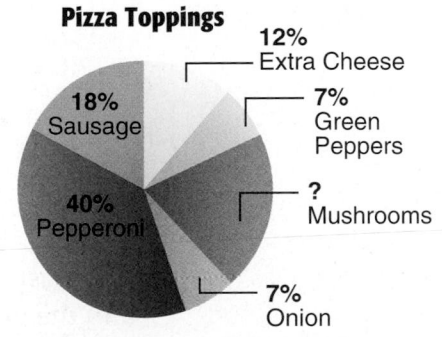

22. **FOOD** A diner sold the following number of sandwiches during one work week: 20, 26, 18, 21, and 27. Construct a box-and-whisker plot of the data. (Lesson 2B)

23. **AGES** The ages of children at a playground were 7, 4, 5, 4, 7, 3, 13, 6, and 8. Name any outliers in the data. (Lesson 1D)

Main Idea
Display and analyze data in a histogram.

 Vocabulary
histogram
relative frequency
cumulative relative frequency

 Get ConnectED

Histograms

BASKETBALL Kylie researched the average ticket prices for NBA basketball games for 30 teams. The frequency table shows the results.

1. What do you notice about the price intervals in the table?

2. How many tickets were at least $20.00 but less than $50.00?

Price Interval ($)	Tally	Frequency
20.00–29.99	I	1
30.00–39.99	IIII IIII I	11
40.00–49.99	IIII IIII	10
50.00–59.99	IIII	5
60.00–69.99	I	1
70.00–79.99	II	2

Data from a frequency table can be displayed as a histogram. A **histogram** is similar to a bar graph and is used to display numerical data that have been organized into equal intervals.

  EXAMPLE **Construct a Histogram**

① **MOVIES** Choose intervals and make a frequency table of the data shown below. Then construct a histogram to represent the data.

Running Time of Movies (minutes)				
135	89	142	219	96
144	104	135	94	155
106	127	134	116	91
118	138	118	110	101

The least value in the data is 89 and the greatest is 219. An interval size of 30 minutes would yield the frequency table at the right.

To construct a histogram, follow these steps.

Step 1 Draw and label a horizontal and vertical axis. Include a title.

Step 2 Show the intervals from the frequency table on the horizontal axis.

Running Time of Movies (minutes)		
Time	Tally	Frequency
81–110	IIII III	8
111–140	IIII III	8
141–170	III	3
171–200		0
201–230	I	1

Step 3 For each time interval, draw a bar whose height is given by its frequency.

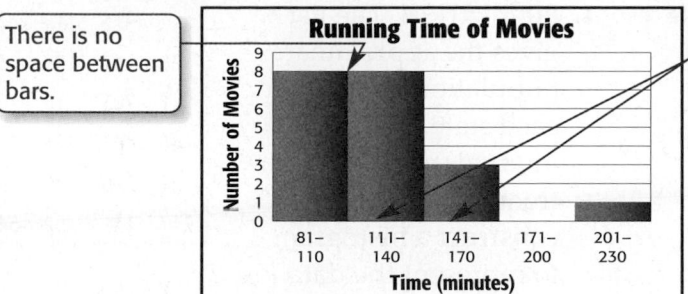

There is no space between bars.

Because all of the intervals are equal, all of the bars have the same width.

Running Time of Movies

Study Tip

Gaps Intervals with a frequency of 0 have a bar height of 0. This is referred to as a gap.

✓ **CHECK Your Progress**

a. SCHOOL The list at the right gives a set of test scores. Choose intervals, make a frequency table, and construct a histogram to represent the data.

Test Scores							
94	85	73	93	75	77	89	80
89	83	79	81	87	85	90	83
88	86	83	91	93	93	92	90
91	88	96	97	98	82	90	100

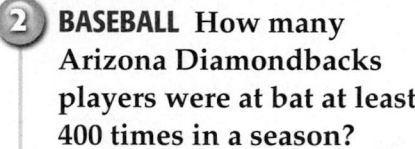 **EXAMPLES** Analyze and Interpret Histograms

2 **BASEBALL** How many Arizona Diamondbacks players were at bat at least 400 times in a season?

One player was at bat 400–499 times, and 3 players were at bat 500–599 times. Two players were at bat 600–699 times. Therefore, $1 + 3 + 2$ or 6 players were at bat at least 400 times.

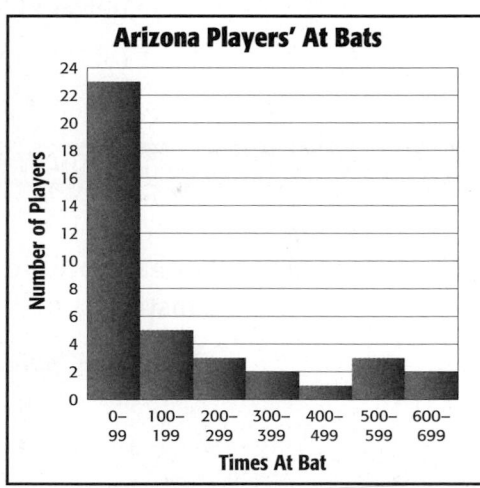

Arizona Players' At Bats

QUICK Review

Percents

To write a decimal as a percent, move the decimal point two places to the right and add a percent sign.
$0.67 = 67\%$

3 **BASEBALL** What percent of the players were at bat 199 times or fewer?

There were $23 + 5 + 3 + 2 + 1 + 3 + 2$ or 39 players at bat. There were $23 + 5$ or 28 players that were at bat 199 times or fewer.

$\frac{28}{39} \approx 0.72$ Divide 28 by 39.

So, about 72% of the players were at bat 199 times or fewer.

✓ **CHECK Your Progress**

b. What was the greatest number of times at bat for any one player?

c. Based on the data above, how many times is an Arizona Diamondbacks player most likely to be at bat?

Example 1

1. POPULATION The list gives the approximate population density for each state. Choose intervals and make a frequency table. Then construct a histogram to represent the data.

U.S. State Population Density (per square mile)									
88	42	189	33	810	6	15	50	10	179
1	703	16	102	175	22	402	36	138	89
45	401	223	103	62	18	165	274	80	75
51	296	170	41	61	138	9	1,003	27	99
217	141	52	542	81	1,135	277	133	66	5

Examples 2 and 3

2. VOLCANOES Use the histogram at the right.

a. What percent of the volcanoes are 8,999 feet or less?

b. How likely is it that any given volcano is at least 15,000 feet tall? Explain your reasoning.

c. What is the height of the tallest volcano?

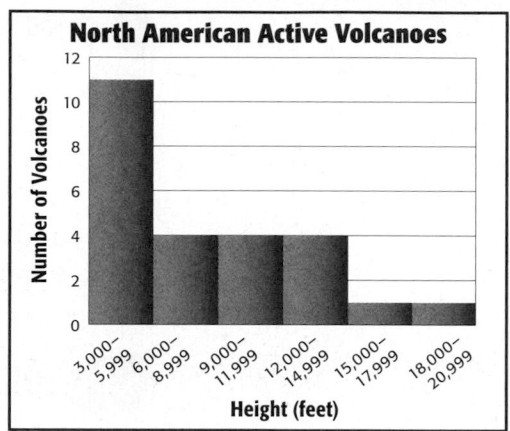

Practice and Problem Solving

● = **Step-by-Step Solutions** begin on page R1.
Extra Practice begins on page EP2.

Example 1 For each problem, choose intervals and make a frequency table. Then construct a histogram to represent the data.

3.

Hours Spent Exercising per Week						
3	0	9	1	4	2	0
3	6	14	4	2	5	3
7	3	0	8	3	10	

4.

Average Speed (mph), Selected Animals						
70	61	50	50	50	45	8
43	42	40	40	40	35	0.17
35	32	32	30	30	30	1.17
30	25	20	9	18	14	200

Examples 2 and 3

5 COUNTRIES Use the histogram below.

a. How many countries have an area less than 401 square kilometers?

b. What percent of the countries have an area of 201–600 square kilometers?

c. How likely is it that any given country will have an area greater than 800 square kilometers? Explain.

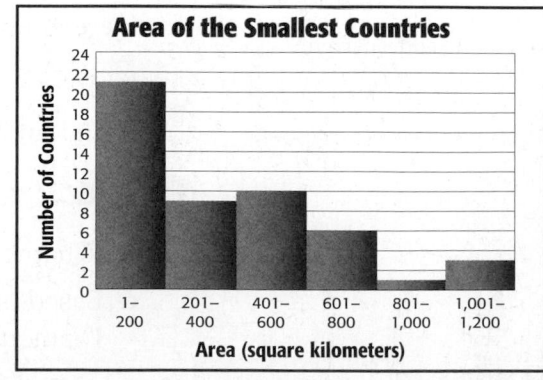

6. ECLIPSES Use the histogram at the right.

   a. What percent of the solar eclipses lasted at least 7 minutes 31 seconds?

   b. How long was the shortest solar eclipse?

   c. How many solar eclipses lasted between 1 second and 5 minutes?

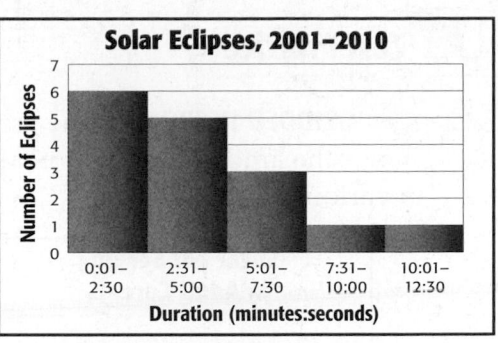

7 BUILDINGS Use the histograms shown.

Tall Buildings in Pittsburgh and Seattle

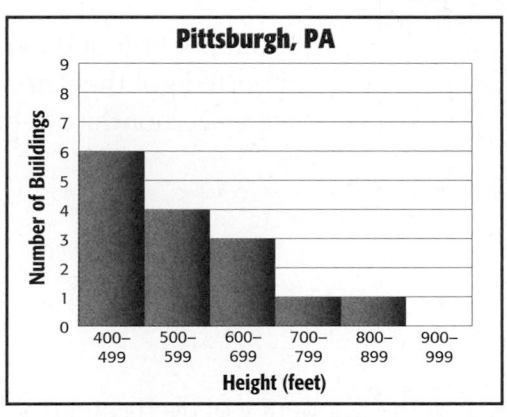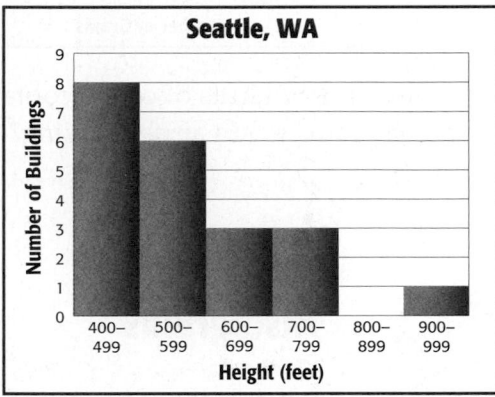

Real-World Link
Total solar eclipses occur about 3 times every 4 years.

   a. Which city has the tallest building?

   b. Determine which city has more buildings that are 800–899 feet tall.

   c. Determine which city has more buildings that are at least 600 feet tall. What percent of the buildings in that city are at least 600 feet tall?

   d. Which city has more tall buildings? by how many?

8. COLLECT THE DATA Conduct a survey of your classmates to determine the number of hours each person spends on the Internet during a typical week. Then construct a histogram of your data.

H.O.T. Problems

9. OPEN ENDED Construct a histogram that has a vertical line of symmetry and two gaps. Then construct a histogram that has a vertical line of symmetry and one gap.

10. CHALLENGE Describe how the histogram at the right would change if larger intervals, such as 0–9 and 10–19, were used. Describe how it would change if smaller intervals, such as 0–2, 3–5, 6–8, and so on, were used.

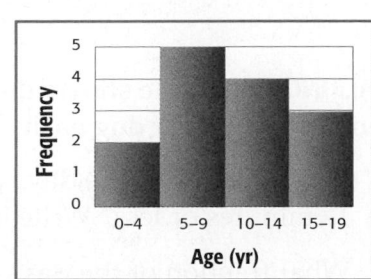

11. WRITE MATH Compare and contrast a bar graph and a histogram.

12. ✎ **GRIDDED RESPONSE** The graph shows the amount of sugar per serving in various adult cereals.

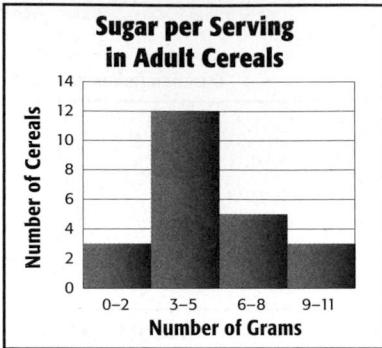

How many kinds of cereal contain 6–8 grams of sugar per serving?

13. **THINK SOLVE EXPLAIN** **SHORT RESPONSE** A group of mothers reported when their children got their first tooth.

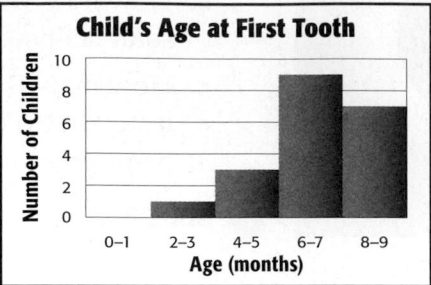

What fraction of the number of children reported got their first tooth when they were six months old or older?

More About   Histograms

Relative frequency is the ratio that compares the frequency of each category to the total. **Cumulative relative frequency** of two or more categories is the ratio of the combined frequencies to the total.

 REAL-WORLD EXAMPLE

SLEEP The table shows the results of a class survey about the number of hours of sleep students get each night. What fraction of students get less than 8 hours of sleep each night?

Find the cumulative relative frequency.

There were 20 students surveyed. And
2 + 2 + 7 or 11 students get less than 8 hours of sleep each night.

So, $\frac{11}{20}$ of students surveyed get less than 8 hours of sleep each night.

Number of Hours	Number of Students
5	2
6	2
7	7
8	8
9	1

WALKING The table shows the number of minutes Penny walked her dog each day this month.

14. What fraction of the days did she walk her dog 12 minutes or less? Write in simplest form.

15. What fraction of the days did she walk her dog less than 16 minutes? Write in simplest form.

Number of Minutes	Number of Days
10	10
12	6
14	4
16	2
18	8

Extend

Main Idea

Use a graphing calculator to make histograms.

Graphing Technology: Histograms

You can make a histogram using a graphing calculator.

 **ACTIVITY**

The Moapa Middle School basketball team listed each player's average points per game. Make a histogram of the data.

Average Points per Game											
15	3	11	7	4	6	18	1	2	21	10	3
9	2	1	12	24	5	13	20	4	12	1	2

STEP 1 Clear any existing data in list L1 by pressing
STAT ENTER ▲ CLEAR ENTER.

Then enter the data in L1. Input each number and press ENTER.

STEP 2 Turn on the statistical plot by pressing 2nd [STAT PLOT] ENTER ENTER.

Select the histogram and L1 as the Xlist by pressing ▼ ► ► ENTER ▼ 2nd [L1] ENTER.

STEP 3 Press WINDOW. To set the viewing window to be [0, 25] scl: 5 by [0, 12] scl: 1, press WINDOW 0 ENTER 25 ENTER 5 ENTER 0 ENTER 12 ENTER 1 ENTER GRAPH.

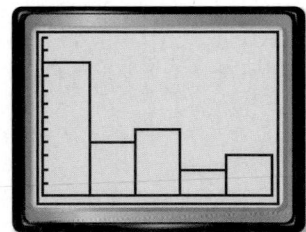

STEP 4 Press GRAPH to create the histogram.

Analyze the Results

1. Press TRACE. Find the frequency of each interval.

2. Explain why the *x*-values for this data set were chosen as 0 to 25.

3. **COLLECT THE DATA** Use the graphing calculator to make a histogram of your classmates' heights in inches. Analyze the graph to make some conclusions about the data.

Main Idea

Display and analyze data in a stem-and-leaf plot.

 Vocabulary

stem-and-leaf plot
leaf
stem
back-to-back stem-and-leaf plot

Stem-and-Leaf Plots

 BIRDS The table shows the average mass in grams of sixteen different species of chicks.

1. Which mass is the lightest?

2. How many of the masses are less than 10 grams?

Chick Mass (g)			
19	6	7	10
11	13	18	25
21	12	5	12
20	21	11	12

In a **stem-and-leaf plot**, the data are organized from least to greatest. The digits of the least place value usually form the **leaves**, and the next place-value digits form the **stems**.

EXAMPLE   Display Data in a Stem-and-Leaf Plot

1. **BIRDS** Display the data from the table above in a stem-and-leaf plot.

Step 1   Choose the stems using digits in the tens place, 0, 1, and 2. The least value, 5, has 0 in the tens place. The greatest value, 25, has 2 in the tens place.

Step 2   List the stems from least to greatest in the *Stem* column. Write the leaves, the ones digits, to the right of the corresponding stems.

Stem	Leaf
0	6 7 5
1	9 0 1 3 8 2 2 1 2
2	5 1 0 1

Step 3   Order the leaves and write a *key* that explains how to read the stems and leaves. Include a title.

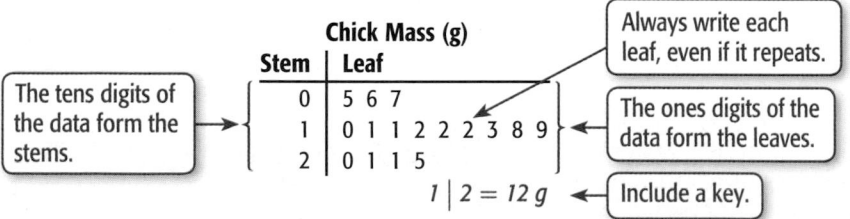

Chick Mass (g)

Stem	Leaf
0	5 6 7
1	0 1 1 2 2 2 3 8 9
2	0 1 1 5

The tens digits of the data form the stems.

Always write each leaf, even if it repeats.

The ones digits of the data form the leaves.

1 | 2 = 12 g   Include a key.

CHECK Your Progress

a. **HOMEWORK** The number of minutes the students in Mr. Blackwell's class spent doing their homework one night is shown. Display the data in a stem-and-leaf plot.

Homework Time (min)				
42	5	75	30	45
47	0	24	45	51
56	23	39	30	49
58	55	75	45	35

 **EXAMPLE** **Analyze Stem-and-Leaf Plots**

 CHESS The stem-and-leaf plot shows the number of chess matches won by members of the Avery Middle School Chess Team. Find the range, median, and mode of the data.

Chess Matches Won

Stem	Leaf
0	8 8 9
1	9
2	0 0 2 4 4 8 9
3	1 1 2 4 5 5 6 6 7 7 8
4	0 0 0 3 8 9
5	2 4
6	1

$3 \mid 2 = 32$ *wins*

Range greatest wins − least wins
$= 61 - 8$ or 53

Median middle value, 35

Mode most frequent value, 40

 CHECK **Your Progress**

b. BIRDS Find the range, median, and mode of the data in Example 1.

Study Tip

Data Sets Remember that measures of central tendency are numbers that describe the center of a data set and include the mean, median, and mode. The range is the difference between the greatest and least numbers in a data set.

Recall that the measures of central tendency can be affected by an outlier.

 **EXAMPLE** **Effect of Outliers**

 SPORTS The stem-and-leaf plot shows the number of points scored by a college basketball player. Which measure of central tendency is most affected by the outlier?

Basketball Points

Stem	Leaf
0	2
1	2 2 3 5 8
2	0 0 1 1 3 4 6 6 6 8 9
3	0 1
4	3

$1 \mid 2 = 12$ *points*

The mode, 26, is not affected by the inclusion of the outlier, 43.

Calculate the mean and median, each without the outlier, 43. Then calculate them including the outlier and compare.

	without the outlier	including the outlier
Mean	$\dfrac{2 + 12 + \cdots + 31}{19} \approx 20.89$	$\dfrac{2 + 12 + 12 + \cdots + 43}{20} = 22$
Median	21	$\dfrac{21 + 23}{2} = 22$

The mean increased by $22 - 20.89$, or 1.11, while the median increased by $22 - 21$, or 1. Since $1.11 > 1$, the mean is more affected.

 CHECK **Your Progress**

c. CHESS Refer to Example 2. If an additional student had 84 wins, which measure of central tendency would be most affected?

Example 1 Display each set of data in a stem-and-leaf plot.

1.

Height of Trees (ft)				
15	25	8	12	20
10	16	15	8	18

2.

Cost of Shoes ($)				
42	47	19	16	21
23	25	25	29	31
33	34	35	39	48

Examples 2 and 3   **CAMP** The stem-and-leaf plot shows the ages of students in a pottery class.

Ages of Students

Stem	Leaf
0	9 9 9
1	0 1 1 1 1 2 2 3 3 4

1 | 0 = 10 years

a. What is the range of the ages of the students?

b. Find the median and mode of the data.

c. If an additional student was 6 years old, which measure of central tendency would be most affected?

Practice and Problem Solving

● = **Step-by-Step Solutions** begin on page R1.
Extra Practice begins on page EP2.

Example 1 Display each set of data in a stem-and-leaf plot.

4.

Quiz Scores (%)			
70	96	72	91
80	80	79	93
76	95	73	93
90	93	77	91

5.

Low Temperatures (°F)				
15	13	28	32	38
30	31	13	36	35
38	32	38	24	20

6.

Floats at Annual Parade			
151	158	139	103
111	134	133	154
157	142	149	159

7.

School Play Attendance			
225	227	230	229
246	243	269	269
267	278	278	278

Examples 2 and 3 **8. CYCLING** The number of Tour de France titles won by eleven countries is shown.

Tour de France Titles
Won by Countries

Stem	Leaf
0	8 8 9
1	0 8
2	
3	6

0 | 8 = 8 titles

a. Find the range of titles won.

b. Find the median and mode of the data.

c. Which measure of central tendency is most affected by the outlier?

9. ELECTRONICS The stem-and-leaf plot shows the costs of various DVD players at an electronics store.

Costs of DVD Players

Stem	Leaf
8	2 5 5
9	9 9
10	0 0 2 5 6 8
11	0 0 5 5 5 9 9
12	5 7 7

11 | 5 = $115

a. What is the range of the prices?

b. Find the median and mode of the data.

c. If an additional DVD player costs $153, which measure of central tendency would be most affected?

10. **HISTORY** Refer to the stem-and-leaf plot below.

Ages of Signers of Declaration of Independence

Stem	Leaf
2	6 6 9
3	0 1 3 3 3 4 4 5 5 5 7 7 8 8 9 9
4	0 0 1 1 1 2 2 2 4 5 5 5 5 6 6 6 6 7 8 9
5	0 0 0 0 2 2 3 3 5 7
6	0 0 2 3 5 9
7	0

3 | 1 = 31 years

a. How many people signed the Declaration of Independence?

b. What was the age of the youngest signer?

c. What is the range of the ages of the signers?

d. Based on the data, can you conclude that the majority of the signers were 30–49 years old? Explain your reasoning.

11. **GYMNASTICS** The scores for 10 girls in a gymnastics event are 9.3, 10.0, 9.9, 8.9, 8.7, 9.0, 8.7, 8.5, 8.8, and 9.3. Analyze a stem-and-leaf plot of the data to draw two conclusions about the scores.

Real-World Link

The saltwater crocodile is the largest living reptile. Some measuring 27–30 feet in length have been recorded in the wild.

12. **REPTILES** The average lengths of certain species of crocodiles are given in the table. Analyze a stem-and-leaf plot of this data to write a convincing argument about a reasonable length for a crocodile.

Crocodile Average Lengths (ft)			
8.1	16.3	16.3	9.8
16.3	16.3	11.4	6.3
13.6	9.8	19.5	16.0

13. **MATH IN THE MEDIA** Find a data set in a newspaper or magazine, on television, or on the Internet that could be represented using a stem-and-leaf plot. Create a stem-and-leaf plot of the data. Then write a sentence that summarizes the data.

14. **TENNIS** Joel and Skylar displayed some statistics about their favorite tennis players in the stem-and-leaf plots below. The stem-and-leaf plots show the number of aces each player had per match in their last tournament.

Joel's Player
Number of Aces

Stem	Leaf
0	7 8 9
1	0 1 2 3 5 8
2	1 4 4 9
3	

2 | 4 = 24 aces

Skylar's Player
Number of Aces

Stem	Leaf
0	8 9 9
1	0 3 4 5
2	1 2 6 7 9
3	0

2 | 6 = 26 aces

a. Which player has a greater range of scores? Explain.

b. Which player should use the mode to emphasize the number of aces? Explain your reasoning.

c. Based on the data, can you conclude that Skylar's player is more likely to average 16 aces in a match? Explain.

H.O.T. Problems

15. FIND THE ERROR Aisha is analyzing the data in the stem-and-leaf plot below. Find her mistake and correct it.

Cut Ribbon Length

Stem	Leaf
2	6 6 9
3	
4	6
5	3 6

2 | 6 = 26 in.

> There are no pieces of ribbon more than 50 inches in length.

16. 🖉 **WRITE MATH** Present the data shown at the right in a line plot and a stem-and-leaf plot. Describe the similarities and differences among the representations. Which representation do you prefer to use? Explain your reasoning.

Fiber in Cereal (g)

5	5	4	3	3
3	1	1	1	2
1	1	1	1	0

✓ Test Practice

17. Denzell's science quiz scores are 11, 12, 13, 21, and 35. Which stem-and-leaf plot **best** represents this data?

A.

Stem	Leaf
1	1
2	1
3	5

3 | 5 = 35

B.

Stem	Leaf
1	3
2	1
3	5

3 | 5 = 35

C.

Stem	Leaf
1	1 2 3
2	1
3	5

3 | 5 = 35

D.

Stem	Leaf
1	1
2	1 1
3	5

3 | 5 = 35

18. The stem-and-leaf plot shows the points scored by the Harding Middle School basketball team.

Points Scored

Stem	Leaf
4	7 8 8
5	0 0 2 3 7 9
6	1 6
7	
8	4

4 | 7 = 47

Which one of the following statements is true concerning how the measures of central tendency are affected by the inclusion of the outlier?

F. The mode is most affected.

G. The median is not affected.

H. The mean is most affected.

I. None of the measures of central tendency are affected.

More About Stem-and-Leaf Plots

A **back-to-back stem-and-leaf plot** uses one stem to compare two sets of data.

 REAL-WORLD EXAMPLE

FOOTBALL Refer to the back-to-back stem-and-leaf plot below.

Fantasy Football Points

The leaves for one set of the data are on one side of the stem.

Eagles' Kicker	Stem	Titans' Kicker
9 8 8 8 7 7 7 5 3 1	0	0 3 4 5 6 6 7 7 7 7
6 4 4 3 2 2	1	0 1 2 2 3 7

2 | 1 = 12 points   1 | 7 = 17 points

The leaves for the other set of data are on the other side of the stem.

a. Which team's kicker had a greater range of points?
The range for the Eagles' kicker is 16 − 1 or 15.
The range for the Titans' kicker is 17 − 0 or 17.

Since 17 > 15, the Titans' kicker had a greater range of points.

b. Which team's kicker scored 11 or more points a greater number of times?
The Eagles' kicker scored 11 or more points 6 times.
The Titans' kicker scored 11 or more points 5 times.

So, the Eagles' kicker scored 11 or more points a greater number of times.

SCORES The quiz scores of two English classes are shown in the back-to-back stem-and-leaf plot.

English Quiz Scores

Class A	Stem	Class B
9 8	6	4 7 9
9 8 6 4 2 2	7	1 7 9 9
8 6 5 1	8	0 1 2 3 5 6
4 3 0	9	1 2

8 | 6 = 68%   6 | 4 = 64%

19. Which measure(s) of central tendency would students in Class B want to use to show that they performed better on the quiz?

20. Which class has a greater range of scores? Explain.

21. Based on the data, can you conclude that Class A has the better English quiz scores? Explain your reasoning.

22. Write a sentence that compares the data.

SANDWICHES The back-to-back stem-and-leaf plot shows nutrition information about sandwiches.

Fat (g) of Various Burgers and Chicken Sandwiches

Chicken	Stem	Burgers
8	0	
9 8 5 5 3 3	1	0 5 9
0	2	0 6
	3	0 3 6

8 | 0 = 8 g   2 | 6 = 26 g

23. What is the greatest number of fat grams in each sandwich?

24. Compare the median number of fat grams for each sandwich.

25. In general, which type of sandwich has a lower amount of fat? Explain.

26. Write a sentence that describes the data.

Problem-Solving Investigation

Main Idea Solve problems by using a graph.

P.S.I. TEAM +

Use a Graph

TESS: I recently purchased a saltwater aquarium. I need to add 1 tablespoon of sea salt for every 5 gallons of water.

Sea Salt Requirements						
Tablespoons of Sea Salt	1	2	3	4	5	6
Capacity of Tank (gallons)	5	10	15	20	25	30

YOUR MISSION: Use a graph to predict the number of tablespoons of salt required for a 50-gallon saltwater fish tank.

Understand	You know the number of gallons of the tank. You need to predict the number of tablespoons of sea salt.	
Plan	Organize the data in a graph so you can easily see any trends.	
Solve	Continue the graph with a dotted line in the same direction until you align horizontally with 50 gallons. Graph a point. Find what value of sea salt corresponds with the point. Ten tablespoons are required for a 50-gallon tank.	

Check   Find the unit rate of tablespoons of sea salt per gallon of water. Multiply the unit rate by the number of gallons to find the number of tablespoons of sea salt.

$$\frac{0.2 \text{ tbsp salt}}{1 \text{ gal water}} \times \frac{50 \text{ gal water}}{1} = 10 \text{ tbsp salt} \checkmark$$

Analyze the Strategy

1. Suppose the tank holds 32 gallons. Predict how much sea salt is required.

= **Step-by-Step Solutions** begin on page R1.
Extra Practice begins on page EP2.

- Use a graph.
- Solve a simpler problem.
- Look for a pattern.
- Choose an operation.

Use a graph to solve Exercises 2–4.

2. **CALORIES** The table shows the average number of Calories burned while sleeping for various hours.

Calories Burned While Sleeping	
Hours	**Calories**
6	386
7	450
8	514
9	579

a. Make a graph of the data.

b. If the trend continues, about how many Calories are burned by sleeping for 10 hours?

3. **ADVERTISING** A local newspaper charges $14.50 for every three lines of a classified ad. Predict the cost of a 7-line ad.

4. **SWIMMING** The table shows the winning Olympic times for the Women's 4 x 100-meter Freestyle Relay in swimming from 1972 to 2008. Make a graph of the data. Predict the winning time in 2012.

Women's 4 × 100-meter Freestyle Olympic Times	
Year	**Time (s)**
1972	235
1976	225
1980	223
1984	224
1988	221
1992	220
1996	219
2000	217
2004	215
2008	213

Use any strategy to solve Exercises 5–9.

5. **WOODWORKING** Two workers can make two chairs in two days. How many chairs can 8 workers working at the same rate make in 20 days?

6. **POSTAGE** The table shows the postage stamp rate from 1975 to 2008. Make a graph of the data. Predict the year the postage rate will reach $0.50.

Postage Stamp Rates	
Year	**Cost ($)**
1975	0.13
1978	0.15
1981	0.20
1985	0.22
1988	0.25
1991	0.29
1995	0.32
1999	0.33
2001	0.34
2002	0.37
2006	0.39
2007	0.41
2008	0.42

7. **ANATOMY** Each human hand has 27 bones. There are 6 more bones in the fingers than in the wrist. There are 3 fewer bones in the palm than in the wrist. How many bones are in each part of the hand?

8. **ALLOWANCE** Tia used half of her allowance to buy a ticket for the class play. Then she spent $1.75 on an ice cream cone. Now she has $2.25 left. How much is her allowance?

9. **WRITE MATH** Explain the importance of constructing an accurate graph when predicting future trends.

Main Idea

Analyze line graphs and scatter plots to make predictions and conclusions.

Vocabulary

scatter plot
line of best fit

Get ConnectED

Scatter Plots and Lines of Best Fit

Explore

• Pour 1 cup of water into a drinking glass.

• Measure the height of the water and record it in a table like the one shown.

• Place 5 marbles in the glass. Measure the height of the water. Record.

• Continue adding marbles, 5 at a time, until there are 20 marbles in the glass. After each time, measure and record the height of the water.

Number of Marbles	Height of Water (cm)
0	
5	
10	
15	
20	

1. By how much did the water's height change after each addition of marbles?

2. Predict the height of the water when 30 marbles are in the drinking glass. Explain how you made your prediction.

3. Test your prediction by placing 10 more marbles in the glass.

4. Graph the ordered pairs you recorded in the table on a coordinate plane.

A **scatter plot** shows the relationship between a set of data with two variables graphed as ordered pairs on a coordinate plane. Like line graphs, scatter plots are useful for making predictions because they show trends in data.

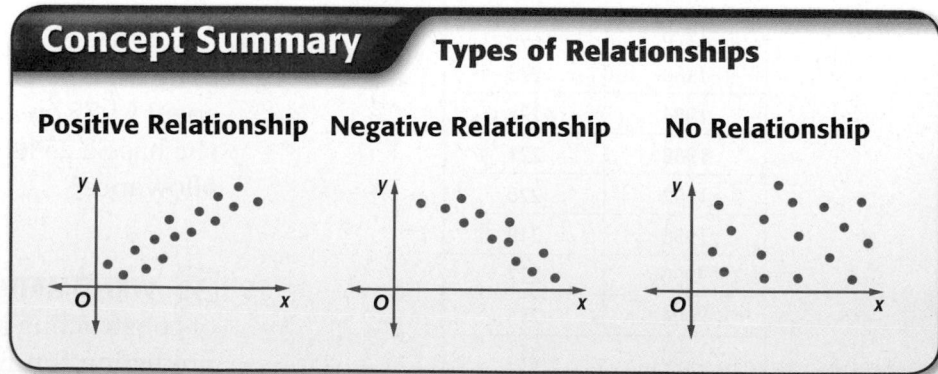

Concept Summary **Types of Relationships**

Positive Relationship **Negative Relationship** **No Relationship**

Study Tip

Scatter Plots In a positive relationship, as the value of x increases, so does the value of y. In a negative relationship, as the value of x increases, the value of y decreases.

EXAMPLE Identify a Relationship

1 Explain whether the scatter plot of the data for school commute times shows a *positive*, *negative*, or *no* relationship.

As the distance increases, the time increases. Therefore, the scatter plot shows a positive relationship.

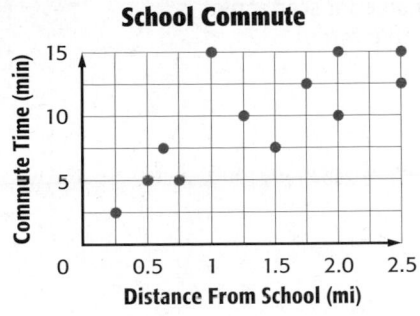

CHECK Your Progress

a. **GRADES** The graph shows the grades on a recent math test. Explain whether the scatter plot shows a *positive*, *negative* or *no* relationship.

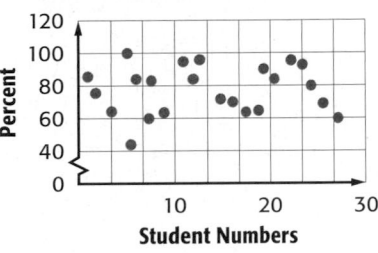

The **line of best fit** is a line that is very close to most of the data points in a scatter plot.

 EXAMPLE Use a Line Graph to Predict

2 **SUMMER CAMP** The graph shows the enrollment at a summer camp for the past several years. Construct a line of best fit. If the trend continues, what will be the enrollment in 2014?

Draw the line that is close to most of the data points. If the trend continues, the enrollment in 2014 will be about 190 campers.

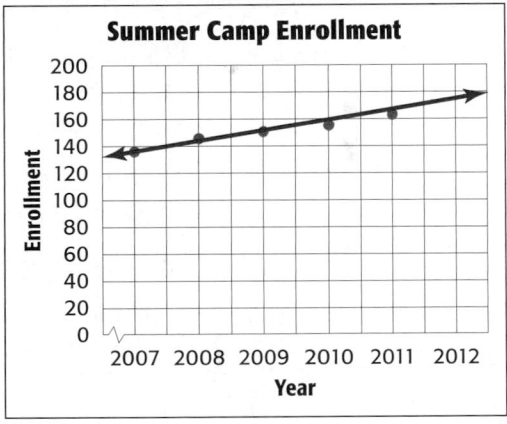

CHECK Your Progress

b. **INTERNET** The graph shows the number of hits on a Web site for the first 5 days. Construct a line of best fit. If the trend continues, predict the day the Web site will have 12,000 hits.

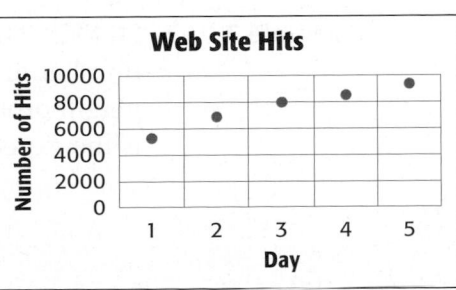

Read Math

Scatter Plots Another name for scatter plot is *scattergram*.

EXAMPLE   Use a Scatter Plot to Predict

3   NASCAR The scatter plot shows the earnings for the winning driver of the Daytona 500 from 1996 to 2008. Predict the winning earnings for the 2012 Daytona 500.

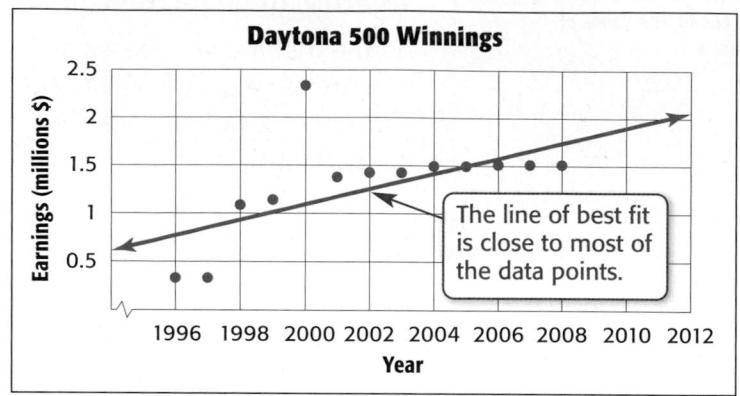

The predicted winning earnings for 2012 will be about $2,000,000.

CHECK Your Progress

c. NASCAR Predict the winning earnings for 2014.

✓ CHECK Your Understanding

Examples 1 and 2

1. POPULATION Fayette County is a fast-growing county in Tennessee. The graph shows its increase in population.

  a. Describe the relationship between the two variables.

  b. Construct a line of best fit.

  c. If the trend continues, what will be the population in 2010?

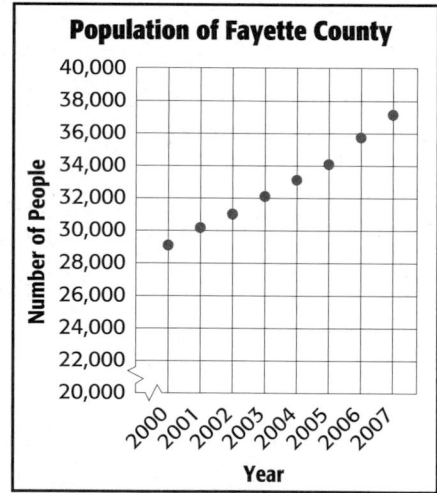

Example 3

2. PICNICS The scatter plot shows the number of people who attended a neighborhood picnic each year. What is the expected attendance for 2010?

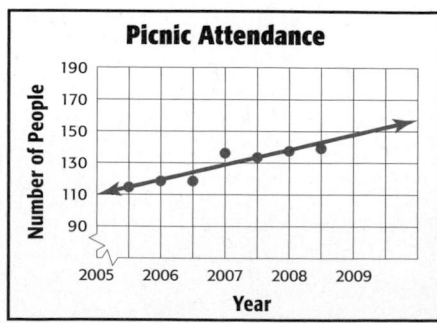

Practice and Problem Solving

 = **Step-by-Step Solutions** begin on page R1.
Extra Practice begins on page EP2.

Examples 1 and 2 **3** **MONUMENTS** Use the graph that shows the time it takes Ciro to climb the Statue of Liberty.

    a. Identify the relationship between the two variables.

    b. Construct a line of best fit.

    c. Predict the time it will take Ciro to climb 354 steps to reach the top.

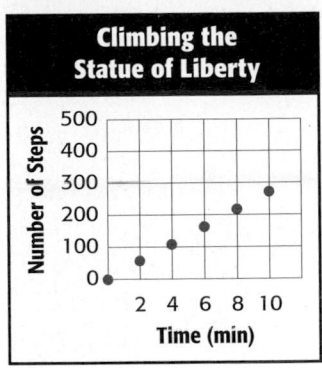

Example 3 **4. SCHOOL** Use the graph that shows the times students spent studying for a test and their test scores.

    a. What score should a student who studies for 50 minutes be expected to earn?

    b. If a student scored 90 on the test, about how much time can you assume the student spent studying?

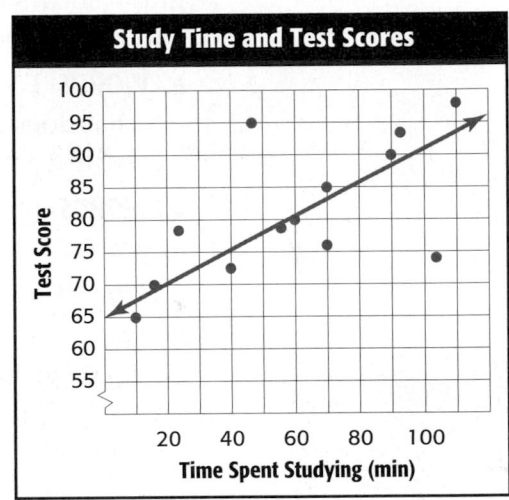

5. SLEEP Use the table that shows the relationship between hours of sleep and scores on a math test.

    a. Display the data in a scatter plot.

    b. Describe the relationship, if any, between the two variables.

    c. Predict the test score for someone that sleeps for 5 hours.

Hours of Sleep	Math Test Score
9	96
8	88
7	76
6	71

6. BASEBALL Use the table at the right.

    a. Make a scatter plot of the data to show the relationship between at bats and hits.

    b. Predict the number of hits if 500 at bats occurred.

    c. Describe the trend in the data.

Player	At Bats	Hits
C. Guzman	403	126
I. Kinsler	398	134
D. Pedroia	395	124
J. Reyes	394	119
I. Suzuki	391	119
M. Young	391	118
O. Cabrera	385	104
D. Lee	382	117

7 **SCHOOLS** Use the graph that shows the number of voters in Clinton County for several years.

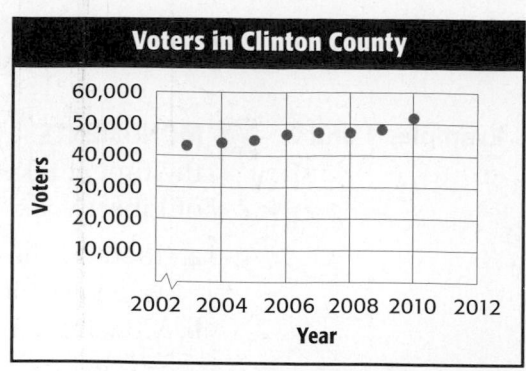

Voters in Clinton County

a. Describe the relationship, if any, between the two sets of data.

b. If the trend continues, what will be the average number of voters in 2011?

8. 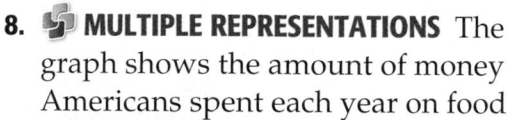 **MULTIPLE REPRESENTATIONS** The graph shows the amount of money Americans spent each year on food.

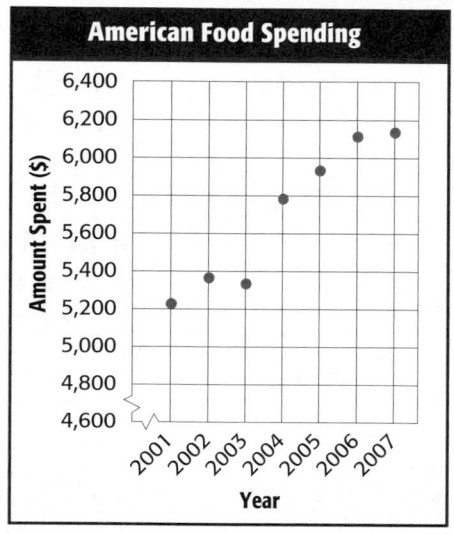

American Food Spending

a. **WORDS** For each graphed point, what does the x-value represent? y-value represent?

b. **WORDS** Describe the trend in the graph.

c. **GRAPHS** Construct a line of best fit.

d. **NUMBERS** Predict the amount Americans will spend on food in 2012.

9. **PETS** What can you conclude about the relationship between pet owner age and number of pets in the scatter plot at the right?

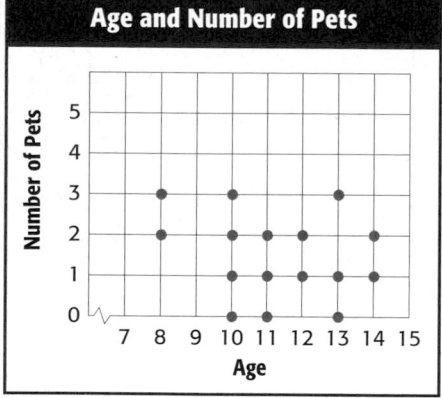

Age and Number of Pets

Real-World Link · · · ·
Research shows that cats have better memories than dogs. A dog's memory lasts no longer than 5 minutes, while a cat's memory can last up to 16 hours.

H.O.T. Problems

10. **OPEN ENDED** Name two sets of data that can be graphed on a scatter plot.

11. **Which One Doesn't Belong?** Identify the term that does not have the same characteristics as the other three. Explain your reasoning.

line plot	mode	bar graph	scatter plot

12. **WRITE MATH** Explain how a graph can be used to make predictions.

13. The number of laps Gaspar has been swimming each day is shown.

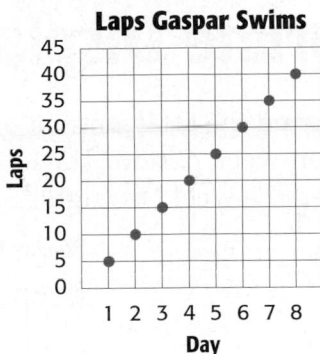

Laps Gaspar Swims

If the trend shown in the graph continues, what is the best prediction for the number of laps he will swim on day 10?

A. 50

B. 65

C. 75

D. 100

14. The number of people at the pool at different times during the day is shown.

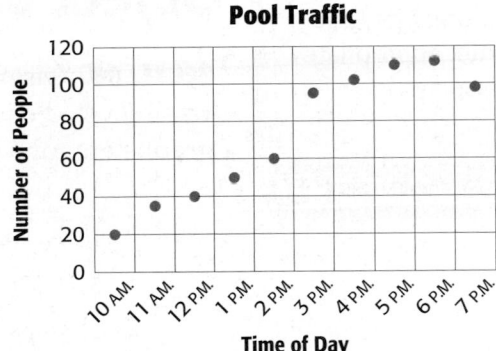

Pool Traffic

If an extra lifeguard is needed when the number of people at the pool exceeds 100, between which hours is an extra lifeguard needed the entire time?

F. 10:00 A.M.–12:00 P.M.

G. 12:00 P.M.–3:00 P.M.

H. 2:00 P.M.–5:00 P.M.

I. 4:00 P.M.–6:00 P.M.

15. **UTILITIES** The table lists monthly heating costs for a household. (Lesson 4A)

Month	Jan	Mar	May	Jul	Sept	Nov
Heating Cost ($)	135	74	41	25	52	94

a. Make a graph of the data.

b. Predict the monthly heating cost for October.

16. **MAMMALS** The stem-and-leaf plot shows the maximum mass in kilograms of several rabbits. (Lesson 3E)

a. Find the range of masses.

b. Find the median and mode of the data.

c. Which measure of central tendency is most affected by the inclusion of the outlier? Explain.

Maximum Mass of Rabbits (kg)

Stem	Leaf
0	8 9
1	0 2 4 6 8
2	7
3	
4	
5	4

$0 \mid 8 = 0.8$ kg

Find the median, upper and lower quartiles, and interquartile range for each data set. (Lesson 2A)

17. stopping distances, in feet: 89, 90, 74, 81, 68

18. swimming pool water temperature, °F: 76, 90, 88, 84, 82, 78

Main Idea

Select, organize, and construct appropriate data displays.

Select an Appropriate Display

RECYCLING Ms. Stevens's class weighed the total amount of paper that was recycled each week during a ten-week period. The following graphs are four ways they displayed the weekly weights of paper.

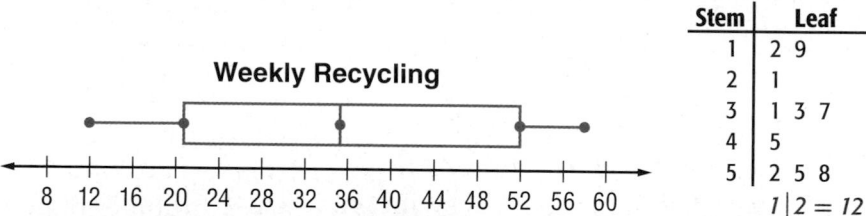

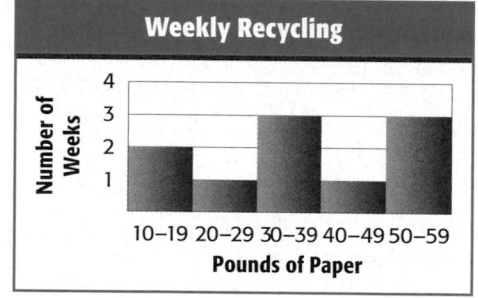

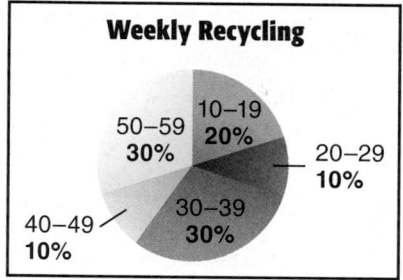

1. Which display(s) shows how many weeks the class collected exactly 31 pounds of paper?

2. Which display(s) most easily shows the number of weeks the class collected between 30 and 39 pounds of paper?

When deciding what type of display to use, ask these questions.

What type of information is given? What do you want the display to show? How will the display be analyzed?

EXAMPLE Select an Appropriate Display

1. **SPORTS** Select an appropriate display to show the number of boys of different age ranges that participate in athletics.

Since the display will show an interval, a histogram would be an appropriate display to represent this data.

CHECK Your Progress

a. Select an appropriate display for the percent of students in each grade at a middle school.

Concept Summary | Statistical Displays

Type of Display	Best Used to…
Bar Graph	show the number of items in specific categories
Box-and-Whisker Plot	show measures of variation for a set of data; also useful for very large sets of data
Circle Graph	compare parts of the data to the whole
Double Bar Graph	compare two sets of categorical data
Double Line Graph	compare change over a period of time for two sets of data
Histogram	show frequency of data divided into equal intervals
Line Graph	show change over a period of time
Line Plot	show frequency of data with a number line
Scatter Plot	show the relationship between a set of data with two variables graphed as ordered pairs
Stem-and-Leaf Plot	list all individual numerical data in condensed form

Real-World Link · · · ·

For every one bushel of corn, it takes 3–5 days to produce 2.8 gallons of ethanol.

REAL-WORLD EXAMPLE — Construct a Display

2 **ETHANOL** Select an appropriate type of display to compare the percent of ethanol production by state. Justify your reasoning. Then construct the display. What can you conclude from your display?

Ethanol Production by State Per Year						
State	Iowa	Nebraska	Illinois	Minnesota	Indiana	Other
Gallons (millions)	3,534	1,665	1,135	1,102	1,074	5,098

You are asked to compare parts to a whole. A circle graph would be an appropriate display.

Indiana, Minnesota, and Illinois produce about the same amount of ethanol.

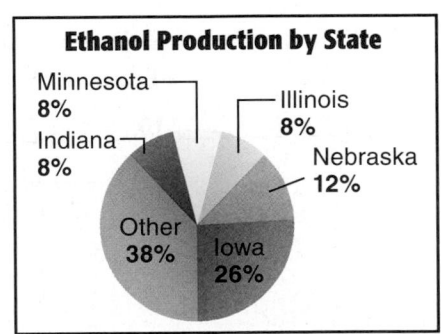

Ethanol Production by State

Minnesota 8%
Indiana 8%
Illinois 8%
Nebraska 12%
Other 38%
Iowa 26%

✓ CHECK Your Progress

b. **MUSICALS** The table lists the ticket prices for school musicals during recent years. Select an appropriate display to predict the price of a ticket in 2012. Justify your reasoning. Then construct the display. What can you conclude from your display?

Ticket Prices	
Year	Price ($)
2008	5.00
2009	5.50
2010	6.50
2011	7.00

Example 1 **Select an appropriate display for each situation. Justify your reasoning.**

1. the temperatures for the past week

2. the number of people who have different kinds of pets

3. the percent of different ways electricity is generated

Example 2 4. **TELEVISION** Use the table at the right. Select an appropriate display to show a relationship between the number of hours of television watched and test scores. Justify your reasoning. Then construct the display. What can you conclude from your display?

Hours of Television Watched	Weekly Test Score
4	78
7	67
1	92
3	81
8	61

5. **SANDWICHES** The prices of sandwiches at a restaurant are $4.50, $5.59, $3.99, $2.50, $4.99, $3.75, $2.99, $3.29, and $4.19. Select an appropriate display to determine how many sandwiches range from $3.00 to $3.99. Justify your reasoning. Then construct the display. What can you conclude from your display?

Practice and Problem Solving

● = **Step-by-Step Solutions** begin on page R1.
Extra Practice begins on page EP2.

Example 1 **Select an appropriate display for each situation. Justify your reasoning.**

6. the number of students that favor chocolate or vanilla as a frosting

7. the median age of members in a community band

8. the resale value of a car over time

9. the percent of people that drink 0, 1, 2, 3, or more than 3 glasses of water a day

Example 2 **Select an appropriate display for each situation. Justify your reasoning. Then construct the display. What can you conclude from your display?**

10.

Favorite Movies	
Type of Movie	Number of People
Comedy	48
Action	17
Drama	5
Horror	2

11.

Temperature (°F)	Depth of Ice (inches)
10	5.1
13	4.5
19	3.8
25	2.8
30	1.1

12.

Age Group	Number of Texts per Day
11–15	25
16–20	23
21–25	17
26–30	10

13.

Number of Push-ups			
56	28	28	54
55	28	25	23
54	53	23	20
57	52	17	10
28	56	51	19

14. GRAPHIC NOVEL Refer to the graphic novel frame below. What are the two best types of displays to use for this data? Explain.

> We are preparing a presentation of our project. Refer to the graphic novel at the start of this chapter.

> Which type of graph should we use?

> Let's show two different types.

15. SURVEY A survey asked teens which subject they felt was most difficult. Of those who responded, 25 said English, 39 said social studies, 17 said English and social studies equally, and 13 said neither subject. Construct an appropriate display of the data.

16. HOME RUNS The number of home runs hit by each player of a high school baseball team is shown in the table. Construct an appropriate display of the data so that each individual value is still represented.

Home Runs						
10	22	15	11	24	12	8
21	13	16	22	10	14	21

17. LAKES The circle graph shows the approximate percent of the total volume of each Great Lake.

   a. Display the data using another type of display.

   b. Write a convincing argument telling which display is most appropriate.

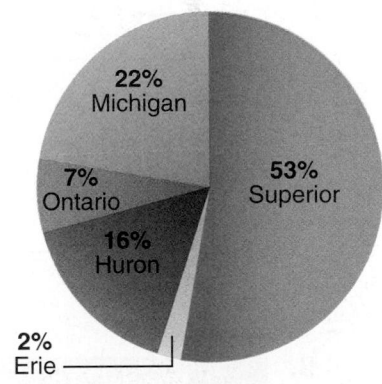

Volume of the Great Lakes

22% Michigan
7% Ontario
16% Huron
2% Erie
53% Superior

18. CUSTOMERS Refer to the situations described below.

Situation A	Situation B
the number of customers ages 12–19 compared to all age groups	the number of customers ages 12, 13, 14, 15, and 16 who made a purchase

   a. Which situation involves data that is best displayed in a bar graph? Explain your reasoning.

   b. Refer to the situation you selected in part **a**. Could you display the data using another type of display? If so, which display? Explain.

H.O.T. Problems

19. **OPEN ENDED** Give an example of a data set that would be best represented in a line graph.

20. **CHALLENGE** Determine if the following statement is *true* or *false*. Explain your reasoning.

    A line plot can be used to display data from a histogram.

21. **CHALLENGE** Which display(s) allows you to easily determine the range of the data set? Explain your reasoning.

22. **REASONING** Determine if the following statement is *always, sometimes,* or *never* true. Justify your response.

    A circle graph can be used to display data from a bar graph.

23. **✏ WRITE MATH** What type of data would be most appropriate to display in a histogram?

Test Practice

24. Moira surveyed 25 of her classmates to find out how many E-mails they received. Which of the following displays gives the most detail about the data?

A.

Number of E-mails Received Each Week

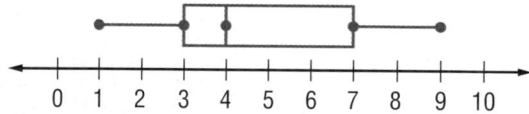

C.

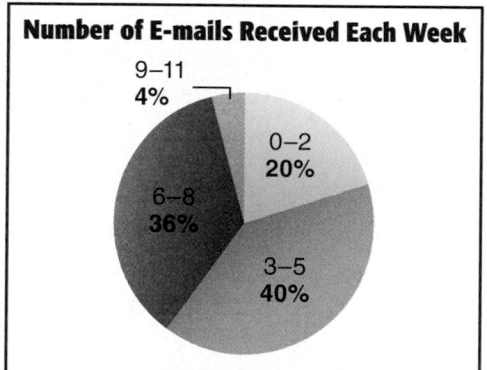

B.

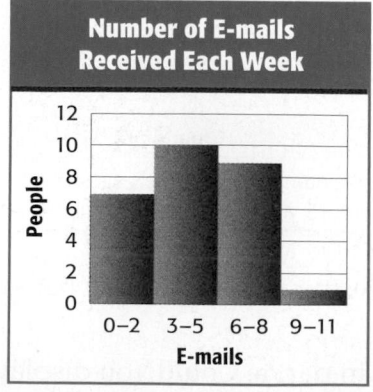

D.

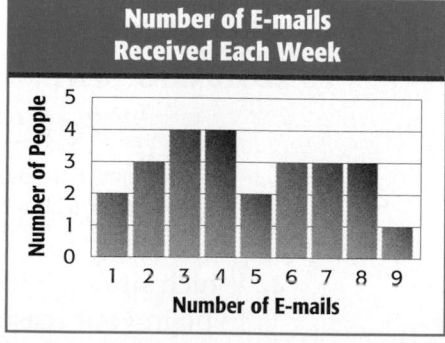

More About Select an Appropriate Display . . . .

Graphs help analyze and interpret data easily, but the way they are drawn may misrepresent the data. The use of different scales can influence conclusions from graphs.

REAL-WORLD EXAMPLE

SCHOOL DANCES The graphs show the attendance at last year's school dances.

Graph A

Graph B

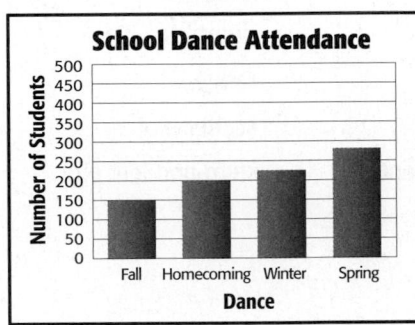

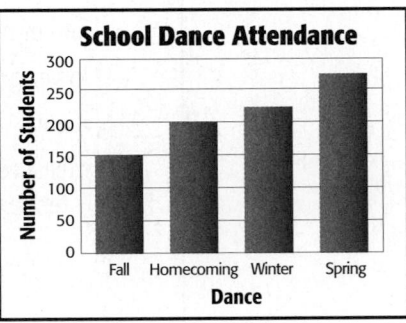

a. **Do the graphs show the same data? If so, explain how they differ.**
The graphs show the same data. However, the graphs differ in that Graph A uses a scale from 0 to 500, and Graph B uses a scale from 0 to 300.

b. **Which graph makes it appear that the attendance increased more rapidly? Explain your reasoning.**
Graph B makes it appear that the attendance increased more rapidly even though the attendance increase is the same.

25. **BUSINESS** The graphs show the admission prices for a museum from 2008 to 2012. Which graph suggests that the price has increased but not significantly? Explain your reasoning.

Graph A

Graph B

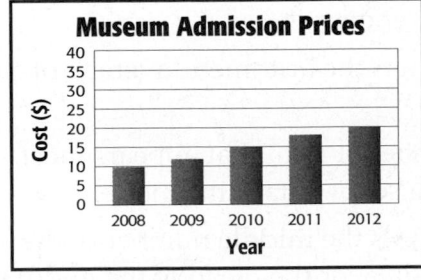

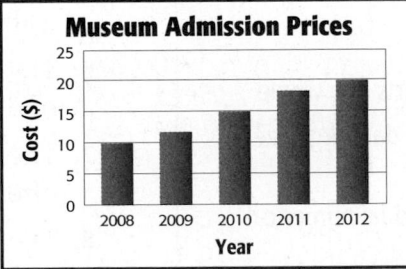

26. **SCHOOL DANCES** Refer to the Example above. Which graph might Student Council use to show that dance attendance has risen? Explain your reasoning.

FOLDABLES®
Study Organizer

Be sure the following Key Concepts are noted in your Foldable.

Key Concepts

Measures of Central Tendency (Lesson 1)
• Measures of central tendency are numbers that describe the center of a data set. They are mean, median, and mode.

Measures of Variation (Lesson 2)
• Measures of variation are used to describe the distribution of a data set.
• The range is the difference between the greatest and the least numbers in the set.
• Box-and-whisker plots use a number line to show the distribution of a set of data.

Statistical Displays (Lesson 3)
• A circle graph shows data as parts of a whole.
• A histogram displays numerical data organized into equal intervals.
• A stem-and–leaf plot lists individual numerical data in a condensed form.

More Statistical Displays (Lesson 4)
• Scatter plots show a relationship between a set of data with two variables.

Key Vocabulary

box-and-whisker plot	measures of variation
circle graph	median
histogram	mode
interquartile range	outlier
line of best fit	quartiles
lower quartile	range
mean	scatter plot
measures of central tendency	stem-and-leaf plot

Vocabulary Check

State whether each sentence is *true* or *false*. If *false*, replace the underlined word or number to make a true sentence.

1. The <u>range</u> is the difference between the greatest and the least values in a set of data.

2. The <u>mode</u> divides a set of data in half.

3. The interquartile range is one of the <u>measures of central tendency</u>.

4. A(n) <u>scatter plot</u> shows a relationship between a set of data with two variables.

5. The median of the bottom half of a data set is the <u>lower quartile</u>.

6. The <u>mean</u> is the arithmetic average of a set of data.

7. The number or item that appears most often in a set of data is the <u>mode</u>.

8. The <u>range</u> is the middle number of the ordered data, or the mean of the middle two numbers.

9. A(n) <u>variation</u> is a piece of data that is more than 1.5 times the value of the interquartile range beyond the quartiles.

Multi-Part Lesson Review

Lesson 1 **Measures of Central Tendency**

Measures of Central Tendency (Lesson 1B)

Find the mean, median, and mode for each data set. Round to the nearest tenth if necessary.

10. number of siblings: 2, 3, 1, 3, 4, 3, 8, 0, 2

11. 89°, 46°, 93°, 100°, 72°, 86°, 74°

12. MONEY Which measure of central tendency best represents the amount of money students spent on clothing?

$21, $75, $48, $52, $65

EXAMPLE 1 Karrie's science fair plants measured 5 inches, 4 inches, 2 inches, 2 inches, and 7 inches. Find the mean, median, and mode of the plant heights.

Mean $\dfrac{5 + 4 + 2 + 2 + 7}{5} = \dfrac{20}{5}$ or 4

Median 2, 2, ④, 5, 7

Mode 2

Lesson 2 **Measures of Variation**

Measures of Variation (Lesson 2A)

Find the measures of variation and any outliers for each data set.

13. number of miles ran each week: 14, 5, 4, 3, 5, 5, 6, 5, 4, 6, 3

14. number of hours spent burning music CDs each week: 2, 3, 2, 4, 1, 4, 1, 2

15. BOWLING The number of times Brittany's friends have been bowling is 6, 3, 7, 2, 5, 6, 1, and 8. Use the measures of variation to describe this data.

EXAMPLE 2 The number of hours Jake spent practicing for a band competition are 11, 9, 3, 9, 4, 9, 5, 5, 6, 10, and 9. Find the measures of variation.

Range 11 − 3 or 8

Median 3, 4, 5, 5, 6, ⑨, 9, 9, 9, 10, 11

Lower Quartile 3, 4, ⑤, 5, 6

Upper Quartile 9, 9, ⑨, 10, 11

Interquartile Range 9 − 5 or 4

Box-and-Whisker Plots (Lesson 2B)

Construct a box-and-whisker plot for the set of data.

16. number of fish caught: 0, 3, 4, 8, 9, 9, 9, 11, 13

17. POSTERS The number of posters various students have in their rooms is 3, 2, 1, 0, 2, 3, 4, 2, 5, 0, and 2. Construct a box-and-whisker plot for the data. What do the lengths of the parts of the plot tell you?

EXAMPLE 3 The lengths of various train rides are 4, 1, 2, 3, 6, 2, 3, 2, 5, 8, and 4 hours. Draw a box-and-whisker plot for the data set.

Length of Train Rides (hr)

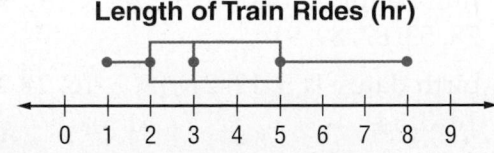

Lesson 3 **Statistical Displays**

Circle Graphs (Lesson 3B)

18. **COLORS** The table shows the shades of blue paint sold. Display the data in a circle graph.

Shade	Percent
Navy	35%
Sky/light blue	30%
Aquamarine	17%
Other	18%

EXAMPLE 4 Which pizza was chosen by about twice as many people as supreme?

Pepperoni was chosen by about twice as many people as supreme.

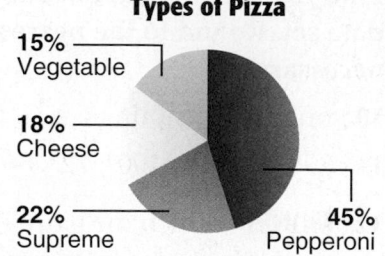

Types of Pizza

15% Vegetable
18% Cheese
22% Supreme
45% Pepperoni

Histograms (Lesson 3C)

Use the histogram at the right.

19. How large is each interval?

20. What percent of the runners ran 75 seconds or slower?

21. What was the most likely time?

22. What was the greatest time?

23. **PLANTS** The heights in inches of various types of plants are listed below. Choose intervals and construct a histogram to represent the data.

1, 1, 2, 4, 4, 5, 6, 7, 7, 8, 9, 10, 11, 12, 12, 12, 13, 14, 17, 18, 18, 19, 21, 23, 24

EXAMPLE 5 Choose intervals and construct a histogram to represent the following 400-meter dash times.

61	71	68	68	69	72	73	61	76	70
64	64	63	82	68	78	74	80	62	75

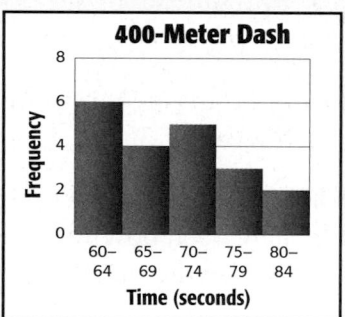

400-Meter Dash

Frequency vs. Time (seconds)
60–64, 65–69, 70–74, 75–79, 80–84

Stem-and-Leaf Plots (Lesson 3E)

Display each set of data using a stem-and-leaf plot.

24. hours worked: 29, 54, 31, 26, 38, 46, 23, 21, 32, 37

25. number of points: 75, 83, 78, 85, 87, 92, 78, 53, 87, 89, 91

26. birth dates: 9, 5, 12, 21, 18, 7, 16, 24, 11, 10, 3, 14

EXAMPLE 6 Display the number of pages read in a stem-and-leaf plot: 12, 15, 17, 20, 22, 22, 23, 25, 27, and 35.

The tens digits form the stems, and the ones digits form the leaves.

Pages Read

Stem	Leaf
1	2 5 7
2	0 2 2 3 5 7
3	5

2 | 3 = 23 pages

More Statistical Displays

PSI: Use a Graph (Lesson 4A)

27. HOCKEY The graph shows the number of points scored in the first twelve hockey games. What is the average number of points scored so far this season?

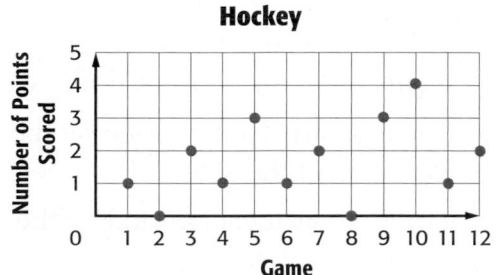

EXAMPLE 7 Use the graph to find the average test score.

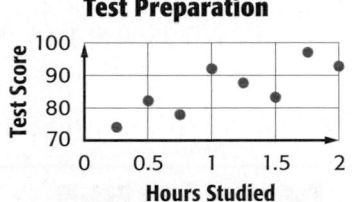

Add the test scores: $73 + 83 + 79 + 91 + 88 + 84 + 96 + 93$ or 687

Divide: $\frac{687}{8}$ or about 85.9

So, the average test score was about 85.9.

Scatter Plots and Lines of Best Fit (Lesson 4B)

28. SPORTS Use the graph to describe the relationship between hours practiced and number of wins.

EXAMPLE 8 Use the scatter plot to describe the relationship between the flower height and water received.

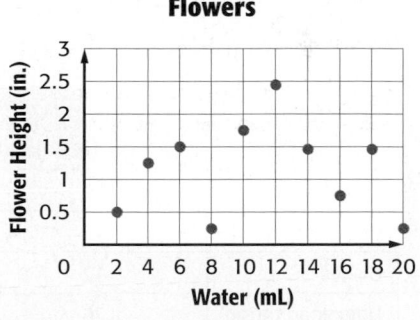

The graph shows no obvious patterns. So, there is no relationship.

Select an Appropriate Display (Lesson 4C)

29. CALORIES Select an appropriate display to show individual students' Calorie intake for lunch in numerical order.

30. SLEEP Is a circle graph an appropriate display to represent the number of calls made each day? Explain.

EXAMPLE 9 Select an appropriate display for the number of Olympic sprinters compared to the total number of athletes. Explain.

A circle graph would be appropriate because you are comparing a part to the whole.

1. **INSECTS** The lengths in inches of several insects are given below. Find the mean, median, and mode of the data. Round to the nearest tenth if necessary.

   0.75, 1.24, 0.95, 8.6, 1.18, 1.3

2. **BASKETBALL** Refer to the histogram.

   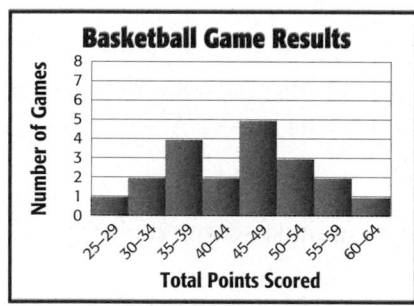

   a. How large is each interval?

   b. What was the most likely point total?

   c. In what percent of games were 50 points or more scored?

3. **TIME** Refer to the table.

Daily Activities	
Activity Type	**Average Time (min)**
Cardio	21
Clean room	15
Dishes	12
Download music	20
E-mail	18
Homework	56
Trash	11

   a. What is the range of the data?

   b. Find the median, upper and lower quartiles, and the interquartile range for the data.

   c. Identify any outliers.

   d. Use the measures of variation to describe the data in the table.

4. **MULTIPLE CHOICE** Refer to the data below. Which of the following statements is true concerning the measures of central tendency?

   41, 45, 42, 38, 77, 44, 36, 43

   A. The mode is most affected by the inclusion of the extreme number.

   B. The median is not affected by the inclusion of the extreme number.

   C. The mean is most affected by the inclusion of the extreme number.

   D. None of the measures of central tendency are affected by the inclusion of the extreme number.

5. **EXERCISE** The graph shows the percent of women who work out after work. Predict the number of women who will work out after work in 2015.

   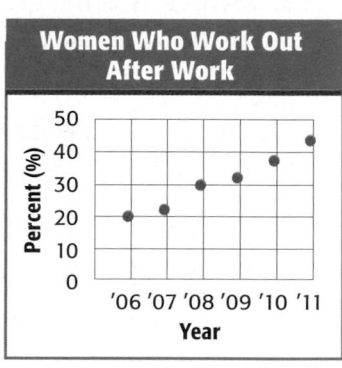

6. **THINK SOLVE EXPLAIN** **EXTENDED RESPONSE** Ronan asked his friends how many CDs they own. Below is the data Ronan collected from his friends.

   0, 0, 3, 5, 6, 8, 9, 15, 15, 15, 18, 20, 20, 28, 31

   Part A Construct a box-and-whisker plot with the data set given above.

   Part B What is the greatest amount of CDs owned by Ronan's friends?

   Part C Explain why the median is not in the middle of the box.

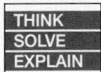

Preparing for Standardized Tests

THINK SOLVE EXPLAIN ## Short Response: Drawing Graphs

When drawing statistical graphs for short-response questions, be sure to include all necessary labels and information in order to receive full credit for that part of the question.

TEST EXAMPLE

The average depths and maximum depths of different lakes around the world are shown below.

Depths of Lakes

Average Depth (m)	149	72	40	72	19	86	50	107	138
Maximum Depth (m)	406	281	84	446	64	224	230	281	267

Construct a scatter plot to display the data.

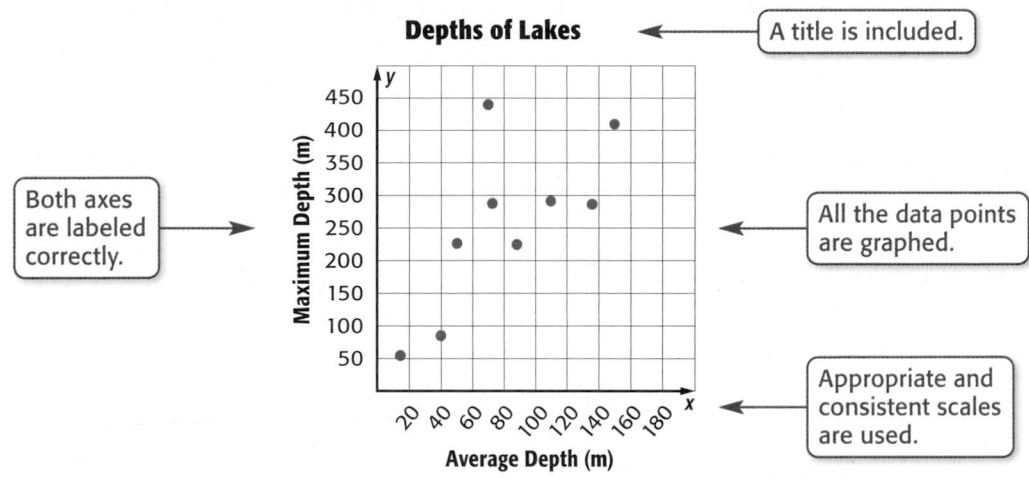

Both axes are labeled correctly.

A title is included.

All the data points are graphed.

Appropriate and consistent scales are used.

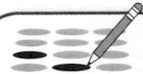

Work on It

Refer to the table of data above. Then construct a box-and-whisker plot showing the average depths of the lakes.

Test Hint

You will often use graphs that you draw to answer other parts of extended-response questions. Double-check your work.

Chapter 9 Preparing for Standardized Tests **551**

Read each question. Then fill in the correct answer on the answer sheet provided by your teacher or on a sheet of paper.

1. Ed's Used Cars bought 5 used cars for $6,400 each. They later bought another used car for $4,600. What measure of central tendency would best represent the cost of each car?

   A. none

   B. mean

   C. median

   D. mode

2. Each spinner is spun once.

    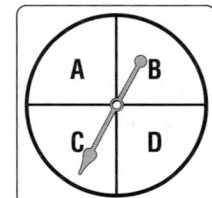

   What is the probability of spinning the number 3 and the letter A?

   F. $\frac{3}{8}$

   G. $\frac{1}{4}$

   H. $\frac{1}{8}$

   I. $\frac{1}{32}$

3. Which of the following equations is equivalent to the equation shown below?

   $$3x + 5 = 7x - 10$$

   A. $10x + 5 = -10$

   B. $4x + 5 = -10$

   C. $7x - 5 = 3x$

   D. $-4x + 5 = -10$

4. **≡≡ GRIDDED RESPONSE** Neela has 11.5 yards of fabric. She will use 20% of the fabric to make a flag. How many yards of fabric will she use?

5. **≡≡ GRIDDED RESPONSE** A patio blueprint has a key that shows 1 inch is equal to 12 feet. If the owner wants the length to be 30 feet, how many inches will the length be on the blueprint?

6. The number of ringtones that twelve middle school students have on their cell phones is 14, 8, 7, 6, 5, 5, 10, 11, 8, 8, 6, and 7. Which of the following statements is NOT supported by these data?

   F. Half of the ringtones are below 7.5 and half are above 7.5.

   G. The range of the data is 9 ringtones.

   H. An outlier of the data is 11 ringtones.

   I. About one fourth of the ringtones that the students have are at or above 9.

7. Which box-and-whisker plot represents the data set 8, 12, 21, 15, 20, 9, 16, 14, and 25?

   A.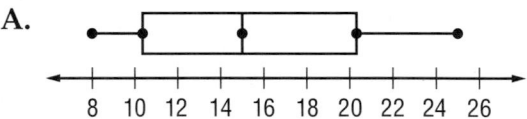

   B.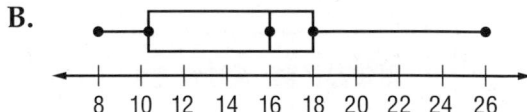

   C.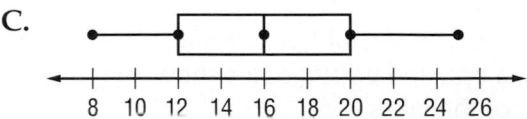

   D.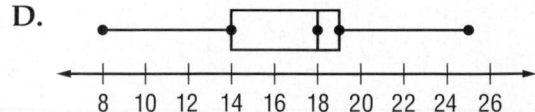

8. Katherine polled 21 classmates to find out the average number of hours each spends watching television each week. Which of the following displays would be most appropriate to show the individual student responses?

F.

Number of Hours Spent Watching Television Each Week

15–19
25%

0–4
8.3%

5–9
16.7%

10–14
50%

G.

Number of Hours Spent Watching Television Each Week

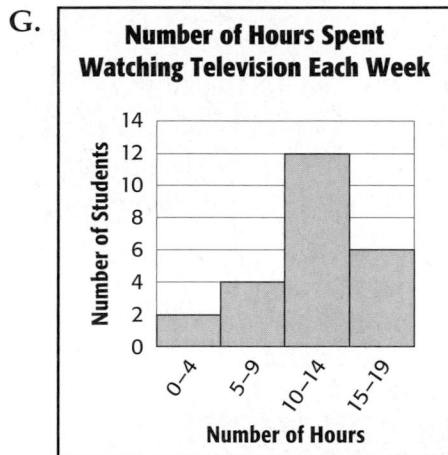

Number of Hours

H. Number of Hours Spent Watching Television Each Week

Stem	Leaf
0	0 3 5 7 8 8
1	0 0 1 2 2 2 3 4 4 4 4 5 6 6 8 8 9

1 | 2 = 12

I.

Number of Hours Spent Watching Television Each Week

0 2 4 6 8 10 12 14 16 18 20

9. The numbers of monthly minutes Gary used on his cell phone for the last eight months are shown below.

Monthly Cell Minutes			
400	550	450	620
550	600	475	425

What is the mode of this data?

A. 550 minutes   **C.** 475 minutes

B. 450 minutes   **D.** 400 minutes

10. THINK SOLVE EXPLAIN **SHORT RESPONSE** Mr. Thompson made 20 liters of punch for a party. The punch contained 5 liters of orange juice. Write and solve a proportion to find the percent of orange juice in the punch.

11. THINK SOLVE EXPLAIN **EXTENDED RESPONSE** The table shows how values of a painting increased over ten years.

Year	Value	Year	Value
2005	$350	2010	$1,851
2006	$650	2011	$2,151
2007	$950	2012	$2,451
2008	$1,200	2013	$2,752
2009	$1,551	2014	$3,052

Part A   Select and create a display that shows the relationship between years and the value of the painting. Justify your reasoning.

Part B   Write a conclusion based on your graph.

Part C   Use the graph to predict what the value of the painting will be in 2018.

NEED EXTRA HELP?										
If You Missed Question...	1	2	3	4	5	6	7	8	9	10	11
Go to Chapter-Lesson...	9-1B	8-2B	4-3D	6-1B	5-2B	9-2A	9-2B	9-4C	9-1B	6-2B	9-4C

CHAPTER 10

Volume and Surface Area

connectED.mcgraw-hill.com

Investigate

 Animations

 Vocabulary

 Multilingual eGlossary

Learn

 Personal Tutor

 Virtual Manipulatives

 Graphing Calculator

 Audio

 Foldables

Practice

 Self-Check Practice

 Worksheets

 Assessment

The ☆BIG Idea

How are volume and surface area related?

 FOLDABLES Study Organizer

Make this Foldable to help you organize your notes.

Volume

Surface Area

Composite Figures

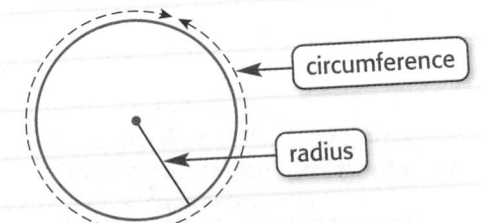

Review Vocabulary

circumference circunferencia the distance around a circle

circumference

radius

Key Vocabulary

English	Español
cone	cono
cylinder	cilindro
prism	prisma
pyramid	pirámide

When Will I Use This?

Your Turn!
You will solve this problem in the chapter.

Are You Ready for the Chapter?

You have two options for checking prerequisite skills for this chapter.

Text Option Take the Quick Check below. Refer to the Quick Review for help.

QUICK Check

Find the area of each triangle.

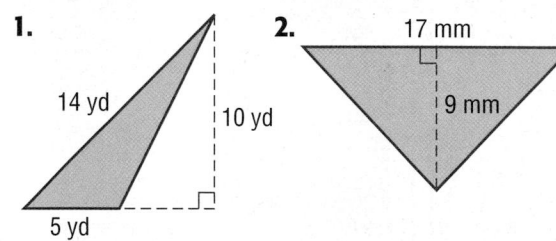

1. 14 yd, 10 yd, 5 yd

2. 17 mm, 9 mm

3. **YARD** Anita's yard is in the shape of a triangle. It has a height of 35 feet and a base of 50 feet. What is the area of the yard?

Find the area of each circle. Use 3.14 for π. Round to the nearest tenth.

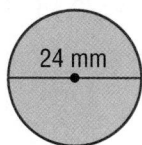

4. 24 mm

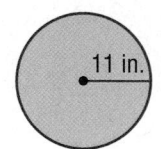

5. 11 in.

6. **PIZZA** Find the area of a circular pizza with a radius of 6 inches.

7. **CUPS** A manufacturer measures a cup's diameter to determine how much material is needed for a lid. The diameter is 10 centimeters. Find the area of the lid.

QUICK Review

EXAMPLE 1

Find the area of the triangle.

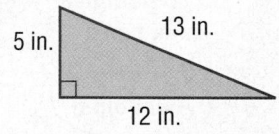

5 in., 13 in., 12 in.

$A = \frac{1}{2}bh$ Area of a triangle

$A = \frac{1}{2}(12)(5)$ Replace b with 12 and h with 5.

$A = \frac{1}{2}(60)$ Multiply.

$A = 30$ Multiply.

The area of the triangle is 30 square inches.

EXAMPLE 2

Find the area of the circle. Use 3.14 for π. Round to the nearest tenth.

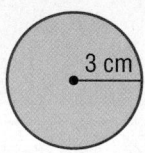

3 cm

Estimate $3.14 \times (3)^2 = 3 \times 9$ or 27

$A = \pi r^2$ Area of a circle

$A \approx 3.14 \times (3)^2$ Replace π with 3.14 and r with 3.

$A \approx 3.14 \times 9$ Evaluate $(3)^2$.

$A \approx 28.26$ Multiply.

Round to the nearest tenth. The area is about 28.3 square centimeters.

Check $28.3 \approx 27$ ✔

 **Online Option** Take the Online Readiness Quiz.

Explore Meaning of Volume

Main Idea
Justify formulas for the volume of prisms.

Vocabulary
prism

Get Connect ED

7.G.6

A cereal box or a box of tissues is shaped like a prism. A **prism** is a three-dimensional figure with at least three rectangular lateral faces and top and bottom faces that are parallel. The top view of a rectangular prism is shown below.

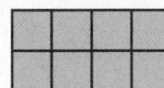
ACTIVITY

STEP 1 Build a prism 1 unit high with the same top view as above. Find its volume.

The prism uses 8 centimeter cubes.

So, the volume is 8 cubic centimeters.

STEP 2 Build a prism 2 units high with the same top view as above. Find its volume.

The prism uses 16 centimeter cubes.

So, the volume is 16 cubic centimeters.

Analyze the Results

1. What is the volume of a prism that is 10 units high with the same top view as above?

2. What is the area of the top view? Justify your reasoning.

3. Write an expression for the volume of a prism with this top view and height h.

4. **MAKE A CONJECTURE** Write a method for finding the volume of any rectangular prism given its dimensions. Justify your answer.

5. Find an object that is a rectangular prism.

   a. Measure the dimensions.

   b. Estimate the volume of the object.

   c. Calculate the volume of the object. Compare to your estimate.

Main Idea
Find the volumes of rectangular and triangular prisms.

Vocabulary
volume
rectangular prism
triangular prism

 7.G.6

 Volume of Prisms

1. If you observed the Great Pyramid in Egypt and the Inner Harbor and Trade Center in Baltimore from directly above, what geometric figures would you see?

2. What is the formula for the area of a square? a rectangle?

The **volume** of a three-dimensional figure is the measure of space it occupies. It is measured in cubic units such as cubic centimeters (cm^3) or cubic inches (in^3).

2 cm
6 cm
6 cm

The bottom layer, or base, has 6 · 6 or 36 cubes.

There are two layers.

It takes 2 layers of 36 cubes to fill the box. So, the volume of the box is 72 cubic centimeters.

The figure above is a rectangular prism. A **rectangular prism** is a prism that has rectangular bases.

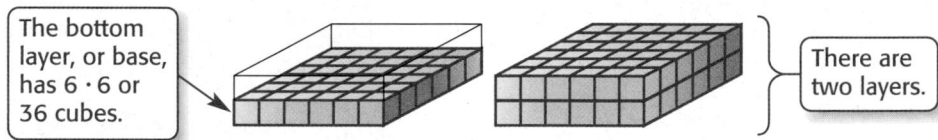

Key Concept — Volume of a Rectangular Prism

Words	The volume V of a rectangular prism is the product of the length ℓ, the width w, and the height h. It is also the area of the base B times the height h.	Model
		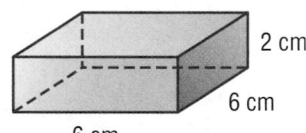 h w ℓ $B = \ell w$
Symbols	$V = \ell wh$ or $V = Bh$	

Study Tip

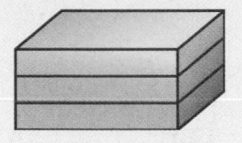

 EXAMPLE Volume of a Rectangular Prism

1. Find the volume of the rectangular prism.

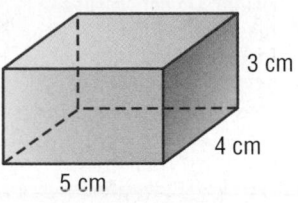

$V = \ell wh$ Volume of a prism

$V = 5 \cdot 4 \cdot 3$ $\ell = 5$, $w = 4$, and $h = 3$

$V = 60$ Multiply.

The volume is 60 cubic centimeters or 60 cm^3.

 CHECK Your Progress

a. Find the volume of the rectangular prism at the right.

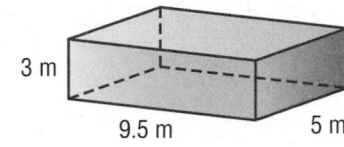

 REAL-WORLD EXAMPLE

2. **MARKETING** A company makes lunch boxes in two different sizes. Which lunch box holds more food?

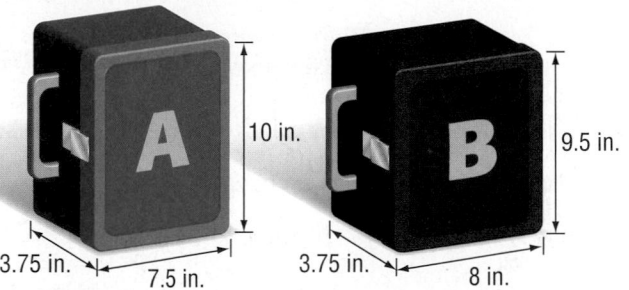

Find the volume of each lunch box. Then compare.

Lunch Box A	**Lunch Box B**
$V = \ell wh$	$V = \ell wh$
$V = 7.5 \cdot 3.75 \cdot 10$	$V = 8 \cdot 3.75 \cdot 9.5$
$V = 281.25 \text{ in}^3$	$V = 285 \text{ in}^3$

Since 285 in^3 > 281.25 in^3, Lunch Box B holds more food.

 CHECK Your Progress

b. **PACKAGING** A concession stand serves peanuts in two different containers. Which container holds more peanuts? Justify your answer.

Box A   Box B

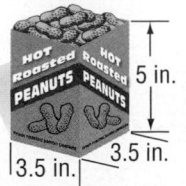

Study Tip

Height Do not confuse the height of the triangular base with the height of the prism.

A **triangular prism** is a prism that has triangular bases. The diagram below shows that the volume of a triangular prism is also the product of the area of the base B and the height h of the prism.

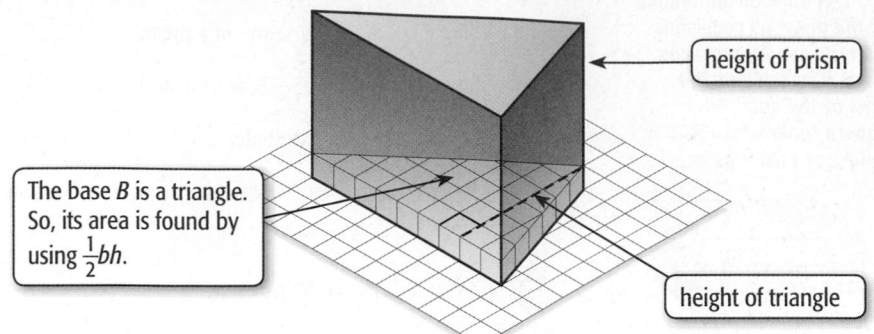

height of prism

The base B is a triangle. So, its area is found by using $\frac{1}{2}bh$.

height of triangle

Key Concept — Volume of a Triangular Prism

Words   The volume V of a triangular prism is the area of the base B times the height h.

Symbols $V = Bh$, where B is the area of the base.

Model

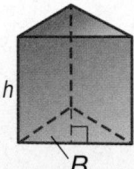

h

B

Study Tip

Base Before finding the volume of a prism, identify the base. In Example 3, the base is a triangle, so you replace B with $\frac{1}{2}bh$.

✐ EXAMPLE   Volume of a Triangular Prism

3   **Find the volume of the triangular prism shown.**

The area of the triangle is $\frac{1}{2} \cdot 6 \cdot 8$, so replace B with $\frac{1}{2} \cdot 6 \cdot 8$.

$V = Bh$ Volume of a prism

$V = \left(\frac{1}{2} \cdot 6 \cdot 8\right)h$ Replace B with $\frac{1}{2} \cdot 6 \cdot 8$.

$V = \left(\frac{1}{2} \cdot 6 \cdot 8\right)9$ The height of the prism is 9.

$V = 216$ Multiply.

The volume is 216 cubic feet or 216 ft³.

6 ft

9 ft

8 ft

✓ CHECK Your Progress

Find the volume of each triangular prism.

c.

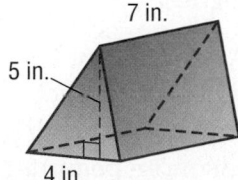

7 in.

5 in.

4 in.

d.

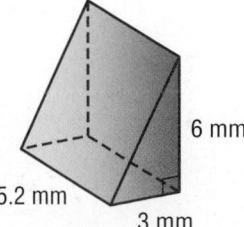

6 mm

5.2 mm

3 mm

Examples 1 and 3 Find the volume of each prism. Round to the nearest tenth if necessary.

1.

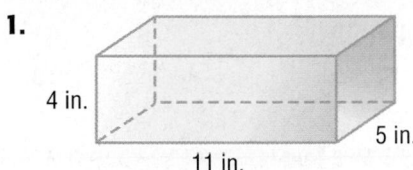

4 in.
5 in.
11 in.

2.

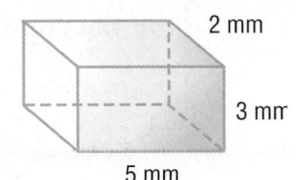

2 mm
3 mm
5 mm

3.

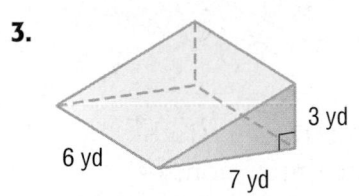

3 yd
6 yd
7 yd

4.

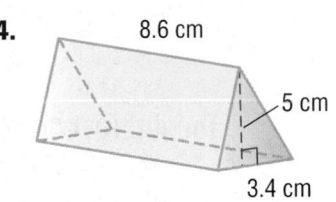

8.6 cm
5 cm
3.4 cm

Example 2 **5. STORAGE** One cabinet measures 3 feet by 2.5 feet by 5 feet. A second measures 4 feet by 3.5 feet by 4.5 feet. Which volume is greater? Explain.

Practice and Problem Solving

● = **Step-by-Step Solutions** begin on page R1.
Extra Practice begins on page EP2.

Examples 1 and 3 Find the volume of each prism. Round to the nearest tenth if necessary.

6.

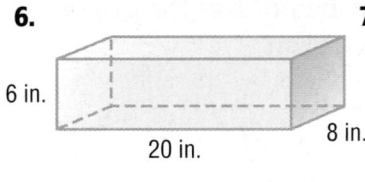

6 in.
20 in.
8 in.

7.

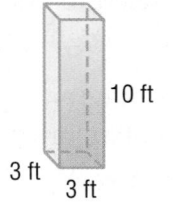

10 ft
3 ft
3 ft

8.

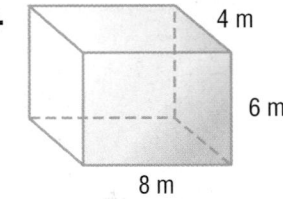

4 m
6 m
8 m

9.

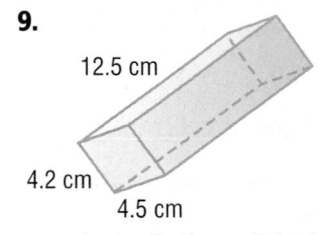

12.5 cm
4.2 cm
4.5 cm

10.

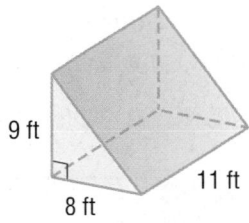

9 ft
11 ft
8 ft

11

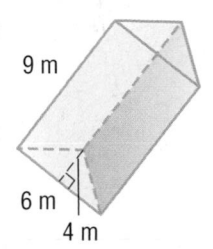

9 m
6 m
4 m

12.

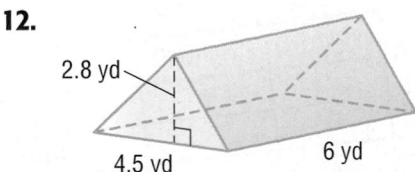

2.8 yd
4.5 yd
6 yd

13.

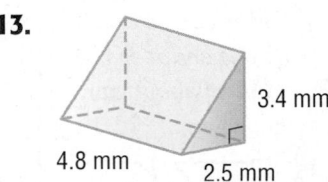

3.4 mm
4.8 mm
2.5 mm

Example 2 **14. PACKAGING** Which container holds more detergent? Justify your answer.

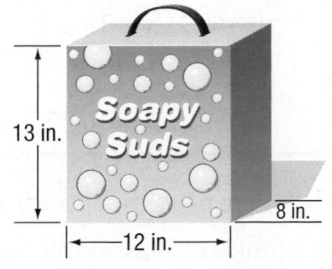

Soapy Suds
13 in.
8 in.
12 in.

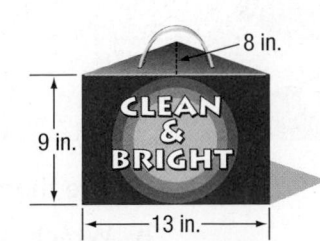

CLEAN & BRIGHT
8 in.
9 in.
13 in.

15. TOYS A toy company makes rectangular sandboxes that measure 6 feet by 5 feet by 1.2 feet. A customer buys a sandbox and 40 cubic feet of sand. Did the customer buy too much or too little sand? Justify your answer.

Find the volume of each prism.

16.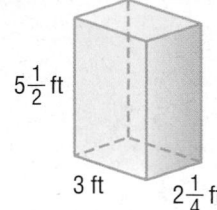
$5\frac{1}{2}$ ft
3 ft $2\frac{1}{4}$ ft

17.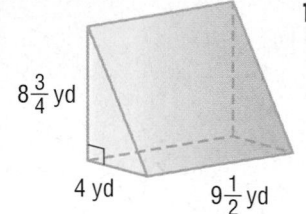
$8\frac{3}{4}$ yd
4 yd $9\frac{1}{2}$ yd

18.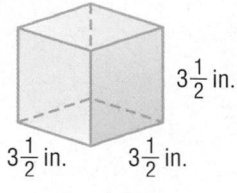
$3\frac{1}{2}$ in.
$3\frac{1}{2}$ in. $3\frac{1}{2}$ in.

19. ARCHITECTURE Use the diagram at the right that shows the approximate dimensions of the Flatiron Building in New York City.

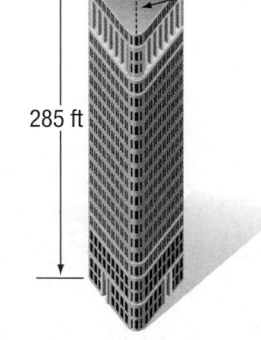
|←87 ft→|
174 ft
285 ft

   a. What is the approximate volume of the Flatiron Building?

   b. The building is 22 stories tall. Estimate the volume of each story.

20. ALGEBRA The base of a rectangular prism has an area of 19.4 square meters and the prism has a volume of 306.52 cubic meters. Write an equation that can be used to find the height h of the prism. Then find the height of the prism.

21 MONEY The diagram shows the dimensions of an office. It costs about $0.11 per year to air condition one cubic foot of space. On average, how much does it cost to air condition the office for one month?

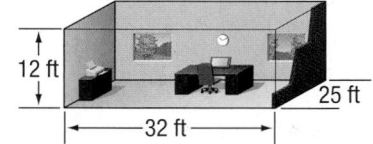
12 ft
25 ft
32 ft

22. GRAPHIC NOVEL Refer to the graphic novel frame below for Exercises **a** and **b**.

Refer to the start of the chapter to read all about our dunk tank.

Length(ft)	Width(ft)	Height(ft)	Surface Area(ft²)	Volume(ft³)
2	12	4	136	96
4	4	8	144	128
4	7	6	160	168
8	5	4	152	160
10	4	3	124	120

   a. Are there any other possibilities for dimensions?

   b. Which dimensions are reasonable for a dunk tank? Explain.

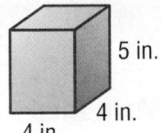

H.O.T. Problems

23. REASONING A rectangular prism is shown.

    a. Suppose the length of the prism is doubled. How does the volume change? Explain your reasoning.

    b. Suppose the length, width, and height are each doubled. How does the volume change?

    c. Which will have a greater effect on the volume of the prism: doubling the height or doubling the width? Explain your reasoning.

24. CHALLENGE How many cubic inches are in a cubic foot?

25. WRITE MATH Compare and contrast finding the volume of a rectangular prism and a triangular prism.

Test Practice

26. A fish aquarium is shown.

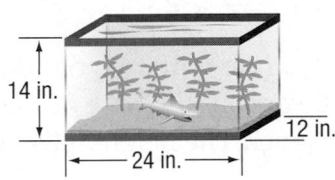

What is the volume of the aquarium?

    A. 168 in^3 **C.** 2,016 in^3

    B. 342 in^3 **D.** 4,032 in^3

27. GRIDDED RESPONSE Box A is 12 inches by 18 inches by 24 inches. Box B is 12 inches by 12 inches by 30 inches. In cubic inches, how much greater is the volume of Box A?

28. The area of the base of a triangular prism is 50 square centimeters. The height of the triangular prism is 8 centimeters. Which represents the volume of the triangular prism?

    F. 58 cm^2

    G. 58 cm^3

    H. 400 cm^2

    I. 400 cm^3

29. The table shows the dimensions of mailing containers.

Container	ℓ (ft)	w (ft)	h (ft)
A	2	2	2
B	1	3	3
C	3	4	0.5
D	3	2	0.5

Which container has the greatest volume?

    A. Container A

    B. Container B

    C. Container C

    D. Container D

30. The volume of the box below is 1.5 cubic inches.

Which of the following are possible dimensions of the box?

    F. 2 in. by 2 in. by 1 in.

    G. 1 in. by 1 in. by 1 in.

    H. 2 in. by 1.5 in. by 0.5 in.

    I. 3 in. by 0.5 in. by 1.5 in.

Main Idea
Find the volumes of cylinders.

 Vocabulary
cylinder
precision

 Get ConnectED

 CCSS 7.G.4, 7.G.6

Volume of Cylinders

Explore Set a soup can on a piece of grid paper and trace around the base, as shown below.

1. Estimate the number of centimeter cubes that would fit at the bottom of the container. Include parts of cubes.

2. If each layer is 1 centimeter high, how many layers would it take to fill the cylinder?

3. **MAKE A CONJECTURE** Write a formula that allows you to find the volume of the container.

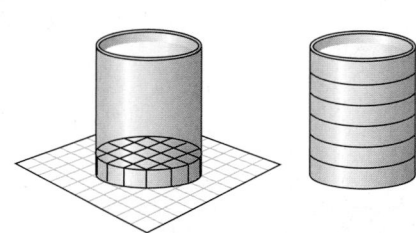

A **cylinder** is a three-dimensional figure with two parallel congruent circular bases. As with prisms, the area of the base of a cylinder tells the number of cubic units in one layer. The height tells how many layers there are in the cylinder.

Key Concept **Volume of a Cylinder**

Words   The volume V of a cylinder with radius r is the area of the base B times the height h.

Model

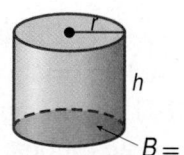

h

$B = \pi r^2$

Symbols $V = Bh$, where $B = \pi r^2$
or $V = \pi r^2 h$

EXAMPLE **Volume of a Cylinder**

1 Find the volume of the cylinder. Round to the nearest tenth.

$V = \pi r^2 h$ Volume of a cylinder

$V = \pi(5)^2(8.3)$ Replace r with 5 and h with 8.3.

Use a calculator.

2nd [π] ✕ 5 x^2 ✕ 8.3 ENTER 651.8804756

The volume is about 651.9 cubic centimeters.

5 cm

8.3 cm

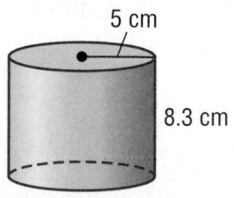

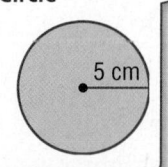

✓ **CHECK** Your Progress

Find the volume of each cylinder. Round to the nearest tenth.

a.

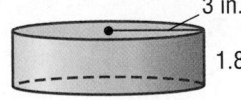

3 in.

1.8 in.

b. 2.4 m

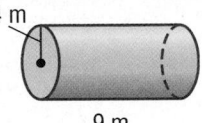

9 m

🏃 🖊️ **REAL-WORLD EXAMPLE**

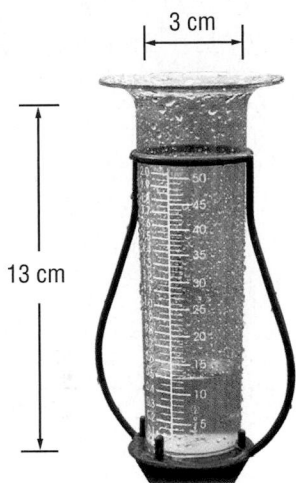

3 cm

13 cm

Study Tip

Circles Recall that the radius is half the diameter.

2 **WEATHER** The decorative rain gauge shown has a height of 13 centimeters and a diameter of 3 centimeters. How much water can the rain gauge hold?

$V = \pi r^2 h$ Volume of a cylinder

$V = \pi (1.5)^2 13$ Replace r with 1.5 and h with 13.

$V \approx 91.9$ Simplify.

The rain gauge can hold about 91.9 cubic centimeters.

✓ **CHECK** Your Progress

c. PAINT Find the volume of a cylindrical paint can that has a diameter of 4 inches and a height of 5 inches. Round to the nearest tenth.

✓ **CHECK** Your Understanding

Example 1 **Find the volume of each cylinder. Round to the nearest tenth.**

1.

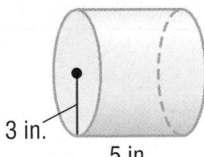

3 in.

5 in

2.

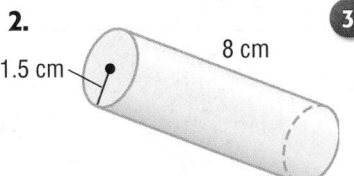

1.5 cm

8 cm

3

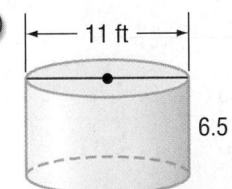

← 11 ft →

6.5 ft

Example 2 **4. CONTAINERS** A can of concentrated orange juice has the dimensions shown at the right. Find the volume of the can of orange juice to the nearest tenth.

15 cm

7 cm

5. CANDLES A scented candle is in the shape of a cylinder. The radius is 4 centimeters and the height is 12 centimeters. Find the volume of the candle. Round to the nearest tenth.

Practice and Problem Solving

= **Step-by-Step Solutions** begin on page R1.
Extra Practice begins on page EP2.

Example 1 **Find the volume of each cylinder. Round to the nearest tenth.**

6.
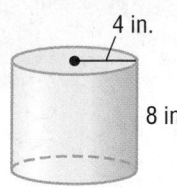
4 in.
8 in.

7.
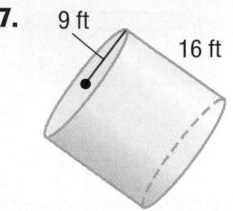
9 ft
16 ft

8.
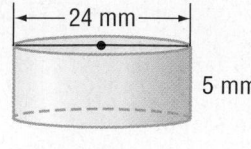
24 mm
5 mm

9.
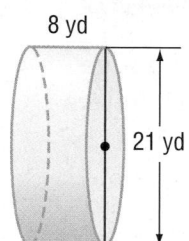
8 yd
21 yd

10.
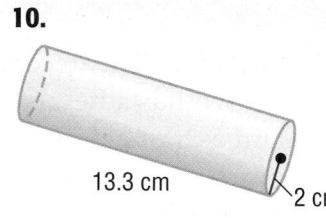
13.3 cm
2 cm

11.
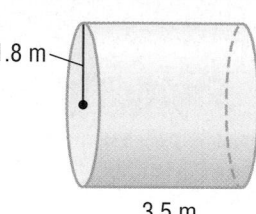
1.8 m
3.5 m

12. diameter = 15 mm
height = 4.8 mm

13 diameter = 4.5 m
height = 6.5 m

14. radius = 6 ft

height = $5\frac{1}{3}$ ft

15. radius = $3\frac{1}{2}$ in.

height = $7\frac{1}{2}$ in.

Example 2 **16. BIRDS** A cylindrical bird feeder has a diameter of 4 inches and a height of 18 inches. How much birdseed can the feeder hold? Round to the nearest tenth.

17. WATER BOTTLE What is the volume of a cylindrical water bottle that has a radius of $1\frac{1}{4}$ inches and a height of 7 inches? Round to the nearest tenth.

Find the volume of each cylinder. Round to the nearest tenth.

18.
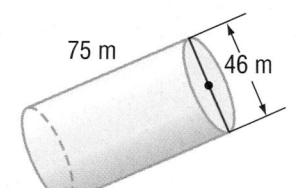
26 ft
40 ft

19.
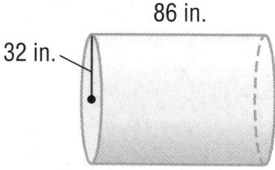
75 m
46 m

20.
86 in.
32 in.

ESTIMATION Match each cylinder with its approximate volume.

21. radius = 4.1 ft, height = 5 ft

22. diameter = 8 ft, height = 2.2 ft

23. diameter = 6.2 ft, height = 3 ft

24. radius = 2 ft, height = 3.8 ft

a. 91 ft³

b. 48 ft³

c. 111 ft³

d. 264 ft³

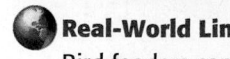

Real-World Link
Bird feeders can attract many species of birds. There are over 800 species of birds in North America.

25. POTTERY A vase in the shape of a cylinder has a diameter of 11 centimeters and a height of 250 millimeters. Find the volume of the vase to the nearest cubic centimeter. Use 3.14 for π. (*Hint*: 1 cm = 10 mm)

26. BAKING Which will hold more cake batter, the rectangular pan or two round pans? Explain.

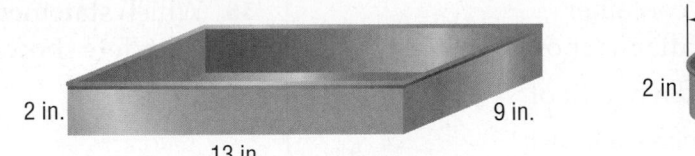

27 ALGEBRA Cylinder A has a radius of 4 inches and a height of 2 inches. Cylinder B has a radius of 2 inches. What is the height of Cylinder B to the nearest inch if both cylinders have the same volume?

28. ⬚ MULTIPLE REPRESENTATIONS The dimensions for four cylinders are shown in the table.

	Radius (cm)	Height (cm)	Volume (cm³)
Cylinder A	1	1	■
Cylinder B	1	2	■
Cylinder C	2	1	■
Cylinder D	2	2	■

  a. SYMBOLS Write an equation to find the volume of each cylinder.

  b. WORDS Compare the dimensions of Cylinder A with the dimensions of Cylinders B, C, and D.

  c. NUMBERS Copy and complete the table.

  d. WORDS Explain how changing the dimensions of a cylinder affects the cylinder's volume.

H.O.T. Problems

29. CHALLENGE Two equally sized sheets of construction paper are rolled; one along the length and the other along the width, as shown. Which cylinder has the greater volume? Explain.

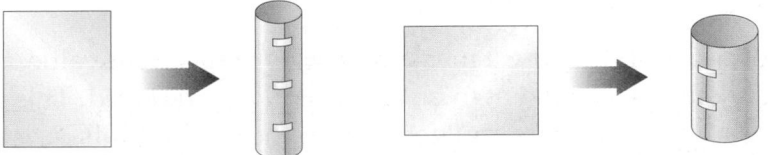

30. OPEN ENDED Draw and label a cylinder that has a larger radius but less volume than the cylinder shown at the right.

8 cm

16 cm

31. NUMBER SENSE What is the ratio of the volume of a cylinder to the volume of a cylinder having twice the height but the same radius?

32. NUMBER SENSE Suppose Cylinder A has the same height but twice the radius of Cylinder B. What is the ratio of the volume of Cylinder B to Cylinder A?

33. ✏ WRITE MATH Explain how the formula for the volume of a cylinder is similar to the formula for the volume of a rectangular prism.

34. The oatmeal container shown has a diameter of $3\frac{1}{2}$ inches and a height of 9 inches. Which is closest to the number of cubic inches it will hold when filled?

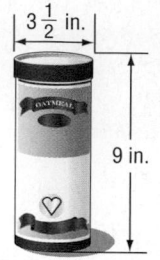

$3\frac{1}{2}$ in.

9 in.

A. 32   C. 75.92
B. 42.78   D. 86.59

35. ✐ **GRIDDED RESPONSE** Jarrod's family stores their sugar in a container like the one shown at the right. They fill a sugar dispenser with a volume of 38.9 cubic inches from the container. If the storage container is full, how many times can the dispenser be completely filled?

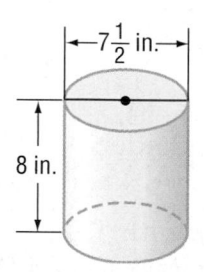

$7\frac{1}{2}$ in.

8 in.

36. Which statement is true about the cylinders shown?

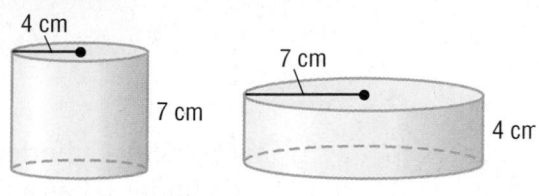

4 cm

7 cm

7 cm

4 cm

Cylinder 1   **Cylinder 2**

F. The volume of Cylinder 1 is greater than the volume of Cylinder 2.

G. The volume of Cylinder 2 is greater than the volume of Cylinder 1.

H. The volumes are equal.

I. The volume of Cylinder 1 is twice the volume of Cylinder 2.

37. THINK SOLVE EXPLAIN **SHORT RESPONSE** Chenoa is using the mold at the right to make candles. What is the volume of candle wax that the mold holds? Use 3.14 for π.

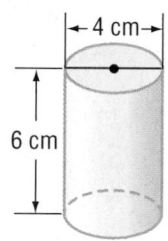

4 cm

6 cm

More About Cylinders

Accuracy describes how close a measurement is to its actual value. **Precision** is the ability of a measurement to be consistently reproduced. This means that every time you measure you get close to the same results.

EXAMPLE

Alley, Ben, and Sophia measured the circumference of a circular tablecloth for trim. Their results are shown in the table. Suppose the radius of the tablecloth is 35 inches. Describe the accuracy and precision of their measurements.

Student	Circumference (in.)
Alley	224
Ben	223.5
Sophia	223.75

Their measurements are precise because they consistently achieved similar results. Their measurements are not accurate because the actual circumference of the circle is $2\pi(35)$ or about 220 inches.

38. Measure the circumference of a cylinder. Compare your results with two classmates who found the circumference of the same cylinder. Describe the precision and accuracy of your results.

Explore Volume of Pyramids and Cones

Main Idea

Justify formulas for the volume of pyramids and cones.

CCSS 7.G.6

 MOVIE THEATERS A movie theater offers two different containers of popcorn: a square prism and a square pyramid. They are both 4 inches tall and have a base area of 16 square inches.

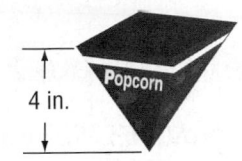

In the following Activity, you will compare their volumes.

🏃 ACTIVITY

① **STEP 1** Draw each net onto card stock. Cut out each and tape together. The prism and pyramid will be open. The pyramid is composed of four congruent isosceles triangles with bases of 4 inches and heights of $4\frac{1}{2}$ inches.

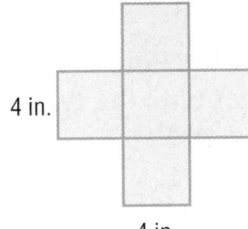

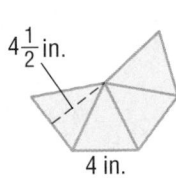

STEP 2 Fill the pyramid with rice. Pour the rice from the pyramid into the prism and repeat until the prism is full. Slide a ruler across the top to level the amount.

Analyze the Results

1. Compare the bases of the prism and pyramid.

2. Compare the heights of the prism and pyramid.

3. How many pyramids of rice did it take to fill the prism?

4. What fraction of the volume of the prism is the volume of the pyramid?

5. Repeat the Activity with a rectangular prism and a rectangular pyramid. The prism and the pyramid should have equal heights and equal bases. What fraction of the volume of the prism is the volume of the pyramid?

6. **MAKE A CONJECTURE** Explain how you could find the volume of a pyramid given a prism with the same base area and height. Justify your answer.

ACTIVITY

2 **MOVIES Suppose the movie theater also offers shaved ice in a cylinder and in a cone. The cylinder and the cone have equal heights and equal base areas. Make a model of each container.**

STEP 1 Draw each net onto card stock. Cut out each and tape together. For the cone, cut out the radius of the larger circle and overlap until the cone is formed. The cylinder and cone will be open.

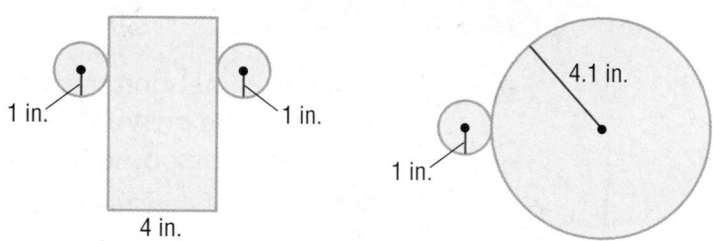

1 in.   1 in.   4 in.   4.1 in.   1 in.

STEP 2 Fill the cone with rice. Pour the rice from the cone into the cylinder and repeat until the cylinder is full.

Analyze the Results

7. Compare the bases and heights of the cylinder and cone.

8. How many cones of rice did it take to fill the cylinder?

9. What fraction of the volume of the cylinder is the volume of the cone?

10. **MAKE A CONJECTURE** If the cylinder of shaved ice has a volume of 27 ounces, what is the volume of the cone of shaved ice? Justify your answer.

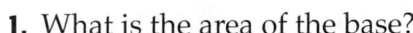
 Main Idea
Find the volume of pyramids.

 Vocabulary
pyramid
lateral face

 **Get ConnectED**

CCSS 7.G.6

Volume of Pyramids

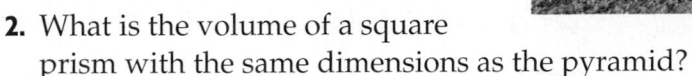 **SAND SCULPTURE** Dion is helping his brother build a sand sculpture at the beach in the shape of a pyramid. The square pyramid has a base with a length and width of 12 inches each and a height of 12 inches.

1. What is the area of the base?

2. What is the volume of a square prism with the same dimensions as the pyramid?

A **pyramid** is a three-dimensional figure with one base and triangular **lateral faces**, or any flat surfaces that are not bases. The lateral edges meet at one vertex. The height of a pyramid is the distance from the vertex perpendicular to the base.

Key Concept **Volume of a Pyramid**

Words The volume V of a pyramid is one third the area of the base B times the height of the pyramid h.

Symbols $V = \frac{1}{3}Bh$

Model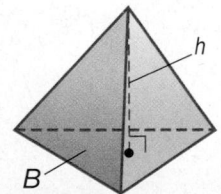

EXAMPLE **Find the Volume of a Pyramid**

1 **Find the volume of the pyramid. Round to the nearest tenth.**

$V = \frac{1}{3}Bh$ Volume of a pyramid

$V = \frac{1}{3}(3.2 \cdot 1.4)2.8$ $B = 3.2 \cdot 1.4, h = 2.8$

$V \approx 4.2$ Simplify.

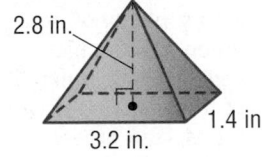
2.8 in.
1.4 in.
3.2 in.

The volume is about 4.2 cubic inches.

CHECK Your Progress

a. Find the volume of a pyramid that has a height of 9 centimeters and a rectangular base with a length of 7 centimeters and a width of 3 centimeters.

 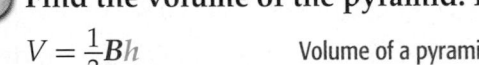

You can also use the formula for the volume of a pyramid to find a missing height.

 **EXAMPLE** **Find the Height of a Pyramid**

2 A triangular pyramid has a volume of 44 cubic meters. It has an 8-meter base and a 3-meter height. Find the height of the pyramid.

QUICK Review

Multiplying Fractions
To find $\frac{1}{3} \cdot \frac{1}{2} \cdot 8 \cdot 3$,
multiply $\frac{1}{3} \cdot \frac{1}{2}$ and $8 \cdot 3$
to get $\frac{1}{6}$ and 24, then
find $\frac{1}{6}$ of 24.

$V = \frac{1}{3}Bh$ Volume of a pyramid

$44 = \frac{1}{3}\left(\frac{1}{2} \cdot 8 \cdot 3\right)h$ $V = 44, B = \frac{1}{2} \cdot 8 \cdot 3$

$44 = 4h$ Multiply.

$\frac{44}{4} = \frac{4h}{4}$ Divide by 4.

$11 = h$ Simplify.

The height of the pyramid is 11 meters.

CHECK Your Progress

b. A triangular pyramid has a volume of 840 cubic inches. It has a base of 20 inches and a height of 21 inches. Find the height of the pyramid.

REAL-WORLD EXAMPLE

3 **SAND SCULPTURES** A model of a sand sculpture is shown at the right. Find the volume of the square pyramid.

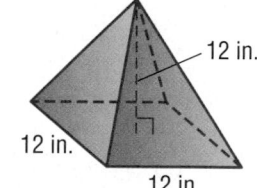

$V = \frac{1}{3}Bh$ Volume of a pyramid

$V = \frac{1}{3}(12 \cdot 12)12$ $B = 12 \cdot 12, h = 12$

$V = 576$ Simplify.

The volume is 576 cubic inches.

CHECK Your Progress

Real-World Link · · · · ·
The U.S. Department of Agriculture recommends that teenagers eat at least three ounces of whole grains, plenty of dark green and orange vegetables, a variety of fruit, calcium-rich foods, lean meats, and a limited intake of oils each day.

c. MODELS Kamilah is making a model of the Food Guide Pyramid for a class project. The model is a square pyramid with a base edge of 4 inches and a height of 5 inches. Find the volume of plaster needed to make the model. Round to the nearest tenth.

d. STADIUMS The Pyramid Arena in Memphis, Tennessee, is a square pyramid 321 feet tall. The base has 600-foot sides. Find the volume of the pyramid.

Example 1 Find the volume of each pyramid. Round to the nearest tenth if necessary.

1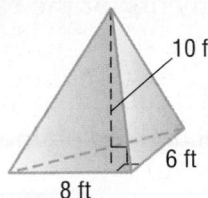
10 ft
6 ft
8 ft

2.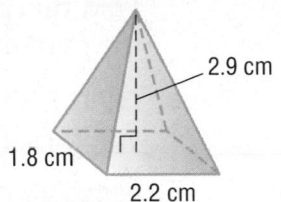
2.9 cm
1.8 cm
2.2 cm

Example 2 Find the height of each pyramid.

3. square pyramid: volume 1,024 cm³, base edge 16 cm

4. triangular pyramid: volume 48 in³, base edge 9 in., base height 4 in.

Example 3 **5. BUILDINGS** The Transamerica Pyramid is the tallest skyscraper in San Francisco. The rectangular base has a length of 175 feet and a width of 120 feet. The height is 853 feet. Find the volume of the building.

Practice and Problem Solving

 = **Step-by-Step Solutions** begin on page R1.
Extra Practice begins on page EP2.

Example 1 Find the volume of each pyramid. Round to the nearest tenth if necessary.

6.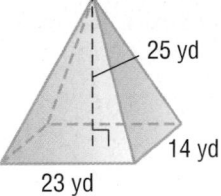
25 yd
14 yd
23 yd

7.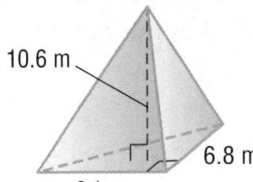
10.6 m
6.8 m
9.1 m

8.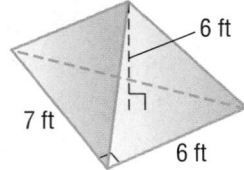
6 ft
7 ft
6 ft

Example 2 Find the height of each pyramid.

9. rectangular pyramid: volume 448 in³, base edge 12 in., base length 8 in.

10. triangular pyramid: volume 270 cm³, base edge 15 cm, height of base 4 cm

11. square pyramid: volume 297 ft³, area of the base 81 ft²

12. hexagonal pyramid: volume 1,320 ft³, area of the base 120 ft²

Example 3 **13 GLASS** A glass pyramid has a height of 4 inches. Its rectangular base has a length of 3 inches and a width of 2.5 inches. Find the volume of glass used to create the pyramid.

14. HISTORY An ancient stone pyramid has a height of 13.6 meters. The edges of the square base are 16.5 meters. Find the volume of the stone pyramid.

15. MATH IN THE MEDIA Find an example of a pyramid in a newspaper or magazine, or on the Internet. Determine the pyramid's volume.

16. MEASUREMENT A rectangular pyramid has a length of 14 centimeters, a width of 9 centimeters, and a height of 10 centimeters. Explain the effect on the volume if each dimension were doubled.

17. **OPEN ENDED** A rectangular pyramid has a volume of 160 cubic feet. Find two possible sets of measurements for the base area and height of the pyramid.

18. **CHALLENGE** A square pyramid and a cube have the same bases and volumes. How are their heights related? Explain.

19. **REASONING** The two figures shown have congruent bases. How does the volume of the two square pyramids in Figure B compare to the volume of the square pyramid in Figure A?

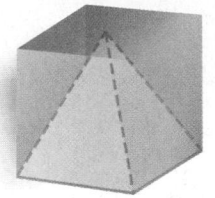
Figure A

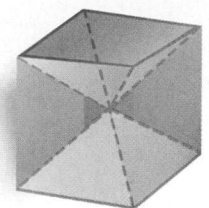
Figure B

20.  **WRITE MATH** Suppose the height of a square pyramid is doubled. Explain what happens to the volume of the pyramid. Support your reasoning with an example.

Test Practice

21. The rectangular pyramid has a volume of 1,560 cubic inches. What is the height of the pyramid?

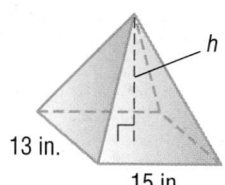

13 in.
15 in.

A. 8 in. **C.** 30 in.

B. 24 in. **D.** 48 in.

22. Find the volume of the rectangular pyramid. Round to the nearest tenth.

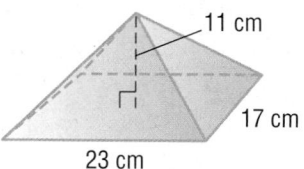

11 cm
17 cm
23 cm

F. 4,301 cm³ **H.** 1,433.7 cm³

G. 2,867.3 cm³ **I.** 716.3 cm³

Find the volume of each cylinder or prism. Round to the nearest tenth.
(Lessons 1B and 1C)

23.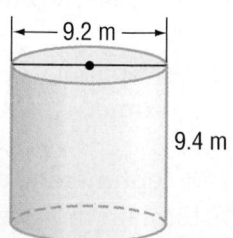

9.2 m
9.4 m

24.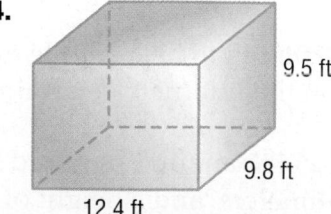

9.5 ft
9.8 ft
12.4 ft

25.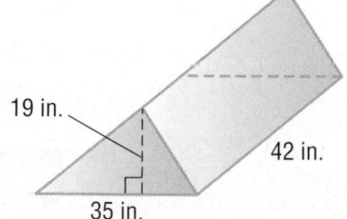

19 in.
42 in.
35 in.

Main Idea
Find the volume of cones.

Vocabulary
cone

 7.G.4

Volume of Cones

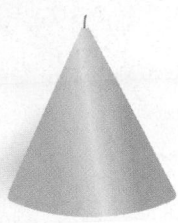

CANDLE MAKING Grace and Elle are making candles to donate for a school fundraiser. The mold they are using is 6 inches tall and has a radius of 3 inches.

1. What would be the approximate volume of the candle if it was a cylinder, but had the same radius and height?

2. **MAKE A CONJECTURE** What fraction of the cylinder is the cone?

A **cone** is a three-dimensional figure with one circular base. A curved surface connects the base and vertex.

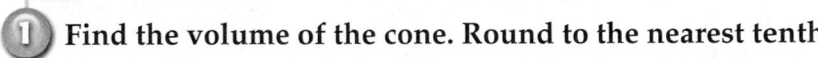
Key Concept — Volume of a Cone

Words The volume V of a cone with radius r is one third the area of the base B times the height h.

Model

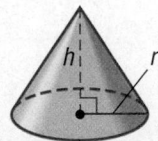

Symbols $V = \frac{1}{3}Bh$ or $V = \frac{1}{3}\pi r^2 h$

Read Math

The $\approx$ symbol is read *is about equal to*.

EXAMPLE — Volume of a Cone

1 Find the volume of the cone. Round to the nearest tenth.

$V = \frac{1}{3}\pi r^2 h$ Volume of a cone

$V = \frac{1}{3} \cdot \pi \cdot 3^2 \cdot 6$ $r = 3, h = 6$

$V \approx 56.5$ Simplify.

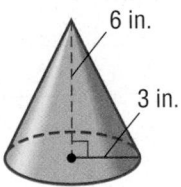
6 in.
3 in.

The volume is about 56.5 cubic inches.

CHECK Your Progress

Find the volume of each cone. Round to the nearest tenth.

a.
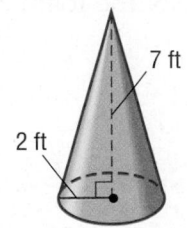
7 ft
2 ft

b.
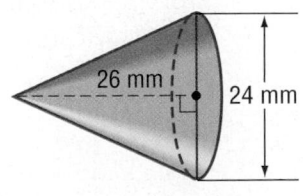
26 mm
24 mm

 REAL-WORLD EXAMPLE

2 **PAPER CUPS** A cone-shaped paper cup is filled with water. The height of the cup is 7 centimeters and the diameter is 6 centimeters. If one cubic centimeter is equal to one milliliter, how many milliliters does the paper cup hold?

$V = \frac{1}{3}\pi r^2 h$ Volume of a cone

$V = \frac{1}{3} \cdot \pi \cdot 3^2 \cdot 7$ $r = 3, h = 7$

$V \approx 65.97$ Simplify.

The paper cup holds about 66 milliliters.

 CHECK Your Progress

c. PIÑATAS April is filling six identical cones for her piñata. Each cone has a radius of 1.5 inches and a height of 9 inches. What is the total volume of the cones? Round to the nearest tenth.

Real-World Link · · · · ·
The piñata originated in Mexico hundreds of years ago. Traditionally, they are made in the shape of stars and human or animal figures.

✓ **CHECK Your Understanding**

Example 1 Find the volume of each cone. Round to the nearest tenth.

1.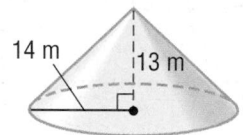
14 m   13 m

2.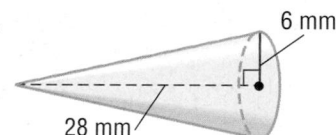
6 mm
28 mm

3 height: 8.4 feet, diameter: 3.5 feet

4. height: 120 millimeters, radius: 45 millimeters

Example 2 **5. FUNNELS** Austin is using the funnel shown to fill a glass bottle with colored sand. If one cup is about 14.4 cubic inches, about how many cups of sand will fill the funnel at one time? Round to the nearest tenth.

5 in.

4 in.

6. FLOWERS Madison is creating a floral centerpiece by attaching artificial flowers to the foam cone shown. What is the volume of the foam cone? Round to the nearest tenth.

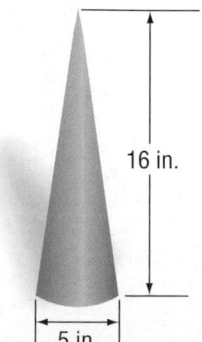

16 in.

5 in.

Practice and Problem Solving

= **Step-by-Step Solutions** begin on page R1.
Extra Practice begins on page EP2.

Example 1   **Find the volume of each cone. Round to the nearest tenth.**

7. 23 mm   14 mm

8. 3.8 ft   1.1 ft

9. 13.4 mm   15.9 mm

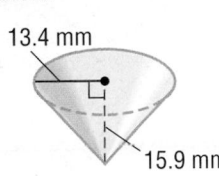

10. height: 3.9 yards, radius: 1.7 yards

11. height: 24 centimeters, diameter: 8 centimeters

12. height: 15 inches, diameter: 5 inches

Example 2   **13. HATS** A party hat like the one at the right is going to be filled with candy. If one cup is about 14.4 cubic inches, about how many cups of candy will fit in the party hat?

14. ICE Isaiah is making cone-shaped ice cubes by using a mold. The radius of the mold is 1.5 inches and the height is 2 inches. If one cubic inch is about 0.55 ounce, how many ounces will ten cone-shaped ice cubes weigh? Round to the nearest tenth.

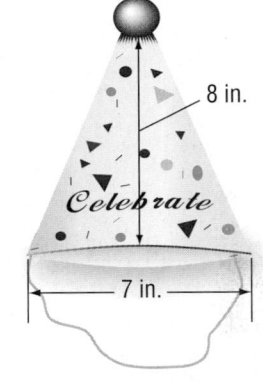

8 in.

Celebrate

7 in.

15. GEOMETRY The volume of a cone with a 30-millimeter radius is 9,420 cubic millimeters. What is the height of the cone to the nearest millimeter?

16. The volume of a cone is 593.46 cubic inches. The radius is 9 inches. Find the height of the cone to the nearest inch.

17 A cylinder has a radius of 5 centimeters and a height of 12 centimeters. What would the height of a cone need to be if it has the same volume and radius? Round to the nearest centimeter.

18. VOLCANOES Mount Rainier, a cone-shaped volcano in Washington, is about 4.4 kilometers tall and about 18 kilometers across its base. Find the volume of Mount Rainier to the nearest whole number.

Find the height of each cone. Round to the nearest tenth.

19.
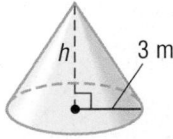
h 3 m
Volume: 42.39 m³

20.
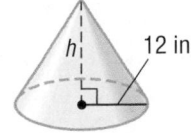
h 12 in.
Volume: 1,205.76 in³

21.
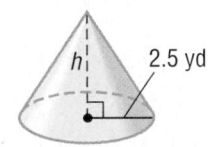
h 2.5 yd
Volume: 19.625 yd³

22. GEOMETRY The volume of a cone with a height of 18 millimeters is 471 cubic millimeters. Find the area of the base to the nearest tenth.

23. **FIND THE ERROR** Aisha is finding the volume of rice that will fill a cone-shaped decorative vase. The vase is 6 inches tall with a 4-inch diameter. Find her mistake and correct it.

$$V = \frac{1}{3}\pi r^2 h$$
$$V = \frac{1}{3}\pi \cdot 4 \cdot 4 \cdot 6$$
$$V \approx 100.5 \text{ in}^3$$

24. **OPEN ENDED** Draw and label two cones with different dimensions but the same volume.

25. **REASONING** Which would have a greater effect on the volume of a cone: doubling its radius or doubling its height? Explain.

26. ⟦▨⟧ **WRITE MATH** Dario and Divya are simplifying $\pi \cdot 5^2$. Dario rounds π to 3.14 and Divya uses the π key on her calculator. Which student's calculation is closer to the exact value? Why?

Test Practice

27. Which is closest to the volume of the cone shown?

    A. 564.4 cm³
    B. 666.7 cm³
    C. 886.5 cm³
    D. 1,238.2 cm³

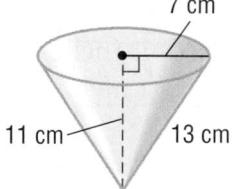

7 cm

11 cm   13 cm

28. Which is closest to the volume of ice cream that the cone can hold?

    F. 47.1 in³
    G. 23.55 in³
    H. 15.7 in³
    I. 11.8 in³

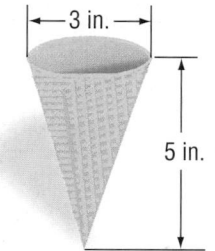

├─ 3 in. ─┤

5 in.

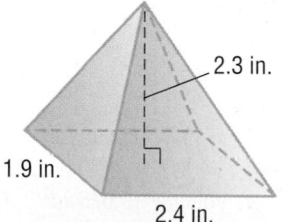
Spiral Review

Find the volume of each pyramid. Round to the nearest tenth. (Lesson 1E)

29.

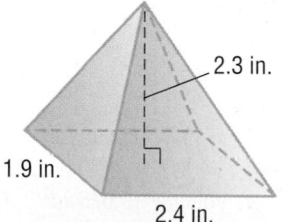

2.3 in.

1.9 in.

2.4 in.

30.

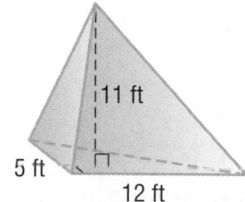

11 ft

5 ft

12 ft

31. Find the volume of a cylinder with a radius of 2.9 meters and a height of 4.9 meters. Round to the nearest tenth. (Lesson 1C)

Mid-Chapter Check

Find the volume of each prism. Round to the nearest tenth if necessary. (Lesson 1B)

1.

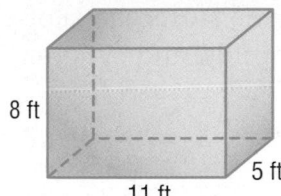

8 ft
5 ft
11 ft

2.

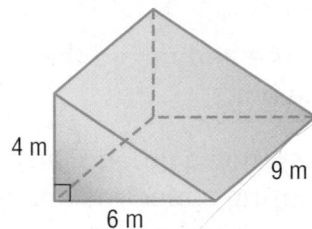

4 m
9 m
6 m

3. CAKE A piece of cake is shaped like a triangular prism. The area of the base is 6 square inches and the height is 3 inches. Find the volume of the piece of cake. (Lesson 1B)

4. Find the volume of the cylinder. Round to the nearest tenth. (Lesson 1C)

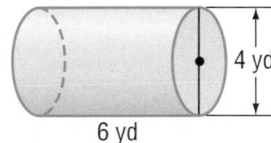

4 yd
6 yd

5. FOOD A can of peanuts has a diameter of 4 inches and a height of 4 inches. Find the volume of the peanut can. Round to the nearest tenth. (Lesson 1C)

Find the volume of each pyramid. Round to the nearest tenth. (Lesson 1E)

6.

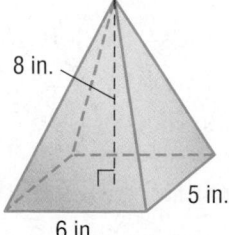

8 in.
5 in.
6 in.

7.

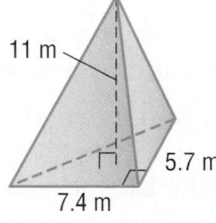

11 m
5.7 m
7.4 m

8. MULTIPLE CHOICE A triangular pyramid has a triangle base of 24 centimeters, triangle height of 16 centimeters, and a pyramid height of 14 centimeters. Which is closest to the volume of the triangular prism in cubic meters? (Lesson 1E)

A. 0.00085 m^3 **C.** 0.00095 m^3

B. 0.0009 m^3 **D.** 0.001 m^3

9. Find the volume of the cone. Round to the nearest tenth. (Lesson 1F)

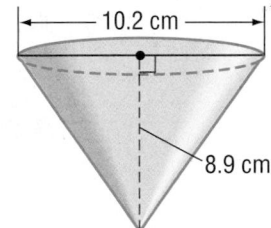

10.2 cm
8.9 cm

10. TRAFFIC CONE An engineer wants to make a traffic cone with a diameter of 12 inches and a height of 16 inches. To make the cone, she needs to know its volume. Find the volume of the traffic cone to the nearest tenth. (Lesson 1F)

11. MULTIPLE CHOICE Find the volume of the cone. Round to the nearest tenth. (Lesson 1F)

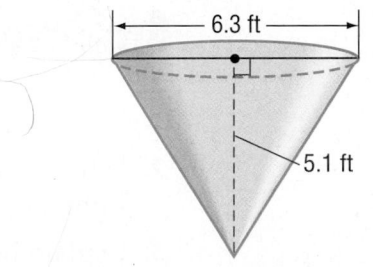

6.3 ft
5.1 ft

F. 212.0 ft^3 **H.** 53.0 ft^3

G. 172.1 ft^3 **I.** 52.0 ft^3

Explore **Nets of Three-Dimensional Figures**

Main Idea

Find the surface area of prisms and cylinders using models and nets.

Vocabulary

net

Get ConnectED

CCSS 7.G.6

Nets are two-dimensional patterns of three-dimensional figures. When you construct a net, you are decomposing the three-dimensional figure into separate figures. You can use a net to find the surface area of three-dimensional figures such as prisms, pyramids, and cylinders.

⚡ ACTIVITY

1. **STEP 1** Use an empty cereal box. Measure and record the length, width, and height of the box.

   STEP 2 Label the top and bottom faces using a green marker. Label the front and back faces using a blue marker and label the left and right faces using a red marker.

   STEP 3 Carefully cut along three edges of the top face and then cut down each vertical edge.

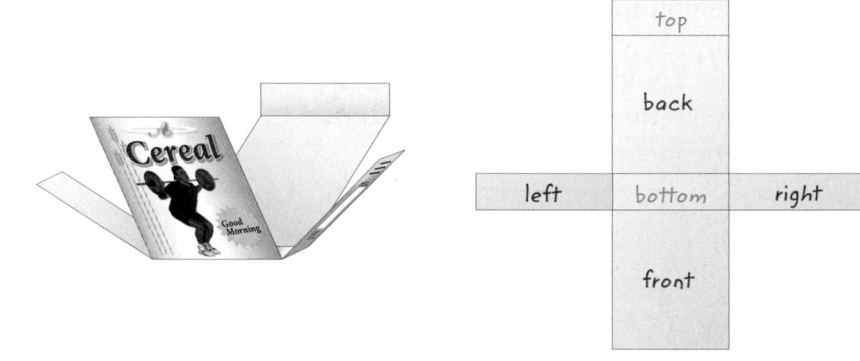

Analyze the Results

1. Name the shape(s) that make up the net.

2. What do you notice about the top and bottom faces, the left and right faces, and the front and back faces?

3. Explain how you could find the total area of the rectangles.

4. **MAKE A CONJECTURE** Write the formula for the surface area of the box using the net. Justify your answer.

2 **STEP 1** Use a soup can. Measure and record the height of the can.

STEP 2 Label the top and bottom faces using a blue marker. Label the curved side using a red marker.

STEP 3 Carefully peel off the label of the soup can and tape it to a piece of paper. Trace the top and bottom faces so they are adjacent to the label.

Analyze the Results

5. Name the shapes that make up the net of the container.

6. Find the diameter of the top of the container and use it to find the perimeter or circumference of that face. Find the area of the top and bottom faces.

7. How is the circumference of the top face of the container related to the rectangle?

8. How could the area of the label be found with the circumference of the top face of the container? Calculate the area of the label.

9. Add the area of the label to the sum of the areas of the two circular bases.

10. **MAKE A CONJECTURE** Write a formula for finding the area of all the surfaces of a cylinder given the measures of its height and the diameter of one of its bases.

Practice and Apply

11. **CAKES** Cake decorators use sheets of icing called fondant. Desa is decorating a rectangular cake that is 24 inches by 18 inches by 4 inches. What is the area of the sheet of fondant that she will need to cover the exposed parts of the cake?

12. **FIRE PIT** James is forging a cylindrical iron fire pit. He wants the pit to be 3 feet in diameter and 2 feet high with the bottom closed and the top open. How much iron will be used to forge the pit? Round to the nearest tenth.

Main Idea
Find the surface areas of prisms.

Vocabulary
surface area

 **7.G.6**

Surface Area of Prisms

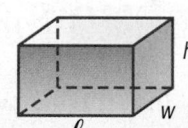 **Explore**

Use the cubes to build a rectangular prism with a length of 8 centimeters. Count the number of squares on the outside of the prism. The sum is the *surface area*.

1. Record the dimensions, volume, and surface area in a table.

2. Build two more prisms using 8 cubes. For each, record the dimensions, volume, and surface area.

3. Describe the prisms with the greatest and least surface areas.

The sum of the areas of all the surfaces, or faces, of a three-dimensional figure is the **surface area**.

> **Key Concept**   **Surface Area of a Rectangular Prism**
>
> **Words**   The surface area *S.A.* of a rectangular prism with base ℓ, width w, and height h is the sum of the areas of its faces.
>
> **Model**
>
> **Symbols** $S.A. = 2\ell h + 2\ell w + 2hw$

EXAMPLES **Surface Area of Rectangular Prisms**

① **Find the surface area of the rectangular prism.**

There are three pairs of congruent faces.

- top and bottom
- front and back
- two sides

Faces	Area
top and bottom	$2(5 \cdot 4) = 40$
front and back	$2(5 \cdot 3) = 30$
two sides	$2(3 \cdot 4) = 24$
sum of the areas	$40 + 30 + 24 = 94$

The surface area is 94 square centimeters.

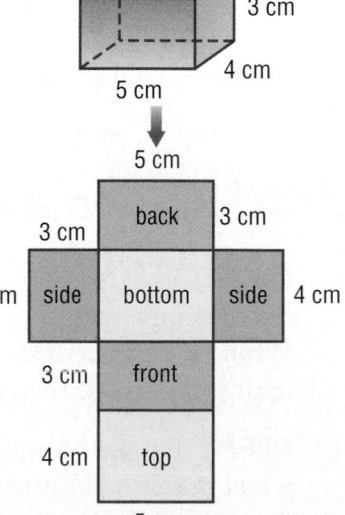

2 Find the surface area of the rectangular prism.

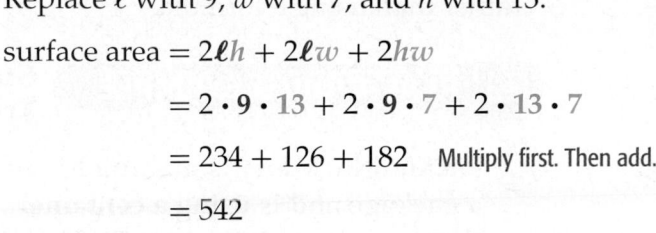

Replace ℓ with 9, w with 7, and h with 13.

surface area $= 2\ell h + 2\ell w + 2hw$

$\qquad = 2 \cdot 9 \cdot 13 + 2 \cdot 9 \cdot 7 + 2 \cdot 13 \cdot 7$

$\qquad = 234 + 126 + 182$ Multiply first. Then add.

$\qquad = 542$

The surface area of the prism is 542 square inches.

 **CHECK Your Progress**

Find the surface area of each rectangular prism.

a.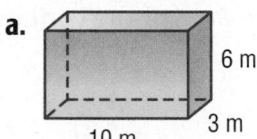

b.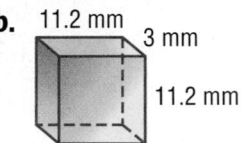

Study Tip

Surface Area When you find the surface area of a three-dimensional figure, the units are square units, not cubic units.

 REAL-WORLD EXAMPLE

3 **PAINTING** Domingo built a toy box 60 inches long, 24 inches wide, and 36 inches high. He has 1 quart of paint that covers about 87 square feet of surface. Does he have enough to paint the toy box? Justify your answer.

Step 1 Find the surface area of the toy box.

Replace ℓ with 60, w with 24, and h with 36.

surface area $= 2\ell h + 2\ell w + 2hw$

$\qquad = 2 \cdot 60 \cdot 36 + 2 \cdot 60 \cdot 24 + 2 \cdot 36 \cdot 24$

$\qquad = 8{,}928 \text{ in}^2$

Step 2 Find the number of square inches the paint will cover.

$1 \text{ ft}^2 = 1 \text{ ft} \times 1 \text{ ft}$ Replace 1 ft with 12 in.

$\qquad = 12 \text{ in.} \times 12 \text{ in.}$ Multiply.

$\qquad = 144 \text{ in}^2$

So, 87 square feet is equal to 87×144 or 12,528 square inches.

Since $12{,}528 > 8{,}928$, Domingo has enough paint.

Study Tip

Consistent Units Since the surface area of the toy box is expressed in inches, convert 87 ft² to square inches so that all measurements are expressed using the same units.

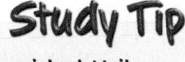 **CHECK Your Progress**

c. **BOXES** The largest corrugated cardboard box ever constructed measured about 23 feet long, 9 feet high, and 8 feet wide. Would 950 square feet of paper be enough to cover the box? Justify your answer.

To find the surface area of a triangular prism, it is more efficient to find the area of each face and calculate the sum of all of the faces rather than using a formula.

 **REAL-WORLD EXAMPLE**

Surface Area of Triangular Prisms

Vocabulary Link

Everyday Use

Base the bottom of an object

Math Use

Base one of two parallel congruent faces of a prism

4 **PACKAGING** Marty is mailing his aunt a package and is using a container that is a triangular prism. The height is 14 inches. The triangular base has a height of 3 inches and a base of 4 inches. How much cardboard is used to create the shipping container?

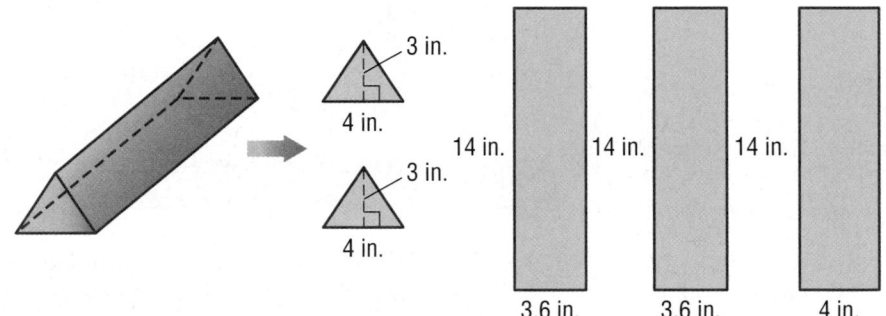

To find the surface area of the container, find the area of each face and add.

The area of each triangle is $\frac{1}{2} \cdot 4 \cdot 3$ or 6.

The area of two of the rectangles is $14 \cdot 3.6$ or 50.4. The area of the third rectangle is $14 \cdot 4$ or 56.

The sum of the areas of the faces is $6 + 6 + 50.4 + 50.4 + 56$ or 168.8 cubic inches.

Study Tip

Area When finding the area of a rectangle, the side lengths can be referred to as length ℓ and width w or base b and height h.

$A = \ell w$ or $A = bh$

CHECK Your Progress

Find the surface area of each triangular prism.

d.

21 mm 15 mm 12 mm 9 mm

e.

5 in. 5 in. 10 in. 5 in. 5 in. 7 in.

f.

2.5 cm 2 cm 4 cm 3 cm

g.

5 m 3 m 6 m 4 m

🔊 **584** Volume and Surface Area

Examples 1 and 2 Find the surface area of each rectangular prism. Round to the nearest tenth if necessary.

1.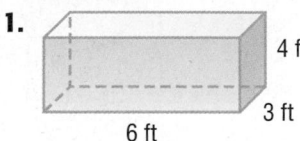
4 ft
3 ft
6 ft

2.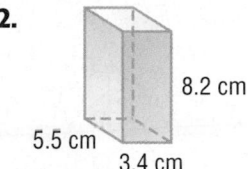
8.2 cm
5.5 cm
3.4 cm

Example 3 **3. GIFTS** Marsha is wrapping a gift. She places it in a box 8 inches long, 2 inches wide, and 11 inches high. If Marsha bought a roll of wrapping paper that is 1 foot wide and 2 feet long, did she buy enough paper to wrap the gift? Justify your answer.

Example 4 **4. INSULATION** The attic shown is a triangular prism. Insulation will be placed inside all walls, not including the floor. Find the surface area that will be covered with insulation.

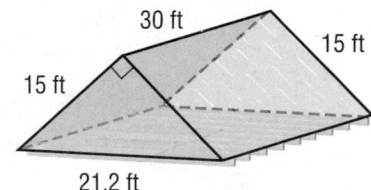
30 ft
15 ft
15 ft
21.2 ft

Practice and Problem Solving

● = **Step-by-Step Solutions** begin on page R1.
Extra Practice begins on page EP2.

Examples 1 and 2 Find the surface area of each rectangular prism. Round to the nearest tenth if necessary.

5.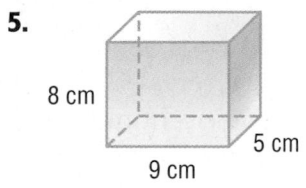
8 cm
5 cm
9 cm

6.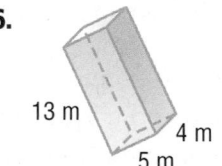
13 m
4 m
5 m

7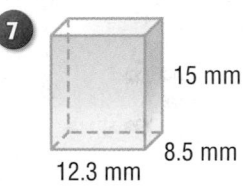
15 mm
8.5 mm
12.3 mm

8.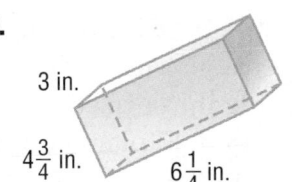
12 ft
1.7 ft
6.4 ft

9.
3 in.
$4\frac{3}{4}$ in.
$6\frac{1}{4}$ in.

10.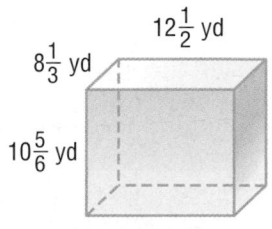
$12\frac{1}{2}$ yd
$8\frac{1}{3}$ yd
$10\frac{5}{6}$ yd

Example 3 **11. BOOKS** When making a book cover, Anwar adds an additional 20 square inches to the surface area to allow for overlap. How many square inches of paper will Anwar use to make a book cover for a book 11 inches long, 8 inches wide, and 1 inch high?

12. FENCES If one gallon of paint covers 350 square feet, will 8 gallons of paint be enough to paint the inside and outside of the fence shown once? Explain.

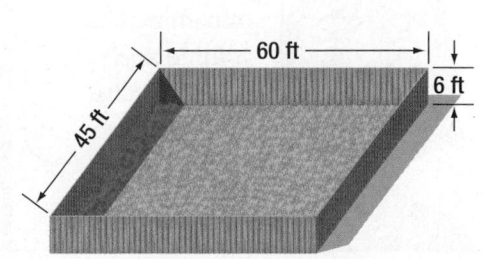
60 ft
6 ft
45 ft

Example 4 Find the surface area of each triangular prism. Round to the nearest tenth if necessary.

13.

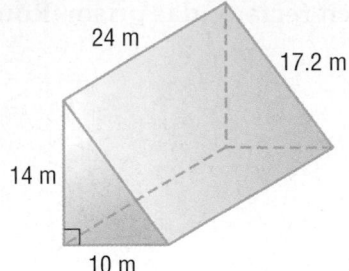

24 m
17.2 m
14 m
10 m

14.

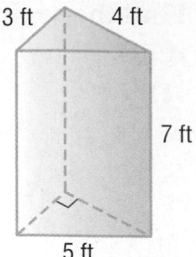

3 ft   4 ft
7 ft
5 ft

15.

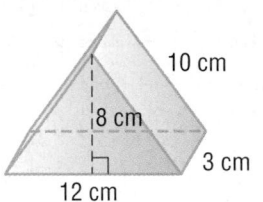

10 cm
8 cm
3 cm
12 cm

16.

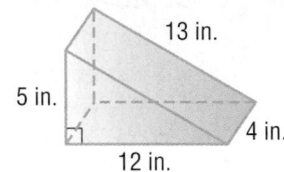

13 in.
5 in.
4 in.
12 in.

17. MUSIC To the nearest tenth, find the approximate amount of plastic covering the outside of the CD case.

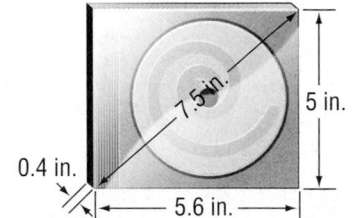

7.5 in.   5 in.
0.4 in.   5.6 in.

18. MEASUREMENT What is the surface area of a rectangular prism that has a length of 6.5 centimeters, a width of 2.8 centimeters, and a height of 9.7 centimeters?

19 ALGEBRA Write a formula for the surface area *S.A.* of a cube in which each side measures *x* units.

20. PACKAGING A company will make a cereal box with whole number dimensions and a volume of 100 cubic centimeters. If cardboard costs $0.05 per 100 square centimeters, what is the least cost to make 100 boxes?

21. GRAPHIC NOVEL Refer to the graphic novel frame below. What dimensions would allow the students to maximize the volume while keeping the surface area at most 160 square feet? Excess metal is permitted. Explain your reasoning.

We are designing a dunk tank. Remember, we want to maximize the volume and minimize the surface area.

22. **REASONING** The bottom and sides of a pool in the shape of a rectangular prism will be painted blue. The length, width, and height of the pool are 18 feet, 12 feet, and 6 feet, respectively. Explain why the number of square feet to be painted is *not* equivalent to the expression $2(18)(12) + 2(18)(6) + 2(12)(6)$.

23. **REASONING** Determine if the following statement is *true* or *false*. Explain your reasoning.

    If you double one of the dimensions of a rectangular prism, the surface area will double.

24. **CHALLENGE** The figure at the right is made by placing a cube with 12-centimeter sides on top of another cube with 15-centimeter sides. Find the surface area.

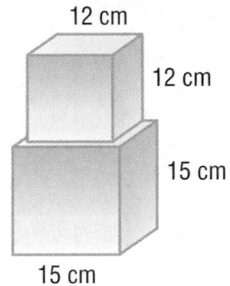
12 cm
12 cm
15 cm
15 cm

25. **WRITE MATH** Explain why the surface area of a three-dimensional figure is measured in square units rather than in cubic units.

Test Practice

26. Which of the following expressions represents the surface area of a cube with side length w?

    A. w^3

    B. $6w^2$

    C. $6w^3$

    D. $2w + 4w^2$

27. How much cardboard is needed to make the box shown?

    F. $37.5\ ft^2$

    G. $24.4\ ft^2$

    H. $8\ ft^2$

    I. $6.1\ ft^2$

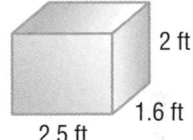
2 ft
1.6 ft
2.5 ft

Find the volume of each cone. Round to the nearest tenth. (Lesson 1F)

28.

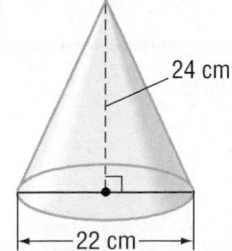
24 cm
22 cm

29.

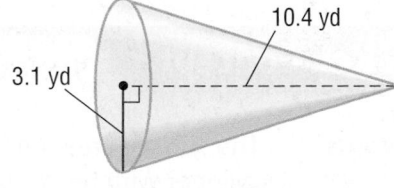
10.4 yd
3.1 yd

30. **LANDSCAPING** Pat is digging circular post holes for a new fence. It is recommended that the holes are at least 10 inches across and 18 inches deep. What is the least amount of dirt she is removing for each hole? Round to the nearest tenth. (Lesson 1D)

Main Idea
Find the surface area of a cylinder.

Get ConnectED

CCSS 7.G.4, 7.G.6

Surface Area of Cylinders

Explore

Step 1 Trace the top and bottom of the can on grid paper. Then cut out the shapes.

Step 2 Cut a long rectangle from the grid paper. The width of the rectangle should be the same as the height of the can. Wrap the rectangle around the can. Cut off the excess paper so that the edges just meet.

1. Make a net of the cylinder.
2. Name the shapes in the net.
3. How is the length of the rectangle related to the circles?

You can put two circles and a rectangle together to make a cylinder.

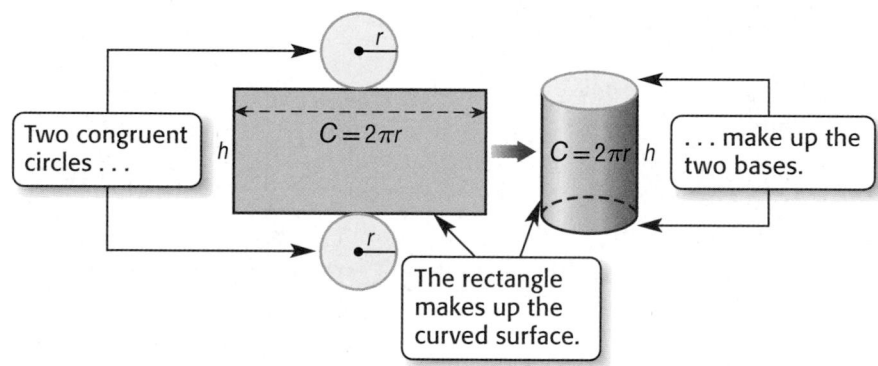

In the diagram above, the length of the rectangle is the same as the circumference of the circle, $2\pi r$. Also, the width of the rectangle is the same as the height of the cylinder.

Key Concept   **Surface Area of a Cylinder**

Words   The surface area *S.A.* of a cylinder with height *h* and radius *r* is the sum of the area of the curved surface and the areas of the circular bases.

Model

Symbols $S.A. = 2\pi rh + 2\pi r^2$

EXAMPLE **Surface Area of a Cylinder**

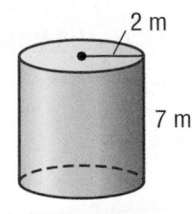

① Find the surface area of the cylinder. Round to the nearest tenth.

$$S.A. = 2\pi rh + 2\pi r^2 \qquad \text{Surface area of a cylinder}$$

$$= 2\pi(2)(7) + 2\pi(2)^2 \quad \text{Replace } r \text{ with 2 and } h \text{ with 7.}$$

$$\approx 113.1 \qquad\qquad \text{Simplify.}$$

The surface area is about 113.1 square meters.

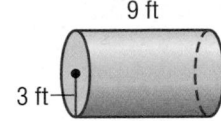

CHECK Your Progress

a. Find the surface area of the cylinder. Round to the nearest tenth.

 REAL-WORLD EXAMPLE

② **CAROUSELS** A circular fence that is 2 feet high is to be built around the outside of a carousel. The distance from the center of the carousel to the edge of the fence will be 35 feet. How much fencing material is needed to make the fence around the carousel?

The radius of the circular fence is 35 feet. The height is 2 feet.

$$S.A. = 2\pi rh \qquad \text{Curved surface of a cylinder}$$

$$= 2\pi(35)(2) \quad \text{Replace } r \text{ with 35 and } h \text{ with 2.}$$

$$\approx 439.8 \qquad \text{Simplify.}$$

So, about 439.8 square feet of material is needed to make the fence.

CHECK Your Progress

b. **DESIGN** Find the area of the label of a can of tuna with a radius of 5.1 centimeters and a height of 2.9 centimeters. Round to the nearest tenth.

 Real-World Link · · · ·
Of the 3,000 to 4,000 wooden carousels carved in America between 1885 and 1930, fewer than 150 operate today.

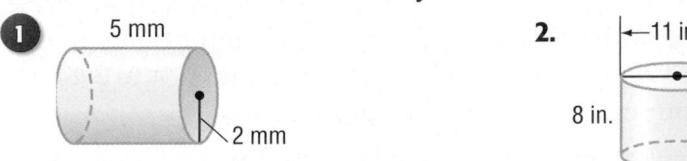 **Your Understanding**

Example 1 Find the surface area of each cylinder. Round to the nearest tenth.

① 5 mm / 2 mm

2. ←11 in.→ / 8 in.

Example 2 3. **STORAGE** The height of a water tank is 10 meters, and it has a diameter of 10 meters. What is the surface area of the tank? Round to the nearest tenth.

Practice and Problem Solving

● = **Step-by-Step Solutions** begin on page R1.
Extra Practice begins on page EP2.

Example 1 **Find the surface area of each cylinder. Round to the nearest tenth.**

4.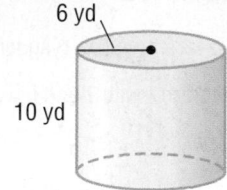
6 yd
10 yd

5.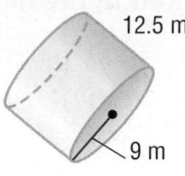
12.5 m
9 m

6.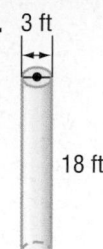
3 ft
18 ft

7.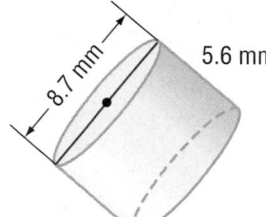
8.7 mm
5.6 mm

8.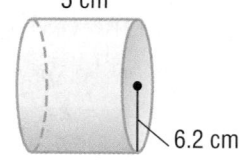
5 cm
6.2 cm

9.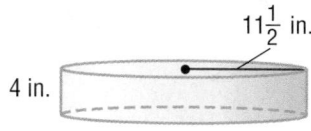
$11\frac{1}{2}$ in.
4 in.

Example 2 **10. CANDLES** A cylindrical candle has a diameter of 4 inches and a height of 7 inches. To the nearest tenth, what is the surface area of the candle?

11. PENCILS Find the surface area of an unsharpened cylindrical pencil that has a radius of 0.5 centimeter and a height of 19 centimeters. Round to the nearest tenth.

ESTIMATION **Estimate the surface area of each cylinder.**

12.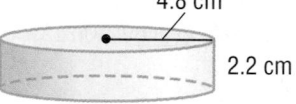
4.8 cm
2.2 cm

13.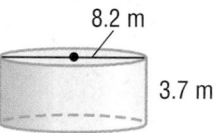
8.2 m
3.7 m

14.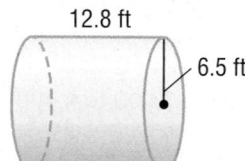
12.8 ft
6.5 ft

15 PACKAGING The mail tube shown is made of cardboard and has plastic end caps. Approximately what percent of the surface area of the mail tube is cardboard?

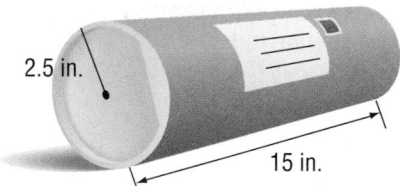
2.5 in.
15 in.

H.O.T. Problems

16. CHALLENGE If the height of a cylinder is doubled, will its surface area also double? Explain your reasoning.

17. **WRITE MATH** Suppose you are simplifying an expression involving π. Will it make a difference if you round π to 3.14 or use the pi key on your calculator? Justify your reasoning.

18. REASONING Which has a greater surface area: a cylinder with radius 6 centimeters and height 3 centimeters or a cylinder with radius 3 centimeters and height 6 centimeters? Explain your reasoning.

19. Stacey has a cylindrical paper clip holder with the net shown.

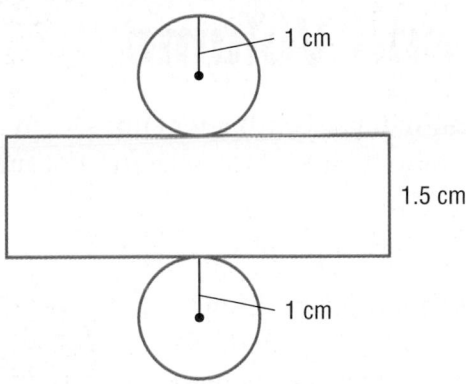

Which is closest to the surface area of the cylindrical paper clip holder?

A. 12.7 cm^2

B. 13.7 cm^2

C. 14.7 cm^2

D. 15.7 cm^2

20. The four containers below each hold about the same amount of liquid. Which container has the greatest surface area?

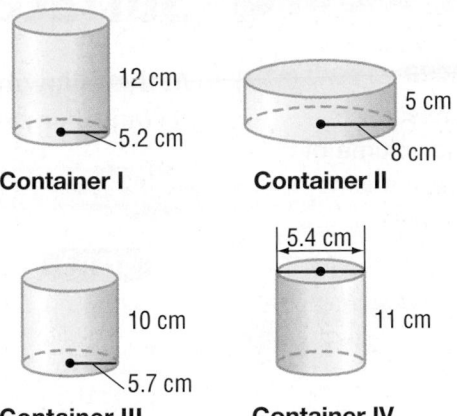

F. Container I

G. Container II

H. Container III

I. Container IV

Spiral Review ....

Find the surface area of each prism. (Lesson 2B)

21.

22.

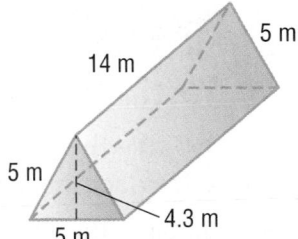

23. TREAT BAGS Mrs. Jones is filling cone-shaped treat bags with candy. Each bag has a height of 6 inches and a radius of 0.75 inch. What is the volume of each bag? Round to the nearest tenth. (Lesson 1F)

Find the volume of each pyramid. Round to the nearest tenth if necessary. (Lesson 1E)

24. rectangular pyramid: base, 12 inches by 5 inches; height, 13 inches

25. rectangular pyramid: base, 28 meters by 4 meters; height, 15 meters

26. hexagonal pyramid: area of base, 212 cm^2; height, 17 cm

27. MEASUREMENT Find the volume of a rectangular prism with a length of 5.5 meters, a width of 4 meters, and a height of 7.2 meters. (Lesson 1B)

Extend

Surface Area and Volume

Main Idea

Compare surface area and volume of rectangular prisms and cylinders.

 **Get ConnectED**

CCSS 7.G.6

In the following Activity, you will use centimeter cubes to create rectangular prisms and compare their surface area and volume.

 **ACTIVITY**

STEP 1 Create a rectangular prism using 8 centimeter cubes. Copy the table below, record the dimensions, and find the surface area and volume of the prism.

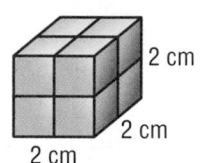
2 cm
2 cm
2 cm

Rectangular Prism	Length	Width	Height	Surface Area	Volume
1	2	2	2	24 cm^2	8 cm^3
2					
3					

STEP 2 Repeat Step 1 for as many different rectangular prisms as you can create with 8 cubes.

Analyze the Results

1. What do you notice about the volume of each rectangular prism?

2. Give the dimensions of the rectangular prism with the greatest surface area. What is the surface area?

3. Give the dimensions of the rectangular prism with the least surface area. What is the surface area?

4. Compare the two rectangular prisms below that have the same volume. What is the volume? Which figure has the lesser surface area?

Figure 1

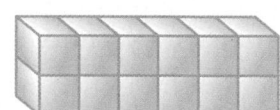
Figure 2

5. **MAKE A CONJECTURE** Suppose you have two prisms with the same volume. One is shaped like a cube and the other is "longer." Which has the lesser surface area? Explain.

Practice and Apply

Compare the two figures that have the same volume. Then determine which one has a greater surface area. Explain.

6.

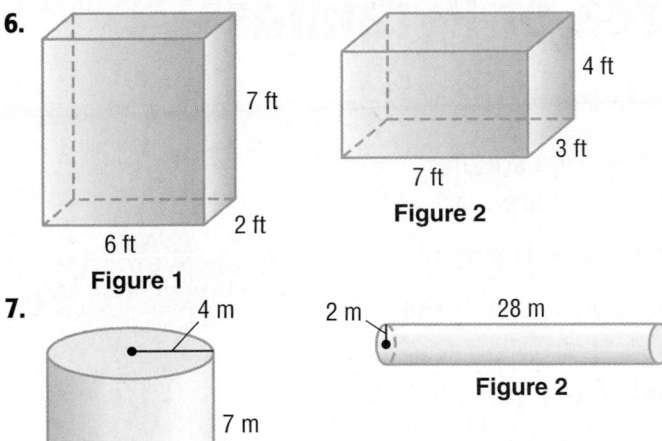

Figure 1
Figure 2

7.

Figure 1

Figure 2

8. PACKAGING Chandler wants to build a box that will hold as many of his collector's coins as possible. He wants the volume of the box to be 20 cubic inches. The dimensions of possible boxes are given in the table below. Find the surface area of each box. Then determine which box has the least amount of surface area.

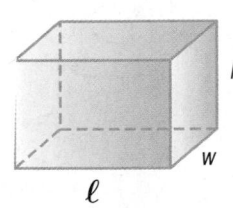

Box	(ℓ)	(w)	(h)	Surface Area	Volume (in³)
1	20	1	1	▪	20
2	4	5	1	▪	20
3	2	2	5	▪	20
4	10	2	1	▪	20

9. CRAFTS Monique sews together pieces of fabric to make gift boxes. If she only uses whole numbers, what are the dimensions of a box with a volume of 50 cubic inches that has the greatest amount of surface area?

10. CONTAINERS Thomas is creating a decorative container to fill with colored sand. What will be the dimensions of the rectangular prism that will hold 100 cubic inches with the least amount of surface area, if he only uses whole numbers? The top of the container is open.

11. BOXES The specifications of a cardboard box indicate that it has the same volume as a rectangular box 4 inches by 10 inches by 12 inches, but with less surface area. What size box would meet these requirements?

12. COOKING Zachariah needs to melt a stick of butter 5 inches by 1 inch by 1 inch in a pan. Explain why cutting the butter into smaller pieces will help the butter melt faster.

Surface Area of Pyramids

Explore

- Copy the following net of a square pyramid onto a piece of paper.

- Cut out the net and tape it together.

1. What shapes make up the net of the square pyramid?

2. What do you notice about the triangular faces of the pyramid?

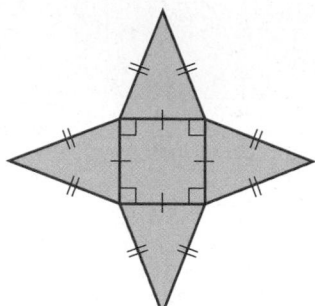

A right square pyramid has a square base and four isosceles triangles that make up the lateral faces. The height of each lateral face is called the **slant height**. The **lateral surface area** of a solid is the sum of the areas of all its lateral faces.

Model of Square Pyramid   **Net of Square Pyramid**

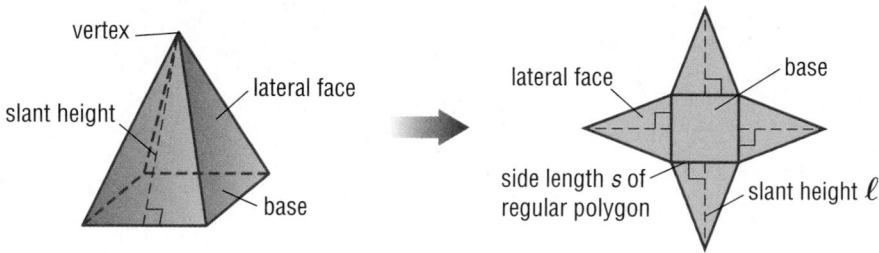

To find the lateral area L of a regular pyramid, refer to the net. The lateral area is the sum of the areas of the triangles.

$L = 4\left(\frac{1}{2}s\ell\right)$ Area of the lateral faces

$L = \frac{1}{2}(4s)\ell$ Commutative Property of Multiplication

$L = \frac{1}{2}P\ell$ The perimeter of the square base P is $4s$.

The total surface area of a regular pyramid is the lateral surface area L plus the area of the base B.

$$S.A. = B + \frac{1}{2}P\ell$$

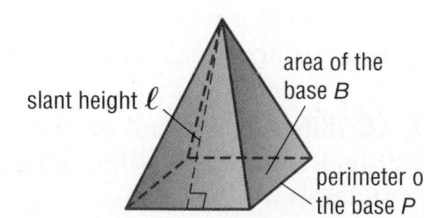

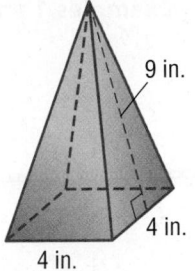

EXAMPLES **Surface Area of a Pyramid**

QUICK Review

Perimeter of a Square
$P = 4s$

(1) Find the total surface area of the pyramid. Round to the nearest tenth.

$S.A. = B + \frac{1}{2}P\ell$ Surface area of a pyramid

$S.A. = 16 + \frac{1}{2}(16 \cdot 9)$ $B = 4 \cdot 4$, $P = 4 \cdot 4$ or 16, $\ell = 9$

$S.A. = 88$ Simplify.

The surface area is 88 square inches.

9 in.

4 in.

4 in.

(2) Find the total surface area of the pyramid with a base area of 111 square meters.

$S.A. = B + \frac{1}{2}P\ell$ Surface area of a pyramid

$S.A. = 111 + \frac{1}{2}(48 \cdot 20)$ $B = 111$, $P = 16 + 16 + 16$ or 48, $\ell = 20$

$S.A. = 591$ Simplify.

The surface area of the pyramid is 591 square meters.

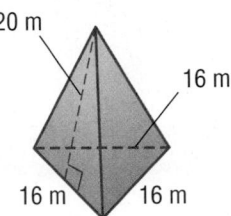

20 m

16 m

16 m   16 m

CHECK Your Progress

a. Find the surface area of a square pyramid that has a slant height of 8 centimeters and a base length of 5 centimeters.

REAL-WORLD EXAMPLE

(3) **GIFT BOXES** Rachel is making gift boxes in the shape of square pyramids for party favors. They have a slant height of 3 inches and base edges 2.5 inches long. How many square inches of card stock are used to make one gift box?

$S.A. = B + \frac{1}{2}P\ell$ Surface area of a pyramid

$S.A. = 6.25 + \frac{1}{2}(10 \cdot 3)$ $B = 2.5^2$ or 6.25, $P = 4(2.5)$ or 10, $\ell = 3$

$S.A. = 21.25$ Simplify.

So, 21.25 square inches of card stock are used to make one gift box.

 Real-World Link · · · · ·
Half of the paper America consumes is used to wrap and decorate consumer products.

CHECK Your Progress

b. **PERFUME** Amado purchased a bottle of perfume that is in the shape of a square pyramid. The slant height of the bottle is 4.5 inches and the base is 2 inches. Find the surface area.

Chapter 10 Lesson 2E Surface Area   **595**

Examples 1 and 2 **Find the total surface area of each pyramid. Round to the nearest tenth.**

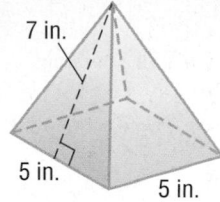
7 in.
5 in.   5 in.

2.
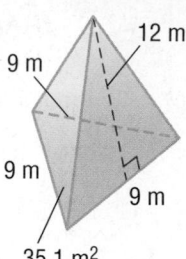
12 m
9 m
9 m
9 m
35.1 m²

Example 3 **3. MONUMENTS** The Washington Monument is an obelisk with a square pyramid top. The slant height of the pyramid is 55.5 feet, and the square base has sides of 34.5 feet. Find the lateral area of the pyramid.

Practice and Problem Solving

 = **Step-by-Step Solutions** begin on page R1.
Extra Practice begins on page EP2.

Examples 1 and 2 **Find the total surface area of each pyramid. Round to the nearest tenth.**

4.
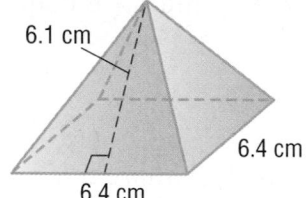
6.1 cm
6.4 cm
6.4 cm

5.
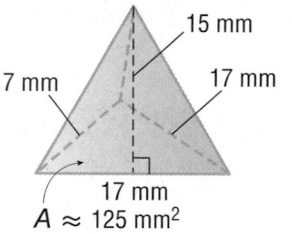
15 mm
17 mm   17 mm
17 mm
$A \approx 125$ mm²

6.
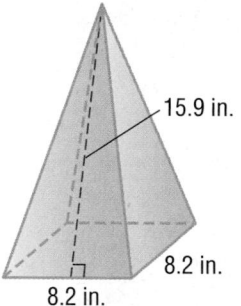
15.9 in.
8.2 in.
8.2 in.

7. The base of a square pyramid has a side length of 27 centimeters. The slant height is 25 centimeters. Find the surface area.

8. A triangular pyramid has a slant height of 0.75 foot. The equilateral triangular base has a perimeter of 1.2 feet and an area of 0.6 square foot. Find the surface area.

9. A square pyramid has a slant height of $4\frac{2}{3}$ feet. The base has side lengths of $2\frac{1}{4}$ feet. Find the surface area.

Example 3 **10. GEMSTONES** The gemstone shown has a base that is a square pyramid with sides 3.4 inches long. The slant height of the pyramid is 3.8 inches. Find the surface area of the gemstone.

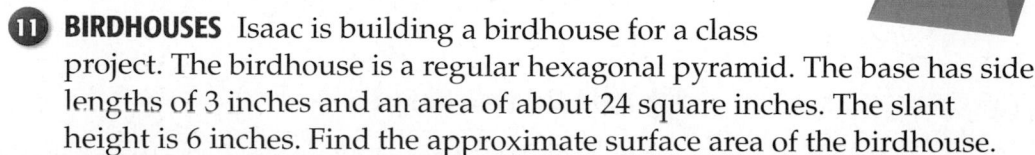

11 **BIRDHOUSES** Isaac is building a birdhouse for a class project. The birdhouse is a regular hexagonal pyramid. The base has side lengths of 3 inches and an area of about 24 square inches. The slant height is 6 inches. Find the approximate surface area of the birdhouse.

12. GEOMETRY A square pyramid has a surface area of 175 square inches. The square base has side lengths of 5 inches. Find the slant height of the pyramid.

13. **CHALLENGE** Suppose you could climb to the top of the Great Pyramid of Giza in Egypt. Which path would be shorter, climbing a lateral edge or the slant height? Justify your response.

14. **OPEN ENDED** Draw a square pyramid and a rectangular pyramid. Explain the differences between the two.

15. **WRITE MATH** Justify the formula for the surface area of a pyramid.

Test Practice

16. The We Entertain company is constructing a tent in the shape of a square pyramid, without a floor, to be used at a party. Find the number of square feet of fabric that will be required.

    A. 1,500 ft²
    B. 1,700 ft²
    C. 2,250 ft²
    D. 2,550 ft²

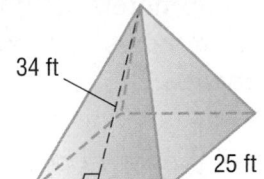

17. Find the surface area of the triangular pyramid shown with a base area of 27.7 square yards.

    F. 39.3 yd²
    G. 117.9 yd²
    H. 171.7 yd²
    I. 213.5 yd²

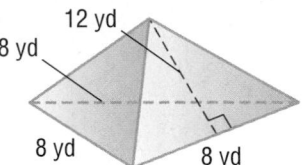

Spiral Review

18. **CRAFTS** LaToya is decorating the side of a cylindrical can. The can has a diameter of 6 inches and a height of 12 inches. To the nearest tenth, what is the minimum amount of construction paper that LaToya needs? (Lesson 2C)

Find the surface area of each prism. (Lesson 2B)

19.
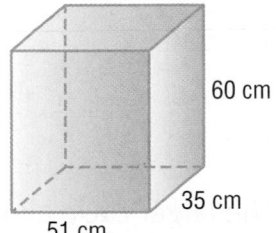

20.
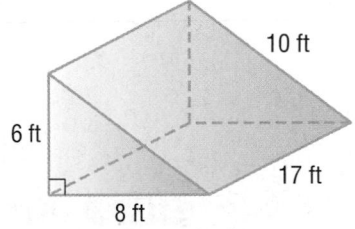

Find the volume of each cone. Round to the nearest tenth. (Lesson 1F)

21.
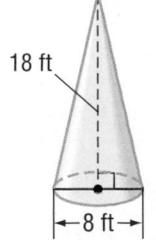

22.
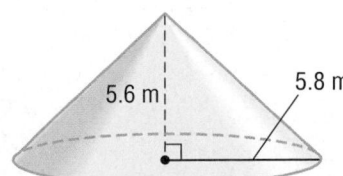

Extend Net of a Cone

Main Idea

Justify the formula for the surface area of a cone by using a net.

CCSS 7.G.4

HATS Corinne is making a party hat for her little sister. It is in the shape of a cone and will be covered with tissue paper. What is the surface area of the cone that will be covered with tissue paper?

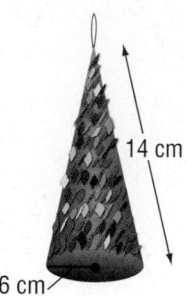

ACTIVITY Construct a Net of a Cone

1 STEP 1 Use a compass to draw two circles slightly touching, one with a radius of 6 centimeters and one with a radius of 14 centimeters.

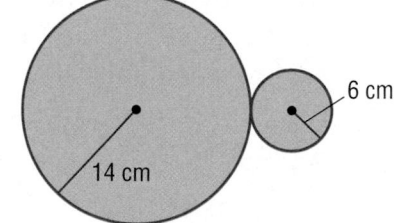

STEP 2 Only a part of the larger circle is needed to make the cone. Use a proportion to find the part.

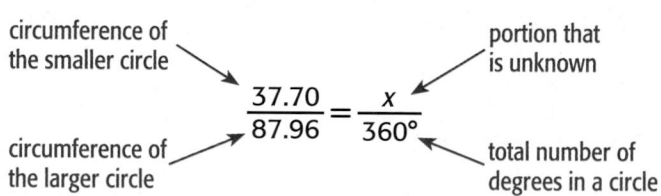

Solve the proportion.

$$\frac{37.70}{87.96} = \frac{x}{360°}$$ Write the proportion.

$$x(87.96) = 13{,}572$$ Cross multiply. $37.70 \cdot 360 = 13{,}572$

$$\frac{x \cdot 87.96}{87.96} = \frac{13{,}572}{87.96}$$ Divide each side by 87.96.

$$x \approx 154$$ Simplify.

You need 154° of the larger circle.

STEP 3 Cut a central angle of 154° from the larger circle and make a cone.

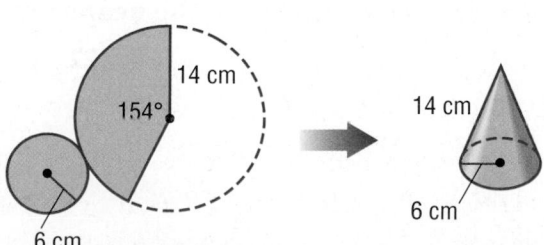

The net shows that the surface area of a cone is the sum of its base B and its lateral area $L.A.$ The base B is a circle. The lateral area $L.A.$ is *part* of a larger circle.

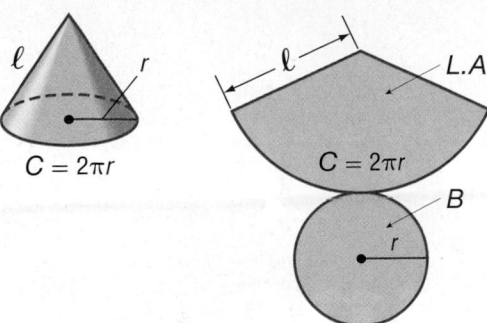

In the following Activity, you will determine the formula for the lateral surface area of a cone.

ACTIVITY Find the Surface Area of a Cone

2 STEP 1 Use a compass to draw a circle. It represents the lateral surface area.

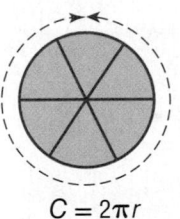

STEP 2 Draw 3 diameter lines that divide the circle equally into 6 sections.

STEP 3 Cut out the sections and form a figure that resembles a parallelogram.

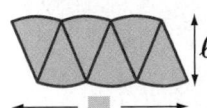

Analyze the Results

1. The circumference of the circle is represented by $2\pi r$. What expression represents the length of the parallelogram in Step 3?

2. **MAKE A CONJECTURE** Use the expression from Exercise 1 to write a formula for the area of the parallelogram, which is the lateral surface area of the cone.

3. **MAKE A CONJECTURE** Write a formula for the total surface area of a cone.

4. Why is only a portion of the larger circle necessary to construct a cone?

5. If the radius of the base is increased while the slant height stays the same, how will that affect the lateral surface area?

6. If a cone's slant height is decreased, which would be affected more: the base or the lateral area? Justify your response.

7. Find the surface area of the party hat that Corinne is covering with tissue paper.

Volume and Surface Area of Composite Figures

BASKETS The Rockwell family uses the basket shown on their staircase. Mrs. Rockwell would like to make a duplicate for extra storage. She needs to determine the surface area to find how much material she will need and the volume to find how much storage it has.

1. What three-dimensional figures make up the basket?

2. What method could you use to find the volume and surface area of the basket?

The basket above is made up of two rectangular prisms. The volume of a composite figure can be found by separating the figure into solids whose volumes you know how to find.

EXAMPLE **Volume of a Composite Figure**

① **BASKETS** Find the volume of the staircase basket above.

Find the volume of each prism.

$V = \ell wh$ $V = \ell wh$

$V = 8 \cdot 6 \cdot 16$ or 768 $V = 8 \cdot 6 \cdot 8$ or 384

The volume is about 768 + 384 or 1,152 cubic inches.

CHECK Your Progress

a. Find the volume of the composite figure.

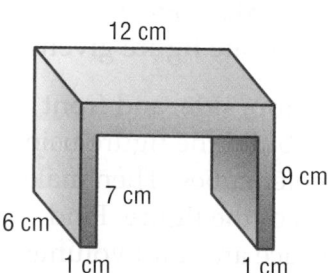

EXAMPLE **Volume of a Composite Figure**

2 **Find the volume of the composite figure. Round to the nearest tenth.**

The figure is made up of a cylinder and a cone.

$$V = \pi r^2 h + \frac{1}{3}\pi r^2 h$$

$$V = \pi \cdot 5^2 \cdot 3 + \frac{1}{3} \cdot \pi \cdot 5^2 \cdot 4$$

$$V \approx 235.6 + 104.7 \text{ or } 340.3$$

The volume of the composite figure is about 340.3 cubic centimeters.

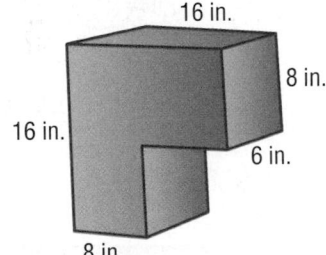

You can also find the surface area of composite figures by finding the areas of the faces that make up the composite figure.

REAL-WORLD EXAMPLE **Surface Area of a Composite Figure**

3 **BASKETS** Find the surface area of the staircase basket in Example 1.

The basket's surface is made up of three different polygons.

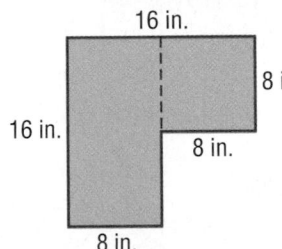

$A = \ell w + \ell w$

$A = (8 \cdot 16) + (8 \cdot 8)$

$A = 128 + 64 \text{ or } 192$

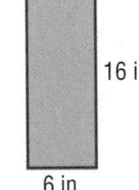

$A = \ell w$

$A = 6 \cdot 16$

$A = 96$

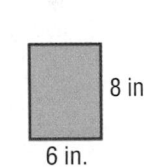

$A = \ell w$

$A = 6 \cdot 8$

$A = 48$

The total surface area of the basket is $2(192) + 2(96) + 4(48)$ or 768 square inches.

 CHECK Your Progress

b. Find the surface area of the composite figure.

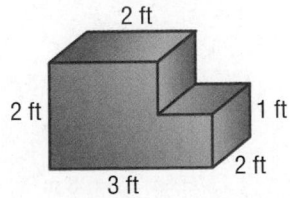

Examples 1 and 2 Find the volume of each composite figure. Round to the nearest tenth if necessary.

1.

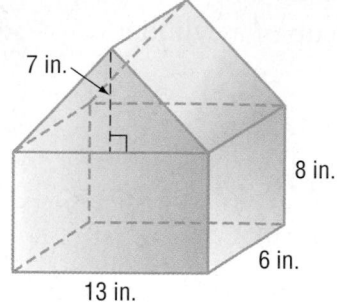

7 in.
8 in.
6 in.
13 in.

2.

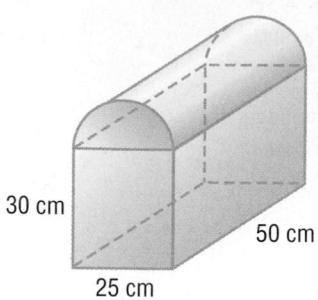

30 cm
50 cm
25 cm

Example 3 Find the surface area of each composite figure.

3.

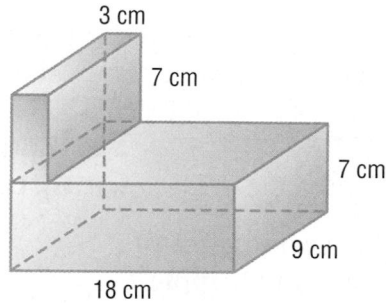

3 cm
7 cm
7 cm
9 cm
18 cm

4.

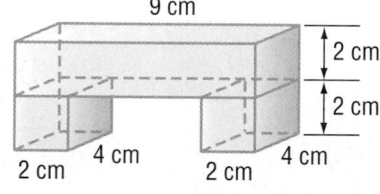

9 cm
2 cm
2 cm
2 cm 4 cm 2 cm 4 cm

5. DIGITAL CAMERA A digital camera is 1 inch by 3 inches by 4 inches. The lens on the front face is 0.5 inch long with a 1-inch diameter. What is the surface area of the digital camera? Round to the nearest tenth.

Practice and Problem Solving

 = **Step-by-Step Solutions** begin on page R1.
Extra Practice begins on page EP2.

Examples 1 and 2 Find the volume of each composite figure. Round to the nearest tenth if necessary.

6.

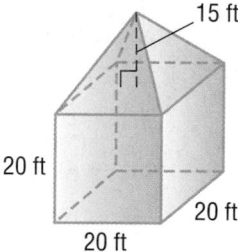

15 ft
20 ft
20 ft
20 ft

7

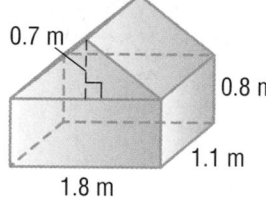

0.7 m
0.8 m
1.1 m
1.8 m

8.

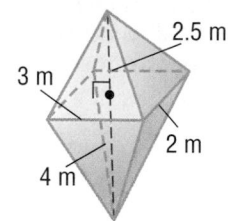

2.5 m
3 m
2 m
4 m

Example 3 Find the surface area of each composite figure. Round to the nearest tenth if necessary.

9.

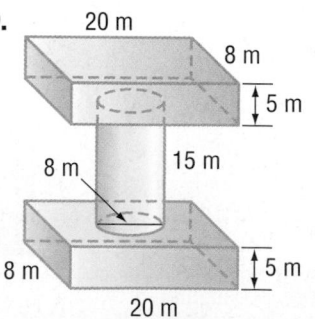

20 m
8 m
5 m
15 m
8 m
8 m
5 m
20 m

10.

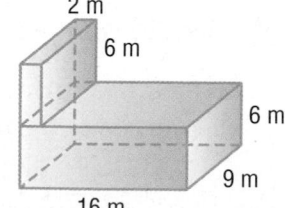

2 m
6 m
6 m
9 m
16 m

11.

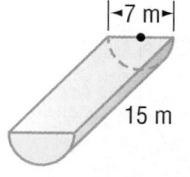

|◄7 m►|
15 m

12. **SWIMMING POOLS** The swimming pool at the right is being filled with water. Find the number of cubic feet that it will take to fill the swimming pool. (*Hint*: The area of a trapezoid is $A = \frac{1}{2}h(b_1 + b_2)$.)

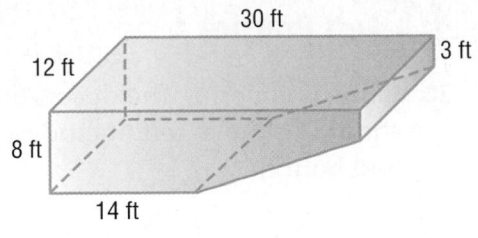

13. **BOXES** Charlotte wants to make the box shown. What is the surface area of the box? Round to the nearest tenth.

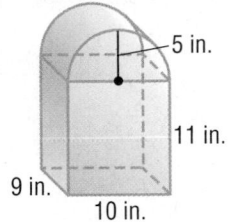

14. **FOOD** A carryout container is shown. The bottom base is a 4-inch square and the top base is a 4-inch by 6-inch rectangle. The height of the container is 5 inches. Find the volume of food that it holds to the nearest tenth.

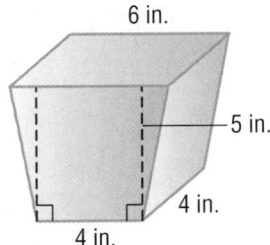

Study Tip

Measurement Conversions
In Exercise 15, remember to convert inches to feet. 12 inches = 1 foot

15. **GEOMETRY** Find the volume of the figure at the right in cubic feet. Round to the nearest tenth.

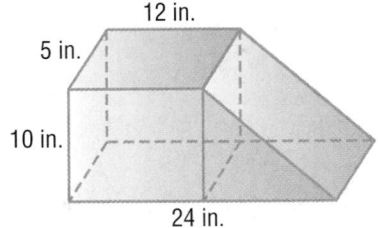

H.O.T. Problems

16. **OPEN ENDED** Draw a composite figure that is made up of a cylinder and two cones. Label its dimensions and find the volume of the figure.

17. **CHALLENGE** Give an example of a composite figure that has a volume between 250 and 300 cubic units.

18. **FIND THE ERROR** Seth is finding the surface area of the composite figure shown. Find his mistake and correct it.

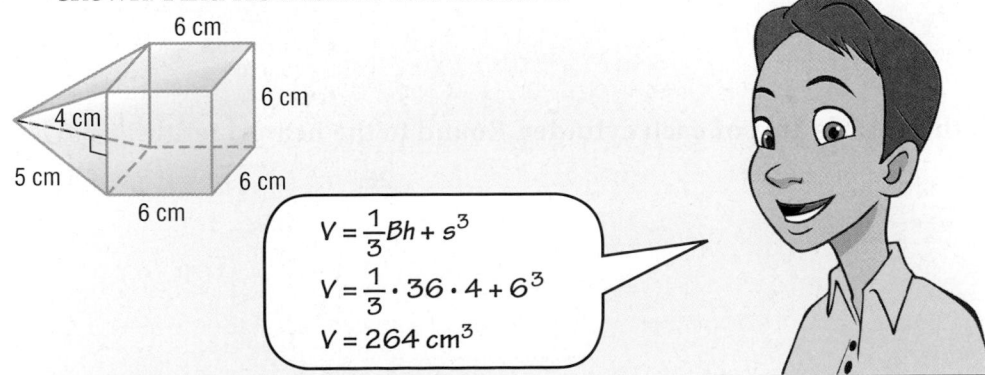

$$V = \frac{1}{3}Bh + s^3$$
$$V = \frac{1}{3} \cdot 36 \cdot 4 + 6^3$$
$$V = 264 \text{ cm}^3$$

19. **WRITE MATH** How is finding the surface area of a half-cylinder different from finding the surface area of a cylinder?

20. Jaime is covering the decorative wall art shown below in felt, including the back and bottom.

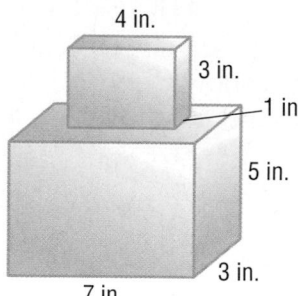

4 in.

3 in.

1 in.

5 in.

3 in.

7 in.

What is the total area to be covered with felt?

A. 23 in²

B. 117 in²

C. 172 in²

D. 1,260 in²

21. Which of the following formulas cannot be used to find the volume of the composite figure?

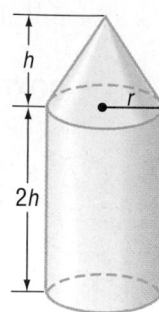

h

r

$2h$

F. $V = 2\pi r^2 h + \frac{1}{3}\pi r^2 h$

G. $V = \frac{7}{3}\pi r^2 h$

H. $V = \pi r^2(2h) + \frac{1}{3}\pi r^2 h$

I. $V = \frac{2}{3}\pi r^2 h$

Spiral Review

22. MONEY Over the weekend, Mr. Lobo spent $534. Of that, about 68% was spent on groceries. About how much money was *not* spent on groceries? Use the *solve a simpler problem* strategy. (Lesson 3A)

Find the surface area of each pyramid. (Lesson 2E)

23.

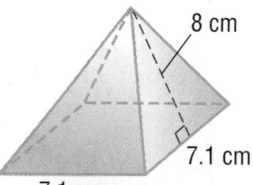

8 cm

7.1 cm

7.1 cm

24.

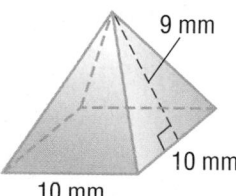

9 mm

10 mm

10 mm

Find the surface area of each cylinder. Round to the nearest tenth. (Lesson 2C)

25.

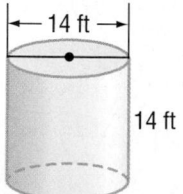

14 ft

14 ft

26.

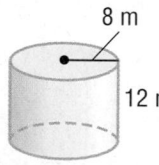

8 m

12 m

27. MEASUREMENT A cylinder has a height of 10 feet and a diameter of 8 feet. What is the volume of the cylinder? Round to the nearest tenth. (Lesson 1C)

FOLDABLES®
Study Organizer

Be sure the following Key Concepts are noted in your Foldable.

Key Concepts

Volume (Lesson 1)
- rectangular prism: $V = Bh$, where $B = \ell w$
- triangular prism: $V = Bh$, where $B = \frac{1}{2}bh$
- cylinder: $V = Bh$, where $B = \pi r^2$

cone

pyramid

$$V = \frac{1}{3}Bh$$

Surface Area (Lesson 2)
- square pyramid

$S.A. = B + \frac{1}{2}P\ell$

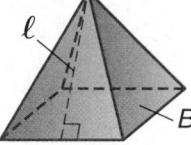

- rectangular prism

$S.A. = 2\ell w + 2\ell h + 2hw$

- cylinder

$S.A. = 2\pi rh + 2\pi r^2$

Composite Figures (Lesson 3)
- A *composite figure* is made up of two or more three-dimensional shapes.

Key Vocabulary

composite figures	pyramid
cone	rectangular prism
cylinder	slant height
lateral surface area	surface area
net	triangular prism
prism	volume

Vocabulary Check

Choose the correct term to complete each sentence.

1. A (rectangular prism, rectangle) is a three-dimensional figure that has three sets of parallel, congruent sides.

2. The (volume, surface area) of a three-dimensional figure is the measure of the space occupied by it.

3. Volume is measured in (square, cubic) units.

4. A (cylinder, prism) is a three-dimensional figure that has two congruent, parallel circles as its bases.

5. The formula for the volume of a (rectangular prism, cone) is $V = \frac{1}{3}Bh$.

6. To find the surface area of a (pyramid, cylinder), you must know the measurements of the height and the radius.

7. The volume of a rectangular prism is found by (adding, multiplying) the length, the width, and the height.

8. The formula for the surface area of a (cylinder, rectangular prism) is $S.A. = 2\pi rh + 2\pi r^2$.

Multi-Part Lesson Review

Lesson 1
Volume

Volume of Prisms (Lesson 1B)

Find the volume of each prism. Round to the nearest tenth if necessary.

9.
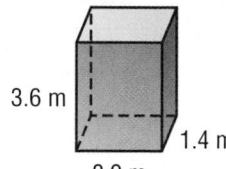
3.6 m
1.4 m
2.9 m

10.
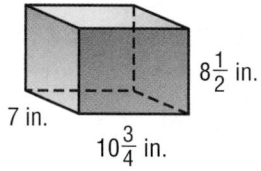
$8\frac{1}{2}$ in.
7 in.
$10\frac{3}{4}$ in.

11. **CEREAL** A box of cereal is 8.5 inches long, 12.5 inches tall, and 3.5 inches wide. What is the maximum amount of cereal the box can contain?

12. **TRUCKS** The dimensions of the bed of a dump truck are length 20 feet, width 7 feet, and height $9\frac{1}{2}$ feet. What is the volume of the bed of the dump truck?

EXAMPLE 1 A local city provides residents with a rectangular container for recycling products. Find the volume of the rectangular container.

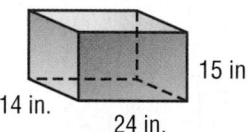
15 in.
14 in.
24 in.

$V = \ell wh$ Volume of a rectangular prism

$V = (24)(14)(15)$ Replace ℓ with 24, w with 14, and h with 15.

$V = 5,040$ Multiply.

The volume of the rectangular container is 5,040 cubic inches.

Volume of Cylinders (Lesson 1C)

Find the volume of each cylinder. Round to the nearest tenth.

13.
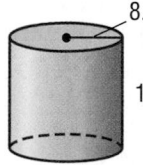
8.7 km
17 km

14.
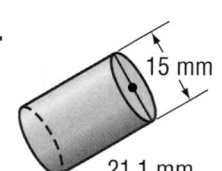
15 mm
21.1 mm

15. **CONTAINERS** A can of soup has a diameter of 3.5 inches and a height of 5 inches. Find the volume of the soup can. Round to the nearest tenth.

16. **COOKIES** Mrs. Delagado stores cookies in a cylinder-shaped jar that has a height of 12 inches and a diameter of 10 inches. Find the volume. Round to the nearest tenth.

EXAMPLE 2 Marquez stores his toys in a cylinder-shaped can like the one shown below. Find the volume of the can. Round to the nearest tenth.

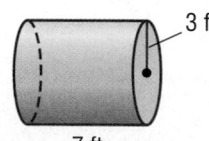
3 ft
7 ft

$V = \pi r^2 h$ Volume of a cylinder

$V = \pi 3^2 (7)$ Replace r with 3 and h with 7.

$V \approx 197.9$ Multiply.

The volume of the cylinder-shaped can is 197.9 cubic feet.

Volume of Pyramids (Lesson 1E)

Find the volume of each pyramid. Round to the nearest tenth.

17.

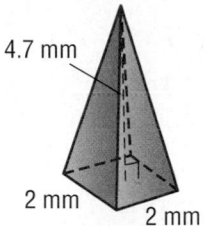

4.7 mm

2 mm 2 mm

18.

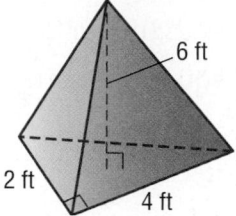

6 ft

2 ft 4 ft

19. CRYSTALS Jess bought a pyramid-shaped crystal to hang in her room. The rectangular base of the crystal measures 9 centimeters by 8 centimeters and has a height of 12 centimeters. What is the volume of the pyramid-shaped crystal?

EXAMPLE 3 Guilia bought a lawn decoration shaped like the pyramid shown below. Find the volume of the pyramid-shaped decoration. Round to the nearest tenth.

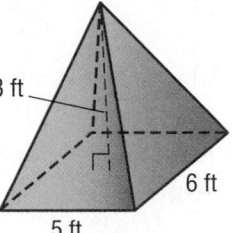

8 ft

6 ft

5 ft

$V = \frac{1}{3}Bh$ Volume of a pyramid

$V = \frac{1}{3}(6 \times 5) \times 8$ Replace B with 6×5 and h with 8.

$V = 80$ Multiply.

The volume of the pyramid-shaped lawn decoration is 80 cubic feet.

Volume of Cones (Lesson 1F)

Find the volume of each cone. Round to the nearest tenth.

20.

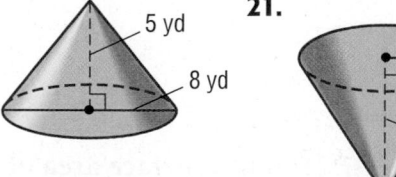

5 yd

8 yd

21.

6 in.

10 in.

22. CONTAINER Mr. Corwin built a cone-shaped storage container with a base 15 inches in diameter and a height of 8 inches. What is the volume of the container? Round to the nearest tenth.

EXAMPLE 4 Find the volume of the cone. Round to the nearest tenth.

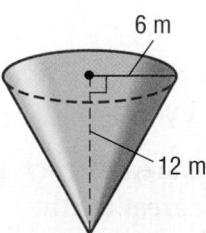

6 m

12 m

$V = \frac{1}{3}\pi r^2 h$ Volume of a cone

$V = \frac{1}{3}\pi 6^2(12)$ Replace r with 6 and h with 12.

$V \approx 452.4$ Simplify.

The volume of the cone is 452.4 cubic meters.

Lesson 2 **Surface Area**

Surface Area of Prisms (Lesson 2B)

Find the surface area of each rectangular prism. Round to the nearest tenth if necessary.

23.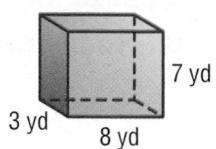
7 yd
3 yd   8 yd

24.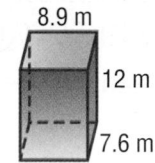
8.9 m
12 m
7.6 m

25. **MOVING** A large wardrobe box is 2.25 feet long, 2 feet wide, and 4 feet tall. How much cardboard was needed to make the box?

EXAMPLE 5 Find the surface area of the rectangular prism.

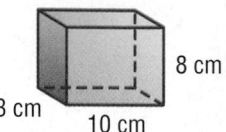
8 cm
3 cm   10 cm

Surface area of rectangular prism

$= 2\ell h + 2\ell w + 2hw$

$= 2(10)(8) + 2(10)(3) + 2(8)(3)$ or 268

The surface area is 268 square centimeters.

Surface Area of Cylinders (Lesson 2C)

Find the surface area of each cylinder. Round to the nearest tenth.

26.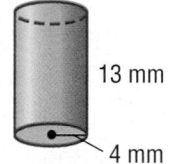
13 mm
4 mm

27.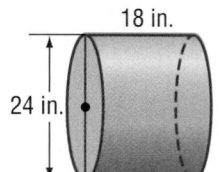
18 in.
24 in.

28. **DESIGN** A can of black beans is $5\frac{1}{2}$ inches high, and its base has a radius of 2 inches. How much paper is needed to make the label on the can? Round to the nearest tenth.

EXAMPLE 6 Find the surface area of the cylinder. Round to the nearest tenth.

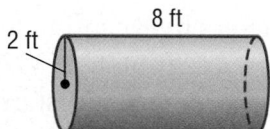
8 ft
2 ft

Surface area of cylinder

$= 2\pi r^2 + 2\pi rh$

$= 2\pi(2)^2 + 2\pi(2)8$ or 125.7

The surface area is about 125.7 square feet.

Surface Area of Pyramids (Lesson 2E)

Find the surface area of each pyramid. Round to the nearest tenth.

29.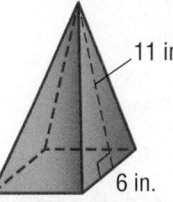
11 in.
6 in.   6 in.

30.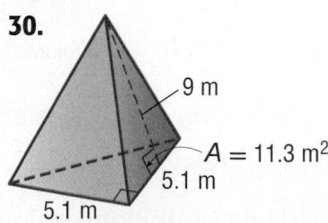
9 m
$A = 11.3$ m²
5.1 m
5.1 m

EXAMPLE 7 Find the surface area of the pyramid. Round to the nearest tenth.

Surface area of pyramid

$= B + \frac{1}{2}P\ell$

$= 25 + \frac{1}{2}(20)(7)$ or 95

The surface area is 95 square meters.

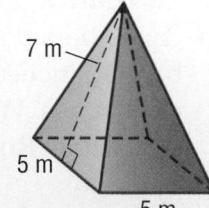
7 m
5 m
5 m

Composite Figures

PSI: Solve a Simpler Problem (Lesson 3A)

31. LAND A rectangular plot of land measures 1,450 feet by 850 feet. A contractor wishes to section off a portion of this land to build an apartment complex. If the complex is 425 feet by 550 feet, how many square feet of land will *not* be sectioned off to build it?

32. TRAVEL Mrs. Whitmore left Chicago at 6:45 A.M. and arrived in St. Louis at 11:15 A.M., driving a distance of approximately 292 miles. Find her approximate average speed.

33. SHOPPING Mercedes spent $175.89 over the weekend. Of the money she spent, 40% was spent on shoes. About how much money was *not* spent on shoes?

EXAMPLE 8 A total of 950 residents voted on whether to build a neighborhood playground. Of those that voted, 70% voted for the playground. How many residents voted for the playground?

Find 10% of 950 and then use the result to find 70% of 950.

10% of 950 = 95

Since there are seven 10s in 70%, multiply 95 by 7.

So, 95 × 7 or 665 residents voted for the playground.

Volume and Surface Area of Composite Figures (Lesson 3C)

Find the volume of each composite figure. Round to the nearest tenth if necessary.

34.

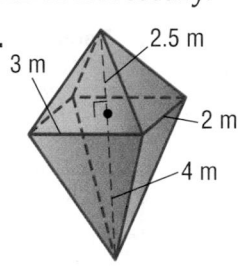
4 ft, 6 ft, 5 ft, 5 ft, 12 ft

35.

2.5 m, 3 m, 2 m, 4 m

36. MAILBOX Berdina is helping her dad paint a mailbox like the one shown. If one quart of paint covers 40 square feet, how many quarts of paint should they buy?

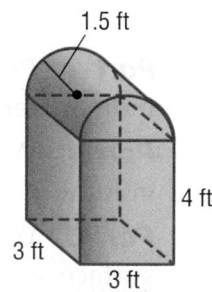
1.5 ft, 4 ft, 3 ft, 3 ft

EXAMPLE 9 Joey built a small greenhouse shown below. What is the volume of the greenhouse?

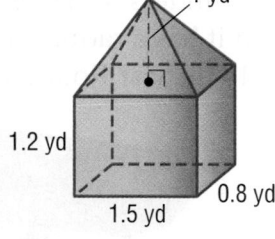
1 yd, 1.2 yd, 1.5 yd, 0.8 yd

Volume of rectangular prism

$= \ell wh$

$= (1.5)(0.8)(1.2)$

$= 1.44 \text{ yd}^3$

Volume of pyramid

$= \frac{1}{3}Bh$

$= \frac{1}{3}(1.2)(1)$

$= 0.4 \text{ yd}^3$

1.44 + 0.4 = 1.84 cubic yards

So, there are 1.84 cubic yards in the greenhouse.

Read each question. Then fill in the correct answer on the answer sheet provided by your teacher or on a sheet of paper.

1. A metal toolbox has a length of 11 inches, a width of 5 inches, and a height of 6 inches. What is the volume of the toolbox?

   A. 22 in³   C. 210 in³

   B. 121 in³   D. 330 in³

2. A bag contains 5 red, 2 yellow, and 8 blue marbles. Xavier removed one blue marble from the bag and did not put it back. He then randomly removed another marble. What is the probability that the second marble removed was blue?

   F. $\frac{8}{14}$ H. $\frac{7}{15}$

   G. $\frac{8}{15}$ I. $\frac{1}{2}$

3. Evelyn has 3 apples to serve to her friends. If Evelyn serves each friend $\frac{1}{3}$ of a whole apple, how many friends can she serve?

   A. 1   C. 9

   B. 3   D. 12

4. ▨ **GRIDDED RESPONSE** Daniel is designing and building a small storage shed. He wants the dimensions of the shed to be one half the dimensions of the shed shown below.

   Storage Shed

   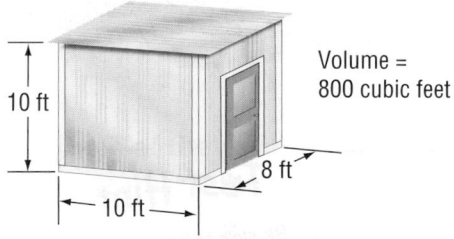

   Volume = 800 cubic feet

   If the dimensions of the shed above are each divided in half, the volume of Daniel's new storage shed will be what fraction of the volume of the original storage shed?

5. ▨ **GRIDDED RESPONSE** Timea ran 3 miles in 19 minutes. At this rate, how many minutes would it take her to run 5 miles?

6. Wilma made a decorative piece shaped like a square pyramid with the dimensions shown.

   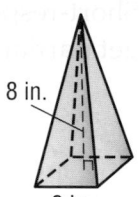

   8 in.

   3 in.

   She wants to double the volume of the piece. Which of the following square pyramid pieces will have a volume that is twice the volume of Wilma's decorative piece?

   F.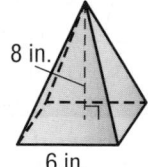
   8 in.
   6 in.

   H.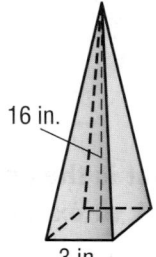
   16 in.
   3 in.

   G.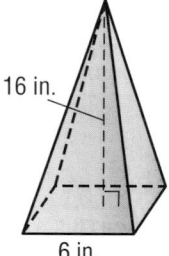
   16 in.
   6 in.

   I.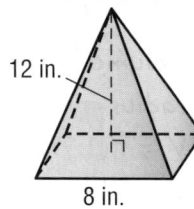
   12 in.
   8 in.

7. A wallet-sized print is about 5 centimeters wide and 8 centimeters long. Grace wants to use the wallet-sized print to make a print that is similar to the wallet-sized print. If the new print will be 20 centimeters long, how wide will the new print be to the nearest centimeter?

   A. 11 centimeters

   B. 12 centimeters

   C. 13 centimeters

   D. 15 centimeters

8. ✎ **GRIDDED RESPONSE** Solve the equation $b - 5 = -8$. What is the value of b?

9. A tube of caulk comes in a cylindrical tube. The tube measures 10 inches long and has a 2-inch diameter. What is the approximate volume of caulk contained in this tube?

F. 3.14 in³   **H.** 62.8 in³
G. 31.4 in³   **I.** 125.7 in³

10. ▨ **SHORT RESPONSE** Compare the surface area of the figures below that have equal volume. Justify your answer.

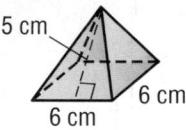

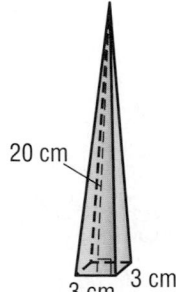

11. Eli made a triangle with an area of 18 square inches. Suppose he makes a similar triangle that has been dilated by a scale factor of 3. What will be the area of the new triangle?

A. 27 in²   **C.** 162 in²
B. 54 in²   **D.** 486 in²

12. Andrea made a tiered cake for a wedding. She wants to cover the outside of each layer marked A, B, and C with white icing.

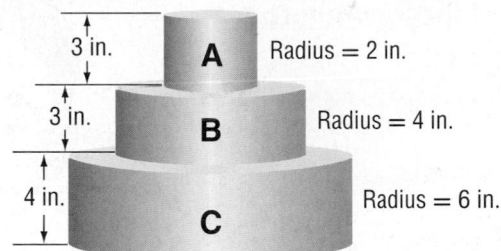

What is the total area of the 3 surfaces to be covered with white icing? The top of each layer will not be covered.

F. 289.4 in²   **H.** 163.5 in²
G. 263.9 in²   **I.** 131.9 in²

13. ▨ **EXTENDED RESPONSE** A ceramic dish company makes small cylindrical dishes with lids that have a radius of 8 centimeters and a height of 6 centimeters. The dishes are shipped in rectangular boxes that are 20 centimeters by 20 centimeters by 16 centimeters. The extra space in the box is filled with packing material to protect the dish.

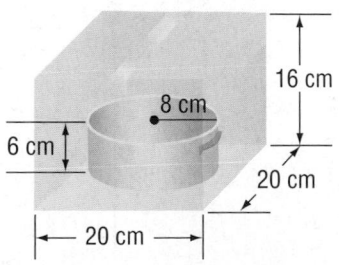

Part A How much space does the dish take up?

Part B How much packing material is needed?

Part C How much material is needed to make the box?

NEED EXTRA HELP?

If You Missed Question...	1	2	3	4	5	6	7	8	9	10	11	12	13
Go to Chapter-Lesson...	10-1B	8-2B	3-3D	10-1B	5-1D	10-1E	5-3A	4-1D	10-1C	10-2E	5-3B	10-2C	10-3C

Measurement and Proportional Reasoning

The
☆BIG Idea

How can proportional reasoning be used to convert measurements?

FOLDABLES®
Study Organizer

Make this Foldable to help you organize your notes.

Convert measurements
Convert between systems
Convert rates
Convert area and volume
Similar solids

Review Vocabulary

ratio table   *tabla de razones* a table with columns filled with pairs of numbers that have the same ratio

Gallons of Paint	1	2	3	4
Drops of Red	6	12	18	24

+1 +1 +1

+6 +6 +6

Key Vocabulary

English	Español
metric system	sistema métrico
similar solids	sólidos semejantes
unit ratio	tasa unitaria

When Will I Use This?

Are You Ready for the Chapter?

You have two options for checking prerequisite skills for this chapter.

Text Option Take the Quick Check below. Refer to the Quick Review for help.

QUICK Check

Multiply. Write in simplest form.

1. $\frac{2}{3} \times \frac{9}{10}$
2. $\frac{1}{2} \times \frac{6}{7}$
3. $\frac{2}{5} \times \frac{5}{8}$
4. $4 \times \frac{7}{8}$
5. $8 \times \frac{3}{16}$
6. $\frac{2}{3} \times 9$

7. **FRUIT** A farmer planted 6 acres of land with orange trees. During the past year, only $\frac{2}{3}$ of the planted acres produced oranges. How many acres produced oranges?

Multiply.

8. 5.8×10
9. $0.9 \times 1,000$
10. 1.04×100
11. $2.4 \times 1,000$
12. 0.03×100
13. 8.15×10

14. **MEASUREMENT** The height of the KVLY Tower in North Dakota in feet can be found by multiplying 2.063 by 1,000. Find the height of the KVLY Tower.

QUICK Review

EXAMPLE 1

Find $\frac{3}{4} \times \frac{2}{9}$.

$\frac{3}{4} \times \frac{2}{9} = \dfrac{\overset{1}{\cancel{3}} \times \overset{1}{\cancel{2}}}{\underset{2}{\cancel{4}} \times \underset{3}{\cancel{9}}}$ Divide 3 and 9 by their GCF, 3.
Divide 2 and 4 by their GCF, 2.

$= \frac{1}{6}$ Simplify.

EXAMPLE 2

Find $5 \times \frac{3}{10}$.

$5 \times \frac{3}{10} = \dfrac{\overset{1}{\cancel{5}}}{1} \times \dfrac{3}{\underset{2}{\cancel{10}}}$ Divide 5 and 10 by their GCF, 5.

$= \frac{3}{2}$ or $1\frac{1}{2}$ Simplify.

EXAMPLE 3

Find 6.3×100.

Method 1   Use paper and pencil.

$$\begin{array}{r} 100 \\ \times\ 6.3 \quad \text{one decimal place} \\ \hline 300 \\ 6000 \\ \hline 630.0 \quad \text{one decimal place} \end{array}$$

Method 2   Use mental math.

$6.3 \times 100 = 6.300$ Move the decimal point two places to the right.

$= 630.0$ or 630

 **Online Option** Take the Online Readiness Quiz.

Studying Math

Power Notes

Power notes are similar to lesson outlines, but they are simpler to organize. Power notes use the numbers 1, 2, 3, and so on.

Do you ever have trouble organizing your notes? Try using power notes.

Power 1: This is the main idea.
> **Power 2:** This provides details about the main idea.
>> **Power 3:** This provides details about Power 2.
>> and so on…

Here's a sample of power notes from this chapter. Note that you can even add drawings or examples to your power notes.

1: The Customary System
  2: Length
    3: Units of Length
      4: inch
      4: foot
        5: 1 foot = 12 inches
        5: 3 feet = 1 yard
      4: yard
      4: mile
    3: Converting Units of Length
      4: from larger to smaller — multiply
      4: from smaller to larger — divide
  2: Weight

> You can have more than one detail under each power.

Practice

Use power notes to make an outline for each concept.

1. metric system (Lesson 1C)

2. convert between systems (Lesson 1D)

3. adding and subtracting integers (Lessons 2B, 2D)

Explore Units of Measure

Main Idea

Determine the appropriate unit of measure. Compare and contrast units of measure.

ELEVATORS Elevators have been around since the mid-1800s. All elevators have a maximum weight limit. Is the maximum weight limit of this elevator more likely to be 3,500 ounces or 3,500 pounds?

In this activity, you will explore different units of measure.

ACTIVITY

 **STEP 1**   Locate the following objects. Estimate their measures. Copy and complete the table.

Object	Estimated Length		Estimated Weight/Mass	
Math textbook	▒ inches	▒ centimeters	▒ pounds	▒ kilograms
Pencil	▒ inches	▒ millimeters	▒ ounces	▒ grams
Board eraser	▒ inches	▒ centimeters	▒ ounces	▒ grams
Classroom door	▒ yards	▒ meters	▒ pounds	▒ kilograms
Chair	▒ feet	▒ centimeters	▒ pounds	▒ kilograms

 **STEP 2**   Find the actual measures of two of the objects in the table.

Analyze the Results

1. Would the maximum weight limit of an elevator be measured in ounces or pounds? Explain your reasoning.

2. Compare and contrast your measurements for each object. Are there any units that are similar?

3. **MATH IN THE MEDIA** Use the Internet or another source to research which countries use the customary system of measurement and which countries use the metric system of measurement.

Practice and Apply

Choose the better unit of measure for each object.

4. kilograms or tons

5. cups or milliliters

6. yards or kilometers

7. milligrams or ounces

8. grams or pounds

9. liters or pints

Write the customary and metric units that you would use to measure each of the following.

10. thickness of a coin

11. amount of water in a pitcher

12. length of a skateboard

13. length of a football field

14. thickness of a pencil

15. vanilla used in a cookie recipe

16. bag of sugar

17. distance between two cities

18. gas in the tank of a car

19. MAPS The map shows part of the eastern coast of central Florida.

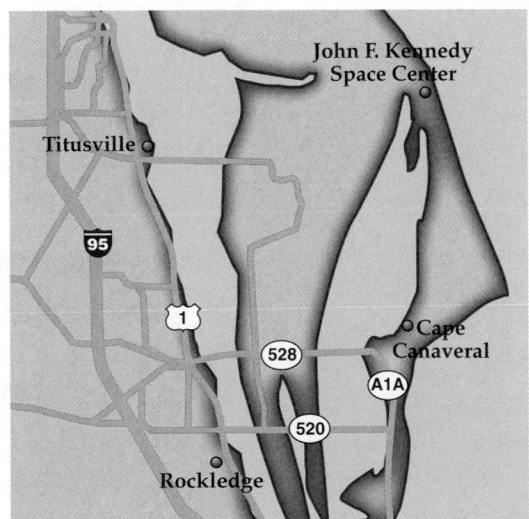

   a. Estimate the map distance in inches between Rockledge and Titusville. Check your measurement with a ruler.

   b. Estimate the map distance in centimeters between Cape Canaveral and the John F. Kennedy Space Center. Check your measurement with a ruler.

20. Which customary unit of length is approximately equal to one meter?

21. Which metric unit of length would you use instead of a mile?

22. HEIGHT Choose two classmates or family members. Which customary and metric units of length would you use to measure each person's height? Estimate the height of each person. Then measure to check the reasonableness of your estimate.

23. FOOD Without looking at their labels, estimate the weight or capacity of three packaged food items in your kitchen. Use customary and metric units. Then compare your estimate to the actual weight or capacity.

Main Idea

Change units of measure in the customary system.

 Vocabulary

unit ratio
dimensional analysis

Convert Customary Units

ANIMALS The table shows the approximate weights in tons of several large land animals.

Animal	Weight (T)
Grizzly bear	1
White rhinoceros	4
Hippopotamus	5
African elephant	8

One ton is equivalent to 2,000 pounds. You can use a *ratio table* to convert each weight from tons to pounds.

1. Copy and complete the ratio table. The first two ratios are done for you.

Tons	1	4	5	8
Pounds	2,000	8,000	▨	▨

To produce equivalent ratios, multiply the quantities in each row by the same number.

2. Then graph the ordered pairs (tons, pounds) from the table. Label the horizontal axis *Weight in Tons* and the vertical axis *Weight in Pounds*. Connect the points. What do you notice about the graph?

Key Concept — Customary Units

Type of Measure	Larger Unit ⟶		Smaller Unit
Length	1 foot (ft)	=	12 inches (in.)
	1 yard (yd)	=	3 feet
	1 mile (mi)	=	5,280 feet
Weight	1 pound (lb)	=	16 ounces (oz)
	1 ton (T)	=	2,000 pounds
Capacity	1 cup (c)	=	8 fluid ounces (fl oz)
	1 pint (pt)	=	2 cups
	1 quart (qt)	=	2 pints
	1 gallon (gal)	=	4 quarts

Each of the relationships on the previous page can be written as a unit ratio. Like a unit rate, a **unit ratio** is one in which the denominator is 1 unit.

$$\frac{3 \text{ ft}}{1 \text{ yd}} \qquad \frac{2{,}000 \text{ lb}}{1 \text{ T}} \qquad \frac{4 \text{ qt}}{1 \text{ gal}}$$

Notice that the numerator and denominator of each fraction above are equivalent, so the value of each ratio is 1. You can multiply by a unit ratio of this type to *convert* or change from larger units to smaller units.

The process of including units of measurement as factors when you compute is called **dimensional analysis**.

Study Tip

Multiplying by 1 Although the number and units changed in Example 1, because the measure is multiplied by 1, the value of the converted measure is the same as the original.

EXAMPLES Convert Larger Units to Smaller Units

1 Convert 20 feet to inches.

Since 1 foot = 12 inches, the unit ratio is $\frac{12 \text{ in.}}{1 \text{ ft}}$.

$20 \text{ ft} = 20 \text{ ft} \cdot \frac{12 \text{ in.}}{1 \text{ ft}}$ Multiply by $\frac{12 \text{ in.}}{1 \text{ ft}}$.

$\quad = 20 \cancel{\text{ ft}} \cdot \frac{12 \text{ in.}}{1 \cancel{\text{ ft}}}$ Divide out common units, leaving the desired unit, inches.

$\quad = 20 \cdot 12 \text{ in. or } 240 \text{ in.}$ Multiply.

So, 20 feet = 240 inches.

2 **GARDENING** Marco mixes $\frac{1}{4}$ cup of fertilizer with soil before planting each bulb. How many fluid ounces of fertilizer does he use per bulb?

$\frac{1}{4} \text{ c} = \frac{1}{4} \cancel{\text{c}} \cdot \frac{8 \text{ fl oz}}{1 \cancel{\text{c}}}$ Since 1 cup = 8 fluid ounces, multiply by $\frac{8 \text{ fl oz}}{1 \text{ c}}$. Then, divide out common units.

$\quad = \frac{1}{4} \cdot 8 \text{ fl oz or } 2 \text{ fl oz}$ Multiply.

So, 2 fluid ounces of fertilizer are used per bulb.

CHECK Your Progress

Complete.

a. 36 yd = ■ ft

b. $\frac{3}{4}$ T = ■ lb

c. $1\frac{1}{2}$ qt = ■ pt

d. **EXERCISE** Jen runs $\frac{1}{8}$ of a mile before tennis practice. How many feet does she run before practice?

To convert from smaller units to larger units, multiply by the reciprocal of the appropriate unit ratio.

 EXAMPLES **Convert Smaller Units to Larger Units**

QUICK Review

Reciprocal The product of a number and its reciprocal is 1. For example, the reciprocal of $\frac{3}{5}$ is $\frac{5}{3}$.

3 **Convert 15 quarts to gallons.**

Since 1 gallon = 4 quarts, the unit ratio is $\frac{4 \text{ qt}}{1 \text{ gal}}$, and its reciprocal is $\frac{1 \text{ gal}}{4 \text{ qt}}$.

$$15 \text{ qt} = 15 \text{ qt} \cdot \frac{1 \text{ gal}}{4 \text{ qt}} \qquad \text{Multiply by } \frac{1 \text{ gal}}{4 \text{ qt}}.$$

$$= 15 \text{ qt} \cdot \frac{1 \text{ gal}}{4 \text{ qt}} \qquad \text{Divide out common units, leaving the desired unit, gallons.}$$

$$= 15 \cdot \frac{1}{4} \text{ gal or } 3.75 \text{ gal} \qquad \text{Multiplying 15 by } \frac{1}{4} \text{ is the same as dividing 15 by 4.}$$

4 **COSTUMES** Umeka needs $4\frac{1}{2}$ feet of fabric to make a costume for a play. How many yards of fabric does she need?

$$4\frac{1}{2} \text{ ft} = 4\frac{1}{2} \text{ ft} \cdot \frac{1 \text{ yd}}{3 \text{ ft}} \qquad \text{Since 1 yard = 3 feet, multiply by } \frac{1 \text{ yd}}{3 \text{ ft}}. \text{ Then, divide out common units.}$$

$$= \frac{\overset{3}{\cancel{9}}}{2} \cdot \frac{1}{\underset{1}{\cancel{3}}} \text{ yd} \qquad \text{Write } 4\frac{1}{2} \text{ as an improper fraction. Then divide out common factors.}$$

$$= \frac{3}{2} \text{ yd or } 1\frac{1}{2} \text{ yd} \qquad \text{Multiply.}$$

So, Umeka needs $1\frac{1}{2}$ yards of fabric.

CHECK Your Progress

Complete.

e. 2,640 ft = ■ mi   **f.** 100 oz = ■ lb   **g.** 3 c = ■ pt

h. **FOOD** A 3-pound pork loin can be cut into 10 pork chops of equal weight. How many ounces is each pork chop?

✓ CHECK Your Understanding

Examples 1 and 2 Complete.

1. 3 lb = ■ oz   **2.** $5\frac{1}{3}$ yd = ■ ft   **3** 6.5 c = ■ fl oz

4. **FISH** A large grouper can weigh $\frac{1}{3}$ ton. How much does a large grouper weigh to the nearest pound?

Examples 3 and 4 Complete.

5. 12 qt = ■ gal   **6.** 28 in. = ■ ft   **7.** 15 pt = ■ qt

8. **VEHICLES** The world's narrowest electric vehicle is about 35 inches wide. How wide is this vehicle to the nearest foot?

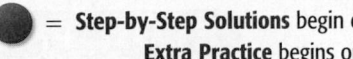

= **Step-by-Step Solutions** begin on page R1.
Extra Practice begins on page EP2.

Examples 1 and 3 Complete.

9. 18 ft = ■ yd

10. 72 oz = ■ lb

11. 2 lb = ■ oz

12. 4 gal = ■ qt

13. $4\frac{1}{2}$ pt = ■ c

14. 3 c = ■ fl oz

15. 2 mi = ■ ft

16. $1\frac{1}{4}$ mi = ■ ft

17. 5,000 lb = ■ T

18. 13 c = ■ pt

19. $2\frac{3}{4}$ qt = ■ pt

20. $3\frac{3}{8}$ T = ■ lb

Examples 2 and 4

21. **PUMPKINS** One of the largest pumpkins ever grown weighed about $\frac{1}{2}$ ton. How many pounds did the pumpkin weigh?

22. **SKIING** Speed skiing takes place on a course that is $\frac{2}{3}$ mile long. How many feet long is the course?

23. **BOATING** A 40-foot power boat is for sale by owner. How long is the boat to the nearest yard?

24. **BLOOD** A total of 35 pints of blood were collected at a local blood drive. How many quarts of blood were collected?

25. **PUNCH** Will a 2-quart pitcher hold the entire recipe of citrus punch given at the right? Explain your reasoning.

Recipe: Citrus Punch Drink
2 cups orange juice
2 cups grapefruit juice
$\frac{1}{4}$ cup apricot nectar
$\frac{1}{3}$ cup pineapple juice
4 cups ginger ale

26. **WEATHER** On Monday, it snowed a total of 15 inches. On Tuesday and Wednesday, it snowed an additional $4\frac{1}{2}$ inches and $6\frac{3}{4}$ inches, respectively. A weather forecaster says that over the last three days, it snowed over $2\frac{1}{2}$ feet. Is this a valid claim? Justify your answer.

MEASUREMENT Complete the following statements.

27. If 16 c = 1 gal, then $1\frac{1}{4}$ gal = ■ c.

28. If 1,760 yd = 1 mi, then 880 yd = ■ mi.

29. If 36 in. = 1 yd, then 2.3 yd = ■ in.

30. **MULTIPLE REPRESENTATIONS** Use the graph at the right.

a. **NUMBERS** What does an ordered pair from this graph represent?

b. **ALGEBRA** Find the slope of the line.

c. **MEASUREMENT** Use the graph to find the capacity in quarts of a 2.5-gallon container. Explain your reasoning.

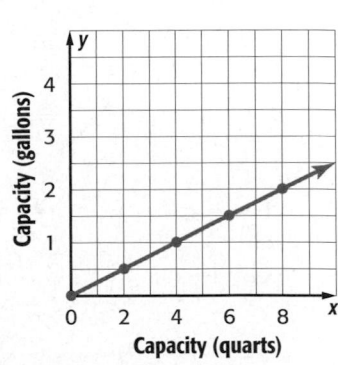

Capacity (gallons)
Capacity (quarts)

31. OPEN ENDED Write a real-world problem in which you would need to convert pints to cups.

CHALLENGE Replace each ● with <, >, or = to make a true sentence. Justify your answers.

32. 16 in. ● $1\frac{1}{2}$ ft

33. $8\frac{3}{4}$ gal ● 32 qt

34. 2.7 T ● 86,400 oz

35. CHALLENGE Give two different measurements that are equivalent to $2\frac{1}{2}$ quarts.

36. 📝 **WRITE MATH** Use multiplication by unit ratios of equivalent measures to convert 5 square feet to square inches. Explain how you solved.

✔ Test Practice

37. Which of the following situations is represented by the graph?

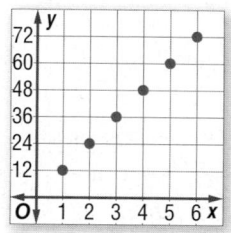

A. conversion of inches to yards

B. conversion of feet to inches

C. conversion of miles to feet

D. conversion of yards to feet

38. How many cups of milk are shown?

F. $\frac{3}{4}$ c

G. $1\frac{1}{4}$ c

H. $2\frac{1}{2}$ c

I. 10 c

39. Which relationship between the given units of measure is true?

A. One foot is $\frac{1}{12}$ of an inch.

B. One yard is $\frac{1}{3}$ of a foot.

C. One yard is $\frac{1}{3}$ of a mile.

D. One inch is $\frac{1}{12}$ of a foot.

40. Which of the following lists the measurements below in order from least to greatest?

> 88 inches, $7\frac{1}{2}$ feet, $2\frac{1}{3}$ yards

F. 88 inches, $7\frac{1}{2}$ feet, $2\frac{1}{3}$ yards

G. $7\frac{1}{2}$ feet, 88 inches, $2\frac{1}{3}$ yards

H. $2\frac{1}{3}$ yards, $7\frac{1}{2}$ feet, 88 inches

I. $2\frac{1}{3}$ yards, 88 inches, $7\frac{1}{2}$ feet

Main Idea

Change metric units of length, capacity, and mass.

 Vocabulary

metric system
meter
liter
gram
kilogram
accuracy

Convert Metric Units

Explore The lengths of two objects are shown below.

Object	Length (millimeters)	Length (centimeters)
Paper clip	45	4.5
CD case	144	14.4

1. Select three other objects. Find and record the width of all three objects to the nearest millimeter and tenth of a centimeter.

2. Compare and contrast the measurements of the objects. Write a rule that describes how to convert from millimeters to centimeters.

3. Measure the length of your classroom in meters. Make a conjecture about how to convert this measure to centimeters. Explain.

The **metric system** is a decimal system of measures. The prefixes commonly used in this system are kilo-, centi-, and milli-.

Prefix	Meaning in Words	Meaning in Numbers
kilo-	thousands	1,000
centi-	hundredths	0.01
milli-	thousandths	0.001

In the metric system, the base unit of *length* is the **meter** (m). Using prefixes, the names of other units of length are formed. Notice that the prefixes tell you how the units relate to the meter.

Unit	Symbol	Relationship to Meter	
kilometer	km	1 km = 1,000 m	1 m = 0.001 km
meter	m	1 m = 1 m	
centimeter	cm	1 cm = 0.01 m	1 m = 100 cm
millimeter	mm	1 mm = 0.001 m	1 m = 1,000 mm

The **liter** (L) is the base unit of *capacity*, the amount of dry or liquid material an object can hold. The **gram** (g) measures *mass*, the amount of matter in an object. The prefixes can also be applied to these units. Whereas the meter and liter are the base units of length and capacity, the base unit of mass is the **kilogram** (kg).

To convert a metric measure of length, mass, or capacity from one unit to another, you can use the relationship between the two units and multiplication by a power of 10.

 EXAMPLES Convert Units in the Metric System

Study Tip

Metric Conversions When converting from a larger unit to a smaller unit, the power of ten being multiplied will be greater than 1.

When converting from a smaller unit to a larger unit, the power of ten will be less than 1.

1 **Convert 4.5 liters to milliliters.**

You need to convert liters to milliliters. Use the relationship $1 L = 1,000 mL$.

$$1 L = 1,000 mL$$ Write the relationship.

$$4.5 \times 1 L = 4.5 \times 1,000 mL$$ Multiply each side by 4.5 since you have 4.5 L.

$$4.5 L = 4,500 mL$$ To multiply 4.5 by 1,000, move the decimal point 3 places to the right.

2 **Convert 500 millimeters to meters.**

You need to convert millimeters to meters. Use the relationship $1 mm = 0.001 m$.

$$1 mm = 0.001 m$$ Write the relationship.

$$500 \times 1 mm = 500 \times 0.001 m$$ Multiply each side by 500 since you have 500 mm.

$$500 mm = 0.5 m$$ To multiply 500 by 0.001, move the decimal point 3 places to the left.

CHECK Your Progress

Complete.

a. $25.4 g = \blacksquare kg$ **b.** $158 mm = \blacksquare m$

 REAL-WORLD EXAMPLE

3 **FLAMINGOS** Use the information at the left to find the average mass of a flamingo in grams.

You are converting kilograms to grams. Since the average mass of a flamingo is 2.95 kilograms, use the relationship $1 kg = 1,000 g$.

$$1 kg = 1,000 g$$ Write the relationship.

$$2.95 \times 1 kg = 2.95 \times 1,000 g$$ Multiply each side by 2.95 since you have 2.95 kg.

$$2.95 kg = 2,950 g$$ To multiply 2.95 by 1,000, move the decimal point 3 places to the right.

So, the average mass of a flamingo is 2,950 grams.

Real-World Link . . . .
The average mass of a flamingo is 2.95 kilograms.

CHECK Your Progress

c. **JUICE** A bottle contains 1.75 liters of juice. How many milliliters does the bottle contain?

Examples 1 and 2   **Complete. Round to the nearest hundredth if necessary.**

1. 3.7 m = ■ cm   **2.** 550 m = ■ km

3. 1,460 mg = ■ g   **4.** 2.34 kL = ■ L

Example 3   **5. SPORTS** How many centimeters does a team of athletes run in a 5-kilometer relay race?

Practice and Problem Solving

 = **Step-by-Step Solutions** begin on page R1.
Extra Practice begins on page EP2.

Examples 1 and 2   **Complete. Round to the nearest hundredth if necessary.**

6. 720 cm = ■ m   **7** 983 mm = ■ m

8. 3.2 m = ■ cm   **9.** 0.03 g = ■ mg

10. 997 g = ■ kg   **11.** 82.1 g = ■ kg

12. 9.1 L = ■ mL   **13.** 130.5 kL = ■ L

Example 3   **14. WATERFALLS** At 979 meters tall, Angel Falls in Venezuela is the highest waterfall in the world. How many kilometers tall is the waterfall?

15. FOOD An 18-ounce jar contains 510 grams of grape jelly. How many kilograms of grape jelly does the jar contain?

16. CYCLING Ramon rode his bike a distance of 8 kilometers. How many centimeters did Ramon ride his bike?

17. BIRDS A gull can fly at a speed of 35 kilometers per hour. How many meters per hour can a gull fly?

Order each set of measures from least to greatest.

18. 0.02 km, 50 m, 3,000 cm   **19.** 660 mL, 0.06 L, 6.6 kL

20. 0.32 kg, 345 g, 35,100 mg   **21.** 2,650 mm, 130 cm, 5 m

Real-World Link
The Mackinac Bridge opened on November 1, 1957. It is the longest two-tower suspension bridge between bases in the Western Hemisphere.

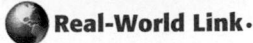

22. ANALYZE TABLES The table shows the lengths of bridges in the United States. Which bridges are about 1 kilometer in length? Justify your answer.

23 CARPENTRY Jacinta needs a 2.5-meter pole for a birdfeeder that she is building. How many centimeters will she need to cut off of a 3-meter pole in order to use it for the birdfeeder?

24. BAKING One recipe for apple pie calls for 0.94 kilogram of apples. Another recipe calls for 950 grams of apples. Which pie requires more apples?

Bridge	Length (m)
Mackinac, MI	1,158
George Washington, NY	1,067
Tacoma Narrows II, WA	853
Oakland Bay, CA	704
Pennybacker, TX	345
Sunshine Skyway, FL	8,712
Golden Gate, CA	2,780

25. FIND THE ERROR Theresa is converting 3.25 kilograms to grams. Find her mistake and correct it.

> 3.25 kg = 0.00325 g

26. CHALLENGE The metric prefix *giga–* refers to something one billion times larger than the base unit. One gigameter equals how many meters?

27.  **WRITE MATH** Explain why you use a power of 10 that is greater than 1 when converting from a larger unit to a smaller unit.

Test Practice

28. The table shows the mass of cell phones. What is the approximate total mass of the cell phones?

Telephone Owner	Mass (g)
Claudio	100.4
Al	70.8
Jane	95.6

   A. 0.27 kilogram   **C.** 27.0 kilograms

   B. 2.7 kilograms   **D.** 270.0 kilograms

29. Which relationship between the given units of measure is correct?

   F. One gram is $\frac{1}{100}$ of a centigram.

   G. One meter is $\frac{1}{100}$ of a centimeter.

   H. One gram is $\frac{1}{1,000}$ of a kilogram.

   I. One milliliter is $\frac{1}{100}$ of a liter.

More About Measurement

Measuring objects as close as possible to the actual value is called **accuracy**.

 **EXAMPLE**

Which is more accurate for measuring the weight of a bag of apples in a grocery store: tenth of a pound or half of a pound?

Measuring to the nearest tenth of a pound is more accurate because it is closer than half a pound to the actual value.

Choose the term that results in a more accurate measurement.

30. In a science experiment, the mass of one drop of solution is found to the nearest 0.01 (gram, kilogram).

31. The length of a bracelet is measured to the nearest (inch, eighth inch).

Main Idea

Convert units of measure between the customary and metric systems.

Get ConnectED

Convert Between Systems

RACES Races are often measured in kilometers. A 5K race is 5 kilometers.

1. How many meters long is the race?

2. One mile is approximately 1.6 kilometers. About how many miles is the race?

To convert measures between customary units and metric units, use the relationships below.

Key Concept	Customary and Metric Relationships		
Type of Measure	**Customary**	→	**Metric**
Length	1 inch (in.)	≈	2.54 centimeters (cm)
	1 foot (ft)	≈	0.30 meter (m)
	1 yard (yd)	≈	0.91 meter (m)
	1 mile (mi)	≈	1.61 kilometers (km)
Weight/Mass	1 pound (lb)	≈	453.6 grams (g)
	1 pound (lb)	≈	0.4536 kilogram (kg)
	1 ton (T)	≈	907.2 kilograms (kg)
Capacity	1 cup (c)	≈	236.59 milliliters (mL)
	1 pint (pt)	≈	473.18 milliliters (mL)
	1 quart (qt)	≈	946.35 milliliters (mL)
	1 gallon (gal)	≈	3.79 liters (L)

EXAMPLES **Convert Between Measurement Systems**

1. **Convert 17.22 inches to centimeters. Round to the nearest hundredth if necessary.**

   Since 2.54 centimeters ≈ 1 inch, multiply by $\frac{2.54 \text{ cm}}{1 \text{ in.}}$.

   $17.22 \approx 17.22 \text{ in.} \cdot \frac{2.54 \text{ cm}}{1 \text{ in.}}$ Multiply by $\frac{2.54 \text{ cm}}{1 \text{ in.}}$. Divide out common units.

   $\approx 43.7388 \text{ cm}$ Simplify.

   So, 17.22 inches is approximately 43.74 centimeters.

2 Convert 828.5 milliliters to cups. Round to the nearest hundredth if necessary.

Since 1 cup ≈ 236.59 milliliters, multiply by $\frac{1\ c}{236.59\ mL}$.

$$828.5\ mL \approx 828.5\ \cancel{mL} \cdot \frac{1\ c}{236.59\ \cancel{mL}} \quad \text{Multiply by } \frac{1\ c}{236.59\ mL} \text{ and divide out common units.}$$

$$\approx \frac{828.5\ c}{236.59} \text{ or } 3.50\ c \quad \text{Simplify.}$$

So, 828.5 milliliters is approximately 3.50 cups.

✓ CHECK Your Progress

Complete. Round to the nearest hundredth if necessary.

a. 7.44 c ≈ ■ mL   **b.** 22.09 lb ≈ ■ kg   **c.** 35.85 L ≈ ■ gal

● Real-World Link · · · · ·

The National Aquatics Center, also called the Water Cube, built for the 2008 Olympics in Beijing consists of more than 100,000 square meters (or 1.08 million square feet) of plastic foils and is the largest, most complicated, plastic-covered structure in the world.

 **REAL-WORLD EXAMPLE**   **Convert Between Systems**

3 **SWIMMING** An Olympic-size swimming pool is 50 meters long. About how many feet long is the pool?

Since 1 foot ≈ 0.30 meter, use the ratio $\frac{1\ ft}{0.30\ m}$.

$$50\ m \approx 50\ m \cdot \frac{1\ ft}{0.30\ m} \quad \text{Multiply by } \frac{1\ ft}{0.30\ m}.$$

$$\approx 50\ \cancel{m} \cdot \frac{1\ ft}{0.30\ \cancel{m}} \quad \text{Divide out common units, leaving the desired unit, feet.}$$

$$\approx \frac{50\ ft}{0.30} \text{ or } 166.67\ ft \quad \text{Divide.}$$

An Olympic-size swimming pool is about 166.67 feet long.

✓ CHECK Your Progress

d. **POOLS** An NCAA regulation-size swimming pool is 25 yards long. About how many meters long is the pool?

✓ CHECK Your Understanding

Examples 1 and 2   Complete. Round to the nearest hundredth if necessary.

1. 3.7 yd ≈ ■ m   **2.** 11.07 pt ≈ ■ mL   **3** 58.14 kg ≈ ■ lb

4. 3.75 c ≈ ■ mL   **5.** 4.725 m ≈ ■ ft   **6.** 680.4 g ≈ ■ lb

Example 3   **7.** **SPORTS** About how many feet does a team of athletes run in a 1,600-meter relay race?

8. **FOOD** Raheem bought 3 pounds of bananas. About how many kilograms did he buy?

Practice and Problem Solving

● = **Step-by-Step Solutions** begin on page R1.
Extra Practice begins on page EP2.

Examples 1 and 2   **Complete. Round to the nearest hundredth if necessary.**

9. 5 in. ≈ ■ cm   **10.** 15 cm ≈ ■ in.   **11.** 2 L ≈ ■ qt   **12.** 650 lb ≈ ■ kg

13. 4 qt ≈ ■ L   **14.** 10 mL ≈ ■ c   **15.** 63.5 T ≈ ■ kg   **16.** 50 mL ≈ ■ fl oz

17. 54 cm ≈ ■ in.   **18.** 17 mi ≈ ■ km   **19.** 32 gal ≈ ■ L   **20.** 350 lb ≈ ■ kg

21. 19 kg ≈ ■ lb   **22.** 3 T ≈ ■ kg   **23.** 6 in. ≈ ■ cm   **24.** 12 in. ≈ ■ m

Example 3   **25. COMPUTERS** A notebook computer has a mass of 2.25 kilograms. About how many pounds does the notebook weigh?

26. TREES A Cabbage Palmetto has a height of 80 feet. What is the estimated height of the tree in meters?

27. BUILDINGS The Willis Tower has a height of 1,451 feet. What is the estimated height of the building in meters?

28. WATER Which is greater, a bottle containing 64 fluid ounces of water or a bottle containing 2 liters of water?

29 FOOD Which box is greater, a 1.5-pound box of raisins or a 650-gram box of raisins?

30. BAKING A bakery uses 900 grams of peaches in a cobbler. About how many pounds of peaches does the bakery use in a cobbler?

Determine which quantity is greater.

31. 3 gal, 10 L   **32.** 14 oz, 0.4 kg   **33.** 4 mi, 6.2 km

34. RATES Velocity is a rate usually expressed in feet per second or meters per second. How can the units help you calculate velocity using the distance a car traveled and the time recorded?

35. GRAPHIC NOVEL Refer to the graphic novel at the start of the chapter. Convert all of the distances from meters to feet. Round to the nearest hundredth if necessary.

H.O.T. Problems

36. NUMBER SENSE One gram of water has a volume of 1 milliliter. What is the volume of the water if it has a mass of 1 kilogram?

37. CHALLENGE The distance from Earth to the Sun is approximately 93 million miles. About how many gigameters is this? Round to the nearest hundredth.

REASONING Order each set of measures from greatest to least.

38. 1.2 cm, 0.6 in., 0.031 m, 0.1 ft

39. 2 lb, 891 g, 1 kg, 0.02 T

40. $1\frac{1}{4}$ c, 0.4 L, 950 mL, 0.7 gal

41. 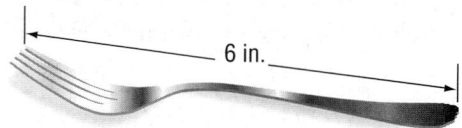 **WRITE MATH** Explain how to order lengths of objects from shortest to longest if lengths are given in both customary and metric units.

✓ Test Practice

42. The diagram shows the length of a fork from the cafeteria.

6 in.

Which of the following measurements is approximately equal to the length of the fork?

A. 2.4 cm

B. 15.2 cm

C. 24 cm

D. 152 cm

43. The table shows the flying speeds of various birds.

Bird	Speed (km/h)
Spur-winged goose	142
Mallard duck	105

About how fast does the Spur-winged goose travel in miles per hour?

F. 229 miles per hour

G. 156 miles per hour

H. 88 miles per hour

I. 71 miles per hour

44. BUILDINGS A skyscraper is 0.484 kilometer tall. What is the height of the skyscraper in meters? (Lesson 1C)

Convert. Round to the nearest tenth if necessary. (Lesson 1B)

45. 17 ft = ■ yd

46. 82 in. = ■ ft

47. 3 mi = ■ ft

48. 66 in. = ■ ft

49. 6 yd = ■ ft

50. 4 yd = ■ in.

Main Idea

Convert units of measure between derived units to solve problems.

 **Vocabulary**

derived unit

Convert Rates

TRAINS Some of the fastest passenger trains are located in Japan and France. The table shows various trains and their speeds.

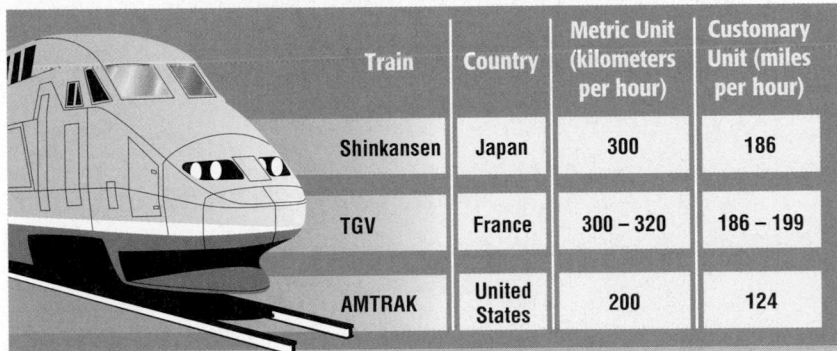

Train	Country	Metric Unit (kilometers per hour)	Customary Unit (miles per hour)
Shinkansen	Japan	300	186
TGV	France	300 – 320	186 – 199
AMTRAK	United States	200	124

1. How many feet are in one mile?

2. How many seconds are in one hour?

You can convert one rate to an equivalent rate by multiplying by a unit ratio or its reciprocal. A **derived unit** is a unit that is derived from a measurement system base unit such as length, mass, or time. Examples include square inches and meters per second.

 REAL-WORLD EXAMPLE **Convert Rates**

① **FISH** A swordfish can swim at a rate of 60 miles per hour. How many feet per hour is this?

You can use 1 mile = 5,280 feet to convert the rates.

$$\frac{60 \text{ mi}}{1 \text{ h}} = \frac{60 \text{ mi}}{1 \text{ h}} \cdot \frac{5{,}280 \text{ ft}}{1 \text{ mi}} \quad \text{Multiply by } \frac{5{,}280 \text{ ft}}{1 \text{ mi}}.$$

$$= \frac{60 \cancel{\text{ mi}}}{1 \text{ h}} \cdot \frac{5{,}280 \text{ ft}}{1 \cancel{\text{ mi}}} \quad \text{Divide out common units.}$$

$$= \frac{60 \cdot 5{,}280 \text{ ft}}{1 \cdot 1 \text{ h}} \quad \text{Simplify.}$$

$$= \frac{316{,}800 \text{ ft}}{1 \text{ h}} \quad \text{Simplify.}$$

A swordfish can swim at a rate of 316,800 feet per hour.

 CHECK Your Progress

a. **BIRDS** A gull can fly at a speed of 22 miles per hour. About how many kilometers per hour can the gull fly?

REAL-WORLD EXAMPLE **Convert Rates**

2 **WALKING** Marvin walks at a speed of 7 feet per second. How many feet per hour is this?

You can use 60 seconds = 1 minute and 60 minutes = 1 hour to convert the rates.

$$\frac{7 \text{ ft}}{1 \text{ s}} = \frac{7 \text{ ft}}{1 \text{ s}} \cdot \frac{60 \text{ s}}{1 \text{ min}} \cdot \frac{60 \text{ min}}{1 \text{ h}}$$ Multiply by $\frac{60 \text{ s}}{1 \text{ min}}$ and $\frac{60 \text{ min}}{1 \text{ h}}$.

$$= \frac{7 \text{ ft}}{1 \text{ s}} \cdot \frac{60 \text{ s}}{1 \text{ min}} \cdot \frac{60 \text{ min}}{1 \text{ h}}$$ Divide out common units.

$$= \frac{7 \cdot 60 \cdot 60 \text{ ft}}{1 \cdot 1 \cdot 1 \text{ h}}$$ Simplify.

$$= \frac{25,200 \text{ ft}}{1 \text{ h}}$$ Simplify.

Marvin walks 25,200 feet in 1 hour.

CHECK Your Progress

b. TRAINS An AMTRAK train travels at 125 miles per hour. Convert the speed to miles per minute.

Study Tip

Unit Rates Make sure the units cancel so that the desired units remain.

 REAL-WORLD EXAMPLE **Convert Derived Units**

3 **IDITAROD** The Iditarod Snow Dog Race is a famous race in Alaska in which mushers and their dog teams compete. The average speed of the team is about 10 miles per hour. What is this speed in feet per second?

We can use 1 mile = 5,280 feet, 1 hour = 60 minutes, and 1 minute = 60 seconds to convert the rates.

$$\frac{10 \text{ mi}}{1 \text{ h}} = \frac{10 \text{ mi}}{1 \text{ h}} \cdot \frac{5,280 \text{ ft}}{1 \text{ mi}} \cdot \frac{1 \text{ h}}{60 \text{ min}} \cdot \frac{1 \text{ min}}{60 \text{ s}}$$ Multiply by distance and time unit ratios.

$$= \frac{10 \text{ mi}}{1 \text{ h}} \cdot \frac{5,280 \text{ ft}}{1 \text{ mi}} \cdot \frac{1 \text{ h}}{60 \text{ min}} \cdot \frac{1 \text{ min}}{60 \text{ s}}$$ Divide out common units.

$$= \frac{10 \cdot 5,280 \cdot 1 \cdot 1 \text{ ft}}{1 \cdot 1 \cdot 60 \cdot 60 \text{ s}}$$ Simplify.

$$= \frac{52,800 \text{ ft}}{3,600 \text{ s}}$$ Simplify.

$$\approx \frac{14.7 \text{ ft}}{1 \text{ s}}$$ Simplify.

The Iditarod mushing teams travel at an average speed of 14.7 feet per second.

CHECK Your Progress

c. RUNNING Charlie runs at a speed of 3 meters per second. About how many miles per hour does Charlie run?

Example 1 ❶ **GO-KARTS** A go-kart's top speed is 607,200 feet per hour. What is the speed in miles per hour?

Example 2 **2. SPORTS** A skydiver is falling at about 176 feet per second. How many feet per minute is the skydiver falling?

Example 3 **3. CYCLING** Lorenzo rides his bike at a rate of 2.2 meters per second. About how many miles per hour can Lorenzo ride his bike?

Practice and Problem Solving

⬤ = **Step-by-Step Solutions** begin on page R1.
Extra Practice begins on page EP2.

Example 1 **4. WATER** Water weighs about 8.34 pounds per gallon. How many ounces per gallon is the weight of the water?

5. BIRDS A peregrine falcon can fly 322 kilometers per hour. How many meters per hour can the falcon fly?

Example 2 **6. RUNNING** The fastest a human has ever run is 27 miles per hour. How many miles per minute did the human run?

Example 3 **7. PLUMBING** A pipe is leaking at 1.5 cups per day. About how many liters per week is the pipe leaking?

Convert each rate. Round to the nearest tenth if necessary.

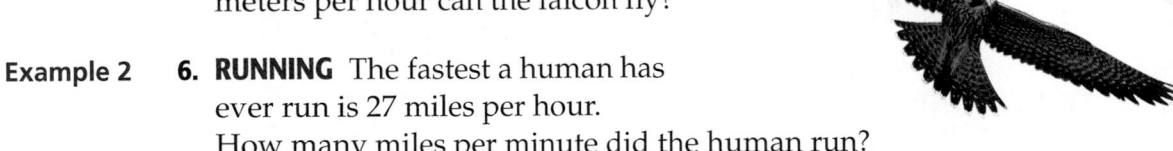

8. $20 \text{ mi/h} = \blacksquare \text{ ft/min}$
9. $16 \text{ cm/min} = \blacksquare \text{ m/h}$
10. $45 \text{ mi/h} = \blacksquare \text{ ft/s}$
11. $26 \text{ cm/s} = \blacksquare \text{ m/min}$
12. $2.5 \text{ qt/min} \approx \blacksquare \text{ L/h}$
13. $13 \text{ lb/gal} \approx \blacksquare \text{ kg/L}$
14. $7 \text{ m/min} \approx \blacksquare \text{ yd/h}$
15. $4.7 \text{ g/cm} \approx \blacksquare \text{ oz/in.}$

16. INTERNET The speed at which a certain computer can access the Internet is 2 megabytes per second. How fast is this in megabytes per hour?

17 INSECTS The table shows the speed and number of wing beats per second for various flying insects.

Flying Insects		
Insect	Speed (miles per hour)	Wing Beats per Second
Housefly	4.4	190
Honeybee	5.7	250
Dragonfly	15.6	38
Hornet	12.8	100
Bumblebee	6.4	130

a. What is the speed of a housefly in feet per second?

b. How many times does a dragonfly's wing beat per minute?

c. How many kilometers can a bumblebee travel in one minute?

d. How many times can a honeybee beat its wings in one hour?

H.O.T. Problems

18. **OPEN ENDED** Give an example of a unit rate used in a real-world situation.

19. **FIND THE ERROR** Divya is converting miles per hour to kilometers per minute. Find her mistake and correct it.

$$\frac{65 \text{ mi}}{1 \text{ h}} = \frac{65 \text{ mi}}{1 \text{ h}} \cdot \frac{1.61 \text{ km}}{1 \text{ mi}} \cdot \frac{60 \text{ min}}{1 \text{ h}}$$

$$= \frac{65 \cdot 1.61 \cdot 60}{1}$$

$$= 6,279 \text{ kilometers per minute}$$

20.  **WRITE MATH** Compare and contrast pounds per gallon to kilograms per liter.

Test Practice

21. Thirty-five miles per hour is the same rate as which of the following?

   A. 150 feet per minute

   B. 1,500 feet per minute

   C. 2,200 feet per minute

   D. 3,080 feet per minute

22. **THINK SOLVE EXPLAIN** **SHORT RESPONSE** An oil tanker empties at 3.5 gallons per minute. Convert this rate to cups per second. Round to the nearest tenth. Show the steps you used.

Spiral Review

23. **PAPER** Standard sized notebook paper is $8\frac{1}{2}$ inches by 11 inches. What are the dimensions in centimeters? (Lesson 1D)

Convert. Round to the nearest hundredth if necessary. (Lessons 1B and 1C)

24. $34 \text{ yd} = \blacksquare \text{ ft}$

25. $1\frac{1}{2} \text{ gal} = \blacksquare \text{ qt}$

26. $4.67 \text{ m} = \blacksquare \text{ cm}$

27. $901 \text{ g} = \blacksquare \text{ kg}$

Main Idea

Convert units of measure between dimensions including area and volume.

Convert Units of Area and Volume

CARPETING Jonathan is carpeting his bedroom. His bedroom is 15 feet long and 12 feet wide. While shopping, he notices carpet is sold in square yards.

1. How many feet are in one yard?

2. How many yards long is the room?

3. How many yards wide is the room?

4. What is the area of the room in square yards?

You can use the formula for the area of a square, $A = s^2$, to find the number of square feet in one square yard.

 **EXAMPLES** **Convert Area Measurements**

1 **Convert one square yard to square feet.**

A square yard is a square with a side length of one yard. You know that one yard is equal to three feet. So, one square yard is a square with side length of three feet.

$A = s^2$ Write the formula.

$A = 3^2$ Replace s with 3.

$A = 9$ Simplify.

So, one square yard is equal to 9 square feet.

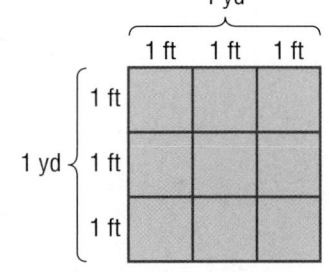

2 **Convert one square meter to square centimeters.**

A square meter is a square with a side length of one meter. You know that one meter is equal to 100 centimeters. So, one square meter is a square with side length of 100 centimeters.

$A = s^2$ Write the formula.

$A = 100^2$ Replace s with 100.

$A = 10,000$ Simplify.

So, one square meter is equal to 10,000 square centimeters.

Complete.

a. $1 \text{ ft}^2 = \blacksquare \text{ in}^2$ **b.** $1 \text{ cm}^2 = \blacksquare \text{ mm}^2$

You can use the formula for the volume of a prism, $V = \ell wh$, to convert cubic units.

> **Study Tip**
>
> **Formulas** Since a cube has three equal dimensions, volume can also be found with the formula $V = s^3$, where s is the length of a side.

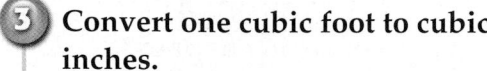 **EXAMPLE** **Convert Volume Measurements**

3 **Convert one cubic foot to cubic inches.**

A cubic foot is a cube with a side length of one foot or 12 inches.

$V = \ell wh$ Write the formula.

$V = 12 \cdot 12 \cdot 12$ Replace ℓ, w, and h with 12.

$V = 1{,}728$ Simplify.

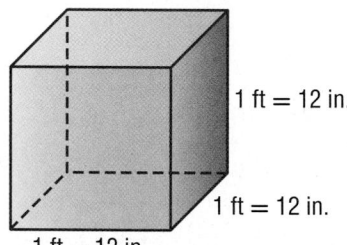

1 ft = 12 in.

1 ft = 12 in.

1 ft = 12 in.

So, one cubic foot is equal to 1,728 cubic inches.

CHECK Your Progress

Complete.

c. $1 \text{ yd}^3 = \blacksquare \text{ ft}^3$ **d.** $1 \text{ cm}^3 = \blacksquare \text{ mm}^3$

The table gives several common measurement conversions for square units and cubic units.

Key Concept	**Measurement Conversions**	
	Customary Units	**Metric Units**
Area	$1 \text{ ft}^2 = 144 \text{ in}^2$	$1 \text{ m}^2 = 10{,}000 \text{ cm}^2$
	$1 \text{ yd}^2 = 9 \text{ ft}^2$	$1 \text{ cm}^2 = 100 \text{ mm}^2$
Volume	$1 \text{ ft}^3 = 1{,}728 \text{ in}^3$	$1 \text{ m}^3 = 1{,}000{,}000 \text{ cm}^3$
	$1 \text{ yd}^3 = 27 \text{ ft}^3$	$1 \text{ cm}^3 = 1{,}000 \text{ mm}^3$

Each relationship in the Key Concept box can be written as a unit ratio. To convert square or cubic units, use the unit ratio or its reciprocal.

REAL-WORLD EXAMPLE **Convert Measurements**

Study Tip

Formulas The formula for finding the area of a rectangle is $A = \ell \cdot w$.

4 CONSTRUCTION A roof is 25 feet by 35 feet. How many square yards is the roof? Round to the nearest tenth.

The area of the roof is 25 feet × 35 feet or 875 square feet. Use the reciprocal of the unit ratio $\dfrac{9\ \text{ft}^2}{1\ \text{yd}^2}$ to find the number of square yards.

$875\ \text{ft}^2 = 875\ \text{ft}^2 \cdot \dfrac{1\ \text{yd}^2}{9\ \text{ft}^2}$ Multiply by $\dfrac{1\ \text{yd}^2}{9\ \text{ft}^2}$.

$= 875\ \cancel{\text{ft}^2} \cdot \dfrac{1\ \text{yd}^2}{9\ \cancel{\text{ft}^2}}$ Divide out common units, leaving the desired unit, yards.

$= \dfrac{875\ \text{yd}^2}{9}$ or $97.2\ \text{yd}^2$ Divide.

The roof is 97.2 square yards.

 CHECK Your Progress

e. CEREAL A cereal box holds 320 cubic inches of cereal. How many cubic feet is this? Round to the nearest tenth.

The metric system also relates length, mass, and capacity.

Key Concept **Length, Mass, and Capacity**

Words	Symbols
1 milliliter has the same volume as 1 cubic centimeter.	1 mL = 1 cc
1 milliliter of water is approximately 1 gram.	1 mL ≈ 1 g

 EXAMPLE **Convert Volume to Capacity**

5 Convert one cubic meter to milliliters.

A cubic meter is a cube with a side length of one meter or 100 centimeters. Use 1 cc = 1 mL to convert the rates.

$V = \ell \cdot w \cdot h$ Write the formula.

$V = 100 \cdot 100 \cdot 100$ Replace ℓ, w, and h with 100.

$V = 1{,}000{,}000$ Simplify.

So, one cubic meter is equal to 1,000,000 milliliters.

 CHECK Your Progress

Complete.

f. $17{,}000\ \text{mm}^3 = $ ■ mL

g. $150\ \text{cc} = $ ■ L

CHECK Your Understanding

Examples 1 and 2 Complete.

1. $3 \text{ ft}^2 = \blacksquare \text{ in}^2$
2. $4 \text{ yd}^2 = \blacksquare \text{ ft}^2$
3. $720 \text{ in}^2 = \blacksquare \text{ ft}^2$
4. $3.2 \text{ m}^2 = \blacksquare \text{ cm}^2$
5. $900 \text{ mm}^2 = \blacksquare \text{ cm}^2$
6. $8 \text{ cm}^2 = \blacksquare \text{ mm}^2$

Examples 3 and 5 Complete.

7. $0.2 \text{ ft}^3 = \blacksquare \text{ in}^3$
8. $4{,}320 \text{ in}^3 = \blacksquare \text{ ft}^3$
9. $1.5 \text{ yd}^3 = \blacksquare \text{ ft}^3$
10. $5{,}600 \text{ mm}^3 = \blacksquare \text{ mL}$
11. $4.1 \text{ m}^3 = \blacksquare \text{ cm}^3$
12. $2 \text{ cm}^3 = \blacksquare \text{ mL}$

Example 4

13. **FENCING** A playground is surrounded by chain-link fencing. The dimensions of the playground are 52 feet by 37 feet. How many square yards does the fencing surround? Round to the nearest tenth if necessary.

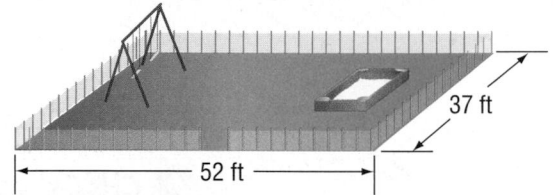

37 ft

52 ft

14. **SCUBA DIVING** Maria is using a cylindrical oxygen tank while scuba diving. It holds 80 cubic inches of air. How many cubic feet of air is she using? Round to the nearest hundredth if necessary.

Practice and Problem Solving

 = **Step-by-Step Solutions** begin on page R1.
Extra Practice begins on page EP2.

Examples 1 and 2 Complete. Round to the nearest hundredth if necessary.

15. $11.5 \text{ ft}^2 = \blacksquare \text{ in}^2$
16. $1{,}396.8 \text{ in}^2 = \blacksquare \text{ ft}^2$
17. $216 \text{ ft}^2 = \blacksquare \text{ yd}^2$
18. $14 \text{ yd}^2 = \blacksquare \text{ ft}^2$
19. $7.5 \text{ m}^2 = \blacksquare \text{ cm}^2$
20. $980 \text{ cm}^2 = \blacksquare \text{ m}^2$
21. $5.4 \text{ cm}^2 = \blacksquare \text{ mm}^2$
22. $597 \text{ mm}^2 = \blacksquare \text{ cm}^2$
23. $1 \text{ mi}^2 = \blacksquare \text{ ft}^2$

Examples 3 and 5 Complete. Round to the nearest hundredth if necessary.

24. $3 \text{ yd}^3 = \blacksquare \text{ ft}^3$
25. $11{,}232 \text{ in}^3 = \blacksquare \text{ ft}^3$
26. $6.06 \text{ ft}^3 = \blacksquare \text{ in}^3$
27. $280.8 \text{ ft}^3 = \blacksquare \text{ yd}^3$
28. $6{,}750 \text{ mm}^3 = \blacksquare \text{ cm}^3$
29. $0.45 \text{ m}^3 = \blacksquare \text{ mL}$
30. $7.7 \text{ cm}^3 = \blacksquare \text{ mm}^3$
31. $973{,}000 \text{ mL} = \blacksquare \text{ m}^3$
32. $1 \text{ yd}^3 = \blacksquare \text{ in}^3$

Example 4

33. **SPORTS** Including the end zones, a football field is 360 feet long by 160 feet wide. What is the area of a football field in square yards?

34. **GARDENING** Tabitha has a small garden. If the garden has an area of 1,512 square inches, what is the area of the garden in square feet?

35. **APPLIANCES** A refrigerator has 25.3 cubic feet of space. How many cubic yards is this?

Real-World Link · · ·
Most parade balloons are filled with 10,000 to 14,000 cubic feet of helium.

36. PARADE A cartoon character was depicted as a balloon in a parade. The balloon contained 2,443 cubic yards of air. How many cubic feet of air was in the balloon?

37. MEDICINE One dose of a certain medicine is 2 teaspoons. One teaspoon is equal to 5 milliliters. How many cubic centimeters is the dose?

38. How many square yards are in one square mile?

39. One square yard is equal to how many square inches?

40. BOXES Two boxes are shown.

a. What is the difference of their volumes, in cubic feet?

b. How many times greater is the volume of Box A than Box B?

c. What conclusion can you draw about the volume of a prism after its dimensions are halved?

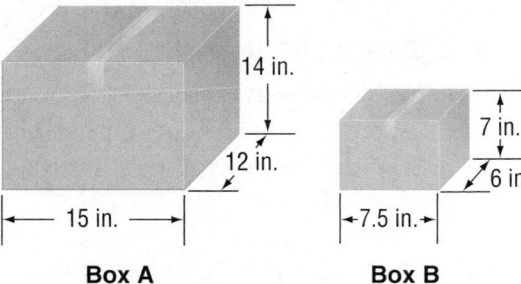

Box A

Box B

41 SWIMMING POOL The world's largest swimming pool is located in Chile and has a volume of 250,000 cubic meters. What is the estimated volume of the pool in cubic yards? (*Hint*: 1 meter ≈ 1.1 yards)

42. DOSAGES A liquid allergy medication comes in a bottle containing 4 fluid ounces. About how many 5 cubic-milliliters doses are in the bottle? (*Hint*: 1 fluid ounce ≈ 29.5 milliliters)

H.O.T. Problems

43. REASONING Alberto measured his bedroom for new carpet. The room is 169 square feet. When he got to the carpet store, all of the prices were given in square yards. Explain how he could convert his calculations to square yards to determine the cost.

44. CHALLENGE The Art Club is tiling a wall 8 feet tall by 10 feet long. Each tile is a 3-inch square and costs $0.59. What is the cost of tiling the wall?

45. FIND THE ERROR Seth is converting the volume of his shed from cubic feet to cubic inches. Find his mistake and correct it.

$$8{,}640\ ft^3 = 8{,}640\ ft^3 \times \frac{1\ ft^3}{1{,}728\ in^3}$$
$$= \frac{8{,}640\ ft^3}{1{,}728\ in^3}$$
$$= 5\ in^3$$

46. WRITE MATH Compare and contrast one mile with one square mile.

47. Which of the following measurements is NOT equivalent to the other three?

   A. 400,000,000 cm²

   B. 4,000 m²

   C. 0.04 km²

   D. 40,000,000,000 mm²

48. **THINK SOLVE EXPLAIN** **SHORT RESPONSE** Two students were asked to design a dog run with an area of 36 square feet. Name a possible length and width of the dog run in yards.

49. Mrs. Westinghouse's classroom measures 13 feet by 14 feet. What is the approximate area of the classroom in square yards?

   F. 2.2 square yards

   G. 20.2 square yards

   H. 60.7 square yards

   I. 182 square yards

50. **THINK SOLVE EXPLAIN** **EXTENDED RESPONSE** A freight container has the dimensions shown below.

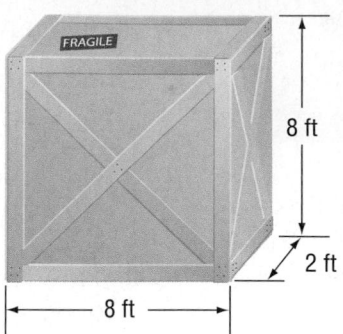

Part A What is the volume of the container in cubic feet?

Part B What is the volume of the container in cubic inches?

Part C What is the volume of the container in cubic meters? Round to the nearest tenth if necessary.

Complete. Round to the nearest tenth if necessary. (Lesson 1E)

51. 24 mi/h = ■ ft/s

52. 39 kg/min = ■ g/s

53. 8.3 m/h ≈ ■ yd/min

54. **WEIGHTLIFTING** The mass of a barbell is 10 pounds. What is its mass in kilograms? Round to the nearest tenth if necessary. (Lesson 1D)

55. **FRUIT** The table shows the amount of fruit Mrs. Roberts bought at the grocery store. Did she buy a greater mass of bananas or peaches? (Lesson 1C)

Fruit	Amount
Bananas	1,151 g
Peaches	1.5 kg

56. **GIRAFFES** The average weight of a giraffe at birth is 110 pounds and the average height is 6 feet. (Lesson 1B)

   a. What is the weight of a giraffe at birth in ounces?

   b. What is the height of a giraffe at birth in inches?

   c. Which is greater, the average height of a giraffe at birth or $1\frac{2}{3}$ yards? Explain your answer.

Mid-Chapter Check

Complete. (Lesson 1B)

1. 42 ft = ■ yd

2. 9 pt = ■ qt

3. 7,600 lb = ■ T

4. $7\frac{1}{2}$ gal = ■ qt

5. **MULTIPLE CHOICE** Which situation is **best** represented by the graph? (Lesson 1B)

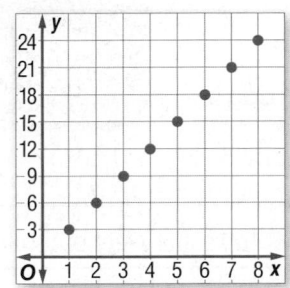

A. conversion of inches to yards

B. conversion of feet to inches

C. conversion of inches to miles

D. conversion of yards to feet

Complete. Round to the nearest hundredth if necessary. (Lessons 1B and 1C)

6. 12.5 mi = ■ yd

7. 4.75 gal = ■ pt

8. 76 cm = ■ m

9. 31.8 kg = ■ g

10. **MEASUREMENT** Bryant was painting lines on the football field and needed to make sure the lines were correctly spaced. He measured the distance to equal 16 feet. What is the distance in yards? (Lesson 1B)

11. **BUILDINGS** The table shows the heights of buildings in the United States. Which building is 0.366 kilometer in height? (Lesson 1C)

Building Location	Height (m)
Empire State Building	381
Bank of America Tower	366
Aon Center	346

Complete. Round to the nearest hundredth if necessary. (Lesson 1D)

12. 11 in. ≈ ■ mm

13. 2.4 T ≈ ■ kg

14. 48 mL ≈ ■ fl oz

15. 30 cm ≈ ■ ft

16. **TRAVEL** A family is driving to Canada, where the speed limit is 100 kilometers per hour. What is the speed limit in miles per hour? (Lesson 1E)

17. **MULTIPLE CHOICE** Which of the following is about the same as 2,088 feet per minute? (Lesson 1E)

F. 0.6264 meter per minute

G. 6.264 meters per minute

H. 389.2 meters per minute

I. 626.4 meters per minute

Complete. (Lesson 1F)

18. 6 ft^2 = ■ in^2

19. 476 in^2 = ■ ft^2

20. 6 yd^3 = ■ ft^3

21. 2 m^3 = ■ cm^3

22. **CONSTRUCTION** A construction worker ordered 5 cubic yards of concrete to pour a sidewalk. How many cubic feet of sidewalk can he pour? (Lesson 1F)

23. What is the volume of the cylinder in cubic centimeters? (Lesson 1F)

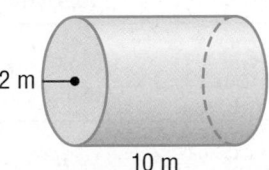

24. Which box has the greater volume? Justify your answer. (Lesson 1F)

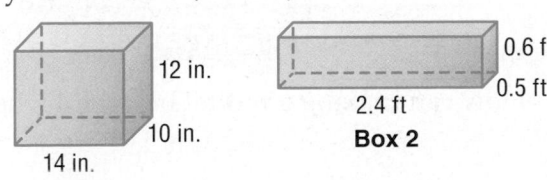

Problem-Solving Investigation

Main Idea Solve problems by making a model.

P.S.I. TEAM ✛

AYITA: I am decorating the school's gymnasium for the spring dance with cubes that will hang from the ceiling. I have 100 square feet of cardboard.

YOUR MISSION: Make a model to find how much cardboard will be needed for each cube if the edge of one cube measures 12 inches.

Understand	You know that each cube is 12 inches long. She has 100 square feet of cardboard.
Plan	Make a cardboard model of a cube with sides 12 inches long. You will also need to determine where to put tabs so that all of the edges are glued together.
Solve	Start with a cube, then unfold it to show the pattern. Five of the edges do not need tabs because they are the fold lines. The remaining 7 edges need a tab. Use $\frac{1}{2}$-inch tabs. 7×12 in. $\times \frac{1}{2}$ in. $\longrightarrow$ 42 in² 7 tabs 6×12 in. $\times 12$ in. $\longrightarrow$ $+\ 864$ in² 6 faces 906 in² total area Convert 906 square inches to square feet. Then divide the total material by the amount of material needed for one cube. $906\ \text{in}^2 \times \dfrac{1\ \text{ft}^2}{144\ \text{in}^2} \approx 6.3\ \text{ft}^2$ $100\ \text{ft}^2 \div 6.3\ \text{ft}^2 \approx 15.9$ So, Ayita has enough cardboard to make 15 cubes.
Check	Make another cube to determine whether all the edges can be glued together using your model.

Analyze the Strategy

1. How can making a model be useful when solving a real-world problem?

= **Step-by-Step Solutions** begin on page R1.
Extra Practice begins on page EP2.

- Make a model.
- Draw a diagram.
- Use logical reasoning.
- Choose an operation.

Use the *make a model* strategy to solve Exercises 2–4.

2. CARS Fiona counted the number of vehicles in the parking lot at a store. She counted a total of 12 cars and motorcycles. If there was a total of 40 wheels, how many cars and motorcycles were there?

3. ART Miguel is making a drawing of his family room for a school project. The room measures 18 feet by 21 feet. He uses a scale of 1 foot $= \frac{1}{2}$ inch. What are the dimensions of the family room on the drawing?

4. MEASUREMENT Francis has a photo that measures 10 inches by $8\frac{1}{2}$ inches. The frame he uses is $1\frac{1}{4}$ inches wide. What is the perimeter of the framed picture?

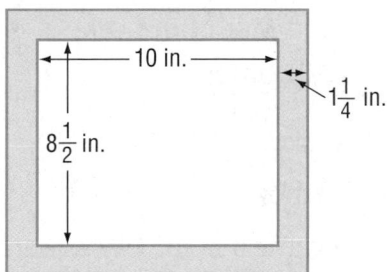

Use any strategy to solve Exercises 5–11.

5. DONATIONS Hickory Point Middle School collected money for a local shelter. The school newspaper reported that more than $5,000 was collected. Is this estimate reasonable? Explain.

Grade	Dollars Collected
Sixth	1,872
Seventh	2,146
Eighth	1,629

6. BIRDHOUSES About how many square inches of the birdhouse will be painted if only the outside of the wood is painted?

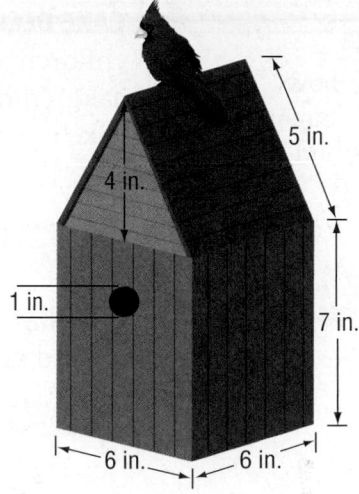

7. MONEY At the beginning of the week, Myra had $45.50. She spent $2.75 each of five days on lunch and bought a sweater for $14.95. Tucker repaid her $10 that he owed her. How much money did she have at the end of the week?

8. BOXES Juliet is placing 20 cereal boxes that measure 8 inches by 2 inches by 12 inches on a shelf that is 3 feet long and 11 inches deep. What is a possible arrangement for the boxes on the shelf?

9. MEASUREMENT A wall measures $15\frac{1}{4}$ feet by $8\frac{3}{4}$ feet and has a window that measures 2 feet by 4 feet. How many square feet of wallpaper are needed to cover the wall?

10. BASEBALL A regulation baseball diamond is a square with an area of 8,100 square feet. Suppose it is on a field that is 172 feet wide and 301 feet long. How much greater is the distance around the whole field than the distance around the diamond?

11. WRITE MATH Write a real-world problem that could be solved by making a model. Then solve the problem.

PART A B C

Explore   Changes in Scale

Main Idea

Determine how changes in dimensions affect area and volume.

 Get ConnectED

CCSS 7.G.6

CAKE DECORATING A cake decorator is making a cake in the shape of a children's board game. The original game is 12 inches by 16 inches. The cake's dimensions will be half as long and half as wide. How does the cake's perimeter and area compare to the game's perimeter and area?

ACTIVITY

1 **What do you need to find?** the change in scale for perimeter and area of the cake

STEP 1 Draw a model of the cake on grid paper where each centimeter represents 1 inch.

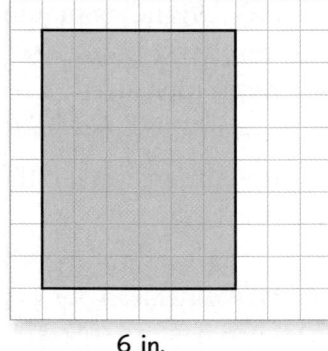

8 in.

6 in.

STEP 2 Find the perimeter of the cake using the grid paper.

8 in. + 8 in. + 6 in. + 6 in. = 28 inches

Compare this to the game's perimeter, 56 inches. Since $\frac{28}{56} = \frac{1}{2}$, the perimeter of the cake is $\frac{1}{2}$ of the perimeter of the board game.

STEP 3 Find the area of the cake.

8 × 6 = 48 square inches

Compare this to the game's area, 192 square inches. Since $\frac{48}{192} = \frac{1}{4}$, the area of the cake is $\frac{1}{4}$ the area of the board game.

Analyze the Results

1. Suppose another cake's dimensions are twice as long and wide as the board game. How do the perimeter and area of the cake compare to the perimeter and area of the game?

2. **MAKE A CONJECTURE** How does changing the dimensions of a figure affect its perimeter? area?

Changing dimensions of a three-dimensional figure also affects the surface area and volume.

ACTIVITY

2 **STEP 1** Find the volume and surface area of one centimeter cube. Then record the data in a table like the one below.

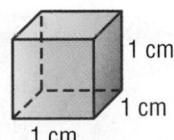

STEP 2 Create a cube with side lengths that are double that of the previous cube. Find the volume and surface area and record the data in the table.

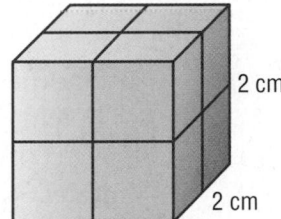

STEP 3 Triple the side lengths of the original centimeter cube. Find the volume and surface area and record the data in the table.

STEP 4 For each cube, write a ratio comparing the side length and the volume. Then write a ratio comparing the side length and the surface area. The first one is done for you.

Side Length (units)	Volume (units³)	Surface Area (units²)	Ratio of Side Length to Volume	Ratio of Side Length to Surface Area
1	$1^3 = 1$	$6 (1^2)$	1 : 1	1 : 6
2				
3				
4				
5				

Practice and Apply

3. **MAKE A CONJECTURE** How does doubling the dimensions of a three-dimensional figure affect the volume of the figure? surface area?

4. **PIZZA** A pizza restaurant advertised the special at the right. The dimensions given are the diameters of each pizza.

   a. What is the area of each pizza? Use $A = 3.14r^2$.

   b. What is the ratio of the area of the personal pizza and the medium pizza?

   c. How does the area of a circle change when the diameter is doubled? tripled?

Main Idea

Solve problems involving similar solids.

Vocabulary

similar solids

Get ConnectED

CCSS 7.G.6

Changes in Dimensions

MODELS Stephen is creating a model of the Washington Monument for history class. The model will be $\frac{1}{100}$ of the monument's actual size.

1. The pyramid that sits atop the monument's obelisk shape has a height of 55.5 feet. What is the height of the pyramid on the model Stephen is creating?

2. **MAKE A CONJECTURE** Write a sentence about the area of the triangular side of the model compared with the actual monument.

Cubes are **similar solids** because they have the same shape and their corresponding linear measures are proportional.

The cubes at the right are similar. The ratio of their corresponding edge lengths is $\frac{8}{4}$ or 2. The scale factor is 2. How are their surface areas related?

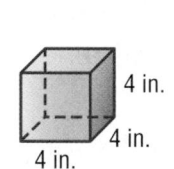

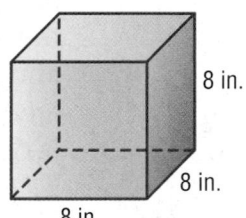

4 in.

4 in.

4 in.

8 in.

8 in.

8 in.

S.A. of Small Cube

$S.A. = 6(4)(4)$

↑

There are 6 faces.

S.A. of Large Cube

$S.A. = 6(\mathbf{2} \cdot 4)(\mathbf{2} \cdot 4)$

$= \mathbf{2} \cdot \mathbf{2}(6)(4 \cdot 4)$

$= \mathbf{2}^2(6)(4 \cdot 4)$

To find the surface area of the large cube, multiply the surface area of the small cube by the *square* of the scale factor, 2^2 or 4. This relationship is true for any similar solids.

> **Key Concept** Surface Area of Similar Solids
>
> If Solid X is similar to Solid Y by a scale factor, then the surface area of X is equal to the surface area of Y times the *square* of the scale factor.

EXAMPLES **Surface Area of Similar Solids**

1 The surface area of a rectangular prism is 78 square centimeters. What is the surface area of a similar prism that is 3 times as large?

$S.A. = 78 \times 3^2$ Multiply by the square of the scale factor.

$S.A. = 78 \times 9$ Square 3.

$S.A. = 702 \text{ cm}^2$ Simplify.

2 **MODELS** Refer to the previous page. The surface area of the exposed portion of the pyramid atop the Washington Monument is 4,012 square feet. What is the surface area in square inches, to the nearest tenth, of the pyramid on Stephen's model?

$S.A. = 4{,}012 \times \left(\dfrac{1}{100}\right)^2$ Multiply by the square of the scale factor.

$S.A. = 4{,}012 \times \dfrac{1}{10{,}000}$ Square $\dfrac{1}{100}$.

$S.A. = 0.4012 \text{ ft}^2$ Simplify.

$S.A. = 0.4012 \; \text{ft} \cdot \text{ft} \times \dfrac{12 \text{ in.}}{1 \text{ ft}} \times \dfrac{12 \text{ in.}}{1 \text{ ft}}$ Convert to inches.

$S.A. \approx 57.8 \text{ in}^2$ Simplify.

The surface area of Stephen's model is 57.8 square inches.

 CHECK Your Progress

a. The surface area of a triangular prism is 34 square inches. What is the surface area of a similar prism that is twice as large?

· · · · · · **b.** **RAISINS** The world's largest box of raisins has a surface area of 352 square feet. If a similar box is smaller than the largest box by a scale factor of $\dfrac{1}{48}$, what is its surface area?

Refer to the cubes on the previous page.

Volume of Small Cube

$V = 4 \cdot 4 \cdot 4$

Volume of Large Cube

$V = (2 \cdot 4)(2 \cdot 4)(2 \cdot 4)$
$= 2 \cdot 2 \cdot 2(4 \cdot 4 \cdot 4)$
$= 2^3(4 \cdot 4 \cdot 4)$

The volumes of similar solids are related by the *cube* of the scale factor.

Key Concept   **Volume of Similar Solids**

If Solid X is similar to Solid Y by a scale factor, then the volume of X is equal to the volume of Y times the *cube* of the scale factor.

EXAMPLE **Volume of Similar Solids**

3 A triangular prism has a volume of 432 cubic yards. If the prism is reduced to one third its original size, what is the volume of the new prism?

$V = 432 \times \left(\frac{1}{3}\right)^3$ Multiply by the cube of the scale factor.

$V = 432 \times \frac{1}{27}$ Cube $\frac{1}{3}$.

$V = 16 \text{ yd}^3$ Simplify.

The volume of the new prism is 16 cubic yards.

✓ CHECK Your Progress

c. A square pyramid has a volume of 512 cubic centimeters. What is the volume of a square pyramid with dimensions one fourth of the original?

 REAL-WORLD EXAMPLE

4 **HOCKEY** The standard hockey puck measures as shown at the right. Find the surface area and volume of the giant puck at the left. Use 3.14 for π.

1.5 in.

1 in.

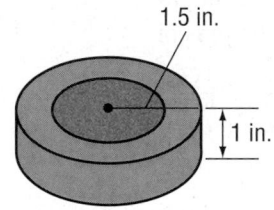

Find the volume and surface area of the standard puck first.

$V = \pi r^2 h$ $S.A. = 2(\pi r^2) + 2\pi rh$

$\approx (3.14)(1.5)^2(1)$ $\approx 2(3.14)(1.5)^2 + 2(3.14)(1.5)(1)$

$\approx 7.065 \text{ in}^3$ $\approx 14.13 + 9.42$

                             $\approx 23.55 \text{ in}^2$

Find the volume and surface area of the giant puck using the scale factor.

$V = V(40)^3$ $S.A. = S.A.(40)^2$

$= (7.065)(40)^3$ $= (23.55)(40)^2$

$= 452{,}160 \text{ in}^3$ $= 37{,}680 \text{ in}^2$

The giant hockey puck has a volume of 452,160 cubic inches and a surface area of 37,680 square inches.

✓ CHECK Your Progress

d. ERASERS The dimensions of a rectangular eraser are 2.4 inches by 4.6 inches by 1 inch. Find the surface area and volume of a similar eraser that is 5 times as large.

Real-World Link · · · ·

The hockey puck that appears to be crashing into the side of the wall at Nationwide Arena in Columbus, Ohio, is about 40 times the actual size of a standard puck.

Example 1   **1.** The surface area of a rectangular prism is 35 square inches. What is the surface area of a similar solid that has been enlarged by a scale factor of 7?

Example 2   **2. MODELS** The surface area of a ship's hull is about 11,000 square meters. What is the surface area, to the nearest tenth, of the hull of a model ship that is smaller by a scale factor of $\frac{1}{100}$?

Example 3   **3.** The volume of a cylinder is about 425 cubic centimeters. What is the volume, to the nearest tenth, of a similar solid that is smaller by a scale factor of $\frac{1}{3}$?

Example 4   **4. ART STUDIO** A sink with a sliding lid in Josh's art studio measures 16 inches by 15 inches by 6 inches. A second sink used just for paintbrushes has a similar shape and is smaller by a scale factor of $\frac{1}{2}$. Find the surface area and volume of the second sink.

Practice and Problem Solving

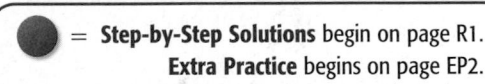

● = **Step-by-Step Solutions** begin on page R1.
Extra Practice begins on page EP2.

Example 1   **5.** The surface area of a rectangular prism is 1,300 square inches. Find the surface area of a similar solid that is larger by a scale factor of 3.

6. The surface area of a triangular prism is 10.4 square meters. What is the surface area of a similar solid that is smaller by a scale factor of $\frac{1}{4}$?

Example 2   **7 FOOD** A cereal box has a surface area of 280 square inches. What is the surface area of a similar box that is larger by a scale factor of 1.4?

8. DISPLAYS A glass display box has a surface area of 378 square inches. How many square inches of glass are used to create a glass display box with dimensions one-half the original?

Example 3   **9.** A cone has a volume of 9,728 cubic millimeters. What is the volume of a similar cone one-eighth the size of the original?

10. A triangular prism has a volume of 350 cubic meters. If the dimensions are tripled, what is the volume of the new prism?

Example 4   **11. ARCHITECTURE** The model of a new apartment building is shown. The architect plans for the building to be 144 times the size of the model. What will be the surface area and volume of the new building when it is completed?

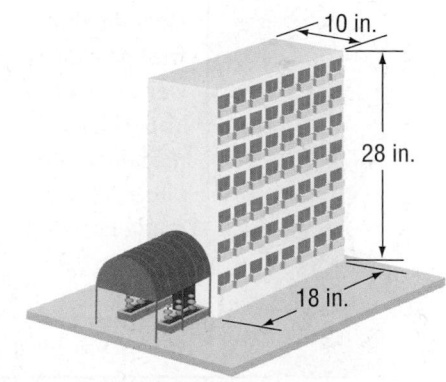

10 in.

28 in.

18 in.

12. PUZZLES The world's largest cube puzzle is in Knoxville, Tennessee. It measures 6 feet on each side. The scale factor between a standard cube puzzle and largest puzzle is $\frac{1}{24}$. Find the surface area and volume of the standard cube puzzle.

13 Two spheres are similar in shape. The scale factor between the smaller sphere and the larger sphere is $\frac{3}{4}$. If the volume of the smaller sphere is 126.9 cubic meters, what is the volume of the larger sphere?

Determine whether each statement is *always*, *sometimes*, or *never* true.

14. Two prisms with equal bases are similar.

15. Similar solids have equal volumes.

16. Two cubes are similar.

17. A prism and pyramid are similar.

18. Find the missing measures for the pair of similar solids.

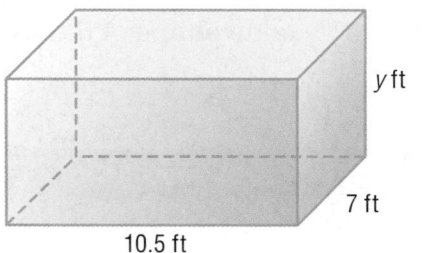
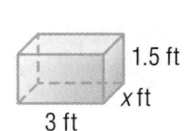

19. Two similar cylinders are shown.

    a. What is the ratio of their radii?

    b. What is the ratio of their surface areas? volumes?

    c. Find the surface area of Cylinder B.

    d. Find the volume of Cylinder A.

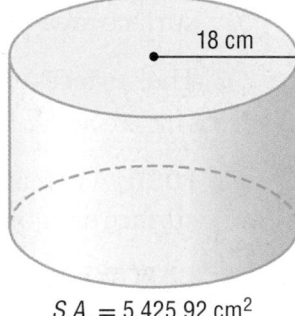
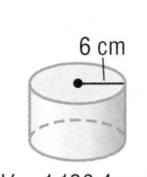

S.A. = 5,425.92 cm²   V = 1,130.4 cm³

Cylinder A   **Cylinder B**

20. GRAPHIC NOVEL Refer to the graphic novel frame below. Use the formula $r = \frac{d}{t}$ to calculate the speeds. Express in miles per hour.

21. **CHALLENGE** A *frustum* is the solid left after a cone is cut by a plane parallel to its base and the top cone is removed.

    a. Is the smaller cone that is removed similar to the original cone? Justify your response.

    b. What is the volume of the smaller cone? the larger cone? Use 3.14 for π.

    c. What is the ratio of the volume of the smaller cone to the volume of the larger cone?

    d. What is the volume of the frustum?

22. **WRITE MATH** Explain what happens to the volume of a prism when its dimensions are tripled.

✔ Test Practice

23. For the similar pyramids, find the ratio of the surface area of the larger pyramid to the smaller pyramid.

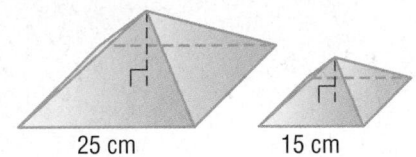

    25 cm   15 cm

    A. $\frac{5}{3}$ C. $\frac{25}{9}$

    B. $\frac{25}{15}$ D. $\frac{10}{6}$

24. Two similar prisms have volumes of 4 cubic meters and 864 cubic meters, respectively. How many times larger is the second prism?

    F. 6 times

    G. 16 times

    H. 96 times

    I. 216 times

25. **ART** Julianna is making a clay figurine of a dog. The dog is 75 centimeters tall. If she uses a scale of 1 centimeter = 10 centimeters, how tall will the clay figurine be? (Lesson 2A)

26. **SPORTS** The table shows the dimensions of the fields used in various sports. (Lesson 1F)

    a. What is the area of the field hockey field in square feet?

    b. What is the difference between the area of the soccer field and the area of the lacrosse field in square feet?

    c. If an acre is 43,560 square feet, about how many acres are all four fields combined?

Sport	Length (yards)	Width (yards)
Field hockey	60	100
Football	$53\frac{1}{3}$	120
Lacrosse	60	110
Soccer	70	115

Food for Thought

Do you love cooking and enjoy trying new recipes? You might want to think about a career in the culinary arts. Research chefs use their culinary skills and their knowledge of food science to develop recipes, menus, and products for restaurant chains and food manufacturers. Research chefs are creative and have a passion for cooking and an understanding of food trends, but they must also have strong computer skills, be knowledgeable about new technologies in cooking and food science, and be able to use mathematics to develop their recipes.

21st Century Careers

Are you interested in a career as a research chef? Take some of the following courses in high school.

- Algebra
- Commercial Foods and Culinary Arts
- Human Nutrition

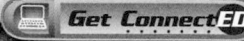 Get ConnectED

Raspberry Lemonade

1 (12 fluid ounce) can frozen raspberry
   lemonade concentrate
710 mL water
4 mL lime juice
354 mL lemon-lime flavored carbonated beverage
235 mL crushed ice
125 g fresh raspberries, garnish
225 g fresh mint, garnish

Banana Split Cake

1 (16-ounce) package vanilla wafers, crushed
1 cup margarine, melted
1 (20-ounce) can crushed pineapple, drained
6 bananas
1 (8-ounce) package cream cheese
2 cups confectioners' sugar
1 (12-ounce) container whipped topping
$\frac{1}{4}$ cup chopped walnuts
8 maraschino cherries

Real-World Math

Use the recipes to solve each problem.

1. How many pounds of crushed pineapple are needed for the cake?

2. How many quarts of sugar are needed for the cake?

3. How many liters of water are used to make the lemonade?

4. How many milligrams of fresh mint are used to make the lemonade?

5. About how many grams of whipped topping are used to make the cake?

6. About how many fluid ounces of carbonated beverage are required for lemonade? Round to the nearest fluid ounce.

FOLDABLES
Study Organizer

Be sure the following Key Concepts are noted in your Foldable.

Key Concepts

Convert Measurements (Lesson 1)

• To convert from larger units to smaller units, multiply by the appropriate unit ratio.

• To convert from smaller units to larger units, multiply by the reciprocal of the appropriate unit ratio.

• When converting metric units, multiply by the appropriate power of 10.

• When converting between customary and metric units, use the appropriate unit ratio.

Similar Solids (Lesson 2)

• Similar solids have the same shape and their corresponding linear measures are proportional.

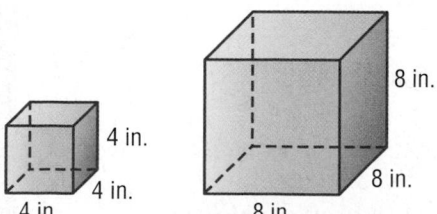

4 in.
4 in.
4 in.

8 in.
8 in.
8 in.

• The areas of similar solids are related by the square of the scale factor. The volumes of similar solids are related by the cube of the scale factor.

Key Vocabulary

derived unit

dimensional analysis

gram

kilogram

liter

meter

metric system

similar solids

unit ratio

Vocabulary Check

Choose the term from the list above that best matches each phrase.

1. the process of including units of measurement as factors when you compute

2. the base unit of capacity

3. a decimal system of measures

4. the base unit of length in the metric system

5. measures mass, which is the amount of matter in an object

6. these have the same shape and their corresponding linear measures are proportional

7. the base unit of mass

8. one in which the denominator is 1 unit

Multi-Part Lesson Review

Lesson 1 Convert Measurements

Convert Customary Units (Lesson 1B)

Complete.

9. 4 qt = ▓ pt

10. 6 gal = ▓ qt

11. 48 oz = ▓ lb

12. 9 c = ▓ pt

13. **RUNNING** Lenora runs 30,000 feet. How many miles does Lenora run?

14. **ESTIMATION** One bushel of apples weighs about 40 pounds. How many bushels of apples would weigh 1 ton?

EXAMPLE 1 Complete: 32 qt = ▓ gal.

Since 1 gallon = 4 quarts, multiply by $\frac{1 \text{ gal}}{4 \text{ qt}}$.

$32 \text{ qt} = 32 \text{ qt} \cdot \frac{1 \text{ gal}}{4 \text{ qt}}$ Multiply by $\frac{1 \text{ gal}}{4 \text{ qt}}$.

$= \overset{8}{\cancel{32}} \text{ qt} \cdot \frac{1 \text{ gal}}{\underset{1}{\cancel{4 \text{ qt}}}}$ Divide out common factors and units.

$= 8 \text{ gal}$ Multiply.

Convert Metric Units (Lesson 1C)

Complete. Round to the nearest hundredth if necessary.

15. 18.25 m = ▓ cm

16. 113.6 g = ▓ kg

17. 24 L = ▓ mL

18. 34 cm = ▓ mm

19. **MASS** Kirk found the mass of his textbook to be 1.02 kilograms. What is the mass of the book in grams?

EXAMPLE 2 Complete: 3.8 km = ▓ m.

Use the relationship 1 km = 1,000 m.

$1 \text{ km} = 1,000 \text{ m}$

$3.8 \times 1 \text{ km} = 3.8 \times 1,000 \text{ m}$

$3.8 \text{ km} = 3,800 \text{ m}$

So, 3.8 kilometers is equal to 3,800 meters.

Convert Between Systems (Lesson 1D)

Complete. Round to the nearest hundredth if necessary.

20. 18.25 ft ≈ ▓ m

21. 113.6 lb ≈ ▓ g

22. 24 L ≈ ▓ gal

23. 46.8 cm ≈ ▓ in.

24. **RUNNING** Justine ran a 10-kilometer race. About how many miles did she run?

25. **BIRDS** The world's largest bird is the ostrich, whose mass can be as much as 156.5 kilograms. What is the approximate weight in pounds?

EXAMPLE 3 Complete: 5.2 mi ≈ ▓ km.

Use the relationship 1 mi ≈ 1.61 km.

$1 \text{ mi} \approx 1.61 \text{ km}$

$5.2 \times 1 \text{ mi} \approx 5.2 \times 1.61 \text{ km}$

$5.2 \text{ mi} \approx 8.372 \text{ km}$

So, 5.2 miles is approximately 8.37 kilometers.

Lesson 1 **Convert Measurements** (continued)

Convert Rates (Lesson 1E)

Convert each rate. Round to the nearest hundredth.

26. $8\frac{1}{4}$ ft/s = ■ yd/s

27. 105.6 L/h = ■ L/min

28. 52 mi/h ≈ ■ km/min

29. 8.4 lb/min = ■ oz/s

30. AIRPLANE An airplane is traveling at an average speed of 245 meters per second. How many kilometers per second is the plane traveling?

31. RIVER A river is flowing at a rate of 45 feet per minute. About how many meters per second is the water flowing?

EXAMPLE 4 Leo drank an average of 0.1 liter of water per hour. About how many gallons of water did he drink per hour?

Since 1 gallon = 3.79 liters, use the unit ratio $\frac{1\text{ gal}}{3.79\text{ L}}$.

$\frac{0.1\text{ L}}{1\text{ h}} \approx \frac{0.1\ \cancel{\text{L}}}{1\text{ h}} \times \frac{1\text{ gal}}{3.79\ \cancel{\text{L}}}$ Multiply by $\frac{1\text{ gal}}{3.79\text{ L}}$. Divide out common units.

$\approx \frac{0.03\text{ gal}}{1\text{ h}}$ Simplify.

So, Leo drank about 0.03 gallon of water per hour.

Convert Units of Area and Volume (Lesson 1F)

Complete. Round to the nearest hundredth if necessary.

32. 0.15 ft³ = ■ in³

33. 0.3 m² = ■ cm²

34. 1.64 yd² = ■ ft²

35. 581 mm³ = ■ cm³

36. AQUARIUM An aquarium holds 3,456 cubic inches of water. How many cubic feet will the aquarium hold?

37. WRAPPING PAPER A gift needed 4.2 square yards of wrapping paper. How many square feet of wrapping paper were used?

EXAMPLE 5 Convert the area of the square to square inches.

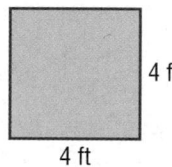

4 ft
4 ft

The area of the square is 4 feet × 4 feet or 16 square feet. Use the ratio $\frac{144\text{ in}^2}{1\text{ ft}^2}$ to find the number of square inches.

$16\text{ ft}^2 = 16\text{ ft}^2 \times \frac{144\text{ in}^2}{1\text{ ft}^2}$ Multiply by $\frac{144\text{ in}^2}{1\text{ ft}^2}$.

$= 16\ \cancel{\text{ft}^2} \times \frac{144\text{ in}^2}{1\ \cancel{\text{ft}^2}}$ Divide out common units, leaving the desired unit, inches.

$= 2{,}304\text{ in}^2$ Simplify.

The area of the square is 2,304 square inches.

Lesson 2 · Similar Solids

PSI: Make a Model (Lesson 2A)

Solve the problems by using the *make a model* strategy.

38. FRAMING A painting 15 inches by 25 inches is bordered by a mat that is 3 inches wide. The frame around the mat is 2 inches wide. Find the area of the picture with the frame and mat.

39. DVDs A video store arranges its bestselling DVDs in their front window. In how many different ways can five bestseller DVDs be arranged in a row?

EXAMPLE 6 The bottom layer of a display of soup cans has 6 cans in it. If each layer has one less can than the one below it and there are 4 layers in the display, how many cans are there in the display?

Based on the model, there are 18 cans.

Changes in Dimensions (Lesson 2C)

40. The two pyramids are similar. The surface area of the larger pyramid is about 108 square meters. Find the surface area of the smaller pyramid if it is $\frac{3}{4}$ the size of the larger pyramid.

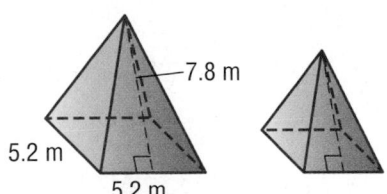

Pyramid 1   Pyramid 2

41. GLUE A cylindrical tube of glue has a volume of 22.0 cubic inches and a surface area of 50.3 square inches. What is the surface area and volume of a similar cylinder that is larger by a scale factor of 2?

42. MODEL A model car has a volume of 1,260 cubic centimeters. What is the volume of a similar car one sixth the size of the original? Round to the nearest tenth.

EXAMPLE 7 The surface area of the box shown is 32 square centimeters. What is the surface area of a similar box that is larger by a scale factor of 4?

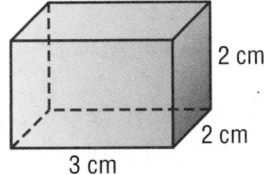

$S.A. = 32 \times 4^2$ or 512

So, the surface area of the similar box that is larger by a scale factor of 4 is 512 square centimeters.

EXAMPLE 8 The volume of the box shown above is 12 cubic centimeters. What is the volume of a similar box that is larger by a scale factor of 4?

$V = 12 \times 4^3$ or 768

So, the volume of the similar box that is larger by a scale factor of 4 is 768 cubic centimeters.

Complete. Round to the nearest hundredth if necessary.

1. 3,600 lb = ■ T

2. 21 pt = ■ qt

3. 28 ft = ■ yd

4. 0.23 g = ■ mg

5. 0.04 L = ■ mL

6. 21 in. = ■ ft

7. MULTIPLE CHOICE The table shows the volume of four containers. What is the approximate total volume of the four containers in gallons?

Container	Volume (oz)
Cup	10
Pitcher	40
Bowl	35
Jug	50

A. 1.5 gal

B. 1.05 gal

C. 0.15 gal

D. 0.11 gal

Complete. Round to the nearest tenth if necessary.

8. 7.62 yd ≈ ■ m

9. 50.8 lb ≈ ■ kg

10. 3,600 mL ≈ ■ qt

11. 19.25 m ≈ ■ ft

12. SCIENCE The population of bacteria in 4 differently sized lab dishes are given. Which dish has an approximate area of 1.4 square feet?

Dish	Bacteria	Area (in²)
1	100	205
2	50	125
3	35	175
4	180	300

13. ROLLER COASTER A roller coaster reaches a speed of 112.7 kilometers per hour. What is the estimated speed in miles per hour?

14. GARDEN The garden below needs to be covered with fertilizer. If the owner has enough fertilizer to cover 200 square yards, will he have enough for the entire garden? Explain your reasoning.

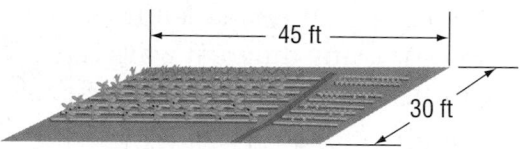

15. PACKING A model of a packing box is below. What is the surface area of a similar box that is larger by a scale factor of 3?

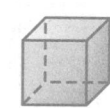

S.A. = 56 in²

16. THINK SOLVE EXPLAIN **EXTENDED RESPONSE** Use the prism below.

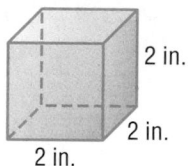

2 in.

2 in.

2 in.

Part A Find the surface area and volume of prisms that are larger by a scale factor of 2, 3, 4, and 5.

Part B Compare the surface area to the volume.

Part C Graph the relationship of the side length to the volume and the side length to the surface area.

Part D Do you think that if the solid were another type of shape, like a pyramid, cylinder, or cone, you would get the same results? Why or why not?

Preparing for Standardized Tests

THINK SOLVE EXPLAIN

Extended Response: Scoring Points

In order to receive all possible points for an extended-response question, an answer must be correct and all work must be shown. An example of a full-credit answer to one part of an extended-response question is shown below.

TEST EXAMPLE

Daniela has a corkboard with a 2-centimeter wide frame around it, as shown at the right. She wants to paint the frame in her school colors. About how many square inches will Daniela need to paint? Use 1 meter ≈ 39.4 inches.

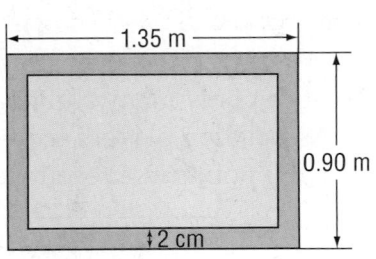

- **Full Credit: 4 points**

$A = \ell \cdot w$
$A_1 = 1.35 \text{ m} \cdot 0.90 \text{ m} = 1.215 \text{ m}^2$
$A_2 = (1.35 \text{ m} - 0.02 \text{ m} - 0.02 \text{ m}) \cdot (0.90 \text{ m} - 0.02 \text{ m} - 0.02 \text{ m})$
$A_2 = 1.31 \cdot 0.86 \text{ m} = 1.1266 \text{ m}^2$
$A_1 - A_2 = 1.215 \text{ m}^2 - 1.1266 \text{ m}^2 = 0.0884 \text{ m}^2$
$1 \text{ m}^2 \approx 1552.36 \text{ in}^2 \rightarrow 0.0884 \text{ m}^2 \cdot \dfrac{1552.36 \text{ in}^2}{1 \text{ m}^2} \approx 137.2 \text{ in}^2$

> All work is shown, answer is correct.

- **Partial Credit: 3 points**
  All work is shown, but an incorrect conversion factor is used, so answer is incorrect.

- **Partial Credit: 2 points**
  The steps shown are incorrect, answer is incorrect.

- **Partial Credit: 1 point**
  No work is shown, answer is incorrect.

Work on It

A company sells cylindrical tins of popcorn in two sizes. The large tin has a diameter of 20 centimeters and a height of 25 centimeters. The small tin has a diameter of 12 centimeters and a height of 18 centimeters. About how many more cubic inches of popcorn does the larger tin hold than the smaller tin? Use 1 centimeter ≈ 0.39 inch.

Test Hint

If you find that you cannot answer every part of an extended-response question, do as much as you can. You may earn partial credit.

Read each question. Then fill in the correct answer on the answer sheet provided by your teacher or on a sheet of paper.

1. Dee moves to a new house. Her old bedroom measured 5 meters by 3 meters. Her new bedroom is 2 meters longer and 1 meter wider. How much more area will she have in her new bedroom?

   A. 13 m²   C. 21 m²

   B. 15 m²   D. 30 m²

2. About how many National Basketball Association players scored at least 1,200 points in a recent year?

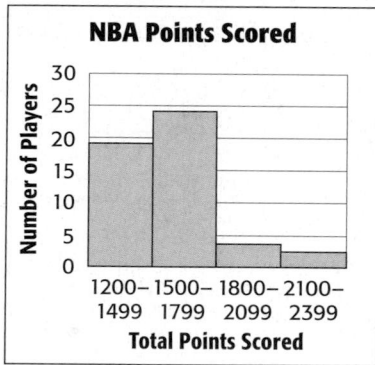

   NBA Points Scored

   F. 7   H. 24

   G. 19   I. 50

3. ≡≡≤ **GRIDDED RESPONSE** The diagram below shows two triangular deck areas along two different houses. The base of Deck B is 4 times the base of Deck A.

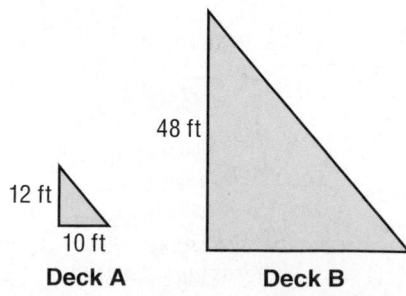

   What is the base, in feet, of Deck B?

4. Courtney ran at a rate of 6.5 miles per hour. At this rate, about how many kilometers can she run in 2 hours?

   A. 10.5 km   C. 21 km

   B. 13 km   D. 24.8 km

5. ≡≡≤ **GRIDDED RESPONSE** Mrs. Black is making 2 pasta salads for a picnic. The first pasta salad requires $4\frac{2}{3}$ cups of pasta and the second pasta salad requires $\frac{1}{3}$ cup more than the first. How many cups of pasta are needed for the second recipe?

6. Amelia carried a package that had a mass of 6.38 kilograms. About how many pounds does the package weigh?

   F. 14.1 lb   H. 9.5 lb

   G. 12.9 lb   I. 6.4 lb

7. The table lists interest rates for savings accounts.

Student Savings and Loan	
Time	Rate
6 months	2.9%
12 months	3.5%
18 months	3.65%
24 months	3.7%

   What is the simple interest earned on $1,500 for 18 months?

   A. $21.75

   B. $52.50

   C. $79.24

   D. $82.13

8. ≡≡≤ **GRIDDED RESPONSE** A building is 55 meters tall. About how tall is the building in feet? (1 meter ≈ 39 inches)

9. A shoe store had to increase prices. The table shows the regular price *r* and the new price *n* of several shoes. Which of the following formulas can be used to calculate the new price?

Shoe	Regular Price (*r*)	New Price (*n*)
A	$25.00	$27.80
B	$30.00	$32.80
C	$35.00	$37.80
D	$40.00	$42.80

F. $n = r - 2.80$ **H.** $n = r \times 0.1$

G. $n = r + 2.80$ **I.** $n = r \div 0.1$

10. **THINK SOLVE EXPLAIN** **SHORT RESPONSE** Nancy wanted to make a model of her room. She measured the room and it was 12 feet by 10 feet. She decided to make a model that was $\frac{1}{16}$ as long and $\frac{1}{16}$ as wide as her room. Find the area of the model room. Explain your answer.

11. Nyomi recorded the amount of time she worked each day last week. If she worked 0.75 of her permitted weekly hours, how many hours is she permitted to work?

Time Worked in One Week	
Day	**Time (h)**
Monday	1
Wednesday	1.25
Friday	1.5
Saturday	3

A. 5 h   **C.** 9 h

B. 6 h   **D.** 12 h

12. What is the volume of the cone in cubic feet? Round to the nearest tenth. Use 3.14 for π.

F. 21.2 ft³

G. 75.8 ft³

H. 84.8 ft³

I. 254.5 ft³

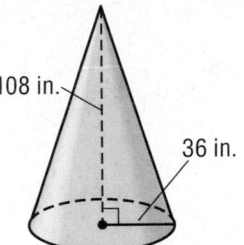

108 in.

36 in.

13. **THINK SOLVE EXPLAIN** **EXTENDED RESPONSE** A movie theater sells popcorn in boxes like the one shown below. The manager wants to sell a new size that is larger by a scale factor of 1.5.

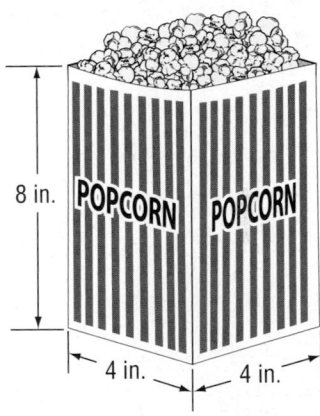

8 in.

POPCORN POPCORN

4 in. 4 in.

Part A What is the volume of the original popcorn box?

Part B What is the volume of a similar popcorn box that is larger by a scale factor of 1.5?

Part C The surface area of the popcorn box is 144 square inches. What would be the surface area of the similar popcorn box that is larger by a scale factor of 1.5?

NEED EXTRA HELP?													
If You Missed Question...	1	2	3	4	5	6	7	8	9	10	11	12	13
Go to Chapter-Lesson...	11-2A	9-3C	5-3A	11-1D	3-2D	11-1D	6-3E	11-1D	7-1B	11-2C	6-2B	11-1F	11-2C

Polygons and Transformations

connectED.mcgraw-hill.com

Investigate

 Animations

 Vocabulary

 Multilingual eGlossary

Learn

 Personal Tutor

 Virtual Manipulatives

Graphing Calculator

Audio

 Foldables

Practice

 Self-Check Practice

Worksheets

 Assessment

The ☆BIG Idea

How does the transformation of a figure affect its size, shape, and position?

FOLDABLES®
Study Organizer

Make this Foldable to help you organize your notes.

Polygons

Review Vocabulary

ordered pair *par ordenado* a pair of numbers used to locate a point in the coordinate plane; an ordered pair is written in the form (x-coordinate, y-coordinate)

The x-coordinate corresponds to a number on the x-axis. → (3, —2) ← The y-coordinate corresponds to a number on the y-axis.

Key Vocabulary

English	Español
dilation	dilatación
quadrilateral	cuadrilatero
reflection	reflexión
rotation	rotación
translation	translación

When Will I Use This?

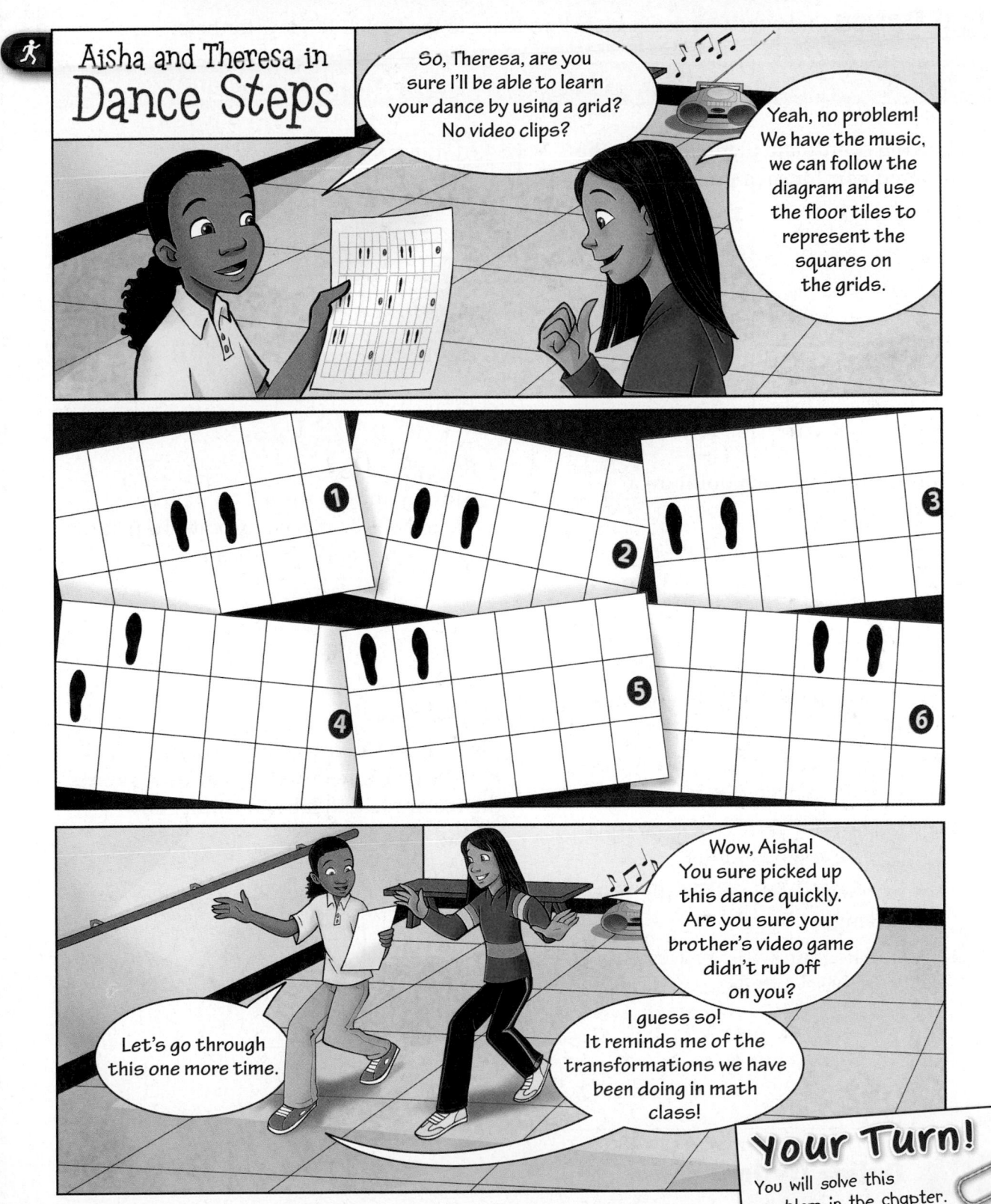

Are You Ready for the Chapter?

You have two options for checking prerequisite skills for this chapter.

Text Option Take the Quick Check below. Refer to the Quick Review for help.

QUICK Check

Use the coordinate plane to name the ordered pair for each point.

1. A **2.** B **3.** C **4.** D

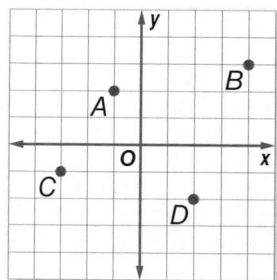

Graph and label each point on a coordinate plane.

5. $G(4, -2)$ **6.** $R(-2, -5)$

7. $P(0, -3)$ **8.** $Q(-3, 3)$

Graph each figure and label the missing vertices.

9. rectangle with vertices: $B(-3, 3)$, $C(-3, 0)$; side length: 6 units

10. triangle with vertices: $Q(-2, -4)$, $R(2, -4)$; height: 4 units

11. square with vertices: $G(5, 0)$, $H(0, 5)$; side lengths: 5 units

12. parallelogram with vertices: $A(0, 0)$, $B(2, 3)$; base: 4 units

QUICK Review

EXAMPLE 1

Graph and label the point $Z(-3, 2)$.

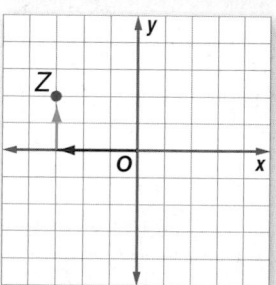

- Start at the origin.
- Move 3 units to the left on the x-axis.
- Then move 2 units up to locate the point.
- Draw a dot and label the dot Z.

EXAMPLE 2

Two vertices of a rectangle are $J(3, 2)$ and $K(1, 2)$. The side length is 4 units. Graph the rectangle and label the other two vertices.

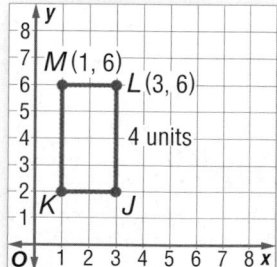

Online Option Take the Online Readiness Quiz.

 **Reading Math**

The Language of Mathematics

Sometimes everyday or scientific usage can give you clues to the mathematical meaning of words. Here are some examples.

> Many of the words you use in math are also used in everyday language.

Usage	Example
Some words are used in English and in mathematics, but have distinct meanings.	leg
Some words are used in science and in mathematics, but the meanings are different.	$x + 4 = -2$ $x = -6$ solution
Some words are used only in mathematics.	hypotenuse

Practice

Explain how the mathematical meaning of each word compares to its everyday meaning.

1. factor

2. rational

Explain how the mathematical meaning of each word compares to its meaning in science.

3. radical

4. variable

Some words are used in English and in mathematics, but the mathematical meaning is more precise. Explain how the mathematical meaning of each word is more precise than the everyday meaning.

5. similar

6. real

Main Idea

Classify and identify angles and find missing measures.

Vocabulary

angle
straight angle
vertical angles
adjacent angles
complementary angles
supplementary angles
alternate interior angles
alternate exterior angles
corresponding angles

Get Connect**ED**

CCSS 7.G.5

Angle Relationships

BIKE RAMPS The angle of descent of a bike ramp is shown.

1. Estimate the measure of the angle of descent.

2. Suppose the angle of descent needs to be 90°. Using your estimate, determine how many degrees the angle of descent would need to be increased.

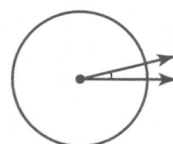

An **angle** has two sides that share a common endpoint called a **vertex**. Angles are measured in units called **degrees**. If a circle were divided into 360 equal-size parts, each part would have an angle measure of 1 degree (1°). **Congruent angles** have the same measure.

EXAMPLE **Naming Angles**

1. **Name the angle at the right in four ways.**

   - Use the vertex as the middle letter and a point from each side. The symbol for angle is ∠.
     ∠LMN or ∠NML

   - Use the vertex only.
     ∠M

   - Use a number.
     ∠1

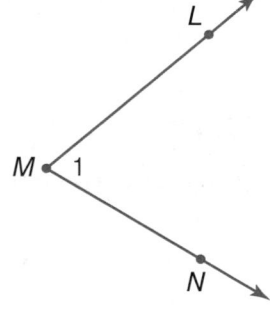

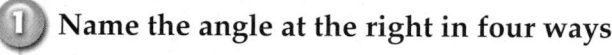

Key Concept **Types of Angles**

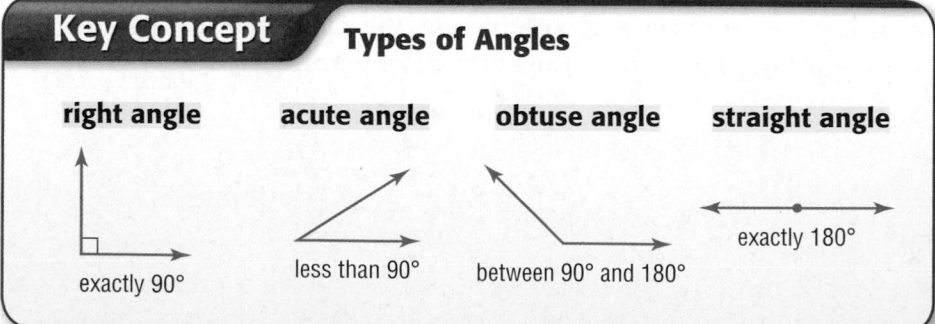

right angle — exactly 90°

acute angle — less than 90°

obtuse angle — between 90° and 180°

straight angle — exactly 180°

  **EXAMPLES** Classify Angles

Classify each angle as *acute, obtuse, right,* or *straight.*

2

3

The angle is less than 90°, so it is an acute angle.

The angle is between 90° and 180°, so it is an obtuse angle.

CHECK Your Progress

a.

b.

c.

Pairs of angles can also be classified by their relationship.

Read Math

Congruent The notation ∠1 ≅ ∠2 is read *angle 1 is congruent to angle 2.*

Read Math

Angle Measure The notation $m\angle 1$ is read *the measure of angle 1.*

Key Concept   **Pairs of Angles**

Words	Models	Symbols
When two lines intersect, they form two pairs of opposite angles, called **vertical angles**, that are congruent.	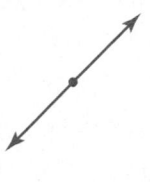 ∠1 and ∠2, ∠3 and ∠4	∠1 ≅ ∠2 ∠3 ≅ ∠4
Two angles that share a vertex and a common side and do not overlap are called **adjacent angles**.		$m\angle ABC = m\angle 5 + m\angle 6$
If the sum of the measures of two angles is 90°, they are called **complementary angles**.		$m\angle 7 + m\angle 8 = 90°$
If the sum of the measures of two angles is 180°, they are called **supplementary angles**.		$m\angle 9 + m\angle 10 = 180°$

EXAMPLES Classify Angles

Identify each pair of angles as *complementary*, *supplementary*, or *neither*.

4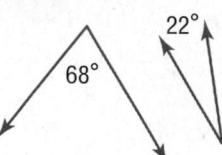

$68° + 22° = 90°$

The angles are complementary.

5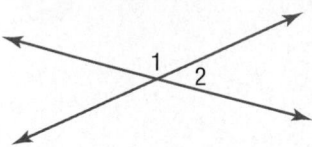

∠1 and ∠2 form a straight angle. So, the angles are supplementary.

CHECK Your Progress

d.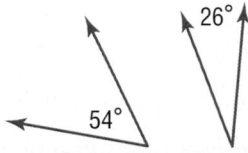

e.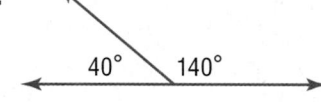

REAL-WORLD EXAMPLE Find a Missing Angle Measure

6 **SIDEWALKS** The two adjacent angles in the sidewalk form a straight line. Find the missing angle measure.

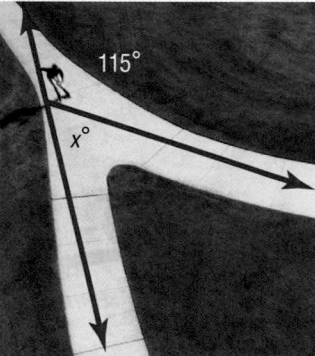

Write an equation.

first angle		second angle		
115	+	x	=	180

Solve the equation.

$$115 + x = 180 \quad \text{Write the equation.}$$
$$\underline{-115 = -115} \quad \text{Subtract 115 from each side.}$$
$$x = 65 \quad \text{Simplify.}$$

So, the value of x is 65.

CHECK Your Progress

f. **ALGEBRA** Find the value of x.

g. Find the measure of ∠3.

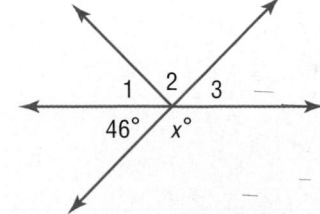

Examples 1–3 Name each angle in four ways. Then classify the angle as *acute*, *right*, *obtuse*, or *straight*.

1.

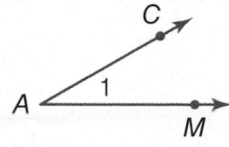

2.

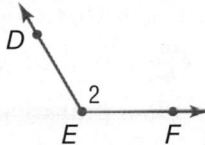

Examples 4 and 5 Identify each pair of angles as *complementary*, *supplementary*, or *neither*.

3.

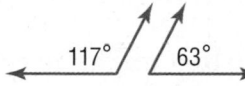

4.

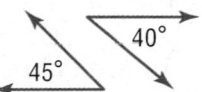

Example 6 **5. BRACES** The picture shows a support brace for a gate. Find the value of *x*.

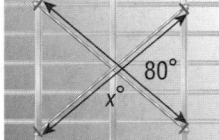

Practice and Problem Solving

● = **Step-by-Step Solutions** begin on page R1.
Extra Practice begins on page EP2.

Examples 1–3 Name each angle in four ways. Then classify the angle as *acute*, *right*, *obtuse*, or *straight*.

6.

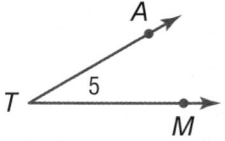

7

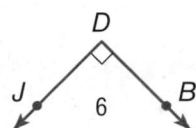

8.

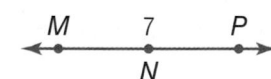

9.

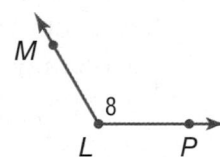

Examples 4 and 5 Identify each pair of angles as *complementary*, *supplementary*, or *neither*.

10.

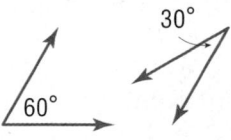

11.

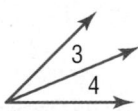

12.

13.

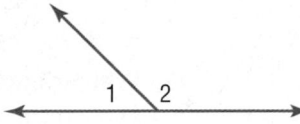

Example 6 **14. LEAVES** What is the measure of the angle between the veins of the leaf, *x*?

15. ALGEBRA If ∠X and ∠Y are complementary and the measure of ∠X is 37°, what is the measure of ∠Y?

16. ALGEBRA What is the measure of $\angle S$ if $\angle S$ and $\angle T$ are supplementary and the measure of $\angle T$ is 109°?

Use the figure at the right to name the following.

17 a pair of congruent angles

18. a pair of supplementary angles

19. a straight angle

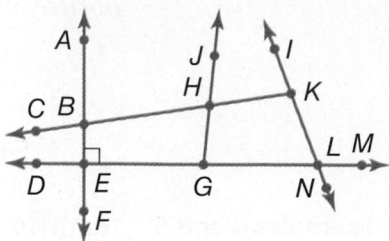

In each diagram describe an angle that might be considered *acute, obtuse, right,* **or** *straight.*

20.

21.

H.O.T. Problems

22. CHALLENGE Determine whether the statement is *true* or *false*. If the statement is true, draw a diagram to support it. If the statement is false, explain why.

> *An obtuse angle and an acute angle are always supplementary.*

Study Tip

Diagrams Sometimes it is helpful to draw a diagram when solving for a variable.

23. CHALLENGE Angles K and O are supplementary. If $m\angle K = 3x + 10$ and $m\angle O = 10x - 12$, find the measure of each angle.

24. REASONING Explain the statement below.

> *If two angles are right angles, they must be supplementary.*

25. Which One Doesn't Belong? Identify the term that does not belong with the other three. Justify your response.

acute	right	complementary	obtuse

26. WRITE MATH Describe the difference between complementary and supplementary angles.

27. Which word best describes the angle marked in the figure?

angle

A. acute   C. right

B. obtuse   D. straight

28. What is the value of x?

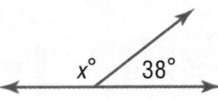

$x°$ $38°$

F. 38

G. 52

H. 142

I. not enough information

More About Angle Relationships

Lines in a plane that never intersect are **parallel lines**. When two parallel lines are intersected by a third line, this line is called a **transversal**.

Key Concept Transversals and Angles

If a pair of parallel lines is intersected by a transversal, these pairs of angles are congruent.

Alternate interior angles are on opposite sides of the transversal and inside the parallel lines.

$$\angle 3 \cong \angle 5, \angle 4 \cong \angle 6$$

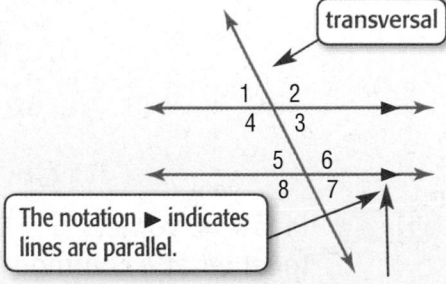

transversal

The notation ▶ indicates lines are parallel.

Alternate exterior angles are on opposite sides of the transversal and outside the parallel lines.

$$\angle 1 \cong \angle 7, \angle 2 \cong \angle 8$$

Corresponding angles are in the same position on the parallel lines in relation to the transversal.

$$\angle 1 \cong \angle 5, \angle 2 \cong \angle 6,$$
$$\angle 3 \cong \angle 7, \angle 4 \cong \angle 8$$

Classify each pair of angles shown in the figure.

29. $\angle 1$ and $\angle 5$

30. $\angle 3$ and $\angle 5$

31. $\angle 6$ and $\angle 4$

32. $\angle 7$ and $\angle 1$

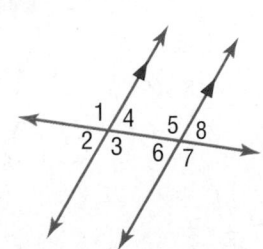

In the figure, if $m\angle 2 = 74°$, find each measure.

33. $m\angle 8$ **34.** $m\angle 6$ **35.** $m\angle 4$ **36.** $m\angle 1$

Explore Triangles

Main Idea

Demonstrate that the sum of the angles in a triangle is 180°.

CONSTRUCTION Natasha is helping her dad build a deck for their backyard. The supports for the deck are triangular. What is the sum of the measures of the three angles of each triangle?

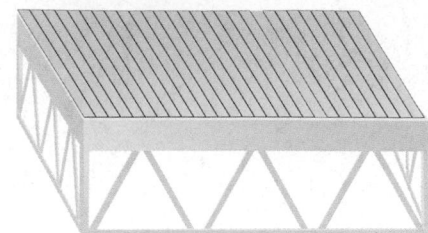

🏃 ACTIVITY

STEP 1 Use a straightedge to draw a triangle with three acute angles. Label the angles *A*, *B*, and *C*. Cut out the triangle.

STEP 2 Cut off the regions around the corners.

STEP 3 Place the corners together at a common meeting point.

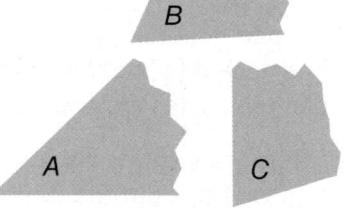

Analyze the Results

1. What kind of angle is formed where the three vertices meet?

2. **MAKE A CONJECTURE** Repeat the activity with another triangle. Make a conjecture about the sum of the measures of the angles of any triangle.

3. *True* or *false*: A triangle can have more than one obtuse angle. Justify your response.

Main Idea

Identify and classify triangles and find missing angle measures.

Vocabulary

triangle
congruent segments
acute triangle
right triangle
obtuse triangle
scalene triangle
isosceles triangle
equilateral triangle

Triangles

Explore

RAMPS Frederick is building a ramp to his house entrance. The incline cannot be greater than 12°.

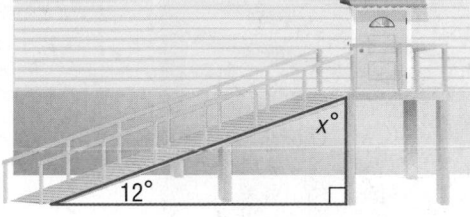

1. What kind of triangle is shown in the figure?

2. Make a prediction about the relationship between the 12° angle and the unknown angle.

A **triangle** is a figure with three sides and three angles. The symbol for triangle is △.

Key Concept — Angles of a Triangle

Words The sum of the measures of the angles of a triangle is 180°.

Model

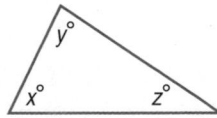

Algebra $x + y + z = 180$

 EXAMPLE **Find a Missing Measure**

 ALGEBRA Find $m\angle Z$.

The sum of the angle measures in a triangle is 180°.

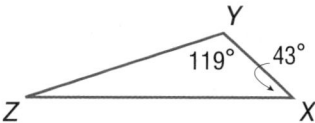

$m\angle Z + 43° + 119° = 180°$ Write the equation.

$m\angle Z + 162° = 180°$ Simplify.

$\underline{\qquad -162° = -162°}$ Subtract 162° from each side.

$m\angle Z \qquad = 18°$

So, $m\angle Z$ is 18°.

 CHECK Your Progress

a. ALGEBRA In △ABC, if $m\angle A = 25°$ and $m\angle B = 108°$, what is $m\angle C$?

 REAL-WORLD EXAMPLE

② **FLAGS** The Alabama state flag is shown. What is the missing measure in the triangle?

To find the missing measure, write and solve an equation.

$$x + 110 + 35 = 180$$ The sum of the measures is 180.

$$x + 145 = 180$$ Simplify.

$$-145 = -145$$ Subtract 145 from each side.

$$x = 35$$

The missing measure is 35°.

 CHECK Your Progress

b. BICYCLES The frame of a bicycle shows a triangle. What is the missing measure?

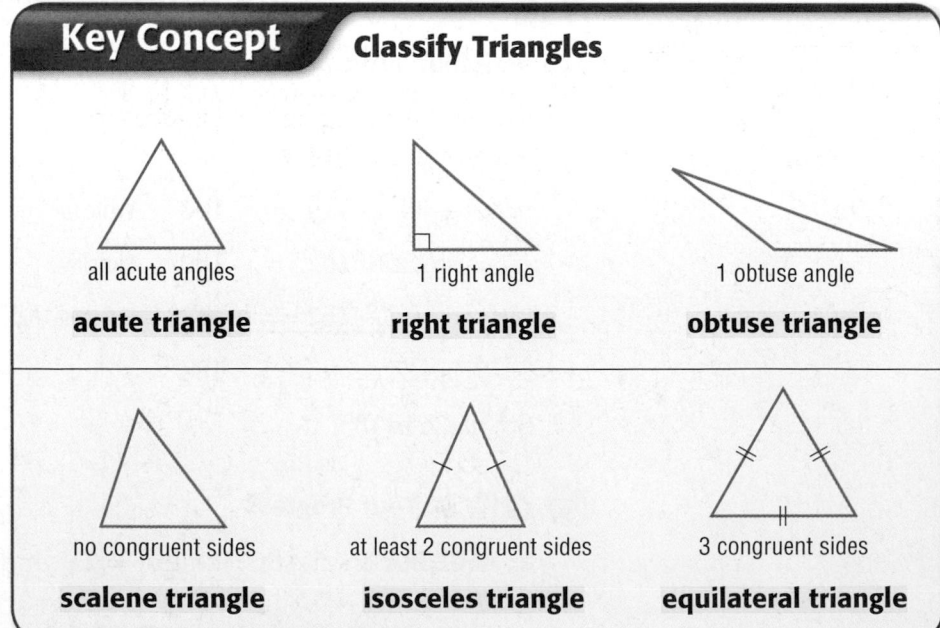

Every triangle has at least two acute angles. One way you can classify a triangle is by using the third angle. Another way to classify triangles is by their sides. Sides with the same length are **congruent segments**.

Key Concept   **Classify Triangles**

all acute angles	1 right angle	1 obtuse angle
acute triangle	**right triangle**	**obtuse triangle**
no congruent sides	at least 2 congruent sides	3 congruent sides
scalene triangle	**isosceles triangle**	**equilateral triangle**

 **Read Math**

Congruent Segments
The tick marks on the sides of the triangle indicate that those sides are congruent.

Real-World Link · · · ·
There are two main types of roofs—flat and pitched. Most houses have pitched, or sloped, roofs.

3 Classify the marked triangle at the right by its angles and by its sides.

The triangle on the side of a house has one obtuse angle and two congruent sides. So, it is an obtuse isosceles triangle.

 **CHECK Your Progress**

c. 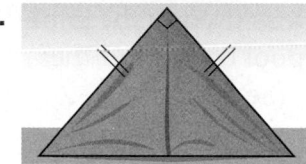 **d.**

EXAMPLES Draw Triangles

4 Draw a triangle with one right angle and two congruent sides. Then classify the triangle.

Draw a right angle. The two segments should be congruent.

Connect the two segments to form a triangle.

The triangle is a right isosceles triangle.

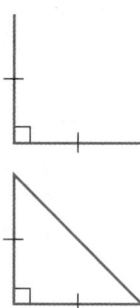

5 Draw a triangle with one obtuse angle and no congruent sides. Then classify the triangle.

Draw an obtuse angle. The two segments of the angle should have different lengths.

Connect the two segments to form a triangle.

The triangle is an obtuse scalene triangle.

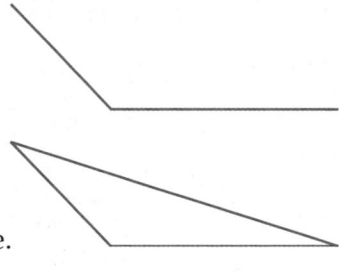

CHECK Your Progress

Draw a triangle that satisfies each set of conditions below. Then classify each triangle.

e. a triangle with three acute angles and three congruent sides

f. a triangle with one right angle and no congruent sides

Example 1   **Find the value of *x*.**

1.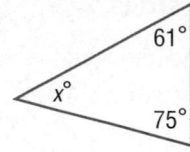

2.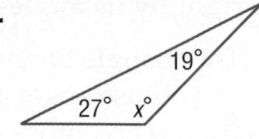

3.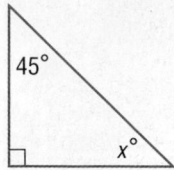

4. ALGEBRA Find *m∠T* in △*RST* if *m∠R* = 37° and *m∠S* = 55°.

Example 2   **5. BILLIARDS** A triangle is used in the game of pool to rack the pool balls. Find the missing measure of the triangle.

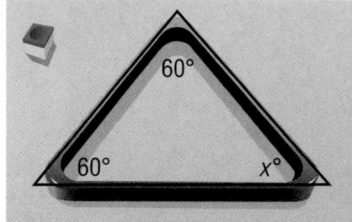

Example 3   **NATURE** Classify the marked triangle in each object by its angles and by its sides.

6.

7.

8.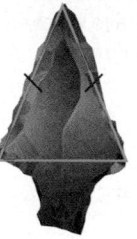

Examples 4 and 5   **DRAWING TRIANGLES** Draw a triangle that satisfies each set of conditions. Then classify each triangle.

9. a triangle with three acute angles and two congruent sides

10. a triangle with one obtuse angle and two congruent sides

Practice and Problem Solving

● = **Step-by-Step Solutions** begin on page R1.
Extra Practice begins on page EP2.

Examples 1 and 2   **Find the value of *x*.**

11.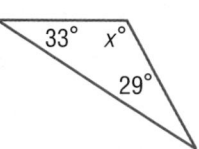

12.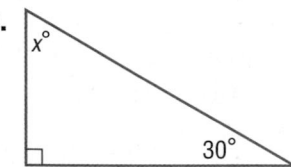

13.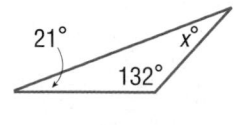

14.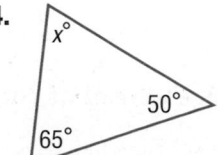

15.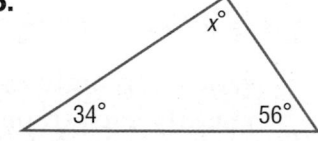

16.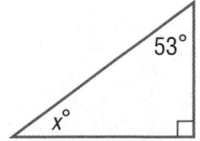

17 ALGEBRA Find *m∠Q* in △*QRS* if *m∠R* = 25° and *m∠S* = 102°.

18. ALGEBRA In △*EFG*, *m∠F* = 46° and *m∠G* = 34°. What is *m∠E*?

Example 3 Classify the marked triangle in each object by its angles and by its sides.

19.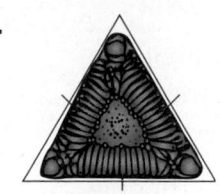

20.

21.

22.

23.

24.

25. ART The sculpture at the right is entitled *Texas Triangles*. It is located in Lincoln, Massachusetts. What type of triangle is shown: *acute, right,* or *obtuse*?

26. ARCHITECTURE Use the photo at the left to classify the side view of the Transamerica building by its angles and by its sides.

Examples 4 and 5 **DRAWING TRIANGLES** Draw a triangle that satisfies each set of conditions. Then classify each triangle.

27. a triangle with three acute angles and no congruent sides

28. a triangle with one obtuse angle and two congruent sides

29. a triangle with three acute angles and three congruent sides

30. a triangle with one right angle and no congruent sides

Find the missing measure in each triangle with the given angle measures.

31. $80°, 20.5°, x°$

32. $75°, x°, 50.2°$

33. $x°, 10.8°, 90°$

34. $45.5°, x°, 105.6°$

35. $x°, 140.1°, 18.6°$

36. $110.2°, x°, 35.6°$

37 **GEOMETRY** Triangle *ABC* is formed by two parallel lines and two transversals. Find the measure of each interior angle *A*, *B*, and *C* of the triangle.

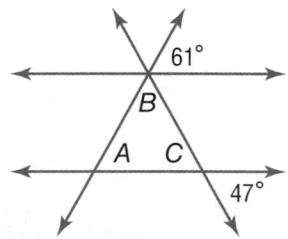

ALGEBRA Find the value of x in each triangle.

38.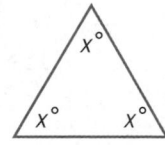

39.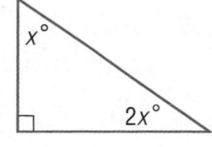

40.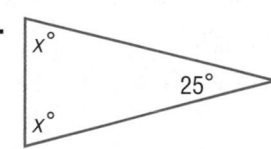

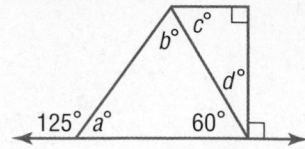

41. CHALLENGE Apply what you know about triangles to find the missing angle measures in the figure.

42. OPEN ENDED Draw an acute scalene triangle. Describe the angles and sides of the triangle.

43. REASONING Determine whether each statement is *sometimes*, *always*, or *never* true. Justify your answer.

  a. It is possible for a triangle to have two right angles.

  b. It is possible for a triangle to have two obtuse angles.

44. **WRITE MATH** An equilateral triangle not only has three congruent sides, but also has three congruent angles. Explain why it is impossible to draw an equilateral triangle that is either right or obtuse.

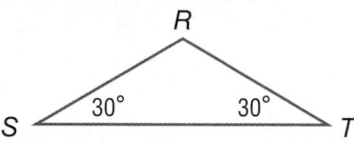

45. How would you find $m\angle R$?

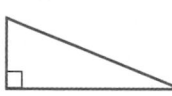

  A. Add 30° to 180°.

  B. Subtract 60° from 180°.

  C. Subtract 30° from 90°.

  D. Subtract 180° from 60°.

46. Which of the following is an acute triangle?

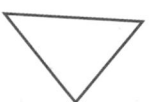

47. **GRIDDED RESPONSE** What is the value of b in the triangle below?

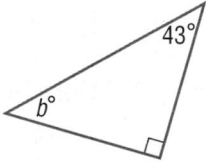

48. Which statement is true about the relationship between the measures of $\angle A$ and $\angle B$, two acute angles of an obtuse triangle?

  A. $m\angle A + m\angle B = 90°$

  B. $m\angle A + m\angle B = 180°$

  C. $m\angle A + m\angle B > 90°$

  D. $m\angle A + m\angle B < 90°$

Spiral Review

Classify each pair of angles as *complementary*, *supplementary*, or *neither*. (Lesson 1A)

49. angle 1: 35°
angle 2: 55°

50. angle 1: 62°
angle 2: 108°

51.

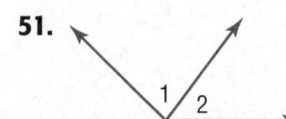

Main Idea

Classify quadrilaterals and find missing angle measures.

 **Vocabulary**

quadrilateral
parallelogram
rectangle
square
rhombus
trapezoid

Get ConnectED

Quadrilaterals

Explore The figure below is a **quadrilateral**, since it has four sides and four angles.

Step 1 Draw a quadrilateral.

Step 2 Pick one vertex and draw the diagonal to the opposite vertex.

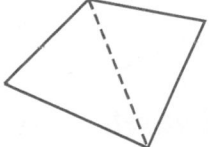

1. What shapes were formed when you drew the diagonal? How many figures were formed?

2. **MAKE A CONJECTURE** Use the relationship among the angle measures in a triangle to find the sum of the angle measures in a quadrilateral. Explain.

3. Find the measure of each angle of your quadrilateral. Compare the sum of these measures to the sum you found in Exercise 2.

The angles of a quadrilateral have a special relationship.

Key Concept **Angles of a Quadrilateral**

Words The sum of the measures of the angles of a quadrilateral is 360°.

Model 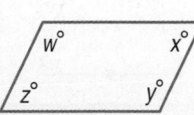 **Symbols** $w + x + y + z = 360$

 EXAMPLE **Find Angle Measures**

1 **Find the value of x in the quadrilateral.**

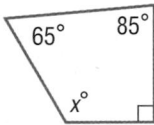

Since the sum of the angle measures in a quadrilateral is 360°, $x + 65 + 85 + 90 = 360$.

$$x + 65 + 85 + 90 = 360 \quad \text{Write the equation.}$$
$$x + 240 = 360 \quad \text{Add 65, 85, and 90.}$$
$$\underline{-240 = -240} \quad \text{Subtract 240 from each side.}$$
$$x = 120$$

So, the value of x is 120.

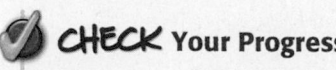 **CHECK** Your Progress

Find the value of x.

a.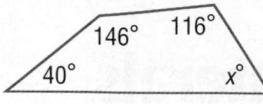

b.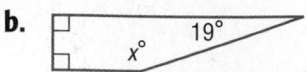

Key Concept		**Classifying Quadrilaterals**
Quadrilateral	**Figure**	**Characteristics**
Parallelogram		• Opposite sides congruent • Opposite sides parallel • Opposite angles congruent
Rectangle		• Opposite sides congruent • Opposite sides parallel • All angles are right angles
Square		• All sides congruent • Opposite sides parallel • All angles are right angles
Rhombus		• All sides congruent • Opposite sides parallel • Opposite angles congruent
Trapezoid		• Exactly one pair of opposite sides parallel

 Read Math

Congruent Angles The red arcs show congruent angles.

 Read Math

Perpendicular Lines The red square corner indicates a perpendicular line. Perpendicular lines intersect to form a right angle.

REAL-WORLD EXAMPLE **Classify Quadrilaterals**

② **QUILTS** Classify the quadrilaterals labeled 1 and 2 in the quilt piece.

Figure 1 is a square. Figure 2 is a rhombus.

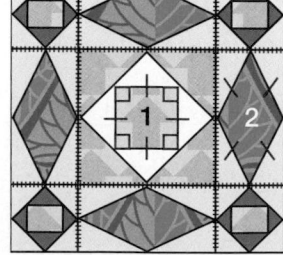

CHECK Your Progress

c. LOGOS Classify the quadrilaterals used in the logo below.

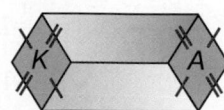

You can use the properties of quadrilaterals to find a missing measure.

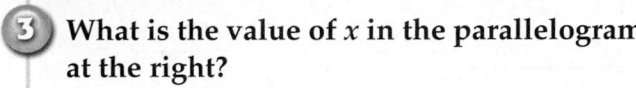 **EXAMPLE** **Find a Missing Measure**

3 **What is the value of x in the parallelogram at the right?**

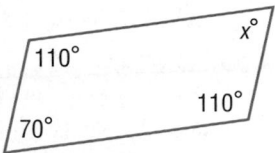

Opposite angles of a parallelogram are congruent. So, $x = 70$.

Check You know the angles in a quadrilateral add to 360°. Since $70° + 110° + 70° + 110° = 360°$, the answer is reasonable. ✔

CHECK Your Progress

d. Find the measure in degrees of $\angle P$ in the rhombus.

e. Find the measure in degrees of $\angle Q$ in the rhombus.

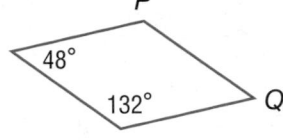

✓ CHECK Your Understanding

Example 1 **Find the value of x in each quadrilateral.**

1.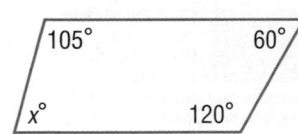

2.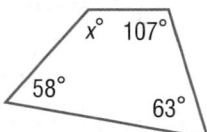

Example 2 **3** **SIGNS** Classify each quadrilateral.

Example 3 **4.** Find the value of x in the parallelogram.

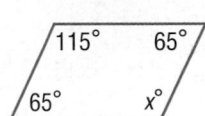

5. In the quadrilateral shown, $\angle ABC$ is congruent to $\angle ADC$. What is the measure of $\angle ABC$?

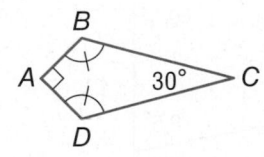

Practice and Problem Solving

 = **Step-by-Step Solutions** begin on page R1.
Extra Practice begins on page EP2.

Examples 1 and 3   **Find the value of x in each quadrilateral.**

6.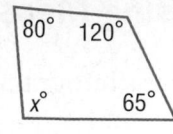
80° 120°
$x°$ 65°

7.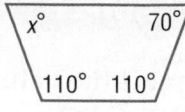
$x°$ 70°
110° 110°

8.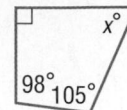
$x°$
98° 105°

9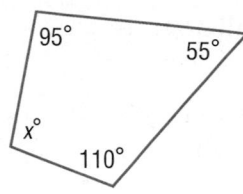
95° 55°
$x°$ 110°

10.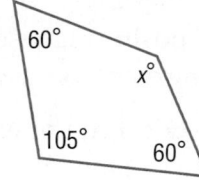
60°
$x°$
105° 60°

11.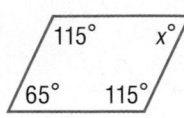
115° $x°$
65° 115°

Example 2   **Classify each quadrilateral.**

12.

13.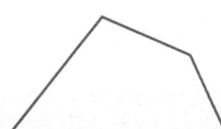

14.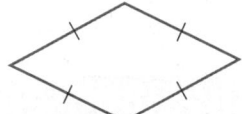

15.

16.

17.

18. FLAGS Many aircraft display the shape of the American flag slightly distorted to indicate motion. Classify each quadrilateral.

19. SIGNS Classify each quadrilateral.

Real-World Link

A tangram is an ancient Chinese puzzle consisting of 7 geometric shapes.

20. TANGRAM Refer to the seven tangram pieces shown at the left. Classify the polygons numbered 3 and 5.

Find the value of x in each quadrilateral.

21.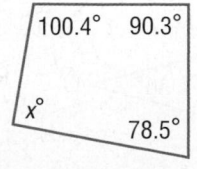
100.4° 90.3°
$x°$ 78.5°

22.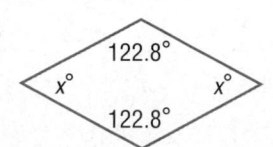
122.8°
$x°$ $x°$
122.8°

23.
$2x°$ $2x°$

$2x°$ $2x°$

24. SORTING Lana sorted a set of quadrilaterals into two categories according to a rule. The shapes that followed the rule were put in Set A and the shapes that did not were put in Set B. What was Lana's rule to sort the quadrilaterals?

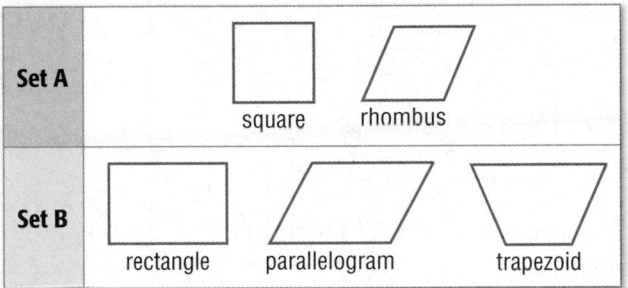

25. FIND THE ERROR Dario is describing the figure below. Find his mistake and correct it.

The figure is a parallelogram with four *congruent* sides.

H.O.T. Problems

26. OPEN ENDED Describe two different real-world items that are shaped as quadrilaterals. Then classify those quadrilaterals.

27. NUMBER SENSE Three of the angle measures of a quadrilateral are congruent. Without calculating, determine if the measure of the fourth angle in each of the following situations is greater than, less than, or equal to 90°. Explain your reasoning.
a. The three congruent angles each measure 89°.
b. The three congruent angles each measure 90°.

CHALLENGE Determine whether each statement is *sometimes*, *always*, or *never* true. Explain your reasoning.

28. A rhombus is a square.

29. A quadrilateral is a parallelogram.

30. A rectangle is a square.

31. A square is a rectangle.

32. ✍ **WRITE MATH** Make a diagram that shows the relationship between each of the following shapes: rectangle, parallelogram, square, rhombus, quadrilateral, and trapezoid. Then use your diagram to explain why some quadrilaterals can be classified in more than one category.

33. [THINK SOLVE EXPLAIN] **SHORT RESPONSE** The drawing shows the shape of Hinto's patio.

Hinto's Patio

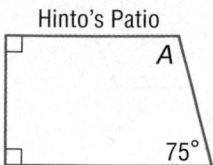

What is the measure of ∠A? Explain how you solved.

34. Which name does NOT describe the quadrilateral shown?

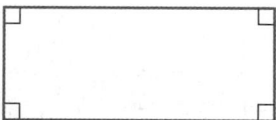

A. quadrilateral

B. rectangle

C. trapezoid

D. parallelogram

35. A parallelogram is shown below. What is the measure of ∠M to the nearest degree?

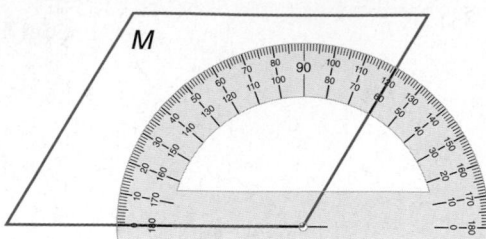

F. 30°   H. 120°

G. 60°   I. 150°

36. Which of the following is NOT true about a square?

A. It has 4 congruent angles.

B. The sum of the measures of the angles is 360°.

C. It is a parallelogram.

D. It has exactly one pair of parallel sides.

Find the value of x in each triangle. (Lesson 1C)

37.

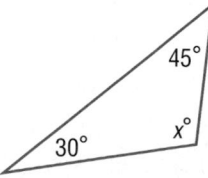

38.

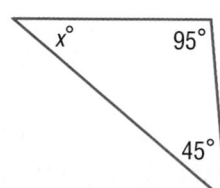

39.

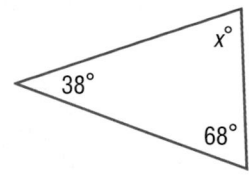

40. **POSTER** A poster has a triangular image with sides that measure 4 inches, 6 inches, and 8 inches. What type of triangle is shown on the poster? (Lesson 1C)

Classify each pair of angles as *complementary*, *supplementary*, or *neither*. (Lesson 1A)

41.

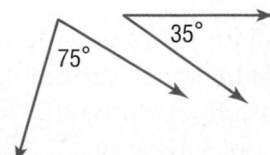

42.

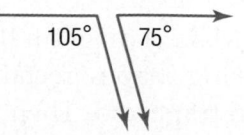

43.

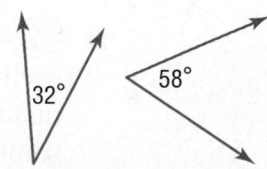

Main Idea

Find the sum of the angle measures of a polygon and the measure of an interior angle of a regular polygon.

 Vocabulary

polygon
pentagon
hexagon
heptagon
octagon
nonagon
decagon
equilateral
equiangular
regular polygon

 **Get ConnectED**

Polygons and Angles

POOLS Prairie Pools designs and builds swimming pools in various shapes and sizes. The shapes of five swimming pool styles are shown in their catalog.

 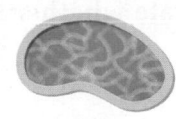 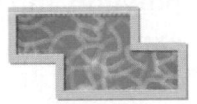  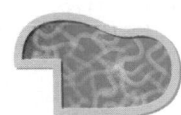

Aquarius   Kidney   Roman   Oval   Rustic

1. In the pool catalog, the Aquarius and Roman styles are listed under Group A. The remaining pools are listed under Group B. Describe one difference between the shapes of the pools in the two groups.

2. Create your own drawing of the shape of a pool that would fit into Group A. Do the same for Group B.

A **polygon** is a simple, closed figure formed by three or more straight line segments. A *simple figure* does not have lines that cross each other. You have drawn a *closed figure* when your pencil ends up where it started.

Polygons	Not Polygons
• Line segments are called sides. • Sides meet only at their endpoints. • Points of intersection are called vertices.	• Figures have sides that cross each other. • Figures are open. • Figures have curved sides.

A polygon can be classified by the number of sides it has.

Words	pentagon	hexagon	heptagon	octagon	nonagon	decagon
Number of Sides	5	6	7	8	9	10
Models						

An **equilateral** polygon has all sides congruent. A polygon is **equiangular** if all of its angles are congruent. A **regular polygon** is equilateral and equiangular, with all sides and all angles congruent.

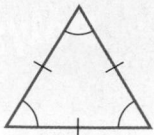 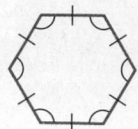

EXAMPLES Classify Polygons

Determine whether each figure is a polygon. If it is, classify the polygon and state whether it is regular. If it is *not* a polygon, explain why.

1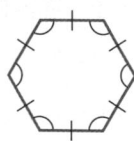

The figure has 6 congruent sides and 6 congruent angles. It is a regular hexagon.

2

The figure is not a polygon since it has a curved side.

CHECK Your Progress

a.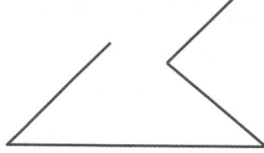

b.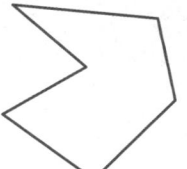

The sum of the measures of the angles of a triangle is 180°. You can use this relationship to find the measures of the angles of polygons.

Study Tip

Diagonals A diagonal is a line segment that joins two nonconsecutive vertices in a polygon.

EXAMPLE Find the Sum of the Angles of a Polygon

3 **Find the sum of the angle measures of the polygon.**

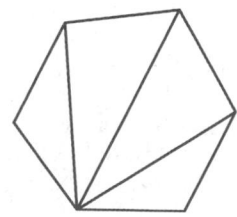

Draw all of the diagonals from one vertex as shown and count the number of triangles formed.

Find the sum of the angle measures in the polygon.

$$4 \times 180° = 720°$$

So, the sum of the angle measures of a hexagon is 720°.

CHECK Your Progress

c. pentagon

d. octagon

Key Concept — Interior Angle Sum of a Polygon

Words The sum of the measures of the angles of a polygon is $(n - 2)180$, where n represents the number of sides.

Symbols $S = (n - 2)180$

 REAL-WORLD EXAMPLE

4 **PICTURE FRAME** Bryan is building a picture frame in the shape of a regular pentagon. Find the measure of each angle of the regular pentagon.

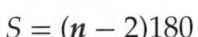

Step 1 Find the sum of the measures of the angles.

$S = (\boldsymbol{n} - 2)180$ Write the formula.

$S = (\boldsymbol{5} - 2)180$ A pentagon has 5 sides. Replace n with 5.

$S = (3)180$ or 540 Simplify.

The sum of the measures of the angles is $540°$.

Step 2 Divide 540 by 5, the number of interior angles, to find the measure of one angle. So, the measure of one angle of a regular pentagon is $540° \div 5$ or $108°$.

Each angle of a regular pentagon measures $108°$.

 CHECK Your Progress

e. SIGNS A stop sign is a regular octagon. What does each angle measure in a regular octagon?

CHECK Your Understanding

Examples 1 and 2 Determine whether each figure is a polygon. If it is, classify the polygon and state whether it is regular. If it is *not* a polygon, explain why.

1.

2.

3.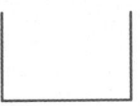

Example 3 Find the sum of the angle measures of each polygon.

4. quadrilateral   **5** heptagon

Example 4 **6. SOCCER** A soccer ball is made up of regular pentagons and hexagons. What is the measure of each angle of a regular hexagon?

Practice and Problem Solving

● = **Step-by-Step Solutions** begin on page R1.
Extra Practice begins on page EP2.

Examples 1 and 2   Determine whether each figure is a polygon. If it is, classify the polygon and state whether it is regular. If it is *not* a polygon, explain why.

7

8.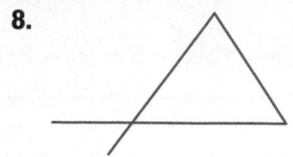

9.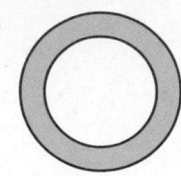

10.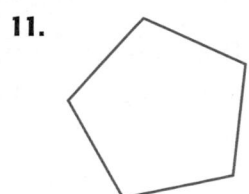

11.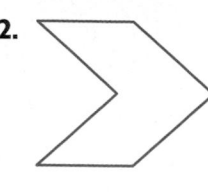

12.

Example 3   Find the sum of the angle measures of each polygon.

13. nonagon

14. decagon

15. 12-sided polygon

16. 20-sided polygon

Example 4   **17. MIRROR** The mirror shown is a regular octagon. What is the measure of each angle of the mirror?

18. GEOMETRY Find the measure of each angle of a regular triangle.

ART The patterns below are made up of polygons. Classify the polygons that are used to create each pattern.

19.

20.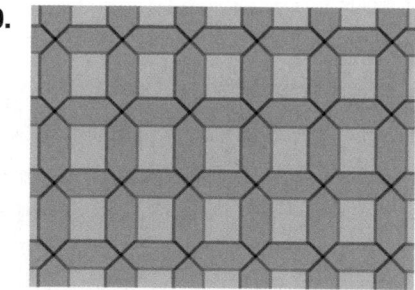

21 **TECHNOLOGY** Use a calculator to find the measure of each of the angles of a regular 20-sided, 50-sided, and 100-sided polygon. What do you notice about the measure of each angle? Explain why the measure of each angle can never be more than 180°.

22. ALGEBRA Each interior angle of a regular polygon measures 156°. Find the number of sides.

Find the value of each variable.

23.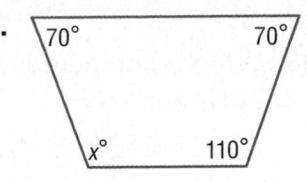

24.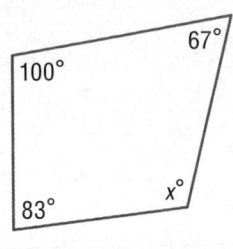

25.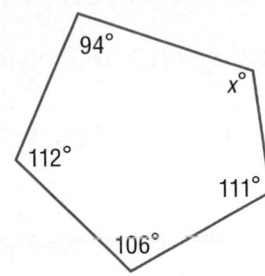

26.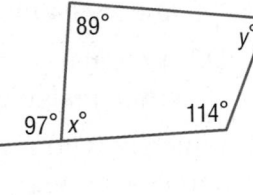

27.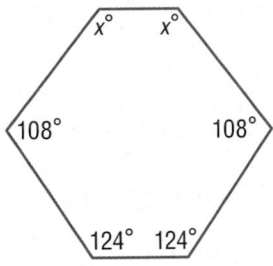

28.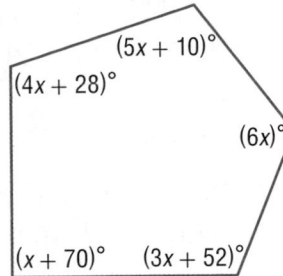

29. **MULTIPLE REPRESENTATIONS** Find the sum of the angles of regular polygons with 3, 4, 5, 6, 7, 8, 9, 10, 12, and 15 sides.

  a. TABLE Record your results in a table.

  b. GRAPH Graph your results as ordered pairs (number of sides, sum of angles). Draw a line through the points.

  c. WORDS Describe the graph. What does the slope represent? Is there a y-intercept? x-intercept?

30. ALPHABET Which letters shown below are *not* considered polygons?

ABCDEFGHIJKLMNOPQRSTUVWXYZ

H.O.T. Problems

OPEN ENDED Sketch each of the following figures. Include tick marks to show congruent parts if needed.

31. equiangular quadrilateral

32. equilateral pentagon

33. equilateral quadrilateral with a right angle

34. CHALLENGE When a side of a polygon is extended, an *exterior angle* is formed. Find the sum of the exterior angles of the pentagon.

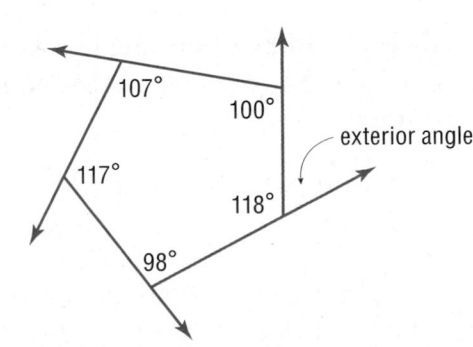

35. **WRITE MATH** Devion drew a regular polygon and measured one of its interior angles. Explain why it is impossible for his angle measure to be 145°.

36. What is the measure of ∠1?

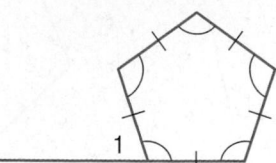

   A. 60°

   B. 72°

   C. 108°

   D. 120°

37. What is the measure of each angle of a regular nonagon?

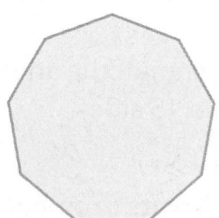

   F. 140°

   G. 145°

   H. 150°

   I. 155°

38. Which statement is true about polygons?

   A. A polygon is classified by the lengths of its sides.

   B. The sides of a polygon overlap.

   C. A polygon is formed by 4 or more line segments.

   D. A regular polygon has equal sides and equal angles.

39. Which polygon is represented by the part shaded gray in the design shown?

   F. hexagon

   G. heptagon

   H. octagon

   I. nonagon

Spiral Review

Classify each quadrilateral using the name that *best* describes it. (Lesson 1D)

40.

41.

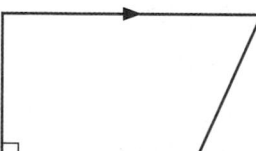

42.

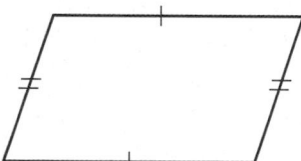

43. ARCHITECTURE A building has a window with 2 pairs of congruent, parallel sides and 4 congruent angles. Name the quadrilateral represented by the window. (Lesson 1D)

Find the value of x. (Lesson 1C)

44.

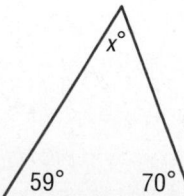

45.

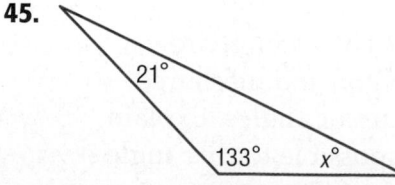

46.

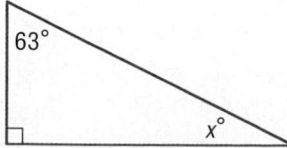

Explore Congruence

Main Idea

Identify the results of a translated figure.

STENCILING Lauren is stenciling her room using a geometric pattern. She wants to make sure that when she slides the stencil, the resulting figures are *congruent*. Congruent figures have the same size and shape.

ACTIVITY

STEP 1 Cut a triangle out of a piece of cardstock. Label its angles 1, 2, and 3.

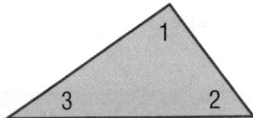

STEP 2 Place the triangle on a piece of notebook paper as shown. Trace the triangle on the paper. Label the vertices *X*, *Y*, and *Z*.

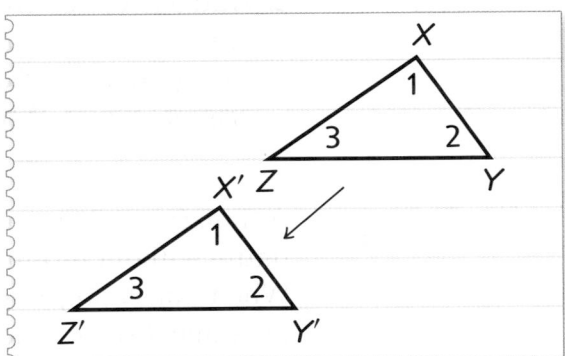

STEP 3 Slide the triangle to another place on the paper without turning it. Trace the triangle again. Label the vertices of this triangle *X'*, *Y'*, and *Z'* so that they correspond to the vertices of the first triangle.

Analyze the Results

1. Use a centimeter ruler to measure the distance between *X* and *X'*, *Y* and *Y'*, and *Z* and *Z'*. What seems to be true about the distance each point in $\triangle XYZ$ moved?

2. Compare and contrast $\triangle XYZ$ and $\triangle X'Y'Z'$. Use the word *congruent* in your comparison.

3. **MAKE A PREDICTION** Suppose you want to slide the triangle to a third position without turning it. What would it look like?

b. Triangle TUV has vertices $T(6, -3)$, $U(-2, 0)$, and $V(-1, 2)$. Find the vertices of $\triangle T'U'V'$ after a translation of 3 units right and 4 units down. Then graph the figure and its translated image.

 REAL-WORLD EXAMPLE

Find Coordinates of a Translation

Read Math

Vertex, Vertices A *vertex* of a figure is a point where two sides of the figure meet. *Vertices* is the plural of *vertex*.

3 **ANIMATION** An animator wants to move a character in a movie **4 units left and 6 units up.** If the character had original coordinates at $A(2, -1)$, $B(4, -1)$, $C(4, -5)$, and $D(2, -5)$, find the new vertices of the character after the translation. Then graph the figure and its translated image.

The vertices can be found by subtracting 4 from the x-coordinates and adding 6 to the y-coordinates.

ABCD	$(x - 4, y + 6)$	A′B′C′D′
$A(2, -1)$	$(2 - 4, -1 + 6)$	$A'(-2, 5)$
$B(4, -1)$	$(4 - 4, -1 + 6)$	$B'(0, 5)$
$C(4, -5)$	$(4 - 4, -5 + 6)$	$C'(0, 1)$
$D(2, -5)$	$(2 - 4, -5 + 6)$	$D'(-2, 1)$

Use the vertices to graph $ABCD$ and $A'B'C'D'$.

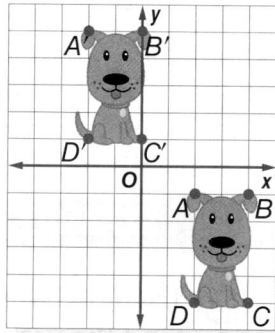

 CHECK Your Progress

c. **ART** A triangular wall decoration has vertices $X(-3, -1)$, $Y(-3, -3)$, and $Z(-1, -2)$. Find the vertices of the decoration after a translation of 5 units right and 3 units up. Then graph the decoration and its translated image.

In Example 3, the figure was translated 4 units left and 6 units up. This translation can be described as $(x, y) \longrightarrow (x - 4, y + 6)$. In Check Your Progress **c**, the figure was translated 5 units right and 3 units up. This translation can be described as $(x, y) \longrightarrow (x + 5, y + 3)$.

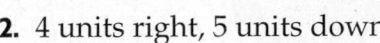
Example 1 **1.** Translate △*ABC* 3 units left and 3 units down. Graph △*A'B'C'*.

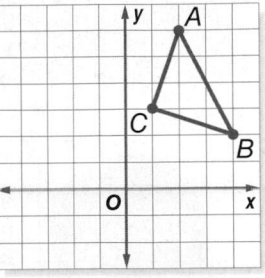

Example 2 **Quadrilateral *DEFG* has vertices *D*(1, 0), *E*(−2, −2), *F*(2, 4), and *G*(6, −3). Find the vertices of *D'E'F'G'* after each translation. Then graph the figure and its translated image.**

    2. 4 units right, 5 units down   **3.** 6 units right

Example 3 **4. MAPS** Julio is in Colorado exploring part of the Denver Zoo as shown. He starts at the Felines exhibit and travels 3 units to the right and 5 units up. At which exhibit is Julio located? If the Felines exhibit is located at (3, 1), what are the coordinates of Julio's new location?

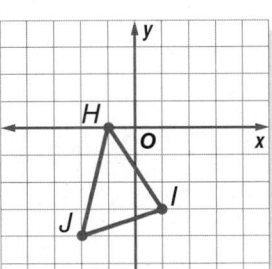

Practice and Problem Solving

    ● = **Step-by-Step Solutions** begin on page R1.
    Extra Practice begins on page EP2.

Example 1 **5.** Translate △*HIJ* 2 units right and 6 units down. Graph △*H'I'J'*.

6. Translate rectangle *KLMN* 1 unit left and 3 units up. Graph rectangle *K'L'M'N'*.

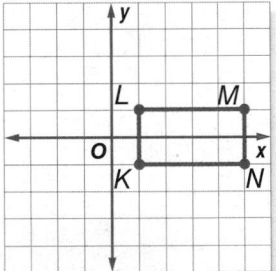

Example 2 **Triangle *PQR* has vertices *P*(0, 0), *Q*(5, −2), and *R*(−3, 6). Find the vertices of *P'Q'R'* after each translation. Then graph the figure and its translated image.**

    7 6 units right, 5 units up   **8.** 8 units left, 1 unit down

    9. 3 units left   **10.** 9 units down

Example 3 **11. GAMES** When playing the game shown at the right, the player can move horizontally or vertically across the board. Describe each of the following player's moves as a translation in words and as an ordered pair.

    a. green player

    b. orange player

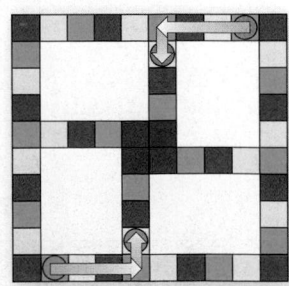

12. GRAPHIC NOVEL Refer to the graphic novel frame below. List the five steps the girls should take and identify any transformations used in the dance steps.

13 Parallelogram *RSTU* is translated 3 units right and 5 units up. Then the translated figure is translated 2 units left. Graph the resulting parallelogram.

14. Triangle *ABC* is translated 2 units left and 3 units down. Then the translated figure is translated 3 units right. Graph the resulting triangle.

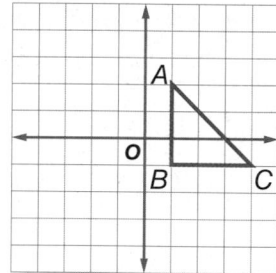

H.O.T. Problems

15. CHALLENGE What are the coordinates of the point (x, y) after being translated m units left and n units up?

16. Which One Doesn't Belong? Identify the transformation that is not the same as the other three. Explain your reasoning.

A	B	C	D
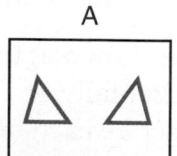	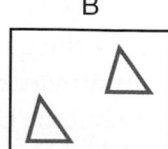	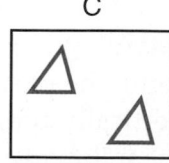	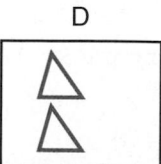

17. ✏️ **WRITE MATH** Triangle *ABC* is translated 4 units right and 2 units down. Then the translated image is translated 7 units left and 5 units up. Describe the final translated image in words.

18. Which graph shows a translation of the letter Z?

A.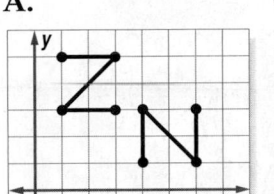

C.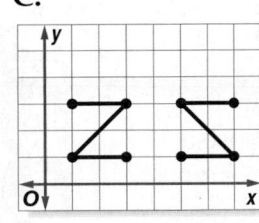

B.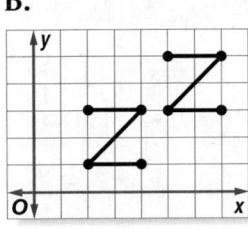

D.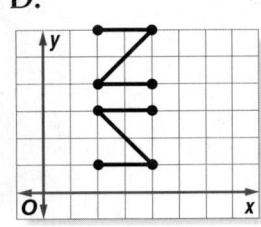

19. Point *A* is translated 4 units left and 3 units up. What are the coordinates of point *A* in its new position?

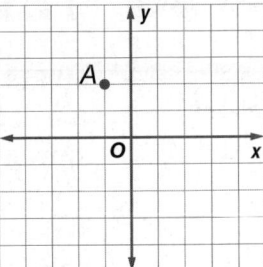

F. (4, 4)

G. (−5, 5)

H. (−5, −1)

I. (−4, 3)

Spiral Review

20. CONSTRUCTION Li-Chih is building a gazebo. He would like the base to be a regular octagon. What is the measure of each interior angle of the octagonal floor? (Lesson 1E)

21. DINNERWARE A dinner plate is in the shape of a regular hexagon. What is the sum of the angles of the dinner plate? (Lesson 1E)

Classify each quadrilateral. (Lesson 1D)

22.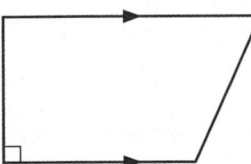

23.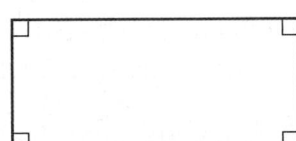

24.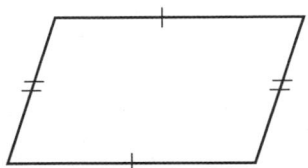

Classify each triangle by its angles and by its sides. (Lesson 1C)

25.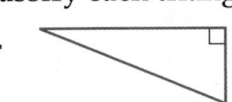

26.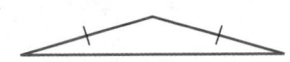

27.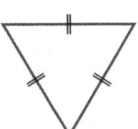

28. ALGEBRA The measure of ∠*U* is 46°. What is the measure of ∠*T* if ∠*U* and ∠*T* are supplementary angles? (Lesson 1A)

Extend

Tessellations

SHIRTS Justin bought a shirt for the new school year. He chose a shirt with patterns in the design. He would like to figure out what types of polygons are used in the shirt's pattern.

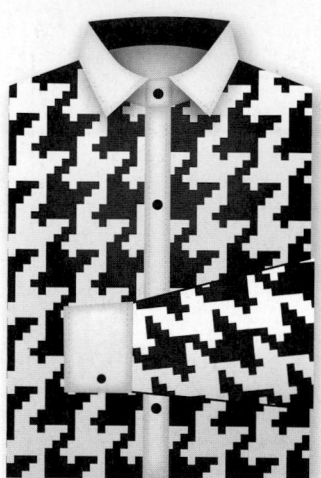

Patterns formed by repeating figures that fill a plane without gaps or overlaps are called **tessellations**. A **regular tessellation** is made from one regular polygon.

ACTIVITY

1 **STEP 1** Cut a regular hexagon out of a piece of cardstock. Use a protractor and ruler to check that the angles and sides are congruent.

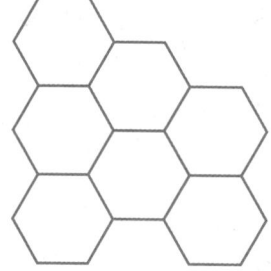

STEP 2 Place the hexagon on a piece of paper and trace the shape. Translate the figure so that its image is adjacent to the original. Trace the image.

STEP 3 Trace the shape so that the entire paper is covered with hexagons with no gaps or overlaps.

Analyze the Results

1. What is the measure of one angle of a regular hexagon? Find the sum of the measures of the angles around each vertex. Explain.

2. What other regular polygons will tessellate? Explain your reasoning.

2 **STEP 1** Draw a square on the back of an index card. Then draw a triangle on the inside of the square and a trapezoid on the bottom of the square as shown in Step 3.

STEP 2 Cut out the square. Then cut out the triangle and slide it from the right side of the square to the left side of the square. Cut out the trapezoid and slide it from the bottom to the top of the square.

STEP 3 Tape the figures together to form a pattern.

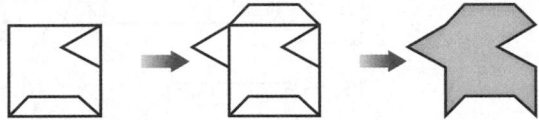

STEP 4 Trace this pattern onto a sheet of paper as shown to create a tessellation.

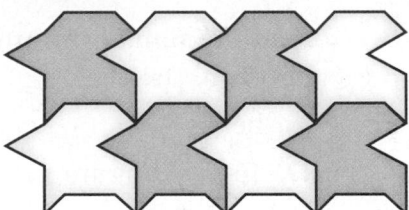

Practice and Apply

Create a tessellation using each pattern.

3. 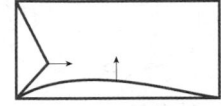   **4.**    **5.**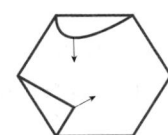

Analyze the Results

6. Design and describe your own tessellation pattern.

7. Explain how congruent figures are used in a tessellation.

8. Was a translation used to create your tessellation? Explain.

9. Some tessellations are formed using combinations of two or three different regular polygons. These are called *semi-regular tessellations*. Identify the regular polygons used in the semi-regular tessellation at the right. Predict the sum of the measures of the angles around each vertex. Check by finding the actual sum.

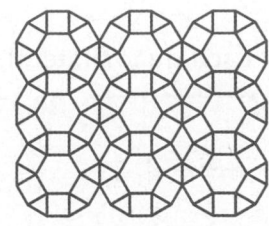

Mid-Chapter Check

Name each angle in four ways. Then classify the angle as *acute, right, obtuse,* **or** *straight.*
(Lesson 1A)

1.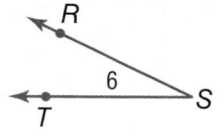

2.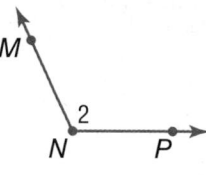

3.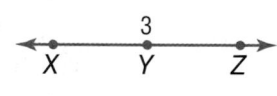

4.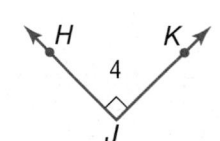

5. ALGEBRA The measure of ∠B is 53°. If ∠B and ∠C are supplementary, what is the measure of ∠C? (Lesson 1A)

Find the value of *x.* (Lesson 1C)

6.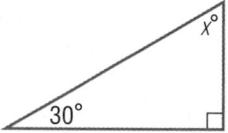

7.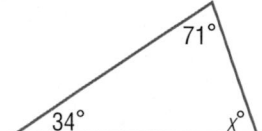

8. MULTIPLE CHOICE Which of the following is NOT a scalene triangle? (Lesson 1C)

A.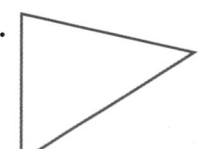

C.

B.

D.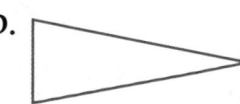

Classify each quadrilateral. (Lesson 1D)

9.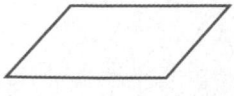

10.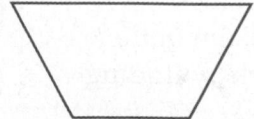

Find the value of *x* **in each quadrilateral.**
(Lesson 1D)

11.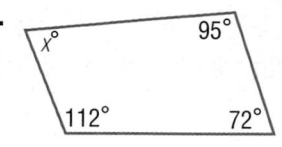

12.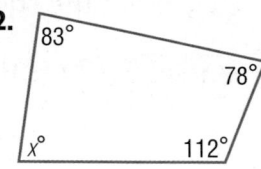

13.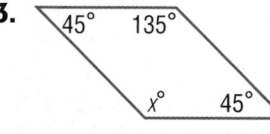

14.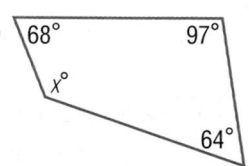

Find the sum of the angle measures of each polygon. (Lesson 1E)

15. hexagon

16. octagon

17. 16-sided figure

18. 22-sided figure

19. MULTIPLE CHOICE Point *D* is translated 5 units right and 2 units down.

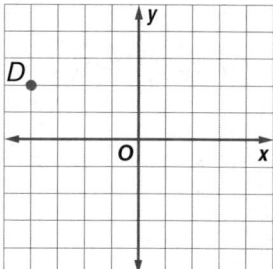

What are the coordinates of point *D* in its new position? (Lesson 2B)

F. (−9, 0)   **H.** (1, 0)

G. (0, −9)   **I.** (0, 9)

20. Translate △*ABC* 4 units right and 1 unit up. Graph △*A′B′C′*. (Lesson 2B)

Explore Line Symmetry

Main Idea

Identify line symmetry.

 Vocabulary

line symmetry
line of symmetry

 Get Connect ED

VISION Scientists have determined that the human eye uses symmetry to see. It is possible to understand what you are looking at even if you do not see all of it.

ACTIVITY

1 **STEP 1** Cut a regular hexagon out of a piece of cardstock. Use a protractor and ruler to check that the angles and sides are congruent.

STEP 2 Fold the hexagon so that one half matches with the other half, and then unfold. Find other lines of symmetry by folding along other lines.

Fold along dotted line.

Figures that match exactly when folded in half have **line symmetry**. The figure at the right has line symmetry. Each fold line is called a **line of symmetry**.

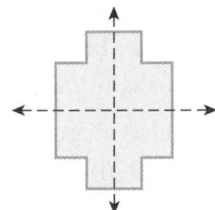

Analyze the Results

1. How many distinct fold lines did you create in the hexagon?

2. What do you notice about the fold lines of the regular hexagon?

Practice and Apply

Determine whether each figure has line symmetry. If so, copy the figure and draw all lines of symmetry.

3.

4.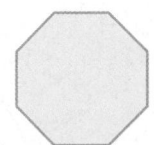

5.

6. **ALPHABET** List all of the capital letters of the alphabet that look exactly the same when folded across a horizontal line.

2 **STEP 1** Use a straightedge and a pencil to draw a quadrilateral on a piece of tracing paper. Label its vertices *A*, *B*, *C*, and *D*.

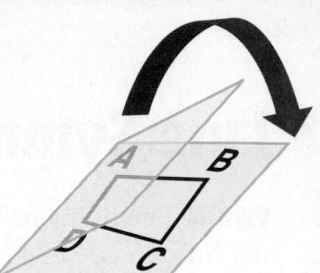

STEP 2 Fold the tracing paper so that quadrilateral *ABCD* is on the inside. Trace the quadrilateral onto the folded side.

STEP 3 Unfold the paper. Label the vertices *A′*, *B′*, *C′*, and *D′* as shown.

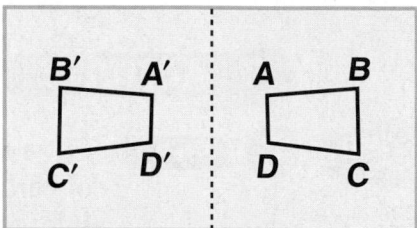

Analyze the Results

7. Use a ruler to measure the distance between *A* and *A′*, *B* and *B′*, *C* and *C′*, and *D* and *D′*. Note the distance between each vertex and the fold line. From your measurements, what can you conclude about the fold line?

8. Use a protractor to measure ∠*A* and ∠*A′*, ∠*B* and ∠*B′*, ∠*C* and ∠*C′*, and ∠*D* and ∠*D′*. What can you conclude about corresponding angles?

9. Compare and contrast quadrilaterals *ABCD* and *A′B′C′D′*. Use the word *congruent* in your comparison.

10. **MAKE A PREDICTION** Write a sentence that would be true for any figure and the resulting figure when the original figure is flipped over a line.

Practice and Apply

Copy each figure. Draw the resulting figure when each figure is flipped over line ℓ.

11.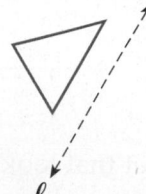

12.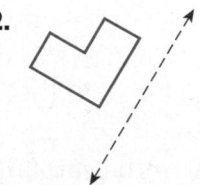

13.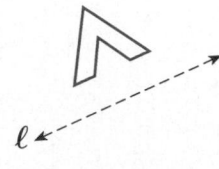

Main Idea

Predict the results of reflections and graph reflections on a coordinate plane.

 Vocabulary
reflection
line of reflection

Reflections in the Coordinate Plane

NATURE The surface of the water in the art shown acts like a mirror by producing an image of the flamingo.

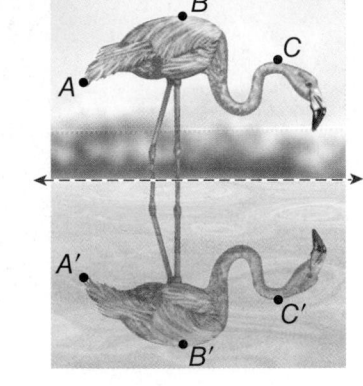

1. Compare the shape and size of the bird on either side of the line of symmetry.

2. Compare the perpendicular distance from the line of symmetry to each of the points shown. What do you observe?

3. The points A, B, and C appear *clockwise* on the bird. How are these points oriented on the other side of the line of symmetry?

The mirror image produced by flipping a figure over a line is called a **reflection**. This line is called the **line of reflection**. A reflection is one type of transformation.

EXAMPLE Draw a Reflection

1. Copy △JKL at the right onto graph paper. Then draw the image of the figure after a reflection over the given line.

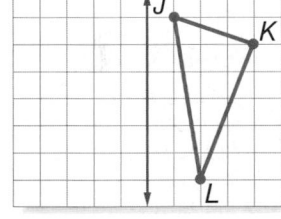

Step 1 Count the number of units between each vertex and the line of reflection.

Step 2 For each vertex, plot a point the same distance away from the line on the other side.

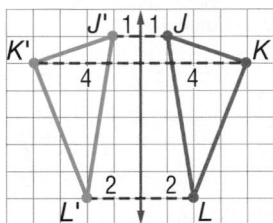

Step 3 Connect the new vertices to form the image of △JKL, △J'K'L'.

CHECK Your Progress

a. Copy the figure onto a piece of graph paper. Then draw the image of the figure after a reflection over the given line.

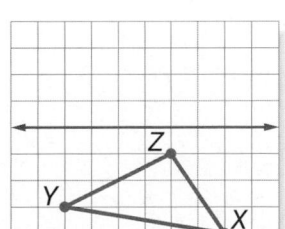

EXAMPLES Reflect a Figure Over an Axis

2 Graph △*PQR* with vertices *P*(−3, 4), *Q*(4, 2), and *R*(−1, 1). Then graph the image of △*PQR* after a reflection over the *x*-axis, and write the coordinates of its vertices.

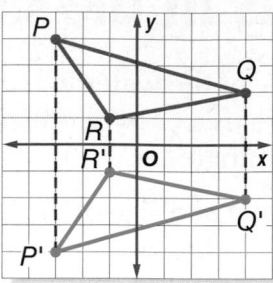

The coordinates of the vertices of the image are *P'*(−3, −4), *Q'*(4, −2), and *R'*(−1, −1). Examine the relationship between the coordinates of each figure.

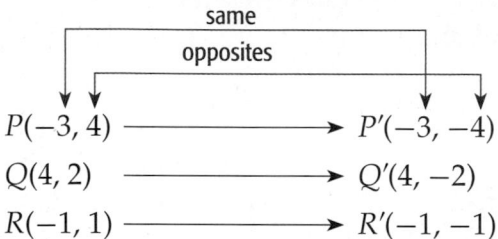

$P(-3, 4) \longrightarrow P'(-3, -4)$

$Q(4, 2) \longrightarrow Q'(4, -2)$

$R(-1, 1) \longrightarrow R'(-1, -1)$

Notice that the *y*-coordinate of a point reflected over the *x*-axis is the opposite of the *y*-coordinate of the original point.

3 Graph quadrilateral *ABCD* with vertices *A*(−4, 1), *B*(−2, 3), *C*(0, −3), and *D*(−3, −2). Then graph the image of *ABCD* after a reflection over the *y*-axis and write the coordinates of its vertices.

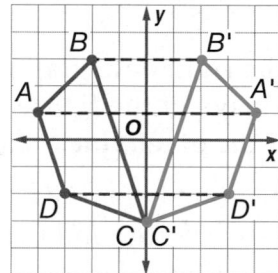

The coordinates of the vertices of the image are *A'*(4, 1), *B'*(2, 3), *C'*(0, −3), and *D'*(3, −2). Examine the relationship between the coordinates of each figure.

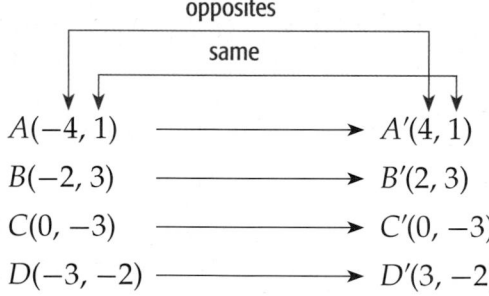

$A(-4, 1) \longrightarrow A'(4, 1)$

$B(-2, 3) \longrightarrow B'(2, 3)$

$C(0, -3) \longrightarrow C'(0, -3)$

$D(-3, -2) \longrightarrow D'(3, -2)$

Notice that the *x*-coordinate of a point reflected over the *y*-axis is the opposite of the *x*-coordinate of the original point.

CHECK Your Progress

Graph △*FGH* with vertices *F*(1, −1), *G*(5, −3), and *H*(2, −4). Then graph the image of △*FGH* after a reflection over the given axis and write the coordinates of its vertices.

b. *x*-axis

c. *y*-axis

  **EXAMPLE** **Use a Reflection**

④ KITES Copy and complete the kite
shown so that the completed figure
has a vertical line of symmetry.

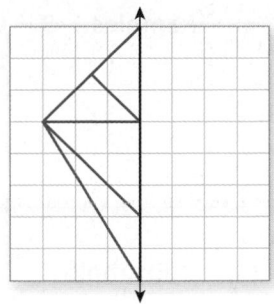

Find the distance from each vertex on
the figure to the line of reflection.

Then plot a point that same distance
away on the opposite side of the line.
Connect vertices as appropriate.

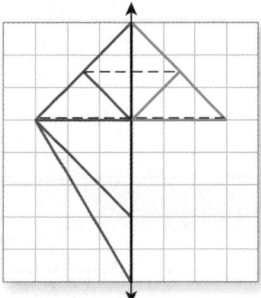 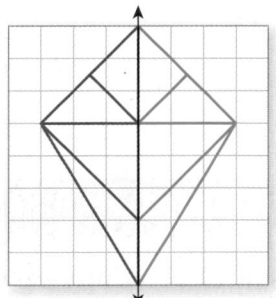

 CHECK Your Progress

d. ART Copy and complete the
animal shown so that the
completed picture has
horizontal line symmetry.

Key Concept **Reflections in the Coordinate Plane**

Words	When a figure is reflected over the *x*-axis, the *x*-coordinate of the image is the same, while the *y*-coordinate of the original figure is multiplied by −1.
Symbols	$(a, b) \longrightarrow (a, -b)$

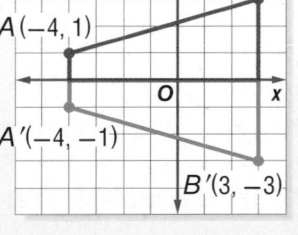

Words	When a figure is reflected over the *y*-axis, the *y*-coordinate of the image is the same, while the *x*-coordinate of the original figure is multiplied by −1.
Symbols	$(a, b) \longrightarrow (-a, b)$

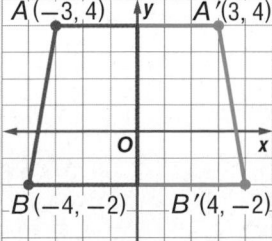

Examples 1–3 Graph the figure with the given vertices. Then graph the image of the figure after a reflection over the x-axis and y-axis and write the coordinates of the image's vertices.

1. △ABC with vertices A(3, 5), B(4, 1), and C(1, 2)

2. △WXY with vertices W(−1, −2), X(0, −4), and Y(−3, −5)

Example 4
3. **HOT TUBS** Copy and complete the hot tub design shown so that the completed design has vertical line symmetry.

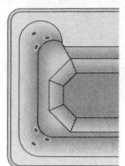

Practice and Problem Solving

● = **Step-by-Step Solutions** begin on page R1.
Extra Practice begins on page EP2.

Examples 1–3 Copy each figure onto graph paper. Then draw the image of the figure after a reflection over the given line.

4.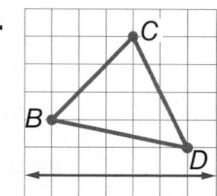

5.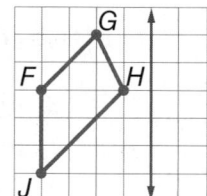

Graph the figure with the given vertices. Then graph the image of the figure after a reflection over the given axis and find the coordinates of the image's vertices.

6. triangle ABC with vertices A(−1, −1), B(−2, −4), and C(−4, −1); x-axis

7. triangle FGH with vertices F(3, 3), G(4, −3), and H(2, 1); y-axis

8. square JKLM with vertices J(−2, 0), K(−1, −2), L(−3, −3), and M(−4, −1); y-axis

9. quadrilateral PQRS with vertices P(1, 3), Q(3, 5), R(5, 2), and S(3, 1); x-axis

Example 4
10. **CARS** The drawing shows the left half of a car. Copy the drawing onto grid paper. Then draw the right side of the car so that the completed drawing has a vertical line of symmetry.

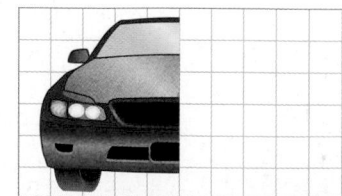

11. **ART** The top half of a Ukrainian decorative egg is shown. Copy the figure onto a piece of paper. Then draw the egg design after it has been reflected over a horizontal line.

12. **ARCHITECTURE** Describe in what ways the symmetry of the Fogong Monastery, shown below at the left, is similar to that of the Eiffel Tower in Paris, France, shown below at the right.

13. **MATH IN THE MEDIA** Find an image that illustrates a reflection in a newspaper or magazine, on television, or on the Internet. Explain why your example is considered a reflection.

14. **FLAGS** Flags of some countries have line symmetry. Of the flags shown below, which flags have line symmetry? Copy and draw all lines of symmetry.

Nigeria   Ghana   Japan   Mexico

15 **MUSIC** Use the photo at the left to determine how many lines of symmetry the body of a violin has.

Real-World Link
A violin is usually around 14 inches long.

For Exercises 16–19, use the graph at the right.

16. Identify the pair(s) of figures for which the x-axis is the line of reflection.

17. For which pair(s) of figures is the line of reflection the y-axis?

18. What type of transformation do figures B and C represent?

19. Describe the possible transformation(s) required to move figure A onto figure D.

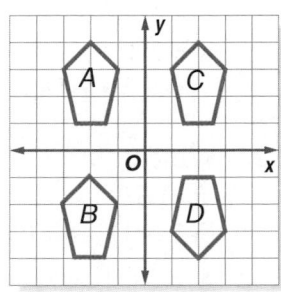

Copy each figure onto graph paper. Then draw the image of the figure after a reflection over the given line.

20.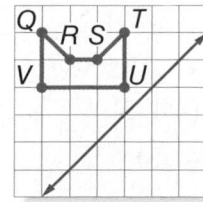

21.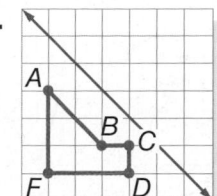

22. **OPEN ENDED** Draw a right triangle *ABC* in the first quadrant of a coordinate plane. Then draw the image after a reflection over the *x*-axis.

23. **FIND THE ERROR** Marisol is finding the new coordinates of the image of a triangle with vertices $A(1, 1)$, $B(4, 1)$, and $C(1, 5)$ after a reflection over the *x*-axis. Find her mistake and correct it.

> The vertices of triangle $A'B'C'$ are $A'(-1, 1)$, $B'(-4, 1)$, and $C'(-1, 5)$.

24. **CHALLENGE** Suppose point *K* with coordinates $(7, -2)$ is reflected so that the coordinates of its image are $(7, 2)$. Without graphing, which axis was this point reflected over? Explain your reasoning.

25. **CHALLENGE** Suppose point *L* with coordinates (x, y) is reflected over the *y*-axis. Then it is translated *a* units right and *b* units up. What are the coordinates of point *L'*?

26. **WRITE MATH** Find the coordinates of the point (x, y) after it has been reflected over the *x*-axis. Then find the coordinates of the point (x, y) after it has been reflected over the *y*-axis. Explain your reasoning.

Test Practice

27. Which of the following is the reflection of △*ABC* with vertices $A(1, -1)$, $B(4, -1)$, and $C(2, -4)$ over the *x*-axis?

A.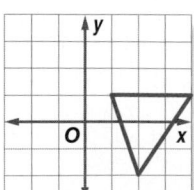

C.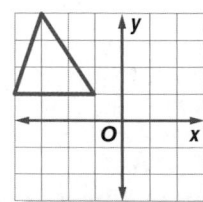

B.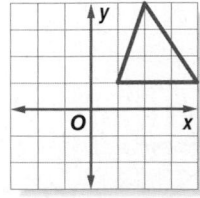

D.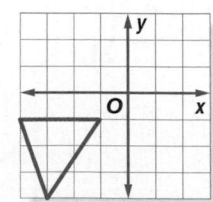

28. Triangle *RST* is reflected over the *x*-axis and translated 4 units to the right.

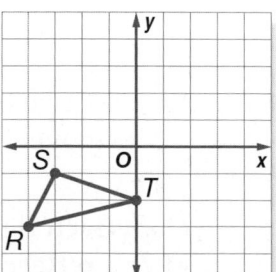

What are the coordinates of point *R* in its new position?

F. $(-4, 3)$

H. $(1, 1)$

G. $(0, 3)$

I. $(4, 2)$

Explore Rotational Symmetry

Main Idea

Identify rotational symmetry.

Vocabulary

rotational symmetry
angle of rotation

Get Connect**ED**

FAMILY BADGES Kamon are Japanese badges used to signify a specific family. If the badge at the right is turned 72°, it will be identical to the original figure.

A figure has **rotational symmetry** if it can be rotated or turned less than 360° about its center so that the figure looks as it does in its original position. The **angle of rotation** is the degree measure that the figure is rotated.

 ACTIVITY

1 **STEP 1** Cut a square out of cardstock. Label the four vertices *A*, *B*, *C*, and *D*. Use a ruler and protractor to make sure the angles and sides are congruent. Lay the square on a piece of patty paper. Trace around the square.

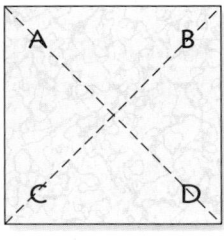

STEP 2 Find the center of the square by drawing the two diagonals shown on the diagram. Place a pin at the center and turn the square clockwise until it matches the original, ignoring the letters.

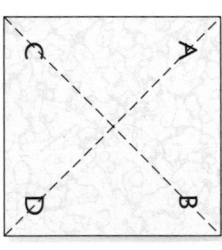

STEP 3 Continue turning the square, noting the angles at which it matches the original, until the vertices are back to their original position.

Analyze the Results

1. How many times did the figure match itself?

2. Describe the relationship between the number of times the figure matched itself and the angle of rotation.

3. **MAKE A PREDICTION** Predict the results of a 90° turn of the figure.

2 **STEP 1** Trace and cut out the figure at the right on a piece of cardstock. Lay a piece of patty paper on the figure.

STEP 2 Place a pin on the red dot. Turn the figure clockwise one quarter turn or 90°. Trace the turned figure.

STEP 3 Turn the figure clockwise another quarter turn to 180° and trace the figure. Repeat another clockwise quarter turn to 270° and trace.

Analyze the Results

4. Did the figure match itself at any of the rotations?

5. Does the figure have rotational symmetry?

6. Compare and contrast the figure using line symmetry and rotational symmetry. Does the figure have either?

7. **MAKE A PREDICTION** Predict the resulting image if you turn the figure 45° counterclockwise. Draw the rotated image.

Practice and Apply

Determine whether each figure has rotational symmetry. Write *yes* or *no*. If *yes*, name its angle(s) of rotation.

8.
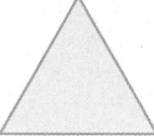

9.

10.

Show each figure after clockwise turns of 45°, 60°, 90°, 180°, and 270°.

11.

12.

13.

14. **DESIGN** Copy and complete the design shown so that the completed figure has rotational symmetry with 90°, 180°, and 270° as its angles of rotation.

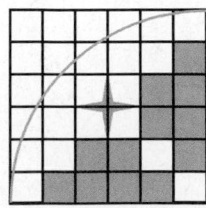

Main Idea

Predict the results of a rotation and graph rotations on a coordinate plane.

 **Vocabulary**

rotation

Rotations in the Coordinate Plane

 **Explore**

Step 1 Draw and label triangle *ABC* with vertices *A*(−4, 1), *B*(−4, 6), and *C*(−1, 1).

Step 2 Attach a piece of tracing paper to the coordinate plane with a fastener at the origin of the graph. Trace the triangle. Label the vertices *A′*, *B′*, and *C′*.

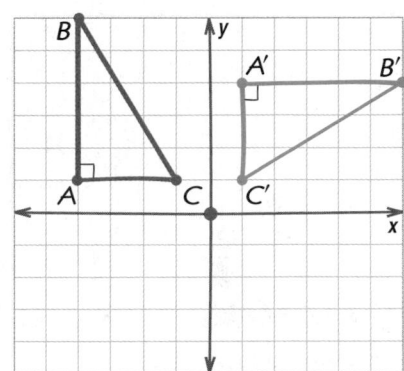

Step 3 Turn the tracing paper clockwise 90° so that the original *y*-axis is on top of the original *x*-axis.

1. Describe the transformation that occurred from triangle *ABC* to triangle *A′B′C′*.

2. What are the coordinates of triangle *A′B′C′*?

A **rotation** is a transformation in which a figure is turned around a fixed point.

Key Concept **Rotations in the Coordinate Plane**

Words When a figure is rotated around a point, neither the size nor the shape of the figure changes.

Models

90° Rotation   180° Rotation   270° Rotation

A rotation can be clockwise or counterclockwise. Unless otherwise indicated, the rotation is about the origin.

EXAMPLE Rotate a Figure Clockwise

1 Triangle *RST* has vertices *R*(1, 3), *S*(4, 4), and *T*(2, 1). Graph the figure and its image after a clockwise rotation of 90° about the origin. Then name the coordinates of the vertices for triangle *R'S'T'*.

Step 1 Graph triangle *RST* on a coordinate plane.

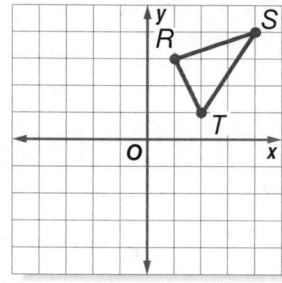

Read Math

Clockwise To rotate the figure the same way the hands on a clock rotate.

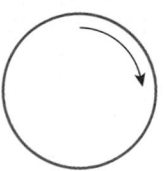

Counterclockwise To rotate the figure the opposite direction of the way the hands on a clock rotate.

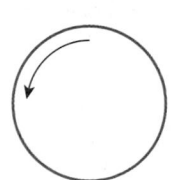

Step 2 Sketch $\overline{RO}$ connecting point *R* to the origin. Sketch another segment, $\overline{R'O}$, so that the angle between points *R*, *O*, and *R'* measures 90° and the segment is congruent to $\overline{RO}$.

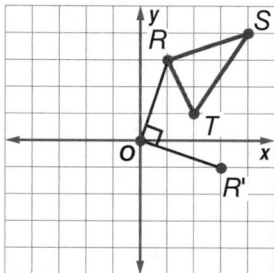

Step 3 Repeat Step 2 for points *S* and *T*. Then connect the vertices to form triangle *R'S'T'*.

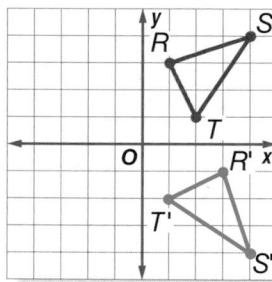

So, the coordinates of the vertices of triangle *R'S'T'* are *R'*(3, −1), *S'*(4, −4), and *T'*(1, −2).

CHECK Your Progress

a. Triangle *XYZ* has vertices *X*(−5, 4), *Y*(−1, 2), and *Z*(−3, 1). Graph the figure and its image after a counterclockwise rotation of 180° about the origin. Then name the coordinates of the vertices for triangle *X'Y'Z'*.

2 **LANDSCAPING** A landscaper is designing a circular garden that has rotational symmetry. The isosceles triangle represents a decorative stone. Find the coordinates of the vertices of the triangles after rotations of 90°, 180°, and 270° counterclockwise.

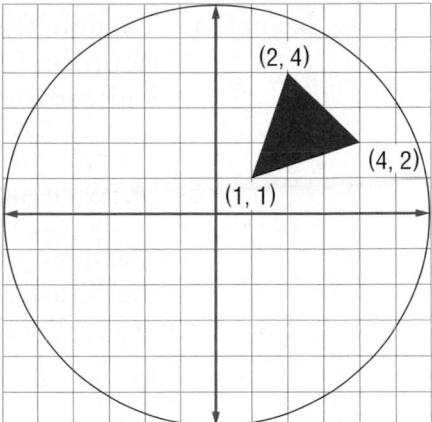

 Real-World Link · · · ·

Landscape designers often use coordinate geometry and algebra as they serve their clients.

Draw each of the three rotations of the isosceles triangle.

List the vertices:
90°: (−1, 1), (−4, 2), (−2, 4)
180°: (−1, −1), (−4, −2), (−2, −4)
270°: (1, −1), (4, −2), (2, −4)

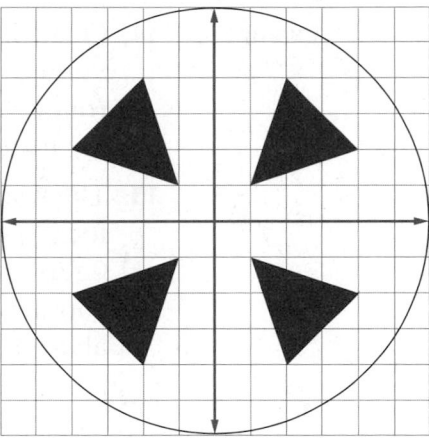

CHECK Your Progress

b. **FLOWERS** The landscaper would like to add marigolds to the circular garden. The marigold plants are located at (2, 1) and (3, 1). Rotate the points 90°, 180°, and 270° clockwise to find the other locations.

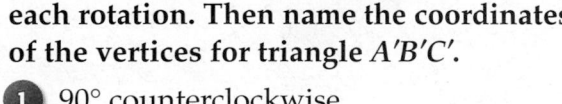 **CHECK** Your Understanding

Example 1 **Graph triangle *ABC* and its image after each rotation. Then name the coordinates of the vertices for triangle *A′B′C′*.**

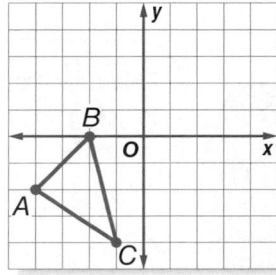

1 90° counterclockwise

2. 180° clockwise

3. 270° counterclockwise

4. 270° clockwise

Example 2 **5.** **GEOMETRY** Parallelogram *EFGH* has coordinates *E*(1, 1), *F*(3, 1), *G*(4, 4), and *H*(2, 4). Find the coordinates of the image after a clockwise rotation of 180°. Then graph the original image to check your work.

Practice and Problem Solving

● = **Step-by-Step Solutions** begin on page R1.
Extra Practice begins on page EP2.

Example 1 Triangle *PQR* has vertices *P*(1, −5), *Q*(2, −1), and *R*(5, −4). Graph the figure and its image after each rotation. Then predict the coordinates of the vertices for triangle *P'Q'R'*.

6. 270° clockwise

7 180° counterclockwise

8. 90° counterclockwise

9. 90° clockwise

Quadrilateral *FGHJ* has vertices *F*(1, 1), *G*(2, 5), *H*(5, 3), and *J*(4, 0). Graph the figure and its image after each rotation. Then name the coordinates of the vertices for quadrilateral *F'G'H'J'*.

10. 90° clockwise

11. 270° clockwise

12. 270° counterclockwise

13. 180° clockwise

Example 2 **14.** Quadrilateral *TUVW* represents a bed in Chantal's bedroom and has vertices *T*(−3, 0), *U*(−3, 2), *V*(−1, 0), and *W*(−1, 2). Chantal would like to rotate her bed 180° clockwise to see if she likes the new placement. What are the coordinates of the final image?

15. The triangle at the right represents the location of three clues in a scavenger hunt. Graph the clue locations after a counterclockwise rotation of 270° about the origin.

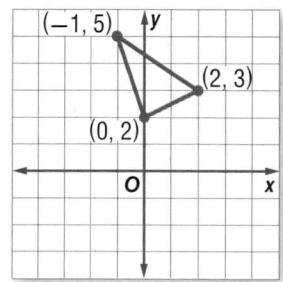

16. GEOMETRY The right isosceles triangle *PQR* has vertices *P*(3, 3), *Q*(3, 1), and *R*(■, ■) and is rotated 90° counterclockwise about the origin. Find the missing vertex of the triangle. Then graph the triangle and its image.

17. GRAPHIC NOVEL Refer to the graphic novel frame below. The last step is shown on grid 6. The girls make a clockwise rotation of 90° and begin the dance again. Expand your grid and mark the ending spot of the second series.

H.O.T. Problems

OPEN ENDED Refer to the information on line symmetry in Lesson 3A.

18. Draw a figure that has line symmetry but not rotational symmetry.

19. Draw a figure that has both line symmetry and rotational symmetry.

20. Is it possible for a figure to have rotational symmetry, but not line symmetry? Justify your response with a drawing or an explanation.

CHALLENGE Triangle *JKL* has vertices *J*(−4, −1), *K*(−1, −2), and *L*(−5, −5). Graph the figure and its image after each rotation about the origin. Then give the coordinates of the vertices for triangle *J'K'L'*.

21. 540° clockwise   **22.** 450° counterclockwise

23. 720° counterclockwise   **24.** 630° counterclockwise

25. ✏ **WRITE MATH** Describe what information is needed to rotate a figure.

Test Practice

26. Which figure shows the letter F after a rotation of 270° clockwise?

A.

C.

B.

D.

27. THINK SOLVE EXPLAIN **EXTENDED RESPONSE** Triangle *XYZ* has vertices *X*(2, −2), *Y*(5, 0), and *Z*(3, −4).

Part A What are the coordinates of point *Y'* after a clockwise rotation of 180°?

Part B Refer to the original triangle. What are the coordinates of *Y'* after a counterclockwise rotation of 90°?

Spiral Review

Graph quadrilateral *ABCD* and its resulting image after each transformation. (Lessons 2B and 3B)

28. reflection over the *y*-axis

29. translation 2 units right, 3 units up

30. reflection over the *x*-axis

31. translation 3 units left, 4 units up

32. reflection over the *x*-axis, then over the *y*-axis

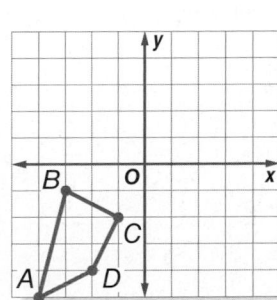

Dilations

Main Idea

Graph dilations on a coordinate plane.

Vocabulary

dilation
center
enlargement
reduction

 Get ConnectED

Explore The figure shown is drawn on 0.5-centimeter grid paper, so each square is 0.5-by-0.5 centimeter. Redraw the figure using squares that are 1-by-1 centimeter. Use point A as your starting point.

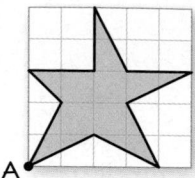

1. Measure and compare corresponding lengths on the original and new figure. Describe the relationship between these measurements. How does this relate to the change in grid size?

The image produced by enlarging or reducing a figure is called a **dilation**. The **center** of the dilation is a fixed point used for measurement when altering the size of the figure.

A dilation image is similar to the original figure. The ratio of a length on the image to a length on the original figure is the scale factor of the dilation.

  EXAMPLE **Draw a Dilation**

 Copy polygon *ABCD* onto graph paper. Then draw the image of the figure after a dilation with center *A* by a scale factor of 2.

Step 1 Draw ray *AB*, or $\overrightarrow{AB}$, extending it to the edges of the grid.

Step 2 Use a ruler to locate point *B′* on $\overrightarrow{AB}$ so that *AB′* = 2(*AB*).

Step 3 Repeat Steps 1 and 2 for points *C′* and *D′*. Then draw polygon *A′B′C′D′* where *A* = *A′*.

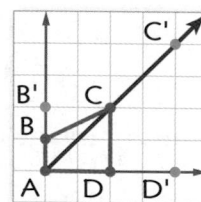 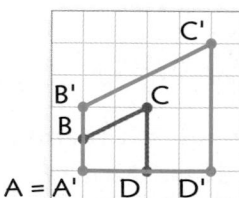

 CHECK Your Progress

a. Draw and label a large triangle *XYZ* on grid paper. Then draw the image of △*XYZ* after a dilation with center *X* and scale factor $\frac{1}{4}$.

 **EXAMPLE** **Graph a Dilation**

Study Tip

Dilations on a Coordinate Plane The ratio of the x- and y-coordinates of the vertices of an image to the corresponding values of the original figure is the same as the scale factor of the dilation.

② Graph △JKL with vertices J(3, 8), K(10, 6), and L(8, 2). Then graph its image △J′K′L′ after a dilation with a scale factor of $\frac{1}{2}$.

To find the vertices of the dilation, multiply each coordinate in the ordered pairs by $\frac{1}{2}$. Then graph both images on the same axes.

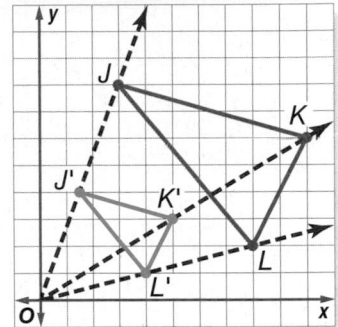

$$J(3, 8) \longrightarrow \left(3 \cdot \frac{1}{2}, 8 \cdot \frac{1}{2}\right) \longrightarrow J'\left(\frac{3}{2}, 4\right)$$

$$K(10, 6) \longrightarrow \left(10 \cdot \frac{1}{2}, 6 \cdot \frac{1}{2}\right) \longrightarrow K'(5, 3)$$

$$L(8, 2) \longrightarrow \left(8 \cdot \frac{1}{2}, 2 \cdot \frac{1}{2}\right) \longrightarrow L'(4, 1)$$

Check for Reasonableness Draw lines through the origin and each of the vertices of the original figure. The vertices of the dilation should lie on those same lines. ✔

 CHECK Your Progress

Find the coordinates of the image of △JKL after a dilation with each scale factor. Then graph △JKL and △J′K′L′.

b. scale factor: 3

c. scale factor: $\frac{1}{3}$

 **Key Concept** **Dilations in the Coordinate Plane**

Words A dilation with a scale factor of a will be:

- an **enlargement**, or an image larger than the original, if $a > 1$,
- a **reduction**, or an image smaller than the original, if $0 < a < 1$, or
- the same as the original figure if $a = 1$.

Model

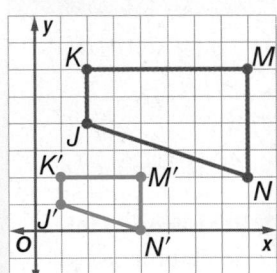

To find the coordinates of the vertices of an image after a dilation with center (0, 0), multiply the x- and y-coordinates by the scale factor.

Symbols $(x, y) \longrightarrow (ax, ay)$

Study Tip

Alternate Form Scale factors can also be written as decimals.

EXAMPLE   Find and Classify a Scale Factor

3 Quadrilateral *V'Z'X'W'* is a dilation of quadrilateral *VZXW*. Find the scale factor of the dilation and classify it as an *enlargement* or a *reduction*.

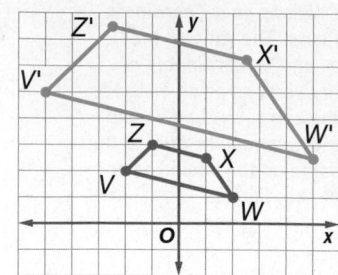

Write a ratio of the *x*- or *y*-coordinate of one vertex of the dilation to the *x*- or *y*-coordinate of the corresponding vertex of the original figure. Use the *y*-coordinates of *V*(−2, 2) and *V'*(−5, 5).

$$\frac{y\text{-coordinate of point } V'}{y\text{-coordinate of point } V} = \frac{5}{2} \qquad \text{Verify by using other coordinates.}$$

The scale factor is $\frac{5}{2}$. Since $\frac{5}{2} > 1$, the dilation is an enlargement.

✓ CHECK Your Progress

d. Triangle *A'B'C'* is a dilation of △*ABC*. Find the scale factor of the dilation and classify it as an *enlargement* or a *reduction*.

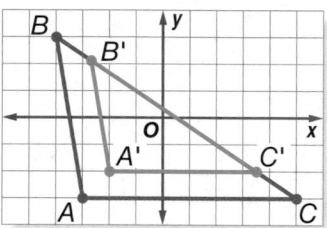

Before Dilation

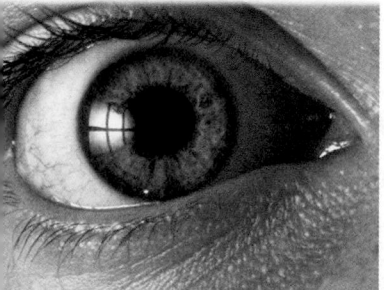

After Dilation

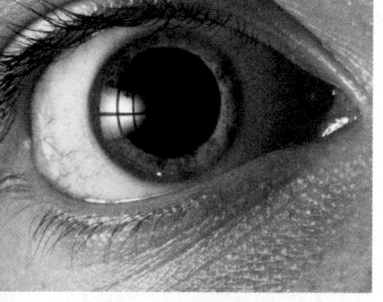

 Real-World Link · · · ·

An optometrist will often dilate the pupils to better examine a patient's retina, the layer of nerve tissue that receives and transmits images to the brain.

🏃 ✎ REAL-WORLD EXAMPLE

4 **EYES** An optometrist dilates a patient's pupils by a factor of $\frac{5}{3}$. If the pupil has a diameter of 5 millimeters before dilation, find the new diameter after the pupil is dilated.

$a = \frac{5}{3}(5)$ Write the equation.

$a \approx 8.33$ Multiply.

The pupil will be about 8.3 millimeters in diameter after dilation.

✓ CHECK Your Progress

e. **COMPUTERS** Padma uses an image of her dog as the wallpaper on her computer desktop. The original image is 5 inches high and 7 inches wide. If her computer scales the image by a factor of $\frac{5}{4}$, what are the dimensions of the dilated image?

✓ **CHECK Your Understanding**

Example 1 Copy △*ABC* on graph paper. Then draw the image of the figure after the dilation with the given center and scale factor.

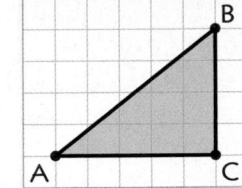

1. center: *A*, scale factor: $\frac{1}{2}$

2. center: *C*, scale factor: $\frac{3}{2}$

Example 2 Triangle *JKL* has vertices *J*(−4, 2), *K*(−2, −4), and *L*(3, 6). Find the vertices of *J′K′L′* after a dilation with the given scale factor. Then graph △*JKL* and △*J′K′L′*.

3. scale factor: 3   **4.** scale factor: $\frac{1}{4}$

Example 3 **5.** On the graph, $\overline{A'B'}$ is a dilation of $\overline{AB}$. Find the scale factor of the dilation and classify it as an *enlargement* or a *reduction*.

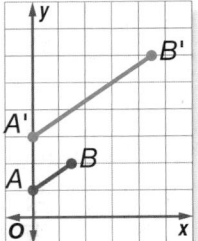

Example 4 **6. GRAPHIC DESIGN** Jacqui designed a 6-inch by $7\frac{1}{2}$-inch logo for her school. The logo is to be reduced by a scale factor of $\frac{1}{3}$ and used to make face paintings. What are the dimensions of the dilated image?

Practice and Problem Solving

 = **Step-by-Step Solutions** begin on page R1.
Extra Practice begins on page EP2.

Example 1 Copy each figure on graph paper. Then draw the image of the figure after the dilation with the given center and scale factor.

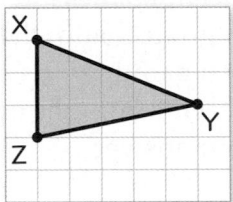

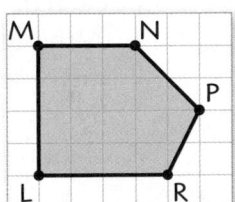

7. center: *X*, scale factor: $\frac{7}{3}$ **8.** center: *Z*, scale factor: $\frac{2}{3}$

9 center: *L*, scale factor: $\frac{3}{4}$ **10.** center: *N*, scale factor: 2

Example 2 Find the vertices of polygon *H′J′K′L′* after polygon *HJKL* is dilated using the given scale factor. Then graph polygon *HJKL* and polygon *H′J′K′L′*.

11. *H*(−1, 3), *J*(3, 2), *K*(2, −3), *L*(−2, −2); scale factor 2

12. *H*(0, 2), *J*(3, 1), *K*(0, −4), *L*(−2, −3); scale factor 3

13. *H*(−6, 2), *J*(4, 4), *K*(7, −2), *L*(−2, −4); scale factor $\frac{1}{2}$

14. *H*(−8, 4), *J*(6, 4), *K*(6, −4), *L*(−8, −4); scale factor $\frac{3}{4}$

Example 3 On each graph, one figure is a dilation of the other. Find the scale factor of each dilation and classify it as an enlargement or a reduction.

15.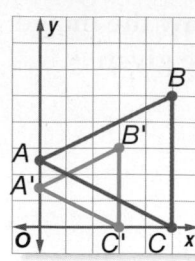

16.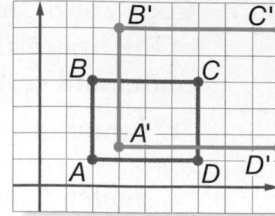

17.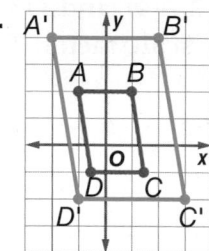

18.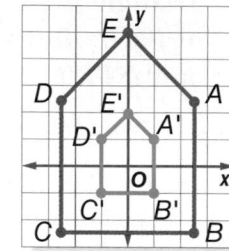

Example 4

19. PUBLISHING To place a picture in his class newsletter, Joquin must reduce the picture by a scale factor of $\frac{3}{10}$. Find the dimensions of the reduced picture if the original is 15 centimeters wide and 10 centimeters high.

20. PROJECTION An overhead projector transforms the image on a transparency so that it is shown enlarged by a scale factor of 3.5 on a screen. If the original image is 3 inches long by 4 inches wide, find the dimensions of the projected image.

Real-World Link

Each Ohio Bicentennial Barn painted by Scott Hagan took approximately 18 hours and 7 gallons of paint to complete.

21 BARN ART Scott Hagan painted the Ohio bicentennial logo on one barn in each of Ohio's 88 counties. Each logo measured about 20 feet by 20 feet. Although Hagan drew each logo freehand, they are amazingly similar. If the original logo on which each painting was based measured 5 inches by 5 inches, what is the scale factor from the original logo to one of Hagan's paintings? Justify your answer.

22. DRAWING Artists use dilations to create the illusion of distance and depth. If you stand on a sidewalk and look in the distance, the parallel sides appear to converge and meet at a point. This is called the vanishing point.

a. Which figure appears to be closer? Explain your reasoning.

b. Draw a figure similar to the one shown at the right. Measure the larger rectangle. Above the horizon, draw a similar figure that is $\frac{7}{5}$ the size of that rectangle.

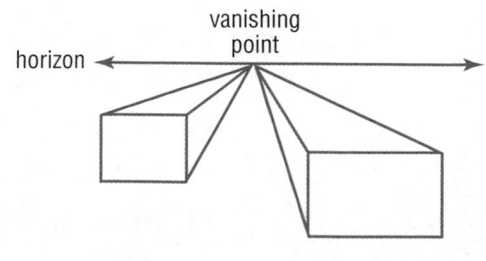

H.O.T. Problems

23. OPEN ENDED Graph a triangle and its image after a dilation with a scale factor greater than 1. Then graph the original figure after a dilation with a scale factor between 0 and 1. Explain your reasoning.

24. CHALLENGE Describe the image of a figure after a dilation with a scale factor of −2. (*Hint:* The original figure and its image are on opposite sides.)

25. ✏ **WRITE MATH** How do polygons that are similar compare to polygons that are congruent? Which transformations produce congruent polygons?

✔ Test Practice

26. Square *A* is similar to square *B*.

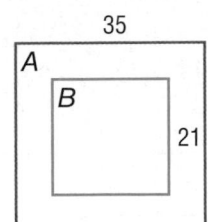

What scale factor was used to dilate square *A* to square *B*?

A. $\frac{1}{7}$

B. $\frac{3}{5}$

C. $\frac{5}{3}$

D. 7

27. Quadrilateral *LMNP* was dilated to form quadrilateral *WXYZ*.

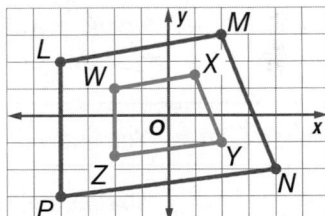

Which number best represents the scale factor used to change quadrilateral *LMNP* into quadrilateral *WXYZ*?

F. 3

G. 2

H. $\frac{1}{2}$

I. $\frac{1}{3}$

Graph △*XYZ* and its image after each rotation. Then give the coordinates of the vertices for △*X′Y′Z′*. (Lesson 4B)

28. 180° clockwise

29. 90° counterclockwise

30. 90° clockwise

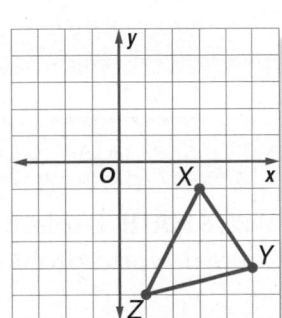

31. GEOMETRY Graph △*JKL* with vertices *J*(−1, −4), *K*(1, 1), and *L*(3, −2) and its reflection over the *x*-axis. Write the ordered pairs for the vertices of the new figure.
(Lesson 3B)

Problem-Solving Investigation

Main Idea Solve problems using the *work backward* strategy.

P.S.I. TEAM +

Work Backward

TENISHA: At the parade yesterday, I saw that the soccer team's float ended at Abbey and Davidson. The float traveled 3 blocks west on Monroe, 2 blocks north on Nelson, 4 blocks west on Main, and 2 blocks north on Davidson.

YOUR MISSION: Work backward to find the starting point of the soccer team's float in the parade.

Understand	You know the translations involved.
Plan	Start at the end of the parade and work backward.
Solve	
Check	Start at the community swimming pool and translate 3 blocks west, 2 blocks north, 4 blocks west, and 2 blocks north. This takes you to where the float ended. So, the starting point of the float is the community swimming pool. ✔

Analyze the Strategy

1. **WRITE MATH** Explain when you would use the *work backward* strategy to solve a real-world problem involving transformations.

= **Step-by-Step Solutions** begin on page R1.
Extra Practice begins on page EP2.

• Work backward.
• Guess, check, and revise.
• Determine reasonable answers.
• Choose an operation.

Use the *work backward* strategy to solve Exercises 2–5.

2. **BEDROOM** Bradley moved his bedroom around as shown. He rotated his bed 90° counterclockwise and translated it 14 units to the right and 4 units down. Graph and find the vertices of the bed before it was moved.

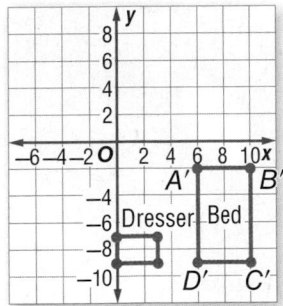

3. **GEOMETRY** The triangle at the right was rotated 180° about vertex S. Draw the original triangle.

4. **CARTOONS** The character below was reflected over the y-axis. Find the original coordinates.

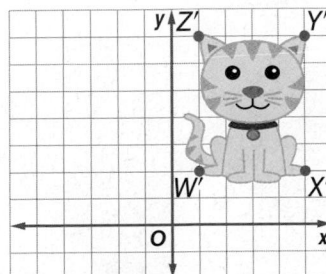

5 **MONEY** Jackie's mom gave her money for her school field trip. Jackie spent half of the money on admission to a museum. She spent $\frac{2}{3}$ of what she had left on lunch. She brought $2.50 back to her mom. How much did Jackie's mom originally give her?

Use any strategy to solve Exercises 6–10.

6. **MONEY** Cole has $2.58 in coins. If he has quarters, dimes, nickels, and pennies, how many of each coin does he have?

7. **FENCING** Mr. Hernandez will build a fence to enclose a rectangular yard for his horse. If the area of the yard to be enclosed is 1,944 square feet and the length of the yard is 54 feet, how much fencing is needed?

8. **ART** Victor draws a right triangle so that one of the acute angles measures 55°. Without measuring, describe how Victor can determine the measure of the other acute angle in the triangle. Then find the angle measure.

9. **GEOMETRY** Refer to quadrilateral *ABCD*.

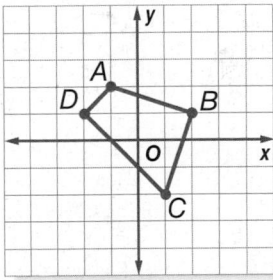

a. Describe the translation that will move point *A* to the point (2, −2). Then graph quadrilateral *A′B′C′D′* using this translation.

b. Find the coordinates of the vertices of quadrilateral *ABCD* after a reflection over the y-axis. Then graph the reflection.

10. **WRITE MATH** Write a transformation problem that could be solved by working backward. Then write the steps you would take to find the solution to your problem.

AN ANIMATION SENSATION!

Have you ever wondered how they make animated movies look so realistic? Computer animators use computer technology and apply their artistic skills to make inanimate objects come alive. If you are interested in computer animation, you should practice drawing, study human and animal movement, and take math classes every year in high school. Tony DeRose, a computer scientist at an animation studio said, "Trigonometry helps rotate and move characters, algebra creates the special effects that make images shine and sparkle, and calculus helps light up a scene."

21st Century Careers

Choose a Major
Are you interested in a career as a computer animator? Take some of the following courses in high school.

- 2-D Animation
- Algebra
- Calculus
- Trigonometry

Get ConnectED

Figure 1

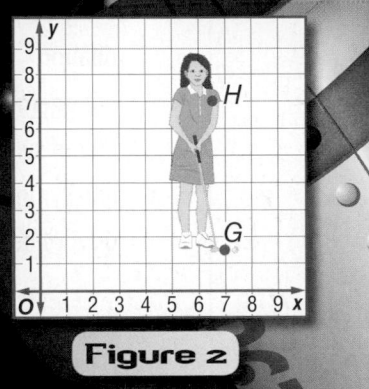

Figure 2

Figure 3

Real-World Math

Use Figures 1–3 to solve each problem.

1. In Figure 1, the car is translated 8 units left and 5 units down so that it appears to be moving. What are the coordinates of A' and B' after the translation?

2. In Figure 1, the car is translated so that A' has coordinates $(-7, 2)$. Describe the translation as an ordered pair. Then find the coordinates of point B'.

3. In Figure 1, the car is reflected over the x-axis in order to make its reflection appear in a pond. What are the coordinates of A' and B' after the reflection?

4. In Figure 2, the artist uses rotation to show the girl's golf swing. Describe the coordinates of G' if the golf club is rotated 90° clockwise about point H.

5. The character in Figure 3 is enlarged by a scale factor of $\frac{5}{2}$. What are the coordinates of Q' and R' after the dilation?

6. The character in Figure 3 is reduced in size by a scale factor of $\frac{2}{3}$. What is the number of units between S' and T', the width of the character's face, after the dilation?

Chapter Study Guide and Review

Be sure the following Key Concepts are noted in your Foldable.

Polygons

Key Concepts

Polygons (Lesson 1)
- The sum of the measures of the angles of a triangle is 180°.
- The sum of the measures of the angles of a quadrilateral is 360°

Translations (Lesson 2)
- When translating a figure, every point is moved the same distance in the same direction.

Reflections (Lesson 3)
- When reflecting a figure, every point in the original figure is the same distance from the line of reflection as its corresponding point on the original figure.

Rotations (Lesson 4)
- A rotation occurs when a figure is rotated, or turned, about a point such as the origin.

Dilations (Lesson 5)
- The image produced by enlarging or reducing a figure is called a dilation.

Key Vocabulary

angle of rotation	reduction
center	reflection
complementary angles	regular polygon
congruent figures	rotation
dilation	rotational symmetry
enlargement	supplementary angles
line of symmetry	transformation
line symmetry	transversal
quadrilateral	triangle

Vocabulary Check

State whether each sentence is *true* or *false*. If *false*, replace the underlined word or number to make a true sentence.

1. The point (3, −2) when translated up 3 units and to the left 5 units becomes <u>(6, −7)</u>.

2. The image produced by enlarging or reducing a figure is called a <u>dilation</u>.

3. Two angles with measures adding up to 180° are called <u>complementary angles</u>.

4. A <u>rotation</u> is a mirror image of the original figure.

5. A <u>reduction</u> is the motion where a figure is moved without being turned.

6. Figures that match exactly when folded in half have <u>line symmetry</u>.

7. <u>Rotational symmetry</u> is the degree measure of the angle through which the figure is rotated.

8. <u>Congruent figures</u> have the same size and same shape, and the corresponding sides and angles have equal measures.

9. The sum of the measures of the angles of a triangle is <u>360°</u>.

Multi-Part Lesson Review

Lesson 1   Polygons

Angle Relationships (Lesson 1A)

10. Name the angle in four ways. Then classify the angle as *acute*, *right*, *obtuse*, or *straight*.

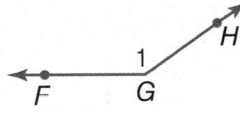

Identify each pair of angles as *complementary*, *supplementary*, or *neither*.

11.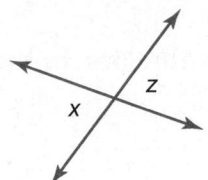

12.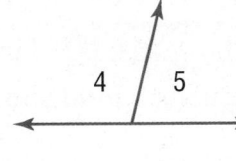

13. **ALGEBRA** Angle Y and ∠Z are complementary, and m∠Z = 35°. What is the measure of ∠Y?

EXAMPLE 1   Name the angle in four ways. Then classify the angle as *acute*, *right*, *obtuse*, or *straight*.

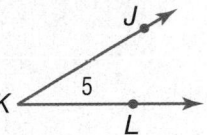

The angle can be named in four ways: ∠JKL, ∠LKJ, ∠K, or ∠5.

The angle is an acute angle because its measure is less than 90°.

EXAMPLE 2   Classify the pair of angles as *complementary*, *supplementary*, or *neither*.

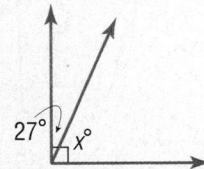

Since the two angles form a right angle, they are complementary.

Triangles (Lesson 1C)

ALGEBRA Find the value of *x*.

14.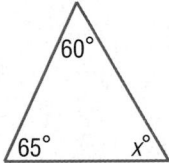

15.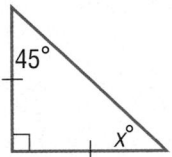

16. **ALGEBRA** Find m∠S in △RST if m∠R = 28° and m∠T = 13°.

17. **SCHOOL SPIRIT** Maddie entered her design for the school flag in the contest. What is the value of *x* in her flag?

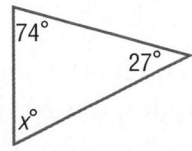

EXAMPLE 3   Find the value of *x*.

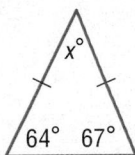

Write and solve an equation.

$$x + 64 + 67 = 180 \quad \text{The sum of the measures is 180.}$$
$$x + 131 = 180 \quad \text{Simplify.}$$
$$\underline{-131 = -131} \quad \text{Subtract 131 from each side.}$$
$$x = 49$$

So, the value of *x* is 49.

Polygons (continued)

Quadrilaterals (Lesson 1D)

Classify the quadrilateral using the name that *best* describes it.

18.    **19.**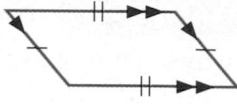

Find the value of *x* in each quadrilateral.

20. 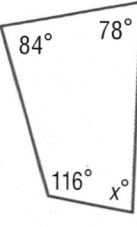   **21.**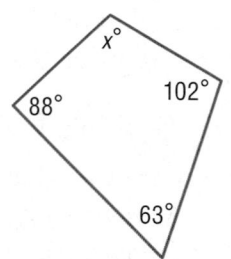

22. SHAPES What type of quadrilateral does not have opposite sides congruent?

23. KITES Identify the quadrilateral outlined.

EXAMPLE 4 Classify the quadrilateral using the name that *best* describes it.

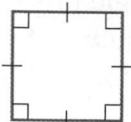

The quadrilateral is a parallelogram with 4 right angles and 4 congruent sides. It is a square.

EXAMPLE 5 Find the value of *x* in the quadrilateral shown.

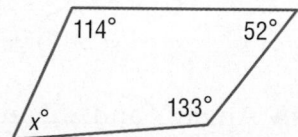

The sum of the angle measures in a quadrilateral is 360°.

$$x + 114 + 52 + 133 = 360$$
$$x + 299 = 360 \quad \text{Add 114, 52, and 133.}$$
$$x = 61 \quad \text{Subtract 299 from each side.}$$

So, the value of *x* is 61.

Polygons and Angles (Lesson 1E)

Determine whether each figure is a polygon. If it is, classify the polygon and state whether it is regular. If it is *not* a polygon, explain why.

24.    **25.**

26. ALGEBRA Find the measure of each angle of a regular 15-gon.

EXAMPLE 6 Determine whether the figure is a polygon. If it is, classify the polygon and state whether it is regular. If it is *not* a polygon, explain why.

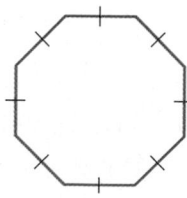

Since the polygon has 8 congruent sides and 8 congruent angles, it is a regular octagon.

Translations

Translations in the Coordinate Plane (Lesson 2B)

Triangle *PQR* has coordinates *P*(4, −2), *Q*(−2, −3), and *R*(−1, 6). Find the coordinates of *P′Q′R′* after each translation. Then graph each translation.

27. 6 units left, 3 units up

28. 4 units right, 1 unit down

29. 3 units left

30. 7 units down

31. Square *ABCD* has vertices *A*(−4, −4), *B*(−1, −4), *C*(−4, −1), and *D*(−1, −1). Find the coordinates of square *A′B′C′D′* after a translation 5 units right and 2 units up. Then graph the figure and its translated image.

EXAMPLE 7 Find the coordinates of △*G′H′I′* after a translation 2 units left and 4 units up.

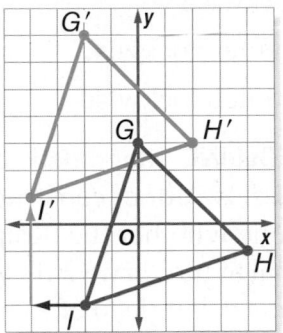

The vertices of △*G′H′I′* are *G′*(−2, 7), *H′*(2, 3), and *I′*(−4, 1).

Reflections

Reflections in the Coordinate Plane (Lesson 3B)

Find the coordinates of each figure after a reflection over the given axis. Then graph the figure and its reflected image.

32. △*RST* with coordinates *R*(−1, 3), *S*(2, 6), and *T*(6, 1); *x*-axis

33. parallelogram *ABCD* with coordinates *A*(1, 3), *B*(2, −1), *C*(5, −1), and *D*(4, 3); *y*-axis

34. rectangle *EFGH* with coordinates *E*(4, 2), *F*(−2, 2), *G*(−2, 5), and *H*(4, 5); *x*-axis

35. square *WXYZ* with coordinates *W*(−3, 0), *X*(−1, 0), *Y*(−1, 2), and *Z*(−3, 2); *y*-axis

EXAMPLE 8 Find the coordinates of △*C′D′E′* after a reflection over the *y*-axis. Then graph its reflected image.

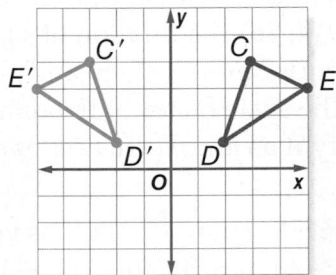

The vertices of △*C′D′E′* are *C′*(−3, 4), *D′*(−2, 1), and *E′*(−5, 3).

Lesson 4 Rotations

Rotations in the Coordinate Plane (Lesson 4B)

Refer to △XYZ in Example 9. Graph the image after each rotation. Then find the coordinates for the vertices of △X'Y'Z'.

36. 270° clockwise

37. 180° counterclockwise

38. 360° counterclockwise

39. PAPER FOLDING Magdalene created the design below out of construction paper. Determine whether the design has rotational symmetry. Write *yes* or *no*. If yes, name its angle(s) of rotation.

EXAMPLE 9 Rotate △XYZ 90° clockwise about the origin and graph its image. Then find the coordinates for the vertices of △X'Y'Z'.

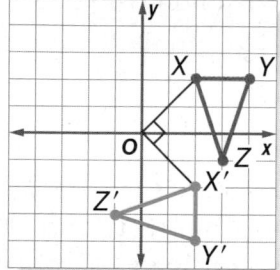

Point X becomes X'(2, −2).

Point Y becomes Y'(2, −4).

Point Z becomes Z'(−1, −3).

Lesson 5 Dilations

Dilations (Lesson 5A)

40. Segment C'D' with endpoints C'(−8, 20) and D'(4, 16) is a dilation of segment CD with endpoints C(−2, 5) and D(1, 4). Find the scale factor of the dilation and classify it as an *enlargement* or a *reduction*.

41. GEOMETRY Triangle ABC has vertices A(−3, −6), B(6, 3), and C(9, −3). Find the vertices of its image for a dilation with a scale factor of $\frac{1}{3}$. Then graph △ABC and its dilation.

EXAMPLE 10 Segment XY has endpoints X(−4, 1) and Y(8, −2). Find the endpoints of its image for a dilation with a scale factor of $\frac{3}{4}$.

Multiply each coordinate in the ordered pair by $\frac{3}{4}$.

$$X(-4, 1) \rightarrow \left(-4 \cdot \frac{3}{4}, 1 \cdot \frac{3}{4}\right) \rightarrow X'\left(-3, \frac{3}{4}\right)$$

$$Y(8, -2) \rightarrow \left(8 \cdot \frac{3}{4}, -2 \cdot \frac{3}{4}\right) \rightarrow Y'\left(6, -1\frac{1}{2}\right)$$

PSI: Work Backward (Lesson 5B)

Solve. Use the *work backward* strategy.

42. **PARKING** Mr. Tetto moved his car and parked it in the 3rd row, 5 places from the end. His car was parked 5 rows away and 2 places closer to the end. Where was Mr. Tetto originally parked?

43. **PHOTO** Elias had a photo of his parents enlarged for their anniversary. He increased the photo 3 times the original size. If the new photo is 9 inches by 12 inches, what was the size of the original?

44. **FINANCIAL LITERACY** Each month, Ben saves half of his paper route paycheck and then spends the rest during the four weeks of the month. He spends $\frac{3}{5}$ of his remaining paycheck on lunches. If he spends $15 on lunches each week, how much is his monthly paycheck?

EXAMPLE 11 Millie has money from her birthday to spend at the art supply store. She spends $\frac{1}{3}$ of the amount on a sketch pad. She then splits the remainder evenly between pencils and paints. If the pencils cost her $10, how much did she have to spend originally?

Work backward to determine the original amount of money Millie had to spend. She spent $10 on pencils and $10 on paints. The pencils and paints account for $\frac{2}{3}$ of the money if she spent $\frac{1}{3}$ on the sketch pad. We can write an equation to find the total amount.

$\frac{2}{3}x = 20$ Write the equation.

$\left(\frac{3}{2}\right)\frac{2}{3}x = \left(\frac{3}{2}\right)20$ Multiplication Property of Equality

$x = 30$ Simplify.

Millie had $30 to spend at the art supply store.

Classify each pair of angles as *complementary*, *supplementary*, or *neither*.

1.

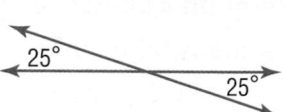

2.

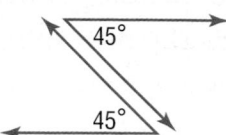

ALGEBRA Find the missing measure in each triangle with the given angle measures.

3. $75°, 25.5°, x°$

4. $x°, 58°, 64°$

5. $23.5°, x°, 109.5$

ALGEBRA Find the missing measure in each quadrilateral.

6.

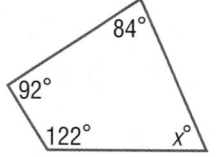

7.

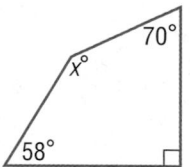

8. MULTIPLE CHOICE Which quadrilateral does NOT have opposite sides congruent?

   A. parallelogram   **C.** trapezoid

   B. square   **D.** rectangle

Determine whether the figure is a polygon. If it is, classify the polygon and state whether it is regular. If it is *not* a polygon, explain why.

9.

10.

Graph $\triangle JKL$ with vertices $J(2, 3)$, $K(-1, 4)$, and $L(-3, -5)$. Then graph its image and write the coordinates of its vertices after each transformation.

11. reflection over the x-axis

12. translation 2 units left and 5 units up

13. Square $ABCD$ is shown. What are the vertices of square $A'B'C'D'$ after a translation 2 units right and 2 units down? Graph the translated image.

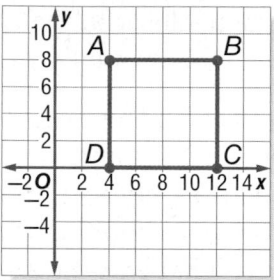

14. Triangle ABC has vertices $A(1, 1)$, $B(-2, 4)$, and $C(-3, -2)$. Find the vertices of its image after a dilation with a scale factor of 2. Then graph $\triangle ABC$ and its dilation.

15. Graph $\triangle ABC$ and its image after a clockwise rotation of 90°. Then name the coordinates of the vertices for $\triangle A'B'C'$.

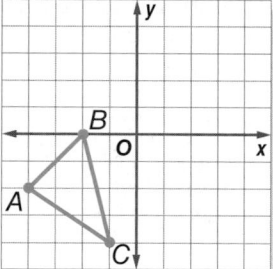

16. THINK SOLVE EXPLAIN EXTENDED RESPONSE Polygon $HJKL$ has vertices $H(-2, 3)$, $J(-3, 1)$, $K(-2, -4)$, and $L(3, 1)$.

   Part A Graph polygon $HJKL$. What are the vertices of polygon $H'J'K'L'$ if it is reflected over the x-axis?

   Part B Find the vertices of polygon $H''J''K''L''$ after polygon $H'J'K'L'$ is dilated using a scale factor of 2. Then graph polygon $H'J'K'L'$ and polygon $H''J''K''L''$.

   Part C Identify at least two things you notice about the relationship between polygon $HJKL$ and polygon $H''J''K''L''$.

Preparing for Standardized Tests

Multiple Choice: Make a Drawing

You are not required to show your work for multiple-choice questions. However, you are allowed to write on the test book to help you solve problems. Doing so will help keep you from making careless errors.

TEST EXAMPLE

Triangle *LMN* has vertices at *L*(−4, −1), *M*(−2, 5), and *N*(1, 2). Triangle *HJK* has one vertex at *H*(0, −3) and one vertex at *J*(2, 3) and represents a slide of triangle *LMN*. Which point would be *K*, the third vertex of triangle *HJK*?

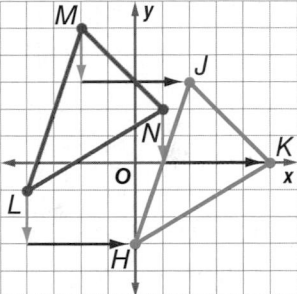

A. (5, 0) **C.** (−1, 2)

B. (5, 2) **D.** (−1, −5)

On the figure provided in the test book, graph vertices *H*(0, −3) and *J*(2, 3), as shown above. You can see that these vertices are 2 units down and 4 units right from the corresponding vertices in triangle *LMN*.

If you translate vertex *N* 2 units down and 4 units right, you get the coordinates of vertex *K*, (5, 0). The correct answer is A.

Work on It

Chloe is creating a design by reflecting the figure below across the *x*-axis. What would be the reflected location of point *Q*?

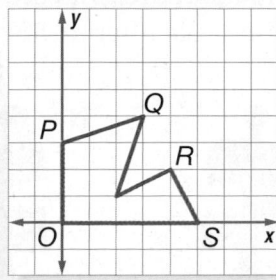

Test Hint

Sometimes it is necessary to make a completely new drawing. Use the white space in the test book to make the drawing.

F. (−2, 4) **H.** (3, −4)

G. (4, 3) **I.** (3, −5)

✓ Test Practice

Read each question. Then fill in the correct answer on the answer sheet provided by your teacher or on a sheet of paper.

1. $\triangle ABC$ is translated 2 units right and 2 units down. What are the coordinates of A'?

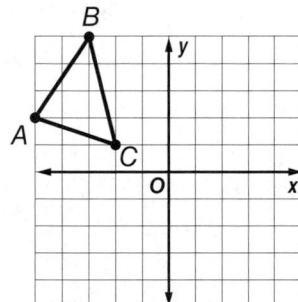

   A. $(0, -1)$

   B. $(-3, 0)$

   C. $(-1, 3)$

   D. $(0, -3)$

2. Seth has $858.60 in his savings account. He plans to spend 15% of his savings on a bicycle. Which of the following represents the amount Seth plans to spend on the bicycle?

   F. $182.79

   G. $171.72

   H. $128.79

   I. $122.79

3. Rectangle M is similar to rectangle N.

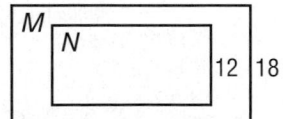

   Which scale factor was used to dilate rectangle M to rectangle N?

   A. $\frac{1}{4}$ C. $\frac{2}{3}$

   B. $\frac{1}{3}$ D. $1\frac{1}{2}$

4. **THINK SOLVE EXPLAIN** **SHORT RESPONSE** The figure shown was transformed from Quadrant I to Quadrant IV. What type of transformation was applied?

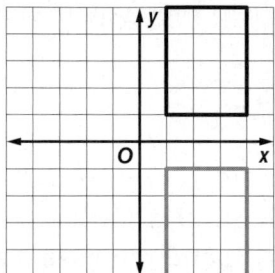

5. **GRIDDED RESPONSE** How many square feet of wrapping paper will Ashton need to cover the box shown?

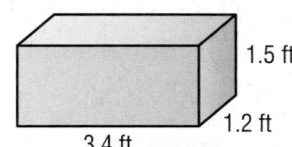

   1.5 ft
   1.2 ft
   3.4 ft

6. Carrie rotated a puzzle piece 180° clockwise around vertex C to see if she could use it.

   Which image represents the position of the puzzle piece after a 180° clockwise rotation?

   F.    H.

   G.    I.

7. The table at the right shows the possible outcomes when tossing two fair coins at the same time.

1st Coin	2nd Coin
H	H
H	T
T	H
T	T

Which of the following must be true?

  A. The probability that both coins have the same outcome is $\frac{1}{4}$.

  B. The probability of getting at least one tail is higher than the probability of getting two heads.

  C. The probability that exactly one coin will turn up heads is $\frac{3}{4}$.

  D. The probability of getting at least one tail is lower than the probability of getting two tails.

8. ✎ **GRIDDED RESPONSE** A manager took an employee to lunch. If the lunch was $48 and she left a 20% tip, what was the total cost in dollars of the lunch?

9. Which of the following groups does NOT contain equivalent fractions and decimals?

  F. $\frac{9}{20}$, 0.45

  G. $\frac{3}{10}$, 0.3

  H. $\frac{7}{8}$, 0.875

  I. $\frac{2}{3}$, 0.7

10. **THINK SOLVE EXPLAIN** **SHORT RESPONSE** Alfonzo drew half of a star on a coordinate plane.

If the drawing was reflected across the y-axis, what would be the reflected location of point B?

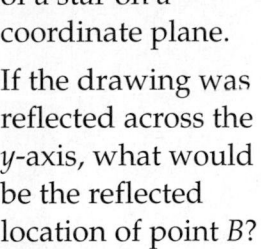

11. Which drawing best represents a reflection over the vertical line segment in the center of the rectangle?

  A.   **C.**

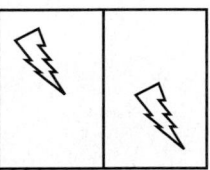 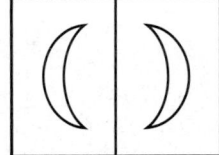

  B.   **D.**

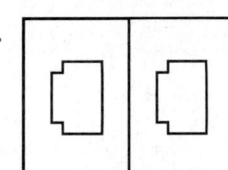 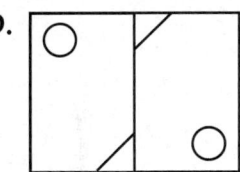

12. **THINK SOLVE EXPLAIN** **EXTENDED RESPONSE** Graph △XYZ with vertices X(2, 1), Y(7, 3), and Z(3, 6).

Part A Translate △XYZ 3 units left and 4 units down. Identify the coordinates of each new vertex.

Part B Find the vertices of △X'Y'Z' after a dilation with a scale factor of $\frac{1}{2}$. Then graph the dilation.

Part C Rotate the dilated figure 270° clockwise around the origin. Draw the rotation.

NEED EXTRA HELP?												
If You Missed Question...	1	2	3	4	5	6	7	8	9	10	11	12
Go to Chapter-Lesson...	12-2B	6-1B	12-5A	12-3B	10-2B	12-4B	8-2B	6-3C	3-1B	12-3B	12-3B	12-5A

Problem-Solving Projects

Are We Similar?

Turn Over a New Leaf

Stand Up and Be Counted!

When will I ever use this?

Have you ever said that? Did you wonder when you would use the math you are learning?

The Problem-Solving Projects apply the math you have learned so far in school. You'll see math in everyday events. Try them!

Be True to Your School

Math Genes

PROJECT 1: Are We Similar?

Las Vegas is famous for its themed hotels. How is the Statue of Liberty at the New York New York hotel similar to the one in New York City? Head to your school or local library to find out. In this project, you will be reading books or stories about the concept of similarity. Don't forget to bring your math thinking cap and your library card. You'll want to check out this adventure!

What You'll Do

Go to a local library and locate books and stories about similarity. Identify careers that use similarity. Brainstorm other places in life that you use similarities. Create a poster to share your findings with the class.

Materials:

- library books about similarity
- poster board

Procedure

1. Take a trip to your local or school library to research information about similarity and the construction of buildings. Look at the different ways buildings are constructed and how the design team incorporates outside influences. As you research, make a list of jobs that use similarity.

2. Select one famous building or statue. Find the actual measurements and draw a smaller version of it using a scale factor. Some ideas are:

- Eiffel Tower
- Leaning Tower of Pisa
- Stonehenge
- Statue of Liberty
- Colosseum
- Golden Gate Bridge

Technology Tips

- You can use a **computer projector** to help enlarge the scale drawing for your poster.

3. Create a poster that contains key facts about similarity and building design. It should include your scale drawing and any calculations you used throughout the project. Share your poster with the class.

Making the Connection

Use the information collected about similarity as needed to help in these investigations.

Art

Create an art project that involves similarity. Where is similarity used in your artwork?

Science

Research objects found in nature that are similar. What makes the objects similar? Does their similarity have an impact on their roles in nature?

Language Arts

Write an essay about a career that uses similarity. Include such information as the type of degree needed, work conditions, and possible industries for that career.

Congratulations!

We hope you enjoyed your experience with construction and buildings! Now that you have gained knowledge about similarities and the jobs that use it, you may find yourself using similarity to earn a living! Keep up the good work and it will take you great places.

Math Genes

What's math have to do with genetics? Well, you're about to find out. You'll research basic genetics and learn how to use a Punnett Square. Then you'll create sample genes for pet traits. You'll make predictions based on the pets' traits to determine the traits of their offspring. Put on your lab coat and grab your math tool kit to begin this adventure.

What You'll Do

Use the Internet to research the Punnett Square and its role in genetics. Then you will use what you learned to practice problems involving a Punnett Square. Next you will collect and record some specific information about pets. Finally, you will analyze and calculate different statistics based on your data.

Materials:

- Internet
- poster board

Procedure

1. Use the Internet to research and find definitions of the following words: alleles, dominant, genotype, heterozygous, homozygous, phenotype, Punnett Square, and recessive.

2. Locate and read different descriptions of the Punnett Square on the Internet. Find four different scenarios that can be represented using a Punnett Square. Print these four scenarios and then explain how the Punnett Square represents them.

3. **Get ConnectED** Create sample genes for pet traits. Go to connectED.mcgraw-hill.com/ to get the recording sheet for the information you will create. Complete the table for 4 different pets and the possible characteristics of each pet. Write 1–2 paragraphs describing how the Punnett Square is used in genetics and how math is involved with the Punnett Square.

4. Create a poster board containing information about the data you have gathered about genetics and pet traits. Include a table displaying your data.

Making the Connection ·····················

Use the information collected about genetics as needed to help in these investigations.

Language Arts

Research the theories of genetics. Write an essay that explains how genes work and what it means for a trait to be dominant or recessive.

Science

Explain how a Punnett Square can be used to determine the probability of inheriting certain features from parents.

Health

Select a health condition or disease and research how genetics may play a part in the disease. Write 1–2 paragraphs explaining how genetics may influence someone's risk of getting the disease and steps that can be taken to reduce the risk factors.

Congratulations!

Great work! Did you find out some interesting things about pet traits? We hope you enjoyed your experience in the amazing world of genetics and saw the large role mathematics plays in understanding this world.

Student Handbook

How to Use the Student Handbook

The Student Handbook is the additional skill and reference material found at the end of books. The Student Handbook can help answer these questions.

What if I need more practice?

You, or your teacher, may decide that working through some additional problems would be helpful. The **Extra Practice** section provides these problems for each lesson so you have ample opportunity to practice new skills.

What if I forget a vocabulary word?

The **English/Spanish Glossary** provides a list of new vocabulary words used throughout the textbook. It provides a definition in English and Spanish.

What if I need to check a homework answer?

The answers to the odd-numbered problems are included in **Selected Answers and Solutions**. Check your answers to make sure you understand how to solve all of the assigned problems. Fully worked out solutions to selected problems are also included in this section.

What if I need to find something quickly?

The **Index** alphabetically lists the subjects covered throughout the entire textbook and the pages on which each subject can be found.

What if I forget a formula?

Inside the back cover of your math book is a **Quick Reference** that lists formulas that are used in the book.

Additional Lessons

Lesson 1

Complex Fractions and Unit Rates

Main Idea

Simplify complex fractions and find unit rates.

New Vocabulary

complex fraction

CCSS 7.RP.1, 7.NS.3

PAINT Tia is mixing paint for an art project. To make green paint, she needs to mix 4 cups of blue paint and 2 cups of yellow paint.

1. Write a ratio in simplest form comparing the amount of blue paint to the amount of yellow paint.

2. Suppose Tia has 1 cup of blue paint. How much yellow paint will she need to make the same shade of green paint? Explain your reasoning.

3. For the previous exercise, write the ratio of blue paint to yellow paint.

Fractions like $\dfrac{1}{\frac{1}{2}}$ are called complex fractions. **Complex fractions** are fractions with a numerator, denominator, or both that are also fractions. Complex fractions are simplified when both the numerator and denominator are integers.

EXAMPLE **Simplify a Complex Fraction**

 Simplify $\dfrac{1}{\frac{1}{2}}$.

Recall that a fraction can also be written as a division problem.

$\dfrac{1}{\frac{1}{2}} = 1 \div \dfrac{1}{2}$ Write the complex fraction as a division problem.

$\quad = \dfrac{1}{1} \times \dfrac{2}{1}$ Multiply by the reciprocal of $\frac{1}{2}$, which is $\frac{2}{1}$.

$\quad = \dfrac{2}{1}$ or 2   Simplify.

So, $\dfrac{1}{\frac{1}{2}}$ is equal to 2.

CHECK Your Progress

Simplify.

a. $\dfrac{2}{\frac{2}{3}}$

b. $\dfrac{6}{\frac{1}{3}}$

c. $\dfrac{14}{\frac{7}{9}}$

d. $\dfrac{3}{\frac{3}{7}}$

EXAMPLE   Simplify a Complex Fraction

2 Simplify $\dfrac{\frac{1}{4}}{2}$.

$\dfrac{\frac{1}{4}}{2} = \dfrac{1}{4} \div 2$ Write the complex fraction as a division problem.

$= \dfrac{1}{4} \times \dfrac{1}{2}$ Multiply by the reciprocal of 2, which is $\frac{1}{2}$.

$= \dfrac{1}{8}$ Simplify.

So, $\dfrac{\frac{1}{4}}{2}$ is equal to $\dfrac{1}{8}$.

Study Tip

To divide by a whole number, first write it as a fraction with a denominator of 1. Then multiply by the reciprocal.

So, $\dfrac{\frac{1}{4}}{2}$ can be written as $\dfrac{1}{4} \div \dfrac{2}{1}$.

✔ CHECK Your Progress

e. $\dfrac{\frac{2}{3}}{7}$

f. $\dfrac{\frac{2}{4}}{2}$

g. $\dfrac{\frac{1}{5}}{\frac{6}{7}}$

h. $\dfrac{\frac{2}{5}}{8}$

i. $\dfrac{\frac{5}{7}}{4}$

j. $\dfrac{\frac{2}{3}}{\frac{3}{4}}$

REAL-WORLD EXAMPLE   Find a Unit Rate

3 **RUNNING** Josiah can jog $1\frac{1}{3}$ miles in $\frac{1}{4}$ hour. Find his average speed in miles per hour.

Write a rate that compares the number of miles to hours.

$\dfrac{1\frac{1}{3}\ \text{mi}}{\frac{1}{4}\ \text{h}} = 1\dfrac{1}{3} \div \dfrac{1}{4}$ Write the complex fraction as a division problem.

$= \dfrac{4}{3} \div \dfrac{1}{4}$ Write the mixed number as an improper fraction.

$= \dfrac{4}{3} \times \dfrac{4}{1}$ Multiply by the reciprocal of $\frac{1}{4}$, which is $\frac{4}{1}$.

$= \dfrac{16}{3}$ or $5\dfrac{1}{3}$ Simplify.

So, Josiah jogs at an average speed of $5\frac{1}{3}$ miles per hour.

✔ CHECK Your Progress

k. **DRIVING** A truck driver drove 350 miles in $8\frac{3}{4}$ hours. What is the speed of the truck in miles per hour?

l. **WALKING** Aubrey can walk $4\frac{1}{2}$ miles in $1\frac{1}{2}$ hours. Find her average speed in miles per hour.

You can also use complex fractions to write fractional percents as fractions.

EXAMPLE

4 Write $33\frac{1}{3}\%$ as a fraction in simplest form.

$$33\frac{1}{3}\% = \frac{33\frac{1}{3}}{100} \qquad \text{Definition of percent}$$

$$= 33\frac{1}{3} \div 100 \qquad \text{Write the complex fraction as a division problem.}$$

$$= \frac{100}{3} \div 100 \qquad \text{Write } 33\frac{1}{3} \text{ as an improper fraction.}$$

$$= \frac{\overset{1}{\cancel{100}}}{3} \times \frac{1}{\underset{1}{\cancel{100}}} \qquad \text{Multiply by the reciprocal of 100, which is } \frac{1}{100}.$$

$$= \frac{1}{3} \qquad \text{Simplify.}$$

So, $33\frac{1}{3}\% = \frac{1}{3}$.

 CHECK Your Progress

m. $4\frac{1}{2}\%$ **n.** $12\frac{1}{2}\%$

o. $10\frac{2}{3}\%$ **p.** $3\frac{1}{3}\%$

QUICK Review

Percent

Percent is a ratio that compares a number to 100.

CHECK Your Understanding

Examples 1–2 Simplify.

1. $\dfrac{18}{\frac{3}{4}}$ 2. $\dfrac{\frac{3}{6}}{4}$ 3. $\dfrac{\frac{1}{3}}{\frac{1}{4}}$

Example 3 **4. READING** Monica reads $7\frac{1}{2}$ pages of a mystery book in 9 minutes. What is her average reading rate in pages per minute?

5. TRIP Patrick drove 220 miles to his grandmother's house. The trip took him $4\frac{2}{5}$ hours. What is his average speed in miles per hour

Example 4 Write each percent as a fraction in simplest form.

6. $6\frac{2}{3}\%$ 7. $7\frac{1}{2}\%$

Examples 1–2 Simplify.

8. $\dfrac{\frac{1}{2}}{3}$

9. $\dfrac{\frac{2}{3}}{11}$

10. $\dfrac{12}{\frac{3}{5}}$

11. $\dfrac{1}{\frac{1}{4}}$

12. $\dfrac{\frac{8}{9}}{6}$

13. $\dfrac{\frac{2}{5}}{9}$

14. $\dfrac{\frac{9}{10}}{9}$

15. $\dfrac{\frac{4}{5}}{10}$

16. $\dfrac{\frac{1}{2}}{\frac{1}{4}}$

17. $\dfrac{\frac{1}{4}}{\frac{7}{10}}$

18. $\dfrac{\frac{1}{12}}{\frac{5}{6}}$

19. $\dfrac{\frac{5}{6}}{\frac{5}{9}}$

Example 3 **20. CANOEING** Richard rowed a canoe $3\frac{1}{2}$ miles in $\frac{1}{2}$ hour. What is his average speed in miles per hour?

21. PLANES A small airplane used $4\frac{3}{8}$ gallons of fuel to fly a $1\frac{1}{4}$ hour trip. How many gallons were used each hour?

22. COOKIES Sally wants to make cookies for her little sister's tea party. Sally is cutting a roll of cookie dough into pieces that are $\frac{1}{2}$ inch thick. If the roll is $11\frac{1}{2}$ inches long, how many cookies can she make?

23. FABRIC Mary Alice is making a curtain for her kitchen window. She bought $2\frac{1}{2}$ yards of fabric. Her total cost was $15. What was the cost per yard?

Example 4 Write each percent as a fraction in simplest form.

24. $56\frac{1}{4}\%$

25. $15\frac{3}{5}\%$

26. $13\frac{1}{3}\%$

27. $2\frac{2}{5}\%$

28. $7\frac{3}{4}\%$

29. $8\frac{1}{3}\%$

30. COOKING A high school Family and Consumer Science class has $12\frac{1}{2}$ pounds of flour with which to make soft taco shells. There are $3\frac{3}{4}$ cups of flour in a pound, and it takes about $\frac{1}{3}$ cup of flour per shell. How many soft taco shells can they make?

31. RUNNING Emma runs $\frac{3}{4}$ mile in 6 minutes. Joanie runs $1\frac{1}{2}$ miles in 11 minutes. Whose speed is greater?

32. MONEY A bank is offering home loans at an interest rate of $5\frac{1}{2}\%$. Write the percent as a fraction in simplest form.

33. STOCKS The value of a certain stock increased by $1\frac{1}{4}\%$. Write the percent as a fraction in simplest form.

34. COSTUMES Mrs. Frasier is making costumes for the school play. The table shows the amount of material needed for each complete costume. She bought $14\frac{3}{4}$ yards of material.

Play Costumes	
Top	$1\frac{1}{4}$ yards
Bottom	$\frac{1}{2}$ yards

  a. How many complete costumes can she make? How much material will be left over?

  b. If she spent a total of $44.25 on fabric, what was the cost per yard? Explain how you solved.

35. INSECTS For a project, Karl measured the wingspan of a butterfly and a moth. His measurements are shown below. How many times larger is the moth than the butterfly?

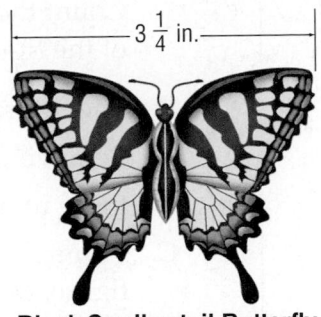

$3\frac{1}{4}$ in.

Black Swallowtail Butterfly

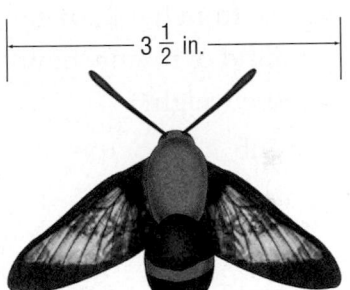

$3\frac{1}{2}$ in.

Hummingbird Moth

Simplify.

36. $\dfrac{5\frac{2}{5}}{3\frac{3}{8}}$

37. $\dfrac{7\frac{1}{2}}{1\frac{2}{5}}$

38. $\dfrac{4\frac{1}{2}}{2\frac{7}{10}}$

39. $\dfrac{4\frac{3}{4}}{2\frac{1}{10}}$

40. $\dfrac{3\frac{2}{3}}{1\frac{8}{9}}$

41. $\dfrac{13\frac{4}{5}}{1\frac{7}{8}}$

H.O.T. Problems

42. OPEN ENDED Write three different complex fractions that can be simplified to $\frac{1}{4}$.

43. CHALLENGE A motorized scooter has tires with a radius of $3\frac{1}{2}$ inches. The tires make one revolution every $\frac{1}{10}$ second. Find the speed of the scooter in miles per hour. Round to the nearest tenth. (*Hint*: The speed of an object spinning in a circle is equal to the circumference divided by the time it takes to complete one revolution)

44. WRITE MATH Explain how complex fractions can be used to solve problems involving ratios.

45. Debra can run $20\frac{1}{2}$ miles in $2\frac{1}{4}$ hours. How many miles per hour can she run?

A. $46\frac{1}{8}$ miles per hour

B. $22\frac{3}{4}$ miles per hour

C. $18\frac{1}{4}$ miles per hour

D. $9\frac{1}{9}$ miles per hour

46. Tina wants to give away 6 bundles of thyme from her herb garden. If she has $\frac{1}{2}$ pound of thyme, how much will each bundle weigh?

F. $\frac{1}{2}$ lb

G. 3 lb

H. $\frac{1}{12}$ lb

I. 12 lb

47. LaShondra is using a model to find the complex fraction below.

$$\frac{\frac{2}{3}}{\frac{1}{12}}$$

| $\frac{1}{12}$ | $\frac{1}{12}$ | $\frac{1}{12}$ | $\frac{1}{12}$ | $\frac{1}{12}$ | $\frac{1}{12}$ | $\frac{1}{12}$ | $\frac{1}{12}$ | $\frac{1}{12}$ | $\frac{1}{12}$ | $\frac{1}{12}$ | $\frac{1}{12}$ |

Which statement shows how to use the model?

A. The figure is divided into twelfths. Count the twelfths that fit within $\frac{2}{3}$ of the figure.

B. The figure is divided into twelfths. Remove $\frac{2}{3}$ of the twelfths, and count those remaining.

C. Count the number of thirds in the figure. Multiply this number by 12.

D. Count the number of rectangles in the figure. Divide this number by 3.

Lesson 2

Graph Proportional Relationships

Main Idea

Identify proportional or nonproportional relationships by graphing on the coordinate plane.

New Vocabulary

ordered pair
coordinate plane
***x*-coordinate**
***y*-coordinate**
origin
***y*-axis**
***x*-axis**
quadrants

CCSS 7.RP.2, 7.RP.2a, 7.RP.2d

PHOTO PRINTERS Katrina wants to purchase a new color photo printer. The table shows the number of color photos one printer can print.

Printer Speed	
Time (min)	Printed Photos
1	14
2	28
3	42
4	56

1. Is the number of photos printed proportional to the number of minutes? Explain your reasoning.

2. How many photos can the printer print in 6 minutes? Explain your reasoning.

The information in the table can be listed as ordered pairs. An **ordered pair** is a pair of numbers, such as (1, 14), used to locate or graph points in the coordinate plane. The **coordinate plane** is formed when two number lines intersect at their zero points.

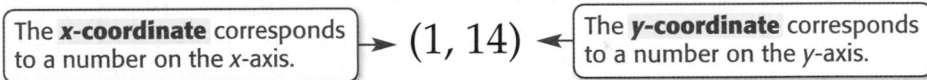

The **x-coordinate** corresponds to a number on the *x*-axis. → (1, 14) ← The **y-coordinate** corresponds to a number on the *y*-axis.

The number lines separate the coordinate plane into four regions called quadrants.

Key Concept — Coordinate Plane

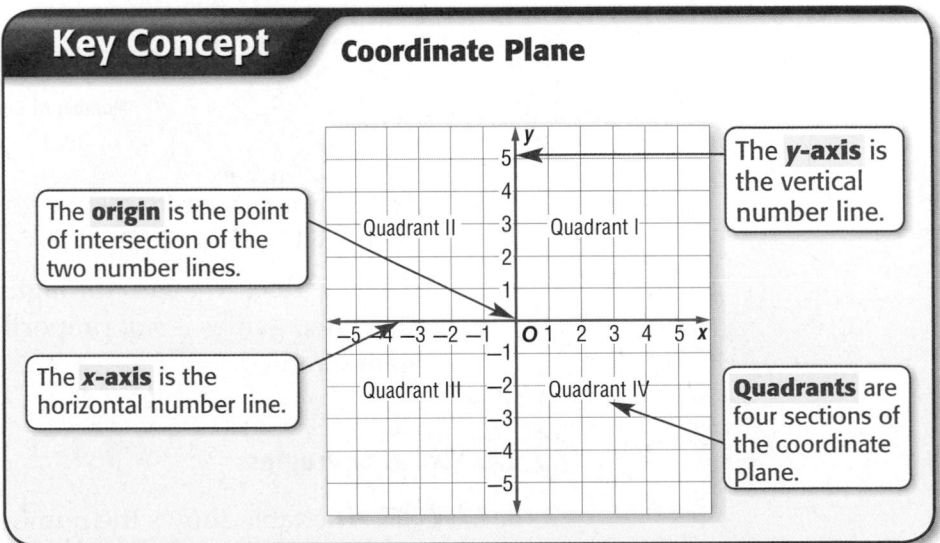

The **origin** is the point of intersection of the two number lines.

The **x-axis** is the horizontal number line.

The **y-axis** is the vertical number line.

Quadrants are four sections of the coordinate plane.

In this lesson, only the first quadrant will be used since we are only working with positive integers.

Another way to determine whether two quantities are proportional is to graph the quantities on the coordinate plane. If the graph of the two quantities is a straight line through the origin, then the two quantities are proportional.

**Identify Proportional
Relationships**

① **VIDEO GAMES** The cost of renting video
games from Games Inc. is shown in the
table. Determine whether the cost is
proportional to the number of games
rented by graphing on the coordinate
plane. Explain your reasoning.

Video Game Rental Rates	
Number of Games	Cost ($)
1	3
2	5
3	7
4	9

Step 1 Write the two quantities as ordered
pairs (number of games, cost).

The ordered pairs are (1, 3), (2, 5), (3, 7), and (4, 9).

QUICKReview

When drawing a graph,
include a title and labels
for the horizontal and
vertical axes.

Step 2 Graph the ordered pairs on the coordinate plane.
Then connect the ordered pairs and extend the line
to the *y*-axis.

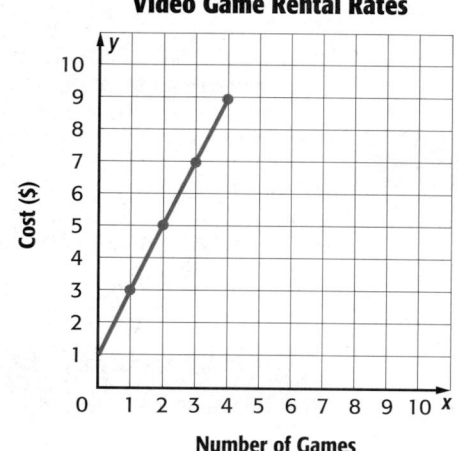

Video Game Rental Rates

Step 3 Determine if the two quantities show a proportional
relationship by looking at the graph.

The line does not pass through the origin. So, the cost of
the video games is not proportional to the number of
games rented.

 CHECK Your Progress

a. EXERCISE The table shows the number of
Calories an athlete burned per minute of
exercise. Determine whether the number
of Calories burned is proportional to the
number of minutes by graphing on the
coordinate plane. Explain your reasoning.

Calories Burned	
Number of Minutes	Number of Calories
1	4
2	8
3	13
4	18

REAL-WORLD EXAMPLE **Identify Proportional Relationships**

2 **ANIMALS** The slowest mammal on Earth is the tree sloth. It moves at a speed of 6 feet per minute. Determine whether the number of feet the sloth moves is proportional to the number of minutes it moves by graphing on the coordinate plane. Explain your reasoning.

Step 1 Make a table to find the number of feet walked for 0, 1, 2, 3, and 4 minutes.

Time (min)	0	1	2	3	4
Distance (ft)	0	6	12	18	24

Step 2 Graph the ordered pairs on the coordinate plane. Then connect the ordered pairs.

Step 3 Determine if the two quantities show a proportional relationship by looking at the graph.

The line passes through the origin and is a straight line. So, the number of feet traveled is proportional to the number of minutes.

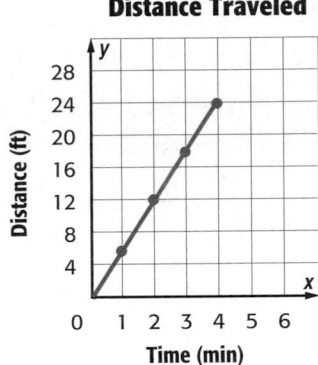

Distance Traveled

CHECK Your Progress

b. JOBS James makes $7 an hour babysitting. Determine whether the amount of money James earns babysitting is proportional to the number of hours he babysits by graphing on the coordinate plane. Explain your reasoning.

REAL-WORLD EXAMPLE **Identify Unit Rates**

3 **ANIMALS** Refer to the graph in Example 2. Explain what the points (0, 0) and (1, 6) represent.

The point (0, 0) represents zero minutes traveled and zero feet traveled. The point (1, 6) represents the sloth walking 6 feet in one minute. This is the unit rate of $\frac{6 \text{ feet}}{1 \text{ minute}}$.

CHECK Your Progress

c. JOBS Refer to the graph you drew in Check Your Progress b. Explain what the points (0, 0) and (1, 7) represent.

✓ CHECK Your Understanding

Example 1

1. MOVIES The cost of 3-D movie tickets is shown in the table. Determine whether the cost is proportional to the number of tickets by graphing on the coordinate plane. Explain your reasoning.

Example 2

2. MUSIC Anna was given a $75 gift card to buy CDs from her favorite store. Each CD costs $15. Determine whether the remaining balance on the gift card is proportional to the number of CDs bought by graphing on the coordinate plane. Explain your reasoning.

Example 3

3. MOVIES Refer to the graph you drew in Exercise 1. Explain what the points (0, 0) and (1, 12) represent.

3D Movie Ticket Prices

Number of Tickets	Cost ($)
1	12
2	24
3	36
4	48

Practice and Problem Solving

Example 1

Determine whether the relationship between the two quantities shown in each table are proportional by graphing on the coordinate plane. Explain your reasoning.

4.

Savings Account

Week	Account Balance ($)
1	125
2	150
3	175
4	200

5.

Cooling Water

Time (min)	Temperature (°F)
5	95
10	90
15	85
20	80

6.

Calories in Fruit Cups

Servings	Calories
1	70
3	210
5	350
7	490

7.

Pizza Recipe

Number of Pizzas	Cheese (oz)
1	8
4	32
7	56
10	80

Example 2

8. MEASUREMENT The perimeter of a square is 4 times as great as the length of any of its sides. Determine whether the perimeter of the square is proportional to the side length. Explain your reasoning.

9. FITNESS A health club charges $35 a month for membership fees. Determine whether the cost of membership is proportional to the number of months. Explain your reasoning.

10. **AIRPLANES** An airplane is flying at an altitude of 4,000 feet and descends at a rate of 200 feet per minute. Determine whether the altitude is proportional to number of minutes. Explain your reasoning.

11. **CONCERT TICKETS** On Saturday morning, 350 concert tickets went on sale. Every 5 minutes 25 tickets were sold. Determine whether the number of remaining tickets is proportional to time in minutes. Explain.

Example 3 Explain what each point represents for the indicated graph and ordered pairs.

12. Exercise 4, points: (0, 100) and (1, 125)

13. Exercise 5, points: (0, 100) and (5, 95)

14. Exercise 6, points: (0, 0) and (1, 70)

15. Exercise 7, points: (0, 0) and (1, 8)

16. **CELL PHONES** Frank and Allie purchased cell phone plans through different providers. Their costs for several minutes are shown.

    a. Whose plan is proportional to the number of minutes the phone is used? Explain.

    b. Refer to Frank's graph. What does the point (1, 0.5) represent?

    c. Refer to Allie's graph. What does the point (3, 4.5) represent?

    d. Who pays a constant rate for cell phone minutes? Explain your reasoning.

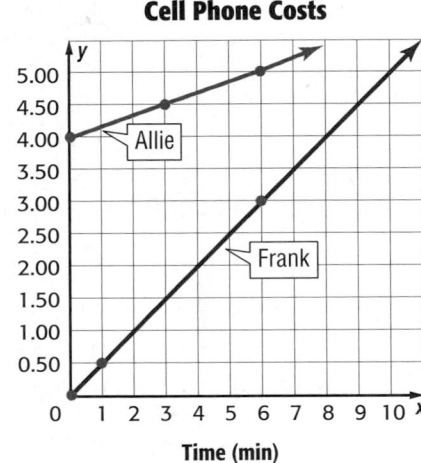

Cell Phone Costs

17. **MULTIPLE REPRESENTATIONS** There are 12 inches in one foot.

    a. TABLES Make a table that shows the number of inches in 0 feet, 1 foot, 2 feet, 3 feet, and 4 feet.

    b. GRAPHS Make a graph of the ordered pairs.

    c. EQUATIONS Write an equation to show the relationship between the total number of inches n and the number of feet f.

    d. WORDS Is the number of inches proportional to the number of feet? Explain your reasoning.

    e. NUMBERS What is the unit rate? Explain how it is represented on the graph.

H.O.T. Problems

18. OPEN ENDED Describe some data and draw a graph that represents a proportional relationship. Explain your reasoning.

19. CHALLENGE Darrell recorded the greenhouse temperatures at certain times of the day in the table. The greenhouse is advertised to maintain temperatures between 65°F and 85°F. Suppose the temperature increases at a constant rate.

Time	Temperature (°F)
1:00 P.M.	66
6:00 P.M.	78.5
8:00 P.M.	83.5

  a. What is the rate of increase as a unit rate?

  b. Create a new table that shows the temperatures recorded hourly from 1:00 P.M. to 8:00 P.M. using the given data in the table.

  c. Create a graph of the time and temperatures from 1:00 P.M. to 8:00 P.M. Is the relationship proportional? Explain.

20. WRITE MATH Write a real-world problem in which you would need to find a unit rate. Then solve your problem. Is the relationship described in your problem proportional? Explain.

✓ Test Practice

21. SHORT RESPONSE Determine whether the relationship between the two quantities shown in the graph is proportional. Explain your reasoning.

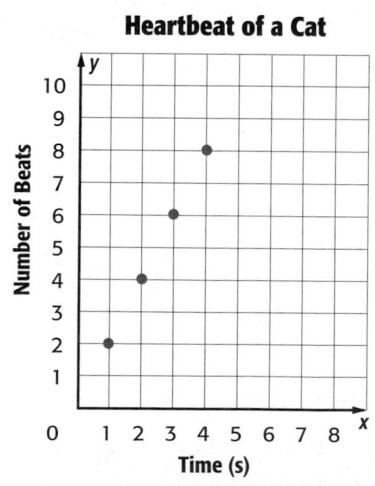

Heartbeat of a Cat

22. Refer to the graph in Exercise 21. Which of the following ordered pairs represent the unit rate?

  A. (0, 0)

  B. (1, 2)

  C. (2, 4)

  D. (3, 6)

Lesson 3

Main Idea

Find the distance between two rational numbers on a number line.

 7.NS.1c

Extend Distance on the Number Line

You can find the distance between two rational numbers on the number line without counting the number of units between them.

ACTIVITY

STEP 1 Copy the table below. Complete the third and fourth columns of the table.

a	b	a + b	a − b	Distance
1	5			
−3	0			
−4	2			
−3	−1			

STEP 2 Use a number line to find the distance between each integer a and b. For example, the distance between 1 and 5 on the number line below is 4 units.

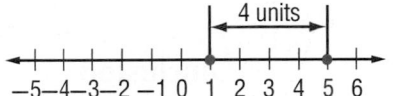

Complete the last column in the table

a	b	a + b	a − b	Distance
1	5			4 units
0	−3			
−4	2			
−3	−1			

Analyze the Results

1. Is there a relationship between the sum of the rational numbers and the distance between them? If so, explain.

2. Is there a relationship between the difference of the rational numbers and the distance between them? If so, explain.

3. For each pair of rational numbers, find b − a. How does b − a compare to a − b and the difference between the points? Use the term *absolute value* in your response.

4. **MAKE A CONJECTURE** Write a statement to describe the relationship regarding the distance between two rational numbers on a number line and their difference. Test your conjecture by finding the distance between −3 and −9 on a number line.

Lesson 4

Extend Use Properties to Multiply

Main Idea

Use properties to prove the rules for multiplying integers.

 7.NS.2a, 7.NS.2c

Properties can be used to justify each statement you make while verifying or proving another statement. In this lesson, you will prove the rules for multiplying integers.

Properties of Mathematics	
Additive Inverse	Multiplicative Property of Zero
Distributive Property	Multiplicative Identity

ACTIVITY

1 **Show that 2(−1) = −2.**
Copy the statements and complete the justifications by writing the correct property from the table above.

Statements	Properties
0 = 2(0)	
0 = 2[1 + (−1)]	
0 = 2(1) + 2(−1)	
0 = 2 + 2(−1)	

Conclusion
In the last statement, 0 = 2 + 2(−1). In order for this to be true, 2(−1) must equal −2. Therefore, 2(−1) = −2.

ACTIVITY

2 **Show that (−2)(−1) = 2.**
Copy the statements and complete the justifications by writing the correct property from the table above.

Statements	Properties
0 = −2(0)	
0 = −2[1 + (−1)]	
0 = −2(1) + (−2)(−1)	
0 = −2 + (−2)(−1)	Simplify.

Analyze the Results

1. Write a conclusion for Activity 2.
2. Prove each of the following statements. Justify each step.

    a. 1(−1) = −1   **b.** (−1)(−1) = 1

Lesson 5

Main Idea

Model factoring linear expressions.

CCSS 7.EE.1, 7.EE.2

Explore Factor Linear Expressions

PENTOMINOES Pentominoes are shapes consisting of five squares. There are twelve different possible ways to arrange the five squares, as shown below.

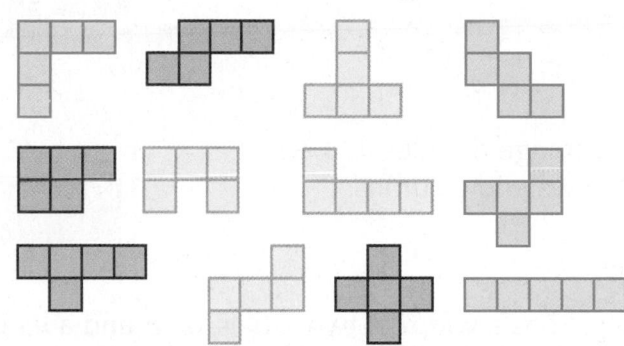

There are twelve shapes, each having an area of 5 square units. So, the total area is 60 square units. It is possible to combine these shapes into a rectangle whose dimensions are 3 units by 20 units.

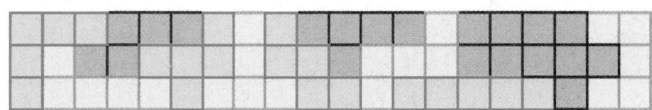

Since $3 \cdot 20 = 60$, 3 and 20 are factors of 60.

Sometimes, you know the product and are asked to find the factors. This process is called *factoring*.

ACTIVITY Factor Linear Expressions

1 Use algebra tiles to factor $2x + 6$.

STEP 1 Model the expression $2x + 6$.

STEP 2 Arrange the tiles into a rectangle with equal rows and columns. The total area of the tiles represents the product. The length and width represent the factors.

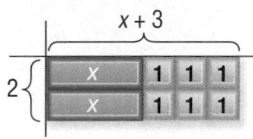

The rectangle has a width of two 1-tiles and a length of one x-tile and three 1-tiles.
So, $2x + 6 = 2(x + 3)$.

ACTIVITY **Factor Linear Expressions**

2 Use algebra tiles to factor 2x − 8.

STEP 1 Model the expression 2x − 8.

STEP 2 Arrange the tiles into equal rows and columns.

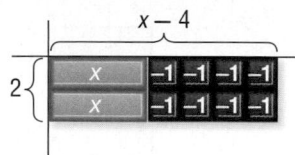

The rectangle has a width of two 1-tiles, or 2, and a length of one x-tile and four −1-tiles, or x − 4.

So, 2x − 8 = 2(x − 4).

Practice and Apply

Use algebra tiles to factor each polynomial.

1. 4x + 6

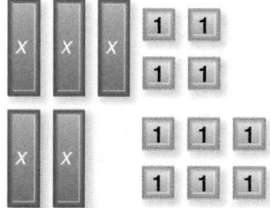

2. 5x + 10

3. 3x + 12

4. 3x − 6

5. 4x − 10

6. 3x − 9

Analyze the Results

7. How is factoring similar to using the Distributive Property?

8. How could you use algebra tiles to factor 3x + 9?

Lesson 6

Factor Linear Expressions

YARD SALE A rectangular yard is being separated into four equal-size sections for different items at a yard sale.

1. How do you find the area of a rectangle?

2. The area of the yard is $(40x + 120)$ square feet. What is the area of each section? Explain your answer.

3. If x is equal to 25 feet, what is the total area of the yard?

A **monomial** is a number, a variable, or a product of a number and one or more variables.

Monomials	Not Monomials
$25, x, 40x$	$x + 4, 40x + 120$

To **factor** a number means to write it as a product of its factors. A monomial can be factored using the same method you would use to factor a number. The greatest common factor (GCF) of two monomials is the greatest monomial that is a factor of both.

EXAMPLES Find the GCF of Monomials

Find the GCF of each pair of monomials.

1 $4x, 12x$

$4x = 2 \cdot 2 \cdot x$ Write the prime factorization of $4x$ and $12x$.
$12x = 2 \cdot 2 \cdot 3 \cdot x$ Circle the common factors.

The GCF of $4x$ and $12x$ is $2 \cdot 2 \cdot x$ or $4x$.

2 $18a, 20ab$

$18a = 2 \cdot 3 \cdot 3 \cdot a$ Write the prime factorization of $18a$ and $20ab$.
$20ab = 2 \cdot 2 \cdot 5 \cdot a \cdot b$ Circle the common factors.

The GCF of $18a$ and $20ab$ is $2 \cdot a$ or $2a$.

CHECK Your Progress

Find the GCF of each pair of monomials.

a. $12, 28c$ **b.** $25x, 15xy$ **c.** $42mn, 14mn$

You can use the Distributive Property and the work backward strategy to express an algebraic expression as a product of its factors. An algebraic expression is in **factored form** when it is expressed as the product of its factors.

$$8x + 4 = 4(2x) + 4(1) \quad \text{The GCF of } 8x \text{ and } 4 \text{ is } 4.$$
$$= 4(2x + 1) \quad \text{Distributive Property}$$

EXAMPLES **Factor Algebraic Expressions**

3 Factor $3x + 9$.

Method 1 **Use a model.**

Model $3x + 9$.

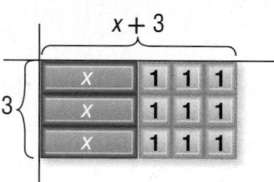

Arrange the tiles into equal rows and columns. The rectangle has a width of three 1-tiles, or 3, and a length of one x-tile and three 1-tiles, or $x + 3$.

Method 2 **Use the GCF.**

$3x = \boxed{3} \cdot x \quad$ Write the prime factorization of $3x$ and 9.

$9 = \boxed{3} \cdot 3 \quad$ Circle the common factors.

The GCF of $3x$ and 9 is 3. Write each term as a product of the GCF and its remaining factors.

$3x + 9 = \mathbf{3}(x) + \mathbf{3}(3)$
$\qquad\quad = \mathbf{3}(x + 3) \quad$ Distributive Property

So, $3x + 9 = 3(x + 3)$.

Study Tip

Factoring Expressions Use algebra tiles to model the expression in Example 4. Since you cannot rearrange the tiles to make a rectangle, the expression cannot be factored.

4 Factor $12x + 7$.

Find the GCF of $12x$ and 7.
$12x = 2 \cdot 2 \cdot 3 \cdot x$
$\quad 7 = 1 \cdot 7$
There are no common factors, so $12x + 7$ *cannot be factored*.

CHOOSE **Your Method**

Factor each expression. If the expression cannot be factored, write *cannot be factored*. Use algebra tiles if needed.

d. $4x + 28$ **e.** $3 + 33x$ **f.** $4x + 35$

Use Expressions to Solve Problems

5 **GARDEN** The garden at the right has a total area of $(15x + 18)$ square feet. Find possible dimensions of the garden.

Factor $15x + 18$.

$15x = 3 \cdot 5 \cdot x$ Write the prime factorization of $15x$ and 18.

$18 = 2 \cdot 3 \cdot 3$ Circle the common factors.

The GCF of $15x$ and 18 is 3. Write each term as a product of the GCF and its remaining factors.

$15x + 18 = 3(5x) + 3(6)$

$\qquad\qquad = 3(5x + 6)$ Distributive Property

So, the dimensions of the garden are 3 feet and $(5x + 6)$ feet.

Check Find the product of 3 and $5x + 6$. $3(5x + 6) = 15x + 18$ ✓

 CHECK Your Progress

g. **FINANCIAL LITERACY** The Reyes family has saved \$480 as a down payment for a new television. If x is the monthly payment for one year, the expression $\$12x + \$480$ represents the total cost of the television. Factor $\$12x + \$480$.

✓ CHECK Your Understanding

Examples 1 and 2   Find the GCF of each pair of monomials.

    1. $32x$, 18   **2.** $15y$, 25   **3.** $45a$, $20ab$

    4. $27s$, $54st$ **5.** $18cd$, $30cd$ **6.** $22mn$, $11kmn$

Examples 3 and 4   Factor each expression. If the expression cannot be factored, write *cannot be factored*. Use algebra tiles if needed.

    7. $36x + 24$ **8.** $6 + 3x$ **9.** $4x + 9$

   10. $13x + 21$ **11.** $2x - 4$ **12.** $14x - 16$

Example 5   **13.** **INCOME** Mr. Phen's monthly income can be represented by the expression $25x + 120$ where x is the number of hours worked. Factor the expression $25x + 120$.

   14. **SPORTS** The area of a high school basketball court is $(50x - 300)$ square feet. Factor $50x - 300$ to find possible dimensions of the basketball court.

Practice and Problem Solving

Examples 1 and 2 Find the GCF of each pair of monomials.

15. 24, 48*m*   **16.** 63*p*, 84   **17.** 40*x*, 60*x*

18. 32*a*, 48*b*   **19.** 30*rs*, 42*rs*   **20.** 54*gh*, 72*g*

21. 36*k*, 144*km*   **22.** 60*jk*, 45*jkm*   **23.** 100*xy*, 75*xyz*

Examples 3 and 4 Factor each expression. If the expression cannot be factored, write *cannot be factored*. Use algebra tiles if needed.

24. $3x + 9$ **25.** $5x + 5$ **26.** $10x - 35$

27. $2x - 15$ **28.** $18x + 6$ **29.** $32 + 24x$

30. $12 + 30x$ **31.** $4x - 7$ **32.** $30x - 40$

Example 5 **33. GEOMETRY** The area of a rectangle is $(4x - 8)$ square units. Factor $4x - 8$ to find possible dimensions of the rectangle.

34. SCHOOL James has $120 in his savings account and plans to save $*x* each month for 6 months. The expression $6*x* + $120 represents the total amount in the account after 6 months. Factor the expression $6x + 120$.

GEOMETRY Write an expression in factored form to represent the total area of each rectangle.

35.

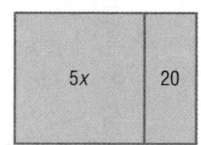

5*x*   20

36.

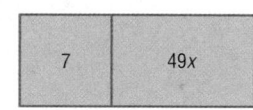

7   49*x*

37.

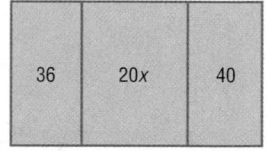

36   20*x*   40

38. SCRAPBOOKING A square scrapbooking page has a perimeter of $(8x + 20)$ inches. What is the length of one side of the page?

39. MUSEUMS Six friends visited a museum to see the new holograms exhibit. The group paid for admission to the museum and $12 for parking. The total cost of the visit can be represented by the expression $6*x* + $12. What was the cost of the visit for one person?

40. LANDSCAPING The diagram represents a flower border that is 3 feet wide surrounding a rectangular sitting area. Write an expression in factored form that represents the area of the flower border.

3*x*

5

3

←3→

Write an expression in factored form that is equivalent to the given expression.

41. $\frac{1}{2}x + 4$ **42.** $\frac{2}{3}x + 6$ **43.** $\frac{3}{4}x - 24$ **44.** $\frac{5}{6}x - 30$

H.O.T. Problems

45. OPEN ENDED Write two monomials whose greatest common factor is $4m$.

46. FIND THE ERROR Enrique is factoring $90x - 15$. Find his mistake and correct it.

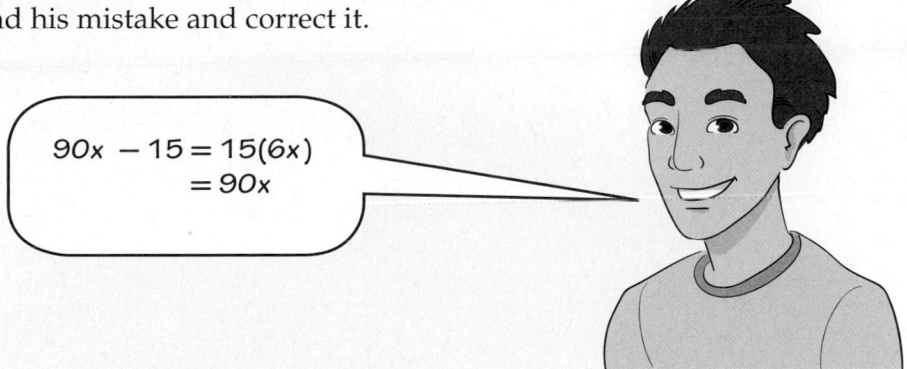

$$90x - 15 = 15(6x)$$
$$= 90x$$

47. WRITE MATH Explain how the GCF is used to factor an expression. Use the term *Distributive Property* in your response.

Test Practice

48. THINK SOLVE EXPLAIN **SHORT RESPONSE** Factor the expression $40x + 15$.

49. Which of the following expressions **cannot** be factored?

A. $6 + 3x$

B. $7x + 3$

C. $15x + 10$

D. $30x + 40$

50. The Venn diagram shows the factors of 12 and $18x$.

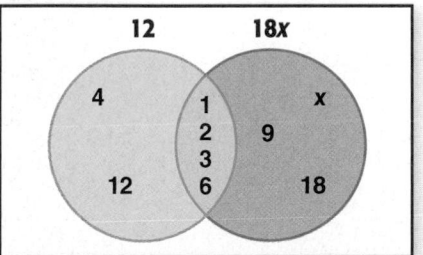

What is the greatest common factor of the two monomials?

F. 2

G. 3

H. 6

I. 36

Lesson 7

Main Idea

Write and solve two-step equations in $p(x + q) = r$ form using bar diagrams and algebra tiles.

 7.EE.4a

Explore More Two-Step Equations

MONEY Emma has two summer jobs. The table shows her earnings each day. She works at each job three days a week and earns a total of $240. How much does she earn each day babysitting?

Job	Earnings ($)
Babysitting	x
Gardening	30

ACTIVITY

1 **What do you need to find?** how much she earns each day babysitting

> **STEP 1** Draw a bar diagram that represents the situation.
>
> ```
> |--------------- $240 ---------------|
> | $x + $30  |  $x + $30  |  $x + $30  |
> earnings each day  earnings each day  earnings each day
> ```
>
> **STEP 2** Write an equation that represents the situation.
> $$3(\$x + \$30) = \$240$$
>
> From the diagram, you can see that one third of her total earnings is equal to $x + $30. So, $x + \$30 = \dfrac{\$240}{3}$ or $80.
> Emma earns $80 − $30 or $50 each day babysitting.

Analyze the Results

1. If Emma worked four days a week and made $360, how much did she earn babysitting each day?

2. If Emma worked five days a week and made $480, how much did she earn babysitting each day?

BAR | DIAGRAM **Draw a bar diagram for each situation. Then solve the problem.**

3. **SCHOOL SUPPLIES** Carolina bought packs of pens and pencils that each cost $2. She bought four packs of pens and spent $14. How many packs of pencils did Carolina buy?

4. **FRUIT** Tyrell bought three apples for 49¢ each , three peaches for 59¢ each, and three bananas for x¢ each. The total cost for all of the fruit was $3.90. How much did each banana cost?

Additional Lessons

Lesson 7 Explore: More Two-Step Equations **781**

ACTIVITY

2 **EXERCISE** Nelson and his brother bought two hamburgers and two soft drinks. The hamburgers cost $6 each. They spent a total of $16. How much did each soft drink cost?

STEP 1 Model the equation using algebra tiles. Use two groups of $x + 6$ tiles.

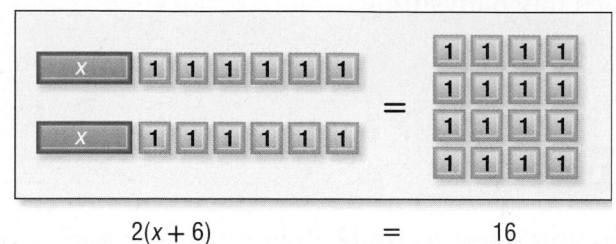

$$2(x + 6) = 16$$

STEP 2 Divide the tiles into two equal groups on each side of the mat. Remove one group from each side.

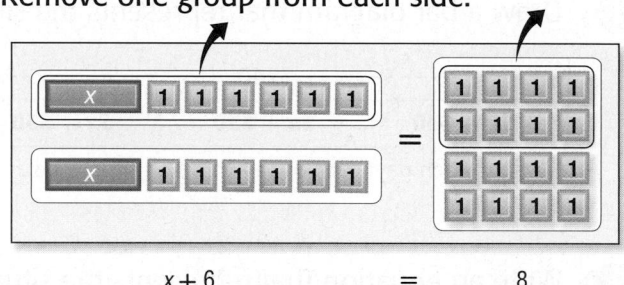

$$x + 6 = 8$$

STEP 3 Remove the same number of 1-tiles from each side.

$$x = 2$$

So, $x = 2$.
Each soft drink costs $2.

Practice and Apply

Solve each equation using algebra tiles.

5. $3(x + 5) = 21$ **6.** $4(x + 2) = 20$ **7.** $5(x + 4) = 30$ **8.** $2(x + 3) = 6$

Solve each real-world work problem. Use a bar diagram or algebra tiles.

9. **SPORTS** Mr. Smith bought five basketballs and five baseballs. Each baseball costs $4. If Mr. Smith spent $50, how much does each basketball cost?

10. **ENTERTAINMENT** Christina's family is attending the school play. Her parents bought two adult tickets and two student tickets that cost $5.00 each. If her parents spent $25 altogether, how much is an adult ticket?

11. **WRITE MATH** Write a word problem that can be represented by the equation $4(x + 15) = 140$. Then solve the problem.

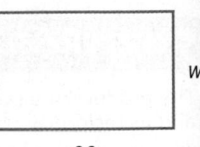
More Two-Step Equations

Main Idea

Solve two-step equations in the form $p(x + q) = r$.

CCSS 7.EE.4a

PLAYGROUNDS A playground is 36 feet long. It has a perimeter of 114 feet.

1. Write an expression that represents the sum of the length and width of the playground.

2. Write an expression that represents twice the sum of the length and the width.

3. Write an equation that represents the perimeter of the playground.

The equation for the perimeter of the playground was written using addition and multiplication. First 36 was added to w and then multiplied by 2. To find the value of w, use inverse operations by dividing the perimeter by 2 and subtracting 36.

The perimeter of the playground is 114 feet.   114
Undo the multiplication. Divide 114 by 2. $114 \div 2 = 57$
Undo the addition. Subtract 36 from 57. $57 - 36 = 21$

An equation like $2(w + 36) = 114$ is in the form $p(x + q) = r$. It contains two factors, p and $(x + r)$, and is considered a two-step equation. Solve these equations using the properties of equality.

EXAMPLE

1. Solve $3(x + 5) = 45$.

Method 1   Use a bar diagram.

Draw a bar diagram. From the diagram, you can see that $x + 5 = 45 \div 3$ or 15. So, $x = 15 - 5$ or 10.

Method 2   Write and solve an equation.

$3(x + 5) =$	45	Write the equation.
$\dfrac{3(x + 5)}{3} =$	$\dfrac{45}{3}$	Divide each side by 3.
$x + 5 =$	15	Simplify.
$x + 5 =$	15	
$-5 =$	-5	Subtract 5 from each side.
$x =$	10	Simplify.

CHOOSE Your Method

a. $2(x + 4) = 20$ b. $5(1 + n) = 30$ c. $3(b - 6) = 12$

EXAMPLE **Solve Two-Step Equations**

2 Solve $\frac{2}{3}(n + 6) = 10$. Check your solution.

$$\frac{2}{3}(n + 6) = 10 \qquad \text{Write the equation.}$$

$$\frac{3}{2} \cdot \frac{2}{3}(n + 6) = \frac{3}{2} \cdot 10 \qquad \text{Multiply each side by the reciprocal of } \frac{2}{3}, \frac{3}{2}.$$

$$(n + 6) = \frac{3}{\underset{1}{\cancel{2}}} \cdot \left(\frac{\overset{5}{\cancel{10}}}{1}\right) \qquad \frac{2}{3} \cdot \frac{3}{2} = 1; \text{ write 10 as } \frac{10}{1}.$$

$$n + 6 = 15 \qquad \text{Simplify.}$$

$$\underline{-\,6 = -\,6} \qquad \text{Subtract 6 from each side.}$$

$$n = 9 \qquad \text{Simplify.}$$

Check $\frac{2}{3}(n + 6) = 10$ $\qquad$ Write the original equation.

$$\frac{2}{3}(9 + 6) = 10 \qquad \text{Replace } n \text{ with 9. Is this sentence true?}$$

$$10 = 10 \checkmark \qquad \text{The sentence is true.}$$

> **QUICK Review**
> The product of a number and its reciprocal is 1.

CHECK Your Progress

d. $5(b + 2) = 30$ $\qquad$ **e.** $-7(6 + d) = 49$ $\qquad$ **f.** $(t + 3)\frac{5}{9} = 40$

REAL-WORLD EXAMPLE

3 **MOVIES** Jamal and two cousins received the same amount of money to go to a movie. Each boy spent $15. Afterward, the boys had $30 altogether. Write and solve an equation to find the amount of money each boy received.

Let m represent the amount of money each boy received.

$$3(m - 15) = 30 \qquad \text{Write the equation.}$$

$$\frac{3(m - 15)}{3} = \frac{30}{3} \qquad \text{Divide each side by 3.}$$

$$m - 15 = 10 \qquad \text{Simplify.}$$

$$\underline{+\,15 = +\,15} \qquad \text{Add 15 to each side.}$$

$$m = 25 \qquad \text{Simplify.}$$

So, each boy received $25.

 CHECK Your Progress

g. **PICNIC** Twelve foot-long sandwiches were cut into equal-size pieces at a school picnic. At the picnic, six pieces were eaten from each sandwich and there were 24 pieces of sandwiches left. Write and solve an equation to find in how many pieces each sandwich was cut.

✓ CHECK Your Understanding

Examples 1 and 2 **Solve each equation. Check your solution.**

1. $2(p + 7) = 18$ **2.** $4(k - 7.3) = 22$ **3.** $-6(1 + m) = -12$

4. $(4 + g)(-11) = 121$ **5.** $\frac{4}{7}(x - 2) = 20$ **6.** $(v + 5)\left(-\frac{1}{9}\right) = 6$

Example 3 **7. STICKERS** Mr. Singh had three sheets of stickers. He gave 20 stickers from each sheet to his students and has 12 total stickers left. Write and solve an equation to find how many stickers were originally on each sheet.

Practice and Problem Solving

Examples 1 and 2 **Solve each equation. Check your solution.**

8. $8(s + 3) = 72$ **9.** $(t + 9)20 = 140$ **10.** $12(0.25 + n) = 3$

11. $(6.3 + w)8 = 28$ **12.** $-7(z - 6) = -70$ **13.** $(t + 8)(-2) = 12$

14. $-4(p + 6.1) = 10$ **15.** $6(y - 1.04) = 21$ **16.** $\frac{8}{11}(t - 10) = 64$

17. $\frac{5}{9}(8 + c) = -20$ **18.** $-\frac{3}{5}\left(r + \frac{1}{5}\right) = \frac{9}{5}$ **19.** $\left(s - \frac{4}{9}\right)\left(-\frac{2}{3}\right) = -\frac{4}{5}$

Example 3 **Write an equation to represent each situation. Then solve the equation.**

20. PERIMETER The length of each side of an equilateral triangle is increased by 5 inches. The perimeter of the triangle is now 60 inches. What was the original length of each side of the equilateral triangle?

21. JEWELRY Anne bought a necklace for each of her three sisters. She paid $7 for each necklace. If she had $9 left, how much money did Anne have initially to spend on each sister?

Solve each equation. Check your solution.

22. $1\frac{3}{5}(t - 6) = -0.4$ **23.** $\left(x + 5\frac{1}{2}\right)0.75 = \frac{5}{8}$

24. SCHOOL SUPPLIES Mrs. Sorenstam bought one ruler, one compass, and one mechanical pencil at the prices shown in the table for each of her 12 students.

a. Suppose Mrs. Sorenstam had 36 cents left after buying the school supplies. Write an equation to find the amount of money Mrs. Sorenstam initially had to spend on each student.

Item	Price ($)
compass	1.49
mechanical pencil	0.59
notebook	0.99
ruler	0.49
protractor	1.00

b. Describe a two-step process you could use to solve your equation. Then solve the equation.

c. Did Mrs. Sorenstam initially have enough money to buy 2 notebooks and 1 protractor for each student? Explain your reasoning.

H.O.T. Problems

25. **OPEN ENDED** Write a real-world situation that can be represented by the equation $2(n + 20) = 110$.

26. **FIND THE ERROR** Marisol is solving the equation $6(x + 3) = 21$. Find her mistake and correct it.

$$6(x + 3) = 21$$
$$\underline{-3 \quad = -3}$$
$$6x = 18$$
$$x = 3$$

27. **CHALLENGE** Solve $p(x + q) = r$ for x. Justify each step.

28. **WRITE MATH** Explain the difference between $px + q = r$ and $p(x + q) = r$. Then write a real-world situation for each type of equation.

 **Test Practice**

29. Which equation has a solution of -4?

   A. $5(n + 2) = 10$

   B. $\frac{1}{5}(n + 2) = 10$

   C. $-5(n + 2) = 10$

   D. $-\frac{1}{5}(n + 2) = 10$

30. Which pair of operations should you use to solve $p(x - q) = r$ for x?

   F. Multiply by p. Subtract q.

   G. Multiply by p. Add q.

   H. Divide by p. Subtract q.

   I. Divide by p. Add q.

31. **THINK SOLVE EXPLAIN** **SHORT RESPONSE** What is the reciprocal of b?

32. Each week, Payat spends a portion of his allowance to download music. The table shows the cumulative amount that he spent on music downloads during a three-week period.

Payat's Downloads	
Week	Total Spent ($)
1	12
2	24
3	36

At the end of Week 3, Payat had $9 left. Which equation can be used to find the amount of Payat's weekly allowance a?

   A. $3(a - 12) = 9$

   B. $9(a - 12) = 3$

   C. $3(a - 12) = 9$

   D. $9(a - 12) = 3$

Lesson 9

Main Idea

Solve and graph two-step inequalities in one variable.

Vocabulary

two-step inequality

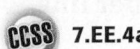

7.EE.4a

Solve Two-Step Inequalities

NEWSPAPERS Kaitlyn is placing an ad in the newspaper for an arts and crafts show. The cost of placing an ad is shown in the table.

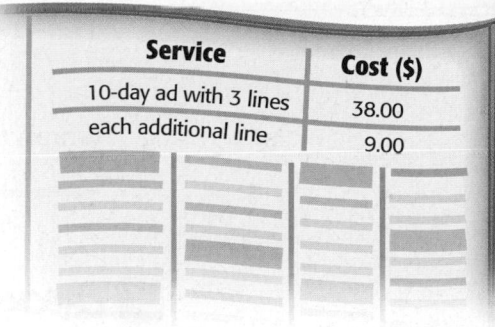

Service	Cost ($)
10-day ad with 3 lines	38.00
each additional line	9.00

1. Write an equation that could be used to find the total cost c of an ad with 4 or more lines ℓ.

2. How much will it cost to place the ad if it is 5 lines long?

3. Suppose Kaitlyn can spend only $60 on the ad. Does she have enough money to place the ad?

A **two-step inequality** is an inequality that contains two operations. To solve a two-step inequality, use inverse operations to undo each operation in reverse order of the order of operations.

EXAMPLE **Solve a Two-Step Inequality**

1. Solve $3x + 4 \geq 16$. Graph the solution set on a number line.

$3x + 4 \geq 16$ Write the inequality.

$\begin{array}{r} 3x + 4 \geq 16 \\ -4 \quad -4 \end{array}$ Subtraction Property of Inequality

$3x \quad\;\; \geq 12$ Simplify.

$\dfrac{3x}{3} \geq \dfrac{12}{3}$ Division Property of Inequality

$x \geq 4$ Simplify.

Graph the solution set.

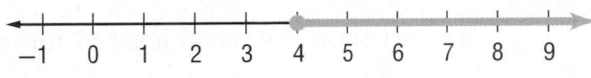

Draw a closed dot at 4 with an arrow to the right.

CHECK Your Progress

Solve each inequality. Graph the solution set on a number line.

a. $2x + 8 > 24$ **b.** $5 + 4x < 33$ **c.** $9 + 2x > 15$

Study Tip

Solving Inequalities
Remember that if multiplying or dividing by a negative number when solving inequalities, reverse the direction of the inequality symbol.

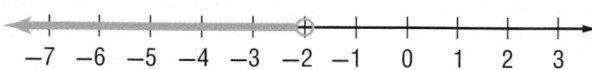 **EXAMPLE** **Solve a Two-Step Inequality**

2 Solve $7 - 2x > 11$. Graph the solution set on a number line.

$7 - 2x > 11$ Write the inequality.

$\underline{-7 \qquad\quad -7}$ Subtraction Property of Inequality

$-2x > 4$ Simplify.

$\dfrac{-2x}{-2} < \dfrac{4}{-2}$ Division Property of Inequality; reverse inequality symbol.

$x < -2$ Simplify. Check your solution.

Graph the solution set.

Draw an open dot at -2 with an arrow to the left.

CHECK Your Progress

Solve each inequality. Graph the solution set on a number line.

d. $\dfrac{x}{2} + 9 \geq 5$ **e.** $16 - 4x > 20$ **f.** $8 - \dfrac{x}{3} \leq 7$

REAL-WORLD EXAMPLE **Two-Step Inequalities**

3 **BASEBALL** Halfway through baseball season, Ichiro Stewart has 34 hits. He averages 2 hits per game. Write and solve an inequality to find how many more games it will take for Stewart to have at least 61 hits, the school record. Interpret the solution.

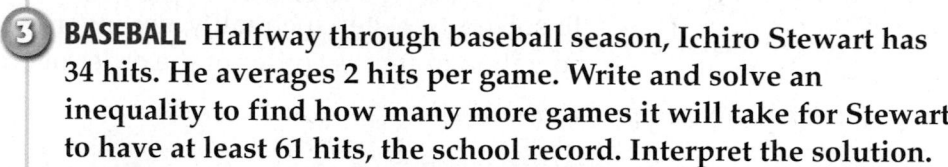

Words	The number of hits plus two hits per game is at least 61.
Variable	Let g represent the number of games he needs to play.
Inequality	$34 \qquad + \qquad 2g \qquad \geq \qquad 61.$

$34 + 2g \geq 61$ Write the inequality.

$\underline{-34 \qquad\quad -34}$ Subtraction Property of Inequality

$2g \geq 27$ Simplify.

$\dfrac{2g}{2} \geq \dfrac{27}{2}$ Division Property of Inequality

$g \geq 13.5$ Simplify.

Stewart should have at least 61 hits after 14 more games.

CHECK Your Progress

g. DVDs Joan has $250. DVDs cost $18.95 each. Write and solve an inequality to find how many DVDs she can buy and still have at least $50. Interpret the solution.

✓ CHECK Your Understanding

Examples 1 and 2 Solve each inequality. Graph the solution set on a number line.

1. $4x + 8 > 40$
2. $9x - 2 < 34$
3. $5x - 7 \geq 43$
4. $\frac{x}{3} + 10 < 12$
5. $11 \leq 7 + \frac{x}{5}$
6. $-3 - x > 4$

Example 3

7. **FINANCIAL LITERACY** A rental car company charges $45 plus an additional $0.20 per mile to rent a car. If Mr. Lawrence does not want to spend more than $100 for his rental car, write and solve an inequality to find how many miles he can drive and not spend more than $100. Interpret the solution.

Practice and Problem Solving

Examples 1 and 2 Solve each inequality. Graph the solution set on a number line.

8. $6x + 14 \geq 20$
9. $5x + 6 > 71$
10. $6x - 3 > 33$
11. $16 > 2x + 4$
12. $4x - 15 \leq 5$
13. $4x - 13 < 11$
14. $\frac{x}{13} + 3 \geq 4$
15. $\frac{x}{5} - 2 > 1$
16. $\frac{x}{4} - 8 \leq 16$
17. $9 \leq \frac{x}{14} + 6$
18. $18.6 \geq \frac{x}{3} - 1.4$
19. $7.5 > 4.5 + \frac{x}{4}$
20. $-7 \leq \frac{x}{10} - 12$
21. $-73 \geq 15 + 11x$
22. $-1 \geq \frac{x}{-10} - 6$
23. $-3x - 11 \geq 34$
24. $-20 > -2x + 4$
25. $143 \leq -12x - 1$

Example 3

26. **ENTERTAINMENT** Tyler needs at least $205 for a new video game system. He has already saved $30. He earns $7 an hour at his part-time job. Write and solve an inequality to find how many hours he will need to work to buy the system. Interpret the solution.

27. **BABYSITTING** Catie is starting a babysitting business. She spent $26 to make signs and flyers to advertise. She charges an initial fee of $5 and then $3 for each hour of service. Write and solve an inequality to find the number of hours she will have to babysit to make a profit. Interpret the solution.

Solve each inequality. Graph the solution set on a number line.

28. $6y > 15 + y$
29. $-5g + 5 \geq -7 - 2g$
30. $10 - 3x \geq 25 + 2x$

Write an inequality for each sentence. Then solve the inequality and graph the solution set on a number line.

31. Three times a number increased by four is less than -62.

32. The quotient of a number and -5 increased by one is at most 7.

33. The quotient of a number and 3 minus two is at least -12.

34. The product of -2 and a number minus six is greater than -18.

H.O.T. Problems

35. **OPEN ENDED** Write a real-world example that could be solved by using the inequality $4x + 8 \geq 32$. Then solve the inequality.

36. **CHALLENGE** In five games, you score 16, 12, 15, 13, and 17 points. How many points must you score in the sixth game to have an average of at least 15 points?

37. **WRITE MATH** Solve $2x + 8 > 18$ and $2x + 8 \leq 18$. How are the inequalities and solutions similar? How are they different?

38. **CHALLENGE** Solve $-x + 6 > -(2x + 4)$. Then graph the solution set on a number line.

Test Practice

39. You want to purchase a necklace for $325. You have already saved $115 and can set aside $22 a week. Which inequality can be used to find the number of weeks it will take to save at least $325?

 A. $22w + 115 \geq 325$

 B. $22w + 115 \leq 325$

 C. $22 + 115w \leq 325$

 D. $22w + 115 < 325$

40. Which inequality represents *six less than three times a number is at least fifteen*?

 F. $3n - 6 \leq 15$

 G. $3n - 6 \geq 15$

 H. $3n - 6 < 15$

 I. $3n - 6 > 15$

41. Which of the following inequalities has the solution set shown below?

 A. $-2x - 5 < 7$

 B. $-2x - 5 > 7$

 C. $-2x - 5 \leq 7$

 D. $-2x - 5 \geq 7$

42. **THINK SOLVE EXPLAIN** **EXTENDED RESPONSE** Dante has 60 baseball cards. This is at least six more than three times as many cards as Anna.

 Part A Write an inequality to represent the situation.

 Part B Solve the inequality from part a. Interpret the solution.

 Part C Graph the solution set on a number line.

Lesson 10

Explore Investigate Online Maps and Scale Drawings

Main Idea

Use online maps to reproduce a scale drawing at a different scale.

 7.G.1

Blueprints and maps are *scale drawings* of the buildings and locations they represent. A scale drawing shows a real object that has been reduced or enlarged.

ACTIVITY Use an Online Map

1 **STEP 1** Use the online map service provided to you by your teacher. Locate your school on a map. Find the scale for the map. Measure the length of the bar in centimeters.

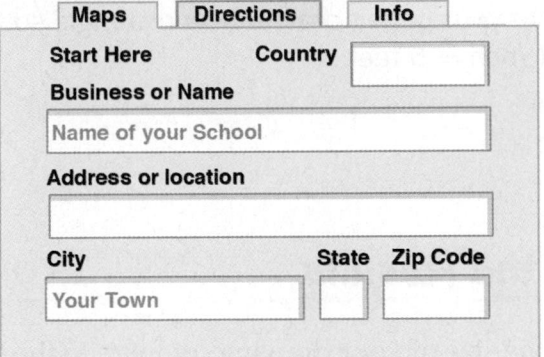

STEP 2 Click on the satellite or aerial view. Use the zoom feature to zoom in until the building shows up on the map.

STEP 3 Find the new scale for the map. Measure the length of the bar in centimeters.

Analyze the Results

1. Copy and complete the table below.

Original View		Zoom View	
Scale Bar		Scale Bar	
Scale Distance		Scale Distance	

2. Describe the appearance of the map as you zoomed in.

3. Write the ratio $\dfrac{\text{scale bar}}{\text{scale distance}}$ for the original view and the zoom view as a decimal. Round to the nearest thousandth.

4. How many times bigger is the zoom view?

ACTIVITY **Reproduce a Scale Drawing**

2 The rectangle shown represents a garden. The scale is 1 inch = 20 feet.

STEP 1 Measure the length and the width of the rectangle.

STEP 2 Use the scale to find the dimensions of the garden.

STEP 3 On your paper, draw a new rectangle so that the scale is 1 inch = 5 feet.

Analyze the Results

5. How does the size of your drawing compare to the size of the drawing shown?

6. Use the scale on your drawing to compute the dimensions of the garden. How do the dimensions compare to the dimensions in Step 2?

Practice and Apply

Recreate each of the drawings shown using the new scale.

7. current scale: 1 in. = 90 ft
   new scale: 1 in. = 15 ft

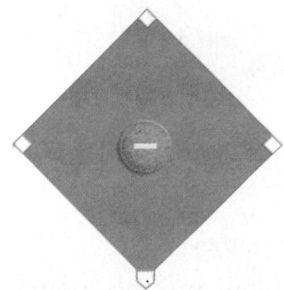

8. current scale: 1 unit = 3 m
   new scale: 1 unit = 6 m

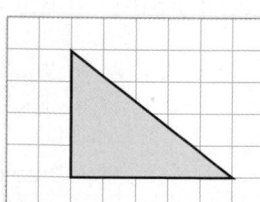

9. A drawing of the Statue of Liberty is 3 inches tall. The scale is 1 inch = 50 feet. How tall would the drawing be if the scale were 0.5 inch = 100 feet?

Explore Draw Triangles

Can you always make a triangle from any three side lengths? Let's see.

Main Idea

Draw triangles using given angles or given side lengths.

 7.G.2

ACTIVITY Use Side Lengths

1 **STEP 1** Use a ruler to cut several plastic straws into lengths that equal 4, 6, 8, 8, 8, 10, 13, 15, 15, and 15 centimeters.

STEP 2 Arrange the three pieces that measure 15 centimeters each to form a triangle.

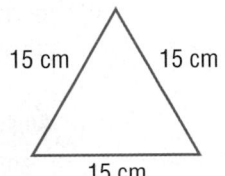
15 cm   15 cm
15 cm

STEP 3 Copy the table below. Try to form figures using the different combinations of side lengths given in the table. Determine whether or not the lengths form a triangle.

Side 1	Side 2	Side 3	Form a Triangle?
6	8	10	
8	8	13	
4	8	15	

STEP 4 Repeat Step 3 ten more times using different combinations of side lengths. Record your results in a table.

Analyze the Results

1. Use the data from Steps 3 and 4. Compare the sum of two sides to the length of a third side. Describe any patterns that are found.

2. Experiment with various side lengths. Can you create another triangle that is the same shape as the triangle with side lengths of 15 centimeters? What are the side lengths of your triangle?

3. Create figures using the side lengths in the table. Describe the figures. How could you change these figures in order to form a triangle?

Figure	Side 1	Side 2	Side 3
A	4 cm	6 cm	13 cm
B	6 cm	8 cm	15 cm

4. **MAKE A CONJECTURE** Could you form a triangle using side lengths 7, 8, 25 centimeters? Justify your reasoning.

ACTIVITY **Use Angle Measures**

2 **STEP 1** Draw two sets of angles measuring 30°, 45°, 60°, and 90° on different pieces of patty paper.

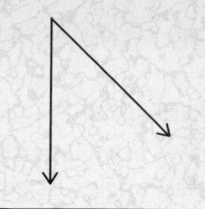

STEP 2 Extend the rays of each angle to the edges of the patty paper.

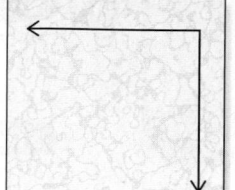 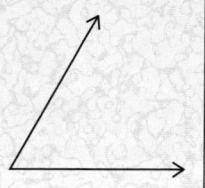

STEP 3 Copy the table below. Try to form the figures described by the angle measures given in the table below. Determine whether or not the figures are triangles.

Angle 1	Angle 2	Angle 3	Form a Triangle?
30°	60°	90°	
30°	45°	60°	
45°	45°	90°	

STEP 4 Repeat Step 3 ten more times using different combinations of angle measures.

Analyze the Results

5. Use the data from Steps 3 and 4. Compare the sum of the angle measures. Describe any patterns that are found.

6. Use the 45°, 60°, and 90° angles. Can you form a triangle using these angle measures? Explain.

7. Draw another 60° angle on a piece of patty paper. Describe the angles and side lengths of the figure you formed using three 60° angles.

8. Draw a set of angles measuring 20°, 70°, and 90°. Do the angle measures form a triangle? Can you create more than one triangle that is the same shape with different side lengths? What are the side lengths of your triangle?

9. **MAKE A CONJECTURE** Based on the Activity 2, what is the sum of the angle measures of any triangle? Explain.

Lesson 12

Main Idea

Identify and draw three-dimensional figures.

New Vocabulary

coplanar
parallel
solid
polyhedron
edge
face
vertex
diagonal
prism
base
pyramid
cylinder
cone
cross section

CCSS 7.G.3

Cross Sections

MONUMENTS A two-dimensional figure, like a rectangle, has two dimensions: length and width. A three-dimensional figure, like a building, has three dimensions: length, width, and height.

1. Name the two-dimensional shapes that make up the sides of the Washington Monument.

2. If you observed the building from directly above, what two-dimensional figure would you see?

3. How are two- and three-dimensional figures related?

The figure at the right shows rectangle *ABCD*. Lines *AB* and *DC* are **coplanar** because they lie in the same plane. They are also **parallel** because they will never intersect, no matter how far they are extended.

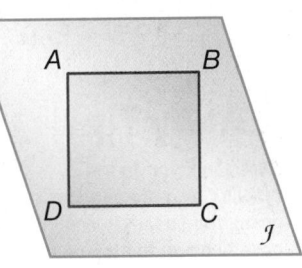

Just as two lines in a plane can intersect or be parallel, there are different ways that planes may be related in space.

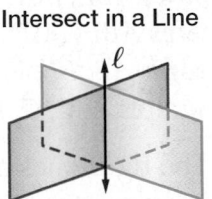

Intersect in a Line

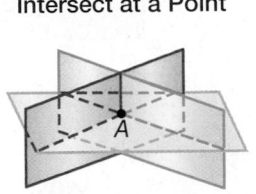

Intersect at a Point

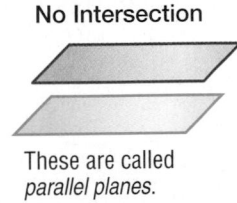

No Intersection

These are called *parallel planes.*

Intersecting planes can also form three-dimensional figures or **solids**. A **polyhedron** is a solid with flat surfaces that are polygons. Some terms associated with three-dimensional figures are *edge, face, vertex,* and *diagonal.*

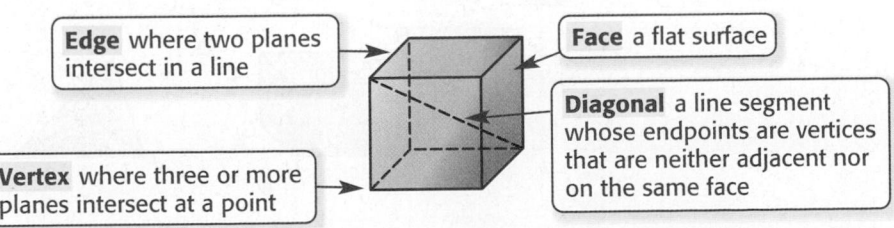

Edge where two planes intersect in a line

Face a flat surface

Diagonal a line segment whose endpoints are vertices that are neither adjacent nor on the same face

Vertex where three or more planes intersect at a point

A **prism** is a polyhedron with two parallel, congruent faces called **bases**. A **pyramid** is a polyhedron with one base that is a polygon and faces that are triangles. Prisms and pyramids are named by the shape of their bases.

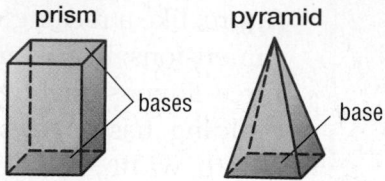

prism   pyramid
bases   base

There are also solids that are not polyhedrons. A **cylinder** is a three-dimensional figure with congruent, parallel bases that are circles connected with a curved side. A **cone** has one circular base and a vertex connected by a curved side.

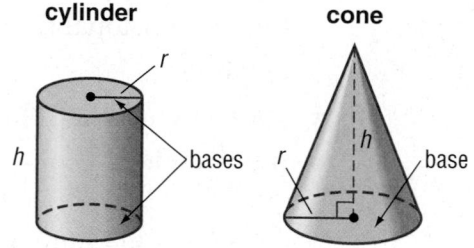

cylinder   cone
h   bases   r   h   base

Study Tip

Common Error In the drawing of a rectangular prism, the bases do not have to be on the top and bottom. Any two parallel rectangles are bases. In a triangular pyramid, any face is a base.

EXAMPLES **Identify Solids**

Identify the figure. Then name the bases, faces, edges, and vertices.

①

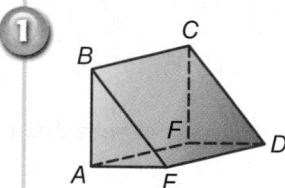

The figure has two parallel congruent bases that are triangles, so it is a triangular prism.
bases *ABE, FCD*
faces *ABE, FCD, BCDE, FAED, ABCF*
edges $\overline{AB}, \overline{BE}, \overline{EA}, \overline{FC}, \overline{CD}, \overline{DF}, \overline{BC}, \overline{ED}, \overline{AF}$
vertices *A, B, C, D, E, F*

②

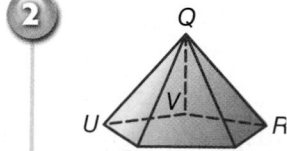

The figure has one base that is a pentagon, so it is a pentagonal pyramid.
base *RSTUV*
faces *RSTUV, QVR, QRS, QST, QTU, QUV*
edges $\overline{QR}, \overline{QS}, \overline{QT}, \overline{QU}, \overline{QV}, \overline{VR}, \overline{RS}, \overline{ST}, \overline{TU}, \overline{UV}$
vertices *Q, R, S, T, U, V*

 CHECK Your Progress

a.

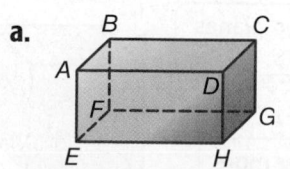

b.

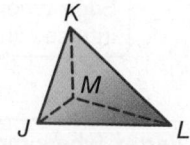

c.

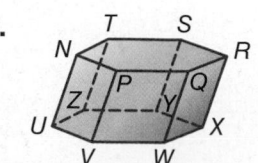

You can use three-dimensional drawings of objects to describe how different parts of the objects are related in space.

REAL-WORLD EXAMPLE **Analyze Drawings**

③ FURNITURE The photo shows a garden bench. Draw and label the top, front, and side views of the bench.

Top   Front   Side

✓ CHECK Your Progress

d. TOOLBOX Draw and label the top, front, and side views of the toolbox shown.

Real-World Link · · · ·
A well-landscaped lawn and garden can increase the value of a home up to 15%.

The intersection of a solid and a plane is called a **cross section** of the solid.

EXAMPLE **Identify Cross Sections**

④ Describe the shape resulting from a vertical, angled, and horizontal cross section of a cylinder.

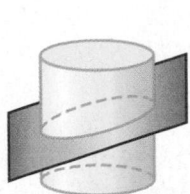

| Vertical Slice | Angled Slice | Horizontal Slice |

The cross section is a rectangle.   The cross section is an oval.   The cross section is a circle.

✓ CHECK Your Progress

e. Describe the shape resulting from a vertical, angled, and horizontal cross section of a square pyramid.

✓ CHECK Your Understanding

Examples 1 and 2 **Identify each figure. Then name the bases, faces, edges, and vertices.**

1.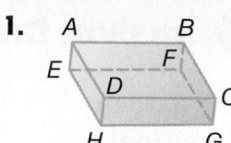

2.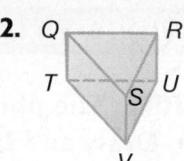

3.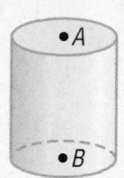

Example 3 **4. AQUARIUMS** Draw and label the top, front, and side views of the aquarium shown.

Example 4 **Describe the shape resulting from each cross section.**

5.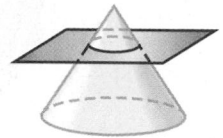

6.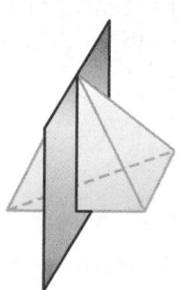

7.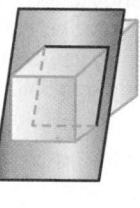

Practice and Problem Solving

Examples 1 and 2 **Identify each figure. Then name the bases, faces, edges and vertices.**

8.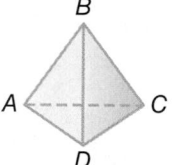

9.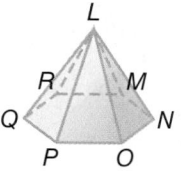

10.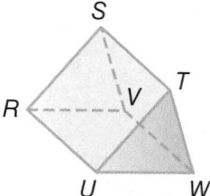

Example 3 **11. BUILDINGS** Draw and label the top, front, and side views of the building.

12. TENT Draw and label the top, front, and side views of the tent.

Example 4 Describe the shape resulting from each cross section.

13.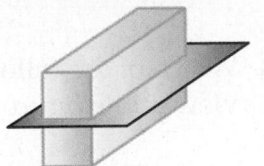

14.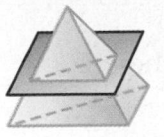

15.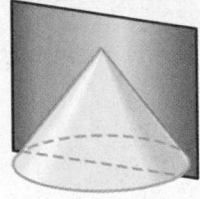

16.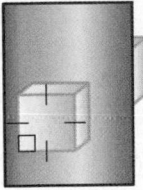

17.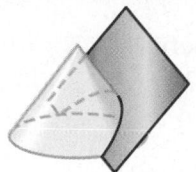

18.

19. State whether the following conjecture is *true* or *false*. If *false*, provide a counterexample.

Two planes in three-dimensional space can intersect at one point.

20. SPORTS A standard basketball is shaped like a *sphere*.

   a. Draw a basketball with a vertical, angled, and horizontal slice.

   b. Describe the cross section made by each slice.

H.O.T. Problems

21. OPEN ENDED Draw the cross sections of a polyhedron, cylinder, or cone. Exchange papers with another student. Identify the three-dimensional figures represented by the cross sections.

22. FIND THE ERROR Brian is identifying the figure below. Find his mistake and correct it.

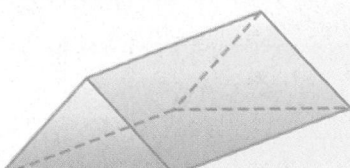

The figure has a triangular base. It is a triangular pyramid.

CHALLENGE Determine whether each statement is *always*, *sometimes*, or *never* true. Explain your reasoning.

23. A prism has 2 bases and 4 faces.　　**24.** A pyramid has parallel faces.

25. WRITE MATH Explain whether a top-front-side view diagram *always* provides enough information to draw a figure. If not, provide a counterexample.

26. Benita received the gift box shown.

Which drawing **best** represents the top view of the gift box?

A.

B.

C.

D.

27. Which of the following is NOT an example of a polyhedron?

   F. cylinder

   G. rectangular prism

   H. octagonal pyramid

   I. triangular prism

28. Which of the following represents a side view of the figure below?

A.    **C.**

B.   **D.**

29. The figure below is a square pyramid.

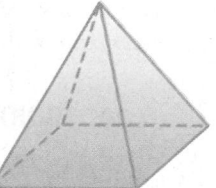

Which of the following is NOT a cross section from the square pyramid?

F.    **H.**

G.   **I.**

Circumference and Area of Circles

Main Idea

Find the circumference and area of circles.

New Vocabulary

circle
center
radius
chord
diameter
circumference
pi

CCSS 7.G.4

ACTIVITY

STEP 1 Measure and record the distance d across the circular part of an object, such as a battery or a can, through its center.

STEP 2 Place the object on a piece of paper. Mark the point where the object touches the paper on both the object and on the paper.

STEP 3 Carefully roll the object so that it makes one complete rotation. Then mark the paper again.

STEP 4 Finally, measure the distance C between the marks.

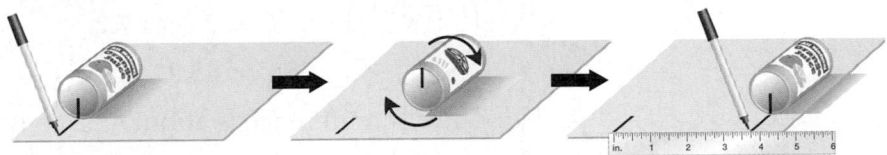

1. What distance does C represent?
2. Find the ratio $\frac{C}{d}$ for this object.
3. Repeat the steps above for at least two other circular objects and compare the ratios of C to d. What do you observe?
4. Graph the data you collected as ordered pairs, (d, C). Then describe the graph.

A **circle** is a set of points in a plane that are the same distance from a given point in the plane, called the **center**. The segment from the center to any point on the circle is called the **radius**. A **chord** is any segment with both endpoints on the circle. A **diameter** is a chord that passes through the center. It is the longest chord.

The diameter of a circle is twice its radius or d = 2r.

The distance around the circle is called the **circumference**. The ratio of the circumference of a circle to its diameter is always 3.1415926…. It is represented by the Greek letter π **(pi)**. The numbers 3.14 and $\frac{22}{7}$ are often used as approximations for π. So, $\frac{C}{d} = \pi$. This can also be written as $C = \pi d$ or $C \approx 3.14d$.

Key Concept · Circumference of a Circle

Words The circumference C of a circle is equal to its diameter d times π, or 2 times its radius r times π.

Model

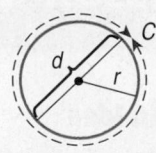

Symbols $C = \pi d$ or $C = 2\pi r$

EXAMPLES   Find the Circumferences of Circles

Find the circumference of each circle. Round to the nearest tenth.

1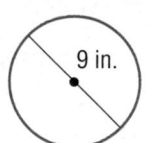

$C = \pi d$ Circumference of a circle

$C = \pi \cdot 9$ Replace d with 9.

$C = 9\pi$ This is the *exact* circumference.

Use a calculator to find 9π.   9 $\boxed{\times}$ $\boxed{\text{2nd}}$ $[\pi]$ $\boxed{\text{ENTER}}$ 28.27433388

The circumference is about 28.3 inches.

1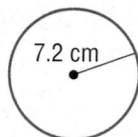

$C = 2\pi r$ Circumference of a circle

$C = 2 \cdot \pi \cdot 7.2$ Replace r with 7.2.

$C \approx 45.2$ Use a calculator.

The circumference is about 45.2 centimeters.

✓ CHECK Your Progress

a.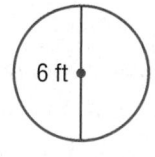

b.

c.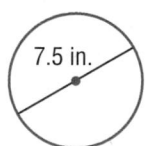

A circle can be decomposed into congruent wedge-like pieces. Then the pieces can be rearranged to form a figure that resembles a parallelogram.

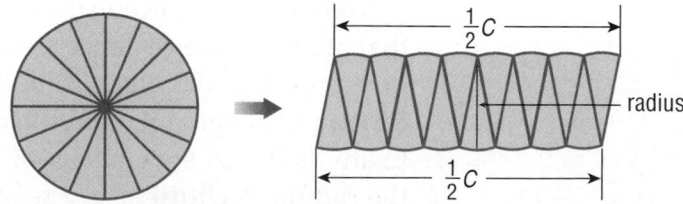

Since the circle has an area that is relatively close to the area of the parallelogram-shaped figure, you can use the formula for the area of a parallelogram to find the formula for the area of a circle.

$A = bh$ Area of a parallelogram

$A = \left(\dfrac{1}{2} \cdot C\right)r$ The base of the parallelogram is one-half the circumference and the height is the radius.

$A = \left(\dfrac{1}{2} \cdot 2\pi r\right)r$ Replace C with $2\pi r$.

$A = \pi \cdot r \cdot r$ or πr^2 Simplify.

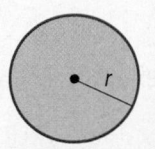

Key Concept · Area of a Circle

Words   The area A of a circle is equal to π times the square of the radius r.

Model

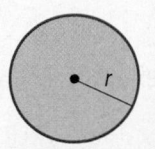

Symbols $A = \pi r^2$

Study Tip

Estimation
To estimate the area of a circle, square the radius and then multiply by 3.

EXAMPLES · Find the Areas of Circles

Find the area of each circle. Round to the nearest tenth.

 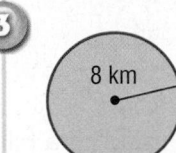

	$A = \pi r^2$	Area of a circle
8 km	$A = \pi \cdot 8^2$	Replace r with 8.
	$A = \pi \cdot 64$	Evaluate 8^2. This is the *exact* area.
	$A \approx 201.1$	Use a calculator.

The area is about 201.1 square kilometers.

 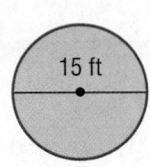

	$A = \pi r^2$	Area of a circle
15 ft	$A = \pi (7.5)^2$	Replace r with half of 15 or 7.5.
	$A = \pi \cdot 56.25$	Evaluate 7.5^2. This is the *exact* area.
	$A \approx 176.7$	Use a calculator.

The area is about 176.7 square feet.

✓ CHECK Your Progress

Find the area of each circle. Round to the nearest tenth.

d. The radius is 11 inches.   **e.** The diameter is 5 meters.

REAL-WORLD EXAMPLE

 STATE PARKS Suppose you walk around the edge of the circular Point State Park fountain in Pittsburgh, Pennsylvania, and estimate its circumference to be 470 feet. Based on your estimate, what is the approximate diameter of the fountain?

	$C = \pi d$	Circumference of a circle
	$470 = \pi d$	Replace C with 470.
	$\dfrac{470}{\pi} = d$	Divide each side by π.
	$149.6 \approx d$	Use a calculator.

The diameter of the fountain is about 150 feet.

✓ CHECK Your Progress

 f. HOME DECOR A catalog states that a circular area rug covers 19.5 square feet. What is the approximate diameter of the rug?

CHECK Your Understanding

Find the circumference of each circle. Round to the nearest tenth.

Examples 1 and 2

1.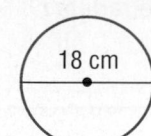

2.

3.

Find the area of each circle. Round to the nearest tenth.

Examples 3 and 4

4.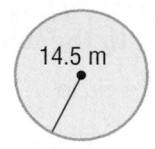

5.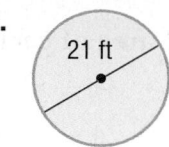

6.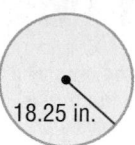

Example 5

7. BRACELETS When Cammie finished making a friendship bracelet, it was 7.9 inches long. What was the diameter of the bracelet?

Practice and Problem Solving

Find the circumference of each circle. Round to the nearest tenth.

Examples 1 and 2

8.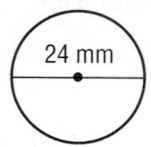

9.

10.

11.

Find the area of each circle. Round to the nearest tenth.

Examples 3 and 4

12.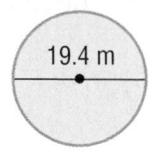

13.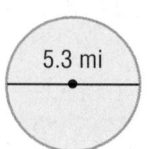

14.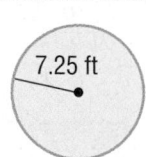

15.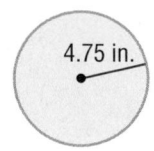

Example 5

16. PETS Simone purchased a circular exercise pen with a radius of 2.5 feet to keep her new puppy safe. Find the area inside the pen.

17. MEASUREMENT A circular table top has a radius of $2\frac{1}{4}$ feet. A decorative trim is placed along the outside edge of the table. How long is the trim?

18. SAFETY A light in a parking lot illuminates a circular area 15 meters across. What is the area of the parking lot covered by the light?

19. BICYCLES Jerrod's mountain bike has a tire diameter of 26 inches. How far will the bike travel in 100 rotations of its tires?

Find the exact circumference and area of each circle.

20. The radius is 3.5 centimeters.

21. The diameter is 8.6 kilometers.

22. The diameter is 9 inches.

23. The radius is 0.6 mile.

24. Find the diameter of a circle if its area is 706.9 square millimeters.

25. **GARDENING** Mr. Townes created a 2-foot wide garden path around a circular garden. The radius of the garden is 7 feet. He wants to cover the path in stones. If Mr. Townes needs 1 bag of stones for every 5 square feet of path, how many bags of stones will he need to cover his garden path?

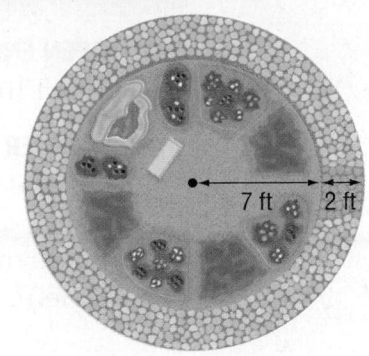
7 ft   2 ft

Another approximate value for π is $\frac{22}{7}$. Use this value to find the circumference and area of each circle.

26. The diameter is 7 feet.

27. The radius is $2\frac{1}{3}$ inches.

28. **BAKING** Joaquim is baking giant cookies for the school bake sale. They will be sold for $20 for one large cookie or $20 for three smaller cookies. Which offer is the better buy? Explain your reasoning.

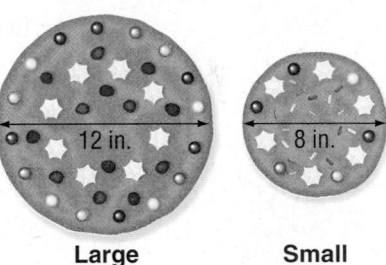

12 in.   8 in.

Large   **Small**

29. **SPORTS** Three tennis balls are packaged one on top of the other in a can. Which measure is greater, the can's height or circumference? Explain.

30. **TREES** During a construction project, barriers are placed around trees. For each inch of trunk diameter, the protection zone should have a radius of $1\frac{1}{2}$ feet. Find the area of this zone for a tree with a trunk circumference of 63 inches.

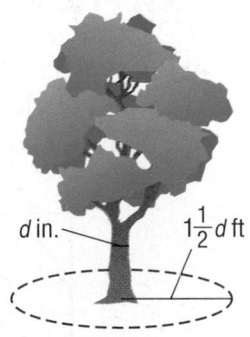

d in. $1\frac{1}{2}d$ ft

31. **GRAPHIC ARTS** Michael is painting a sign for a new coffee shop. On the sign, he drew a circle with a radius of 2 feet. He then drew another circle with a radius 1.5 times larger. How much greater is the area of the larger circle?

H.O.T. Problems

32. OPEN ENDED Draw and label a circle that has a circumference between 5 and 10 centimeters. Justify your answer.

33. NUMBER SENSE If the radius of a circle is halved, how will this affect its circumference and its area? What happens to the circumference and area if the radius is doubled or tripled? Explain your reasoning. (*Hint*: Find the circumference and area for each circle and organize the data in a table.)

CHALLENGE Find the area of each shaded region.

34.

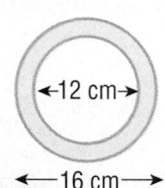

35.

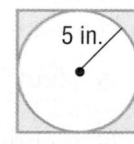

36.

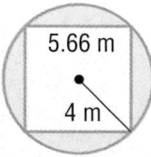

37. WRITE MATH Explain how the circumference and area of a circle are related or different.

Test Practice

38. In the figure below, the radius of the inscribed circle is 8 inches. What is the perimeter of square *WXYZ*?

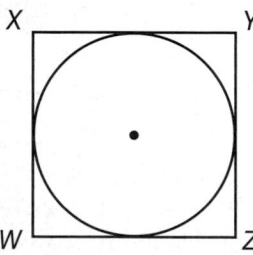

A 16π in.

B 64 in.

C 32 in.

D 64π in.

39. Using the two circles shown below, what is $\dfrac{\text{circumference of circle } x}{\text{circumference of circle } y}$?

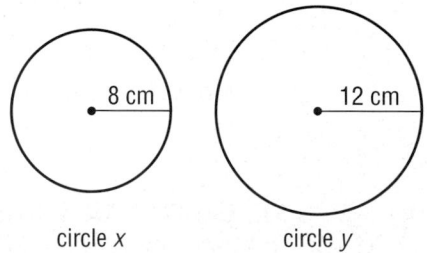

circle *x*   circle *y*

F $\dfrac{3\pi}{4}$

G $\dfrac{4\pi}{3}$

H $\dfrac{2}{3}$

I $\dfrac{4}{3}$

Lesson 14

Extend Multiple Samples of Data

When you draw a conclusion about a population from a sample of data, you are drawing an **inference** about that population. Sometimes, drawing inferences about a population from only one sample of data is not as accurate as using multiple samples of data.

Main Idea

Analyze the variation in multiple samples of data.

New Vocabulary

inferences

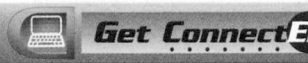

 7.SP.2

ACTIVITY

1. Do you know which letter is the most frequently occurring letter in words in the English language? The table gives fifteen randomly selected words from a dictionary.

Randomly Selected Words from the English Language Dictionary		
airport	juggle	sewer
blueberry	lemon	standard
costume	mileage	thread
doorstop	percentage	vacuum
instrument	print	whale

STEP 1 Find the frequency of each letter that appears in the table. Copy and complete the table below. Label as Sample 1.

Letter	Frequency	Letter	Frequency	Letter	Frequency
a	8	j		s	
b	2	k		t	
c		l		u	
d		m		v	
e		n		w	
f		o		x	
g		p		y	
h		q		z	
i		r			

STEP 2 Randomly select another fifteen words from a dictionary and record the frequency of letters in a separate table. Label the second table as Sample 2.

STEP 3 Repeat Step 2 to obtain a third sample. Label the third table as Sample 3.

Analyze the Results

1. The *relative frequency* of a letter occurring is the ratio of the number of times the letter occurred to the total number of times all letters occurred. Find the relative frequency for the letter *e* for each sample. Round to the nearest hundredth.

2. Find the mean and median relative frequency of the letter *e* for the three samples. Round to the nearest tenth if necessary.

3. Use your answers from Exercise 2 to predict the relative frequency of the letter *e* for words in the English language.

4. Research the Internet to find the relative frequency for the letter *e* for words in the English language. Compare to your results.

5. WRITE MATH Write a few sentences describing inferences you can make about the frequency of letters in words in the English language using your three samples. Then compare to the actual frequencies. Note any differences.

You can also simulate multiple samples of data.

ACTIVITY

2 A fast-food restaurant randomly hands out prizes with each combo meal. There are three different prizes: a restaurant gift certificate (G), a movie theater pass (M), and an amusement park pass (A). The restaurant gives out the gift certificate 40% of the time, the movie pass 40% of the time, and the amusement park pass 20% of the time.

STEP 1 Create a spinner with five equal sections. Label two sections G. Label another two sections M and label one section A.

STEP 2 Each spin of the spinner represents a customer receiving a prize. Spin the spinner 20 times. Record the number of times each prize was received. Label this table as Sample 1.

Prize	Number of Times
G	
M	
A	

STEP 3 Repeat Step 2 two more times to obtain Sample 2 and Sample 3.

Analyze the Results

6. Explain how the set up of the spinner accurately reflects the surveyed data.

7. Which prize was received the most often in each sample?

8. Find the relative frequency of the number of times that each prize was received for each sample. Round to the nearest hundredth if necessary.

9. Find the mean and median relative frequency of the number of times that each prize was received. Round to the nearest tenth if necessary.

10. Use your answers from Exercise 9 to predict the number of times that each prize will be received if the restaurant gives out 200 prizes.

11. WRITE MATH Write a few sentences describing which prize might be received the most often over time. How far off might your prediction be? Explain your reasoning.

Lesson 15

Main Idea

Analyze the visual overlap of two numerical data distributions.

New Vocabulary

visual overlap

7.SP.3

Explore Visual Overlap of Data Distributions

You can compare two numerical data sets by comparing the shape of their distributions. The **visual overlap** of two distributions with similar variation is a visual demonstration that compares their centers to their *variation*, or spread.

ACTIVITY

1 The tables show the number of text messages sent and received daily for two different age groups.

Text Messages Ages 12–15			
70	90	80	90
85	75	85	80
90	80	75	95
100	85	95	85

Text Messages Ages 16–19			
85	75	80	70
75	80	65	75
85	70	90	80
70	75	60	65

STEP 1 Create a double dot plot to display the data.

Text Messages Sent and Received

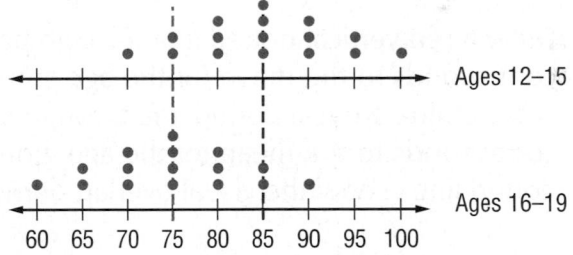

Ages 12–15

Ages 16–19

60 65 70 75 80 85 90 95 100

STEP 2 Find the mean number of text messages for each age group.

Ages 12–15

$$\frac{1,360}{16} = 85$$

Ages 16–19

$$\frac{1,200}{16} = 75$$

STEP 3 Draw a red vertical dotted line through both dot plots that corresponds to the mean for the age group, 12–15 years. Draw a blue vertical dotted line through both dot plots that corresponds to the mean for the age group, 16–19 years. The dotted lines show the visual overlap between the centers.

Analyze the Results

1. Which age group had a greater mean number of text messages?

2. What is the difference between the means of the distributions?

3. The mean absolute deviation of each distribution is 6.25 text messages. Write the ratio of the difference between the means to the mean absolute deviation. Express the ratio as a decimal.

4. **WRITE MATH** Look at the distributions. Describe any overlap in data values.

ACTIVITY

2 The double dot plot below compares the number of text messages by a third age group to the age group, 12–15 years, from Activity 1.

STEP 1 You know the mean number of text messages for the age group, 12–15 years, is 85 text messages. Find the mean number of text messages for the age group, 24–27 years.

Text Messages Sent and Received

Ages 24–27

$$\frac{1,040}{16} = 65$$

Ages 12–15

50 55 60 65 70 75 80 85 90 95 100

Ages 24–27

STEP 2 Draw a red vertical dotted line through both dot plots that corresponds to the mean for the age group, 12–15 years. Draw a blue vertical dotted line through both dot plots that corresponds to the mean for the age group, 24–27 years. The dotted lines show the visual overlap between the centers.

Analyze the Results

5. What is the difference between the means of the distributions?

6. The mean absolute deviation of each distribution is 6.25 text messages. Write the ratio of the difference between the means to the mean absolute deviation. Express the ratio as a decimal.

7. Compare the ratio in Exercise 6 to the ratio in Exercise 3.

8. Compare the variations (mean absolute deviations) for the distributions in each activity.

9. **WRITE MATH** By comparing the distributions in each activity, compare how the ratio, $\frac{\text{difference in means}}{\text{mean absolute deviation}}$, tells you how much visual overlap there is between two distributions with similar variation.

Lesson 16

Compare Populations

Main Idea

Compare two populations using the measures of center and variation.

New Vocabulary

double box-and-whisker plot

CCSS 7.SP.4

BLOGGING Kacey surveyed the students in her math class to find out how many times they blogged this month. The box-and-whisker plot shows the results.

How Many Times Have You Posted A Blog This Month?

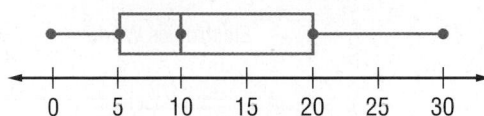

1. Find the median and interquartile range.

2. If Kacey randomly asked a different group of students the same question, would you expect the results to be the same? Explain.

A **double box-and-whisker plot** consists of two box-and-whisker plots graphed on the same number line. You can draw inferences about two populations in a double box-and-whisker plot by comparing their centers and variations.

REAL-WORLD EXAMPLE **Compare Two Populations**

① **BLOGGING** Kacey surveyed a different group of students in her science class. The double box-and-whisker plot shows the results for both classes. Compare their centers and variations. Write an inference you can draw about the two populations.

How Many Times Have You Posted A Blog This Month?

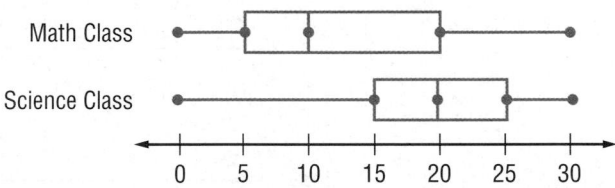

Neither line plot is symmetric. Use the median to compare the centers. Use the interquartile range to compare the variations.

	Math Class	Science Class
Median	10	20
Interquartile range	20 – 5, or 15	25 – 15, or 10

Overall, the students in Kacey's science class posted more blogs this month than the students in her math class. The median for the science class is twice the median for the math class. There is a greater spread of data around the median for the math class than the science class.

Study Tip

If a distribution is *not* symmetric, use the median and interquartile range. If the distribution is symmetric, use the mean and mean absolute deviation.

CHECK Your Progress

a. MP3 PLAYERS The double box-and-whisker plot shows the costs of MP3 players at two different stores. Compare the centers and variations of the two populations. Write an inference you can draw about the two populations.

Cost of MP3 Players ($)

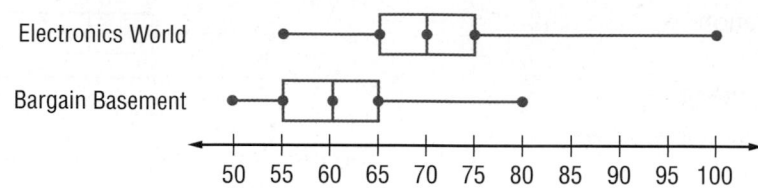

Another name for a line plot is a dot plot. A *dot plot* uses dots to show the frequency of data values in a distribution. A *double dot plot* consists of two dot plots that are drawn on the same number line.

> **REAL-WORLD EXAMPLE** **Compare Two Populations**

2 **WEATHER** The double dot plot below shows the daily high temperatures for two cities for thirteen days. Compare the centers and variations of the two populations. Write an inference you can draw about the two populations.

Daily High Temperatures (°F)

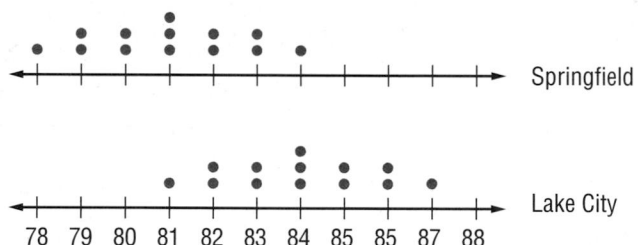

Both dot plots are symmetric. Use the mean to compare the centers and use the mean absolute deviation, rounded to the nearest tenth, to compare the variations.

	Springfield	Lake City
Mean	81	84
Mean Absolute Deviation	1.4	1.4

While both cities have the same variation, or spread of data about each of their means, Lake City has a greater mean temperature than Springfield.

Study Tip

Mean Absolute Deviation To find the mean absolute deviation, find the absolute values of the differences between each value and the mean. Then find the average of those differences.

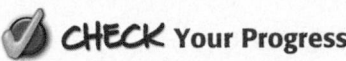

Additional Lessons

b. E-MAIL The double dot plot shows the number of new E-mails in each of Pedro's and Annika's inboxes for sixteen days. Compare the centers and variations of the two populations. Write an inference you can draw about the two populations.

Number of E-mails in Inbox

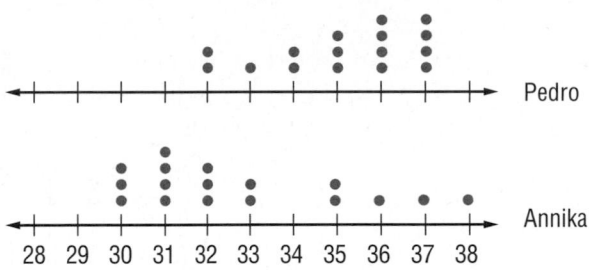

You can compare two populations when only one is symmetric. The mean and median of a symmetric distribution are very similar, if not the same value. So, you can use the median and interquartile range for both populations.

REAL-WORLD EXAMPLE

3 **ZIP LINING** The double box-and-whisker plot shows the daily participants for two zip line companies for one month. Compare the centers and variations of the two populations. Write an inference you can draw about the two populations.

Number of Daily Participants

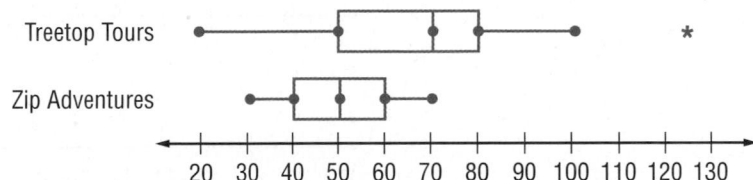

The distribution for Zip Adventures is symmetric, while the distribution for Treetop Tours is not symmetric. Use the median and the interquartile range to compare the populations.

	Treetop Tours	Zip Adventures
Median	70	50
Interquartile Range	30	20

Overall, Treetop Tours has a greater number of daily participants. However, Treetop Tours also has a greater variation, so it is more difficult to predict how many participants they may have each day. Zip Adventures has a greater consistency in their distribution.

 **CHECK** Your Progress

c. RUNNING The double dot plot shows Kareem's and Martin's race times for a three-mile race. Compare the centers and variations of the two populations. Write an inference you can draw about the two populations.

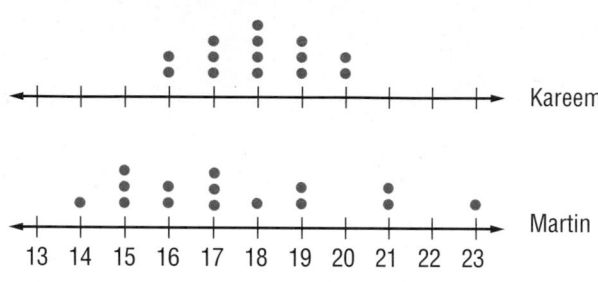

Race Times (min)

 **CHECK Your Understanding**

Examples 1 and 2

1. SCHOOL The double dot plot below shows the quiz scores out of 20 points for two different class periods. Compare the centers and variations of the two populations. Round to the nearest tenth. Write an inference you can draw about the two populations.

Quiz Scores (points)

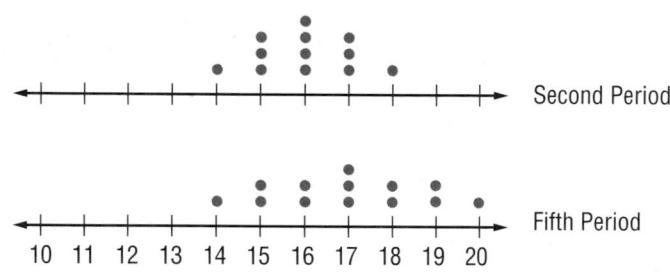

Example 3

2. SPEEDING The double box-and-whisker plot shows the speeds of cars recorded on two different roads in Hamilton County. Compare the centers and variations of the two populations. Write an inference you can draw about the two populations.

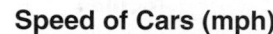

Speed of Cars (mph)

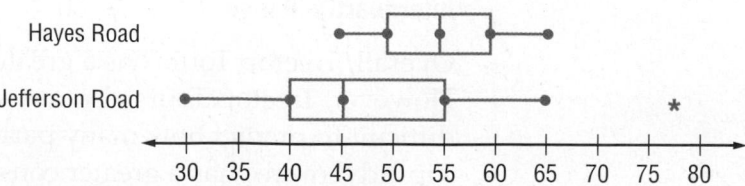

Examples 1 and 2

3. RESTAURANTS Jordan randomly asked customers at two different restaurants how long they waited for a table before they were seated. The double box-and-whisker plot shows the results. Compare their centers and variations. Write an inference you can draw about the two populations.

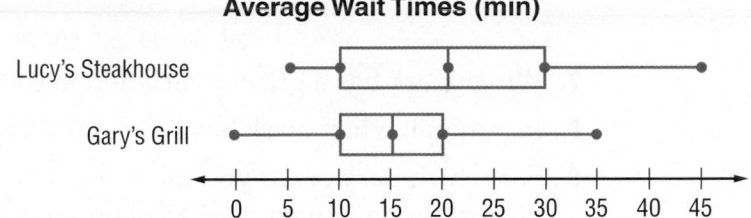

Average Wait Times (min)

Lucy's Steakhouse

Gary's Grill

0 5 10 15 20 25 30 35 40 45

4. HEIGHTS The double dot plot below shows the heights in inches for the girls and boys in Franklin's math class. Compare the centers and variations of the two populations. Round to the nearest tenth. Write an inference you can draw about the two populations.

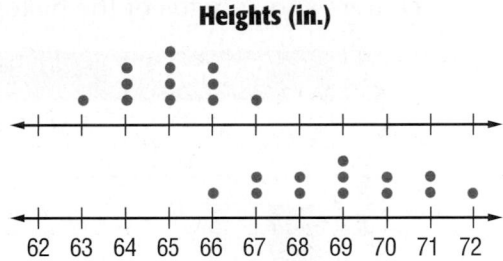

Heights (in.)

62 63 64 65 66 67 68 69 70 71 72

Example 3

5. FOOTBALL The double box-and-whisker plot shows the number of points scored by the football team for two seasons. Compare the centers and variations of the two populations. Write an inference you can draw about the two populations.

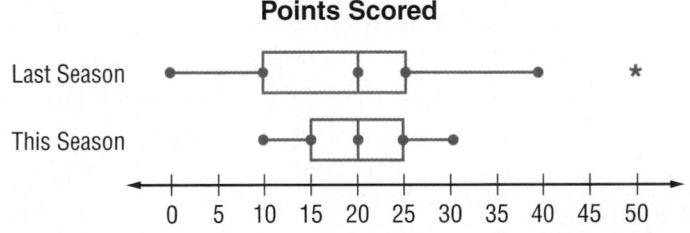

Points Scored

Last Season

This Season

0 5 10 15 20 25 30 35 40 45 50

6. FLYING The double dot plot below shows the times, in hours, for flights of two different airlines flying out of the same airport. Compare the centers and variations of the two populations. Write an inference you can draw about the two populations.

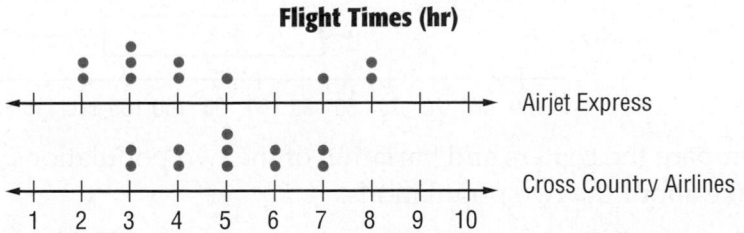

Flight Times (hr)

Airjet Express

Cross Country Airlines

1 2 3 4 5 6 7 8 9 10

PARKS For Exercises 7–9, use the double box-and-whisker plot below that shows the number of daily visitors to two different parks.

Number of Daily Visitors

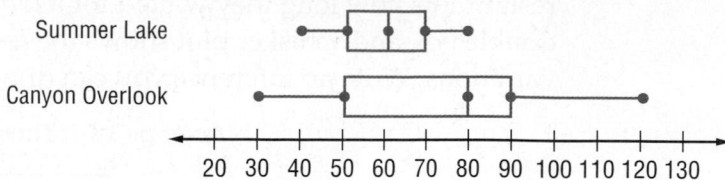

7. Which park has a greater variation in the number of daily visitors?

8. In general, which park has more daily visitors? Justify your response.

9. For which park could you more accurately predict the number of daily visitors on any given day? Explain your reasoning.

H.O.T. Problems

10. **CHALLENGE** The histograms below show the number of tall buildings for two cities. Explain why you cannot describe the specific location of the centers and spreads of the histograms.

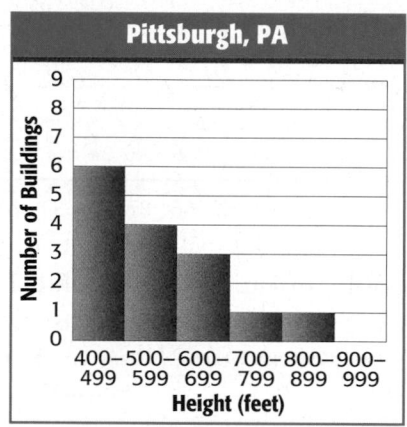
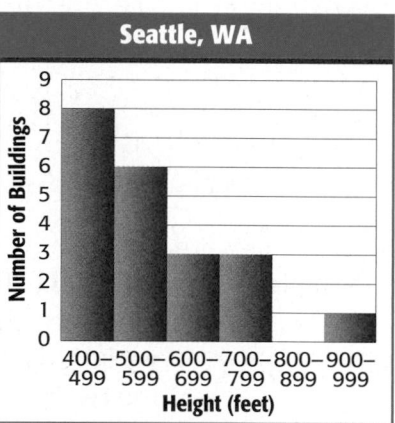

11. **WRITE MATH** Marcia recorded the daily temperatures for two cities for 30 days. The two populations have similar centers, but City A has a greater variation than City B. For which city can you more accurately predict the daily temperature? Explain.

Test Practice

12. **SHORT RESPONSE** The double box-and-whisker plot below shows the speeds in miles per hour of several wood and steel roller coasters.

Speed (miles per hour) of Roller Coasters

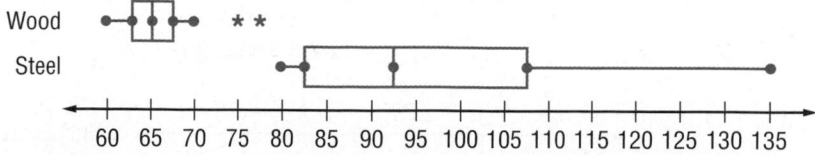

Compare the centers and variation of the two populations. Write an inference you can make about the two populations.

Lesson 17

Main Idea
Use a simulation to generate frequencies for a compound event.

Vocabulary
simulation

 7.SP.8c

Extend Simulate Compound Events

A **simulation** is a way of modeling a problem situation. Simulations often model events that would be difficult or impractical to perform. In this lab you will use different methods of simulations to predict outcomes of compound events.

You can use a spinner to simulate a compound event.

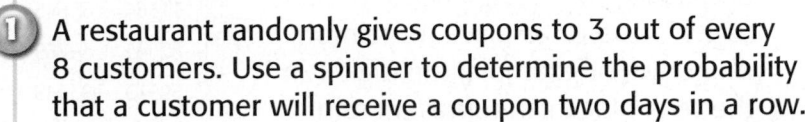 ACTIVITY

1. A restaurant randomly gives coupons to 3 out of every 8 customers. Use a spinner to determine the probability that a customer will receive a coupon two days in a row.

   STEP 1 Create a spinner with eight equal sections. Label three of the sections C for "receive a coupon". Label the other sections as D for "do not receive a coupon".

   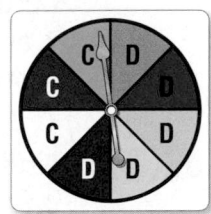

   STEP 2 Every two spins of the spinner represents one trial. Spin the spinner and record whether or not the customer receives a coupon on the first day. Spin the spinner again and record whether or not the customer receives a coupon on the second day.

   STEP 3 Perform a total of 40 trials. Record the number of times a coupon was given to a customer on both days.

Analyze the Results

1. Describe what the outcome of spinning a C on the first spin and spinning a D on the second spin represents in this activity.

2. Based on your results, what is the experimental probability that a customer will receive a coupon two days in a row?

3. How might your results change if you simulated 80 customers?

4. One out of every ten coupons has a red circle symbol. If a customer receives a coupon with a red circle, their meal is free. Use the spinner above to create a second spinner and simulate the probability that a customer will receive a free meal. Perform 30 trials.

You can also use a random number table to simulate a compound event. A random number table has random digits in rows that can be grouped in different combinations as needed.

ACTIVITY

2 **WEATHER** There is a 10% chance of rain for a city on Sunday and a 20% chance of rain on Monday. Use a random number table to find the probability that it will rain on both days.

The digits in the random number table below can be separated into two-digit numbers because the situation involves two days. The red lines show the separation.

STEP 1 Using the digits 0 through 9, assign one digit in the tens place for rain on Sunday and assign two different digits in the ones place for rain on Monday. For example, the digit 1 in the tens place can represent rain occurring on Sunday and the digits 1 and 2 in the ones place can represent rain occurring on Monday.

48587	49460	89640	30270
19507	87835	99812	52353
11364	35645	90087	64254
87045	39769	77995	14316
69913	93449	68497	31270
81827	32901	82033	43714
33386	99637	25725	31900
41575	86692	40882	44123
77351	12790	62795	77307

STEP 2 Find the numbers in the table that have a 1 in the tens place *and* either a 1 or 2 in the ones place. Those numbers are 11 and 12 and are noted in red in the table.

STEP 3 Find the probability using the numbers found in Step 2. There were 3 instances of the random numbers 11 and 12 occurring. So, the probability is $\frac{3}{90}$ or $3\frac{1}{3}$ %.

Analyze the Results

5. What would the number 19 in the random number table represent in this activity? the number 61? the number 55?

6. What would be the probability if the digit 8 was used for rain occurring on Sunday and the digits 8 and 9 were used for rain occurring on Monday?

7. How would your results for Exercise 5 be different if you only used the first 3 rows of numbers in the random number table? the last 5 rows?

8. **WRITE MATH** Explain how using more numbers from a random number table affects the predicted probability.

9. **DESIGN A SIMULATION** The probability of the Perry Panthers winning a basketball game is 60%. Design a simulation using the random number table above to find the simulated probability that the team will win four games in a row. Explain how you used the random number table. State the number of trials you performed.

Extra Practice

Multi-Part Lesson 1-1: Expressions

PART A PAGES 25–28

Write each power as a product of the same factor.

1. 13^4 **2.** 9^6 **3.** 1^7

4. 12^2 **5.** 5^8 **6.** 15^4

Evaluate each expression.

7. 5^6 **8.** 17^3 **9.** 2^{12} **10.** 3^5

11. 1^4 **12.** 5^3 **13.** 10^2 **14.** 2^8

15. 8^2 **16.** 7^4 **17.** 20^3 **18.** 42^3

Write each product in exponential form.

19. $2 \cdot 2 \cdot 2 \cdot 2 \cdot 2$ **20.** $3 \cdot 3$ **21.** $1 \cdot 1 \cdot 1 \cdot 1 \cdot 1 \cdot 1$

22. $18 \cdot 18 \cdot 18 \cdot 18$ **23.** $9 \cdot 9 \cdot 9 \cdot 9 \cdot 9 \cdot 9 \cdot 9 \cdot 9$ **24.** $10 \cdot 10 \cdot 10 \cdot 10 \cdot 10 \cdot 10$

PART B PAGES 29–32

Evaluate each expression.

1. $14 - (5 + 7)$ **2.** $(32 + 10) - 5 \times 6$ **3.** $(50 - 6) + (12 + 4)$

4. $12 - 2 \cdot 3$ **5.** $16 + 4 \times 5$ **6.** $(5 + 3) \times 4 - 7$

7. $2 \times 3 + 9 \times 2$ **8.** $6 \cdot (8 + 4) \div 2$ **9.** $7 \times 6 - 14$

10. $8 + (12 \times 4) \div 8$ **11.** $13 - 6 \cdot 2 + 1$ **12.** $(80 \div 10) \times 8$

13. $14 - 2 \cdot 7 + 0$ **14.** $156 - 6 \times 0$ **15.** $30 - 14 \cdot 2 + 8$

16. $3 \times 4 - 3^2$ **17.** $10^2 - 5$ **18.** $3 + (10 - 5 + 1)^2$

19. $(4 + 3)^2 \div 7$ **20.** 8×10^3 **21.** $10^4 \times 6$

22. 4.5×10^3 **23.** 1.8×10^2 **24.** $3 + 5(1.7 + 2.3)$

PART C PAGES 33–37

Evaluate each expression if $a = 3$, $b = 4$, $c = 12$, and $d = 1$.

1. $a + b$ **2.** $c - d$ **3.** $a + b + c$ **4.** $b - a$

5. $c - ab$ **6.** $a + 2d$ **7.** $b + 2c$ **8.** ab

9. $a + 3b$ **10.** $6a + c$ **11.** $\dfrac{c}{d}$ **12.** abc

13. $2(a + b)$ **14.** $\dfrac{2c}{b}$ **15.** $144 - abc$ **16.** $2ab$

17. $\dfrac{b}{2}$ **18.** a^2 **19.** $c^2 - 100$ **20.** $a^3 + 3$

21. $2b^2$ **22.** $b^3 + c$ **23.** $\dfrac{a^2}{d}$ **24.** $5a^2 + 2d^2$

25. $\dfrac{4d^2}{b}$ **26.** $\dfrac{15}{a}$ **27.** $3a^2$ **28.** $10d^3$

29. $\dfrac{(2c + b)}{b}$ **30.** $\dfrac{(b^2 + 2d)}{a}$ **31.** $\dfrac{(2c + ab)}{c}$ **32.** $\dfrac{(3.5c + 2)}{11}$

Multi-Part Lesson 1-1 (continued)

PART D

PAGES 38–41

Use the Distributive Property to evaluate each expression.

1. $3(4 + 5)$

2. $(2 + 8)6$

3. $4(9 - 6)$

4. $8(6 - 3)$

5. $5(200 - 50)$

6. $20(3 + 6)$

7. $(20 - 5)8$

8. $50(8 + 2)$

9. $15(1{,}000 - 200)$

10. $3(2{,}000 + 400)$

11. $12(1{,}000 + 10)$

12. $7(1{,}000 - 50)$

Evaluate each expression mentally. Justify each step.

13. $(5 + 17) + 25$

14. $13 + (22 + 17)$

15. $(8 + 18) + 92$

16. $(11 + 32) + 9$

17. $4 + (15 + 76)$

18. $(25 + 56) + 75$

19. $(4 \cdot 21) \cdot 25$

20. $5 \cdot (40 \cdot 8)$

21. $(2 \cdot 38) \cdot 50$

22. $(12 \cdot 7) \cdot 5$

23. $25 \cdot (12 \cdot 4)$

24. $(15 \cdot 9) \cdot 2$

Multi-Part Lesson 1-2: Patterns

PART A

PAGES 42–43

1. MONEY The table shows the amount of money Georgia has in her piggy bank for each week. If the pattern continues, how much money will she have saved in Week 8?

Week 1	Week 2	Week 3	Week 4
$6.75	$7.25	$7.75	$8.25

2. GEOMETRY Draw the next two figures in the pattern.

3. ALGEBRA What are the next two numbers in the pattern?

564, 549, 534, 519, 504, _____, _____

PARTS B C

PAGES 44–50

Describe the relationship between the terms in each arithmetic sequence. Then write the next three terms in each sequence.

1. $5, 9, 13, 17, \ldots$

2. $3, 5, 7, 9, \ldots$

3. $10, 15, 20, 25, \ldots$

4. $90, 93, 96, 99, \ldots$

5. $8, 14, 20, 26, \ldots$

6. $4.5, 5.4, 6.3, 7.2, \ldots$

7. $0.3, 0.4, 0.5, \ldots$

8. $2.3, 3.4, 4.5, 5.6, \ldots$

9. $8.9, 9.1, 9.3, 9.5, \ldots$

10. $3, 11, 19, 27, \ldots$

11. $350, 375, 400, 425, \ldots$

12. $620, 635, 650, 665, \ldots$

13. $2, 7, 12, 17, \ldots$

14. $10, 17, 24, 31, \ldots$

15. $9, 90, 171, 252, \ldots$

16. $2.6, 2.8, 3.0, 3.2, \ldots$

17. $4.1, 4.6, 5.1, 5.6, \ldots$

18. $6.6, 7.7, 8.8, 9.9, \ldots$

19. $19.5, 21, 22.5, 24, \ldots$

20. $14.5, 14.8, 15.1, 15.4, \ldots$

21. $0.1, 0.4, 0.7, 1.0, \ldots$

Multi-Part Lesson 1-3: Square Roots

PARTS A B

PAGES 52–56

Find the square of each number.

1. 4	**2.** 19	**3.** 13
4. 25	**5.** 9	**6.** 2
7. 14	**8.** 24	**9.** 40
10. 50	**11.** 100	**12.** 250

Find each square root.

13. $\sqrt{324}$	**14.** $\sqrt{900}$	**15.** $\sqrt{2,500}$
16. $\sqrt{576}$	**17.** $\sqrt{8,100}$	**18.** $\sqrt{676}$
19. $\sqrt{100}$	**20.** $\sqrt{784}$	**21.** $\sqrt{1,024}$
22. $\sqrt{841}$	**23.** $\sqrt{2,304}$	**24.** $\sqrt{3,025}$

PART C

PAGES 57–61

Estimate each square root to the nearest whole number.

1. $\sqrt{27}$	**2.** $\sqrt{112}$	**3.** $\sqrt{249}$
4. $\sqrt{88}$	**5.** $\sqrt{1,500}$	**6.** $\sqrt{612}$
7. $\sqrt{340}$	**8.** $\sqrt{495}$	**9.** $\sqrt{264}$
10. $\sqrt{350}$	**11.** $\sqrt{834}$	**12.** $\sqrt{3,700}$

13. ALGEBRA Estimate $\sqrt{a - b}$ to the nearest whole number if $a = 16$ and $b = 4$.

14. ALGEBRA Estimate the value of $\sqrt{x + y}$ to the nearest whole number if $x = 64$ and $y = 25$.

Multi-Part Lesson 2-1: Integers and the Coordinate Plane

PARTS A B

PAGES 76-80

Write an integer for each situation.

1. seven degrees below zero	**2.** a loss of 3 pounds	**3.** a loss of 20 yards
4. a profit of $25	**5.** 112°F above 0	**6.** 2,830 feet above sea level

Graph each set of integers on a number line.

7. $\{-2, 0, 2\}$	**8.** $\{1, 3, 5\}$	**9.** $\{-2, -5, 3\}$	**10.** $\{7, -1, 4\}$

ALGEBRA Evaluate each expression.

11. $|1|$	**12.** $|-8|$	**13.** $|0|$
14. $|-82|$	**15.** $|64|$	**16.** $|-128|$
17. $|-22| + 5$	**18.** $|-40| - 8$	**19.** $|-18| + |10|$
20. $|-7| + |-1|$	**21.** $|98| - |-5|$	**22.** $|-49| - |-10|$

Multi-Part Lesson 2-1 (continued)
PART C

PAGES 81–85

Write the ordered pair for each point graphed at the right.
Then name the quadrant or axis on which each point
is located.

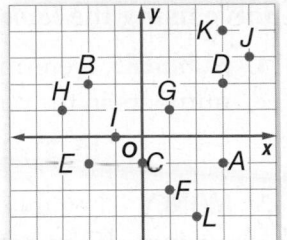

1. A 2. B 3. C

4. D 5. E 6. F

7. G 8. H 9. I

10. J 11. K 12. L

Graph and label each point on a coordinate plane.

13. $N(-4, 3)$ 14. $K(2, 5)$ 15. $W(-6, -2)$ 16. $X(5, 0)$

17. $Y(4, -4)$ 18. $M(0, -3)$ 19. $Z(-2, 0.5)$ 20. $S(-1, -3)$

21. $A(0, 2)$ 22. $C(-2, -2)$ 23. $E(0, 1)$ 24. $G(1, -1)$

Multi-Part Lesson 2-2: Add and Subtract Integers
PARTS A B

PAGES 86–92

Add.

1. $-4 + 8$ 2. $14 + 16$ 3. $-7 + (-7)$

4. $-9 + (-6)$ 5. $-18 + 11$ 6. $-36 + 40$

7. $42 + (-18)$ 8. $-42 + 29$ 9. $18 + (-32)$

10. $-96 + 6$ 11. $-2 + (-5)$ 12. $-6 + (-32)$

13. $3 + 98$ 14. $-120 + (-2)$ 15. $-120 + (-6)$

16. $5 + (-2)$ 17. $6 + 3$ 18. $-6 + 6$

PARTS C D

PAGES 93–98

Subtract.

1. $3 - 7$ 2. $-5 - 4$ 3. $-6 - 2$

4. $8 - 13$ 5. $6 - (-4)$ 6. $12 - 9$

7. $-2 - 23$ 8. $63 - 78$ 9. $0 - (-14)$

10. $15 - 6$ 11. $18 - 20$ 12. $-5 - 8$

ALGEBRA Evaluate each expression if $k = -3$, $p = 6$, $n = 1$, and $d = -8$.

13. $55 - k$ 14. $p - 7$ 15. $d - 15$

16. $n - 12$ 17. $-51 - d$ 18. $k - 21$

19. $n - k$ 20. $-99 - k$ 21. $p - k$

22. $d - (-1)$ 23. $k - d$ 24. $n - d$

Multi-Part Lesson 2-3: Multiply and Divide Integers

PART A

PAGES 100–101

Solve using the *look for a pattern* strategy.

1. **NUMBERS** Determine the next three numbers in the pattern below.

   15, 21, 27, 33, 39, …

2. **TIME** Determine the next two times in the pattern below.

   2:30 A.M., 2:50 A.M., 3:10 A.M., 3:30 A.M., …

3. **MONEY** The table shows Abigail's savings. If the pattern continues, what will be the total amount in week 6?

Week	Total ($)
1	$400
2	$800
3	$1,200
4	$1,600
5	$2,000
6	■

4. **SCIENCE** A single rotation of Earth takes about 24 hours. Copy and complete the table to determine the number of hours in a week.

Number of Days	Number of Hours
1	24
2	48
3	72
4	■
5	■
6	■
7	■

PARTS B C

PAGES 102–108

Multiply.

1. $5(-2)$
2. $6(-4)$
3. $4(21)$
4. $-11(-5)$
5. $-6(5)$
6. $-50(0)$
7. $-5(-5)$
8. $-4(8)$
9. $(-6)^2$
10. $(-2)^2$
11. $(-4)^3$
12. $(-5)^3$

ALGEBRA Evaluate each expression if $a = -5, b = 2, c = -3,$ and $d = 4$.

13. $-2d$
14. $6a$
15. $3ab$
16. $-12d$
17. $-4b^2$
18. $-5cd$
19. a^2
20. $13ab$

PART D

PAGES 109–113

Divide.

1. $4 \div (-2)$
2. $16 \div (-8)$
3. $-14 \div (-2)$
4. $\frac{32}{8}$
5. $18 \div (-3)$
6. $-18 \div 3$
7. $8 \div (-8)$
8. $0 \div (-1)$
9. $-25 \div 5$
10. $\frac{-14}{-7}$
11. $-32 \div 8$
12. $-56 \div (-8)$
13. $-81 \div 9$
14. $-42 \div (-7)$
15. $121 \div (-11)$
16. $-81 \div (-9)$

ALGEBRA Evaluate each expression if $a = -2, b = -7, x = 8,$ and $y = -4$.

17. $-64 \div x$
18. $\frac{16}{y}$
19. $x \div 2$
20. $\frac{a}{2}$
21. $ax \div y$
22. $\frac{bx}{y}$
23. $2y \div 1$
24. $\frac{x}{ay}$
25. $-y \div a$
26. $x^2 \div y$
27. $\frac{ab}{1}$
28. $\frac{xy}{a}$

Multi-Part Lesson 3-1: Rational Numbers

PARTS (A) (B)

PAGES 127–132

Write each fraction or mixed number as a decimal. Use bar notation if the decimal is a repeating decimal.

1. $\frac{16}{20}$ **2.** $\frac{30}{120}$ **3.** $1\frac{7}{8}$ **4.** $\frac{1}{6}$

5. $\frac{11}{40}$ **6.** $5\frac{13}{50}$ **7.** $\frac{55}{300}$ **8.** $1\frac{1}{2}$

9. $\frac{5}{9}$ **10.** $2\frac{3}{4}$ **11.** $\frac{9}{11}$ **12.** $4\frac{1}{9}$

Write each decimal as a fraction or mixed number in simplest form.

13. 0.26 **14.** 0.75 **15.** 0.4 **16.** 0.1

17. 4.48 **18.** 9.8 **19.** 0.91 **20.** 11.15

PART (C)

PAGES 133–138

Replace each ● with <, >, or = to make a true sentence. Use a number line if necessary.

1. $-\frac{1}{5}$ ● $-\frac{3}{5}$ **2.** $-\frac{7}{8}$ ● $-\frac{5}{8}$ **3.** $-\frac{1}{6}$ ● $-\frac{5}{6}$ **4.** $-\frac{3}{4}$ ● $-\frac{1}{4}$

5. $-2\frac{1}{4}$ ● $-2\frac{2}{8}$ **6.** $-4\frac{3}{7}$ ● $-4\frac{2}{7}$ **7.** $-1\frac{4}{9}$ ● $-1\frac{8}{9}$ **8.** $-3\frac{4}{5}$ ● $-3\frac{2}{5}$

9. $\frac{7}{9}$ ● $\frac{3}{5}$ **10.** $\frac{14}{25}$ ● $\frac{3}{4}$ **11.** $\frac{8}{24}$ ● $\frac{20}{60}$ **12.** $\frac{5}{12}$ ● $\frac{4}{9}$

13. $\frac{18}{24}$ ● $\frac{10}{18}$ **14.** $\frac{4}{6}$ ● $\frac{5}{9}$ **15.** $\frac{11}{49}$ ● $\frac{12}{42}$ **16.** $\frac{5}{14}$ ● $\frac{2}{6}$

Order each set of numbers from least to greatest.

17. 70%, 0.6, $\frac{2}{3}$ **18.** 0.8, $\frac{17}{20}$, 17% **19.** $\frac{61}{100}$, 0.65, 61.5%

20. $0.\overline{42}$, $\frac{3}{7}$, 42% **21.** 2.15, 2.105, $2\frac{7}{50}$ **22.** $7\frac{1}{8}$, 7.81, 7.18

Multi-Part Lesson 3-2: Add and Subtract Fractions

PART (A)

PAGES 139–143

Add or subtract. Write in simplest form.

1. $\frac{5}{11}+\frac{9}{11}$ **2.** $\frac{5}{8}-\frac{1}{8}$ **3.** $\frac{7}{10}+\frac{7}{10}$

4. $\frac{9}{12}-\frac{5}{12}$ **5.** $\frac{2}{9}+\frac{1}{9}$ **6.** $\frac{1}{4}+\frac{3}{4}$

7. $\frac{17}{21}+\left(-\frac{13}{21}\right)$ **8.** $-\frac{8}{13}+\left(-\frac{11}{13}\right)$ **9.** $\frac{13}{28}-\frac{9}{28}$

10. $\frac{15}{16}+\frac{13}{16}$ **11.** $-\frac{4}{35}-\left(-\frac{17}{35}\right)$ **12.** $\frac{3}{8}+\left(-\frac{5}{8}\right)$

13. $\frac{8}{15}-\frac{2}{15}$ **14.** $-\frac{3}{10}+\frac{7}{10}$ **15.** $\frac{5}{6}-\frac{7}{6}$

16. $\frac{7}{24}+\frac{7}{24}$ **17.** $-\frac{29}{9}-\left(-\frac{26}{9}\right)$ **18.** $\frac{3}{7}-\frac{4}{7}$

Multi-Part Lesson 3-2 (continued)
PARTS B C

Add or subtract. Write in simplest form.

1. $\frac{1}{4} - \frac{3}{12}$ **2.** $\frac{3}{7} + \frac{6}{14}$ **3.** $\frac{1}{4} + \frac{3}{5}$

4. $\frac{4}{9} + \frac{1}{2}$ **5.** $\frac{5}{7} - \frac{4}{6}$ **6.** $\frac{3}{4} - \frac{1}{6}$

7. $\frac{3}{5} + \frac{3}{4}$ **8.** $\frac{2}{3} - \frac{1}{8}$ **9.** $\frac{9}{10} + \frac{1}{3}$

10. $-\frac{3}{4} + \frac{7}{8}$ **11.** $\frac{3}{8} + \frac{7}{12}$ **12.** $\frac{3}{5} - \frac{2}{3}$

13. $\frac{2}{5} + \left(-\frac{2}{7}\right)$ **14.** $-\frac{3}{5} - \left(-\frac{5}{6}\right)$ **15.** $-\frac{7}{12} - \frac{3}{4}$

Evaluate each expression if $a = \frac{2}{3}$ and $b = \frac{7}{12}$.

16. $\frac{1}{5} + a$ **17.** $a - \frac{1}{2}$ **18.** $b + \frac{7}{8}$

19. $\frac{7}{8} - a$ **20.** $a + b$ **21.** $a - b$

PART D

Add or subtract. Write in simplest form.

1. $2\frac{1}{3} + 1\frac{1}{3}$ **2.** $5\frac{2}{7} - 2\frac{3}{7}$ **3.** $6\frac{3}{8} + 7\frac{1}{8}$

4. $2\frac{3}{4} - 1\frac{1}{4}$ **5.** $5\frac{1}{2} - 3\frac{1}{4}$ **6.** $2\frac{2}{3} + 4\frac{1}{9}$

7. $7\frac{4}{5} + 9\frac{3}{10}$ **8.** $3\frac{3}{4} + 5\frac{5}{8}$ **9.** $10\frac{2}{3} + 5\frac{6}{7}$

10. $17\frac{2}{9} - 12\frac{1}{3}$ **11.** $6\frac{5}{12} + 12\frac{5}{12}$ **12.** $7\frac{1}{4} + 15\frac{5}{6}$

13. $6\frac{1}{8} + 4\frac{2}{3}$ **14.** $7 - 6\frac{4}{9}$ **15.** $8\frac{1}{12} + 12\frac{6}{11}$

16. $7\frac{2}{3} + 8\frac{1}{4}$ **17.** $12\frac{3}{11} + 14\frac{3}{13}$ **18.** $21\frac{1}{3} + 15\frac{3}{8}$

Multi-Part Lesson 3-3: Multiply and Divide Fractions
PARTS A B

Multiply. Write in simplest form.

1. $\frac{2}{3} \times \frac{3}{5}$ **2.** $\frac{1}{6} \times \frac{2}{5}$ **3.** $\frac{4}{9} \times \frac{3}{7}$ **4.** $\frac{5}{12} \times \frac{6}{11}$

5. $\frac{3}{8} \times \frac{8}{9}$ **6.** $\frac{2}{5} \times \frac{5}{8}$ **7.** $\frac{7}{15} \times \frac{3}{21}$ **8.** $\frac{5}{6} \times \frac{15}{16}$

9. $\frac{2}{3} \times \frac{3}{13}$ **10.** $\frac{4}{9} \times \frac{1}{6}$ **11.** $3 \times \frac{1}{9}$ **12.** $5 \times \frac{6}{7}$

13. $\frac{3}{5} \times 15$ **14.** $3\frac{1}{2} \times 4\frac{1}{3}$ **15.** $\frac{4}{5} \times 2\frac{3}{4}$ **16.** $6\frac{1}{8} \times 5\frac{1}{7}$

17. $2\frac{2}{3} \times 2\frac{1}{4}$ **18.** $\frac{7}{8} \times 16$ **19.** $5\frac{1}{5} \times 2\frac{1}{2}$ **20.** $7 \times \frac{1}{14}$

21. $22 \times \frac{3}{11}$ **22.** $8\frac{2}{3} \times 1\frac{1}{2}$ **23.** $4 \times 6\frac{1}{2}$ **24.** $\frac{1}{2} \times 10\frac{2}{3}$

25. $\frac{2}{3} \times 21\frac{1}{3}$ **26.** $\frac{7}{8} \times \frac{8}{7}$ **27.** $21 \times \frac{1}{2}$ **28.** $11 \times \frac{1}{4}$

Multi-Part Lesson: 3-3 (continued)
PART C
PAGES 166–167

Use the *draw a diagram* strategy to solve the following problems.

1. **TESTS** The scores on a test are found by adding or subtracting points as shown below. If Salazar's score on a 15-question test was 86 points, how many of his answers were correct, incorrect, and blank?

Answer	Points
Correct	+8
Incorrect	−4
Blank	−2

2. **GAMES** Six members of a video game club are having a tournament. In the first round, every player will play a video game against every other player. How many games will be in the first round of the tournament?

3. **FAMILY** At Latrice's family reunion, $\frac{4}{5}$ of the people are 18 years of age or older. Half of the remaining people are under 12 years old. If 20 children are under 12 years old, how many people are at the reunion?

PART D
PAGES 168–173

Divide. Write in simplest form.

1. $\frac{2}{3} \div \frac{3}{2}$
2. $\frac{3}{5} \div \frac{2}{5}$
3. $\frac{7}{10} \div \frac{3}{8}$
4. $\frac{5}{9} \div \frac{2}{5}$
5. $4 \div \frac{2}{3}$
6. $8 \div \frac{4}{5}$
7. $9 \div \frac{5}{9}$
8. $\frac{2}{7} \div 2$
9. $\frac{1}{14} \div 7$
10. $15 \div \frac{3}{5}$
11. $\frac{9}{14} \div \frac{3}{4}$
12. $\frac{7}{8} \div 10$
13. $16 \div \frac{3}{4}$
14. $\frac{3}{8} \div 2\frac{1}{2}$
15. $5\frac{1}{2} \div 2\frac{1}{2}$
16. $3\frac{1}{4} \div 5\frac{1}{2}$
17. $12\frac{5}{6} \div 2\frac{1}{6}$
18. $7\frac{1}{2} \div 3\frac{1}{2}$

Multi-Part Lesson 3-4: Monomials
PART A
PAGES 176–180

Find each product or quotient. Express using exponents.

1. $2^3 \cdot 2^4$
2. $5^6 \cdot 5$
3. $t^4 \cdot t^2$
4. $y^5 \cdot y^3$
5. $(-3x^3)(-2x^2)$
6. $b^{12} \cdot b$
7. $3^5 \cdot 3^8$
8. $(-2y^3)(5y^7)$
9. $(6a^5)(-3a^6)$
10. $\frac{7^9}{7^6}$
11. $\frac{2^5}{2^2}$
12. $\frac{11^{10}}{11}$
13. $\frac{16x^3}{4x^2}$
14. $\frac{25y^5}{5y^2}$
15. $\frac{-48y^3}{-8y}$
16. $\frac{12y^5}{3y^2}$
17. $\frac{39x^7y^5}{3x^3y}$
18. $\frac{21a^7b^2}{7ab^2}$

Multi-Part Lesson 3-4 (continued)
PART B

PAGES 181–184

Write each expression using a positive exponent.

1. 4^{-4}
2. $(-3)^{-2}$
3. 7^{-6}

4. $(-5)^{-8}$
5. $(-2)^{-4}$
6. c^{-6}

7. $3a^{-2}$
8. b^{-9}
9. $-8c^{-8}$

Write each fraction as an expression using a negative exponent other than −1.

10. $\frac{1}{3^5}$
11. $\frac{1}{5^2}$
12. $\frac{1}{243}$
13. $\frac{1}{9}$

PART C

PAGES 185–189

Express each number in standard form.

1. 4.5×10^3
2. 2×10^4
3. 1.725896×10^6
4. 9.61×10^2
5. 1×10^7
6. 8.256×10^8
7. 3.25×10^2
8. 3.1×10^{-4}
9. 2.51×10^{-2}
10. 6×10^{-1}
11. 2.15×10^{-3}
12. 3.14×10^{-6}

Express each number in scientific notation.

13. 720
14. 7,560
15. 892

16. 1,400
17. 91,256
18. 51,000

19. 0.012
20. 0.0002
21. 0.054

22. 0.231
23. 0.0000056
24. 0.000123

Multi-Part Lesson 4-1: Addition and Subtraction Equations
PART A

PAGES 202–203

Use the *work backward* strategy to solve each problem.

1. **DVDS** Jeffrey rented 2 times as many DVDs as Paloma last month. Paloma rented 4 fewer than Robbie, but 4 more than Sanjay. Robbie rented 9 DVDs. How many DVDs did each person rent?

2. **MONEY** Holly spent $13.76 on a birthday present for her mom. She also spent $3.25 on a snack for herself. If she now has $7.74, how much money did she have initially?

3. **NUMBERS** A number is divided by 2. Then 4 is added to the quotient. Next, the sum of these numbers is multiplied by 3. The result is 21. Find the number.

4. **TIME** A portion of a shuttle bus schedule is shown below. What is the earliest time after 9 A.M. when the bus departs?

5. **FOOD** After four days, 0.5 pound of lunch meat was left in the refrigerator. If half this amount was eaten on each of the previous four days, how much lunch meat was initially in the refrigerator?

Departs	Arrives
8:55 A.M.	9:20 A.M.
?	10:08 A.M.
10:31 A.M.	10:56 A.M.
11:19 A.M.	11:44 A.M.

Multi-Part Lesson 4-1 (continued)

PARTS B C D

PAGES 204–213

Solve each equation. Check your solution.

1. $r - 3 = 14$
2. $t + 3 = 21$
3. $s + 10 = 23$

4. $7 + a = -10$
5. $14 + m = 24$
6. $-9 + n = 13$

7. $s - 2 = -6$
8. $6 + f = 71$
9. $x + 27 = 30$

10. $k - 9 = -3$
11. $j + 12 = 11$
12. $-42 + v = -42$

13. $s + 1.3 = 18$
14. $x + 7.4 = 23.5$
15. $p + 3.1 = 18$

16. $w - 3.7 = 4.63$
17. $m - 4.8 = 7.4$
18. $x - 1.3 = 12$

19. $y + 3.4 = 18$
20. $7.2 + g = 9.1$
21. $z - 12.1 = 14$

22. $v - 18 = 13\frac{7}{10}$
23. $w - \frac{1}{10} = \frac{8}{25}$
24. $r + 6\frac{7}{10} = 1\frac{1}{5}$

Multi-Part Lesson 4-2: Multiplication and Division Equations

PARTS A B

PAGES 214–219

Solve each equation. Check your solution.

1. $2m = 18$
2. $-42 = 6n$
3. $72 = 8k$

4. $-20r = 20$
5. $420 = 5s$
6. $325 = 25t$

7. $-14 = -2p$
8. $18q = 36$
9. $40 = 10a$

10. $100 = 20b$
11. $416 = 4c$
12. $45 = 9d$

13. $0.5m = 3.5$
14. $1.8 = 0.6x$
15. $0.4y = 2$

16. $1.86 = 6.2z$
17. $-8x = 24$
18. $8.34 = 2r$

19. $1.67t = 10.02$
20. $243 = 27a$
21. $0.9x = 4.5$

22. $\frac{r}{7} = -8$
23. $\frac{w}{7} = 8$
24. $\frac{y}{12} = -6$

25. $\frac{c}{-4} = 10$
26. $\frac{s}{9} = 8$
27. $\frac{m}{8} = 5$

PARTS C D

PAGES 220–226

Find the multiplicative inverse of each number.

1. $\frac{2}{3}$
2. $\frac{5}{4}$
3. 1
4. 10

5. $\frac{1}{7}$
6. $\frac{9}{16}$
7. $1\frac{1}{3}$
8. $3\frac{3}{4}$

9. $7\frac{3}{8}$
10. $6\frac{2}{5}$
11. $33\frac{1}{3}$
12. $66\frac{2}{3}$

Solve each equation. Check your solution.

13. $\frac{a}{13} = 2$
14. $\frac{8}{9}x = 24$
15. $\frac{3}{8}r = 36$
16. $\frac{3}{4}t = \frac{1}{2}$

17. $16 = \frac{h}{4}$
18. $\frac{m}{8} = 12$
19. $\frac{5}{8}n = 45$
20. $10 = \frac{b}{10}$

21. $\frac{1}{7}x = 7$
22. $5 = \frac{1}{5}y$
23. $\frac{4}{3}m = 28$
24. $\frac{2}{3}z = 20$

25. $\frac{c}{9} = 81$
26. $\frac{m}{9} = 9$
27. $16 = \frac{4}{9}f$
28. $\frac{15}{8}x = 225$

Multi-Part Lesson 4-3: Multi-Step Equations
PARTS A B

PAGES 228–234

Solve each equation. Check your solution.

1. $3x + 6 = 6$
2. $2r - 7 = -1$
3. $-10 + 2d = 8$
4. $2b + 4 = -8$
5. $5w - 12 = 3$
6. $5t - 4 = 6$
7. $2q - 6 = 4$
8. $2g - 3 = -9$
9. $15 = 6y + 3$
10. $3s - 4 = 8$
11. $18 - 7f = 4$
12. $13 + 3p = 7$
13. $7.5r + 2 = -28$
14. $4.2 + 7z = 2.8$
15. $-9m - 9 = 9$
16. $32 + 0.2c = 1$
17. $5t - 14 = -14$
18. $-0.25x + 0.5 = 4$
19. $5w - 4 = 8$
20. $4d - 3 = 9$
21. $2g - 16 = -9$
22. $4k + 13 = 20$
23. $7 = 5 - 2x$
24. $8z + 15 = -1$
25. $92 - 16b = 12$
26. $14e + 14 = 28$
27. $1.1j + 2 = 7.5$

PARTS C D

PAGES 235–239

Express each equation as another equivalent equation. Justify your answer.

1. $x - 6 = -3x + 10$
2. $2x + 7 = x - 6$
3. $5x - 3 = 18 + 2x$

Solve each equation. Check your solution.

4. $11x - 7 = 5 - x$
5. $2 - 8x = 10x + 20$
6. $17 - 3x = 2 + 2x$
7. $4x - 5 = 2x + 11$
8. $\frac{x}{2} + 4 = \frac{5}{2}x - 6$
9. $7 - 4x = x + 12$
10. $-1 + 8x = 4x + 5$
11. $-6 - 5x = 12 - x$
12. $6x + 11 = 4x + 10$
13. $3 + \frac{x}{6} = -2 + x$
14. $-9x + 7 = 25 - 3x$
15. $-42 + 3x = 12 - 3x$
16. $5x + 3.6 = 3x - 1.2$
17. $0.6x + 15 = -1.4x + 12$
18. $2x - 8 = 0.75x - 5$

Multi-Part Lesson 4-4: Inequalities
PART A B

PAGES 242–248

Solve each inequality. Graph the solution set on a number line.

1. $y + 3 > 7$
2. $c - 9 < 5$
3. $x + 4 \geq 9$
4. $y - 3 < 15$
5. $t - 13 \geq 5$
6. $x + 3 < 10$
7. $y - 6 \geq 2$
8. $x - 3 \geq -6$
9. $a + 3 \leq 5$
10. $c - 2 \leq 11$
11. $a + 15 \geq 6$
12. $y + 3 \geq 18$
13. $y + 16 \geq -22$
14. $x - 3 \geq 17$
15. $y - 6 > -17$
16. $y - 11 < 7$
17. $a + 5 \geq 21$
18. $c + 3 > -16$
19. $x - 12 \geq 12$
20. $x + 5 \geq 5$
21. $y - 6 > 31$

Multi-Part Lesson 4-4 (continued)
PART C PAGES 249–253

Solve each inequality. Graph the solution set on a number line.

1. $5p \geq 25$

2. $4x < 12$

3. $15 \leq 3m$

4. $\dfrac{d}{3} > 15$

5. $8 < \dfrac{r}{7}$

6. $9g < 27$

7. $4p \geq 24$

8. $5p > 25$

9. $-4 > \dfrac{-k}{3}$

10. $\dfrac{-z}{5} > 2$

11. $-3x \leq 9$

12. $-5x > -35$

13. $\dfrac{a}{-6} < 1$

14. $\dfrac{x}{-5} \leq -2$

15. $-2x < 16$

Multi-Part Lesson 5-1: Proportions
PARTS A B PAGES 265–271

Find each unit rate. Round to the nearest hundredth if necessary.

1. $240 for 4 days

2. 250 people in 5 buses

3. 500 miles in 10 hours

4. $18 for 24 pounds

5. 32 people in 8 cars

6. $4.50 for 3 dozen

7. 245 tickets in 5 days

8. 12 classes in 4 semesters

9. 60 people in 4 rows

10. 48 ounces in 3 pounds

11. 20 people in 4 groups

12. 1.5 pounds for $3.00

13. 45 miles in 60 minutes

14. $5.50 for 10 disks

15. 360 miles on 12 gallons

16. $8.50 for 5 yards

17. 24 cups for $1.20

18. 160 words in 4 minutes

19. $60 for 5 books

20. $24 for 6 hours

PART C PAGES 272–275

1. POPCORN Fun Center rents popcorn machines for $20 per hour. In addition to the hourly charge, there is a rental fee of $32. Is the number of hours you rent the popcorn machine proportional to the total cost? Explain.

2. BAKING Mrs. Govin is making cakes for the school bake sale. She needs 2 cups of sugar for every cake she makes. Is the number of cakes Mrs. Govin makes proportional to the number of cups of sugar? Explain.

3. MUSIC At a local music store, CDs cost $11.99 including tax. Is the number of CDs purchased proportional to the cost of the CDs? Explain.

4. SAVINGS Jean has $280 in her savings account. Starting next week, she will deposit $30 in her account every week. Is the amount of money in her account proportional to the number of weeks? Explain.

Multi-Part Lesson 5-1 (continued)

PARTS D E

PAGES 276–281

Solve each proportion.

1. $\dfrac{x}{15} = \dfrac{4}{5}$ **2.** $\dfrac{a}{11} = \dfrac{24}{8}$ **3.** $\dfrac{19}{p} = \dfrac{16}{32}$ **4.** $\dfrac{5}{t} = \dfrac{0.5}{0.3}$

5. $\dfrac{5}{19} = \dfrac{c}{57}$ **6.** $\dfrac{3.6}{3} = \dfrac{b}{2.5}$ **7.** $\dfrac{18}{4.5} = \dfrac{8}{f}$ **8.** $\dfrac{36}{7} = \dfrac{a}{21}$

9. $\dfrac{9}{8} = \dfrac{36}{a}$ **10.** $\dfrac{b}{126} = \dfrac{3}{14}$ **11.** $\dfrac{n}{6} = \dfrac{1}{4}$ **12.** $\dfrac{7}{9} = \dfrac{c}{54}$

13. $\dfrac{2}{3} = \dfrac{a}{12}$ **14.** $\dfrac{7}{8} = \dfrac{c}{16}$ **15.** $\dfrac{3}{7} = \dfrac{21}{d}$ **16.** $\dfrac{2}{5} = \dfrac{18}{x}$

17. $\dfrac{3}{5} = \dfrac{n}{21}$ **18.** $\dfrac{5}{12} = \dfrac{b}{5}$ **19.** $\dfrac{4}{36} = \dfrac{2}{y}$ **20.** $\dfrac{16}{8} = \dfrac{y}{12}$

Multi-Part Lesson 5-2: Scale Drawings and Models

PART A

PAGES 282–283

Use the *draw a diagram* strategy to solve the problem.

1. PICTURE FRAMES Mr. Francisco has 4 picture frames that he wants to hang on the wall. In how many different ways can he hang the picture frames in a row on the wall?

2. PONDS Carter is filling the pond in his backyard. After 2 minutes and 20 seconds, the pond is only $\frac{1}{7}$ full. If the pond can hold 280 gallons, how much longer will it take to fill the pond?

3. MARCHING BAND The marching band is in formation on the field. In the first row, there are 10 band members. Each additional row has 6 more members in it than the previous row. If there are a total of 6 rows, how many band members are there?

PARTS B C

PAGES 284–291

On a map, the scale is 1 inch = 50 miles. For each map distance, find the actual distance.

1. 5 inches   **2.** 12 inches

3. $2\frac{3}{8}$ inches   **4.** $\frac{4}{5}$ inch

5. $2\frac{5}{6}$ inches   **6.** 3.25 inches

7. 4.75 inches   **8.** 5.25 inches

On a scale drawing, the scale is $\frac{1}{2}$ inch = 2 feet. Find the dimensions of each room in the scale drawing.

9. 14 feet by 18 feet   **10.** 32 feet by 6 feet

11. 3 feet by 5 feet   **12.** 20 feet by 30 feet

13. A photograph was enlarged from 5.5 inches by 7 inches to 11 inches by 14 inches. Find the scale factor of the enlargement.

Multi-Part Lesson 5-3: Similarity and Proportional Reasoning

PART A PAGES 293–298

Find the value of x in each pair of similar figures.

1.
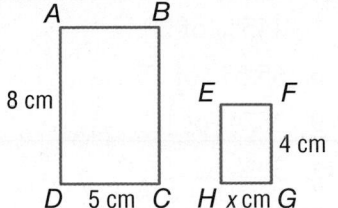

2.
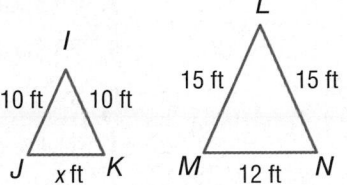

3.
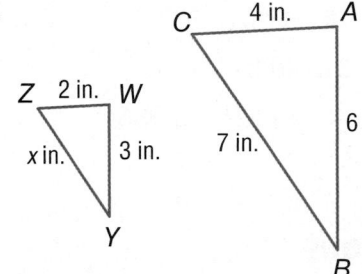

4.
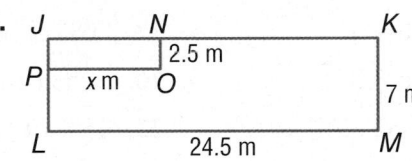

PARTS B C PAGES 299–303

For each pair of similar figures, find the perimeter of the second figure.

1. $P = 15$ mm $\qquad P = ?$ mm

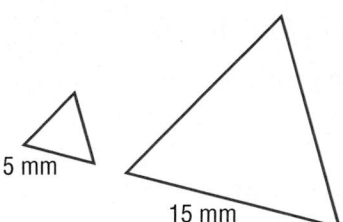

2. $P = 24$ cm $\qquad P = ?$ cm

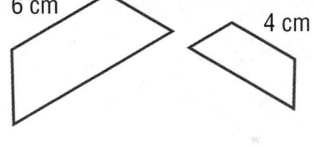

3. $P = 42$ in. $\qquad P = ?$ in.

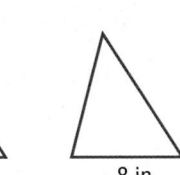

4. $P = 20$ in. $\qquad P = ?$ in.

Multi-Part Lesson 6-1: Percents

PARTS A B PAGES 318–324

Find each number. Round to the nearest tenth if necessary.

1. 5% of 40   **2.** 10% of 120   **3.** 12% of 150   **4.** 12.5% of 40

5. 75% of 200   **6.** 13% of 25.3   **7.** 250% of 44   **8.** 0.5% of 13.7

9. 600% of 7   **10.** 1.5% of 25   **11.** 81% of 134   **12.** 43% of 110

13. 61% of 524   **14.** 100% of 3.5   **15.** 20% of 58.5   **16.** 45% of 125.5

17. 23% of 500   **18.** 80% of 8   **19.** 90% of 72   **20.** 32% of 54

Multi-Part Lesson 6-1 (continued)
PART C
PAGES 325–330

Estimate.

1. 28% of 48
2. 99% of 65
3. 445% of 20
4. 9% of 81
5. 73% of 240
6. 65.5% of 75
7. 48.2% of 93
8. 39.45% of 51
9. 287% of 122
10. 53% of 80
11. 414% of 72
12. 59% of 105
13. 50% of 37
14. 18% of 90
15. 300% of 245
16. 1% of 48
17. 70% of 300
18. 35% of 35
19. 60.5% of 60
20. $5\frac{1}{2}$% of 100
21. 40.01% of 16
22. 80% of 62
23. 45% of 119
24. 14.81% of 986

Multi-Part Lesson 6-2: Proportions and Equations
PARTS A B
PAGES 331–336

Find each number. Round to the nearest tenth if necessary.

1. What number is 25% of 280?
2. 38 is what percent of 50?
3. 54 is 25% of what number?
4. 24.5% of what number is 15?
5. What number is 80% of 500?
6. 12% of 120 is what number?
7. Find 68% of 50.
8. What percent of 240 is 32?
9. 99 is what percent of 150?
10. Find 75% of 1.
11. What number is $33\frac{1}{3}$% of 66?
12. 50% of 350 is what number?
13. What percent of 450 is 50?
14. What number is $37\frac{1}{2}$% of 32?
15. 95% of 40 is what number?
16. Find 30% of 26.
17. 9 is what percent of 30?
18. 52% of what number is 109.2?
19. What number is 65% of 200?
20. What number is 15.5% of 45?

PART C
PAGES 337–341

Write an equation for each problem. Then solve. Round to the nearest tenth if necessary.

1. Find 45% of 50.
2. 75 is what percent of 300?
3. 16% of what number is 2?
4. 75% of 80 is what number?
5. 5% of what number is 12?
6. Find 60% of 45.
7. 90 is what percent of 95?
8. $28\frac{1}{2}$% of 64 is what number?
9. Find 46.5% of 75.
10. What number is 55.5% of 70?
11. 80.5% of what number is 80.5?
12. $66\frac{2}{3}$% of what number is 40?
13. Find 122.5% of 80.
14. 250% of what number is 75?

Multi-Part Lesson 6-2 (continued)

PART D

PAGES 342–343

Solve each problem using the *reasonable answers* strategy.

1. **SKIING** Emil skied for 13.5 hours and estimated that he spent 30% of his time on the ski lift. Did he spend about 4, 6, or 8 hours on the ski lift?

2. **CLASS TRIP** The class trip at Wilson Middle School costs $145 per student. A fundraiser earns 38% of this cost. Will each student have to pay about $70, $80, or $90?

3. **GAS MILEAGE** Holden's car gets 38 miles per gallon and has 2.5 gallons of gasoline left in the tank. Can he drive for 85, 95, or 105 more miles before he runs out of gas?

4. **DINING** At a restaurant, the total cost of a meal is $87.50. Dawn wants to leave a 20% tip. Should she leave a total of $95, $105, or $115?

Multi-Part Lesson 6-3: Applying Percents

PARTS A B

PAGES 345–350

Find each percent of change. Round to the nearest whole percent if necessary. State whether the percent of change is an *increase* or *decrease*.

1. 450 centimeters to 675 centimeters

2. 77 million to 200.2 million

3. 500 albums to 100 albums

4. 350 yards to 420 yards

5. 3.25 meters to 2.95 meters

6. $65 to $75

7. 180 dishes to 160 dishes

8. 450 pieces to 445.5 pieces

9. 700 grams to 910 grams

10. 55 women to 11 women

11. 412 children to 1,339 children

12. 464 kilograms to 20 kilograms

13. 24 hours to 86 hours

14. 16 minutes to 24 minutes

PART C

PAGES 351–354

Find the total cost to the nearest cent.

1. $45 sweater; 6% tax

2. $29 shirt; 7% tax

3. $145 coat; 6.25% tax

4. $12 meal; 4.5% tax

5. $105 skateboard; $7\frac{1}{2}$% tax

6. $12,500 car; $3\frac{3}{4}$% tax

7. $49.95 gloves; $5\frac{1}{4}$% tax

8. $525 stereo; 6% tax

9. **DINNER** The Lombardo family went out to dinner and the cost of their meal was $62. If they left a 20% tip, how much did they leave?

10. **HAIR CUTS** The total cost of Mrs. Draper's haircut including tip was $38. If she gave her hair dresser a 20% tip, how much was her haircut?

11. **LAWN CARE** Jordan earned $35 for taking care of his neighbor's lawn. Suppose the neighbor gave him a 15% tip. What would be the total amount Jordan earned?

Multi-Part Multi-Part Lesson 6-3 (continued)
PART D
PAGES 355–358

Find the sale price to the nearest cent.

1. $18.99 CD; 15% discount
2. $199 ring; 10% discount
3. $19 purse; 25% discount
4. $899 computer; 20% discount
5. $599 TV; 12% discount
6. $210 watch; 20% discount
7. $40 game; 30% off; 5% tax
8. $6 notebook; 50% off; 7% tax
9. $80 jacket; 15% off; 5.5% tax
10. $10 binder; 60% off; 6% tax

PARTS E F
PAGES 359–363

Find the simple interest earned to the nearest cent for each principal, interest rate, and time.

1. $2,000, 8%, 5 years
2. $500, 10%, 8 months
3. $750, 5%, 1 year
4. $175.50, $6\frac{1}{2}$%, 18 months
5. $236.20, 9%, 16 months
6. $89, $7\frac{1}{2}$%, 6 months
7. $800, 5.75%, 3 years
8. $225, $1\frac{1}{2}$%, 2 years
9. $12,000, $4\frac{1}{2}$%, 40 months

Find the simple interest paid to the nearest cent for each loan, interest rate, and time.

10. $750, 18%, 2 years
11. $1,500, 19%, 16 months
12. $300, 9%, 1 year
13. $4,750, 19.5%, 30 months
14. $2,345, 17%, 9 months
15. $689, 12%, 2 years
16. $390, 18.75%, 15 months
17. $1,250, 22%, 8 months
18. $3,240, 18%, 14 months

Multi-Part Multi-Part Lesson 7-1: Rates and Functions
PARTS A B
PAGES 377–384

Copy and complete each function table. Then identify the domain and range.

1.

x	2x	y
0		
1		
2		
3		

2.

x	3x + 1	y
1		
2		
3		
4		

3.

x	x − 2	y
3		
4		
5		
6		

4.

x	x + 0.1	y
2		
3		
4		
5		

Multi-Part Lesson 7-1 (continued)

PARTS **C** **D**

PAGES 385–390

Graph the function represented by the table.

1.

Total Cost of Tennis Balls	
Number of Tennis Balls	Total Cost ($)
3	6
4	8
5	10
6	12

2.

Convert Gallons to Quarts	
Gallon	Quarts
1	4
2	8
3	12
4	16

Graph each equation.

3. $y = 3x$

4. $y = 2x + 3$

5. $y = -x$

6. $y = 0.5x + 2$

7. $y = -x + 3$

8. $y = 0.25x + 6$

Multi-Part Lesson 7-2: Slope

PARTS **A** **B**

PAGES 391–395

Find the rate of change for each table.

1.

Age (yr)	Height (in.)
9	54
10	56
11	58
12	60

2.

Time (h)	Temperature (°C)
0	0
4	3
8	6
12	9

3. MOVIE RENTALS The graph shows the cost of renting movies. Use the graph to find the rate of change.

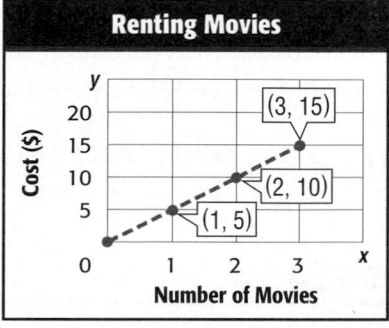

PARTS **C** **D**

PAGES 396–400

1. Find the slope of the line at the right and describe what happens when the amount of flour decreases.

2. The table below shows the number of apples y per basket x. Find the slope of the line.

Baskets	2	4	6	8
Apples	10	20	30	40

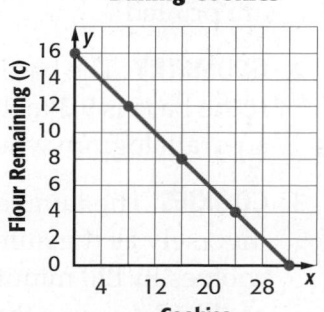

Multi-Part Lesson 7-3: Variation

PART A

PAGES 402–403

CLUBS Use the table that shows the math club membership from 2007 to 2012.

1. Make a graph of the data.

2. Describe how the number of math club memberships changed from 2007 to 2012.

3. What is a reasonable prediction for the membership in 2013 if this membership trend continues?

Math Club Membership	
Year	Number of Students
2007	20
2008	21
2009	30
2010	34
2011	38
2012	45

PARTS B C

PAGES 404–410

TRAVEL Use the graph that shows distance traveled.

1. The number of miles traveled varies directly with the number of hours traveled. What is the rate of speed in miles per hour?

2. Going at the rate shown, what distance would one travel in 39 hours?

3. **GAS MILEAGE** Dustin's car can travel about 100 miles on 3 gallons of gas. Assuming that the distance traveled remains constant to the amount of gas used, how many gallons of gas would be needed to travel 650 miles?

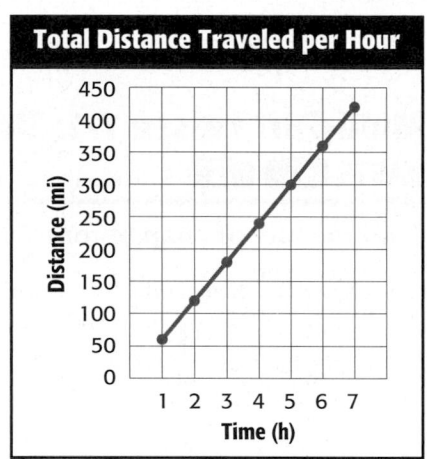

Total Distance Traveled per Hour

4. **MONEY** Determine whether the linear function shown is a direct variation. If so, state the constant of variation.

Years, x	2	3	4	5
Savings, y	$2,154	$3,231	$4,308	$5,385

PARTS D E

PAGES 411–415

1. **PROGRAMS** The cost to print programs for football season varies inversely as the number of programs printed. If 1,250 programs are printed, the cost is $2.00 each. Find the cost per program if 2,000 programs are printed.

2. **GEOMETRY** The base b of a parallelogram varies inversely as the height h. If the base is 9.3 meters when the height is 1.4 meters, what is the base of a parallelogram when the height is 3.2 meters?

3. **COOKIES** The number of bakers needed to make 100 cookies varies inversely as the number of hours needed. Three bakers can make the cookies in 120 minutes. How long will it take 4 bakers to make the cookies? Assume that they all work at the same rate.

Multi-Part Lesson 8-1: Probability

PART A

PAGES 429–434

Use the spinner at the right to find each probability. Write as a fraction in simplest form.

1. P(even number)
2. P(prime number)
3. P(factor of 12)
4. P(composite number)
5. P(greater than 10)
6. P(neither prime nor composite)

A package of balloons contains 5 green, 3 yellow, 4 red, and 8 pink balloons. Suppose you reach in the package and choose one balloon at random. Find the probability of each event. Write as a fraction in simplest form.

7. P(red balloon)
8. P(yellow balloon)
9. P(pink balloon)
10. P(orange balloon)
11. P(red or yellow balloon)
12. P(*not* green balloon)

PART B

PAGES 435–439

For each situation, find the sample space using a tree diagram.

1. rolling 2 number cubes
2. choosing an ice cream cone from waffle, plain, or sugar and a flavor of ice cream from chocolate, vanilla, or strawberry
3. making a sandwich from white, wheat, or rye bread, cheddar or Swiss cheese, and ham, turkey, or roast beef
4. tossing a penny twice
5. choosing one math class from Algebra and Geometry and one foreign language class from French, Spanish, or Latin

PART C

PAGES 440–443

Use the Fundamental Counting Principle to find the total number of outcomes in each situation.

1. choosing a local phone number if the exchange is 398 and each of the four remaining digits is different
2. choosing a way to drive from Lodi to Akron if there are 5 roads that lead from Lodi to Miami, 3 roads that connect Miami to Niles, and 4 highways that connect Niles to Akron
3. tossing a quarter, rolling a number cube, and tossing a dime
4. spinning the spinners shown below

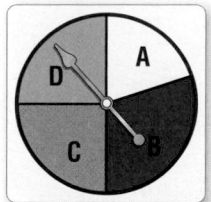 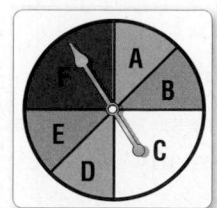 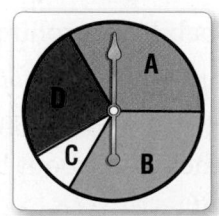

Multi-Part Lesson 8-1 (continued)
PART D

PAGES 444–447

1. **RACES** Eight runners are competing in a 100-meter sprint. In how many ways can the gold, silver, and bronze medals be awarded?

2. **LOCKERS** Five-digit locker combinations are assigned using the digits 1–9. In how many ways can the combinations be formed if no digit can be repeated?

3. **SCHEDULES** In how many ways can the classes math, language arts, science, and social studies be ordered on student schedules as the first four classes of their day?

4. **TOYS** At a teddy bear workshop, customers can select from black, brown, gold, white, blue, or pink for their bear's color. If a father randomly selects two bear colors, what is the probability that he will select a white bear for his son and a pink bear for his daughter? The father cannot pick the same color for both bears.

5. **WRITING** If you randomly select three of your last seven writing assignments to submit to an essay contest, what is the probability that you will select your first, fourth, and sixth essays in that order?

Multi-Part Lesson 8-2: Compound Events
PARTS A B

PAGES 449–455

Two socks are drawn from a drawer which contains one red sock, three blue socks, two black socks, and two green socks. Once a sock is selected, it is not replaced. Find each probability.

1. P(a black sock and then a green sock)   2. P(two blue socks)

There are three quarters, five dimes, and twelve pennies in a bag. Once a coin is drawn from the bag, it is not replaced. If two coins are drawn at random, find each probability.

3. P(a quarter and then a penny)   4. P(a nickel and then a dime)

Multi-Part Lesson 8-3: Predictions
PARTS A B

PAGES 458–463

The frequency table shows the results of a fair number cube rolled 40 times.

1. Find the experimental probability of rolling a 4.

2. Find the theoretical probability of *not* rolling a 4.

3. Find the theoretical probability of rolling a 2.

4. Find the experimental probability of *not* rolling a 6.

5. Suppose the number cube was rolled 500 times. Based on the results in the table, about how many times would it land on 5?

Face	Frequency
1	5
2	9
3	2
4	8
5	12
6	4

Multi-Part Lesson 8-3 (continued)

PART C

PAGES 464–465

Use the *act it out* strategy to solve each problem.

1. **STAIRS** Lynnette lives on a certain floor of her apartment building. She goes up two flights of stairs to put a load of laundry in a washing machine on that floor. Then she goes down five flights to borrow a book from a friend. Next, she goes up 8 flights to visit another friend who is ill. How many flights up or down does Lynette now have to go to take her laundry out of the washing machine?

2. **LOGIC PUZZLE** Suppose you are on the west side of a river with a fox, a duck, and a bag of corn. You want to take all three to the other side of the river, but...

   • your boat is only large enough to carry you and either the fox, duck, or bag of corn.

   • you cannot leave the fox alone with the duck.

   • you cannot leave the duck alone with the corn.

   • you cannot leave the corn alone on the east side of the river because some wild birds will eat it.

   • the wild birds are afraid of the fox.

   • you cannot leave the fox, duck, and the corn alone.

   • you can bring something across the river more than once.

   If there is no other way to cross the river, how do you get everything to the other side?

PARTS D E

PAGES 466–471

1. **SURVEYS** The table shows the results of a survey of students' favorite cookies. Predict how many of the 424 students at Scobey High School prefer chocolate chip cookies.

Cookie	Number
Chocolate chip	49
Peanut butter	12
Oatmeal	10
Sugar	8
Raisin	3

2. **VACATION** The circle graph shows the results of a survey of teens and where they would prefer to spend a family vacation. Predict how many of 4,000 teens would prefer to go to an amusement park.

Vacation Survey

5% Foreign country
7% Other
18% Mountains
45% Amusement park
25% Beach

3. **TRAVEL** In 2000, about 29% of the foreign visitors to the U.S. were from Canada. If a particular hotel had 150,000 foreign guests in one year, how many would you predict were from Canada?

Multi-Part Lesson 8-3 (continued)
PART F

PAGES 472–476

Determine whether the conclusions are valid. Justify your answer.

1. To award prizes at a hockey game, four tickets with individual seat numbers printed on them are picked from a barrel. Since Elvio's section was not selected for any of the four prizes, he assumes that they forgot to include the entire section in the drawing.

2. To evaluate the quality of the televisions coming off the assembly line, the manufacturer takes one every half hour and tests it. About 1 out of every 10,000 is found to have a minor mechanical problem. The company assumes from this data that about 1 out of every 10,000 televisions they produce will be returned for mechanical problems after being purchased.

3. To determine whether most students participate in after-school activities, the principal of Humberson Middle School randomly surveyed 75 students from each grade level. Of these, 34% said they participate in after school activities. The principal concluded that about a third of the students at Humberson Middle School participate in after school activities.

Multi-Part Lesson 9-1: Measures of Central Tendency
PARTS A B C

PAGES 490–497

Find the mean, median, and mode for each data set. Round to the nearest tenth if necessary.

1. number of siblings: 2, 5, 8, 9, 7, 6, 3, 5, 1, 4

2. ages of children living on a street: 1, 5, 9, 1, 2, 6, 8, 2

3. points scored each game: 82, 79, 93, 91, 95, 95, 81

4. number of coins saved: 117, 103, 108, 120

5. E-mails sent this month: 256, 265, 247, 256

6. typing speed in words per minute: 47, 54, 66, 54, 46, 66

Multi-Part Lesson 9-2: Measures of Variation
PART A

PAGES 498–503

Find the range, median, lower and upper quartiles, interquartile range, and any outliers for each set of data.

1. ages of players on a team: 15, 12, 21, 18, 25, 11, 17, 19, 20

2. ages of cousins: 2, 24, 6, 13, 8, 6, 11, 4

3. dollars in an account: 189, 149, 155, 290, 141, 152

4. daily attendance at a fair: 451, 501, 388, 428, 510, 480, 390

5. number of calls made: 22, 18, 9, 26, 14, 15, 6, 19, 28

6. text messages recieved: 245, 218, 251, 255, 248, 241, 250

7. ages of people in a restaurant: 46, 45, 50, 40, 49, 42, 64

8. points earned in a game: 128, 148, 130, 142, 164, 120, 152, 202

Multi-Part Lesson 9-2 (continued)
PART B

PAGES 504–509

Draw a box-and-whisker plot for each set of data.

1. 2, 3, 5, 4, 3, 3, 2, 5, 6

2. 6, 7, 9, 10, 11, 11, 13, 14, 12, 11, 12

3. 15, 12, 21, 18, 25, 11, 17, 19, 20

4. 2, 24, 6, 13, 8, 6, 11, 4

ZOOS Use the box-and-whisker plot.

Area (acres) of Major Zoos in the United States

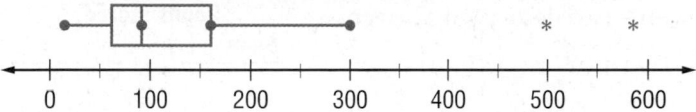

5. How many outliers are in the data?

6. Describe the distribution of the data. What can you say about the areas of the major zoos in the United States?

Multi-Part Lesson 9-3: Statistical Displays
PARTS A B

PAGES 513–519

Display each set of data in a circle graph.

1.

Car Sales	
Style	**Percent**
Sedan	55%
SUV	30%
Pickup truck	15%

2.

Favorite Flavor of Ice Cream	
Flavor	**Number**
Vanilla	12
Chocolate	15
Strawberry	8

PARTS C D

PAGES 520–525

For each problem, choose intervals and make a frequency table. Then construct a histogram to represent the data.

1.

Cost of a Movie Ticket at Selected Theaters			
$5.25	$6.50	$3.50	$3.75
$7.50	$9.25	$10.40	$4.75
$10.00	$4.50	$8.75	$7.25
$3.50	$6.70	$4.20	$7.50

2.

Highest Recorded Wind Speeds for Selected U.S. Cities (mph)					
52	55	81	46	73	57
75	54	58	76	46	58
60	91	53	53	51	56
80	60	73	46	49	47

ARCHITECTURE Use the histogram.

3. How many buildings are represented in the histogram?

4. Which interval represents the most number of buildings?

5. What percent of the buildings are taller than 70 feet?

6. What is the height of the tallest building?

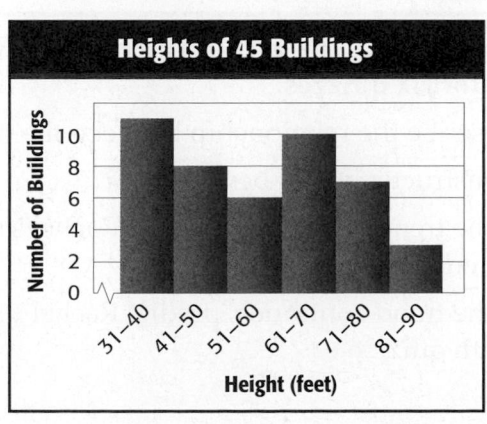

Multi-Part Lesson 9-3 (continued)

PART E

PAGES 526–531

Display each set of data in a stem-and-leaf plot.

1. 37, 44, 32, 53, 61, 59, 49, 69

2. 3, 26, 35, 8, 21, 24, 30, 39, 35, 5, 38

3. 15.7, 7.4, 0.6, 0.5, 15.3, 7.9, 7.3

4. 172, 198, 181, 182, 193, 171, 179, 186, 181

5. 55, 62, 81, 75, 71, 69, 74, 80, 67

6. 121, 142, 98, 106, 111, 125, 132, 109, 117, 126

7. 17, 54, 37, 86, 24, 69, 77, 92, 21

8. 7.3, 6.1, 8.9, 6.7, 8.2, 5.4, 9.3, 10.2, 5.9, 7.5, 8.3

BASKETBALL Use the stem-and-leaf plot shown.

9. What is the greatest value?

10. In which interval do most of the values occur?

11. What is the median value?

Points Scored

Stem	Leaf
7	2 2 3 5 9
8	0 1 1 4 6 6 8 9
9	3 4 8

$9 \mid 4 = 94$ points

Multi-Part Lesson 9-4: More Statistical Displays

PART A

PAGES 532–533

1. **BAMBOO** The table shows the amount of growth of bamboo in inches in a certain number of years.

   a. Make a graph of the data.

   b. If the trend continues, about how high will the bamboo grow in 4 years?

Bamboo Growth	
Time (yr)	Height (in.)
0	20
1	30
2	40
3	50

2. **WALKING** The table shows how long it takes Lacey to walk a given distance.

   a. Make a graph of the data.

   b. If the trend continues, predict how far Lacey will go in 50 seconds.

Walking	
Time (s)	Distance (ft)
0	0
5	2.5
10	5
15	7.5
20	10
25	12.5
30	15

PART B

PAGES 534–539

QUIZZES Refer to the graph which shows Rachel's quiz scores for six quizzes.

1. Describe the relationship between the two variables.

2. Construct a line of best fit.

3. If the trend continues, predict Rachel's score on the seventh quiz.

4. If the trend continues, predict Rachel's score on the tenth quiz.

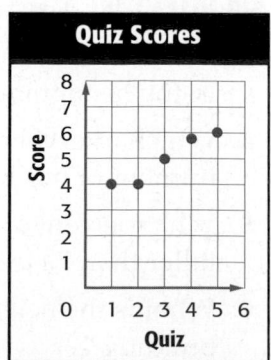

Multi-Part Lesson 9-4 (continued)
PART C

PAGES 540–545

Select an appropriate display for each situation. Justify your reasoning.

1. the average daily high temperatures for Wichita, KS recorded for two weeks

2. the favorite new release movies recorded in a survey of seventh grade students

3. the birth years of a population arranged by intervals

4. **SCHOOLS** The table shows the results of a school survey. Select an appropriate display for the data. Then construct the display. What can you conclude from your display?

Should the Name of Deer Valley Middle School be Changed?	
Vote	**Number of Votes**
Yes	15
No	21
Don't Know	8

Multi-Part Lesson 10-1: Volume
PARTS A B

PAGES 557–563

Find the volume of each prism. Round to the nearest tenth if necessary.

1.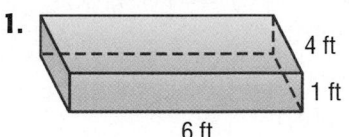
   4 ft, 1 ft, 6 ft

2.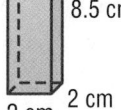
   8.5 cm, 2 cm, 2 cm

3.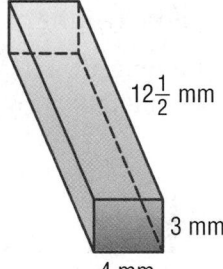
   $12\frac{1}{2}$ mm, 3 mm, 4 mm

4.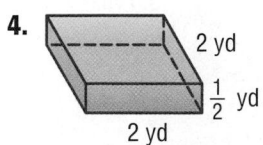
   2 yd, $\frac{1}{2}$ yd, 2 yd

5.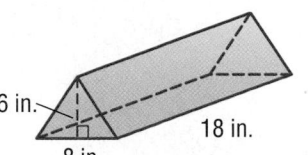
   6 in., 8 in., 18 in.

6.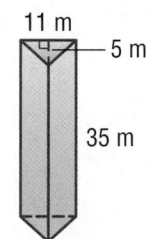
   11 m, 5 m, 35 m

7.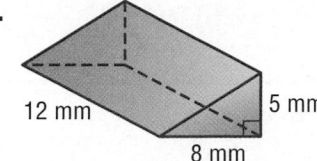
   12 mm, 5 mm, 8 mm

8.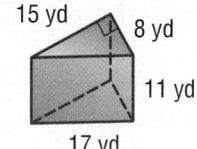
   15 yd, 8 yd, 11 yd, 17 yd

9.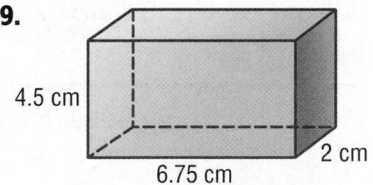
   4.5 cm, 2 cm, 6.75 cm

10. Find the volume of a rectangular prism with a length of 9 feet, a width of 5 feet, and a height of 12 feet.

11. Find the volume of a triangular prism with a base area of 416 square feet and height of 22 feet.

Multi-Part Lesson 10-1 (continued)

PART C

PAGES 564–568

Find the volume of each cylinder. Round to the nearest tenth.

1.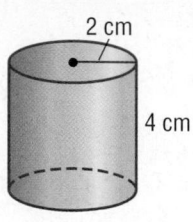
2 cm
4 cm

2.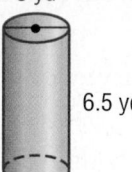
3 yd
6.5 yd

3.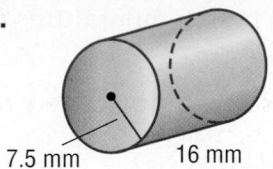
7.5 mm   16 mm

4.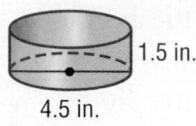
1.5 in.
4.5 in.

5. radius = 6 in.
height = 3 in.

6. radius = 8 ft
height = 10 ft

7. radius = 6 km
height = 12 km

8. radius = 8.5 cm
height = 3 cm

9. diameter = 16 yd
height = 4.5 yd

10. diameter = 3.5 mm
height = 2.5 mm

11. radius = 40.5 m
height = 65.1 m

12. radius = 0.5 cm
height = 1.6 cm

13. diameter = $8\frac{3}{4}$ in.
height = $5\frac{1}{2}$ in.

14. Find the volume of a cylinder with a diameter of 6 inches and height of 24 inches. Round to the nearest tenth.

15. How tall is a cylinder that has a volume of 2,123 cubic meters and a radius of 13 meters? Round to the nearest tenth.

16. A cylinder has a volume of 310.2 cubic yards and a radius of 2.9 yards. What is the height of the cylinder? Round to the nearest tenth.

17. Find the height of a cylinder with a diameter of 25 centimeters and volume of 8,838 cubic centimeters. Round to the nearest tenth.

PARTS D E

PAGES 569–574

Find the volume of each pyramid. Round to the nearest tenth if necessary.

1.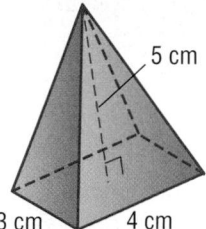
5 cm
3 cm   4 cm

2.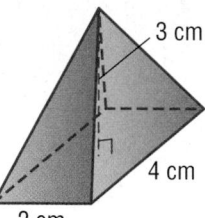
3 cm
4 cm
2 cm

3.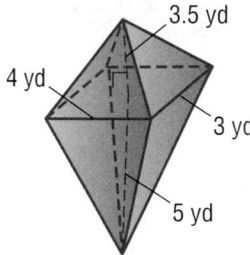
3.5 yd
4 yd
3 yd
5 yd

PART F

PAGES 575–578

Find the volume of each cone. Round to the nearest tenth if necessary.

1.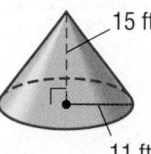
15 ft
11 ft

2.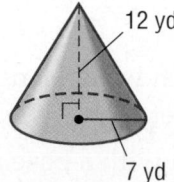
12 yd
7 yd

3.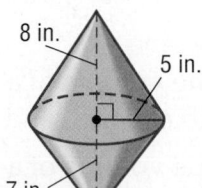
8 in.
5 in.
7 in.

Multi-Part Lesson 10-2: Surface Area

PARTS A B

PAGES 580–587

Find the surface area of each rectangular prism. Round to the nearest tenth if necessary.

1.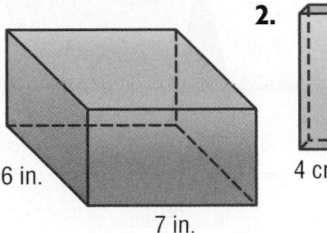
   4 in.
   6 in.
   7 in.

2.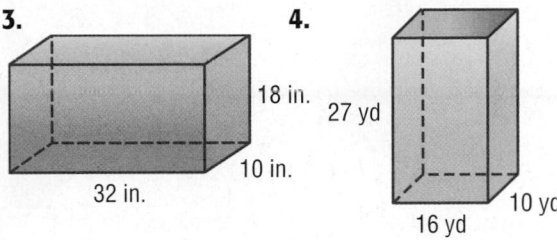
   15 cm
   4 cm
   4 cm

3. 18 in.
   10 in.
   32 in.

4. 27 yd
   10 yd
   16 yd

5. length = 10 m
   width = 6 m
   height = 7 m

6. length = 20 mm
   width = 15 mm
   height = 25 mm

7. length = 8 ft
   width = 6.5 ft
   height = 7 ft

8. length = 20.4 cm
   width = 15.5 cm
   height = 8.8 cm

9. length = 8.5 mi
   width = 3 mi
   height = 5.8 mi

10. length = $7\frac{1}{4}$ ft
    width = 5 ft
    height = $6\frac{1}{2}$ ft

11. length = $15\frac{2}{3}$ yd
    width = $7\frac{1}{3}$ yd
    height = 9 yd

12. length = $4\frac{1}{2}$ in.
    width = 10 in.
    height = $8\frac{3}{4}$ in.

13. length = 12.2 mm
    width = 7.4 mm
    height = 7.4 mm

14. Find the surface area of an open-top box with a length of 18 yards, a width of 11 yards, and a height of 14 yards.

15. Find the surface area of a rectangular prism with a length of 1 yard, a width of 7 feet, and a height of 2 yards.

PARTS C D

PAGES 588–593

Find the surface area of each cylinder. Round to the nearest tenth if necessary.

1.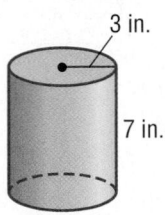
   3 in.
   7 in.

2.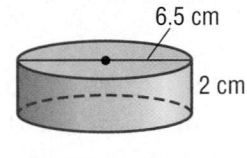
   6.5 cm
   2 cm

3.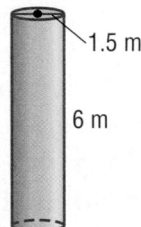
   1.5 m
   6 m

4.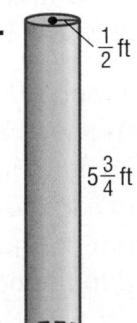
   $\frac{1}{2}$ ft
   $5\frac{3}{4}$ ft

5. height = 6 cm
   radius = 3.5 cm

6. height = 16.5 mm
   diameter = 18 mm

7. height = 22 yd
   radius = 10.5 yd

8. height = 10.2 mi
   diameter = 4 mi

9. height = 8.6 cm
   diameter = 8.2 cm

10. height = 32.7 m
    radius = 21.5 m

11. height = $2\frac{2}{3}$ yd
    diameter = 6 yd

12. height = $12\frac{3}{4}$ ft
    radius = $7\frac{1}{4}$ ft

13. height = $5\frac{1}{2}$ in.
    diameter = 3 in.

Multi-Part Lesson 10-2 (continued)
PARTS Ⓔ Ⓕ

PAGES 594–599

Find the total surface area of each pyramid. Round to the nearest tenth if necessary.

1.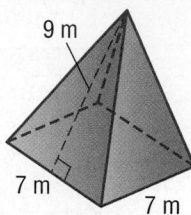

9 m

7 m   7 m

2.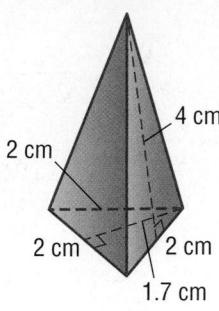

4 cm

2 cm

2 cm   2 cm

1.7 cm

3.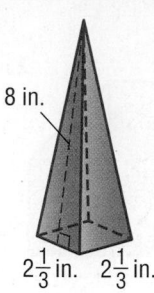

8 in.

$2\frac{1}{3}$ in. $2\frac{1}{3}$ in.

4. **GIFT BOXES** Mr. Parker is making gift boxes in the shape of square pyramids for party favors. They have a slant height of 2.5 inches and base edges of 2 inches. How many square inches of card stock does he use to make one gift box?

5. **BIRDHOUSES** Bonita is building a birdhouse. The birdhouse is a regular octagonal pyramid. The base has side lengths of 4 inches and an area of about 77 square inches. The slant height is 8 inches. If the entire birdhouse is made from plywood, how much plywood will Bonita need?

Multi-Part Lesson 10-3: Composite Figures
PART Ⓐ

PAGES 602–603

Use the *solve a simpler problem* strategy to solve each problem.

1. **EARNINGS** Cedric makes $51,876 each year. If he is paid once every two weeks and actually takes home about 67% of his wages after taxes, how much does he take home each paycheck? Round to the nearest cent if necessary.

2. **CARS** Jonah plans to decorate the rims on his tires by putting a strip of shiny metal around the outside edge of each rim. The diameter of each tire is 17 inches, and each rim is 2.75 inches from the outside edge of each tire. If he plans to cut the four individual pieces for each tire from the same strip of metal, how long of a strip should he buy? Round to the nearest tenth.

BIOLOGY About five quarts of blood are pumped through the average human heart in one minute.

3. At this rate, how many quarts of blood are pumped through the average human heart in one year? (Use 365 days = 1 year.)

4. If the average heart beats 72 times per minute, how many quarts of blood are pumped with each beat? Round to the nearest tenth.

5. About how many total gallons of blood are pumped through the average human heart in one week?

Multi-Part Lesson 10-3 (continued)
PARTS B C
PAGES 604–610

Find the volume of each composite figure. Round to the nearest tenth if necessary.

1.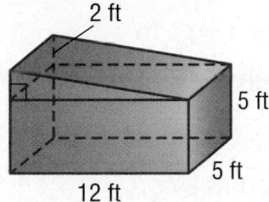

2.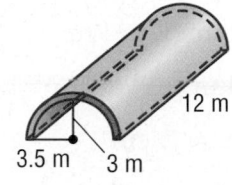

3.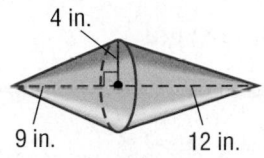

Find the surface area of each composite figure. Round to the nearest tenth if necessary.

4.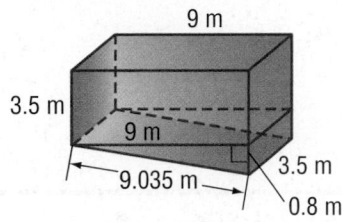

5.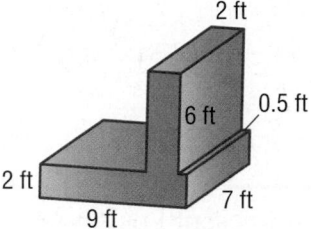

6. Find the surface area of the inside, outside, and end surfaces of the pipe at the right.

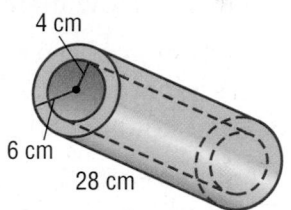

Multi-Part Lesson 11-1: Convert Measurements
PARTS A B
PAGES 624–630

Complete.

1. $4,000 \text{ lb} = \blacksquare \text{ T}$	**2.** $5 \text{ T} = \blacksquare \text{ lb}$	**3.** $5 \text{ lb} = \blacksquare \text{ oz}$
4. $12,000 \text{ lb} = \blacksquare \text{ T}$	**5.** $\frac{1}{4} \text{ lb} = \blacksquare \text{ oz}$	**6.** $12 \text{ pt} = \blacksquare \text{ c}$
7. $3 \text{ gal} = \blacksquare \text{ pt}$	**8.** $24 \text{ fl oz} = \blacksquare \text{ c}$	**9.** $8 \text{ pt} = \blacksquare \text{ c}$
10. $10 \text{ pt} = \blacksquare \text{ qt}$	**11.** $2\frac{1}{4} \text{ c} = \blacksquare \text{ fl oz}$	**12.** $6 \text{ lb} = \blacksquare \text{ oz}$
13. $10 \text{ gal} = \blacksquare \text{ qt}$	**14.** $4 \text{ qt} = \blacksquare \text{ fl oz}$	**15.** $4 \text{ pt} = \blacksquare \text{ c}$

PART C
PAGES 631–634

Complete.

1. $400 \text{ mm} = \blacksquare \text{ cm}$	**2.** $4 \text{ km} = \blacksquare \text{ m}$	**3.** $660 \text{ cm} = \blacksquare \text{ m}$
4. $0.3 \text{ km} = \blacksquare \text{ m}$	**5.** $30 \text{ mm} = \blacksquare \text{ cm}$	**6.** $84.5 \text{ m} = \blacksquare \text{ km}$
7. $\blacksquare \text{ m} = 54 \text{ cm}$	**8.** $18 \text{ km} = \blacksquare \text{ cm}$	**9.** $\blacksquare \text{ mm} = 45 \text{ cm}$
10. $4 \text{ kg} = \blacksquare \text{ g}$	**11.** $632 \text{ mg} = \blacksquare \text{ g}$	**12.** $4,497 \text{ g} = \blacksquare \text{ kg}$
13. $\blacksquare \text{ mg} = 0.51 \text{ kg}$	**14.** $0.63 \text{ kg} = \blacksquare \text{ g}$	**15.** $\blacksquare \text{ kg} = 563 \text{ g}$
16. $662 \text{ m} = \blacksquare \text{ km}$	**17.** $5,283 \text{ mL} = \blacksquare \text{ L}$	**18.** $0.24 \text{ cm} = \blacksquare \text{ mm}$
19. $380 \text{ kL} = \blacksquare \text{ L}$	**20.** $10.8 \text{ g} = \blacksquare \text{ mg}$	**21.** $83,000 \text{ mL} = \blacksquare \text{ L}$

Multi-Part Lesson 11-1 (continued)

PART D

PAGES 635–638

Complete. Round to the nearest hundredth if necessary.

1. 2 ft ≈ ■ m

2. 37 cm ≈ ■ in.

3. 2.3 lb ≈ ■ kg

4. 2 L ≈ ■ gal

5. 5,280 mi ≈ ■ km

6. 4 yd ≈ ■ m

7. 3.6 lb ≈ ■ g

8. 271 km ≈ ■ mi

9. 500 m ≈ ■ ft

10. 1,200 kg ≈ ■ T

11. 16 in. ≈ ■ cm

12. 2.4 c ≈ ■ mL

13. 108 lb ≈ ■ kg

14. 2,000 mL ≈ ■ qt

15. 100 m ≈ ■ yd

16. 56 in. ≈ ■ cm

17. 32.8 ft ≈ ■ m

18. 609 yd ≈ ■ m

19. 21.78 mi ≈ ■ km

20. 48 lb ≈ ■ g

21. 2.3 T ≈ ■ kg

22. 8.5 c ≈ ■ mL

23. 33 gal ≈ ■ L

24. 1.8 qt ≈ ■ mL

PART E

PAGES 639–642

1. The winner in a recent Los Angeles Marathon ran the 26-mile race in 2.23 hours. How many yards per minute did he run?

2. A swimming pool can be filled at the rate of 25 liters per minute using a special pump. About how many hours will it take to fill a pool that holds 5,000 gallons of water?

3. A water plant can process 25 million gallons of water a day. About how many liters of water per hour does it process?

Convert each rate. Round to the nearest hundredth if necessary.

4. 55 mi/h = ■ ft/s

5. 8 gal/h = ■ L/min

6. 12 m/min = ■ ft/s

7. 50 g/cm = ■ oz/in.

8. 15 qt/h = ■ gal/min

9. 2.6 oz/50 lb = ■ g/kg

10. 50 mi/h = ■ ft/h

11. 5 ft/s = ■ ft/h

12. 36 m/h = ■ ft/min

PART F

PAGES 643–648

Complete. Round to the nearest hundredth if necessary.

1. $8 \text{ yd}^2 = \blacksquare \text{ ft}^2$

2. $12 \text{ m}^2 = \blacksquare \text{ cm}^2$

3. $450 \text{ mm}^2 = \blacksquare \text{ cm}^2$

4. $1,000 \text{ in}^2 = \blacksquare \text{ ft}^2$

5. $18 \text{ ft}^2 = \blacksquare \text{ in}^2$

6. $100 \text{ ft}^2 = \blacksquare \text{ yd}^2$

7. $6.5 \text{ cm}^2 = \blacksquare \text{ mm}^2$

8. $8 \text{ cm}^2 = \blacksquare \text{ mm}^2$

9. $8.2 \text{ ft}^2 = \blacksquare \text{ in}^2$

10. $12 \text{ m}^2 \approx \blacksquare \text{ ft}^2$

11. $6.5 \text{ yd}^2 \approx \blacksquare \text{ m}^2$

12. $50 \text{ cm}^2 \approx \blacksquare \text{ in}^2$

Complete. Round to the nearest hundredth if necessary.

13. $3,000,000 \text{ cm}^3 = \blacksquare \text{ m}^3$

14. $100 \text{ mm}^3 = \blacksquare \text{ cm}^3$

15. $100 \text{ in}^3 = \blacksquare \text{ ft}^3$

16. $5 \text{ ft}^3 = \blacksquare \text{ in}^3$

17. $100,000,000 \text{ cm}^3 = \blacksquare \text{ m}^3$

18. $4,000 \text{ in}^3 = \blacksquare \text{ ft}^3$

19. $13.5 \text{ yd}^3 = \blacksquare \text{ ft}^3$

20. $8.4 \text{ cm}^3 = \blacksquare \text{ mm}^3$

21. $1,000 \text{ in}^3 = \blacksquare \text{ ft}^3$

Multi-Part Lesson 11-2: Similar Solids

PART A
PAGES 650–651

Use the *make a model* strategy to solve each problem.

1. **ARCHITECTURE** An architect is designing a large skyscraper for a local firm. The skyscraper is to be 1,200 feet tall, 500 feet long, and 400 feet wide. If his model has a scale of 80 feet = 1 inch, find the volume of the model.

2. **STACKING BOXES** Box A has twice the volume of Box B. Box B has a height of 10 centimeters and a length of 5 centimeters. Box A has a width of 20 centimeters, a length of 10 centimeters, and a height of 5 centimeters. What is the width of Box B?

3. **TRAVEL** On Monday, Mara drove 400 miles as part of her journey to see her sister. She drove 60% of this distance on Tuesday. If the distance she drove on Tuesday represents one third of her total journey, how many more miles does she still need to drive?

4. **PIZZA** On Monday, there was a whole pizza in the refrigerator. On Tuesday, Enrico ate $\frac{1}{3}$ of the pizza. On Wednesday, he ate $\frac{1}{3}$ of what was left. On Thursday, he ate $\frac{1}{2}$ of what remained. What fraction of the pizza is left?

5. **GARDENS** Mr. Hawkins has a circular garden in his backyard. He wants to build a curved brick pathway around the entire garden. The garden has a radius of 18 feet. The distance from the center of the garden to the outside edge of the brick pathway will be 21.5 feet. Find the area of the brick pathway. Round to the nearest tenth.

PARTS B C
PAGES 652–659

1. The surface area of a triangular prism is 18.5 square inches. What is the surface area of a similar solid that has been enlarged by a scale factor of 2.5?

2. The volume of a cylinder is 630 cubic centimeters. What is the volume of a similar cylinder that has been reduced by a scale factor of $\frac{1}{3}$? Round to the nearest tenth.

3. The volume of water in the first floor touch tank at the Florida Aquarium in Tampa is 2,016 cubic feet. Betty is making a model and will use a scale of $\frac{1}{5}$. What will be the volume of water in her model?

4. The volume of the Great Pyramid in Egypt is approximately 33,764 cubic meters. What is the volume, to the nearest tenth, of a scale model that is smaller by a scale factor of $\frac{1}{12}$?

5. The surface area of Spaceship Earth at Epcot Center is 150,000 square feet. What would be the surface area of a model that is decreased by a scale of $\frac{1}{100}$?

Multi-Part Lesson 12-1: Polygons

PART A

PAGES 674–679

Name each angle in four ways. Then classify each angle as *acute*, *right*, *obtuse*, or *straight*.

1.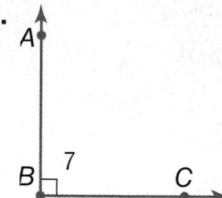

2.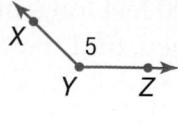

3.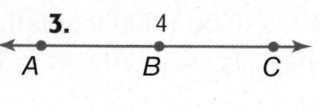

4.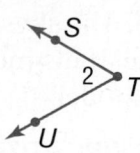

Classify each pair of angles as *complementary*, *supplementary*, or *neither*.

5.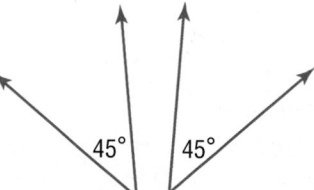

6.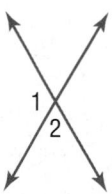

7.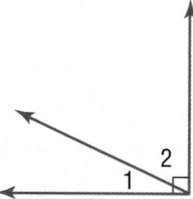

Find the value of *x* in each figure.

8.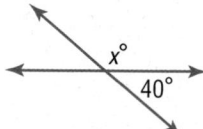

9.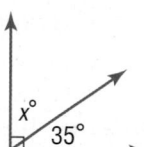

10.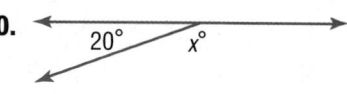

PARTS B C

PAGES 680–686

Find the value of *x*.

1.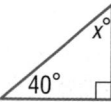

2.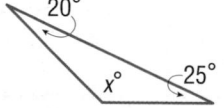

3.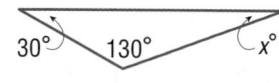

Classify each triangle by its angles and by its sides.

4.

5.

6.

PART D

PAGES 687–692

Classify each quadrilateral.

1.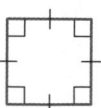

2.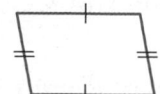

3.

Find the value of *x* in each quadrilateral.

4.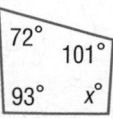

5.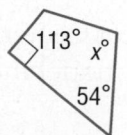

6.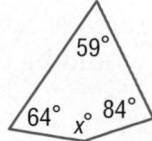

Multi-Part Lesson 12-1 (continued)
PART E

PAGES 693–698

Determine whether each figure is a polygon. If it is, classify the polygon and state whether it is regular. If it is not a polygon, explain why.

1.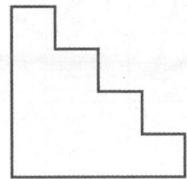

2.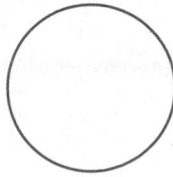

3.

Find the sum of the angle measures of each polygon.

4. dodecagon (12-gon)   **5.** 17-gon   **6.** 21-gon

Multi-Part Lesson 12-2: Translations
PARTS A B C

PAGES 699–707

1. Translate △ABC 2 units right and 1 unit down.

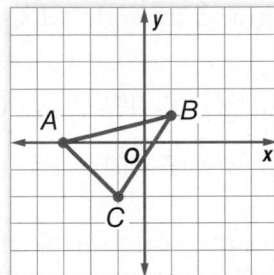

2. Translate quadrilateral RSTU 4 units left and 3 units down.

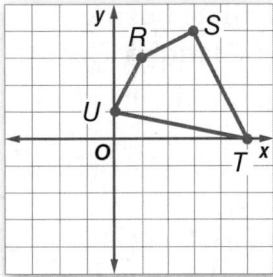

Triangle *TRI* has vertices $T(1, 1)$, $R(4, -2)$, and $I(-2, -1)$. Find the vertices of *T'R'I'* after each translation. Then graph the figure and its translated image.

3. 2 units right, 1 unit down

4. 5 units left, 1 unit up

5. 3 units right

6. 2 units up

Multi-Part Lesson 12-3: Reflections
PARTS A B

PAGES 709–716

Graph each figure and its reflection over the *x*-axis. Then find the coordinates of the vertices of the reflected image.

1. quadrilateral *QUAD* with vertices $Q(-1, 4)$, $U(2, 2)$, $A(1, 1)$, and $D(-2, 2)$

2. triangle *ABC* with vertices $A(0, -1)$, $B(4, -3)$, and $C(-4, -5)$

Graph each figure and its reflection over the *y*-axis. Then find the coordinates of the vertices of the reflected image.

3. parallelogram *PARL* with vertices $P(3, 5)$, $A(5, 4)$, $R(5, 1)$, and $L(3, 2)$

4. pentagon *PENTA* with vertices $P(-1, 3)$, $E(1, 1)$, $N(0, -2)$, $T(-2, -2)$, and $A(-3, 1)$

Multi-Part Lesson 12-4: Rotations

PARTS Ⓐ Ⓑ

PAGES 717–723

Graph triangle *ABC* and its image after each
rotation. Then give the coordinates of the vertices
for triangle *A′B′C′*.

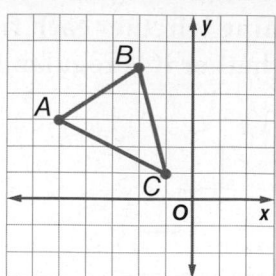

1. 90° counterclockwise

2. 180° clockwise

3. 270° counterclockwise

4. 270° clockwise

Multi-Part Lesson 12-5: Dilations

PART Ⓐ

PAGES 724–729

Find the coordinates of the vertices of triangle *A′B′C′* after triangle *ABC*
is dilated using the given scale factor. Then graph triangle *ABC* and
its dilation.

1. $A(-1, 0)$, $B(2, 1)$, $C(2, -1)$; scale factor 2   **2.** $A(4, 6)$, $B(0, -2)$, $C(6, 2)$; scale factor $\frac{1}{2}$

3. $A(1, -1)$, $B(1, 2)$, $C(-1, 1)$; scale factor 3   **4.** $A(2, 0)$, $B(0, -4)$, $C(-2, 4)$; scale factor $\frac{3}{2}$

On each graph, one figure is of the other. Find the scale factor of each dilation and
classify each dilation as an *enlargement* or as a *reduction*.

5. 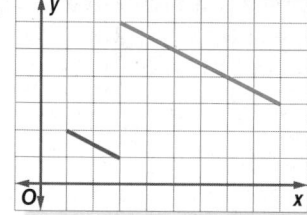   **6.** 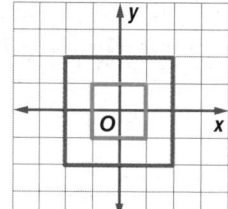   **7.**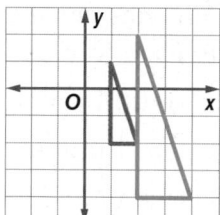

PART Ⓑ

PAGES 730–731

1. TIME The table shows the amount of time it
takes Leo to complete different activities
before going to school. If he needs to be at
school at 9 A.M., what time should he leave
his house?

Activity	Time (minutes)
Walking the dog	10
Showering and getting ready	20
Eating breakfast	10
Getting to school	25

2. AGES Anne is 3 years younger than her sister
Maddie. Maddie is half as old as her uncle
Bobby. If Bobby is 32, how old is Anne?

3. NUMBER SENSE A number is multiplied by 5 and then 7 is added to the
product. The result is 22. What is the number?

Selected Answers and Solutions

Chapter 1 Expressions and Patterns

Page 24   Chapter 1   Are You Ready?
1. 105.8   **3.** 11.3   **5.** 82.96   **7.** 2.37   **9.** $72.94
11. 29.4   **13.** 5.3   **15.** 10.2   **17.** 4.46

Pages 26–28   Lesson 1-1A
1. $9 \cdot 9 \cdot 9$

3 The base is 5 and the exponent is 5. $5^5 = 5 \cdot 5 \cdot 5 \cdot 5 \cdot 5$

5. 1,000   **7.** 117,649 people   **9.** 1^4
11. $1 \cdot 1 \cdot 1 \cdot 1 \cdot 1$ **13.** $7 \cdot 7 \cdot 7 \cdot 7 \cdot 7 \cdot 7 \cdot 7$
15. $9 \cdot 9 \cdot 9 \cdot 9$ **17.** 64   **19.** 121   **21.** 1

23 $10^6 = 10 \cdot 10 \cdot 10 \cdot 10 \cdot 10 \cdot 10$
$= 1,000,000$

25. 3^2 **27.** 1^8 **29.** $4 \cdot 4 \cdot 4 \cdot 4 \cdot 4$ **31.** 1,296

33 a. Each edge has 3 cubes, so there are $3 \cdot 3$, or 9 cubes in one layer. There are 3 layers, so there are $9 \cdot 3$, or 27 cubes. Written as one multiplication problem this is $3 \cdot 3 \cdot 3$, or 3^3.
b. Sample answer: A number taken to the third power is the same as the volume of a cube, or the amount of space inside a cube.
35. 2,147,483,648   **37.** $6^3, 15^2, 3^5, 2^8$ **39.** Sample answer: $4^5 = 1,024$ **41.** 1,000; 1,000 cannot be expressed as a square: $11^2 = 121$, $19^2 = 361$, $24^2 = 576$.   **43.** Sample answer: Using exponents is a more efficient way to describe and compare numbers.   **45.** F

Pages 30–32   Lesson 1-1B
1. 11; Sample answer: Subtract first since $5 - 2$ is in parentheses. Then add 8.   **3.** 11; Sample answer: Multiply 2 by 6 first since multiplication comes before addition or subtraction. Then subtract and add in order from left to right.   **5.** 400; Sample answer: Evaluate 10^2 first since it is a power. Then multiply by 4.   **7.** 11; Sample answer: Subtract first since $6 - 3$ is in parentheses. Then multiply the difference by 2 and multiply 3 by 4 since multiplication comes before addition or subtraction. Finally, add 17 and 6 and subtract 12 in order from left to right.

9 $3(0.05) + 2(0.25) + 2(0.10) + 7(0.01)$
$= 0.15 + 0.50 + 0.20 + 0.07$
$= 0.92$
He has $0.92.
11. 3; Sample answer: Add first since $3 + 4$ is in parentheses. Then subtract.

13 1; Sample answer: Subtract first since $11 - 2$ is in parentheses. Then divide.
15. 8; Sample answer: Divide first since division comes before addition or subtraction. Then subtract 1 and add 7 in order from left to right.
17. 5; Sample answer: Multiply first since multiplication comes before addition or subtraction. Then subtract the product from 18 and add 5.   **19.** 30,000; Sample answer: Evaluate 10^4 first since it is a power. Then multiply by 3.
21. 386; Sample answer: Evaluate 7^2 first since it is a power. Then multiply by 8 since multiplication comes before subtraction. Finally, subtract.
23. 75; Sample answer: Evaluate 9^2 since it is a power. Then divide 14 by 7 and multiply the quotient by 3 since multiplication and division occur from left to right. Finally, subtract.
25. 22; Sample answer: Add first since $6 + 5$ is in parentheses. Then subtract 6 from 8 since $8 - 6$ is in parentheses. Finally, multiply.   **27.** 23; Sample answer: Add first since $4 + 7$ is in parentheses. Then multiply the sum by 3. Next, multiply 5 by 4 and divide the product by 2 since multiplication and division occur in order from left to right. Finally, subtract.   **29.** $3(0.25) + 5(0.50)$; $3.25
31. 19; Sample answer: Evaluate 3^3 first since it is a power. Then add 8. Next, subtract 6 from 10 since $10 - 6$ is in parentheses. Then square the difference since 4^2 is a power. Finally, subtract 16 from 35.
33. 64; Sample answer: Subtract first since $4 - 3.2$ is in parentheses. Multiply 7 by 9 next since multiplication occurs from left to right. Then subtract 0.8 from the product, 63, and add 1.8 since addition and subtraction occur from left to right.

35 The total cost is equal to $0.15, the cost of one text message over 250, times the number of text messages over 250 plus $5, the cost of the plan.
$0.15(275 - 250) + $5
$\quad = $0.15(25) + $5 \qquad$ Subtract.
$\quad = $3.75 + $5 \qquad$ Multiply.
$\quad = $8.75 \qquad$ Add.
37. $2 \times 8 + 2^4$; Sample answer: The expression when evaluated is equal to 32. The other three expressions are equal to 0.   **39.** Sample answer: There are 24 cookies to be divided among 12 people. If 4 people do not want cookies, how many cookies will each person receive?; $12 - 4 = 8$ and $24 \div 8 = 3$; So, each person will get 3 cookies.
41. H   **43.** Find the value of 3^2. Then multiply by 5 and then add 7; 52

Pages 36–37   Lesson 1-1C

1. 2   **3.** 22

5 $\dfrac{mn}{4} = \dfrac{2(6)}{4}$ Replace *m* with 2 and *n* with 6.

   $= \dfrac{12}{4}$ Multiply.

   $= 3$ Divide.

9. 17   **11.** 17   **13.** 2   **15.** 3   **17.** 96   **19.** 71   **21.** $17

23 The total cost is equal to the $50 fee plus $0.17 times *m*, the number of miles.

$50 + 0.17m = 50 + 0.17(150)$ Replace *m* with 150.

$ = 50 + 25.50$ Multiply.

$ = 75.50$ Add.

The cost is $75.50.

25. 9.1   **27.** 37.85   **29.** Sample answer: $5x - 37$ if $x = 8$ **31.** Sample answer: The fee to rent a bicycle is $10 plus $5 for each hour. The expression $5x + 10$ represents the total cost for renting a bicycle for *x* hours.   **33.** A   **35.** $4(8) + 3(5)$; $47

Pages 40–41   Lesson 1-1D

1. $7(4) + 7(3)$; 49   **3.** $3(9 + 6)$; 45

5 $4(12 + 5)$; $68; Sample answer: The expression $12 + 5$ represents the cost of one ticket and one hot dog. The expression $4(12 + 5)$ represents the cost of four tickets and four hot dogs. Since $4 \times 12 = 48$ and $4 \times 5 = 20$, find $48 + 20$, or 68, to find the total cost of four tickets and four hot dogs.

7. Sample answer: Rewrite $44 + (23 + 16)$ as $44 + (16 + 23)$ using the Commutative Property of Addition. Rewrite $44 + (16 + 23)$ as $(44 + 16) + 23$ using the Associative Property of Addition. Find $44 + 16$, or 60, mentally. Then find $60 + 23$, or 83, mentally.   **9.** $2(6) + 2(7)$; 26   **11.** $4(3 + 8)$; 44   **13.** Sample answer: Rewrite $(8 + 27) + 52$ as $(27 + 8) + 52$ using the Commutative Property of Addition. Rewrite $(27 + 8) + 52$ as $27 + (8 + 52)$ using the Associative Property of Addition. Find $8 + 52$, or 60, mentally. Then find $60 + 27$, or 87, mentally.   **15.** Sample answer: Rewrite $91 + (15 + 9)$ as $91 + (9 + 15)$ using the Commutative Property of Addition. Rewrite $91 + (9 + 15)$ as $(91 + 9) + 15$ using the Associative Property of Addition. Find $91 + 9$, or 100, mentally. Then find $100 + 15$, or 115, mentally.   **17.** Sample answer: Rewrite $(4 \cdot 18) \cdot 25$ as $(18 \cdot 4) \cdot 25$ using the Commutative Property of Multiplication. Rewrite $(18 \cdot 4) \cdot 25$ as $18 \cdot (4 \cdot 25)$ using the Associative Property of Multiplication. Find $4 \cdot 25$, or 100, mentally. Then find $100 \cdot 18$, or 1,800, mentally.

19 Sample answer: Rewrite $15 \cdot (8 \cdot 2)$ as $15 \cdot (2 \cdot 8)$ using the Commutative Property of Multiplication. Rewrite $15 \cdot (2 \cdot 8)$ as $(15 \cdot 2) \cdot 8$ using the Associative Property of Multiplication. Find $15 \cdot 2$, or 30, mentally. Then find $30 \cdot 8$, or 240, mentally.

21. Sample answer: Rewrite $5 \cdot (30 \cdot 12)$ as $5 \cdot (12 \cdot 30)$ using the Commutative Property of Multiplication. Rewrite $5 \cdot (12 \cdot 30)$ as $(5 \cdot 12) \cdot 30$ using the Associative Property of Multiplication. Find $5 \cdot 12$, or 60, mentally. Then find $60 \cdot 30$, or 1,800, mentally.   **23.** $5(20 + 7)$; 135 million; Sample answer: The expression $20 + 7$ represents the number of millions of people who visit Paris each year. The expression $5(20 + 7)$ represents the number of millions of people who visit Paris over a five-year period. Since $5 \times 20 = 100$ and $5 \times 7 = 35$, find $100 + 35$, or 135, to find the number of millions of people who visit Paris over a five-year period.

25. $7(9 - 3)$; 42   **27.** $9(7 - 3)$; 36   **29.** $y + 5$ **31.** $32b$ **33.** $2x + 6$ **35.** $6c + 6$ **37.** Sample answer: $6 \times 0 = 0$ **39.** false; $(18 + 35) \times 4 = 212$; $18 + 35 \times 4 = 158$ **41.** Sample answer: Using properties with real numbers helps to understand mathematical relationships.   **43.** G

Pages 42–43   Lesson 1-2A

1. 486 posts   **3.** 19 h   **5.** 24,900 mi   **7.** 6 tins

9 Find how many passengers can ride in one run. So, multiply 8 by 4. Then divide the number of passengers that can ride in one hour by the number of passengers that can ride in one run. $1,056 \div 32 = 33$. So, 33 runs.

11. 5:10 P.M.

Pages 46–49   Lesson 1-2B

1. 9 is added to each term; 36, 45, 54   **3.** 0.1 is added to each term; 1.4, 1.5, 1.6

5 First, find how many inches the plant's height increases each month. Look at the pattern in the table.

Month	1	2	3
Height (in.)	3	6	9

+3   +3

The plant's height increases by 3 inches each month. Let *n* represent the number of months. So, $3n$ represents the total height after *n* months.

$3n$ Write the expression.

$3(12)$ Replace *n* with 12.

36 Multiply.

So, the plant's height is 36 inches.

7. 6 is added to each term; 25, 31, 37   **9.** 12 is added to each term; 67, 79, 91   **11.** 5 is added to each term; 53, 58, 63   **13.** 0.8 is added to each term; 5.6, 6.4, 7.2   **15.** 1.5 is added to each term; 10.5, 12.0, 13.5   **17.** 4 is added to each term; 20.6, 24.6, 28.6   **19.** $7n$; 42 laps

21 25 is added to each term; $95 + 25 = 120$; $120 + 25 = 145$; $145 + 25 = 170$. So, the next three terms are 120, 145, and 170.

23. 12, 7, 2　**25.** 8.5, 8, 7.5　**27a.** Sample answer: For each text message over 250, the cost increases by $0.15.　**27b.** 0.15$n$ + 5　**27c.** $8.75; $12.50; $16.25　**29.** 1,400　**31.** 20　**33.** 7,425

35 **a.** Each figure is 8 less than the previous figure.
b. In figure 3, there are 48 rectangles. 48 − 8 = 40; 40 − 8 = 32. So, in the next two figures there will be 40 and 32 rectangles.
37. + 2, + 4, + 6, + 8, … ; 30, 42, 56　**39.** Sample answer: The total amount earned forms the sequence 6.5, 13, 19.5, 26, … . Since each term is 6.5 more than the previous term, the sequence is arithmetic.　**41.** F　**43.** each term is multiplied by 3; 162, 486, 1,458

Pages 55–56　Lesson 1-3B

1. 36

3 17 × 17 = 289

5. 3　**7.** 11　**9.** 24 in. by 24 in.　**11.** 1　**13.** 121
15. 400　**17.** 1,156　**19.** 4　**21.** 10　**23.** 16　**25.** 25
27. 40 ft

29 18 × 18 = 324, so the length of the side of the chessboard including the border is 18 inches. Subtract an inch for the border on both sides of the region containing the small squares. The length of one side of that region is 16 inches.
31a. Yes; Sample answer: A pen that measures 10 feet by 10 feet has the same perimeter, but its area is 100 square feet, which is greater than 84 square feet.　**31b.** A square that measures 10 feet on each side; the perimeter is the same, and the area is 100 square feet, or 16 square feet greater than that of the original pen.　**33.** $\sqrt{25} \cdot \sqrt{4}$; Sample answer: $\sqrt{25}$ is 5 and $\sqrt{4}$ is 2. So, 5 × 2 = 10. The $\sqrt{25 \cdot 4} = \sqrt{100}$ and the $\sqrt{100} = 10$. So, 10 = 10.　**35.** D　**37.** C

Pages 58–60　Lesson 1-3C

1. 6

3 90 is between the perfect squares 81 and 100. Find the square root of each number.

$$81 < 90 < 100$$
$$\sqrt{81} < \sqrt{90} < \sqrt{100}$$
$$9 < \sqrt{90} < 10$$

So, $\sqrt{90}$ is between 9 and 10. Since 90 is closer to 81 than to 100, the best whole number estimate is 9.
5. 10 ft　**7.** 3　**9.** 6

11 89 is between 81 or 9^2 and 100, or 10^2. 89 is 8 away from 81 and 11 away from 100. So it is closer to 81. Therefore, $\sqrt{89}$ to the nearest whole number is 9.
13. 12　**15.** 8 in.　**17.** about 1.5　**19.** 30　**21.** 40
23. 4　**25.** 11　**27.** 8

29 **a.** Estimate the square root of 1,008.

1,008 is between the perfect squares 961 and 1,024. Find the square root of each number.

$$961 < 1,008 < 1,024$$
$$\sqrt{961} < \sqrt{1,008} < \sqrt{1,024}$$
$$31 < \sqrt{1,008} < 32$$

So, $\sqrt{1,008}$ is between 31 and 32. Since 1,008 is closer to 1,024 than to 961, the best whole number estimate is 32. The approximate length of the postage stamp is 32 millimeters.
b. Use the relationship 1 mm = 0.1 cm.

$$1 \text{ mm} = 0.1 \text{ cm}$$
$$32 \times 1 \text{ mm} = 32 \times 0.1 \text{ cm}$$
$$32 \text{ mm} = 3.2 \text{ cm}$$

One side is about 3 cm.

31. $\sqrt{81}$; Sample answer: It is not irrational.
33. Sample answer: 71 is between the perfect squares 64 and 81. Since 71 is closer to 64 than 81, $\sqrt{71}$ is closer to $\sqrt{64}$, or 8.　**35.** 0.7　**37.** 1.7
39. Sample answer: It cannot be written as a fraction.　**41.** G　**43.** 6　**45.** 9　**47.** 5

Pages 64–67　Chapter Study Guide and Review

1. false, equivalent expressions　**3.** false, factors
5. false, square　**7.** false, two　**9.** 243　**11.** 324
13. 18　**15.** 36　**17.** 5.75h$ + 8.95s$; $62
19. 68; Sample answer: Rewrite 14 + (38 + 16) as 14 + (16 + 38) using the Commutative Property of Addition. Rewrite 14 + (16 + 38) as (14 + 16) + 38 using the Associative Property of Addition. Find 14 + 16, or 30, mentally. Then find 30 + 38, or 68, mentally.

21.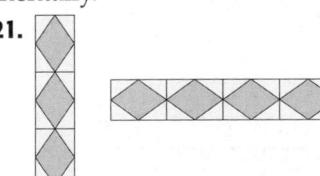

23. 21　**25.** Each term is found by adding 0.8 to the previous term; 6.6, 7.4, 8.2　**27.** 4.50n$
29. 9　**31.** 80 ft

33. 10　**35.** 9　**37.** 7　**39.** 8　**41.** 51

Chapter 2　Integers

Page 74　Chapter 2　Are You Ready?

1. 6　**3.** 17　**5.** 24　**7.** 7　**9.** (1, 1)　**11.** (8, 1)
13. (1, 5)　**15.** (7, 6)

Pages 79–80　Lesson 2-1B

1. −11　**3.** 16　**5.** −15
7.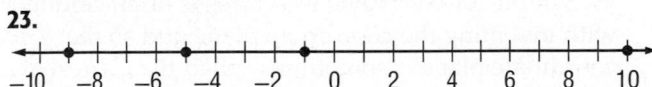
$$-10\ -8\ -6\ -4\ -2\ \ 0\ \ 2\ \ 4\ \ 6\ \ 8\ \ 10$$

9 1 + |7| = 1 + 7
$ = 8$
11. 9　**13.** −53　**15.** −2　**17.** 12　**19.** −7
21.
$$-3\ -2\ -1\ \ 0\ \ 1$$

23.
$$-10\ \ -8\ \ -6\ \ -4\ \ -2\ \ \ 0\ \ \ 2\ \ \ 4\ \ \ 6\ \ \ 8\ \ \ 10$$

25. 10 **27.** 2 **29.** 14 **31.** 25
33 $|27| \div 3 - |-4|$
$= 27 \div 3 - 4$ $|27| = 27; |-4| = 4$
$= 9 - 4$ Divide.
$= 5$ Subtract.
37. Sometimes; it is always true if A and B are both positive or if A or B are negative, but not if both A and B are negative.   **39.** $-|7 + 3|$; All others are positive.   **41.** C   **43.** C

Pages 83–85   Lesson 2-1C

1. $(-2, -4)$; III
3 From the origin, move no units horizontally, and up 3 units. $(0, 3)$; the point lies on the y-axis.

5–8.

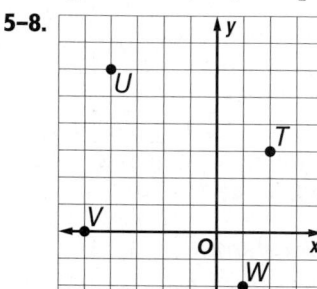

9a. Sea Cliffs   **9b.** I   **11.** $(5, 4)$; I   **13.** $(4, -3)$; IV
15 From the origin, move 3 units left, and up 5 units. $(-3, 5)$; the point lies in Quadrant II.
17. $(0, -4)$; y-axis   **19.** $(-4, -5)$; III   **21.** $(1, 0)$; x-axis

22–33.

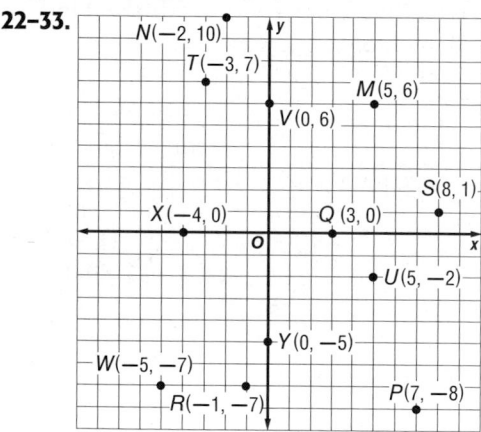

35–37.

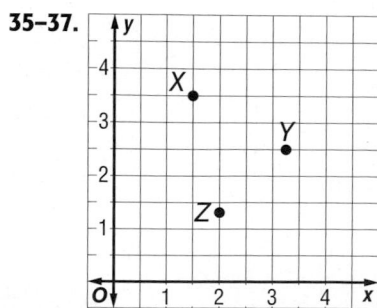

35 From the origin, move one and a half units to the right. Then move up three and a half units.
39. Sample answer: René Descartes is often credited with inventing the coordinate plane and so the coordinate plane is sometimes called the Cartesian

plane, in his honor.   **41.** She switched the x- and y-coordinates. She should have gone to the left 3 units and up 4 units.   **43.** Sample answer: Point A is 1 unit to the right and 2 units down from the origin, in Quadrant IV. Point B is 2 units to the left and 1 unit up from the origin, in Quadrant II.
45. H   **47.** G

Pages 91–92   Lesson 2-2B

1. -14 **3.** 7   **5.** -4
7 $-17 + 20 + (-3)$
$= (-17) + (-3) + 20$ Commutative (+)
$= [(-17) + (-3)] + 20$ Associative (+)
$= (-20) + 20$ Add.
$= 0$
9. -1 **11.** -38 **13.** 16   **15.** 13   **17.** 5   **19.** -4
21. -34 **23.** 3   **25.** $-14 + 3$; -11; The shark is 11 meters below sea level.

27 In a bank account, withdrawing $20 would be -20 and depositing $84 would be $+84$.
$152 + (-20) + 84 = (152 + 84) + (-20)$
$= 236 + (-20)$
$= 216$
The balance is $216.
29. green T-shirt: profit of $1; white T-shirt: profit of $3; black T-shirt: profit of $3   **31a.** Additive Inverse Property   **31b.** Commutative Property (+)
33. $x + (-4)$ **35.** $n + 6$ **37.** $-8 + (-3) = -11$
39. G   **41.** $(-2, 4)$; II   **43.** $(-3, -1)$; III   **45.** 75
47. -13 **49** -12

Pages 97–98   Lesson 2-2D

1. -3
3 $-4 - 8$
$= -4 + (-8)$ To subtract 8, add -8.
$= -12$ Simplify.
5. 24   **7.** -2 **9.** -21 **11.** 22   **13.** -10 **15.** -14
17. -14 **19.** -30 **21.** 23   **23.** 31   **25.** 104   **27.** 6
29. 0   **31.** 0   **33.** 15   **35.** 11
37 **a.** $2{,}407 - (-8) = 2{,}407 + 8$ or $2{,}415$ ft
b. $2{,}842 - (-282) = 2{,}842 + 282$ or $3{,}124$ ft
c. $345 - (-282) = 345 + 282$ or 627 ft
d. $0 - (-8) = 0 + 8$ or 8 ft
39. 16   **41.** Sample answer: $-5 - 11 = -5 + (-11) = -16$; Add 5 and 11 and keep the negative sign.
43. true   **45.** A   **47.** 7   **49.** -13 **51.** Quadrant IV   **53.** 8

Pages 100–101   Lesson 2-3A

1. 24 free throws
3 Each month, the amount of money Peter has increases by $35. Extend the pattern.

Months	1	2	3	4	5	6	7	8
Money Saved	$50	$85	$120	$155	$190	$225	$260	$295

So, it will take Peter 8 months to save $295.

5. Sample answer: 3 quarters, 2 nickels, and 1 penny
7. 13 toothpicks   **9.** Each term is found by adding the two previous terms. The next two terms would be 34 + 55 or 89 and 55 + 89 or 144.

Pages 106–108   Lesson 2-3C

1. −60   **3.** −28   **5.** 45   **7.** 64   **9.** −12
11. 100(−3) = −300; Tamera's investment is now worth $300 less than it was worth before the price of the stock dropped.

13 $fgh = (-1)(7)(-10)$
$\qquad = (-7)(-10)$
$\qquad = 70$

15. −220   **17.** −70   **19.** −50   **21.** 80   **23.** −125
25. 81   **27.** 6   **29.** 24

31 5(−650); −3,250; Ethan burns 3,250 Calories each week.

33. −84   **35.** −160   **37.** −108

39 Replace a with −6 and b with −4.
$-2a + b = -2(-6) + (-4)$
$\qquad\qquad = 12 + (-4)$
$\qquad\qquad = 8$

41. 5 black T-shirts   **43a.** Sample answer: Evaluate −7 + 7 first. Since −7 + 7 = 0, and any number times 0 is 0, the value of the expression is 0.
43b. Sample answer: First, use the Distributive Property to rewrite the expression as −15(−26 + 25). Then evaluate −26 + 25. Since −26 + 25 = −1 and −15 × (−1) = 15, the value of the expression is 15.
45. Sample answer: The product of three integers is positive when exactly two of the integers are negative or when all three integers are positive.   **47.** 64
49a. 419 points   **49b.** 47 points   **51.** −19   **53.** −4

Pages 111–113   Lesson 2-3D

1. −4   **3.** −6   **5.** 5

7 $15 ÷ y = 15 ÷ (-5)$ Replace y with −5.
$\qquad\quad = -3$ Divide. The quotient is negative.

9. −48.3°C   **11.** −7   **13.** −9   **15.** 10   **17.** −7
19. −9   **21.** The integers have the same sign. The quotient is positive. So, $\dfrac{-54}{-6} = 9$.   **23.** 5   **25.** −12
27. −3   **29.** 2   **31.** −1   **33.** −10°F   **35.** 8   **37.** 1

39 mean
$= \dfrac{-70 + (-40) + (-25) + (-50) + (-80)}{5}$
$= \dfrac{-265}{5}$
$= -53$ ft

41. −32 ÷ (−4) has a positive quotient; the others have negative quotients.   **43.** First, evaluate each 2^2 since they are powers. Rewrite the expression as $-2 \cdot (4 + 2) ÷ 4$. Then add since 4 + 2 is in parentheses. Rewrite the expression as $-2 \cdot 6 ÷ 4$. Next, multiply and then divide since multiplication and division occur in order from left to right. Since $-2 \cdot 6 = -12$ and $-12 ÷ 4 = -3$, the value of the expression is −3.   **45.** B   **47.** 60   **49.** 81   **51.** 18

Pages 116–119   Chapter Study Guide and Review

1. false; negative   **3.** false; opposite   **5.** false; y-coordinate   **7.** true   **9.** false; positive   **11.** 350
13. −48   **15.** 32   **17.** 33
19–22.

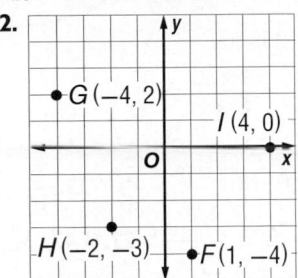

23. (−4, 1)   **25.** −13   **27.** 0   **29.** −15   **31.** 0
33. 600 lb   **35.** −3   **37.** 4   **39.** −5   **41.** −1
43. 2,980 ft   **45.** $46,400   **47.** −12   **49.** 35   **51.** 28
53. −35   **55.** 5   **57.** −2

Chapter 3   Rational Numbers

Page 126   Chapter 3   Are You Ready?

1. $\dfrac{3}{4}$ **3.** $\dfrac{47}{100}$ **5.** $\dfrac{7}{20}$

7.
```
 0   1   2   3   4   5   6
```

9.
```
 0   1   2   3   4   5   6
```

11. 0.57   **13.** 0.94   **15.** 0.72

Pages 131–132   Lesson 3-1B

1. 0.4   **3.** 7.5

5 Divide 1 by 8.

$$
\begin{array}{r}
0.125 \\
8\overline{)1.000} \\
-8 \\
\hline
20 \\
-16 \\
\hline
40 \\
-40 \\
\hline
0
\end{array}
$$

So, $\dfrac{1}{8}$ is 0.125.

7. $0.\overline{5}$ **9.** $-\dfrac{11}{50}$ **11.** $4\dfrac{3}{5}$ **13.** 0.8   **15.** −4.16   **17.** 0.3125
19. −0.66   **21.** 5.875   **23.** $-0.\overline{4}$ **25.** $-0.1\overline{6}$ **27.** $5.\overline{3}$
29. $-\dfrac{1}{5}$ **31.** $\dfrac{11}{20}$ **33.** $5\dfrac{24}{25}$

35 $30.5 = 30 + 0.5$
$\qquad = 30 + \dfrac{5}{10}$
$\qquad = 30 + \dfrac{1}{2} = 30\dfrac{1}{2}$

39. $\dfrac{22}{3}$ **41.** $-\dfrac{16}{5}$ **43.** Sample answer: $\dfrac{3}{5}$
45. Sample answer: $3\dfrac{1}{7} \approx 3.14286$ and $3\dfrac{10}{71} \approx 3.14085$; Since 3.1415926… is between $3\dfrac{1}{7}$ and $3\dfrac{10}{71}$, Archimedes was correct.   **47.** I   **49.** G

Pages 136–138   Lesson 3-1C

1. > **3.** < **5.** Elliot; 3 out of 4 has an average of 0.75; 7 out of 11 has an average of about 0.64. **7.** 0.02, 0.1, $\frac{1}{8}$, $\frac{2}{3}$ **9.** < **11.** =

13 The LCM of 7 and 8 is 56.

$\frac{4}{7} = \frac{4 \times 8}{7 \times 8} = \frac{32}{56}$ and $\frac{5}{8} = \frac{5 \times 7}{8 \times 7} = \frac{35}{56}$

Since $\frac{32}{56} < \frac{35}{56}$, $\frac{4}{7} < \frac{5}{8}$.

15. > **17.** > **19.** < **21.** Jim; $\frac{10}{16} > \frac{4}{15}$ **23.** $\frac{8}{10}$, 0.805, 0.81 **25.** $-1.4, -1.25, -1\frac{1}{25}$ **27.** $3.47, 3\frac{4}{7}, 3\frac{3}{5}$

29. > **31.** = **33.** >

35 Write the mixed numbers as decimals and then compare the decimals.

$1\frac{7}{12}$ gallons ● $1\frac{5}{8}$ gallons

$7 \div 12 = 0.58\overline{3}$ $5 \div 8 = 0.625$

Since $1.58\overline{3} < 1.625$, then $1\frac{7}{12}$ gallons $< 1\frac{5}{8}$ gallons.

37. 6 c, $6\frac{1}{3}$ c, 6.5 c **39.** $\frac{1}{5}$ g, 1.5 g, 5 g **41a.** 3 cubes: $18\frac{6}{8}$ in., rod: $12\frac{7}{8}$ in. **41b.** Yes; $69\frac{1}{8} < 69\frac{6}{8}$.

43. Sample answer: $\frac{63}{32}$ is closest to 2 because the difference of $\frac{63}{32}$ and 2 is the least. **45.** C **47.** C **49.** A

Pages 141–143   Lesson 3-2A

1. $\frac{4}{5}$

3 $-\frac{3}{4} + \left(-\frac{3}{4}\right) = \frac{-3 + (-3)}{4}$ Add the numerators.

$= -\frac{6}{4}$ Simplify.

$= -1\frac{1}{2}$

5. $-\frac{3}{5}$ **7.** $\frac{19}{50}$

9 $\frac{4}{5} + \frac{3}{5} = \frac{4 + 3}{5}$ Add the numerators.

$= \frac{7}{5}$ Simplify.

$= 1\frac{2}{5}$

11. $-\frac{1}{2}$ **13.** $-1\frac{2}{3}$ **15.** $\frac{3}{5}$ **17.** $\frac{3}{7}$ **19.** $\frac{5}{12}$

21 $\frac{17}{28} - \frac{11}{28} = \frac{6}{28}$ or $\frac{3}{14}$

23a. $\frac{33}{100}$ **23b.** $\frac{67}{100}$ **25.** $\frac{1}{8}$ **27.** 2 **29.** $1\frac{1}{7}$

31 $\frac{13}{16} - \frac{5}{16} = \frac{8}{16}$ or $\frac{1}{2}$ in.

33. Sample answer: $\frac{11}{18}$ and $\frac{5}{18}$; $\frac{11}{18} - \frac{5}{18} = \frac{6}{18}$, which simplifies to $\frac{1}{3}$. **35.** To add or subtract fractions with the same denominator, add or subtract the numerators. Write the result using the common denominator. **37.** I **39.** 4 **41.** > **43.** <

45. $\frac{3}{8}$ **47.** $\frac{19}{1,000}$

Pages 148–151   Lesson 3-2C

1. $\frac{7}{9}$ **3.** $\frac{5}{8}$

5 $\frac{1}{6} + \frac{3}{8}$

$= \frac{1 \times 4}{6 \times 4} + \frac{3 \times 3}{8 \times 3}$ Rename using the LCD, 24.

$= \frac{4}{24} + \frac{9}{24}$ Add the fractions.

$= \frac{13}{24}$ Simplify.

7. $-\frac{1}{4}$ **9.** Addition; $\frac{11}{16}$ in.; Sample answer: To find how much shorter the total height of the photo is now, add $\frac{5}{16}$ and $\frac{3}{8}$. **11.** $\frac{7}{10}$

13 $\frac{5}{6} - \left(-\frac{2}{3}\right) = \frac{5}{6} - \left(-\frac{2 \times 2}{3 \times 2}\right)$ Rename using the LCD, 6.

$= \frac{5}{6} - \left(-\frac{4}{6}\right)$ Subtract.

$= \frac{5}{6} + \frac{4}{6}$ Add the numerators.

$= \frac{9}{6}$ Simplify.

$= 1\frac{3}{6}$ or $1\frac{1}{2}$ Simplify.

15. $-\frac{2}{3}$ **17.** $1\frac{13}{24}$ **19.** $\frac{4}{9}$ **21.** $-\frac{26}{45}$ **23.** Addition; $1\frac{11}{20}$ ft; Sample answer: To find the smallest width to make the shelf, add $\frac{4}{5}$ and $\frac{3}{4}$. **25.** Subtraction; $\frac{3}{8}$ lb; Sample answer: To find how much more turkey Makayla bought, subtract $\frac{1}{4}$ from $\frac{5}{8}$. **27.** $\frac{23}{28}$ **29.** $\frac{7}{12}$

31. $1\frac{1}{4}$ **33.** $2\frac{2}{3}$ **35.** $\frac{2}{15}$

37 $\frac{1}{2} + \frac{3}{4}$

$= \frac{1 \times 2}{2 \times 2} + \frac{3 \times 1}{4 \times 1}$ Rename using the LCD, 4.

$= \frac{2}{4} + \frac{3}{4}$ Add the fractions.

$= \frac{5}{4}$ or $1\frac{1}{4}$ Simplify.

39. $\frac{1}{12}$

41 Find the total amount of time used by the second and third students. Find the difference between the available $\frac{2}{3}$ hour and the time used by the two students. This will show the amount of time remaining for the final student.

$\frac{2}{3} - \left(\frac{1}{6} + \frac{1}{4}\right) = \frac{2 \times 4}{3 \times 4} - \left(\frac{1 \times 2}{6 \times 2} + \frac{1 \times 3}{4 \times 3}\right)$

$= \frac{8}{12} - \left(\frac{2}{12} + \frac{3}{12}\right)$

$= \frac{8}{12} - \frac{5}{12}$

$= \frac{3}{12}$ or $\frac{1}{4}$ hr

43. LaTasha; $\frac{1}{8}$ mile; LaTasha's position on the track is $\frac{1}{4}$ of a mile from the start. Colin's position on the track is $\frac{5}{8} - \frac{1}{2}$, or $\frac{1}{8}$, of a mile from the start. So, LaTasha is farther ahead on the track by $\frac{1}{4} - \frac{1}{8}$ or $\frac{1}{8}$ of a mile. **45.** Sample answer: $\frac{1}{4} + \frac{1}{6} + \frac{1}{12} = \frac{6}{12}$ or $\frac{1}{2}$

47. filling the $\frac{3}{4}$-measuring cup once; Filling the $\frac{2}{3}$-measuring cup twice gives her $\frac{1}{3}$ cup more flour than she needs. Filling the $\frac{3}{4}$-measuring cup twice gives her $\frac{1}{2}$ cup more flour than she needs. Filling

the $\frac{2}{3}$-measuring cup once gives her $\frac{1}{3}$ cup less flour than she needs. Filling the $\frac{3}{4}$-measuring cup once gives her $\frac{1}{4}$ cup less flour than she needs. Since $\frac{1}{4} < \frac{1}{3} < \frac{1}{2}$, filling the $\frac{3}{4}$-measuring cup once will bring Felicia closest to having the amount of flour she needs. **49.** H **51.** $\frac{4}{5}$ **53.** $\frac{2}{3}$ **55.** $\frac{4}{7}$
57. 1 **59.** < **61.** = **63.** Vijay; Sample answer: $\frac{3 \times 7}{8 \times 7} = \frac{21}{56}, \frac{4 \times 8}{7 \times 8} = \frac{32}{56}$; Since 32 > 21, $\frac{32}{56} > \frac{21}{56}$.

Pages 154–156   Lesson 3-2D
1. $9\frac{6}{7}$ **3.** $4\frac{2}{3}$
5 $2\frac{5}{4} - 1\frac{3}{4} = 1\frac{2}{4}$
$\qquad = 1\frac{1}{2}$
7. $4\frac{5}{8}$ **9.** $3\frac{3}{20}$ gallons **11.** $7\frac{5}{7}$ **13.** $2\frac{1}{7}$
15 $11\frac{3}{4} - 4\frac{1}{3} = 11\frac{9}{12} - 4\frac{4}{12}$ Rename using the LCD, 12. Then subtract.
$\qquad = 7\frac{5}{12}$
17. $18\frac{17}{24}$ **19.** $3\frac{1}{2}$ **21.** $2\frac{11}{20}$ **23.** $5\frac{7}{8}$ **25.** $7\frac{1}{6}$
27. $17\frac{7}{8}$ in.; Addition, since the necklace is $10\frac{5}{8}$ in. longer than the bracelet. **29.** $3\frac{1}{4}$ in.; Subtraction, since cutting her hair makes it shorter. **31.** $15\frac{1}{4}$
33. $1\frac{3}{4}$
35 Find the sum of the sides.
$\qquad 2\frac{3}{8} + 2\frac{3}{8} + 2\frac{3}{8} = 6\frac{9}{8}$ or $7\frac{1}{8}$ yd
37. $5\frac{1}{2} - 3\frac{7}{8} = 1\frac{5}{8}$ **39.** Sample answer: Since the garden is a rectangle, the length of one side added to the length of the other side would equal half the length of the perimeter, or 6 ft. If one side of the garden is $2\frac{5}{12}$ ft long, find 6 ft $- 2\frac{5}{12}$ ft, or $3\frac{7}{12}$ ft. **41.** H
43. $\frac{9}{20}$ **45.** $\frac{46}{63}$ **47.** $\frac{3}{4}$ ft

Pages 163–165   Lesson 3-3B
1. $\frac{2}{9}$ **3.** $\frac{2}{3}$ **5.** $1\frac{1}{2}$ **7.** 32 pounds **9.** $\frac{4}{15}$
11. $-4\frac{4}{5}$ **13.** $\frac{1}{9}$ **15.** $\frac{1}{20}$
17 $\frac{2}{5} \times \frac{15}{16} = \frac{2 \times 15}{5 \times 16}$
$\qquad = \frac{\overset{1}{\cancel{2}}}{\underset{1}{\cancel{5}}} \times \frac{\overset{3}{\cancel{15}}}{\underset{8}{\cancel{16}}}$ Divide 2 and 16 by their GCF, 8, and divide 5 and 15 by their GCF, 5.
$\qquad = \frac{1 \times 3}{1 \times 8}$ Multiply.
$\qquad = \frac{3}{8}$
19. $-\frac{2}{3}$ **21.** $\frac{3}{16}$ **23.** 14 c **25.** 30 **27.** 50 **29.** $\frac{1}{16}$
31. $-\frac{8}{27}$ **33.** $7\frac{1}{20}$ mi **35.** $\frac{8}{21}$ **37.** $\frac{11}{48}$ **39.** one pint
41. one centimeter **43.** $15\frac{3}{4}$ **45.** $28\frac{3}{4}$

47 Alano needs to find $1\frac{1}{2}$ times the amount of each ingredient.
Broccoli: $1\frac{1}{2} \times 1\frac{1}{4} = \frac{3}{2} \times \frac{5}{4}$
$\qquad = \frac{3 \times 5}{2 \times 4}$
$\qquad = \frac{15}{8}$
$\qquad = 1\frac{7}{8}$ c
Cooked pasta: $1\frac{1}{2} \times 3\frac{3}{4} = \frac{3}{2} \times \frac{15}{4}$
$\qquad = \frac{45}{8}$
$\qquad = 5\frac{5}{8}$ c
Salad dressing: $1\frac{1}{2} \times \frac{2}{3} = \frac{3}{2} \times \frac{2}{3}$
$\qquad = \frac{3 \times 2}{2 \times 3}$
$\qquad = \frac{6}{6}$ or 1 c
Cheese: $1\frac{1}{2} \times 1\frac{1}{3} = \frac{3}{2} \times \frac{4}{3}$
$\qquad = \frac{3 \times 4}{2 \times 3}$
$\qquad = \frac{12}{6}$ or 2 c
49. Never; Sample answer: improper fractions are always greater than 1, so their product will be greater than 1. **51.** Sample answer: Addition requires common denominators for completion, whereas multiplication does not. Multiplication can utilize cancelling. **53.** H **55a.** $\frac{5}{8}$ mi **55b.** $\frac{13}{16}$ mi
57. >

Pages 166–167   Lesson 3-3C
1. $2\frac{10}{27}$ foot
3 Draw a model of the full trip. Since the denominator is 5, divide it into 5 equal parts.

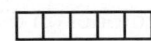

They have covered $\frac{4}{5}$ of trip, so color in 4 sections.

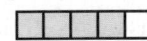

Since they have traveled one mile, each colored section is $\frac{1}{4}$ mile. Therefore, they have $\frac{1}{4}$ mile more remaining.
5. $\frac{3}{40}$; Sample answer: $\frac{1}{5} \cdot \frac{3}{8} = \frac{3}{40}$ **7.** 28 games
9. Mary, Isabela, Anna, Rachana **11.** Sample answer: There are 4 books in the front window of a bookstore. How many different ways can you arrange the books in the front window?; 24 ways.

Pages 170–173   Lesson 3-3D
1. $\frac{3}{8}$ **3.** $3\frac{1}{2}$

5 $\frac{1}{2} \div 7\frac{1}{2} = \frac{1}{2} \div \frac{15}{2}$ Rename $7\frac{1}{2}$ as $\frac{15}{2}$.

$= \frac{1}{\cancel{2}} \cdot \frac{\cancel{2}}{15}$ Divide 2 and 2 by their GCF, 2. Then multiply.

$= \frac{1}{15}$

7. $1\frac{1}{5}$ **9.** 56 slices   **11.** $\frac{7}{16}$ **13.** $1\frac{1}{3}$ **15.** -12 **17.** $\frac{2}{3}$

19. 12 portions   **21.** $\frac{4}{15}$ **23.** $-\frac{2}{3}$

25 $3\frac{4}{5} \div 1\frac{1}{3} = \frac{19}{5} \div \frac{4}{3}$ Rename each mixed number.

$= \frac{19}{5} \times \frac{3}{4}$ Multiply.

$= \frac{57}{20}$ Simplify.

$= 2\frac{17}{20}$

27. $-7\frac{4}{5}$ **29.** 36 servings   **31a.** $3\frac{34}{35}$ **31b.** $1\frac{40}{99}$

33. $2\frac{1}{2}$

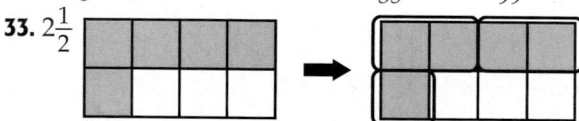

Sample answer: The model on the left shows that $\frac{5}{8}$ of a rectangle with 8 sections is 5 sections. $\frac{1}{4}$ of 8 sections is 2 sections. The model on the right shows those 5 sections divided into $2\frac{1}{2}$ groups of 2 sections.

35. $3\frac{1}{4}$

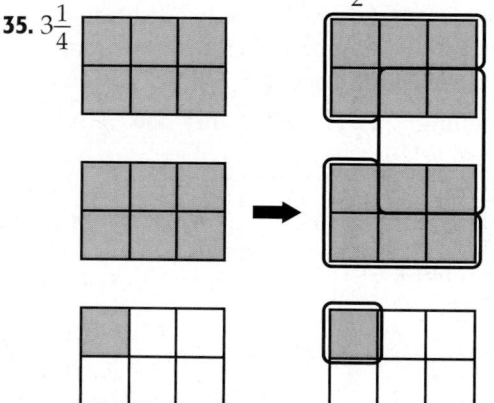

Sample answer: The model on the left shows that $2\frac{1}{6}$ of a rectangle with 6 sections is 13 sections. $\frac{2}{3}$ of 6 sections is 4 sections. The model on the right shows those thirteen sections being divided into $3\frac{1}{4}$ groups of four sections.

37. $7\frac{1}{3}$ **39.** -1

41 $12\frac{1}{2} \div \frac{3}{4} = \frac{25}{2} \div \frac{3}{4}$

$= \frac{25}{2} \times \frac{4}{3}$

$= \frac{25 \times 2}{1 \times 3}$

$= \frac{50}{3}$ or $16\frac{2}{3}$ times larger

43a. $2\frac{3}{8}$ times as many   **43b.** $7\frac{4}{5}$ times as many

45. 6:30 P.M.; Sample answer: $105 \div 35 = 3$. The storm will travel 105 miles in 3 sets of half hours, or $1\frac{1}{2}$ hours. Adding $1\frac{1}{2}$ hours to 5:00 P.M. will make it 6:30 P.M.   **47.** $\frac{10}{3}$ **49.** Yes; sample answer: If the

first proper fraction is larger than the second proper fraction, then the resulting quotient will be a whole number or mixed number.   **51.** H   **53.** $\frac{1}{16}$

Pages 179–180   **Lesson 3-4A**

1. 6^9 **3.** $\left(\frac{2}{3}\right)^9$ **5.** $-12c^8$

7 $\frac{4^7}{4^3} = 4^{7-3}$ The common base is 4.

$= 4^4$ Subtract the exponents.

9. $2t^3$ **11.** 5^{10} **13.** $\left(\frac{1}{5}\right)^5$ **15.** 7^8 **17.** n^3 **19.** $18j^4k^{13}$

21. $14h^{11}$ **23.** $24x^{12}$ **25.** $42p^{16}$ **27.** 8^5 **29.** c^3 **31.** x

33. $10n^5$ **35.** 4^5 or 1,024 fish

37 **a.** $\frac{10^{12}}{10^6} = 10^{12-6}$ The common base is 10. Division is used to find how many times greater one number is than another.

$= 10^6$ Subtract the exponents.

One trillion is 10^6 times greater than one million.

b. $\frac{10^{18}}{10^9} = 10^{18-9}$ The common base is 10. Division is used to find how many times greater one number is than another.

$= 10^9$ Subtract the exponents.

One quintillion is 10^9 times greater than one billion.

39. Equal; sample answer: Using the quotient of powers, $\frac{4^{200}}{4^{199}} = 4^{200-199}$, or 4^1, which is 4.   **41.** A

43. $42x^9$ ft^2 **45.** $6\frac{2}{3}$ **47.** 4   **49.** $-9\frac{2}{3}$ **51.** 8 ft

53. $0.38, \frac{7}{16}, 44\%$

Pages 183–184   **Lesson 3-4B**

1. $\frac{1}{5^2}$

3 $t^{-10} = \frac{1}{t^{10}}$ Definition of a negative exponent

5. 3^{-4} **7.** 7^{-2} **9.** h^2 **11.** r **13.** 10^{-6} **15.** $\frac{1}{5^3}$

17. $\frac{1}{(-3)^3}$ **19.** $\frac{1}{10^4}$ **21.** $\frac{1}{a^{10}}$ **23.** $\frac{1}{q^4}$ **25.** $\frac{1}{x^2}$

27. 5^{-5} **29.** k^{-2} **31.** 9^{-2} or 3^{-4} **33.** 2^{-4} or 4^{-2}

35. g^{-2} or $\frac{1}{g^2}$ **37.** $15v^{-5}$ or $\frac{15}{v^5}$ **39.** k^{-1} or $\frac{1}{k}$

41. $9c^{-4}$ or $\frac{9}{c^4}$

43 $100\text{cm} = 1\text{m}$

$\frac{100\text{cm}}{100} = \frac{1\text{m}}{100}$

$1\text{cm} = \frac{1}{100}\text{m}$

$\frac{1}{100} = \frac{1}{10^2}$ or 10^{-2}

45. 10^{-2} **47.** 10^{-5} **49.** $8^3, 8^0, 8^{-8}$; Sample answer: 8^{-8} is $\frac{1}{8}$ to the 8th power. This is a very small number between 0 and 1. 8^0 is 1 and 8^3 is 8 multiplied by itself 3 times.   **51.** Sample answer: A base of 10 raised to a negative exponent represents a number between 0 and 1.   **53.** H   **55.** B

57. x^6 **59.** $2n^7$ **61.** $2\frac{3}{10}$ T/mo

Pages 187–189   Lesson 3-4C

1. 375,400   **3.** 0.00015

5 $4,510,000 = 4.51 \times 1,000,000$ The decimal point moves 6 places.

$= 4.51 \times 10^6$ The exponent is positive.

7. 9.2×10^{-5} **9.** 3×10^5 **11.** >   **13.** 61,000
15. 0.33   **17.** 0.09014   **19.** 2,505   **21.** 0.001
23. 4.99×10^5 **25.** 6×10^{-3} **27.** 5×10^7
29. 7.8×10^{-5} **31.** 3×10^9 **33a.** golf   **33b.** golf
35. <   **37.** =   **39.** 3.2×10^{-8}

41 a. The Mount St. Helens eruption is one power of ten or 10 times greater than the Ngauruhoe eruption. **b.** Since the powers of ten for both eruptions are the same, we can compare the eruptions by comparing 4 and 2. The Hekla eruption was 2 times greater than the Ngauruhoe eruption.

43. 48,396

45a. $\dfrac{(4.2 \times 10^5)(1.5 \times 10^{-2})}{(2.5 \times 10^{-2})}$; 2.52×10^5; 252,000

45b. $\dfrac{(7.8 \times 10^{-2})(8.5 \times 10^0)}{(1.6 \times 10^{-1})(2.5 \times 10^5)}$; 1.6575×10^{-5}; 0.000016575

47. Sample answer: The distance between Earth and the Sun is 1.55×10^8 km or 155,000,000 km.

49. D   **51.** 5.21×10^4; 52,100   **53.** 9.2×10^{-2}; 0.092

Pages 190–193   Chapter Study Guide and Review

1. terminating   **3.** numerators   **5.** unlike
7. multiply   **9.** scientific notation   **11.** $\frac{7}{10}$ **13.** $\frac{1}{20}$
15. <   **17.** English   **19.** $\frac{2}{3}$ **21.** $1\frac{2}{9}$ **23.** 1   **25.** $3\frac{7}{15}$
27. $2\frac{4}{15}$ **29.** $7\frac{11}{12}$ h   **31.** $-2\frac{3}{5}$ **33.** $9\frac{3}{8}$ **35.** 45 cookies
37. -6 **39.** $\frac{3}{4}$ **41.** 3^1 or 3 times   **43.** $\frac{1}{3^5}$ **45.** 5^{-2}
47. x^{-4} **49.** 1.96×10^{-5}

Chapter 4 Equations and Inequalities

Page 200   Chapter 4   Are You Ready?

1. $p + 3$ **3.** $j + 10$ **5.** 17   **7.** 1   **9.** 29   **11.** 6   **13.** 15
15. 50   **17.** 38 years old

Pages 202–203   Lesson 4-1A

1. when you are given the final result and asked to find an earlier amount   **3.** Sample answer: In the first four games, Hannah scored a total of 83 points. In the fourth game, she scored 19 points. In game three, she scored 27 points and in the second game, she scored 22 points. How many points did she score in the first game? To solve, first subtract 19 from 83, which is 64. Then subtract 27 from 64 to get 37. Finally, subtract 22 from 37. So, Hannah scored 15 points in her first game.

5 Work backward. Start with -25 and reverse all operations. Subtract -7 to get -18. Add 6 to get -12. Divide by -3. The original number is 4.

7. 19,200 tennis balls   **9.** Brie is 14 years old.

11.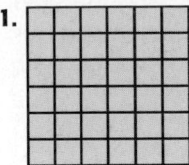

13. Raquel's car gets 1,774,074 more inches per gallon than an aircraft carrier.   **15.** 1 ten-dollar bill, 2 five-dollar bills, and 7 one-dollar bills

Pages 211–213   Lesson 4-1D

1. 2   **3.** -2
5. $n + 120 = 364$; 244 ft

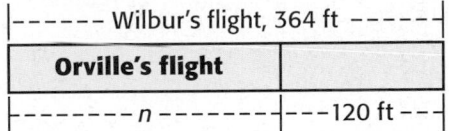

7. 5   **9.** 7   **11.** 7   **13.** -3

15 $\begin{aligned} r + 6 &= -3 \\ -6 &= -6 \\ \hline r &= -9 \end{aligned}$

17. 17   **19.** 7   **21.** $7 = w + 2$; 5 h   **23.** $15 = t - 3$; 18 years old   **25.** 18.4   **27.** 6.4   **29.** -0.68 **31.** $\frac{1}{12}$
33. $\frac{7}{18}$ **35.** $-\frac{1}{12}$ **37.** $d - 5 = 18$; \$23
39. $35 + 45 + x = 180$; 100°

41 $\begin{aligned} -1 + -3 + s + 2 &= 0 \\ -2 + s &= 0 \\ +2 \quad\quad &= +2 \\ \hline s &= +2 \end{aligned}$

43a. $s - 65 = 13$; 78 mph   **43b.** $d + 52 = 176$; 124 ft

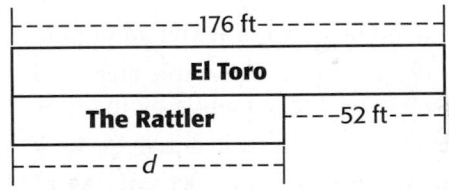

43c. The solution of each equation is 170; Colossos is 170 feet tall.   **45.** The value of y decreases by 2.
47. Sample answer: An office building is 50 stories tall. Its height is 25 stories less than an apartment building. What is the height of the apartment building?   **49.** I   **51.** 7

Pages 218–219   Lesson 4-2B

1. 3   **3.** -3 **5.** 81   **7.** $6h = 48$; 8 h   **9.** 7   **11.** -3
13. -5

15 $\begin{aligned} \frac{m}{10} &= 7 \\ 10 \cdot \frac{m}{10} &= 10 \cdot 7 \\ m &= 70 \end{aligned}$

17. -36 **19.** 80   **21.** $15w = 300$; 20 weeks

23 $\begin{aligned} 205 &= \frac{d}{3} \\ 3 \cdot 205 &= 3 \cdot \frac{d}{3} \\ 615 \text{ miles} &= d \end{aligned}$

25. $20.88h = 145$; ≈ 6.94 h   **27.** Sample answer: Raul divided by $+6$ and should have divided by -6; $x = -12$.   **29.** True; Sample answer: Multiply each side of the equation by $\frac{1}{5}$ instead of dividing each side by 5.   **31.** Sample answer: Three coins of the same value are worth 75 cents. How much is each coin worth?   **33.** C   **35a.** $16.50x = 99$

35b. 6 months; Sample answer: To find x, divide each side by 16.50.

Pages 224–226   Lesson 4-2D

1. $\frac{5}{8}$ **3.** $\frac{5}{29}$ **5.** 2   **7.** -8.2 **9.** $\frac{4}{5}$ **11.** $\frac{3}{4}n = 24$;

32 pieces of fruit   **13.** $\frac{6}{5}$ or $1\frac{1}{5}$ **15.** $\frac{6}{1}$ or 6   **17.** $\frac{1}{3}$

19. $\frac{8}{41}$ **21.** 5   **23.** -6 **25.** -3 **27.** $\frac{6}{5}$ or $1\frac{1}{5}$

29
$$\frac{7}{8}k = \frac{5}{6}$$
$$\left(\frac{8}{7}\right)\frac{7}{8}k = \left(\frac{8}{7}\right)\frac{5}{6}$$
$$k = \frac{20}{21}$$

31. $6\frac{2}{3}$ **33.** $0.5t = 3$; 6 months   **35.** $140 = \frac{7}{15}x$; 300 ft

37 If one serving is $\frac{3}{4}$ cup, you can find how many servings are in $16\frac{1}{2}$ cups by writing an equation where x is the number of servings.
$$\frac{3}{4}x = 16\frac{1}{2}$$
$$\frac{3}{4}x = \frac{33}{2}$$
$$\left(\frac{4}{3}\right)\frac{3}{4}x = \left(\frac{4}{3}\right)\frac{33}{2}$$
$$x = 22$$
There are 22 servings in $16\frac{1}{2}$ cups.

39. $1.75m = 3.5$; 2 movies   **41.** $\frac{3}{5}$, 5; The other pairs of numbers are reciprocals.   **43.** Sample answer: If you multiply each side of an equation by the same nonzero number, two sides remain equal; $\frac{2}{5}x = 7$.

45. I   **47.** 16 hours   **49.** 5   **51.** 450   **53.** -8 **55.** 4
57. -65

Pages 232–234   Lesson 4-3B

1. 2

3
$$-6r + 1 = -17$$
$$\underline{\quad -1 = \quad -1}$$
$$-6r = -18$$
$$\frac{-6r}{-6} = \frac{-18}{-6}$$
$$r = 3$$

5. 3   **7.** $14c + 23 = 65$; 3 CDs   **9.** 3   **11.** -4 **13.** 4
15. 9   **17.** 8   **19.** 24   **21.** $2c + 10 = 14$; 2 cups
23. 2.25   **25.** 2.1   **27.** 11

29 a.
$$F = 1.8C + 32 \qquad \text{Write the equation.}$$
$$16 = 1.8C + 32 \qquad \text{Substitute 16 for } F.$$
$$16 - 32 = 1.8C + 32 - 32 \qquad \text{Subtract 32.}$$
$$\frac{-16}{1.8} = \frac{1.8C}{1.8} \qquad \text{Divide each side by 1.8.}$$
$$-8.9 \approx C \qquad \text{Simplify.}$$
$$-9°C \qquad \text{Round to the nearest degree.}$$

b.
$$F = 1.8C + 32 \qquad \text{Write the equation.}$$
$$F = 1.8(-11) + 32 \qquad \text{Substitute } -11 \text{ for } C.$$

$$F = -19.8 + 32 \qquad \text{Multiply 1.8 and } -11.$$
$$F = 12.2 \qquad \text{Simplify.}$$
Subtract Alaska's record low temperature from Hawaii's record low temperature.
$$12.2 - (-80) = 92.2°F$$

31. 22 subscriptions   **33.** D   **35.** $1\frac{1}{2}$ **37.** 20
39. $3.25n = 22.75$; 7 E-cards   **41.** -29 **43.** -24

Pages 238–239   Lesson 4-3D

1. Sample answer: $2x + 8 = 40$; subtracted $2x$ from both sides   **3.** Sample answer: $5x - 7 = 43$; subtracted x from both sides   **5.** 52

7
$$3 - x = 4 - 3x$$
$$\underline{+ 3x = \quad + 3x}$$
$$3 + 2x = 4$$
$$\underline{-3 \qquad = -3}$$
$$2x = 1$$
$$\frac{2x}{2} = \frac{1}{2}$$
$$x = \frac{1}{2}$$

9. 6.9   **11.** Sample answer: $2x + 14 = 20$; subtracted $4x$ from both sides   **13.** Sample answer: $3x = 36$; subtracted $3x$ from both sides and added 3 to both sides   **15.** Sample answer: $2x = 20$; subtracted $2x$ from both sides and added 15 to both sides   **17.** 6
19. 24   **21.** 4.2   **23.** -10 **25.** $\frac{1}{4}$ **27.** -6

29 Manny bought supplies for \$48 and 3 buckets at a certain price, b. Jin bought 7 buckets at a certain price, b. Manny and Jin each spent the same amount of money, so set the expressions equal to each other. Then solve.
$$3b + 48 = 7b$$
$$\underline{- 3b \qquad - 3b}$$
$$48 = 4b$$
$$12 = b$$
The cost of one bucket is \$12.

31. Adding $2b$ to both sides and then subtracting 14 from both sides of the first equation will result in the second equation.   **33.** -10 **35.** B
37a. $15c = 12c + 0.99$ **37b.** 0.33; A baseball card costs 33 cents.   **37c.** Sergio could have bought 9 baseball cards if he bought 2 packs of gum.

39. -3 **41.** $-\frac{3}{5}$ **43.** 8 guests

Pages 246–248   Lesson 4-4B

1. $c < 4$ **3.** $y < 20$

5 $c + 4 \geq 17$ Write the inequality.
$\underline{\quad -4 \quad -4}$ Subtract 4 from each side.
$\quad c \geq 13$ Simplify.

7. $h \geq 6$

9. $-16 \leq n$

11. $5 < n$ **13.** $y \geq -7$ **15.** $x > -1$

17. $-18 > t$

$-20\ -19\ -18\ -17\ -16\ -15\ -14\ -13\ -12\ -11\ -10$

19. $-5 < a$

$-9\ -8\ -7\ -6\ -5\ -4\ -3\ -2\ -1\ \ 0\ \ 1$

21. $n \leq \frac{1}{2}$

$-2\frac{1}{2}\ -2\ -1\frac{1}{2}\ -1\ -\frac{1}{2}\ \ 0\ \ \frac{1}{2}\ \ 1\ \ 1\frac{1}{2}\ \ 2\ \ 2\frac{1}{2}$

23. $n + 4 > 13; n > 9$ **25.** $n - 8 < 10; n < 18$
27. $p + 17 \leq 26; p \leq 9$

(29) $785 + m \leq 1,500$ Write the inequality.
$\underline{\ -785\qquad\ -785\ }$ Subtract 785 from each side.
$\quad\quad m \leq\ \ 715$ Simplify.
Lalo has 715 minutes remaining.
31a. $42 + x \geq 74; x \geq 32$ **31b.** $74 + y \geq 110; y \geq 36$
33. Sample answer: $x + 3 < 25$ **35.** C   **37.** C
39. $25x = 18x + 21; 3$ appetizers   **41.** -2 **43.** 7

Pages 252–253   Lesson 4-4C

1. $x > 3$ **3.** $h \geq 7$ **5.** $y < -4$

(7) $\frac{t}{-4} < -11$ Write the inequality.

$-4\left(\frac{t}{-4}\right) > -4(-11)$ Multiply each side by -4 and reverse the symbol.

$\quad\quad t > 44$ Simplify.

9. $y \leq -1$

$-5\ -4\ -3\ -2\ -1\ \ 0\ \ 1\ \ 2\ \ 3\ \ 4\ \ 5$

11. $n \leq -0.5$

$-2.5\ -2\ -1.5\ -1\ -0.5\ \ 0\ \ 0.5\ \ 1\ \ 1.5\ \ 2\ \ 2.5$

13. $4x > 100; x > 25$ **15.** $x \geq 9$ **17.** $4 < t$ **19.** $h > 81$
21. $y > -5$ **23.** $n < 2$ **25.** $t > -12$ **27.** $s < 96$
29. $x < 4$

$0\ \ 1\ \ 2\ \ 3\ \ 4\ \ 5\ \ 6\ \ 7\ \ 8\ \ 9\ \ 10\ \ 11$

31. $y \geq 2$

$-5\ -4\ -3\ -2\ -1\ \ 0\ \ 1\ \ 2\ \ 3\ \ 4\ \ 5$

33. $t \leq -25$

$-27\quad -25\quad -23\quad -21\quad -19$

35. $0.5x > 15; x > 30$

(37) $5n - 7 < -52$ Write the inequality. The word *times* means multiply; the word *decrease* means subtract.

$\underline{\ +7\qquad +7\ }$ Add 7 to each side.

$\quad 5n < -45$ Simplify.

$\quad \dfrac{5n}{5} < -\dfrac{45}{5}$ Divide each side by 5.

$\quad\quad n < -9$ Simplify.

39. Sample answer: $7y > 42$ and $\frac{1}{2}y > 3$

41. Sample answer: When you are not multiplying or dividing by a negative number.   **43.** G
45. $k > 29$ **47.** $x < -33$ **49.** 4 times

Pages 254–257   Chapter Study Guide and Review

1. true   **3.** true   **5.** false; less than   **7.** false;
subtract 3 from each side   **9.** true   **11.** false; $\frac{3}{2}$
13. 26   **15.** 3   **17.** -13 **19.** -2 **21.** $\frac{1}{4}$
23. $c - 6 = 18; 24$ cookies   **25.** 4   **27.** -9 **29.** 12
31. $14w = 98; 7$ weeks   **33.** 1.5   **35.** 20   **37.** -5
39. 6   **41.** -1 **43.** $6d + 5 = 155; 25$ DVDs   **45.** 1
47. 12 rentals   **49.** $n \leq -2$ **51.** $g > 3\frac{2}{3}$ **53.** $p < 18$
55. $d < -7$ **57.** $6x < 18.75; 3$ packs

Chapter 5 Proportions and Similarity

Page 264   Chapter 5   Are You Ready?

1. $\frac{2}{15}$ **3.** $\frac{1}{51}$ **5.** $\frac{15}{17}$ **7.** no; $\frac{2}{8} \neq \frac{8}{14}$ **9.** no; $\frac{2}{7} \neq \frac{10}{15}$

Pages 268–271   Lesson 5-1B

1. 6 mi per gal

(3) 5 pounds for $\$2.49 = \dfrac{\$2.49}{5 \text{ lb}}$ Write the rate as a fraction.

$= \dfrac{\$2.49 \div 5}{5 \text{ lb} \div 5}$ Divide the numerator and the denominator by 8.

$= \dfrac{\$0.498}{1 \text{ lb}}$ Simplify.

$\$0.498$ rounds to $\$0.50$. So, the cost is $\$0.50$ per pound.
5. Music Place   **7.** 60 mi/h   **9.** 30 people per class

(11) 45.5 meters in 13 seconds $= \dfrac{45.5 \text{ m}}{13 \text{ s}}$

Write the rate as a fraction.

$= \dfrac{45.5 \div 13}{13 \div 13}$ Divide the numerator and the denominator by 13.

$= \dfrac{3.5 \text{ m}}{1 \text{ s}}$ Simplify.

3.5 meters per second
13. $\$0.14/\text{oz}$ **15.** Sample answer: about $\$0.50$ per pair   **17.** Susana; 1.78 m/s $>$ 1.66 m/s $>$ 1.23 m/s   **19a.** Soft drink C; Soft drinks A and B have about 3 milligrams of sodium per ounce, and soft drink C has 6 mg per ounce.   **19b.** Soft drink A; 1.83 g/oz $<$ 1.88 g/oz $<$ 4.29 g/oz   **21.** $\$4.98$
23. $\$130.50$ **25.** Sample answer: $\$1.25$ per qt; $\$2.49 \div 2 \approx \$2.50 \div 2$ or $\$1.25$ **27.** Sample answer: $\$0.06$ per oz; $\$1.13 \div 20 \approx \$1.20 \div 20$ or $\$0.06$

(29) a. 1 hr, 18 minutes, 27 seconds is $(1 + 18 \div 60 + 27 \div 3600)$ hr or 1.3075 hr.

26.2 miles in 1.3075 hours $= \dfrac{26.2 \text{ miles}}{1.3075 \text{ hr}}$

Write the rate as a fraction.

$= \dfrac{26.2 \div 1.3075}{1.3075 \div 1.3075}$ Divide the numerator and the denominator by 1.3075.

$= \dfrac{20.04}{1}$ Simplify.

The average speed was 20.04 mph.   **b.** Divide the distance by the average speed to find the time.

$$\frac{30 \text{ miles}}{20.04 \text{ mph}} \approx 1.497 \text{ hours or 1 hr 29 min 49 s}$$

31a. The bear's heart beats 120 times in 2 minutes when it is active.   **31b.** The bear's heart beats 18 times in 1.5 minutes when it is hibernating.   **31c.** the bear's heart rate in beats per minute   **31d.** active: 60 beats per minute; hibernating: 12 beats per minute   **31e.** when it is active; Sample answer: The active line increases faster than the hibernating line when read from left to right.   **35.** Sometimes; a ratio that compares two measurements with different units is a rate, such as $\frac{2 \text{ miles}}{10 \text{ minutes}}$. A ratio that compares two numbers or two measurements with like units is not a rate, such as $\frac{2 \text{ cups}}{3 \text{ cups}}$.   **37.** Sample answer: The rate 55 miles per hour is a measure of the number of miles traveled per unit hour.   **39.** H   **41.** H

Pages 273–275   Lesson 5-1C

1 Make a table and compare the values. The ratios can be simplified to 225 liters. So the rates are proportional.; Sample answer:

Time (days)	1	2	3	4
Water (L)	225	450	675	900

The time to water ratio for 1, 2, 3, and 4 days is $\frac{1}{225}$, $\frac{2}{450}$ or $\frac{1}{225}$, $\frac{3}{675}$ or $\frac{1}{225}$, and $\frac{4}{900}$ or $\frac{1}{225}$. Since these ratios are all equal to $\frac{1}{225}$, the number of days the supply lasts is proportional to the amount of water the elephant drinks.

3. no; Sample answer:

Number of Teachers	4	5	6	7
Number of Students	28	56	84	112

The ratio of students to teachers for 4, 5, 6, and 7 teachers is $\frac{28}{4}$ or 7, $\frac{56}{5}$ or 11.2, $\frac{84}{6}$ or 14, and $\frac{112}{7}$ or 16. Since these ratios are not all equal, the number of students at the school is not proportional to the number of teachers.

5. no; Sample answer:

Rental Time (h)	1	2	3	4
Cost ($)	37	62	87	112

The cost to time ratio for 1, 2, 3, and 4 hours is $\frac{37}{1}$ or 37, $\frac{62}{2}$ or 31, $\frac{87}{3}$ or 29, and $\frac{112}{4}$ or 28. Since these ratios are not all equal, the cost of a rental is not proportional to the number of hours you rent the boat.

7. yes; Sample answer:

Time (days)	5	10	15	20
Length (in.)	7.5	15	22.5	30

The length to time ratio for 5, 10, 15, and 20 days is $\frac{7.5}{5}$ or 1.5, $\frac{15}{10}$ or 1.5, $\frac{22.5}{15}$ or 1.5, and $\frac{30}{20}$ or 1.5. Since these ratios are all equal to 1.5 ft per day, the length of vine is proportional to the number of days of growth.

9a. yes; Sample answer:

Number of Hours Worked on Sunday	1	2	3	4
Number of Coupons Given Away on Sunday	52	104	156	208

The coupons to hours ratios for 1, 2, 3, and 4 hours of work on Sunday are $\frac{52}{1}$ or 52, $\frac{104}{2}$ or 52, $\frac{156}{3}$ or 52, and $\frac{208}{4}$ or 52. Since these ratios are all equal to 52 coupons per hour, the number of coupons given away is proportional to the number of hours worked on Sunday.   **9b.** no; Sample answer:

Number of Hours Worked on Sunday	1	2	3	4
Total Number of Coupons Given Away that Weekend	468	520	572	624

The coupons to hours ratios for 1, 2, 3, and 4 hours of work on Sunday are $\frac{468}{1}$ or 468, $\frac{520}{2}$ or 260, $\frac{572}{3}$ or about 190, and $\frac{624}{4}$ or 156. Since these ratios are not all equal, the total number of coupons given away is not proportional to the number of hours worked on Sunday.

11 **a.** Make a table and compare the side length to the perimeter. The measures are proportional to the perimeter. Sample answer:

Side length (units)	1	2	3	4
Perimeter (units)	4	8	12	16

The side length to perimeter ratio for side lengths of 1, 2, 3, and 4 units is $\frac{1}{4}$, $\frac{2}{8}$ or $\frac{1}{4}$, $\frac{3}{12}$ or $\frac{1}{4}$, and $\frac{4}{16}$ or $\frac{1}{4}$. Since these ratios are all equal to $\frac{1}{4}$, the measure of the side length of a square is proportional to the square's perimeter.   **b.** Make a table and compare the side length to the area. The measures are not proportional to the area. Sample answer:

Side length (units)	1	2	3	4
Area (units^2)	1	4	9	16

The side length to area ratio for side lengths of 1, 2, 3, and 4 units is $\frac{1}{1}$ or 1, $\frac{2}{4}$ or $\frac{1}{2}$, $\frac{3}{9}$ or $\frac{1}{3}$, and $\frac{4}{16}$ or $\frac{1}{4}$. Since these ratios are not all equal, the measure of the side length of a square is not proportional to the square's area.

13. It is not proportional because the ratio of laps : time is not consistent; $\frac{4}{1} \neq \frac{6}{2} \neq \frac{8}{3} \neq \frac{10}{4}$.

15. always; Sample answer: All the cars would have 4 tires.   **17.** D   **19.** G

Pages 278–280   Lesson 5-1D

1 $\dfrac{1.5}{6} = \dfrac{10}{p}$

$1.5p = 10(6)$ Find the cross products.
$1.5p = 60$ Multiply.
$\dfrac{1.5p}{1.5} = \dfrac{60}{1.5}$ Divide each side by 1.5.
$p = 40$ Simplify.

3. 16.4   **5.** $m = 9.5h$; \$19; \$42.75   **7.** 6   **9.** 3.5   **11.** 3.75
13. 13.5   **15.** $\dfrac{14}{483} = \dfrac{x}{600}$; about 17.4 gal   **17.** $\dfrac{4}{5} = \dfrac{x}{30}$; 24 people

19 shoulder width → $\dfrac{16.2}{64} = \dfrac{18.5}{h}$ ← shoulder width
height → ← height

$16.2h = 18.5(64)$ Find the cross products.
$16.2h = 1,184$ Multiply.
$\dfrac{16.2h}{16.2} = \dfrac{1,184}{16.2}$ Divide each side by 16.2.
$h \approx 73$ inches   Simplify.

21a. 81.9   **21b.** 144.7   **21c.** 110.2   **21d.** 48.0
23. 256 c; Sample answer: The ratio of cups of mix to cups of water is 1 : 8, which means that the proportion $\dfrac{1}{8} = \dfrac{32}{x}$ is true and can be solved.   **25.** 18
27. Sample answer: By writing an equation to represent a proportional relationship, you can use this equation to find any other similar quantities. The result is one calculation involving multiplication, rather than the two calculations that would result by writing and solving a new proportion.   **29.** G
31. Yes; the constant rate of change is $\dfrac{15}{1}$ or \$15 per hour.   **33.** 500 kB/min   **35.** 7.2 m/s

Pages 282–283   Lesson 5-2A   PSI

1. Sample answer: It helps to visualize the parts in relation to the whole.   **3.** 1,536 tiles
5 If Mr. Sanchez can only change the width of his flower bed, then the additional 12 meters needs to be split up between the two sides that are the widths. Therefore, add half of 12, or 6 meters to the width. This changes the width to 11 meters.
7. 16   **9.** 100 lb   **11.**

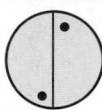

Pages 287–290   Lesson 5-2B

1 Use a centimeter ruler to find the distance between the two cities. The distance between the two cities on the map is 2 centimeters. Write and solve a proportion using the scale. Let d represent the actual distance between the cities.

$\dfrac{\text{map} \to 1\ \text{cm}}{\text{actual} \to 25\ \text{km}} = \dfrac{2\ \text{cm} \leftarrow \text{map}}{d\ \text{km} \leftarrow \text{actual}}$
$1 \times d = 2 \times 25$
$d = 50$

So, the actual distance between Carlsbad and Artesia is 50 kilometers.
3. 130 km   **5a.** $16\frac{2}{3}$ in.   **5b.** $1\frac{1}{3}$ in.   **7.** $\frac{2}{3}$
9. about 10.3 inches wide   **11.** 81 mi   **13.** 37.8 mi

15 $\dfrac{0.5\ \text{cm}}{1.5\ \text{m}} = \dfrac{x}{36\ \text{m}}$
$0.5 \cdot 36 = 1.5 \cdot x$
$18 = 1.5x$
$\dfrac{18}{1.5} = \dfrac{1.5x}{1.5}$
$12 = x$

The model is 12 cm tall.
To find the scale factor, first convert cm to m: 0.5 cm = 0.005 m. Next make a ratio and simplify.
$\dfrac{0.005}{1.5} = \dfrac{5}{1,500}$ or $\dfrac{1}{300}$
17. $11\frac{3}{4}$ in.; $\frac{2}{1}$ **19.** $109\frac{3}{8}$ ft   **21a.** $\frac{1}{3}$

21b.

Height of Thomas Jefferson (feet)	1	2	3	4	5	6
Height of Statue (feet)	3	6	9	12	15	18

21c. $3x$ **21d.** about $6\frac{1}{3}$ ft or 6 ft 4 in.
23 **a.** Count to find that it is 4 units from the lake to the cabin.
b. To find the actual distance, use the key that states 1 unit = 75 yards to set up a proportion. Then solve the proportion.
$\dfrac{1\ \text{unit}}{75\ \text{yards}} = \dfrac{4\ \text{units}}{x\ \text{yards}}$ Set up the proportion.
$1 \cdot x = 75 \cdot 4$ Find the cross products.
$x = 300$ Simplify.
The children will travel 300 yards from the lake to the cabin.
25. Sample answer: The scale is the ratio comparing the measurements, including the units, on the model to the measurements on the actual figure. Once the units have been converted to the same unit, the scale factor is the ratio written without units as a fraction in simplest form. For example, if the scale of the model to the actual figure is 1 in. = 4.5 ft, then the scale factor would be $\dfrac{1}{4.5 \times 12}$ or $\dfrac{1}{54}$.
27. D   **29.** F   **31.** 5 ft   **33.** 25   **35.** 63   **37.** −5

Pages 296–299   Lesson 5-3A

1. rectangle $PQRS$ **3.** 45 mm
5 Find which triangle is similar to $\triangle FGH$.

Triangle PMN Triangle CAB
$\dfrac{MP}{GF} = \dfrac{1}{2}$ $\dfrac{AC}{GF} = \dfrac{4}{2}$ or 2
$\dfrac{PN}{FH} = \dfrac{4}{5}$ $\dfrac{CB}{FH} = \dfrac{10}{5}$ or 2
not similar   similar
Triangle CAB is similar to triangle FGH.
7. 25 mi   **9.** $\dfrac{3}{x} = \dfrac{4.5}{18}$; 12 ft
11 **a.** If the kids follow the path to the cabin, to the mess hall, and finally to the campfire, they will

walk 17 units. We are given the key that 1 square =
75 yards. Write a proportion to find the total
distance walked.

$$\begin{array}{l}\text{units} \rightarrow \\ \text{yards} \rightarrow\end{array} \dfrac{1}{75} = \dfrac{17}{x} \begin{array}{l}\leftarrow \text{units} \\ \leftarrow \text{yards}\end{array}$$

$1 \cdot x = 17 \cdot 75$ Find cross products.

$x = 1{,}275$ Simplify.

The kids walk 1,275 yards from the time they leave
the lake until they get to the campfire.

b. By the time they return to the cabin, they will
have walked a total of 30 units.

$$\begin{array}{l}\text{units} \rightarrow \\ \text{yards} \rightarrow\end{array} \dfrac{1}{75} = \dfrac{30}{x} \begin{array}{l}\leftarrow \text{units} \\ \leftarrow \text{yards}\end{array}$$

$1 \cdot x = 30 \cdot 75$ Find cross products.

$x = 2{,}250$ Simplify.

The kids walk 2,250 yards total from the time they
leave the lake until they return to the cabin.
13a. true; Sample answer: All squares have 4
90° angles and equal sides. Therefore, all squares
are proportional to other squares. **13b.** false;
Sample answer: Rectangles have different
lengths than widths that may not be proportional
to another rectangle. **15.** 20 cm **17.** yes; Sample
answer: The ratios for the corresponding sides are
$\dfrac{3}{4.5}, \dfrac{5}{7.5},$ and $\dfrac{2.8}{4.2}$. All three ratios simplify to $\dfrac{2}{3}$. Since
the ratios are all equal, the corresponding sides are
proportional and the triangles are similar. **19.** 4; 12
21. Sample answer: When two triangles have equal
corresponding angles formed by proportional
corresponding sides, the triangles are similar.

Pages 301–302   Lesson 5-3B

1. $x = 38\left(\dfrac{18}{12}\right)$ Multiply by the scale factor.

$x = \dfrac{\overset{19}{\cancel{38}}}{1}\left(\dfrac{18}{\underset{6}{\cancel{12}}}\right)$ Cancel common factors.

$x = 57$ Simplify.

The perimeter is 57 millimeters.
3. 78.75 in² **5.** 14.55 in. **7.** 49 ft²

9. $x = 1{,}134\left(\dfrac{1}{3}\right)^2$ Multiply the square of the scale factor.

$x = 1{,}134\left(\dfrac{1}{9}\right)$ Evaluate the power.

$x = 126$ Multiply.

The area of the putting green will be 126 square feet.
11. Sample answer:

$P = 30$ m $P = 36$ m
$A = 50$ m² $A = 72$ m²

13. Robert is thinking of size in terms of area and
Denise is thinking of size in terms of perimeter.
15. H **17.** 27 yd **19.** 1 in. = 60 mi

Pages 306–309   Chapter Study Guide and Review

1. proportional **3.** proportion **5.** unit rate **7.** scale
9. nonproportional **11.** 90 mi per day

13. No; the ratios are not equal. **15.** 4 **17.** 3.5
19. 12.5 **21.** 3.2 mi **23.** 25 minutes **25.** 21 km
27. 1 in. = 24 in. or 1 in. = 2 ft **29.** *RUTS* **31.** 4.2 ft
33. 150 yd²

Chapter 6 Percent

Page 316   Chapter 6   Are You Ready?

1. 48 **3.** 1,512 **5.** $54.75 **7.** 0.17 **9.** 1.57 **11.** 0.085
13. 0.075 **15.** 8% **17.** 580% **19.** 72.5%

Pages 322–324   Lesson 6-1B

1. 4 **3.** 110.5 **5.** 23 **7.** $3.25 **9.** $194.40 **11.** 45.9
13. 14.7

15 Method 1 Write the percent as a fraction.

$175\% = \dfrac{175}{100}$ or $\dfrac{7}{4}$

$\dfrac{7}{4}$ of $10 = \dfrac{7}{4} \times 10$

$= \dfrac{7}{4} \times \dfrac{10}{1}$ or $17\dfrac{1}{2}$

Method 2 Write the percent as a decimal.

$175\% = \dfrac{175}{100}$ or 1.75

1.75 of $10 = 1.75 \times 10 = 17.5$
17. 62.5 **19.** $290 **21.** 3.5 **23.** 97.8 **25.** 92.5
27. about 19.5 million **29.** 3.3 **31.** 990 **33.** 520
35. 0.24 **37.** $241.50 **39.** $297 **41.** about 30
43. about 15

45 There are 20 answers on the test. If 25% of the
answers are choice B, there are 0.25×20 or
5 answers that are choice B. Subtracting these 5
from the total number of 20 answers leaves
15 answers that are not choice B.
47a. 66.7% **47b.** 33.3% **47c.** 50% **47d.** 33.3%
51. C **53.** I **55.** 124

Pages 327–330   Lesson 6-1C

1. Sample answer: 5; $\dfrac{1}{2} \cdot 10 = 5$; $0.1 \cdot 10 = 1$ and
$5 \cdot 1 = 5$

3 Method 1: Use a fraction to estimate.
38% is about 40% or $\dfrac{2}{5}$.
38% of $62 \approx \dfrac{2}{5} \cdot 60$ or 24
Method 2: Use 10% to estimate.
10% of 60 is 6.
38% is about $4 \cdot 10\%$.
$4 \cdot 6 = 24$
So, 38% of 62 is about 24.
5. Sample answer: 105; $(1 \cdot 70) + \left(\dfrac{1}{2} \cdot 70\right) = 105$

7. Sample answer: about 48 teenagers; $\dfrac{3}{5} \cdot 80 = 48$;
$0.1 \cdot 80 = 8$ and $8 \times 6 = 48$ **9.** Sample answer:
35; $\dfrac{1}{2} \cdot 70 = 35$; $0.1 \cdot 70 = 7$ and $5 \cdot 7 = 35$

11 Method 1 Use a fraction to estimate.
21% is about 20%, or $\dfrac{1}{5}$.
20% of $90 = \dfrac{1}{5} \times 90$ or 18.
So, 21% of 90 is about 18.

Method 2 Use 10% to estimate.
10% of 90 is 0.1×90 or 9.
21% is about $2 \times 10\%$.
$2 \times 9 = 18$.
So, 21% of 90 is about 18.

13. Sample answer: 18; $\frac{3}{5} \cdot 30 = 18$; $0.1 \cdot 30 = 3$ and $6 \cdot 3 = 18$ **15.** Sample answer: 12.5; $\frac{1}{4} \cdot 50 = 12.5$; $0.1 \cdot 50 = 5$ and $2.5 \cdot 5 = 12.5$ **17.** Sample answer: 180; $\frac{9}{10} \cdot 200 = 180$; $0.1 \cdot 200 = 20$ and $9 \cdot 20 = 180$ **19.** Sample answer: 100; $\frac{2}{3} \cdot 150 = 100$; $0.1 \cdot 150 = 15$ and $6.6 \cdot 15 \approx 100$ **21.** Sample answer: about \$6; $\frac{3}{20} \cdot \$40 = \$6$ **23.** Sample answer: $(1 \cdot 50) + \left(\frac{3}{10} \cdot 50\right) = 65$ **25.** Sample answer: $0.01 \cdot 400 = 4$ and $\frac{1}{2} \cdot 4 = 2$ **27.** Sample answer: $0.01 \cdot 500 = 5$ and $\frac{2}{5} \cdot 5 = 2$ **29.** Sample answer: about 96 mi; $0.01 \cdot 12{,}000 = 120$ and $\frac{4}{5} \cdot 120 = 96$ **31.** Sample answer: $\frac{2}{3} \cdot 9 = 6$ **33.** Sample answer: $\frac{1}{3} \cdot 90 = 30$ **35.** Sample answer: $\frac{1}{5} \cdot 100 = 20$

37 **a.** Estimate 13% of 24 hours.
13% is about 10% or $\frac{1}{10}$.
13% of $24 \approx \frac{1}{10} \cdot 24$ or 2.4
So, 13% of 24 is about 2.4. Avery spends about 2.4 hours on homework.
b. Estimate 33% of 24.
33% is about $\frac{1}{3}$.
33% of $24 \approx \frac{1}{3} \cdot 24$ or 8
So, 33% of 24 is about 8. Avery spends about 8 hours sleeping.
Estimate 19% of 24.
19% is about 20% or $\frac{1}{5}$. 24 is about 25.
19% of $24 \approx \frac{1}{5} \cdot 25$ or 5
So, 19% of 24 is about 5. Avery spends about 5 hours doing the activities in the "other" category. Avery spends $8 - 5$ or 3 hours more sleeping than doing the activities in the "other" category.
c. There are $24 \cdot 60$ or 1,440 minutes in one day. Estimate 8% of 1,440.
8% is about 10% or $\frac{1}{10}$.
8% of $1{,}440 \approx \frac{1}{10} \cdot 1{,}440$ or 144
So, 8% of 1,440 is about 144. Avery spends about 144 minutes each day on extracurricular activities.
39a. Sample answer:

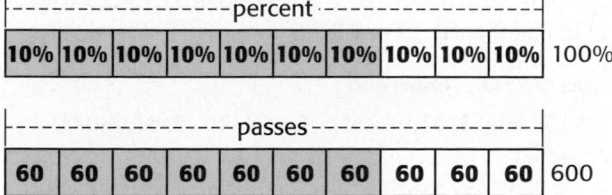

39b. 420 **39c.** greater; Both the number of passes and the percent were rounded up.
39d. Peyton Manning; Sample answer: 65% of 515 must be greater than 65% of 404.

41a. Sample answer: about 260 canned foods; $200 + 0.3 \cdot 200$ **41b.** Sample answer: about 780 canned foods; $600 + 0.3 \cdot 600$ **43a.** Sample answer: \$75; $(100\% \text{ of } 60) + (25\% \text{ of } 60) = 60 + 15 = 75$
43b. Sample answer: less than; 125% < 130%
45. He incorrectly changed 1.5% to 1.5, which is 150%. It should be about 3. **47.** Sample answer: One way to find 22% of 136 is to find $\frac{1}{5} \cdot 140 = 28$. Another way to find 22% of 136 is to first find $(0.1 \cdot 140)$ and then multiply by 2. The result is 28.
49. H

Pages 335–336 **Lesson 6-2B**

1. 36% **3.** 0.7

5
$$\frac{9}{w} = \frac{12}{100} \quad \text{Write the proportion.}$$
$$9 \cdot 100 = 12 \cdot w \quad \text{Find the cross products.}$$
$$900 = 12w \quad \text{Simplify.}$$
$$\frac{900}{12} = \frac{12w}{12} \quad \text{Divide each side by 12.}$$
$$75 = w$$
So, 9 is 12% of 75.

7. 3 c **9.** 7.5% **11.** 8.6 **13.** 375 **15.** 24 points
17. 40% **19.** 4.1 **21.** 192

23
$$\frac{0.6}{300} = \frac{n}{100} \quad \text{Write the proportion.}$$
$$0.6 \cdot 100 = 300 \cdot n \quad \text{Find the cross products.}$$
$$60 = 300n \quad \text{Simplify.}$$
$$\frac{60}{300} = \frac{300n}{300} \quad \text{Divide each side by 300.}$$
$$0.2 = n$$
So, 0.6 is 0.2% of 300.

25. 75 students; 70 of the students bought their lunch. Some amount, x, left the cafeteria. The new percent proportion would be $\frac{70 - x}{100 - x} = \frac{60}{100}$. So, $x = 25$. $100 - 25 = 75$ **27.** Sample answer: A runner won 15% of the races he ran. If he won 3 races, how many races did he run? **29.** F **31.** 30
33a. 16% **33b.** 24% **33c.** 60%

Pages 339–341 **Lesson 6-2C**

1. $p = 0.88 \cdot 300$; 264

3 75 is what percent of 150?
$$75 = n \cdot 150 \quad \text{Write the percent equation.}$$
$$\frac{75}{150} = \frac{150n}{150} \quad \text{Divide each side by 150.}$$
$$0.5 = n \quad \text{Simplify.}$$
Since n represents the decimal form, the percent is 50%. So, 75 is 50% of 150.

5. $3 = 0.12 \cdot w$; 25 **7.** 39 loaves **9.** $p = 0.39 \cdot 65$; 25.4 **11.** $p = 0.53 \cdot 470$; 249.1 **13.** $26 = n \cdot 96$; 27.1% **15.** $30 = n \cdot 64$; 46.9% **17.** $84 = 0.75 \cdot w$; 112 **19.** $64 = 0.8 \cdot w$; 80 **21.** 4,400 games

23. 1 lobster　**25.** $p = 0.004 \cdot 82.1$; 0.3
27. $230 = n \cdot 200$; 115%

29 **a.** Let n represent the percent of viewers watching comedy, jokes, and bloopers.

$52,540,000 = n \cdot 142,000,000$	Write the percent equation.
$\dfrac{52,540,000}{142,000,000} = \dfrac{142,000,000n}{142,000,000}$	Divide each side by 142 million.
$0.37 = n$	Simplify.

37% of viewers watch comedy, jokes, and bloopers.
b. Let n represent the percent of viewers watching news stories.

$44,020,000 = n \cdot 142,000,000$	Write the percent equation.
$\dfrac{44,020,000}{142,000,000} = \dfrac{142,000,000n}{142,000,000}$	Divide each side by 142 million.
$0.31 = n$	Simplify.

31% of viewers watch news stories.
c. Subtract to find the number of viewers who do not watch movie previews.
$142 - 39.76 = 102.24$
Write an equation.

$$\text{part} = \text{percent} \cdot \text{whole}$$
$$102.24 = n \cdot 142$$
$$\frac{102.24}{142} = \frac{142n}{142}$$
$$0.72 = n$$

So, 72% of viewers do not watch movie previews.
31. Sample answer: If the percent is less than 100%, then the part is less than the base; if the percent equals 100%, then the part equals the base; if the percent is greater than 100%, then the part is greater than the base.　**33.** B　**35.** D　**37.** 170.73
39. Soccer; 21 is 15% of 140

Pages 342–343　Lesson 6-2D　PSI

1. Find 10% of the number, then half of the 10%, and add together.　**3.** $3.50　**5.** 500; $60\% \cdot 830 \approx 500$
7. Sample answer: 2 quarters, 1 dime, 4 nickels, 3 pennies

9 25% off is one fourth less than the original price. $41 is close to $40. One fourth of $40 is $10. Taking $10 off the original price leaves us with $30 as the sale price.
11. $120; Sample answer: $10 \cdot 12 = $120
13. 70 yd^2; $15 \cdot 18 + 18 \cdot 20 = 630$ ft^2. Then convert 630 ft^2 to square yards. $630 \div (3 \cdot 3) = 70$ yd^2.

Pages 348–350　Lesson 6-3B

1. −20%, decrease　**3.** 19%, increase　**5a.** about 3.8%, increase　**5b.** about −2.9%, decrease　**7.** 40%, increase　**9.** −71%, decrease

11 The amount of change is $11.70 − $15.60 = −$3.90. Since the percent of change is negative, this is a percent of decrease.

percent of change $= \dfrac{\text{amount of change}}{\text{original amount}}$	
$= \dfrac{-3.90}{15.60}$	Substitution
$= -0.25$	Simplify.
$= -25\%$	Write 0.25 as a percent.

The percent of change is −25%.
13. −5%, decrease　**15.** 13%, increase　**17.** −38%, decrease　**19.** 2%, increase

23 Write the percent of change equation. Substitute the values you know, and solve to find the original amount. Let x = amount of change.

percent of change $= \dfrac{\text{amount of change}}{\text{original amount}}$	
$0.20 = \dfrac{x}{25,900,000,000}$	Substitute.

$(25,900,000,000)(0.20) = 25,900,000,000x$
Multiply each side by 25,900,000,000.

$5,180,000,000 = x$	Simplify.

The amount of change was 5.18 billion.
Amount of change = new amount − original amount
Let n = new amount.

5.18 billion $= n - 25.9$ billion	Substitute.
5.18 billion $+ 25.9$ billion $= n$	Add 25.9 billion to each side.
31.08 billion $= n$	Simplify.

The projected sales are $31.08 billion or about $31 billion.
25. $60 sound system; 10 is a greater part of 60 than of 90.　**27.** Sample answer: Compare the original amounts. The ratio of the amount of change to a smaller original amount would be larger than the ratio of the same amount of change to a larger original amount.　**29.** 6,500 comments
31. 200　**33.** $0.3w = 17$; 56.7　**35.** $p = 0.64 \cdot 150$; 96
37. $5 = 0.2 \cdot w$; 25　**39a.** Sample answer: $1.5 = n \cdot 24$; 6.25%　**39b.** about 35%

Pages 353–354　Lesson 6-3C

1. $3.10　**3.** $32.20　**5.** $37.73
7 $100\% + 7\% = 107\%$　Add the percent of tax to 100%.
The total cost is 107% of the regular price.
107% of $1,500 = 1.07×1500　Write 107% as a decimal.
$= $1,605$　Multiply.
The total cost of the computer is $1,605.
9. $14.95　**11.** $7.99　**13.** $96.26　**15.** Yes; Sample answer: $84 was earned.

17 Combine the tax and the tip before calculating the total price: $0.0625 + 0.20 = 0.2625$. Multiply the tax and tip by the price of the bill to find the tax and tip: $28.35 \times 0.2625 = $7.44. Add tax and tip to bill. $28.35 + $7.44 = $35.79
19. $134.82　**21.** $54, $64.80; The percent gratuity is 20%. All of the other pairs have a gratuity of 15%.
23. B　**25.** 50%, increase　**27.** 73%, increase
29. Sample answer: 10% of $11.90 is $1.19. 20% of $11.90 is $2 \times 1.19 or $2.38.

Pages 357–358　Lesson 6-3D

1. $157.50　**3.** $1,395.65　**5.** $98.90　**7.** $1,080.00
9. $4.90

11 First, find the amount of the discount.

20% of $7.50 = 0.20 · 7.50   Write 20% as a decimal.
            = $1.50   The discount is $1.50.

Next, subtract the discount from the regular price.

$7.50 − $1.50 = $6.00

5.75% of $6.00 = 0.0575 · 6.00   Apply the tax.
              = $0.35   Write 5.75% as a decimal.

Add the tax to the sale price.

$6.00 + $0.35 = $6.35

The cost of the ticket, including tax, is $6.35.

13. $64.80   **15.** $7.50   **17.** $180.00

19 a.

Camera Model	Regular Price ($)	Discount	Sale Price
A	97.99	15%	$83.30
B	102.50	20%	$82.00
C	75.99	14%	$65.35
D	150.50	10%	$135.45

b. Sample answer: Subtract the sale price from the regular price to find the amount of the discount. Divide the amount of the discount by the regular price and multiply by 100. Round to the nearest percent if needed.   **c.** Camera B with a 20% discount

21. Sample answer: The regular price of a CD is $17.95. The total sale price is $13.46 and the total cost including tax is $14.23.   **23.** Sample answer: One method is to find 30% of the regular price, and then subtract this amount from the regular price. Another method is to find 70% of the regular price. The second method is more efficient because after mental math is used to find the percent, it can be done in one step rather than in two.   **25.** F

27. about $2   **29.** −37%, decrease

Pages 361–362   Lesson 6-3E

1. $38.40   **3.** $5.80

5 $I = prt$ Formula for simple interest

$I = \$4,500 · 0.09 · 3.5$ Replace p with $4,500, r with 0.09, and t with 3.5.
$I = \$1,417.50$

$1,417.50 interest is paid in 3.5 years.

7. $1,219.00   **9.** $21.38   **11.** $123.75   **13.** $45.31

15. $14.06   **17.** $1,353.13

19 a. $I = prt$ Formula for simple interest

$630 = 4,200 · r · 3$ Substitute.
$630 = 12,600r$ Simplify.

$\dfrac{630}{12,600} = \dfrac{12,600r}{12,600}$ Divide each side by 12,600.

$0.05 = r$ Simplify.

Write 0.05 as a percent. The interest rate is 5%.

b. Find the interest earned on $4,200 at a rate of 6% over 4 years.

$I = prt$ Formula for simple interest
$I = 4,200 · 0.06 · 4$ Substitute.
$I = \$1,008$ Multiply.

Add the interest earned to the principal. Is the total at least $5,000?

$4,200 + $1,008 = $5,208

Ramon would earn at least $5,000.

21. $825.60, $852.02, $879.28   **23.** C   **25.** $717.40

27. $8,440   **29.** $28.75

Pages 366–369   Chapter Study Guide and Review

1. true   **3.** true   **5.** false; 100%   **7.** true   **9.** false; original   **11.** 39   **13.** 135   **15.** Sample answer: 20; $\frac{1}{4}$ · 80 = 20   **17.** Sample answer: 30; $\frac{3}{4}$ · 40 = 30   **19.** Sample answer: 8; 0.1 · 80 = 8   **21.** 5%   **23.** 10.7%   **25.** $27.49   **27.** 39 = 0.65 · w; 60   **29.** 333   **31.** 93%, increase   **33.** 10%, increase   **35.** $13.80   **37.** $52.12   **39.** $11.25   **41.** $47.50   **43.** $101.25

Chapter 7 Linear Functions

Page 376   Chapter 7   Are You Ready?

1. 22   **3.** 42   **5.** 8   **7.** 14   **9.** 28   **11.** 63   **13.** 30

15. C   **17.** D   **19.** B

Pages 380–384   Lesson 7-1B

1.

x	3x	y
1	3 × 1	3
2	3 × 2	6
3	3 × 3	9
4	3 × 4	12

Domain: {1, 2, 3, 4};
Range: {3, 6, 9, 12}

3.

x	8x	y
1	8 · 1	8
2	8 · 2	16
3	8 · 3	24
4	8 · 4	32

Domain: {1, 2, 3, 4};
Range: {8, 16, 24, 32}

5.

x	2x	y
0	2 × 0	0
1	2 × 1	2
2	2 × 2	4
3	2 × 3	6

Domain: {0, 1, 2, 3};
Range: {0, 2, 4, 6}

7 In the rule column, show that you will multiply 9 times each of the x-values. In the y column, show each product.

x	9x	y
1	9 × 1	9
2	9 × 2	18
3	9 × 3	27
4	9 × 4	36

List the x-values as the domain. Domain: {1, 2, 3, 4}
List the y-values as the range. Range: {9, 18, 27, 36}

9.

x	60x	y
5	60 · 5	300
10	60 · 10	600
15	60 · 15	900
20	60 · 20	1,200

Domain: {5, 10, 15, 20};
Range: {300, 600, 900, 1,200}

11a. $t = 35m$ **11b.** 525 times; Sample answer:
Replace m with 15 in the equation $t = 35m$ to find
the number of times a cricket will have chirped
after 15 minutes at this temperature.

13.

x	1.3x	y
1	1.3 × 1	1.3
2	1.3 × 2	2.6
3	1.3 × 3	3.9
4	1.3 × 4	5.2

Domain: {0, 1, 2, 3};
Range: {1.3, 2.6, 3.9, 5.2}

15.

x	$\frac{2}{3}x$	y
2	$\frac{2}{3} \times 2$	$1\frac{1}{3}$
3	$\frac{2}{3} \times 3$	2
4	$\frac{2}{3} \times 4$	$2\frac{2}{3}$
5	$\frac{2}{3} \times 5$	$3\frac{1}{3}$

Domain: {2, 3, 4, 5};
Range: $\left\{1\frac{1}{3}, 2, 2\frac{2}{3}, 3\frac{1}{3}\right\}$

17 **a.** Use the distance formula, $D = rt$ to write the
equation. The rate is 8, the distance is m, and the
time is s. Therefore, the equation is $m = 8s$.
b. Use the distance formula, $D = rt$ to write the
equation. The rate is 19, the distance is m and the
time is s. Therefore, the equation is $m = 19s$.
c. 480 mi; 1,140 mi; Sample answer: Replace s with
60 in the equation $m = 8s$ and in the equation
$m = 19s$ to find the number of miles Jupiter
travels in 1 minute and the number of miles
Earth travels in 1 minute, respectively.
19. Divya added 3 to each input rather than
multiplying by 3. The range should be {6, 12, 18, 24}.
21a. $y = 0.25x$ **21b.** 2.5; 3; 3.5 **23.** Sample answer:
You start with an input number, perform the
operations in the function rule, and the result is the
output number. **25.** G **27.** function; Sample
answer: each element of the domain is paired with
exactly one element of the range **29.** not a
function; Sample answer: the domain value 12 is
paired with two range values 4 and 16 **31.** not a
function; Sample answer: the graph does not pass
the vertical line test. **33.** −100

35.

x	f(x) = x − 3	f(x)
2	2 − 3	−1
4	4 − 3	1
6	6 − 3	3
8	8 − 3	5

37.

x	f(x) = 2x + 6	f(x)
0	2(0) + 6	6
1	2(1) + 6	8
3	2(3) + 6	12
5	2(5) + 6	16

39. −100

Pages 388–389 **Lesson 7-1C**

1.

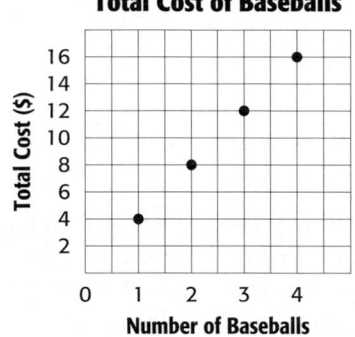

Total Cost of Baseballs

3 Graph $y = x - 1$. Select any four values for
the input x. For example, try 2, 1, 0, and −1.
Substitute these values for x to find the output y.
Graph the points, and draw a line through the
points.

x	x − 1	y	(x, y)
−1	−1 − 1	−2	(−1, −2)
0	0 − 1	−1	(0, −1)
1	1 − 1	0	(1, 0)
2	2 − 1	1	(2, 1)

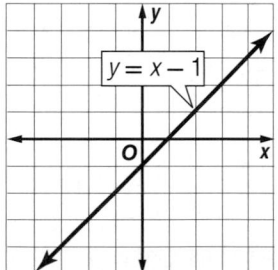

5.

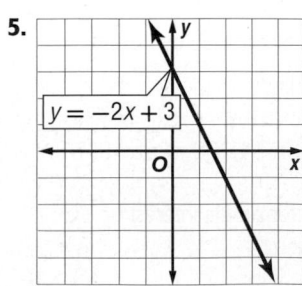

7.

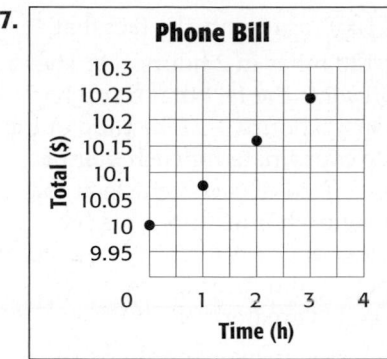

Phone Bill

9.

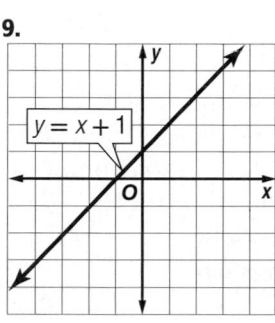

$y = x + 1$

11.

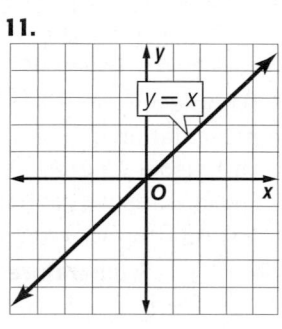

$y = x$

13.

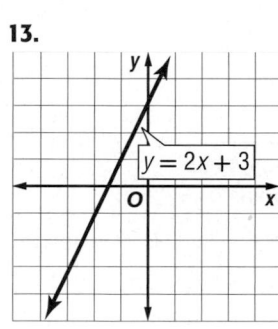

$y = 2x + 3$

15.

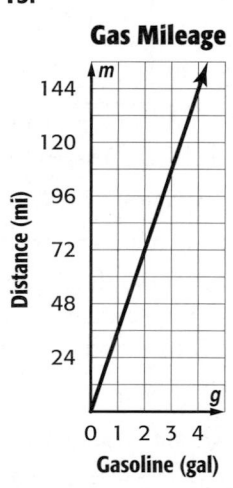

Gas Mileage

17.

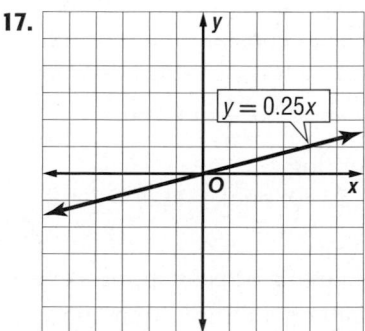

$y = 0.25x$

19 Graph $y = 0.5x - 1$. Select any four values for the input x. We chose 4, 2, 0, and -2 in order to have whole number values for y. Substitute these values for x to find the output y. Graph the points, and draw a line through the points.

x	0.5x − 1	y	(x, y)
−2	0.5(−2) − 1	−2	(−2, −2)
0	0.5(0) − 1	−1	(0, −1)
2	0.5(2) − 1	0	(2, 0)
4	0.5(4) − 1	1	(4, 1)

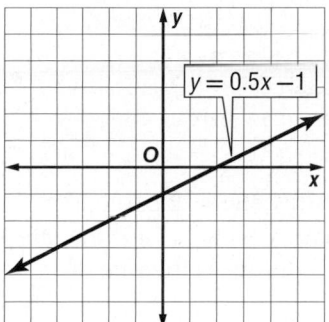

$y = 0.5x - 1$

23. Sample answer: {(1, 2), (1, 3), (2, 4)}; This set is not a function because 1 is mapped to two y-values, 2 and 3. **25.** F

Pages 393–395   Lesson 7-2B

1 Find the unit rate to determine the constant rate of change.

$$\frac{\text{change in temperature}}{\text{change in time}} = \frac{3°F}{2 \text{ hours}}$$

$$= \frac{1.5°F}{1 \text{ hour}}$$

So, the temperature rises by 1.5°F per hour.

3. 6 m per s   **5.** about $0.03 per min

7 To find the rate of change, pick any two points on the line, such as (18, 216) and (6, 72).

$$\frac{\text{change in length}}{\text{change in height}} = \frac{(216 - 72) \text{ inches}}{(18 - 6) \text{ inches}}$$

$$= \frac{144 \text{ inches}}{12 \text{ inches}} \text{ or } \frac{12 \text{ inches}}{1 \text{ inch}}$$

The length increases 12 inches for every inch the height increases. So, the rate of change is 12 inches of length for every 1 inch of height.

9 a. Write and solve a proportion to determine how many laps equal 1 mile.

$$\frac{20 \text{ laps}}{5 \text{ miles}} = \frac{x \text{ laps}}{1 \text{ mile}}$$

$$5x = 20$$

$$\frac{5x}{5} = \frac{20}{5}$$

$$x = 4$$

So, 4 laps equal 1 mile. From the information on the clipboard, it takes Seth 57.1 seconds to complete 4 laps, or 1 mile.

Find the unit rate to determine the constant rate of change.

$$\frac{\text{change in miles}}{\text{change in seconds}} = \frac{1 \text{ mile}}{57.1 \text{ seconds}}$$

$$\approx \frac{0.0175 \text{ mile}}{1 \text{ second}}$$

So, the constant rate of change is about 0.02 mile per second.

b.   **Go-Kart Racing Times**

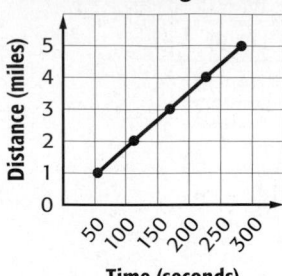

11. Sample answer: At Sam's Shoes, 2 pairs of sandals cost $30 and 4 pairs cost $60. Find the rate of change.   **13.** 390   **15.** nonlinear; the rate of change is not constant

Pages 398–399   Lesson 7-2C

1

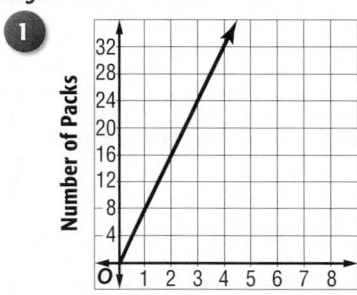

Number of Boxes

The slope of the line is equal to the rate at which the number of fruit snacks increases as the number of boxes increases or $\frac{8 \text{ fruit snacks}}{1 \text{ box}}$. 8 packs per box; The slope represents how many packages of fruit snacks are in each box.

3.

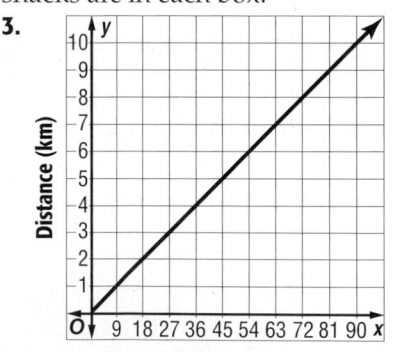

Time (min)

$\frac{1}{9}$; Adriano traveled 1 kilometer for every 9 minutes.

5.

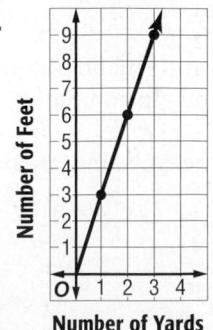

Number of Yards

$\frac{3}{1}$ or 3; There are 3 feet per yard.

7 **a.** The point (2, 120) represents the fact that Car A had traveled 120 miles in 2 hours.   **b.** The point (1.5, 67.5) represents the fact that Car B had traveled 67.5 miles in 1.5 hours.   **c.** The ratio of the y-coordinate to the x-coordinate for each pair of points represents the speed of each vehicle at that point.   **d.** The slope of each line represents the speed of each vehicle.   **e.** Car A is traveling faster. Since its slope is steeper, and its graph is rising more rapidly, its speed is greater.

9. Marisol found $\frac{\text{run}}{\text{rise}}$. Her answer should be $\frac{3}{2}$.
11. $\frac{4}{1}$ or 4   **13.** $9 per hour

Pages 402–403   Lesson 7-3A

1. Sample answer: Graphs provide a visual representation of a situation involving comparisons. A graphical model can sometimes show conclusively what is often difficult to interpret from looking at lists alone.

3a.   **Temperature Conversions**

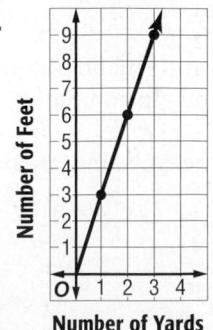

3b. Sample answer: 77°F   **5.** 158, 318   **7.** Wednesday

9 If the helicopter can carry 2,400 pounds and the crates weigh 75 pounds each, divide 2,400 by 75 to find the amount of crates the helicopter can carry; 2,400 ÷ 75 = 32 crates
11. 18

Pages 408–410   Lesson 7-3C

1. 30 cakes per hour   **3.** yes; 36

5 Since the points on the graph lie in a straight line, the rate of change is a constant. The constant ratio is what Shelley earns for each dog she walks.
$$\frac{\text{pay (\$)} \rightarrow 10}{\text{number of dogs} \rightarrow 4} \text{ or } \$2.50, \frac{5}{2} \text{ or } \$2.50$$
Shelley earns $2.50 per dog.
7. no   **9.** yes; 0.2   **11.** $0.80/book   **13.** $y = \frac{1}{5}x$; 3
15. $y = \frac{7}{8}x$; 16

17 Since we are given that a direct variation exists, we can find the constant of variation from one point on the graph. Using the point (36, 12) we find that the constant of variation is $\frac{36}{12}$ or $\frac{3}{1}$. Use this fact to set up and solve a proportion to find the measure of an object in yards if it is 78 feet long.
$$\frac{3}{1} = \frac{78}{x} \qquad \text{Set up the proportion.}$$
$$3 \cdot x = 1 \cdot 78 \qquad \text{Find the cross products.}$$

$$3x = 78 \qquad \text{Simplify.}$$
$$\frac{3x}{3} = \frac{78}{3} \qquad \text{Divide each side by 3.}$$
$$x = 26 \text{ yards} \quad \text{Simplify.}$$

An object that is 78 feet long is 26 yards long.

19a. 10 gal per minute; yes

19b.

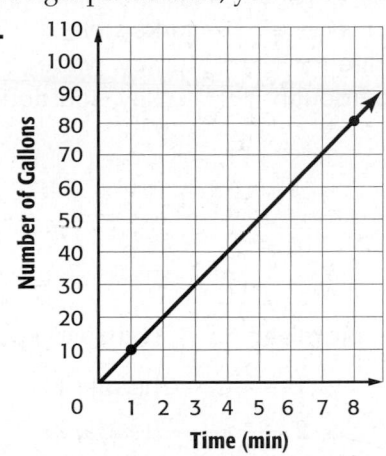

19c. Sample answer: $g = 10m$ **21.** Sample answer: $x = 9$, $y = 5\frac{1}{2}$; $x = 36$; $y = 22$ **25.** G   **27.** a speed of 45 mph; the distance driven on Day 1

29. no; Sample answer: Possible times and distances are shown in the table. Since $\frac{\text{distance}}{\text{time}}$ ratios are not equal, the relationship is not proportional.

Time (h)	1	2	3	4
Distance (mi)	145	190	235	280

Pages 414–415   Lesson 7-3E

1 Determine if the relationship is an inverse variation by checking that $xy = k$ for some constant k.
$48(360) = 17{,}280$; $36(480) = 17{,}280$; $24(720) = 17{,}280$; $12(1{,}440) = 17{,}280$
The relationship is an inverse variation.

Piano Strings

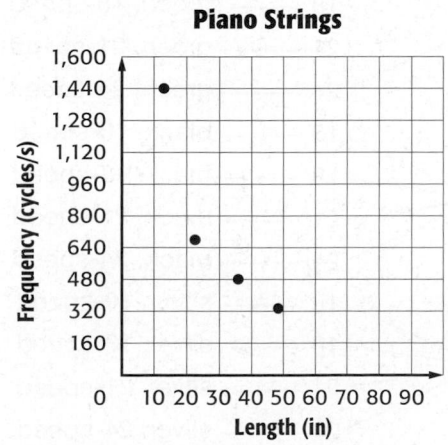

3.

Length and Width

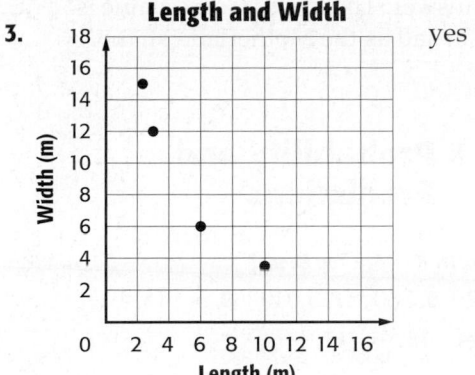

yes

5. 24 h

7 Check if it is a direct variation by looking at the ratio of each pair of numbers.
$\frac{88}{2}$ or 44, $\frac{44}{4}$ or 11. The relationship is not a direct variation. Next, check to see if the relationship is an inverse variation. $88(2) = 176$, $44(4) = 176$; The relationship is an inverse variation.

9. Sample answer: $x = 12$, $y = 2$; $x = 6$, $y = 4$; $x = 8$, $y = 3$.   **11.** B   **13.** A   **15.** Sample answer: 200 stamps

Pages 418–421   Chapter Study Guide and Review

1. slope   **3.** inverse variation   **5.** function
7. function table   **9.** constant rate of change
11. rate of change

13.

x	2x	y
1	2 × 1	2
2	2 × 2	4
3	2 × 3	6
4	2 × 4	8
5	2 × 5	10

Domain: {1, 2, 3, 4, 5};
Range: {2, 4, 6, 8, 10}

15.   Texting Rate

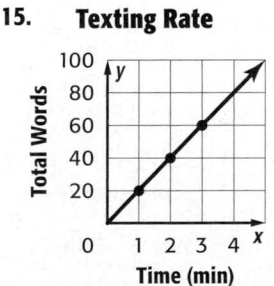

17.

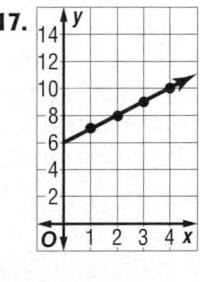

19. 2 laps per minute   **21.** 3 ft per yd

23.   Rainfall

Slope $= \frac{2}{1}$ or 2; for every 1 hour of rainfall, 2 inches of rain will accumulate

25. Sample answer: The Crazy Horse statue is about twice as tall as the Motherland statue. **27.** yes; 500

Chapter 8 Probability and Predictions

Page 428   Chapter 8   Are You Ready?

1. 105   **3.** 52   **5.** 360   **7.** 5,040   **9.** 5   **11.** 4
13. 315 songs   **15.** $\frac{1}{6}$ **17.** $\frac{1}{3}$

Pages 431–434   Lesson 8-1A

1. $\frac{1}{8}$, 0.125, 12.5%   **3.** $\frac{1}{8}$, 0.125, 12.5%

5 12 red + 6 orange = 18 possibilities out of 30 marbles. So, P(red or orange) is $\frac{18}{30}$ or $\frac{3}{5}$.

7. $\frac{23}{30}$ **9.** 1   **11.** $\frac{1}{20}$, 0.05, 5%   **13.** $\frac{3}{10}$, 0.3, 30%

15. $\frac{19}{20}$, 0.95, 95%   **17.** $\frac{2}{7}$ **19.** $\frac{1}{6}$ **21.** $\frac{5}{7}$

23 Since you are finding the probability the student is NOT in Room 10, find the total number of students in the other rooms. 24 + 20 + 16 + 14 = 74 students in the favorable outcomes group. The total number of students is 74 + 10 or 84. P(*not* Room 10) = $\frac{74}{84}$ or $\frac{37}{42}$

25. 60%   **27.** about 70%   **29.** Sample answer: The complementary event is the chance of no rain. Its probability is 60%.

31 There is one favorable outcome, and 5 total outcomes. Therefore, the probability that any given song would be hip-hop is $\frac{1}{5}$.

33. Sample answer: There are currently 6 + 4 + 8 or 18 marbles in the bag. A total of 27 − 18 or 9 marbles should be added to the bag. To do so without changing the probability of randomly selecting one marble of each color, add 3 red marbles, 2 blue marbles, and 4 green marbles. Then, P(red) = $\frac{9}{27}$ or $\frac{1}{3}$, P(blue) = $\frac{6}{27}$ or $\frac{2}{9}$, and P(green) = $\frac{12}{27}$ or $\frac{4}{9}$ **35.** C   **37.** B   **39a.** $\frac{1}{4}$ **39b.** $\frac{2}{5}$ **39c.** $\frac{1}{2}$

Pages 436–439   Lesson 8-1B

1 Sample answer: Write heads. Show that it leads to heads and tails. Repeat this for tails. The tree diagram is shown.

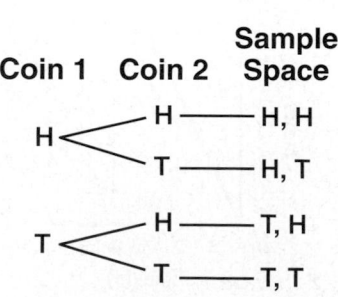

3.

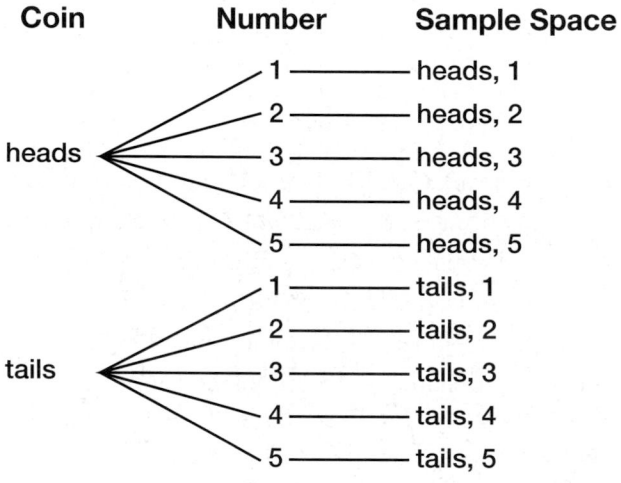

5.

7.

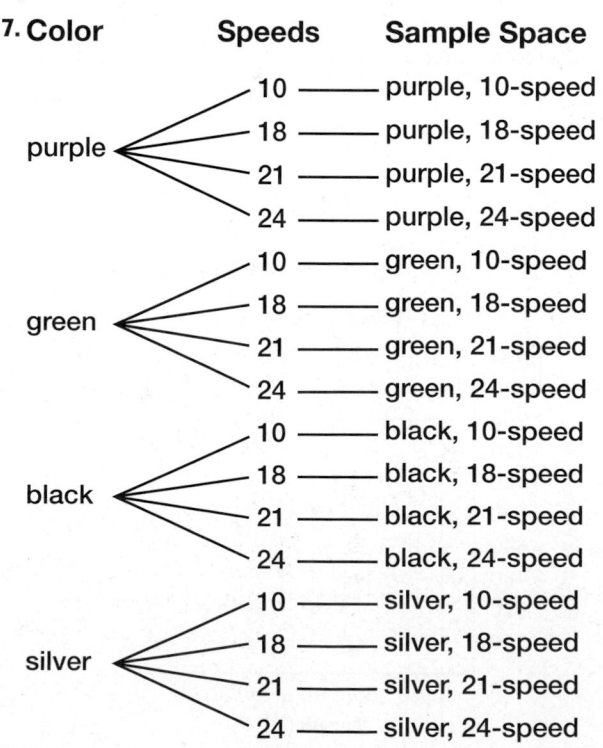

9.

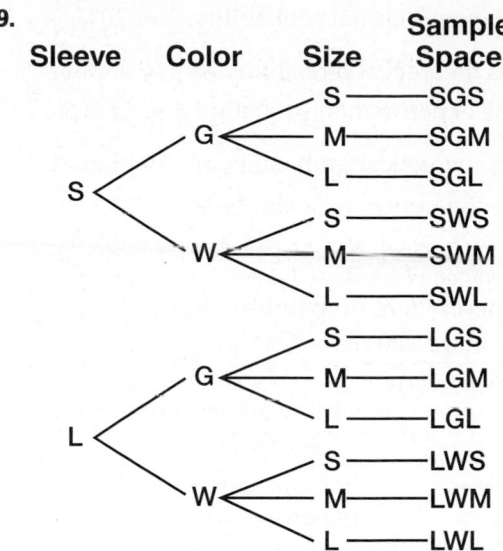

| Sleeve | Color | Size | Sample Space |

11.

Quarter Dime Nickel Sample Space

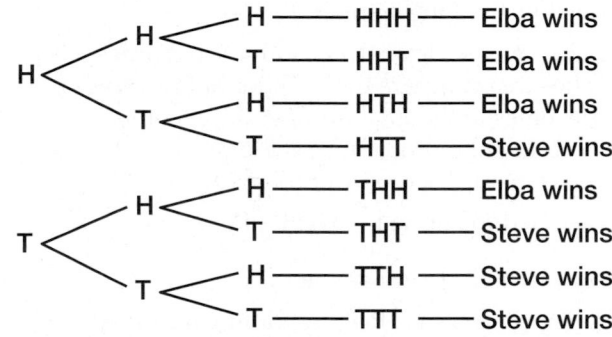

				Sample Space	
H	H	H	HHH	Elba wins
H	H	T	HHT	Elba wins
H	T	H	HTH	Elba wins
H	T	T	HTT	Steve wins
T	H	H	THH	Elba wins
T	H	T	THT	Steve wins
T	T	H	TTH	Steve wins
T	T	T	TTT	Steve wins

There are 8 equally likely outcomes with 4 favoring Elba. So, the probability that Elba wins is $\frac{4}{8}$ or $\frac{1}{2}$.

13 Draw a tree diagram.   **a.** One of eight outcomes has three boys. Therefore, P(all three children will be boys) $= \frac{1}{8}$.

b. Six outcomes include at least one boy and one girl. Therefore, P(at least one boy and one girl) $= \frac{6}{8}$ or $\frac{3}{4}$.   **c.** Three outcomes have two boys and one girl. Therefore, P(two boys and one girl) $= \frac{3}{8}$.

d. Four outcomes have at least two girls. Therefore, P(at least two girls) $= \frac{4}{8}$ or $\frac{1}{2}$.   **e.** There is only 1 outcome in which the first two born are boys and the last born is a girl. Therefore, P(the first two born are boys and the last born is a girl) $= \frac{1}{8}$.

15a. 16   **15b.** $\frac{1}{16}$

15c.

Shoes   Socks   Sample Space

Shoes	Socks	Sample Space
black	green	black, green
	yellow	black, yellow
	black	black, black
	white	black, white
yellow	green	yellow, green
	yellow	yellow, yellow
	black	yellow, black
	white	yellow, white

17. The first outcome in the I bracket should be IC.
19. Sample game: Each player tosses a coin 10 times. If it comes up heads, player 1 receives 1 point. If it comes up tails, player 2 receives 1 point. The player with the most points at the end of 20 tosses wins.
21. 12   **23.** 2:11

Pages 441–443   *Lesson 8-1C*

1 There are two outcomes for tossing the quarter, two for tossing the dime and two for tossing the nickel. The Fundamental Counting Principle says that you can multiply to find out the total number of outcomes: $2 \cdot 2 \cdot 2 = 8$ outcomes.

3. 72; $\frac{1}{72}$ or about 1.4%; unlikely   **5.** 140   **7.** 12

9. 16   **11.** 18 possible bread choices; $\frac{1}{18}$; very unlikely

13 The total number of T-shirt choices is the product of the number of designs and the number of color choices, or $32 \cdot 11$ or 352 choices. The advertisement is not true because there are 365 days in a year and 352 is less than 365.

15. 2; 4; 8; 2^n; Sample answer: I used a pattern to determine the number of outcomes for n coins. One coin: 2^1 outcomes, two coins: $2 \cdot 2$ or 2^2 outcomes, three coins: $2 \cdot 2 \cdot 2$ or 2^3 outcomes, n coins: 2^n outcomes.   **17.** Sample answer: Both are methods used to find the total number of possible outcomes. The Fundamental Counting Principle is a computational procedure therefore is a much faster method of obtaining the total number of outcomes than drawing a tree diagram. Tree diagrams allow you to see the specific outcomes. The Fundamental Counting Principle only gives the number of outcomes.   **19.** 6   **21.** $\frac{1}{2}$ **23.** $\frac{1}{8}$

Pages 445–447   *Lesson 8-1D*

1. 5,040

3 There are 5 choices for the first DVD. There are 4 choices that remain for the second DVD. Since $5 \cdot 4 = 20$, there is a $\frac{1}{20}$ chance of selecting Season 1 followed by Season 2.

5 There are 4 choices for first place. There are 3 choices that remain for second place. There are 2 choices that remain for third place. There is 1 choice that remains for fourth place. Since $4 \cdot 3 \cdot 2 \cdot 1 = 24$, there are 24 possible ways in which the four bands can be ordered.
7. 720   **9.** $\frac{1}{90}$ **11.** $\frac{1}{12}$
13 Find consecutive whole numbers, starting with 1, that have a product of 720. Since $1 \cdot 2 \cdot 3 \cdot 4 \cdot 5 \cdot 6 = 720$, there are 6 members in the Coughlin family.
15. Sample answer: The number of ways you can order 3 books on a shelf is $3 \cdot 2 \cdot 1$ or 6.   **17.** 20 things taken 4 at a time; Sample answer: To find $P(20, 4)$, multiply $20 \cdot 19 \cdot 18 \cdot 17$.   **19.** C   **21.** 56

Pages 452–455   Lesson 8-2B

1 There is a $\frac{1}{2}$ chance of landing on tails and a $\frac{1}{6}$ chance of rolling a 3. $\frac{1}{2} \times \frac{1}{6} = \frac{1}{12}$

3. $\frac{1}{6}$ **5.** $\frac{5}{24}$ **7.** $\frac{1}{48}$ **9.** $\frac{3}{8}$ **11.** $\frac{5}{24}$ **13.** $\frac{92}{287}$

15 There are a total of $5 + 7 + 4 + 4$ or 20 students. $P(\text{green}) = \frac{4}{20}$ or $\frac{1}{5}$; $P(\text{brown after green}) = \frac{7}{19}$ $P(\text{green then brown}) = \frac{1}{5} \cdot \frac{7}{19}$ or $\frac{7}{95}$
17. $\frac{7}{76}$ **19.** $\frac{12}{19}$ **21a.** 11.76%   **21b.** 24.64%

23. $\frac{4}{13}$; dependent event; The first coin is not replaced.

25 The probability of the first player choosing a tile with the same number of dots on each side is $\frac{7}{28}$. If each player were to draw such a tile, both the number of favorable outcomes and the number of possible outcomes would reduce by one for each person, resulting in the following product: $\frac{7}{28} \cdot \frac{6}{27} \cdot \frac{5}{26} \cdot \frac{4}{25}$ or $\frac{1}{585}$
27. Sample answer: Raul used the method for dependent events. Spinning the spinner twice represents two independent events. The probability of getting an even number is $\frac{2}{5}$ each time. So, the correct answer is $\frac{2}{5} \cdot \frac{2}{5}$ or $\frac{4}{25}$.   **29.** Both independent events and dependent events are compound events. Outcomes of independent events do not affect each other. Outcomes of dependent events affect each other.
31. I   **33.** G   **35.** $\frac{2}{3}$

Pages 460–462   Lesson 8-3A

1 **a.** The experimental probability is $\frac{\text{number of times heads appears}}{\text{number of tosses}} = \frac{28}{50}$ or $\frac{14}{25}$.
b. The theoretical probability is $\frac{\text{favorable outcomes}}{\text{possible outcomes}} = \frac{1}{2}$.   **c.** The experimental probability, $\frac{14}{25}$ or 56%, is close to its theoretical probability of $\frac{1}{2}$ or 50%.

3a. $\frac{1}{5}$; The experimental probability, $\frac{1}{5}$ or 20%, is close to its theoretical probability of $\frac{1}{6}$ or about 17%.
3b. $\frac{9}{10}$; The experimental probability, $\frac{9}{10}$ or 90%, is close to its theoretical probability of $\frac{5}{6}$ or about 83%.

5 **a.** The theoretical probability is $\frac{\text{favorable outcomes}}{\text{possible outcomes}} = \frac{1}{3}$.
b. The experimental probability $P(A) = \frac{\text{number of times landing on A}}{\text{number of spins}} = \frac{6}{25}$. The experimental probability $P(C) = \frac{\text{number of times landing on C}}{\text{number of spins}} = \frac{13}{50}$.
c.

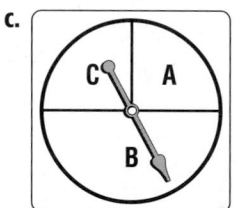

Sample answer: The experimental probability of landing on section A is $\frac{6}{25}$, or about $\frac{1}{4}$. The experimental probability of landing on section B is $\frac{50}{100}$, or $\frac{1}{2}$. The experimental probability of landing on section C is $\frac{13}{50}$, or about $\frac{1}{4}$. So, section B should be one half of the spinner and sections A and C should each be one fourth of the spinner.

7. 72   **9.** Sample answer: Both probabilities are ratios that compare the number of favorable outcomes to the total number of outcomes. Experimental probability is the result of an experiment. Theoretical probability is what is expected to happen.   **11.** H   **13.** $\frac{2}{7}$ **15.** $\frac{1}{3}$

Pages 464–465   Lesson 8-3C   PSI

1. Sample answer: Results could vary greatly.
3. No; Sample answer: the experiment produces about 1–2 correct answers, so using a spinner with 4 sections is not a good way to answer a 5-question multiple-choice quiz.   **5.** 30

7 There were 160 students surveyed and 40% of the students preferred a dunking booth. 40% of 160 is 64. So, 65 students is the most reasonable choice for the answer.

9. 31   **11.** $70; Sample answer: Round $95 to $100. 10% of $100 is $10, so 20% would be $20. $100 − $20 = $80, which is closest to $70.

Pages 469–471   Lesson 8-3E

1 **a.** 33% of the teens said they would save their money. Let n be the number of students that would save their money.
$n = 0.33 \cdot 60,000$ Write the percent equation with 33% as a decimal.
$n = 19,800$ Simplify.
So, about 19,800 teens surveyed said they would save their money.   **b.** Write a proportion. Let p be the number of students that would buy a music CD.

$$\frac{p}{28{,}000{,}000} = \frac{32}{100} \qquad \text{Percent proportion}$$
$$100p = 32 \cdot 28{,}000{,}000 \qquad \text{Find cross products.}$$
$$100p = 896{,}000{,}000 \qquad \text{Simplify.}$$
$$\frac{100p}{100} = \frac{896{,}000{,}000}{100} \qquad \text{Divide each side by 100.}$$
$$p = 8{,}960{,}000$$

About 8.96 million teens would buy a music CD if they were given $20.

3a. about 1,419 people   **3b.** about 581 people

5 Since 27 MP3s is a part of 238 MP3s, and you are looking for a percentage, answer c matches this situation.

7. b   **11.** Sample answer: Randomly select a part of the group to get a sample. Find their preferences, and use the results to find the percent of the total group. It makes sense to use a sample when surveying the population of a city.   **13.** G   **15a.** 44%
15b. 86%   **15c.** Sample answer: Use a percent equation. Divide the total number of students who play video games, $42 + 32 + 12$ or 86, by the total number of students, 100. $p = \frac{86}{100} = 0.86$ or 86%
17. $\frac{1}{3}$ **19.** 40,320

Pages 474–476   **Lesson 8-3F**

1. The conclusion is not valid. This is a biased sample, since people in other states would spend much more than those in Alaska. The sample is a convenience sample since all the people are from the same state.

3 This is a simple random survey because each person is randomly chosen for the survey, so the sample is valid; 28% chose Rolling Meadows. $0.28 \times 150 = 42$; 42 people.   **5.** The conclusion is not valid. This is a biased sample, since only art club members were surveyed. This is a convenience sample.

7 The conclusion is valid. This is an unbiased, simple random sample because randomly selected Californians were surveyed.

9. This is a simple random sample, so the sample is valid; about 205 people.   **11.** Sample answer: If the questions are not asked in a neutral manner, the people may not give their true opinions. For example, the question, "You really don't like Brand X, do you?" might not get the same answer as the question, "Do you prefer Brand X or Brand Y?" Also, the question, "Why would anyone like rock music?" might not get the same answer as the question, "What do you think about rock music?"   **13.** Sample answer: A survey uses an actual sample to predict the results in a population.   **15.** F   **17.** 16 pizzas

Pages 477–481   **Chapter Study Guide and Review**

1. true   **3.** false; multiplication   **5.** false; complement
7. false; outcome   **9.** $\frac{1}{6}$ **11.** $\frac{13}{18}$ **13.** 8%

15.

Pizza	Beverage	Sample Space
pepperoni	water	pepperoni, water
	juice	pepperoni, juice
	milk	pepperoni, milk
mushroom	water	mushroom, water
	juice	mushroom, juice
	milk	mushroom, milk
cheese	water	cheese, water
	juice	cheese, juice
	milk	cheese, milk

17. 36   **19.** 60   **21.** 120   **23.** $\frac{1}{42}$ **25.** $\frac{2}{15}$ **27.** $\frac{4}{45}$
29. $\frac{8}{25}$ **31.** 12   **33.** about 208 freshmen
35. 162 teens   **37.** about 204 students

Chapter 9 Statistical Displays

Page 488   **Chapter 9**   **Are You Ready?**

1. Rihanna   **3.** 25%

Pages 493–496   **Lesson 9-1B**

1. 5; 3.5; 0   **3.** 17.5; 14.5; 14   **5.** mean; Sample answer: no extreme values; $24.38

7 mean:
$$\frac{25 + 23 + 23 + 27 + 39 + 27 + 23 + 31 + 26 + 28}{10} = \frac{272}{10}$$
or 27.2
median: 23, 23, 23, 25, (26, 27) 27, 28, 31, 39;
$$\downarrow$$
$$26.5$$
mode: There are more 23 year olds than players of any other age. So, the mode is 23.
9. 27.4; 24; no mode   **11.** mean   **13.** median; Sample answer: There are no big gaps in the middle of data.   **15.** 78

19 mean:
$$\frac{3 + 5 + 8 + 2 + 1 + 1 + 7 + 3 + 9 + 2 + 3 + 10}{12} = \frac{54}{12}$$
or 4.5
21a. Sample answer: 3, 4, 5, 8   **21b.** Sample answer: 8, 9, 10, 10, 11, 12   **23.** 35, 41, 44   **25.** A   **27.** H

Pages 500–503   **Lesson 9-2A**

1a. 3.5   **1b.** 8.9; 7.6; 9.5   **1c.** 1.9   **1d.** none
3 a. To find the range, subtract the least number of golf courses from the greatest: $1{,}465 - 437$ or 1,028.
b. Put the data in order from least to greatest.
437, 456, 513, 650, 893, 954, 1,018, 1,038, 1,117, 1,465
Median: Since there is an even number of states represented, average the middle two numbers to find the median: $\frac{893 + 954}{2} = \frac{1{,}847}{2}$ or 923.5.

Upper Quartile: The middle number of the top half of the data is 1,038.
Lower Quartile: The middle number of the lower half of the data is 513. **c.** To find the inner quartile range, subtract the lower quartile from the upper quartile. 1,038 − 513 or 525. **d.** To check for outliers, first multiply the inner quartile range by 1.5: 525 × 1.5 = 787.5. Add 787.5 to 1,038 and subtract it from 513. No data falls outside this range, so there are no outliers.
5. ranges 50 and 45; medians both 50; upper quartiles both 60; lower quartiles 45 and 35; interquartile ranges 15 and 25; Sample answer: Both sets of data have the same median while the first set is more spread out than the second. **7.** median: 357.5, upper quartile: 422, lower quartile: 298, interquartile range: 124 **9.** median: $24.27, upper quartile: $28.99, lower quartile: $19.99, interquartile range: $9.00
11 Lucy − median: 7, upper quartile: 9, lower quartile: 5, interquartile range: 4; Dena − median: 7, upper quartile: 9, lower quartile: 6, interquartile range: 3; They have the same median and upper quartile, but Lucy's interquartile range is greater, meaning her mileage is more spread out than Dena's.
13. Sample answer: 6, 30, 33, 41, 45, 71 **15.** D **17.** H **19.** Sample answer: The mean price of the items listed is $3; however, they are calculating the mean price by including the kid's cone, a low priced item.

Pages 506–509 Lesson 9-2B

1a. 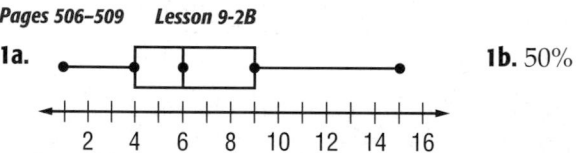 **1b.** 50%

1c. Sample answer: The length of the box-and-whisker plot shows that the depths of the earthquakes are not concentrated around a certain depth.

3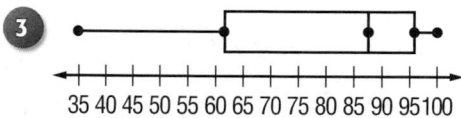

Step 1 Order the numbers from least to greatest. Then draw a number line that covers the range of data. 35, 55, 61, 65, 74, 88, 92, 96, 97, 99, 100

Step 2 Find the median, the extremes, and the upper and lower quartiles. Mark these points above the number line. Median: 88; Extremes: 35, 100; UQ: 97; LQ: 61

Step 3 Draw the box so that it includes the quartile values. Draw a vertical line through the median value. Extend the whiskers from each quartile to the extreme data points.

5.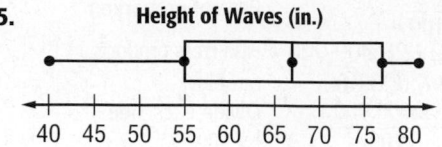

7a.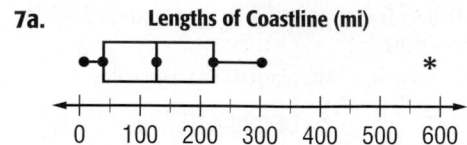

7b. 25% **7c.** 127 mi **7d.** Sample answer: The length of the box-and-whisker plot shows that the number of miles of coastline for the top 25% of the states varies greatly. The number of miles of coastline for the bottom 25% of the states is concentrated.

9. Half of the fruits have between 50 and 130 Calories, and half of the vegetables have between 25 and 110 Calories. At least one fruit has almost 450 Calories and at least one vegetable has about 350 Calories. So on the whole, the food energy of fruits and vegetables does not vary greatly.

11 a.

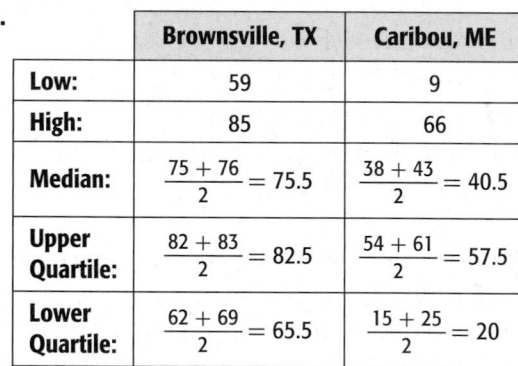

	Brownsville, TX	Caribou, ME
Low:	59	9
High:	85	66
Median:	$\frac{75+76}{2} = 75.5$	$\frac{38+43}{2} = 40.5$
Upper Quartile:	$\frac{82+83}{2} = 82.5$	$\frac{54+61}{2} = 57.5$
Lower Quartile:	$\frac{62+69}{2} = 65.5$	$\frac{15+25}{2} = 20$

Brownsville: 59, 85, 75.5, 82.5, 65.5; Caribou: 9, 66, 40.5, 57.5, 20

11b.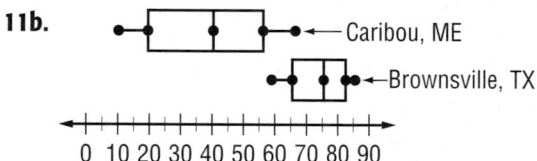

11c. The Brownsville data have a smaller interquartile range, so the data are more tightly clustered around the median. The Caribou data are more spread out. **13.** Sample answer: {60, 60, 60, 60, 60, 70, 75, 80, 85, 85, 85, 85}; median = 72.5; LQ = lower extreme = 60; UQ = upper extreme = 85

15. Sample answer: {28, 30, 52, 68, 90, 92};

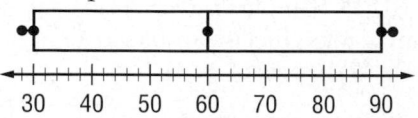

17. A **19.** I **21.** range: 12; median: 7; UQ: 9.5; LQ: 4.5; IQR: 5 **23.** mean: 5.6; median: 5.5; mode: 2 **25.** mean: 11.6; median: 11; mode: 10

1. Blood Types in U.S.

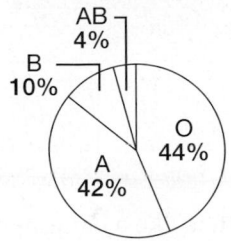

3a. blue   **3b.** 40

5.   Orange Production

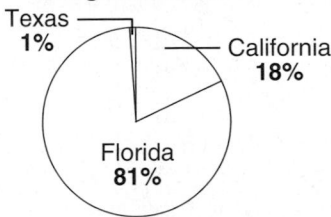

7.   Favorite Games

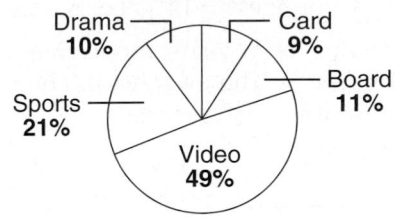

9 **a.** Look at the title of the graph. It asks, "Do Americans Favor Common North American Currency?" The section that says "Yes" represents the percent of Americans who do favor a common North American currency. The number there is 43%. So, 43% of Americans favor common North American currency.   **b.** Let p represent the part of the American people who would say "Don't Know". Write an equation to find the number. $p = 0.04 \cdot 298{,}000{,}000$ or about 12 million people do not know if they favor a common North American currency.   **c.** Find the number of people opposed to a common currency. $0.53 \cdot 298{,}000{,}000 = 157{,}940{,}000$ people opposed. Find the number of people in favor of a common currency. $0.43 \cdot 298{,}000{,}000 = 128{,}140{,}000$ people in favor. Find the difference: $157{,}940{,}000 - 128{,}140{,}000 = 29{,}800{,}000$ So, about 30 million people more are opposed to a common currency than are in favor of it.   **11.** 40

13. circle graph   Tanya's Day

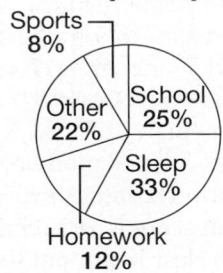

15 **a.** $366 + 126 + 10 = 502$ **15b.** No; a 50% increase in 126 students is 189 students and 189 students is not equal to or close to 366 students. So, it is not reasonable to say that 50% more students said they could make a difference. Since 300% of 126 is 378, it is reasonable to say that 300% more students said they could make a difference than those who said they cannot make a difference.   **17.** 12.5%; English is half of the circle. Since Science is half of English, Science is half of 50% or 25%. Math is half of Science, which is half of 25% or 12.5%.   **19.** No; Sample answer: The sum of the percents is greater than 100. The people surveyed must have been able to choose more than one fruit juice.   **21.** 16   **23.** 13

1. Sample answer:

State Population Density (per square mile)		
Density	Tally	Frequency
0–199	𝍫𝍫𝍫𝍫 𝍫𝍫𝍫 |||	38
200–399	𝍫	5
400–599	|||	3
600–799			1
800–999			1
1,000–1,999	||	2

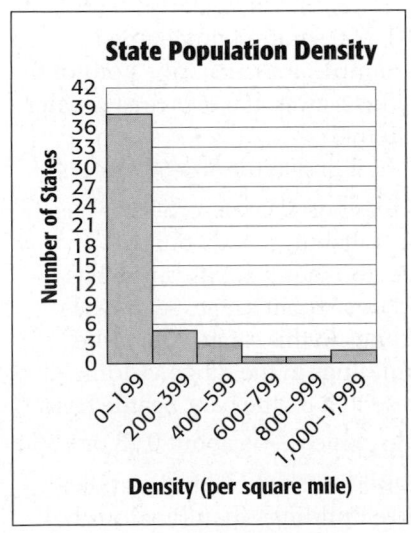

3. Sample answer:

Hours Spent Exercising per Week

Hours	Tally	Frequency
0–2	IIII I	6
3–5	IIII III	8
6–8	III	3
9–11	II	2
12–14	I	1

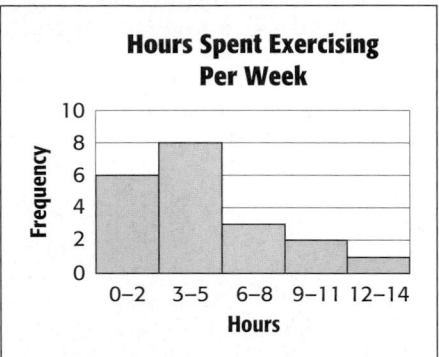

Hours Spent Exercising Per Week

(histogram: Frequency vs Hours; bars at 0–2 = 6, 3–5 = 8, 6–8 = 3, 9–11 = 2, 12–14 = 1)

5 **a.** Look at the horizontal scale of the graph. The countries represented in the first two bars have areas less than 401 square kilometers. Look at the vertical scale of the graph. The first bar represents 21 countries. The second bar represents 9 countries. Add these to find a total of 30 countries with areas less than 401 square kilometers.
b. Adding the number of countries in the second and third bars, you find that 19 countries have between 201 and 600 square kilometers. Adding all the bars shows a total of 50 countries represented. Find what percent 19 is of 50. Since $\frac{19}{50}$ is about 0.38 or 38%, about 38% of the countries have an area of 201–600 square kilometers.
c. Not very likely. Sample answer: Only 4 out of the 50 countries in the histogram have an area greater than 800 square kilometers.
7 **a.** Seattle has a building in the 900–999 feet range. Pittsburgh does not. Therefore, Seattle has the tallest building. **b.** Pittsburgh has a building in that range, while Seattle does not. **c.** Pittsburgh has $3 + 1 + 1$ or 5 buildings in that range. Seattle has $3 + 3 + 1$ or 7 buildings in this range. Therefore, Seattle has more buildings in the range. Adding all of the buildings represented by the bars, Seattle has a total of 21 buildings. Since $\frac{7}{21}$ is about 0.33 or 33%, about 33% of the buildings are 600 feet or taller.
d. Seattle has 6 more buildings than Pittsburgh.

R28 Selected Answers and Solutions

9. Sample answer:

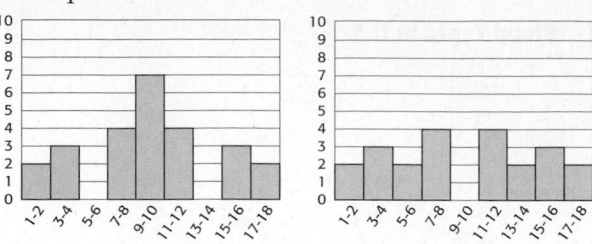

11. Sample answer: A histogram is like a bar graph because it displays data. A histogram is different because it uses intervals of numbers instead of one specific category. **13.** $\frac{4}{5}$ **15.** $\frac{2}{3}$

Pages 528–530   Lesson 9-3E

1. Height of Trees (ft)

Stem	Leaf
0	8 8
1	0 2 5 5 6 8
2	0 5 2 | 0 = 20 ft

3 **a.** range: oldest − youngest = 14 − 9 or 5 years
b. median: middle value, or 11 years; mode: most frequent value, 11 years **c.** The mean would be most affected because it would decrease.

5. Low Temperatures (°F)

Stem	Leaf
1	3 3 5
2	0 4 8
3	0 1 2 2 5 6 8 8 8 1 | 3 = 13°F

7. School Play Attendance

Stem	Leaf
22	5 7 9
23	0
24	3 6
25	
26	7 9 9
27	8 8 8 26 | 7 = 267

9a. $45 **9b.** $108 and $115 **9c.** mean

11 Gymnastics Scores

Stem	Leaf
8	5 7 7 8 9
9	0 3 3 9
10	0 8 | 9 = 8.9

Sample answer: There is only one person who scored a perfect 10. An average score is about 9.1.
15. Sample answer: Aisha read the plot incorrectly. There are two pieces of ribbon that are more than 50 inches at 53 and 56 inches. **17.** C **19.** median or mode; Sample answer: Class B has a median of 80. Class A has a median of 79. 80 > 79; Class B has a mode of 79. Class A has a mode of 72. 79 > 72
21. no; Sample answer: Class A's mean score is 80% and Class B's mean score is 79%. Since 80% and 79% are close, each class has about the same average.

23. Chicken: 20 g; Burger: 36 g   **25.** chicken; Sample answer: Chicken sandwiches have 8–20 grams of fat. Burgers have 10–36 grams of fat.

Pages 532–533   Lesson 9-4A   PSI

1. about $6\frac{1}{2}$ tablespoons   **3.** Sample answer: $34

5 If two workers make 2 chairs in 2 days, it takes each worker 2 days to make 1 chair. In 20 days, a worker could make 10 chairs. Eight workers could therefore make 80 chairs in 20 days.
7. bones in wrist: 8; bones in fingers: 14; bones in palm: 5   **9.** Sample answer: Because the graph is being used to predict, straight axes, equally spaced scales, and accurate plotting of points are necessary to get the most accurate prediction.

Pages 536–539   Lesson 9-4B

1a. The graph shows a positive relationship. That is, as the years pass, the population increases.

1b.

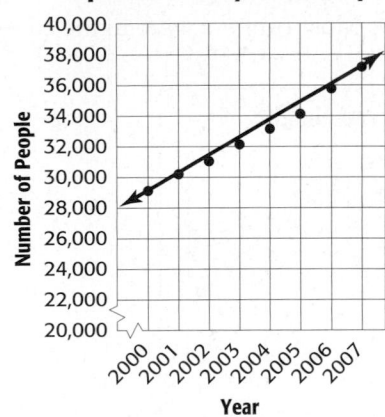

Population of Fayette County

1c. Sample answer: 40,000

3 **a.** positive relationship

3b.

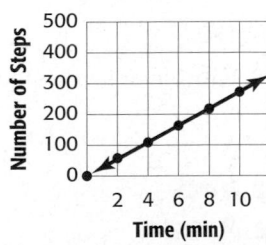

Climbing the Statue of Liberty

3c. Ciro climbs about 50–60 steps per minute. If the trend continues, it should take Ciro about 13 minutes to climb 354 steps.

5a.

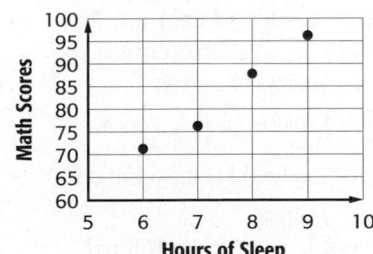

How Sleep Affects Math Scores

5b. positive relationship; Sample answer: As the hours of sleep increases, the scores increase.
5c. Sample answer: 65

7 **a.** The two sets of data show a positive relationship. As the years increase, so do the number of voters.   **7b.** Use a straightedge to make a line that comes close to all of the points. Look across to the vertical axis where the line is above the year 2011, halfway from 2010 to 2012. It looks as though it would be about 56,000 voters.
9. Sample answer: The data do not appear to fall along a straight line. So, age and number of pets do not seem to be related.   **11.** mode; Sample answer: The other three are ways to display data.   **13.** A

15a.

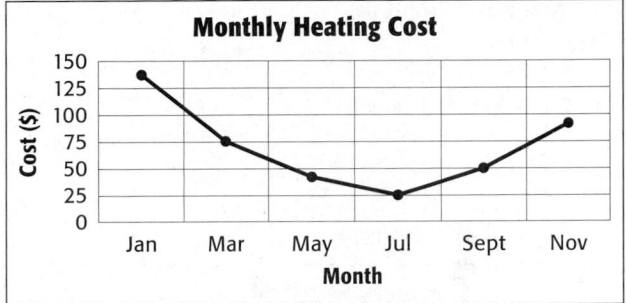

Monthly Heating Cost

15b. Sample answer: $75
17. median: 81; UQ: 89.5; LQ: 71; IQR: 18.5

Pages 542–545   Lesson 9-4C

1. line graph; compares values over a length of time
3. circle graph; compares parts to a whole

5 A histogram would be appropriate because it shows data divided into equal intervals. To construct a histogram of this data, use the intervals $2.00–$2.99, $3.00–$3.99, $4.00–$4.99, and $5.00–$5.99. Put these intervals on the horizontal axis. On the vertical axis, show the scale 0 to 3.5. Be sure to label the axis and title the graph.

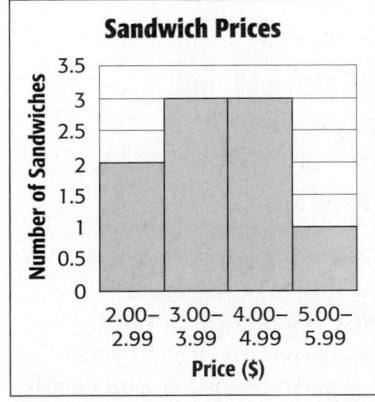

Sandwich Prices

Sample answer: Most sandwiches cost between $3.00–$4.99.

7 A box-and-whisker plot would be the most appropriate display because it shows the median along with other measures of variation.
9. circle graph; compares parts to whole

11. scatter plot; compares two sets of data; as the temperature increases, the depth of ice decreases.

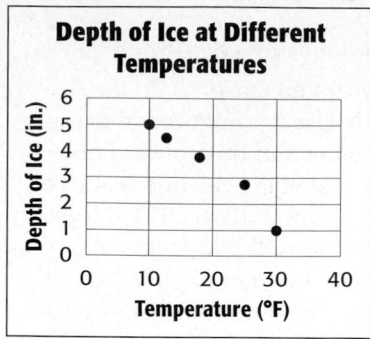

Depth of Ice at Different Temperatures

13. stem-and-leaf plot; displays individual data; The range of push-ups is $57 - 10$ or 47.

Number of Push-Ups

Stem	Leaf
1	0 7 9
2	0 3 3 5 8 8 8 8
3	
4	
5	1 2 3 4 4 5 6 6 7

$1 \mid 0 = 10$ push-ups

15.

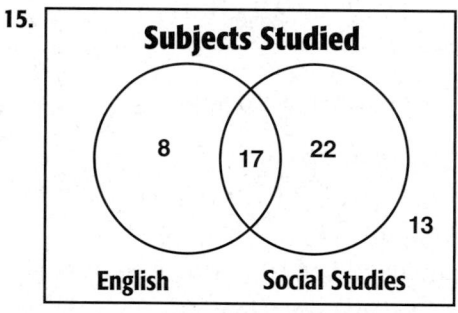

Subjects Studied

8   17   22

13

English   Social Studies

17a.

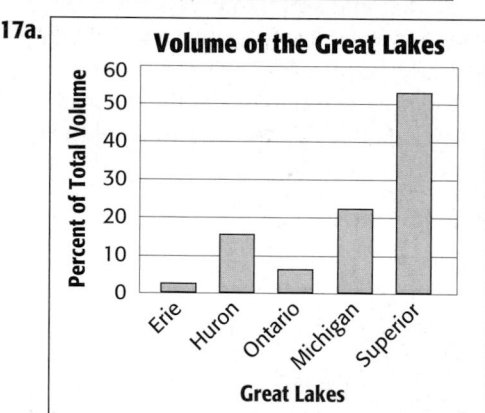

Volume of the Great Lakes

17b. Sample answer: The circle graph is most appropriate because it shows how each lake compares to the whole.   **19.** Sample answer: The population of polar bears over the last 10 years.
21. Sample answer: A box-and-whisker plot clearly shows the least and greatest values. Then the range can be found by subtracting the least value from the greatest.   **23.** Sample answer: Numerical data that can be displayed using consecutive intervals are best displayed in a histogram.   **25.** Graph A; Sample answer: The scale used makes the increase appear less significant.

1. true   **3.** false; measures of variation   **5.** true
7. true   **9.** false; outlier   **11.** 80; 86; no mode
13. range: 11; median: 5; LQ: 4; UQ: 6; IR: 2; 14 is an outlier   **15.** Sample answer: The range of the data is 7. The median is 5.5. The interquartile range is 4, the lower quartile is 2.5, the upper quartile is 6.5, and there are no outliers. Half of Brittany's friends have been bowling between 2.5 and 6.5 times. One fourth have been bowling less than 2.5 times. One fourth have gone more than 6.5 times.   **17.** Sample answer: The length of the left half of the box-and-whisker plot is shorter than the length of the right half of the plot. This means that the lower numbers of posters are more concentrated, and the higher numbers of posters are more spread out.

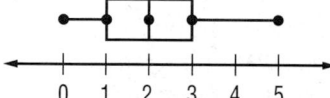

19. 5 seconds   **21.** Most runners were in the 60–64 interval.

23.

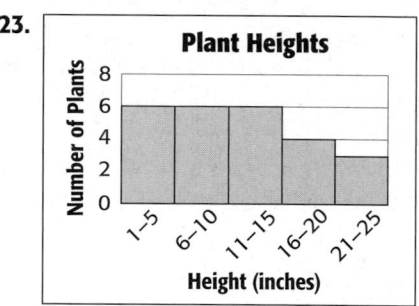

Plant Heights

25.   **Points Scored**

Stem	Leaf
5	3
6	
7	5 8 8
8	3 5 7 7 9
9	1 2

$7 \mid 5 = 75$ points

27. 1.7 points per game   **29.** Sample answer: stem-and-leaf plot

Chapter 10 Volume and Surface Area

1. 25 yd^2 **3.** 875 ft^2 **5.** 379.9 in^2 **7.** 78.5 cm^2

1. 220 in^3 **3.** 63 yd^3 **5.** second cabinet; 37.5 ft^3 < 63 ft^3 **7.** 90 ft^3 **9.** 236.3 cm^3

11 $V = Bh$ Volume of a prism

$V = \left(\frac{1}{2} \cdot 6 \cdot 4\right)h$ Replace B with $\frac{1}{2} \cdot 6 \cdot 4$.

$V = \left(\frac{1}{2} \cdot 6 \cdot 4\right)9$ The height of the prism is 9.

$V = 108$ Multiply.

The volume is 108 cubic meters or 108 m^3.

13. 20.4 mm^3 **15.** 40 ft^3 > 36 ft^3, so too much was bought. **17.** $166\frac{1}{4}$ yd^3 **19a.** 2,157,165 ft^3
19b. 98,053 ft^3

㉑ First, find the volume of the office.

$V = \ell wh$	Volume of a prism
$V = 32 \cdot 25 \cdot 12$	$\ell = 32, w = 25, h = 12$
$V = 9{,}600$	Multiply.

The volume of the office is 9,600 cubic feet. Next, find the cost to air condition the office for a year. Find the product of the cost per cubic foot and the number of cubic feet. The cost to air condition the space is 9,600 ft^3 · \$0.11/ft^3 or \$1,056 per year. To find the average cost per month, divide the yearly cost by 12. On average, it costs \$1,056 ÷ 12 or \$88 to air condition the office for one month.
23a. Sample answer: There is a direct relationship between the volume and the length. Since the length is doubled, the volume is also doubled.
23b. The volume is eight times greater.
23c. neither; Sample answer: Doubling the height will result in a volume of 4 · 4 · 10 or 160 in^3; doubling the width will result in a volume of 4 · 8 · 5 or 160 in^3. **25.** Sample answer: They are similar in that the volume is the product of the area of the base and the height of the prism. They are different in the formulas used to find the area of the base of the figure. **27.** 864 **29.** B

Pages 565–568 **Lesson 10-1C**

1. 141.4 in^3

③

$V = \pi r^2 h$	Volume of a cylinder
$V = \pi(5.5)^2 6.5$	Replace r with 5.5 and h with 6.5.
$V = 617.7$ ft^3	Simplify.

5. 603.2 cm^3 **7.** 4,071.5 ft^3 **9.** 2,770.9 yd^3 **11.** 35.6 m^3

⑬

$r = \frac{1}{2}d$	Relationship between diameter and radius
$r = \frac{1}{2}(4.5)$ or 2.25	Use $d = 4.5$. Simplify.
$V = \pi r^2 h$	Volume of a cylinder
$V = \pi(2.25)^2 6.5$	Replace r with 2.25 and h with 6.5.
$V = 103.4$	Simplify.

The volume of the cylinder is about 103.4 m^3.
15. 288.6 in^3 **17.** 34.4 in^3 **19.** 124,642.7 m^3 **21.** d
23. a **25.** 2,375 cm^3

㉗ Let V_A be the volume of cylinder A, and V_B be the volume of cylinder B.

$V_A = V_B$	Both cylinders have the same volume.
$\pi r^2 h = \pi r^2 h$	Volume of a cylinder
$\pi(4)^2 2 = \pi(2)^2 h$	For Cylinder A, $r = 4$ and $h = 2$. For Cylinder B, $r = 2$.
$32\pi = 4\pi h$	Simplify.
$\dfrac{32\pi}{4\pi} = \dfrac{4\pi h}{4\pi}$	Divide each side by 4π.
$8 = h$	Simplify.

The height of Cylinder B is 8 inches.

29. Sample answer: The shorter cylinder, because the radius is larger and that is the squared value in the formula. **31.** 1 to 2 **33.** Sample answer: In both, the volume equals the area of the base times the height. **35.** 9 **37.** 75.36 cm^3

Pages 573–574 **Lesson 10-1E**

①

$V = \frac{1}{3}Bh$	Volume of a cone
$V = \frac{1}{3}\left(\frac{1}{2} \cdot 8 \cdot 6\right)10$	$B = \frac{1}{2} \cdot 8 \cdot 6, h = 10$
$V = 80$	Simplify.

The volume of the cone is 80 ft^3.
3. 12 cm **5.** 5,971,000 ft^3 **7.** 109.3 m^3 **9.** 14 in.
11. 11 ft

⑬

$V = \frac{1}{3}Bh$	Volume of a pyramid
$V = \frac{1}{3}(3 \cdot 2.5)4$	$B = 3 \cdot 2.5, h = 4$
$V = 10$	Simplify.

The volume of glass used to create the pyramid was 10 in^3.
17. Sample answer: first set: area of the base, 40 ft^2; height of the pyramid, 12 ft; second set: area of the base, 30 ft^2; height of the pyramid, 16 ft. **19.** The volumes are the same. **21.** B **23.** 624.9 m^3
25. 13,965 in^3

Pages 576–579 **Lesson 10-1F**

1. 2,668.3 m^3

③

$V = \frac{1}{3}\pi r^2 h$	Volume of a cone
$V = \frac{1}{3}\pi(1.75)^2 8.4$	$r = \frac{1}{2} \cdot 3.5$ or 1.75, $h = 8.4$
$V = 26.9$	Simplify.

The volume of the cone is 26.9 ft^3.
5. 1.8 c **7.** 4,720.8 mm^3 **9.** 2,989.8 mm^3
11. 402.1 cm^3 **13.** about 7 c **15.** 10 mm

⑰ Let h be the height of the cylinder, H be the height of the cone, and r be the radius of both.

$\pi r^2 h = \frac{1}{3}\pi r^2 H$	The volume of the cylinder is equal to the volume of the cone.
$\pi(5)^2(12) = \frac{1}{3}\pi(5)^2 H$	$r = 5, h = 12$
$300\pi = \frac{1}{3}(25)\pi H$	Simplify.
$900\pi = 25\pi H$	Multiply both sides by 3.
$\dfrac{900\pi}{25\pi} = \dfrac{25\pi H}{25\pi}$	Divide each side by 25π.
$36 = H$	Simplify.

The height of the cone is 36 cm.
19. 4.5 m **21.** 3.0 yd **23.** Aisha used the incorrect radius; 25.1 in^3 **25.** Sample answer: Depending on the length of the radius and the height, generally doubling the radius has more effect as it is squared in the formula. **27.** A **29.** 3.5 in^3 **31.** 129.5 m^3

Pages 585–587 **Lesson 10-2B**

1. 108 ft^2 **3.** Yes; the surface area of the box is 252 in^2. The surface area of the paper is 288 in^2. Since 252 in^2 < 288 in^2, she has enough paper.

5. 314 cm²

7 Replace ℓ with 12.3, w with 8.5, and h with 15.

$S.A. = 2\ell w + 2\ell h + 2wh$

$= 2 \cdot 12.3 \cdot 8.5 + 2 \cdot 12.3 \cdot 15 + 2 \cdot 8.5 \cdot 15$

$= 209.1 + 369 + 255$ Multiply first. Then add.

$= 833.1$

The surface area of the prism is 833.1 mm².

9. 125.4 in²   **11.** 207 in²   **13.** 1,128.8 m²

15. 192 cm²   **17.** 64.5 in²

19 Remember that each edge of a cube has the same measure. Replace ℓ with x, w with x, and h with x.

$S.A. = 2\ell w + 2\ell h + 2wh$

$= 2 \cdot x \cdot x + 2 \cdot x \cdot x + 2 \cdot x \cdot x$

$= 2x^2 + 2x^2 + 2x^2$ Multiply first. Then add like terms.

$= 6x^2$

The formula for the surface area of a cube is $S.A. = 6x^2$.

21. 4 ft by 7 ft by 6 ft; The dimensions of the dunk tank are sufficient for a person to fall in and get wet. **23.** False; Sample answer: A rectangular prism with a length of 9 units, a width of 7 units, and height of 13 units has a surface area of $2(9 \times 13) + 2(9 \times 7) + 2(13 \times 7)$ or 542 square units. If you double the length, the surface area is $2(18 \times 13) + 2(18 \times 7) + 2(13 \times 7)$ or 902 square units. So, 2×542 is 1,084. $1,084 \neq 902$ **25.** Sample answer: Surface area measures the area of the faces, and area is measured in square units.   **27.** G   **29.** 104.7 yd³

Pages 589–591   **Lesson 10-2C**

1 $S.A. = 2\pi rh + 2\pi r^2$ Surface area of a cylinder

$= 2\pi(2)(5) + 2\pi(2)^2$ Replace r with 2 and h with 5.

≈ 88.0 Simplify.

The surface area is about 88.0 mm².

3. 471.2 m²   **5.** 1,215.8 m²   **7.** 272.0 mm²

9. 1,120.0 in²   **11.** 61.3 cm²   **13.** Sample answer: $2 \cdot 3 \cdot 4^2 + 2 \cdot 3 \cdot 4 \cdot 4$ or 192 m²

15 Find the surface area of the tube.

$S.A. = 2\pi rh + 2\pi r^2$ Surface area of a cylinder

$= 2\pi(2.5)(15) + 2\pi(2.5)^2$ Replace r with 2.5 and h with 15.

$= 274.9$ Simplify.

Find the curved surface of the tube.

$S.A. = 2\pi rh$ Curved surface of a cylinder

$= 2\pi(2.5)(15)$ Replace r with 2.5 and h with 15.

$= 235.6$ Simplify.

Find the percent of the tube that is cardboard.

$235.6 = N \cdot 274.9$ Percent Equation

$\dfrac{235.6}{274.9} = \dfrac{N \cdot 274.9}{274.9}$ Divide each side by 274.9.

$0.857 = N$ Simplify.

Changing the decimal to a percent, about 85.7% of the mail tube is cardboard.

---

17. Sample answer: Yes, it could make a difference. As a rule, calculating with more decimal places produces an answer closer to the exact value. **19.** D   **21.** 23.08 ft²   **23.** 3.5 in³   **25.** 560 m³ **27.** 158.4 m³

Pages 596–597   **Lesson 10-2E**

1 $S.A. = B + \frac{1}{2}P\ell$ Surface area of a pyramid

$= 25 + \frac{1}{2}(20 \cdot 7)$ $B = 5 \cdot 5$ or 25, $P = 4(5)$ or 20, $\ell = 7$

$= 95$ Simplify.

The surface area of the pyramid is 95 in².

3. 3,829.5 ft²   **5.** 507.5 mm²   **7.** 2,079 cm²   **9.** 26.1 ft²

11 $S.A. = B + \frac{1}{2}P\ell$ Surface area of a pyramid

$= 24 + \frac{1}{2}(18 \cdot 6)$ $B = 24$, $P = 6(3)$ or 18, $\ell = 6$

$= 78$ Simplify.

The surface area of the birdhouse is 78 in².

13. It would be shorter to climb up the slant height. Sample answer: The bottom of the slant height is closer to the center of the base of the pyramid. The bottom of the lateral edge is farther from the center of the base of the pyramid.   **15.** The formula is based on finding the area of each base and then adding them together.   **17.** H   **19.** 13,890 cm² **21.** 301.6 ft³

Pages 602–603   **Lesson 10-3A**   **PSI**

1. Sample answer: Finding the areas of the separate geometric figures and then adding is easier than trying to find the area of the irregular figure as a whole.   **3.** 114 ft²   **5.** 2 h

7 First find the number of minutes in $\frac{1}{3}$ hour. One third of 60 is 20. Make a chart.

Monday	45	45 min
Tuesday	45 + 20	65 min or 1 hr 5 min
Wednesday	1 hr 5 min + 20 min	1 hr 25 min
Thursday	1 hr 25 min + 20 min	1 hr 45 min
Friday	1 hr 45 min + 20 min	1 hr 65 min or 2 hr 5 min
Saturday	2 hr 5 min + 20 min	2 hr 25 min

2 hours 25 minutes

9. 76 students

Pages 608–610   **Lesson 10-3C**

1. 897 in³   **3.** 870 cm³   **5.** 39.6 in²

7 The solid is composed of a rectangular prism and a triangular prism. Let B_1 be the area of the base of the rectangular prism, $1.8 \cdot 1.1$ or 1.98. Let B_2 be the area of the base of the triangular prism,

$\frac{1}{2} \cdot 1.8 \cdot 0.7$ or 0.63.

$V = B_1 h_1 + B_2 h_2$ Volume of rectangular prism + volume of triangular prism.

$= 1.98(0.8) + 0.63(1.1)$ Replace B_1 with 1.98, B_2 with 0.63, h_1 with 0.8, h_2 with 1.1.

$= 1.584 + 0.693$

$= 2.3$ Simplify.

The volume of the solid, to the nearest tenth, is 2.3 cubic meters.

9. 1,476.5 m²   **11.** 308.4 m²   **13.** 727.9 in²

15 The solid is composed of a rectangular prism and a triangular prism. Convert inches to feet before calculating.

Find the volume of the rectangular prism.

$V = \ell w h$ Volume of a rectangular prism

$\approx (1)(0.42)(0.83)$ Replace ℓ with 1, w with 0.42, and h with 0.83.

≈ 0.35 Simplify.

The volume of the rectangular prism is about 0.35 cubic feet.

Find the volume of the triangular prism.

$V = \frac{1}{2}bh\ell$ Volume of a triangular prism

$\approx \frac{1}{2}(1)(0.83)(0.42)$ Replace b with 1, h with 0.83, and ℓ with 0.42.

≈ 0.17 Simplify.

The volume of the triangular prism is about 0.17 cubic feet.

The volume of the entire solid is about $0.35 + 0.17$ or 0.5 cubic feet.

17. Sample answer:

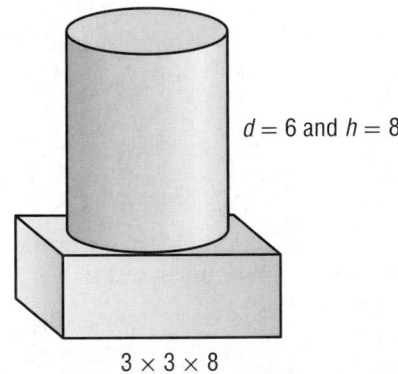

$d = 6$ and $h = 8$

$3 \times 3 \times 8$

19. Sample answer: The area of the rectangular face on the half cylinder must be found, making it a composite figure.   **21.** I   **23.** 164.01 cm²   **25.** 923.6 ft²   **27.** 502.7 ft³

Pages 611–615   *Chapter Study Guide and Review*

1. rectangular prism   **3.** cubic   **5.** cone
7. multiplying   **9.** 14.6 m³   **11.** 371.9 in³
13. 4,042.4 km³   **15.** 48.1 in³   **17.** 6.3 mm³
19. 288 cm³   **21.** 377.0 in³   **23.** 202 yd³   **25.** 43 ft²
27. 2,261.9 in²   **29.** 168 in²   **31.** 998,750 ft²
33. \$106   **35.** 13 m³

Chapter 11 Measurement and Proportional Reasoning

Page 622   *Chapter 11*   *Are You Ready?*

1. $\frac{3}{5}$ **3.** $\frac{1}{4}$ **5.** $1\frac{1}{2}$ **7.** 4 acres   **9.** 900   **11.** 2,400
13. 81.5

Pages 628–630   *Lesson 11-1B*

1. 48

3 Since 1 cup = 8 fluid ounces, the unit ratio is $\frac{16\ \text{fl oz}}{1c}$.

$6.5\ c = 6.5\ c \cdot \frac{8\ \text{fl oz}}{1c}$ Multiply by $\frac{8\ \text{fl oz}}{1c}$. Divide out common units, leaving the desired unit, fluid ounces.

$= 6.5 \cdot \frac{8\ \text{fl oz}}{1}$

$= 6.5 \cdot 8\ \text{fl oz}$ or 52 fl oz   Simplify.

So, 6.5 cups = 52 fluid ounces.

5. 3   **7.** $7\frac{1}{2}$ **9.** 6   **11.** 32   **13.** 9   **15.** 10,560   **17.** $2\frac{1}{2}$
19. $5\frac{1}{2}$

21 Since 1 ton = 2,000 pounds, the unit rate is $\frac{2,000\ \text{lb}}{1\ \text{T}}$.

$\frac{1}{2}\ \text{T} = \frac{1}{2}\ \text{T} \cdot \frac{2,000\ \text{lb}}{1\ \text{T}}$ Multiply $\frac{2,000\ \text{lb}}{1\ \text{T}}$ by $\frac{1}{2}$ since you have $\frac{1}{2}$ T.

$= \frac{1}{2} \cdot \frac{2,000\ \text{lb}}{1}$ Divide out common units, leaving the desired unit, pounds.

$= \frac{2,000\ \text{lb}}{2}$ or 1,000 lb   Simplify.

So, $\frac{1}{2}$ ton = 1,000 pounds.

23. 13 yd

25 No; $2 + 2 + \frac{1}{4} + \frac{1}{3} + 4 = 8\frac{7}{12}$ c punch and a 2-qt pitcher holds $2\ \text{qt} \cdot \frac{2\ \text{pt}}{1\ \text{qt}} \cdot \frac{2\ \text{c}}{1\ \text{pt}} = 2 \cdot 2 \cdot 2$ c or 8 c. Since $8\ \text{c} < 8\frac{7}{12}$ c, the pitcher will not hold all of the punch.

27. 20   **29.** 82.8   **31.** Sample answer: Annabelle is making brownies. The recipe calls for 2 cups of sour cream. She has 2 pints of sour cream. Does she have enough sour cream to make the brownies?
33. >; Sample answer: $8\frac{3}{4}$ gal is equivalent to 35 qt, Since 35 qt > 32 qt, $8\frac{3}{4}$ gal > 32 qt.   **35.** Sample answer: 5 pt; 80 oz   **37.** B   **39.** D

Pages 633–634   *Lesson 11-1C*

1. 370   **3.** 1.46   **5.** 500,000 cm

7 1 mm = 0.001 m   Write the relationship.

$983 \times 1\ \text{mm} = 983 \times 0.001\ \text{m}$ Multiply each side by 983 since you have 983 mm. To multiply 983 by 0.001, move the decimal point 3 places to the left.

$983\ \text{mm} = 0.983\ \text{m}$

So, 983 mm ≈ 0.98 m.

9. 30   **11.** 0.08   **13.** 130,500   **15.** 0.51 kg   **17.** about 35,000 meters per hour   **19.** 0.06 L, 660 mL, 6.6 kL
21. 130 cm, 2,650 mm, 5 m

23 First subtract to find how many meters Jacinta needs to cut off the pole. She needs to cut off $3 - 2.5$ or 0.5 meter. Next, convert 0.5 meter to centimeters.

$1\text{ m} = 100\text{ cm}$	Write the relationship.
$0.5 \times 1\text{ m} = 0.5 \times 100\text{ cm}$	Multiply each side by 0.5 since you have 0.5 m.
$0.5\text{ m} = 50\text{ cm}$	

25. Sample answer: Theresa divided 3.25 by 1,000. She should have multiplied; 3,250 g.   **27.** Sample answer: There are a greater number of smaller units.   **29.** H   **31.** eighth inch

Pages 636–638   Lesson 11-1D

1. 3.37

3 Since 1 pound is approximately equal to 0.4536 kg, multiply by $\dfrac{1\text{ lb}}{0.4536\text{ kg}}$.

$58.14\text{ kg} \approx 58.14\text{ kg} \cdot \dfrac{1\text{ lb}}{0.4536\text{ kg}}$	Multiply by $\dfrac{1\text{ lb}}{0.4536\text{ kg}}$.
$= 58.14 \cdot \dfrac{1\text{ lb}}{0.4536}$	Divide out common units, leaving the desired unit, pounds.
$\approx 128.17\text{ lb}$	Simplify.

So, $58.14\text{ kg} \approx 128.17\text{ lb}$.

5. 15.75   **7.** 5,333 ft   **9.** 12.7   **11.** 2.11   **13.** 3.79
15. 57,607.2   **17.** 21.26   **19.** 121.28   **21.** 41.89
23. 15.24   **25.** 4.96 lb   **27.** 435.3 m

29 First, convert 1.5 pounds to grams in order to compare like units. Since 1 pound is approximately equal to 453.6 g, multiply by $\dfrac{453.6\text{ g}}{1\text{ lb}}$.

$1.5\text{ lb} \approx 1.5\text{ lb} \cdot \dfrac{453.6\text{ g}}{1\text{ lb}}$	Multiply by $\dfrac{453.6\text{ g}}{1\text{ lb}}$.
$= 1.5 \cdot \dfrac{453.6\text{ g}}{1}$	Divide out common units, leaving the desired unit, grams.
$\approx 680.4\text{ g}$	Simplify.

So, $1.5\text{ lb} \approx 680.4\text{ g}$.
Next, compare the two measures. 680.4 is greater than 650, so the 1.5-pound box is greater than the 650-gram box.

31. 3 gal   **33.** 4 mi   **35.** 0.83 ft; 1.67 ft; 2.5 ft; 3.33 ft
37. about 149.73 gigameters   **39.** 0.02 T, 1 kg, 2 lb, 891 g   **41.** Sample answer: When ordering lengths with different units of measures, you must change all of the measurements to the same unit and then order the lengths.   **43.** H   **45.** 5.7   **47.** 15,840   **49.** 18

Pages 641–642   Lesson 11-1E

1 You can use 1 mile = 5,280 feet to convert.

$\dfrac{607,200\text{ ft}}{1\text{ hr}} = \dfrac{607,200\text{ ft}}{1\text{ hr}} \cdot \dfrac{1\text{ mi}}{5,280\text{ ft}}$	Multiply by $\dfrac{1\text{ mi}}{5,280\text{ ft}}$.
$= \dfrac{607,200}{1\text{ hr}} \cdot \dfrac{1\text{ mi}}{5,280}$	Divide out common units.
$= \dfrac{607,200 \cdot 1\text{ mi}}{1 \cdot 5,280\text{ hr}}$	Simplify.
$= \dfrac{115\text{ mi}}{1\text{ hr}}$	Simplify.

The go-kart's top speed is 115 mph.

3. 5 mi/h   **5.** 322,000 m/h   **7.** 2.48 L/week   **9.** 9.6
11. 15.6   **13.** 1.6   **15.** 0.4

17 a. You can use 1 mile = 5,280 feet, 1 hour = 60 minutes, and 1 minute = 60 seconds.

$\dfrac{4.4\text{ m}}{1\text{ hr}} = \dfrac{4.4\text{ mi}}{1\text{ hr}} \cdot \dfrac{5,280\text{ ft}}{1\text{ mi}} \cdot \dfrac{1\text{ hr}}{60\text{ min}} \cdot \dfrac{1\text{ min}}{60\text{ sec}}$	Multiply by distance and time unit ratios.
$= \dfrac{4.4}{1} \cdot \dfrac{5,280\text{ ft}}{1} \cdot \dfrac{1}{60} \cdot \dfrac{1}{60\text{ sec}}$	Divide out common units.
$= \dfrac{4.4 \cdot 5,280 \cdot 1 \cdot 1\text{ ft}}{1 \cdot 1 \cdot 60 \cdot 60\text{ sec}}$	Simplify.
$= \dfrac{23,232\text{ ft}}{3,600\text{ sec}}$	Simplify.
$= \dfrac{6.453\text{ ft}}{1\text{ sec}}$	Simplify.

The speed of the housefly is about 6.45 feet per second.

b. You can use 1 minute = 60 seconds.

$\dfrac{38\text{ beats}}{1\text{ sec}} = \dfrac{38\text{ beats}}{1\text{ sec}} \cdot \dfrac{60\text{ sec}}{1\text{ min}}$	Multiply by $\dfrac{60\text{ sec}}{1\text{ min}}$.
$= \dfrac{38\text{ beats}}{1} \cdot \dfrac{60}{1\text{ min}}$	Divide out common units.
$= \dfrac{38 \cdot 60\text{ beats}}{1 \cdot 1\text{ min}}$	Simplify.
$= \dfrac{2,280\text{ beats}}{1\text{ min}}$	Simplify.

A dragonfly's wings beat 2,280 times per minute.

c. You can use 1 hour = 60 minutes and 1 mile = 1.61 kilometers.

$\dfrac{6.4\text{ mi}}{1\text{ hr}} \approx \dfrac{6.4\text{ mi}}{1\text{ hr}} \cdot \dfrac{1.61\text{ km}}{1\text{ mi}} \cdot \dfrac{1\text{ hr}}{60\text{ min}}$	Multiply by distance and time unit ratios.
$= \dfrac{6.4}{1} \cdot \dfrac{1.61\text{ km}}{1} \cdot \dfrac{1}{60\text{ min}}$	Divide out common units.
$= \dfrac{6.4 \cdot 1.61 \cdot 1\text{ km}}{1 \cdot 1 \cdot 60\text{ min}}$	Simplify.
$= \dfrac{10.304}{60\text{ min}}$	Simplify.
$\approx \dfrac{0.17\text{ km}}{1\text{ min}}$	Simplify.

A bumblebee can travel 0.17 kilometer in 1 minute.

d. You can use 1 minute = 60 seconds and 1 hour = 60 minutes.

$\dfrac{250\text{ beats}}{1\text{ sec}} = \dfrac{250\text{ beats}}{1\text{ sec}} \cdot \dfrac{60\text{ min}}{1\text{ hr}} \cdot \dfrac{60\text{ sec}}{1\text{ min}}$	Multiply by distance and time unit ratios.
$= \dfrac{250\text{ beats}}{1} \cdot \dfrac{60}{1\text{ hr}} \cdot \dfrac{60}{1}$	Divide out common units.
$= \dfrac{250 \cdot 60 \cdot 60\text{ beats}}{1 \cdot 1 \cdot 1\text{ hr}}$	Simplify.
$= \dfrac{900,000\text{ beats}}{1\text{ hr}}$	Simplify.

A honeybee's wings beat 900,000 times an hour.

19. Divya should have multiplied by the reciprocals of the unit ratios. The correct answer is about 1.74 km/min.   **21.** D   **23.** 21.6 cm by 27.9 cm   **25.** 6   **27.** 0.90

Pages 646–648   Lesson 11-1F

1. 432   **3.** 5   **5.** 9   **7.** 345.6   **9.** 40.5   **11.** 4,100,000
13. 213.8 yd^2 **15.** 1,656

17 Use 1 square yard = 9 square feet

$216 \text{ ft}^2 = 216 \text{ ft}^2 \cdot \dfrac{1 \text{ yd}^2}{9 \text{ ft}^2}$ Multiply by $\dfrac{1 \text{ yd}^2}{9 \text{ ft}^2}$.

$\quad\quad = 216 \cdot \dfrac{1 \text{ yd}^2}{9}$ Divide out common units, leaving the desired unit, square yards.

$\quad\quad = 24 \text{ yd}^2$ Simplify.

19. 75,000   **21.** 540   **23.** 27,878,400   **25.** 6.5   **27.** 10.4
29. 450,000   **31.** 0.97   **33.** 6,400 yd²   **35.** 0.94 yd³
37. 10 cc   **39.** 1,296 in²

41 First, find the number of cubic yards in a cubic meter.

$V = \ell wh$ Write the formula.
$\quad = (1.1 \text{ yd})(1.1 \text{ yd})(1.1 \text{ yd})$ Replace ℓ, w, and h with 1.1.
$\quad = 1.331 \text{ yd}^3$ Simplify.

Next, convert 250,000 m³ to cubic yards.

$250{,}000 \text{ m}^3 \approx 250{,}000 \text{ m}^3 \cdot \dfrac{1.331 \text{ yd}^3}{1 \text{ m}^3}$ Multiply.

$\quad\quad = 250{,}000 \cdot \dfrac{1.331 \text{ yd}^3}{1}$ Divide out common units, leaving the desired unit, square yards.

$\quad\quad \approx 332{,}750 \text{ yd}^3$ Simplify.

The swimming pool has a volume of 332,750 yd³.
43. Sample answer: He would have to divide 169 square feet by 9 since there are 3 feet in each yard, and he was converting to square yards.   **45.** He should have multiplied by the reciprocal; 14,929,920 in³.   **47.** B   **49.** G   **51.** 35.2   **53.** 0.2   **55.** peaches

---

Pages 650–651   **Lesson 11-2A**   **PSI**

1. Sample answer: Making a model helps you to see what is happening in the problem which can help you solve it.
3. 9 in. × $10\frac{1}{2}$ in.   **5.** yes; $1{,}900 + 2{,}100 + 1{,}600 > 5{,}000$

7 $45.50 - 5(2.75) - 14.95 + 10.00$ Write an expression.
$\quad = 26.80$ Simplify.
At the end of the week, Myra has $26.80.

9. $125\frac{7}{16}$ ft²   **11.** Sample answer: How many feet of chrome edging do you need to finish a tabletop that is 54 inches long and 30 inches wide? 14 ft

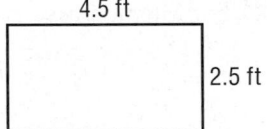
4.5 ft / 2.5 ft

Pages 657–659   **Lesson 11-2C**

1. 1,715 in²   **3.** 15.7 cm³   **5.** 11,700 in²

7 $S.A. = 280 \times \left(\dfrac{1.4}{1}\right)^2$ Multiply by the square of the scale factor.
$\quad = 280 \times \dfrac{1.96}{1}$ Square $\dfrac{1.4}{1}$.
$\quad = 548.8$ Simplify.
The surface area of the larger box is 548.8 square inches.

---

9. 19 mm³   **11.** 251,712 ft²; 8,709,120 ft³

13 $V = 126.9 \times \left(\dfrac{4}{3}\right)^3$ Multiply by the cube of the scale factor.
$\quad V = 126.9 \times \dfrac{64}{27}$ Cube $\dfrac{4}{3}$.
$\quad V = 300.8$ Simplify.

The volume of the larger sphere is 300.8 m³.
15. sometimes   **17.** never   **19a.** 3 : 1   **19b.** surface area, 9 : 1; volume, 27 : 1   **19c.** 602.88 cm²
19d. 30,520.8 cm³   **21a.** Yes, the ratios $\frac{3}{6}$ and $\frac{1.5}{3}$ are equal.   **21b.** 7.065 in³; 56.52 in³   **21c.** 1:8
21d. 49.455 in³   **23.** C   **25.** 7.5 cm

Pages 662–665   **Chapter Study Guide and Review**

1. dimensional analysis   **3.** metric system   **5.** gram
7. kilogram   **9.** 8   **11.** 3   **13.** 5.7 mi   **15.** 1,825
17. 24,000   **19.** 1,020 grams   **21.** 51,528.96
23. 18.43   **25.** about 345 lb   **27.** 1.76   **29.** 2.24
31. 0.23 m/s   **33.** 3,000   **35.** 0.58   **37.** 37.8 ft²
39. 120 ways   **41.** $S.A. = 201.2$ in²; $V = 176$ in³

Chapter 12 Polygons and Transformations

Page 672   **Chapter 12**   **Are You Ready?**

1. $(-1, 2)$ **3.** $(-3, -1)$

5–8.
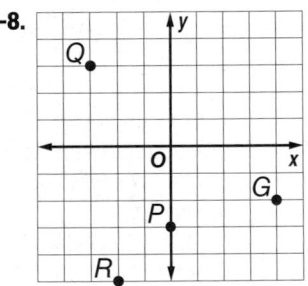

9. Sample answer:
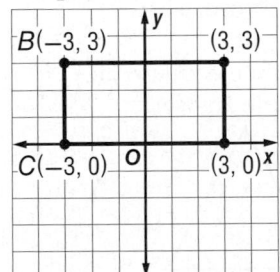

11.
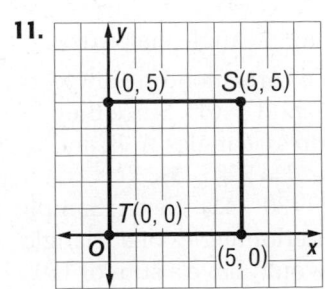

Pages 677–679   Lesson 12-1A

1. ∠CAM, ∠MAC, ∠A, ∠1; acute   **3.** supplementary
5. 100

7 Use the vertex as the middle letter and a point from each side: ∠JDB, ∠BDJ
Use the vertex only: ∠D
Use a number: ∠6
The square at the vertex indicates that this is a right triangle.
9. ∠MLP, ∠PLM, ∠L, ∠8; obtuse   **11.** neither
13. supplementary   **15.** 53°

17 Sample answer: ∠BED and ∠BEG because ∠BEG makes a 90° angle as the two lines intersect. Since ∠BEG is 90°, then ∠BED must also be 90°.
19. Sample answer: ∠CBH   **21.** Sample answer: stairs and landings form obtuse angles.
23. $m\angle K = 52°$ and $m\angle O = 128°$
25. Complementary; the other terms name specific angles, while complementary is a relationship between two angles.   **27.** B   **29.** corresponding
31. alternate interior   **33.** 74°   **35.** 74°

Pages 684–686   Lesson 12-1C

1. 44   **3.** 45   **5.** 60°   **7.** right, scalene
9. acute isosceles; Sample answer:

11. 118   **13.** 27   **15.** 90

17

$$25 + 102 + m\angle Q = 180$$ The sum of the measures of the angles of a triangle is 180°.

$$127 + m\angle Q = 180$$ Simplify.

$$127 + m\angle Q - 127 = 180 - 127$$ Subtract 127 from each side.

$$m\angle Q = 53°$$ Simplify.

Therefore, $m\angle Q = 53°$.
19. acute, equilateral   **21.** obtuse, isosceles
23. obtuse, isosceles   **25.** right
27. acute scalene;   **29.** acute equilateral;
Sample answer: Sample answer:

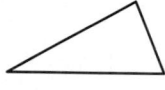   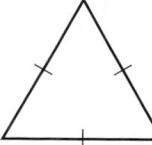

31. 79.5   **33.** 79.2   **35.** 21.3

37 ∠C is a vertical angle to the angle marked as 47°. So, $m\angle C = 47°$. ∠A is a corresponding angle to the angle marked 61°. So, $m\angle A = 61°$. Since they form a triangle, the measures of angles A, B, and C have a sum of 180°. So, $m\angle B = 72°$.   **39.** 30
41. $a = 55$; $b = 65$; $c = 60$; $d = 30$ **43a.** Never; Sample answer: The sum of the interior angles of a triangle is 180°. Two right angles would have a sum of 180°.

The third angle would have to equal 0°, which is not possible.   **43b.** Never; Sample answer: The sum of the interior angles of a triangle is 180°. If a triangle has two obtuse angles, the sum of these two angles, not including the third angle, would already be greater than 180°.   **45.** B   **47.** 47
49. complementary   **51.** neither

Pages 689–692   Lesson 12-1D

1. 75

3 The one way sign has opposite sides that are congruent and four right angles. It is a rectangle. The men working sign has all four sides congruent and four right angles. It is a square.
5. 120°   **7.** 70

9

$$x + 95 + 55 + 110 = 360$$ The sum of the angles of a quadrilateral is 360°.

$$x + 260 = 360$$ Simplify.

$$x + 260 - 260 = 360 - 260$$ Subtract 260 from each side.

$$x = 100$$ Simplify.

11. 65   **13.** square   **15.** trapezoid
17. rhombus   **19.** The first figure is a rectangle. The second figure is a trapezoid.   **21.** 90.8   **23.** 45
25. Sample answer: The figure is a rectangle. It has four congruent angles, not sides.   **27a.** Greater than; Sample answer: 360° ÷ 4 = 90°, so the average angle measure is 90°. Since the three congruent angles are all less than 90°, the remaining angle must be greater than 90° to make sure that all four angles add to 360°.   **27b.** Equal to; Sample answer: 360° ÷ 4 = 90°, so the average angle measure is 90°. Since the three congruent angles are equal to 90°, the remaining angle must also be equal to 90° to make sure that all four angles add to 360°.   **29.** sometimes; Sample answer: When the quadrilateral has opposite sides parallel and congruent, it is a parallelogram.
31. always; Sample answer: A square has all the characteristics of a rectangle. A square is a special kind of rectangle with all sides congruent.   **33.** 105; Sample answer: 360° − (90° + 90° + 75°) = 105°
35. H
37. 105   **39.** 74   **41.** neither   **43.** complementary

Pages 695–698   Lesson 12-1E

1. no; curved   **3.** no; open
5 Draw a heptagon, a 7-sided polygon. Draw all the diagonals from one vertex. This makes 5 triangles. 5 × 180° = 900° So, the sum of the angle measures of a heptagon is 900°.

7 The figure has 16 straight sides that meet only at their endpoints, so it is a polygon. The figure is not equiangular or equilateral, so it is not regular.
9. no; curved   **11.** yes; pentagon, regular   **13.** 1,260°
15. 1,800°   **17.** 135°   **19.** triangles and hexagons

21 20 sides, 162°; 50 sides, 172.8°; 100 sides, 176.4°;

Polygon	Sum of the Angles	Measure of One Angle
20-sided	18 × 180° = 3,240	3,240 ÷ 20 = 162°
50-sided	48 × 180° = 8,640	8,640 ÷ 50 = 172.8°
100-sided	98 × 180° = 17,640	17,640 ÷ 100 = 176.4°

Sample answer: As the number of sides increases, the angle measure gets closer to 180°, but will never be 180° because that is a straight angle.

23. 110° **25.** 117° **27.** 128°

29a.

Number of Sides of Polygon	Sum of Interior Angles
3	180
4	360
5	540
6	720
7	900
8	1,080
9	1,260
10	1,440
12	1,800
15	2,340

29b.

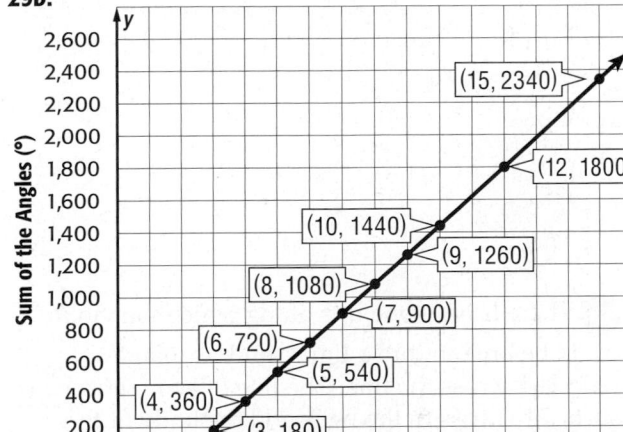

29c. Sample answer: The sum of the angles increases at a constant rate; there is no y- or x-intercept.

31. Sample answer: **33.** Sample answer:

 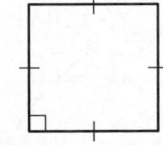

35. Sample answer: Regular decagons have equal angles measuring 144° and regular 11-sided polygons have angles measuring 147.27°

37. F **39.** G **41.** trapezoid **43.** rectangle **45.** 26

Pages 703–705   Lesson 12-2B

1.

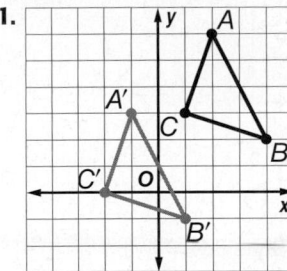

3. $D'(7, 0)$, $E'(4, -2)$, $F'(8, 4)$, $G'(12, -3)$

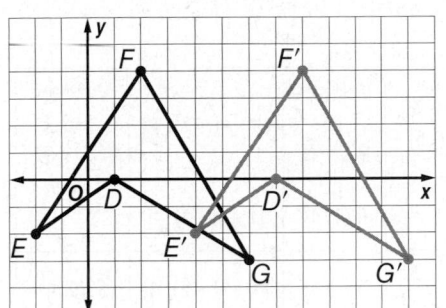

5.

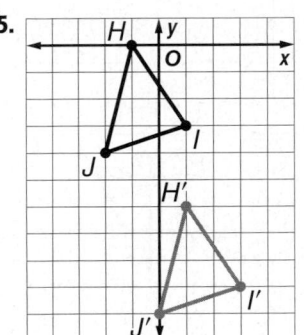

7 $P'(6, 5)$, $Q'(11, 3)$, $R'(3, 11)$

Vertices of △PQR	(x + 6, y + 5)	Vertices of △P'Q'R'
P(0, 0)	(0 + 6, 0 + 5)	P'(6, 5)
Q(5, −2)	(5 + 6, −2 + 5)	Q'(11, 3)
R(−3, 6)	(−3 + 6, 6 + 5)	R'(3, 11)

Use the vertices of △PQR and △P'Q'R' to graph both triangles.

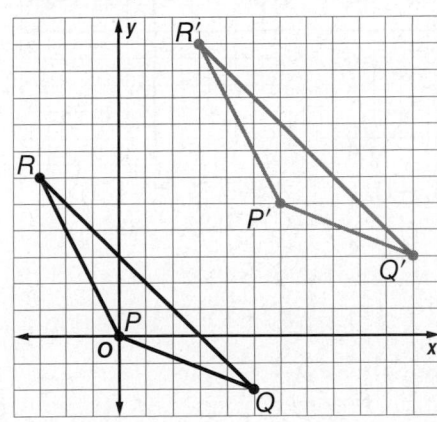

9. $P'(-3, 0), Q'(2, -2), R'(-6, 6)$

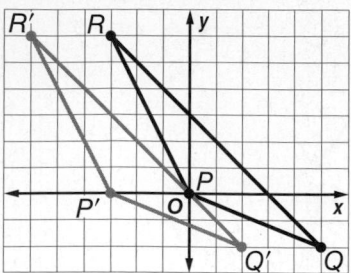

11a. 3 units right and 1 unit up; $(x + 3, y + 1)$
11b. 3 units left and 1 unit down; $(x - 3, y - 1)$

 13

Vertices of *RSTU*	$(x + 3, y + 5)$	Vertices of $R'S'T'U'$	$[x + (-2), y]$	Vertices of $R''S''T''U''$
$R(-2, -1)$	$[-2 + 3, (-1) + 5]$	$R'(1, 4)$	$[1 + (-2), 4]$	$R''(-1, 4)$
$S(1, -1)$	$(1 + 3, -1 + 5)$	$S'(4, 4)$	$[4 + (-2), 4]$	$S''(2, 4)$
$T(0, -3)$	$(0 + 3, -3 + 5)$	$T'(3, 2)$	$[3 + (-2), 2]$	$T''(1, 2)$
$U(-3, -3)$	$(-3 + 3, -3 + 5)$	$U'(0, 2)$	$[0 + (-2), 2]$	$U''(-2, 2)$

Use the verticews of $R''S''T''U''$ to graph the parallelogram.

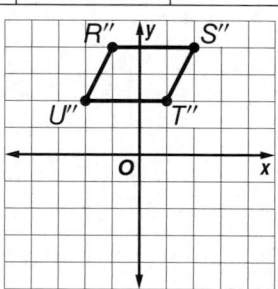

15. $(x - m, y + n)$ **17.** The final image is 3 units left and 3 units up from the original figure.   **19.** G
21. 720°   **23.** rectangle   **25.** right, scalene
27. acute, equilateral

Pages 714–716   Lesson 12-3B

1.

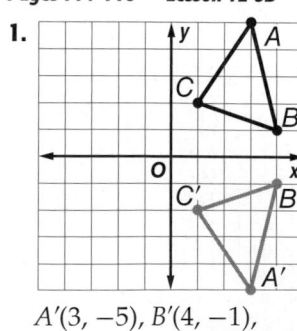 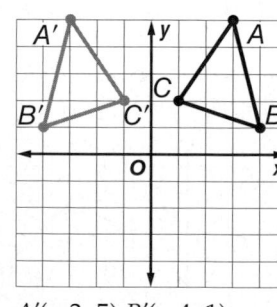

$A'(3, -5), B'(4, -1),$ $A'(-3, 5) B'(-4, 1),$
$C'(1, -2)$ $C'(-1, 2)$

3.

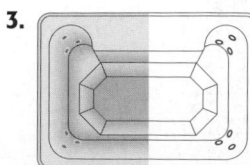

5.

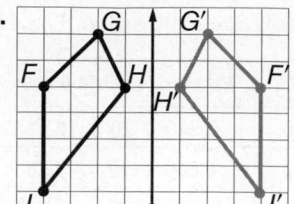

7 When a figure is reflected over the y-axis, the y-coordinate of the image is the same, while the x-coordinate is multiplied by -1. Therefore, triangle $F'G'H'$ has vertices $F'(-3, 3), G'(-4, -3),$ and $H'(-2, 1)$.

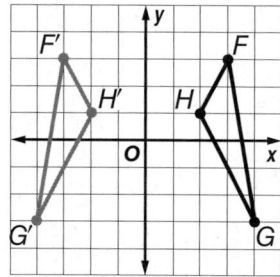

$F'(-3, 3), G'(-4, -3),$
$H'(-2, 1)$

9.

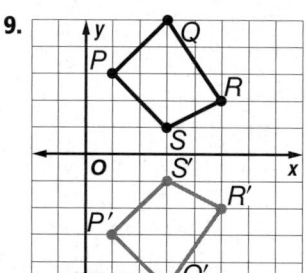

$P'(1, -3), Q'(3, -5),$
$R'(5, -2), S'(3, -1)$

11.

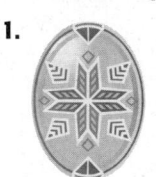

15 There is only one line along which the violin could be broken into a figure and its image reflected across the line. The line of symmetry runs vertically through the neck and the body of the violin.
17. figures A and C **19.** Sample answer: A reflection over the y-axis followed by a reflection over the x-axis.

21.

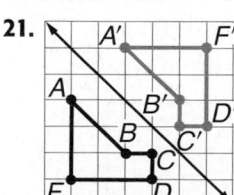

23. Sample answer: Marisol found the results of a reflection over the y-axis. They should be $A'(1, -1),$ $B'(4, -1)$ and $C'(1, -5)$.   **25.** $(-x + a, y + b)$ **27.** B

Step 1 Sketch △ABC on a coordinate plane.

Step 2 Sketch $\overline{AO}$ connecting point A to the origin. Sketch another segment, $\overline{A'O}$, so that the angle between point A, O, and A′ measures 90° and the segment is congruent to $\overline{AO}$.

Step 3 Repeat Step 2 for points B and C. Then connect the vertices to form △A′B′C′.

A′(2, −4), B′(0, −2), C′(4, −1)

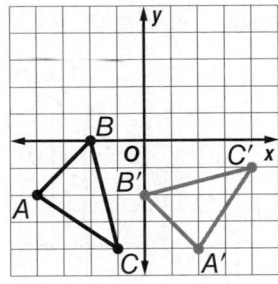

3. A′(−2, 4), B′(0, 2), C′(−4, 1)

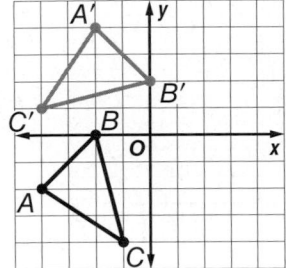

5. E′(−1, −1), F′(−3, −1), G′(−4, −4), H′(−2, −4)

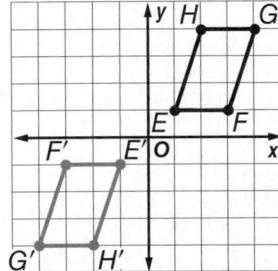

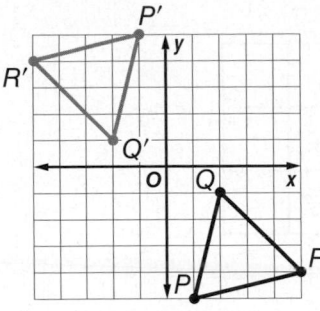

9. P′(−5, −1), Q′(−1, −2), R′(−4, −5)

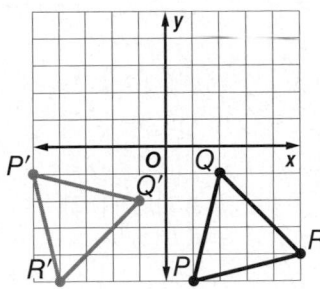

11. F′(−1, 1), G′(−5, 2), H′(−3, 5), J′(0, 4)

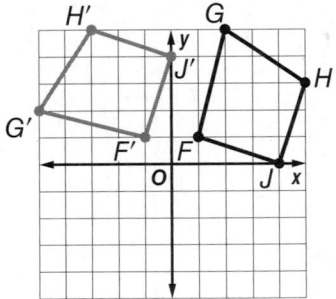

13. F′(−1, −1), G′(−2, −5), H′(−5, −3), J′(−4, 0)

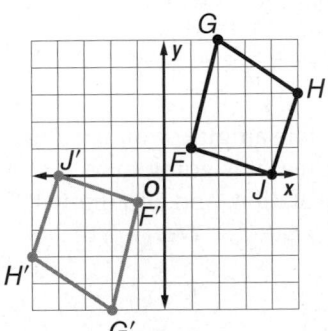

7

Step 1 Sketch △PQR on a coordinate plane.

Step 2 Sketch $\overline{PO}$, connecting point P to the origin. Sketch another segment, $\overline{P'O}$, so that the angle between point P, O, and P′ measures 180° and the segment is congruent to $\overline{PO}$.

Step 3 Repeat Step 2 for points Q and R. Then connect the vertices to form △P′Q′R′.

The vertices for △P′Q′R′ are P′(−1, 5), Q′(−2, 1), R′(−5, 4).

15.

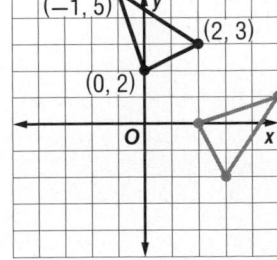

17. Sample answer:

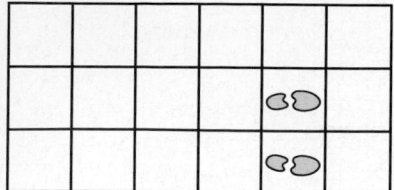

19. Sample answer: 180°

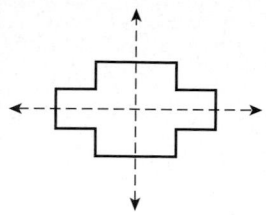

21.

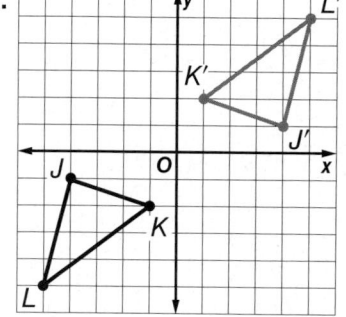

$J'(4, 1)$, $K'(1, 2)$, and $L'(5, 5)$

23.

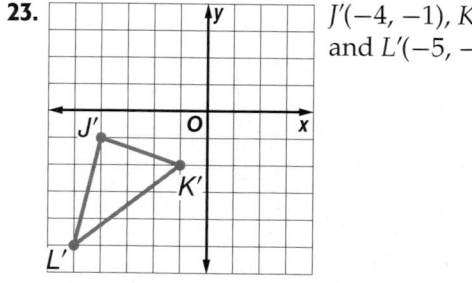

$J'(-4, -1)$, $K'(-1, -2)$, and $L'(-5, -5)$

25. Sample answer: To rotate a figure, you need to know the number of degrees it is going to be rotated, the direction of rotation, and about what point the figure is being rotated. **27a.** $(-5, 0)$ **27b.** $(0, 5)$

29.

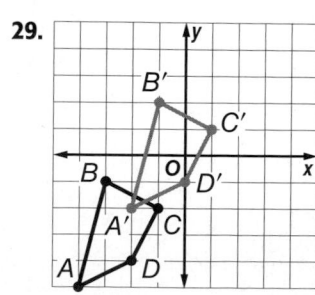

31.

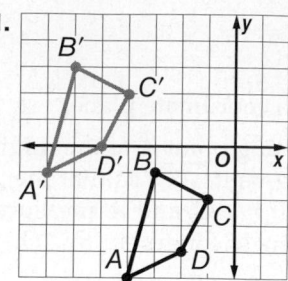

Pages 727–729   Lesson 12-5A

1.

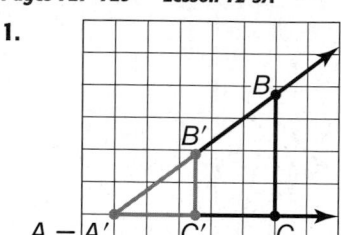

3. $J'(-12, 6)$, $K'(-6, -12)$, $L'(9, 18)$

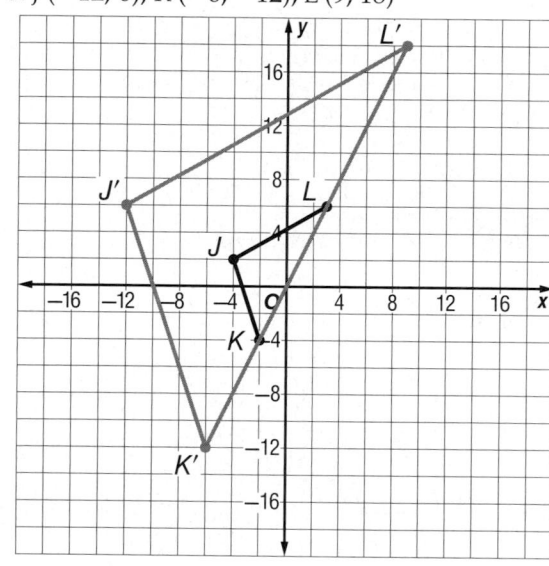

5. 3; enlargement

7.

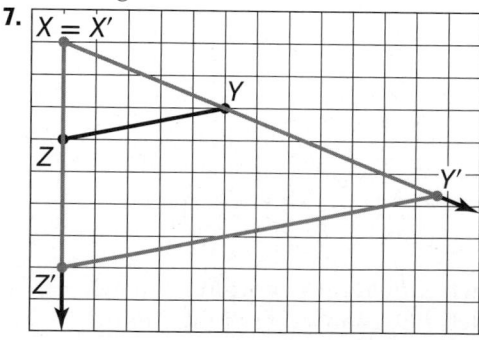

9 Step 1: Draw ray LM, or $\overrightarrow{LM}$, extending it to the edges of the grid. Step 2: Use a ruler to locate point M' on $\overrightarrow{LM}$ so that $LM' = \frac{3}{4}(LM)$. Step 3: Repeat Steps 1 and 2 for points N', P', and R'. Then draw polygon $L'M'N'P'R'$ where $L = L'$.

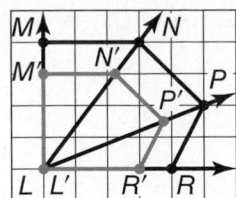

11. $H'(-2, 6)$, $J'(6, 4)$, $K'(4, -6)$, $L'(-4, -4)$

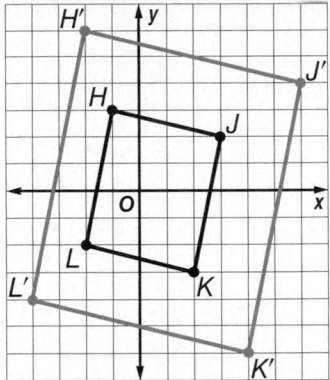

13. $H'(-3, 1)$, $J'(2, 2)$, $K'\left(3\frac{1}{2}, -1\right)$, $L'(-1, -2)$

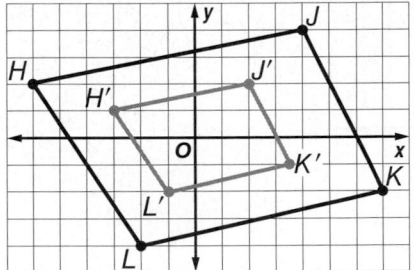

15. $\frac{3}{5}$; reduction   **17.** 2; enlargement

19. 4.5 cm by 3 cm

21 The scale factor is a ratio of the two sizes expressed in the same units. First, change 20 feet to inches; 20 feet times 12 inches per foot is 240 inches. So, the ratio is $\frac{240 \text{ in.}}{5 \text{ in.}}$ or 48.

23. Sample answer:

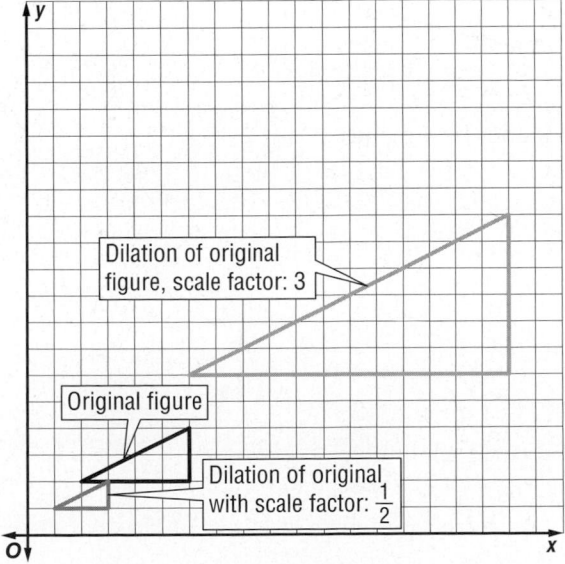

To create the enlarged image, multiply the vertices of the original triangle by a scale factor of 3. To create the reduced image, multiply the vertices of the original triangle by $\frac{1}{2}$.

25. Sample answer: Corresponding angles are equal in similar and congruent polygons. Corresponding sides of similar polygons are proportional, while corresponding sides of congruent polygons are equal; translations, reflections, and rotations   **27.** H

29. $X'(1, 3)$, $Y'(4, 5)$, and $Z'(5, 1)$

31. $J'(-1, 4)$, $K'(1, -1)$, and $L'(3, 2)$

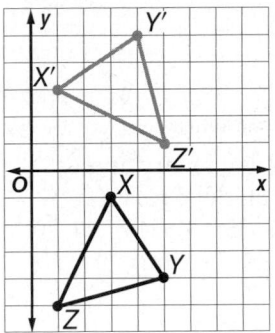

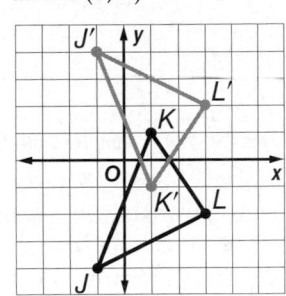

Pages 730–731   Lesson 12-5B   PSI

1. Sample answer: The work backward strategy could be used with hiking, where you may need to retrace your steps to return to the starting point.
3. Sample answer: Translating the triangle will affect the y-coordinates by increasing them, and depends on how many units it was translated up. The rotation is shown.

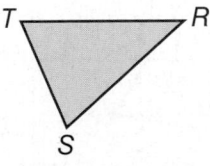

5 Work backward.

$2.50 = \frac{1}{3}x$ Write an equation for the question, "$2.50 is $\frac{1}{3}$ of what?" If she spent $\frac{2}{3}$, she has $\frac{1}{3}$ left.

$3(2.50) = 3 \cdot \frac{1}{3}x$ Multiply each side by 3.

$7.50 = x$ Simplify.

$2(7.50) = 15$ $7.50 is half of what she started with, so multiply by two to find the original amount.

Jackie's mom gave her $15.00.

7. 180 ft

9a. 3 units right and 4 units down

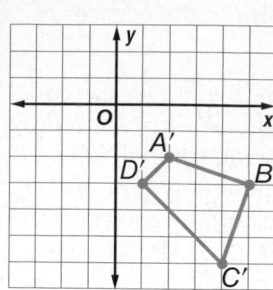

9b. $A'(1, 2)$, $B'(-2, 1)$, $C'(-1, -2)$, $D'(2, 1)$

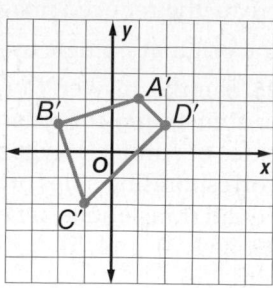

33. $A'(-1, 3)$, $B'(-2, -1)$, $C'(-5, -1)$, and $D'(-4, 3)$

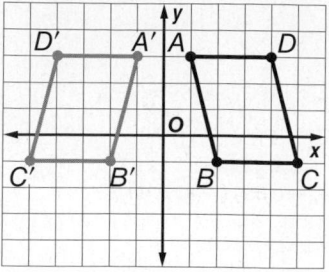

35. $W'(3, 0)$, $X'(1, 0)$, $Y'(1, 2)$, and $Z'(3, 2)$

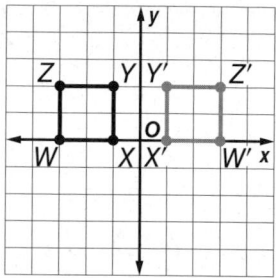

37. $X'(-2, -2)$, $Y'(-4, -2)$, and $Z'(-3, 1)$

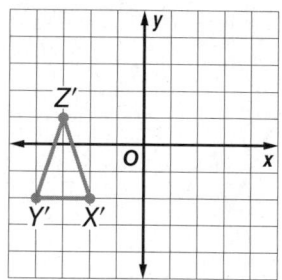

39. yes; 45°, 90°, 135°, 180°, 225°, 270°, 315°

41. $A'(-1, -2)$, $B'(2, 1)$, and $C'(3, -1)$

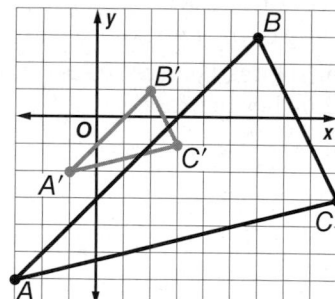

43. 3 inches by 4 inches

Pages 734–739   Chapter Study Guide and Review

1. false; $(-2, 1)$ **3.** false; supplementary angles
5. false; translation   **7.** false; angle of rotation
9. false; 180°   **11.** neither   **13.** 55°   **15.** 45   **17.** 79
19. parallelogram   **21.** 107   **23.** rhombus
25. regular nonagon
27. $P'(-2, 1)$, $Q'(-8, 0)$, and $R'(-7, 9)$

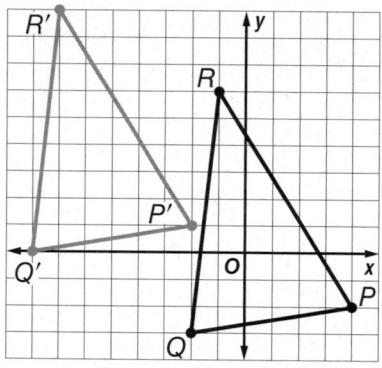

29. $P'(1, -2)$, $Q'(-5, -3)$, and $R'(-4, 6)$

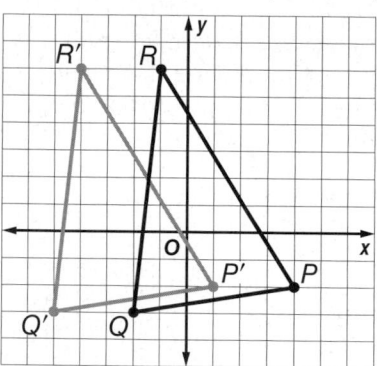

31. $A'(1, -2)$, $B'(4, -2)$, $C'(1, 1)$, and $D'(4, 1)$

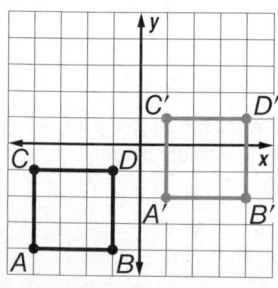

Photo Credits

Cover Tim Pannell/Masterfile; v Jupiter Images/Brand X Pictures/Getty Images; x–xi Digital Vision/Getty Images; xii Mike Parry/Minden Pictures; xv Fritz Liedtke/Alamy; xvi–xvii Ullamaija Hanninen/Getty Images; xviii–xix The McGraw-Hill Companies; xx–xxi Digital Vision/Getty Images; 1 (l) Streeter Lecka/Getty Images, (r) Robert Landau/CORBIS; 4 Grant Faint/Getty Images; 6 John Rawsterne/Alamy; 7 Goodshoot/PunchStock; 9 DreamPictures/Getty Images; 10 JOE KLAMAR/AFP/Getty Images; 13 David Madison/Photographer's Choice RF/Getty Images; 15 Photodisc Collection/Getty Images; 16 Digital Vision/Getty Images; 18 Comstock Images/Alamy; 19 Digital Vision/Getty Images; 21 Ryan McVay/Getty Images; 27 Stefano Bianchetti/CORBIS; 30 Compassionate Eye Foundation/Rob Daly/OJO Images Ltd/Getty Images; 34 SW Productions/Photodisc/Getty Images; 39 MARTIN BUREAU/AFP/Getty Images; 42 Plush Studios/Bill Reitzel/Getty Images; 54 Lawrence Manning/CORBIS; 59 DEA/M. BERTINETTI/Getty Images; 62 (inset) Justin Sullivan/Getty Images; 62–63 (bkgd) Pete Atkinson/Getty Images; 77 Chase Jarvis/CORBIS; 83 Mike Parry/Minden Pictures; 93 altrendo travel/Getty Images; 96 NASA/Photodisc/Getty Images; 100 John Evans; 101 Prisma/SuperStock; 104 Joanne Schmaltz/Getty Images; 106 Ralph White/CORBIS; 111 Lothar Lenz/zefa/CORBIS; 114 (inset) Francis Specker/epa/CORBIS; 114–115 (bkgd) NASA; 130 Comstock/SuperStock; 133 Inga Spence/age fotostock; 136 United States Mint; 146 Thomas Northcut/Photodisc/Getty Images; 148 Susumu Nishinaga/Photo Researchers; 150 Kathryn Harris/4Goodnesscake/Flickr/Getty Images; 152 Diane Macdonald/Stockbyte/Getty Images; 162 Garry Black/Masterfile; 166 KS Studios; 167 Michael Newman/PhotoEdit; 171 Phyllis Greenberg/Animals Animals-Earth Scenes; 174 (inset) Masterfile; 174–175 (bkgd) Gallo Images/Getty Images; 178 Jose Luis Pelaez Inc/Getty Images; 182 Prisma/SuperStock; 185 Charles O'Rear/CORBIS; 186 Chris A Crumley/Alamy Images; 188 M. Doukas-U.S. Forest Service; 202 John Evans; 205 David Young-Wolff/PhotoEdit; 209 Stephen Frink/zefa/CORBIS; 211 Eddie Adams/Sygma/CORBIS; 212 Streeter Lecka/Stringer/Getty Images; 216 Jupiterimages; 222 Stewart Cohen/Pam Ostrow; 228 (l) Getty Images, (r) Ryan McVay/Getty Images; 232 John Eder/Stone/Getty Images; 236 Burke/Triolo/Brand X Pictures; 237 Michael Downes/The Oklahoman/AP Images; 240 (inset) National Geographic/Getty Images; 240–241 (bkgd) Masterfile Royalty-Free; 243 (l) Oleksiy Maksymenko/Alamy, (r) The McGraw-Hill Companies; 244 (l to r, t to b) Hola Images/Getty Images, (2) BananaStock/age fotostock, (3) Barbara Penoyar/Getty Images, (4) The McGraw-Hill Companies, (5) RubberBall Productions/Getty Images, (6) PunchStock, (7) Doug Menuez/Getty Images, (8) Brand X Pictures/PunchStock; 246 Getty Images; 265 Dave & Les Jacobs/Getty Images; 266 David Young-Wolff/PhotoEdit; 268 Doug Menuez/Photodisc/Getty Images; 270 Don Emmett/AFP/Getty Images; 272 CORBIS Premium RF/Alamy; 274 age fotostock/SuperStock; 276 Stockbyte/Getty Images; 277 Jeff Greenberg/PhotoEdit; 281 age fotostock/SuperStock; 282 Jose Luis Pelaez, Inc/Getty Images; 289 Bilderbuch/Design Pics/CORBIS; 295 Getty Images; 301 Fritz Liedtke/Alamy; 303 DEA/M.BERTINETTI/Getty Images; 304 (inset) AP Photo/Paramount's Kings Island, Rick Norton; 304–305 (bkgd) Gunter Marx/Alamy; 318 Tim Platt/Getty Images; 320 DLILLC/CORBIS; 327 Ariel Skelley/CORBIS; 329 George Gojkovich/Getty Images; 332 BARTEK WRZESNIOWSKI/AFP/Getty Images; 334 Donna Ikenberry/Animals Animals-Earth Scenes; 335 Digital Vision/Getty Images; 339 Arend/Smith/AlaskaStock.com; 342 Adrian Peacock/Digital Vision/Getty Images; 346 Tom Carter/PhotoEdit; 348 Ryan McVay/Getty Images; 352 Big Cheese Photo/SuperStock; 355 Wolfgang Kaehler/Alamy; 360 Mark Scheuern/Alamy;

364 (inset) Picture Contact/Alamy; 364–365 (bkgd) Masterfile Royalty-Free; 379 Eric and David Hosking/CORBIS; 381 IT Stock Free/Alamy; 389 Ullamaija Hanninen/Getty Images; 391 Ingram Publishing/SuperStock; 396 Paul Carstairs/Alamy; 402 Gabe Palmer/Alamy Images; 406 C Squared Studios/Getty Images; 409 Tetra Images/Getty Images; 413 IT Stock Free; 416 (inset) Philippe Psaila/Photo Researchers, Inc.; 416–417 (bkgd) Tony West/Alamy; 429 McGraw-Hill Companies; 430 MARK RALSTON/AFP/Getty Images; 437 (t) United States Mint, (b) Big Cheese Photo/JupiterImages; 438 Tom Hauck/Getty Images; 439 McGraw-Hill Companies; 442 O'Brien Productions/CORBIS; 445 Shenval/Alamy Images; 447 Ryan McVay/Getty Images; 451 McGraw-Hill Companies; 454 David Muir/Masterfile; 455 (l) United States Mint, (r) United States Mint; 456 (inset) Digital Vision; 458 (l) McGraw-Hill Companies; 464 Laura Sifferlin; 465 Glencoe/McGraw-Hill; 467 (l) McGraw-Hill Companies, (r) Glencoe/McGraw-Hill; 471 McGraw-Hill Companies; 472 CORBIS; 484 McGraw-Hill Companies; 491 Brand X Pictures; 493 Noel Hendrickson/Masterfile; 499 Bob Daemmrich/PhotoEdit; 502 Masterfile; 504 Doug Pensinger/Getty Images; 507 Royalty-Free-CORBIS; 508 Eric Baccega/age fotostock; 510–511 (bkgd) Brian Sytnyk/Masterfile; 511 (inset) Mark Burnett/Alamy; 515 Stephen Dalton/Animals Animals-Earth Scenes; 523 Digital Vision/PunchStock; 526 Barbara Peacock/Taxi/Getty Images; 529 Brandon D. Cole/CORBIS; 532 Getty Images/Digital Vision; 534 The McGraw-Hill Companies; 538 traumlichtfabrik/Flickr/Getty Images; 541 Karen Whylie/Masterfile; 558 (l) Digital Vision/Getty Images, (r) Andre Jenny/Alamy; 559 Brand X Pictures/PunchStock; 565 Tony Freeman/PhotoEdit; 566 Duncan Usher/Foto Natura/Minden Pictures; 571 Photodisc/Getty Images; 572 (l) Mitch Hrdlicka/Getty Images, (r) SuperStock, Inc./SuperStock; 576 (l) Jeffrey Coolidge/Iconica/Getty Images, (r) The McGraw-Hill Companies; 589 W.A.Hamilton/Alamy; 595 Paul Taylor/Getty Images; 600–601 (bkgd) John Lamb/Getty Images; 601 (inset) Hemis/Alamy; 602 Jim Esposito Photography LLC/Photodisc/Getty Images; 606 The McGraw-Hill Companies; 607 Image Source/Getty Images; 624 Jon Feingersh/Getty Images; 625 United States Mint; 626 John Lambert/Brand X/CORBIS; 632 Comstock/PunchStock; 633 Bruce Edwards/America 24-7/Getty Images; 635 STAN HONDA/AFP/Getty Images; 636 View Stock/Alamy; 641 Tom Vezo/Minden Pictures; 647 Al Messerschmidt/Getty Images; 650 Laura Sifferlin; 654 Brand X Pictures/PunchStock; 655 The McGraw-Hill Companies; 656 Mark Ransom; 658 Monica Flatford/Knoxville Convention Center; 660 (inset) Pixtal/SuperStock; 660–661 (bkgd) Mark Tomalty/Masterfile; 674 Michael Clark/Getty Images; 676 Chase Jarvis/Getty Images; 678 (l) Charles Smith/CORBIS, (r) Image Source/Getty Images; 682 The McGraw-Hill Companies; 683 Jeff Greenberg/PhotoEdit; 685 (l) Damir Frkovic/Masterfile, (r) DeCordova Museum and Sculpture Park; 689 (l) David Pollack/CORBIS, (r) Mark Gibson; 713 DON EMMERT/AFP/Getty Images; 715 (l) Allen Wallace/Photonica/Getty Images, (c) Liu Liqun/CORBIS, (r) Daryl Benson/Masterfile; 721 newstream rf/Alamy; 726 Adam Hart-Davis/Photo Researchers; 728 BARNpix.com/Alamy Images; 730 Image Source/PunchStock; 732 Jon Feingersh/Getty Images; 744 (t) Robert Landau/CORBIS, (c) Royalty-Free/CORBIS, (b) Photofusion Picture Library/Alamy; 745 (l) Streeter Lecka/Getty Images, (r) Comstock/JupiterImages; 746 Robert Landau/CORBIS; 747 John A. Karachewski; 748 Royalty-Free/CORBIS; 749 Ron Chapple/Thinkstock/Getty Images; 750 Photofusion Picture Library/Alamy; 751 Andrea Pistolesi/Getty Images; 752 Streeter Lecka/Getty Images; 753 J. Meric/Getty Images; 754 Comstock/JupiterImages; 755 Gary Randall/Getty Images; EP0 Jose Luis Pelaez, Inc/age fotostock.

Glossary/Glosario

Click on the eGlossary link to find out more about these words in the following 13 languages.

Arabic	Cantonese	Hmong	Spanish	Urdu
Bengali	English	Korean	Tagalog	Vietnamese
Brazilian Portuguese	Haitian Creole	Russian		

English

Spanish
(Español)

Aa

absolute value   The distance the number is from zero on a number line.

valor absoluto   Distancia a la que se encuentra un número de cero en la recta numérica.

acute angle   An angle with a measure greater than 0° and less than 90°.

ángulo agudo   Ángulo que mide más de 0° y menos de 90°.

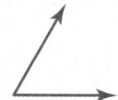

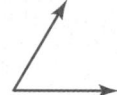

acute triangle   A triangle having three acute angles.

triángulo acutángulo   Triángulo con tres ángulos agudos.

Addition Property of Equality   If you add the same number to each side of an equation, the two sides remain equal.

propiedad de adición de la igualdad   Si sumas el mismo número a ambos lados de una ecuación, los dos lados permanecen iguales.

additive inverse   Two integers that are opposites. The sum of an integer and its additive inverse is zero.

inverso aditivo   Dos enteros opuestos.

adjacent angles   Angles that have the same vertex, share a common side, and do not overlap.

ángulos adyacentes   Ángulos que comparten el mismo vértice y un común lado, pero no se sobreponen.

algebra   A branch of mathematics that involves expressions with variables.

álgebra   Rama de las matemáticas que trata de las expresiones con variables.

algebraic expression   A combination of variables, numbers, expressions, and at least one operation.

expresión algebraica   Combinación de variables, números y por lo menos una operación.

alternate exterior angles   Angles that are on opposite sides of the transversal and outside the parallel lines.

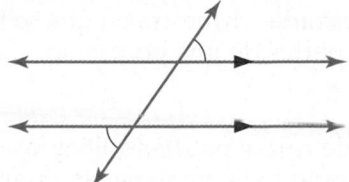

alternate interior angles   Angles that are on opposite sides of the transversal and inside the parallel lines.

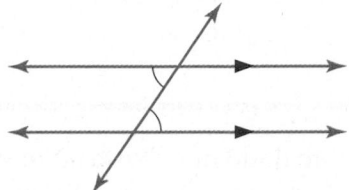

angle   Two rays with a common endpoint form an angle. The rays and vertex are used to name the angle.

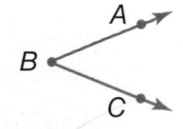

∠ABC, ∠CBA, or ∠B

Angle-Angle Similarity   Two triangles are similar if two of their corresponding angles are equal.

angle of rotation   The degree measure of the angle through which the figure is rotated.

arithmetic sequence   A sequence in which the difference between any two consecutive terms is the same.

ángulos alternos externos   Ángulos en lados opuestos de la trasversal y afuera de las rectas paralelas.

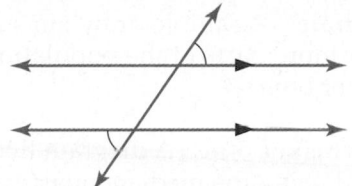

ángulos alternos internos   Ángulos en lados opuestos de la trasversal y dentro de las rectas paralelas.

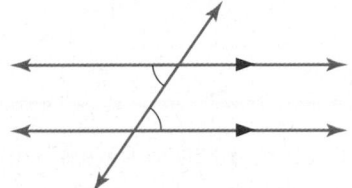

ángulo   Dos rayos con un extremo común forman un ángulo. Los rayos y el vértice se usan para nombrar el ángulo.

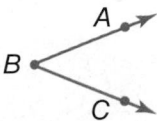

∠ABC, ∠CBA o ∠B

semejanza ángulo-ángulo   Dos triángulos son semejantes si dos de sus ángulos correspondientes son iguales.

ángulo de rotación   Medida en grados del ángulo sobre el cual se rota una figura.

sucesión aritmética   Sucesión en la cual la diferencia entre dos términos consecutivos es constante.

back-to-back stem-and-leaf plot   A display that uses one stem to compare two sets of data.

bar notation   In repeating decimals, the line or bar placed over the digits that repeat. For example, 2.6̄3̄ indicates that the digits 63 repeat.

diagrama de tallo y hojas   Representación que utiliza un tallo para comparar dos conjuntos de datos.

notación de barra   Línea o barra que se coloca sobre los dígitos que se repiten en decimales periódicos. Por ejemplo, 2.6̄3̄ indica que los dígitos 63 se repiten.

base In a power, the number used as a factor. In 10^3, the base is 10. That is, $10^3 = 10 \times 10 \times 10$.

base En una potencia, el número usado como factor. En 10^3, la base es 10. Es decir, $10^3 = 10 \times 10 \times 10$.

biased sample A sample drawn in such a way that one or more parts of the population are favored over others.

muestra sesgada Muestra en que se favorece una o más partes de una población.

box-and-whisker plot A diagram that is constructed using the median, quartiles, and extreme values.

diagrama de caja y patillas Diagrama que se construye usando la mediana, los cuartiles y los valores extremos.

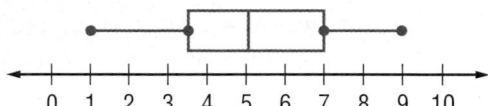

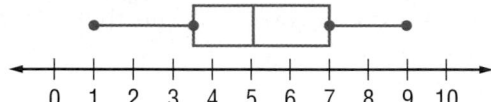

Cc

center A fixed point used for measurement when altering the size of a figure.

centro Punto dado que se usa en medición cuando se altera el tamaño de una figura.

circle graph A graph that shows data as parts of a whole. In a circle graph, the percents add up to 100.

gráfica circular Gráfica que muestra los datos como partes de un todo. En una gráfica circular los porcentajes suman 100.

Area of Oceans

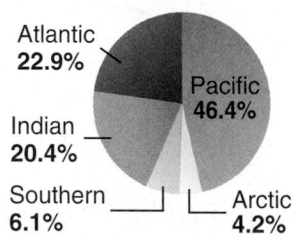

Área de superficie de los océanos

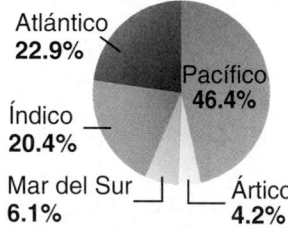

coefficient The numerical factor of a term that contains a variable.

coeficiente El factor numérico de un término que contiene una variable.

combination An arrangement or listing of objects in which order is not important.

combinación Arreglo o lista de objetos en la cual el orden no es importante.

common denominator A common multiple of the denominators of two or more fractions. 24 is a common denominator for $\frac{1}{3}$, $\frac{5}{8}$, and $\frac{3}{4}$ because 24 is the LCM of 3, 8, and 4.

común denominador El múltiplo común de los denominadores de dos o más fracciones. 24 es un denominador común para $\frac{1}{3}$, $\frac{5}{8}$ y $\frac{3}{4}$ porque 24 es el mcm de 3, 8 y 4.

complementary angles Two angles are complementary if the sum of their measures is 90°.

ángulos complementarios Dos ángulos son complementarios si la suma de sus medidas es 90°.

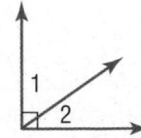

∠1 and ∠2 are complementary angles.

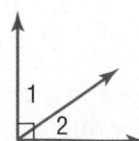

∠1 y ∠2 son complementarios.

complementary events The events of one outcome happening and that outcome not happening. The sum of the probabilities of an event and its complement is 1 or 100%. In symbols, $P(A) + P(not\ A) = 1$.

eventos complementarios Los eventos de un resultado que ocurre y ese resultado que no ocurre. La suma de las probabilidades de un evento y su complemento es 1 ó 100. En símbolos $P(A) + P(no\ A) = 1$.

composite figure A figure that is made up of two or more three-dimensional figures.

figura compuesta Figura formada por dos o más figuras tridimensionales.

compound event An event consisting of two or more simple events.

evento compuesto Un evento que consiste en dos o más eventos simples.

cone A three-dimensional figure with a curved surface and a circular base.

cono Figura tridimensional con una superficie curva y una base circular.

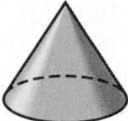

congruent angles Angles that have the same measure.

ángulos congruentes Ángulos que tienen la misma medida.

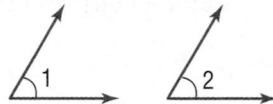

∠1 and ∠2 are congruent angles.

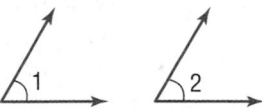

∠1 y ∠2 son congruentes.

congruent figures Figures that have the same size and same shape and corresponding sides and angles with equal measure.

figuras congruentes Figuras que tienen el mismo tamaño y la misma forma y los lados y los ángulos correspondientes tienen igual medida.

congruent segments Sides with the same length.

segmentos congruentes Lados con la misma longitud.

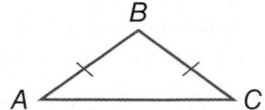

Side $\overline{AB}$ is congruent to side $\overline{BC}$.

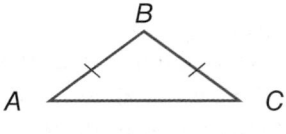

$\overline{AB}$ es congruente a $\overline{BC}$.

constant of variation The constant ratio in a direct variation.

constante de variación Razón constante en una variación directa.

constant rate of change The rate of change in a linear relationship.

razón constante de cambio Tasa de cambio en una relación lineal.

continuous data Data that take on any real number value. It can be determined by considering what numbers are reasonable as part of the domain.

datos continuos Datos que asumen cualquier valor numérico real. Se pueden determinar al considerar qué números son razonables como parte del dominio.

convenience sample A sample which consists of members of a population that are easily accessed.

muestra de conveniencia Muestra que incluye miembros de una población fácilmente accesibles.

coordinate plane A plane in which a horizontal number line and a vertical number line intersect at their zero points. Also called a coordinate grid.

plano de coordenadas Plano en el cual se han trazado dos rectas numéricas, una horizontal y una vertical, que se intersecan en sus puntos cero. También conocido como sistema de coordenadas.

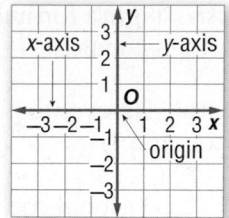

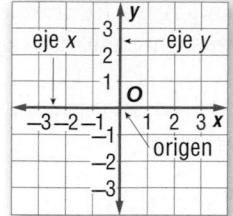

corresponding angles The angles of similar figures that are in the same relative postion.

ángulos correspondientes Ángulos de figuras semejantes que estan en la misma posición.

corresponding angles Angles in the same position on parallel lines in relation to a transversal.

ángulos correspondientes Ángulos que están en la misma posición sobre rectas paralelas en relación con la transversal.

corresponding sides The sides of similar figures that are in the same relative postion.

lados correspondientes Lados de figuras semejantes que estan en la misma posición.

cross product The product of the numerator of one ratio and the denominator of the other ratio. The cross products of any proportion are equal.

producto cruzado Producto del numerador de una razón por el denominador de la otra razón. Los productos cruzados de cualquier proporción son iguales.

cube root One of three equal factors of a number. If $a^3 = b$, then a is the cube root of b. The cube root of 125 is 5 since $5^3 = 125$.

raíz cúbica Uno de tres factores iguales de un número. Si $a^3 = b$, entonces a es la raíz cúbica de b. La raíz cúbica de 125 es 5, dado que $5^3 = 125$.

cubed The product in which a number is a factor three times. Two cubed is 8 because $2 \times 2 \times 2 = 8$.

al cubo El producto de un número por sí mismo, tres veces. Dos al cubo es 8 porque $2 \times 2 \times 2 = 8$.

cumulative frequency A ratio of combined frequencies of two or more categories to the total.

frecuencia cumulativa Razón de las frecuencias combinadas de dos o más categorías al total.

cylinder A three-dimensional figure with two parallel congruent circular bases.

cilindro Figura tridimensional que tiene dos bases circulares congruentes y paralelas.

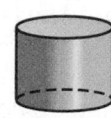

Dd

decagon A polygon having ten sides.

decágono Un polígono con diez lados.

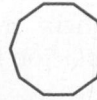

degrees   The most common unit of measure for angles. If a circle were divided into 360 equal-sized parts, each part would have an angle measure of 1 degree.

dependent events   Two or more events in which the outcome of one event affects the outcome of the other event(s).

dependent variable   The variable in a relation with a value that depends on the value of the independent variable.

derived unit   A unit that is derived from a measurement system base unit, such as length, mass, or time.

dilation   The image produced by enlarging or reducing a figure.

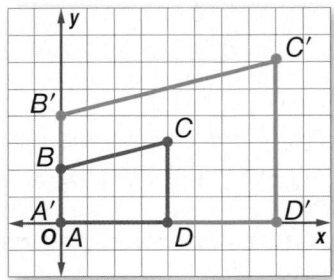

dimensional analysis   The process of including units of measurement when you compute.

direct variation   The relationship between two variable quantities that have a constant ratio.

discount   The amount by which the regular price of an item is reduced.

discrete data   When solutions of a function are only integer values. It can be determined by considering what numbers are reasonable as part of the domain.

disjoint events   Events that cannot happen at the same time.

Division Property of Equality   If you divide each side of an equation by the same nonzero number, the two sides remain equal.

domain   The set of input values for a function.

grados   La unidad más común para medir ángulos. Si un círculo se divide en 360 partes iguales, cada parte tiene una medida angular de 1 grado.

eventos dependientes   Dos o más eventos en que el resultado de un evento afecta el resultado de otro u otros eventos.

variable dependiente   La variable en una relación cuyo valor depende del valor de la variable independiente.

unidad derivada   Unidad que se deriva de una unidad básica de un sistema de medidas, como la longitud, la masa o el tiempo.

homotecia   Imagen producida al ampliar o reducir una figura.

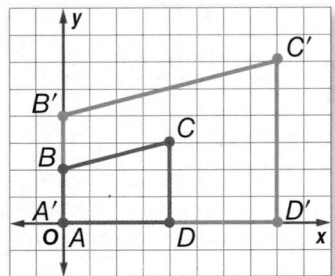

análisis dimensional   Proceso que incluye las unidades de medida al hacer cálculos.

variación directa   Relación entre las cantidades de dos variables que tienen una tasa constante.

descuento   Cantidad que se le rebaja al precio regular de un artículo.

datos discretos   Cuando las soluciones de una función son solo valores enteros. Se pueden determinar considerando qué números son razonables como parte del dominio.

eventos disjuntos   Eventos que no pueden ocurrir al mismo tiempo.

propiedad de igualdad de la división   Si divides ambos lados de una ecuación entre el mismo número no nulo, los lados permanecen iguales.

dominio   El conjunto de valores de entrada de una función.

enlargement   An image larger than the original.

ampliación   Imagen más grande que la original.

equation   A mathematical sentence that contains an equals sign, =, stating that two quantities are equal.

ecuación   Enunciado matemático que contiene el signo de igualdad = indicando que dos cantidades son iguales.

equiangular   In a polygon, all of the angles are congruent.

equiangular   En un polígono, todos los ángulos son congruentes.

equilateral   In a polygon, all of the sides are congruent.

equilátero   En un polígono, todos los lados son congruentes.

equilateral triangle   A triangle having three congruent sides.

triángulo equilátero   Triángulo con tres lados congruentes.

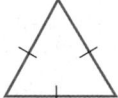

equivalent equations   Two or more equations with the same solution.

ecuaciones equivalentes   Dos o más ecuaciones con la misma solución.

equivalent expressions   Expressions that have the same value.

expresiones equivalentes   Expresiones que tienen el mismo valor.

equivalent ratios   Two ratios that have the same value.

razones equivalentes   Dos razones que tienen el mismo valor.

evaluate   To find the value of an expression.

evaluar   Calcular el valor de una expresión.

experimental probability   An estimated probability based on the relative frequency of positive outcomes occurring during an experiment. It is based on what *actually* occurred during such an experiment.

probabilidad experimental   Probabilidad estimada que se basa en la frecuencia relativa de los resultados positivos que ocurren durante un experimento. Se basa en lo que *en realidad* ocurre durante dicho experimento.

exponent   In a power, the number that tells how many times the base is used as a factor. In 5^3, the exponent is 3. That is, $5^3 = 5 \times 5 \times 5$.

exponente   En una potencia, el número que indica las veces que la base se usa como factor. En 5^3, el exponente es 3. Es decir, $5^3 = 5 \times 5 \times 5$.

exponential form   Numbers written with exponents.

forma exponencial   Números escritos usando exponentes.

Ff

factors   Two or more numbers that are multiplied together to form a product.

factores   Dos o más números que se multiplican entre sí para formar un producto.

fair game   A game where each player has an equally likely chance of winning.

juego justo   Juego donde cada jugador tiene igual posibilidad de ganar.

formula An equation that shows the relationship among certain quantities.

function A relationship which assigns exactly one output value for each input value.

function notation A form where the input x is any real number and the output $f(x)$ is read as *the function of x* or *f of x*.

function rule The operation performed on the input of a function.

function table A table used to organize the input numbers, output numbers, and the function rule.

Fundamental Counting Principle Uses multiplication of the number of ways each event in an experiment can occur to find the number of possible outcomes in a sample space.

fórmula Ecuación que muestra la relación entre ciertas cantidades.

función Relación que asigna exactamente un valor de salida a cada valor de entrada.

notación funcional Forma donde la entrada x es cualquier número real y la salida $f(x)$ se lee como *la función de x* o *f de x*.

regla de función Operación que se efectúa en el valor de entrada.

tabla de funciones Tabla que organiza las entradas, la regla y las salidas de una función.

Principio Fundamental de Contar Este principio usa la multiplicación del número de veces que puede ocurrir cada evento en un experimento para calcular el número de posibles resultados en un espacio muestral.

Gg

geometric probability Using the principles of length and area to find the probability of an event.

gram A unit of mass in the metric system equivalent to 0.001 kilogram. The amount of matter an object can hold.

graph The process of placing a point on a number line or on a coordinate plane at its proper location.

gratuity Also known as a tip. It is a small amount of money in return for a service.

probabilidad geométrica Empleo de los principios de longitud y área para calcular la probabilidad de un evento.

gramo Unidad de masa en el sistema métrico que equivale a 0.001 de kilogramo. La cantidad de materia que puede contener un objeto.

graficar Proceso de dibujar o trazar un punto en una recta numérica o en un plano de coordenadas en su ubicación correcta.

gratificación También conocida como propina. Es una cantidad pequeña de dinero en retribución por un servicio.

Hh

heptagon A polygon having seven sides.

heptágono Polígono con siete lados.

hexagon A polygon having six sides.

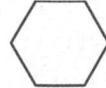

hexágono Polígono con seis lados.

histogram A type of bar graph used to display numerical data that have been organized into equal intervals.

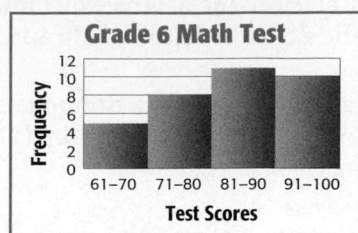

histograma Tipo de gráfica de barras que se usa para exhibir datos que se han organizado en intervalos iguales.

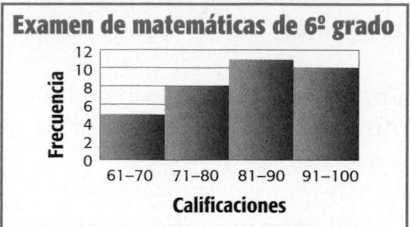

image The position of a figure after a transformation.

imagen Posición de una figura después de una transformación.

independent events Two or more events in which the outcome of one event does not affect the outcome of the other event(s).

eventos independientes Dos o más eventos en los cuales el resultado de uno de ellos no afecta el resultado de los otros eventos.

independent variable The variable in a function with a value that is subject to choice.

variable independiente Variable en una función cuyo valor está sujeto a elección.

indirect measurement Finding a measurement using similar figures to find the length, width, or height of objects that are too difficult to measure directly.

medición indirecta Hallar una medición usando figuras semejantes para calcular el largo, ancho o altura de objetos que son difíciles de medir directamente.

inequality An open sentence that uses <, >, ≠, ≤, or ≥ to compare two quantities.

desigualdad Enunciado abierto que usa <, >, ≠, ≤ o ≥ para comparar dos cantidades.

integer Any number from the set {..., −4, −3, −2, −1, 0, 1, 2, 3, 4, ...}, where ... means continues without end.

entero Cualquier número del conjunto {..., −4, −3, −2, −1, 0, 1, 2, 3, 4, ...}, donde ... significa que continúa sin fin.

interquartile range A range where half of the data lie between the lower quartile and upper quartile.

rango intercuartílico Rango donde la mitad de los datos se hallan entre el cuartil inferior y el cuartil superior.

inverse variation A relationship where the product of x and y is a constant k. As x increases in value, y decreases in value, or as y decreases in value, x increases in value.

variación inversa Relación en la cual el producto de x y y es una constante k. A medida que aumenta el valor de x, disminuye el valor de y o a medida que disminuye el valor de y, aumenta el valor de x.

irrational number A number that cannot be expressed as the ratio of two integers.

número irracional Número que no se puede expresar como el razón de dos enteros.

isosceles triangle A triangle having at least two congruent sides.

triángulo isósceles Triángulo que tiene por lo menos dos lados congruentes.

kilogram   The base unit of mass in the metric system. One kilogram equals 1,000 grams.

kilogramo   Unidad básica de masa del sistema métrico. Un kilogramo equivale a 1,000 gramos.

lateral face   One side of a three-dimensional figure. It is any flat surface that is not a base.

cara lateral   Lado de una figura tridimensional. Cualquier superficie plana que no sea la base.

lateral surface area   The sum of the areas of all of the lateral faces of a solid.

área de superficie lateral   Suma de las áreas de todas las caras de un sólido.

leaf   The digits of the least place value of data in a stem-and-leaf plot.

hoja   En un diagrama de tallo y hojas, los dígitos del menor valor de posición.

least common denominator (LCD)   The least common multiple of the denominators of two or more fractions. You can use the LCD to compare fractions.

mínimo común denominador (mcd)   El menor de los múltiplos de los denominadores de dos o más fracciones. Puedes usar el mínimo común denominador para comparar fracciones.

like fractions   Fractions that have the same denominators.

fracciones semejantes   Fracciones que tienen los mismos denominadores.

line graph   A type of statistical graph using lines to show how values change over a period of time.

gráfica lineal   Tipo de gráfica estadística que usa segmentos de recta para mostrar cómo cambian los valores durante un período de tiempo.

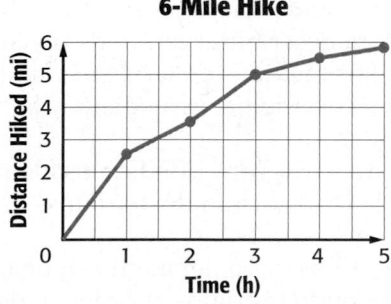

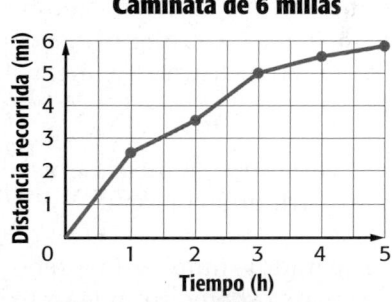

line of best fit   A line that is very close to most of the data points in a scatter plot.

recta de mejor ajuste   Recta que está muy cercana a la mayoría de los puntos de los datos en un diagrama de dispersión.

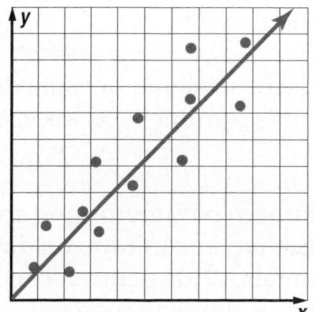

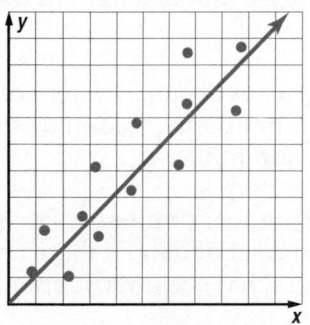

line of reflection   The line over which a figure is reflected.

line of symmetry   A line that divides a figure into two halves that are reflections of each other.

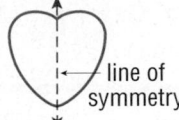
line of symmetry

line symmetry   Figures that match exactly when folded in half have line symmetry.

linear function   A function for which the graph is a straight line.

liter   The base unit of capacity in the metric system. The amount of dry or liquid material an object can hold.

lower quartile   The median of the lower half of a set of data.

eje de reflexión   La línea sobre la cual se refleja una figura.

eje de simetría   Recta que divide una figura en dos mitades que son reflexiones entre sí.

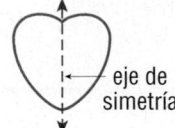
eje de simetría

simetría lineal   Exhiben simetría lineal las figuras que coinciden exactamente al doblarse una sobre otra.

función lineal   Función cuya gráfica es una recta.

litro   Unidad básica de capacidad del sistema métrico. La cantidad de materia líquida o sólida que puede contener un objeto.

cuartil inferior   Mediana de la mitad inferior de un conjunto de datos.

Mm

mean   The sum of the data divided by the number of items in the data set.

measures of central tendency   Numbers that are used to describe the center of a set of data. The most common measures are mean, median, and mode.

measures of variation   A measure used to describe the distribution of data.

median   The middle number in a set of data when the data are ordered from least to greatest. If the data have an even number of items, the median is the mean of the two numbers closer to the middle.

meter   The base unit of length in the metric system.

metric system   A decimal system of measures. The prefixes commonly used in this system are kilo-, centi-, and milli-.

mode   The number or numbers that appear most often in a set of data. If there are two or more numbers that occur most often, all of them are modes.

media   La suma de los datos dividida entre el número total de artículos en el conjunto de datos.

medidas de tendencia central   Números que se usan para describir el centro de un conjunto de datos. Las medidas de tendencia central más comunes son la media, la mediana y la moda.

medidas de variación   Medida usada para describir la distribución de los datos.

mediana   El número del medio en un conjunto de datos cuando los datos se ordenan de menor a mayor. Si los datos tienen un número par de artículos, la mediana es la media de los dos números más cercanos al medio.

metro   Unidad fundamental de longitud del sistema métrico.

sistema métrico   Sistema decimal de medidas. Los prefijos más comunes son kilo-, centi- y mili-.

moda   El número o números que aparece con más frecuencia en un conjunto de datos. Si hay dos o más números que ocurren con más frecuencia, todosellos son modas.

monomial A number, variable, or product of a number and one or more variables.

monomio Número, variable o producto de un número y una o más variables.

Multiplication Property of Equality If you multiply each side of an equation by the same nonzero number, the two sides remain equal.

propiedad de multiplicación de la igualdad Si multiplicas ambos lados de una ecuación por el mismo número no nulo, lo lados permanecen iguales.

multiplicative inverse Two numbers with a product of 1. For example, the multiplicative inverse of $\frac{2}{3}$ is $\frac{3}{2}$.

inverso multiplicativo Dos números cuyo producto es 1. Por ejemplo, el inverso multiplicativo de $\frac{2}{3}$ es $\frac{3}{2}$.

Nn

negative exponent Any nonzero number to the negative n power. It is the multiplicative inverse of its nth power.

Exponente negativo Cualquier número que no sea cero a la potencia negative de n. Es el inverso multiplicativo de su enésimo potencia.

negative integer An integer that is less than zero. Negative integers are written with a $-$ sign.

entero negativo Número menor que cero. Se escriben con el signo $-$.

net A two-dimensional figure that can be used to build a three-dimensional figure.

red Figura bidimensional que sirve para hacer una figura tridimensional.

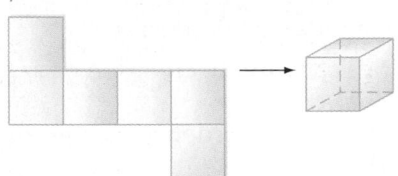

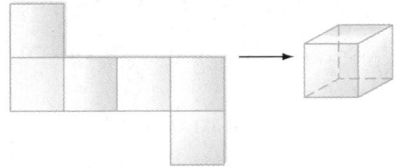

nonagon A polygon having nine sides.

enágono Polígono que tiene nueve lados.

nonlinear function A function for which the graph is *not* a straight line.

nonlinear function Función cuya gráfica *no* es una línea recta.

nonproportional The relationship between two ratios with a rate or ratio that is not constant.

no proporcional Relación entre dos razones cuya tasa o razón no es constante.

numerical expression A combination of numbers and operations.

expresión numérica Combinación de números y operaciones.

Oo

obtuse angle Any angle that measures greater than 90° but less than 180°.

ángulo obtuso Cualquier ángulo que mide más de 90° pero menos de 180°.

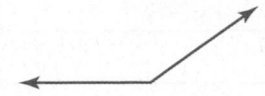

obtuse triangle   A triangle having one obtuse angle.

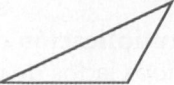

octagon   A polygon having eight sides.

odds in favor   A ratio that compares the number of ways an event *can* occur to the number of ways the event *cannot* occur.

opposites   Two integers are opposites if they are represented on the number line by points that are the same distance from zero, but on opposite sides of zero. The sum of two opposites is zero.

order of operations   The rules to follow when more than one operation is used in a numerical expression.

1. Evaluate the expressions inside grouping symbols.
2. Evaluate all powers.
3. Multiply and divide in order from left to right.
4. Add and subtract in order from left to right.

ordered pair   A pair of numbers used to locate a point in the coordinate plane. An ordered pair is written in the form (x-coordinate, y-coordinate).

origin   The point at which the x-axis and the y-axis intersect in a coordinate plane. The origin is at (0, 0).

outcome   Any one of the possible results of an action. For example, 4 is an outcome when a number cube is rolled.

outlier   A data value that is either much *greater* or much *less* than the median.

triángulo obtusángulo   Triángulo que tiene un ángulo obtuso.

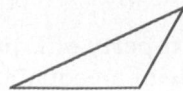

octágono   Polígono que tiene ocho lados.

posibilidades a favor   Razón que compara el número de maneras en que *puede* ocurrir un evento al número de maneras en que *no puede* ocurrir el evento.

opuestos   Dos enteros son opuestos si, en la recta numérica, están representados por puntos que equidistan de cero, pero en direcciones opuestas. La suma de dos opuestos es cero.

orden de las operaciones   Reglas a seguir cuando se usa más de una operación en una expresión numérica.

1. Primero, evalúa las expresiones dentro de los símbolos de agrupación.
2. Evalúa todas las potencias.
3. Multiplica y divide en orden de izquierda a derecha.
4. Suma y resta en orden de izquierda a derecha.

par ordenado   Par de números que se utiliza para ubicar un punto en un plano de coordenadas. Se escribe de la siguiente forma: (coordenada x, coordenada y).

origen   Punto en que el eje x y el eje y se intersecan en un plano de coordenadas. El origen está ubicado en (0, 0).

resultado   Cualquiera de los resultados posibles de una acción. Por ejemplo, 4 puede ser un resultado al lanzar un cubo numerado.

valor atípico   Valor de los datos que es mucho *mayor* o mucho *menor* que la mediana.

Pp

parallel lines   Lines in a plane that never intersect.

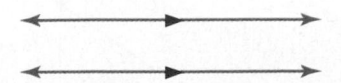

rectas paralelas   Rectas en un plano que nunca se intersecan.

parallelogram A quadrilateral with opposite sides parallel and opposite sides congruent.

pentagon A polygon having five sides.

percent equation An equation that describes the relationship between the part, whole, and percent.

$$\text{part} = \text{percent} \cdot \text{whole}$$

percent of change A ratio that compares the change in a quantity to the original amount.

$$\text{percent of change} = \frac{\text{amount of change}}{\text{original amount}}$$

percent of decrease A negative percent of change.

percent of increase A positive percent of change.

percent proportion One ratio or fraction that compares part of a quantity to the whole quantity. The other ratio is the equivalent percent written as a fraction with a denominator of 100.

$$\frac{\text{part}}{\text{whole}} = \frac{\text{percent}}{100}$$

perfect squares Numbers with square roots that are whole numbers. 25 is a perfect square because the square root of 25 is 5.

permutation An arrangement, or listing, of objects in which order is important.

perpendicular lines Lines that meet or cross each other to form right angles.

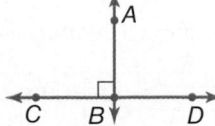

polygon A simple closed figure formed by three or more straight line segments.

paralelogramo Cuadrilátero cuyos lados opuestos son paralelos y congruentes.

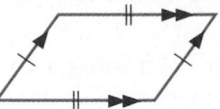

pentágono Polígono que tiene cinco lados.

ecuación porcentual Ecuación que describe la relación entre la parte, el todo y el por ciento.

$$\text{parte} = \text{por ciento} \cdot \text{todo}$$

porcentaje de cambio Razón que compara el cambio en una cantidad a la cantidad original.

$$\text{porcentaje de cambio} = \frac{\text{cantidad del cambio}}{\text{cantidad original}}$$

porcentaje de disminución Porcentaje de cambio negativo.

porcentaje de aumento Porcentaje de cambio positivo.

proporción porcentual Razón o fracción que compara parte de una cantidad a toda la cantidad. La otra razón es el porcentaje equivalente escrito como fracción con 100 de denominador.

$$\frac{\text{parte}}{\text{todo}} = \frac{\text{porcentaje}}{100}$$

cuadrados perfectos Números cuya raíz cuadrada es un número entero. 25 es un cuadrado perfecto porque la raíz cuadrada de 25 es 5.

permutación Arreglo o lista de objetos en la cual el orden es importante.

rectas perpendiculares Rectas que al encontrarse o cruzarse forman ángulos rectos.

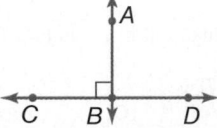

polígono Figura cerrada simple formada por tres o más segmentos de recta.

population The entire group of items or individuals from which the samples under consideration are taken.

población El grupo total de individuos o de artículos del cual se toman las muestras bajo estudio.

positive integer An integer that is greater than zero. They are written with or without a + sign.

entero positivo Entero que es mayor que cero; se escribe con o sin el signo +.

powers Numbers expressed using exponents. The power 3^2 is read *three to the second power,* or *three squared.*

potencias Números que se expresan usando exponentes. La potencia 3^2 se lee *tres a la segunda potencia o tres al cuadrado.*

precision The ability of a measurement to be consistently reproduced.

precisión Capacidad que tiene una medición de poder reproducirse consistentemente.

principal The amount of money deposited or borrowed.

capital Cantidad de dinero que se deposita o se toma prestada.

prism A three-dimensional figure with at least three rectangular lateral faces and top and bottom faces parallel.

prisma Figura tridimensional que tiene por lo menos tres caras laterales rectangulares y caras paralelas superior e inferior.

probability The chance that some event will happen. It is the ratio of the number of favorable outcomes to the number of possible outcomes.

probabilidad La posibilidad de que suceda un evento. Es la razón del número de resultados favorables al número de resultados posibles.

properties Statements that are true for any number or variable.

propiedades Enunciados que son verdaderos para cualquier número o variable.

proportion An equation stating that two ratios or rates are equivalent.

proporción Ecuación que indica que dos razones o tasas son equivalentes.

proportional The relationship between two ratios with a constant rate or ratio.

proporcional Relación entre dos razones con una tasa o razón constante.

pyramid A three-dimensional figure with at least three lateral faces that are triangles and only one base.

pirámide Figura tridimensional que tiene por lo menos tres caras laterales triangulares que son triángulos y una sola base.

Qq

quadrant One of the four regions into which the two perpendicular number lines of the coordinate plane separate the plane.

cuadrante Una de las cuatro regiones en que dos rectas numéricas perpendiculares dividen el plano de coordenadas.

quadrilateral   A closed figure having four sides and four angles.

cuadrilátero   Figura cerrada que tiene cuatro lados y cuatro ángulos.

quartile   A value that divides the data set into four equal parts.

cuartil   Valor que divide el conjunto de datos en cuatro partes iguales.

Rr

radical sign   The symbol used to indicate a nonnegative square root, $\sqrt{}$.

signo radical   Símbolo que se usa para indicar una raíz cuadrada no negativa, $\sqrt{}$.

random   Outcomes occur at random if each outcome occurs by chance. For example, rolling a number on a number cube occurs at random.

azar   Los resultados ocurren aleatoriamente si cada resultado ocurre por casualidad. Por ejemplo, sacar un número en un cubo numerado ocurre al azar.

range   The set of output values for a function.

rango   Conjunto de valores de salida para una función.

range   The difference between the greatest and least data value.

rango   La diferencia entre el número mayor y el menor en un conjunto de datos.

rate   A ratio that compares two quantities with different kinds of units.

tasa   Razón que compara dos cantidades que tienen distintas unidades de medida.

rate of change   A rate that describes how one quantity changes in relation to another. A rate of change is usually expressed as a unit rate.

tasa de cambio   Tasa que describe cómo cambia una cantidad con respecto a otra. Por lo general, se expresa como tasa unitaria.

rational number   Numbers that can be written as the ratio of two integers in which the denominator is not zero. All integers, fractions, mixed numbers, and percents are rational numbers.

número racional   Números que pueden escribirse como la razón de dos enteros en los que el denominador no es cero. Todos los enteros, fracciones, números mixtos y porcentajes son números racionales.

real numbers   A set made up of rational and irrational numbers.

números reales   Conjunto de números racionales e irracionales.

reciprocal   The multiplicative inverse of a number.

recíproco   El inverso multiplicativo de un número.

rectangle   A parallelogram having four right angles.

rectángulo   Paralelogramo con cuatro ángulos rectos.

rectangular prism   A solid figure that has two parallel and congruent bases that are rectangles.

prisma rectangular   Figura sólida con dos bases paralelas y congruentes que son rectángulos.

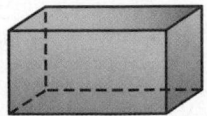

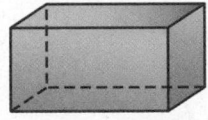

reduction   An image smaller than the original.

reducción   Imagen más pequeña que la original.

reflection   The mirror image produced by flipping a figure over a line.

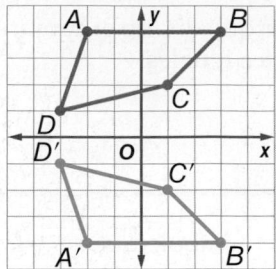

reflexión   Imagen especular producida al voltear una figura sobre una línea.

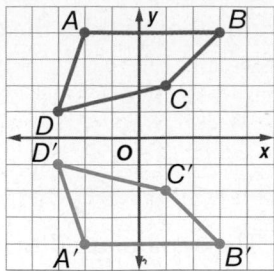

regular polygon   A polygon that has all sides congruent and all angles congruent.

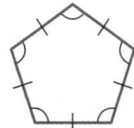

polígono regular   Polígono con todos los lados y todos los ángulos congruentes.

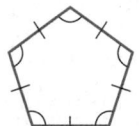

regular tessellation   A tessellation made from one regular polygon.

teselado regular   Teselado formado a partir de un polígono regular.

relation   Any set of ordered pairs.

relación   Cualquier conjunto de pares ordenados.

relative frequency   A ratio that compares the frequency of each category to the total.

frecuencia relativa   Razón que compara la frecuencia de cada categoría al total.

repeating decimals   A decimal with a pattern in its digits that repeats forever. Examples are 0.181818… and 0.83333… .

decimales periódicos   Decimal con un patrón en los dígitos que se repiten infinitamente. Por ejemplo, 0.181818… y 0.83333… .

rhombus   A parallelogram having four congruent sides.

rombo   Paralelogramo que tiene cuatro lados congruentes.

right angle   An angle that measures exactly 90°.

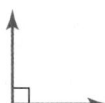

ángulo recto   Ángulo que mide exactamente 90°.

right triangle   A triangle having one right angle.

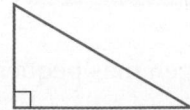

triángulo rectángulo   Triángulo que tiene un ángulo recto.

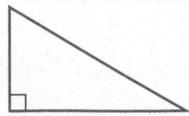

rotation When a figure is rotated around a point to change the figure's orientation with respect to that point.

rotación Cuando una figura se gira alrededor de un punto para cambiar su orientación con respecto al punto.

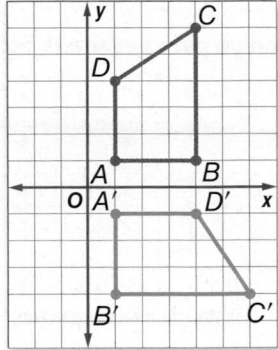

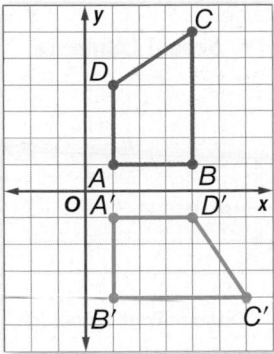

rotational symmetry A type of symmetry a figure has if it can be rotated less than 360 degrees about its center and still look like the original.

simetría rotacional Tipo de simetría que tiene una figura si se puede girar menos que 360 grados en torno al centro y aún sigue viéndose como la figura original.

Ss

sales tax An additional amount of money charged on items that people buy.

impuesto sobre las ventas Cantidad de dinero adicional que se cobra por los artículos que se compran.

sample A randomly selected group chosen for the purpose of collecting data.

muestra Grupo escogido al azar o aleatoriamente que se usa con el propósito de recoger datos.

sample space The set of all possible outcomes of a probability experiment.

espacio muestral Conjunto de todos los resultados posibles de un experimento probabilístico.

scale The scale that gives the ratio that compares the measurements of a drawing or model to the measurements of the real object.

escala Razón que compara las medidas de un dibujo o modelo a las medidas del objeto real.

scale drawing A drawing that is used to represent objects that are too large or too small to be drawn at actual size.

dibujo a escala Dibujo que se usa para representar objetos que son demasiado grandes o demasiado pequeños como para dibujarlos de tamaño natural.

scale factor A scale written as a ratio without units in simplest form.

factor de escala Escala escrita como una razón sin unidades en forma simplificada.

scale model A model used to represent objects that are too large or too small to be built at actual size.

modelo a escala Réplica de un objeto real, el cual es demasiado grande o demasiado pequeño como para construirlo de tamaño natural.

scalene triangle A triangle having no congruent sides.

triángulo escaleno Triángulo sin lados congruentes.

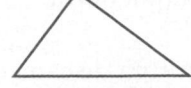

scatter plot In a scatter plot, two sets of related data are plotted as ordered pairs on the same graph.

diagrama de dispersión Diagrama en que dos conjuntos de datos relacionados aparecen graficados como pares ordenados en la misma gráfica.

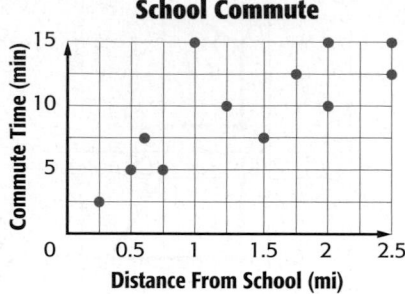

School Commute

Tiempo para llegar a la escuela

scientific notation When a very large or very small number is written as the product of a factor and a power of 10.

notación científica Cuando un número muy grande o muy pequeño se escribe como el producto de un factor y una potencia de 10.

sequence An ordered list of numbers, such as 0, 1, 2, 3 or 2, 4, 6, 8.

sucesión Lista ordenada de números, como 0, 1, 2, 3 ó 2, 4, 6, 8.

Side-Angle-Side Similarity (SAS) Two triangles are similar if two of the corresponding sides are proportional and the included angles are congruent.

semejanza lado-ángulo-lado Dos triángulos son semejantes si dos de los lados correspondientes son proporcionales y los ángulos incluidos son iguales.

Side-Side-Side Similarity (SSS) Two triangles are similar if the corresponding sides are proportional.

semejanza lado-lado-lado Dos triángulos son semejantes si los lados correspondientes son proporcionales.

similar figures Figures that have the same shape but not necessarily the same size.

figuras semejantes Figuras que tienen la misma forma, pero no necesariamente el mismo tamaño.

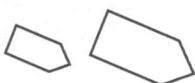

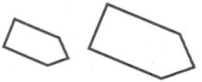

similar solids Solids with the same shape. Their corresponding linear measures are proportional.

sólidos semejantes Sólidos con la misma forma. Sus medidas lineales correspondientes son proporcionales.

simple event One outcome or a collection of outcomes.

eventos simples Un resultado o una colección de resultados.

simple interest The amount paid or earned for the use of money. The formula for simple interest is $I = prt$.

interés simple Cantidad que se paga o que se gana por el uso del dinero. La fórmula para calcular el interés simple es $I = prt$.

simple random sample An unbiased sample where each item or person in the population is as likely to be chosen as any other.

muestra aleatoria simple Muestra de una población que tiene la misma probabilidad de escogerse que cualquier otra.

slant height The height of each lateral face.

altura oblicua Altura de cada cara lateral.

slope The rate of change between any two points on a line. It is the ratio of vertical change to horizontal change. The slope tells how steep the line is.

slope-intercept form An equation written in the form $y = mx + b$, where m is the slope and b is the y-intercept.

square The product of a number and itself. 36 is the square of 6.

square A parallelogram having four right angles and four congruent sides.

squared The product of a number and itself. 36 is the square of 6.

square root The factors multiplied to form perfect squares.

standard form Numbers written without exponents.

stem The digits of the greatest place value of data in a stem-and-leaf plot.

stem-and-leaf plot A system where data are organized from least to greatest. The digits of the least place value usually form the leaves, and the next place-value digits form the stems.

Stem	Leaf
1	2 4 5
2	
3	1 2 3 3 9
4	0 4 6 7

$4 \mid 7 = 47$

straight angle An angle that measures exactly 180°.

Subtraction Property of Equality If you subtract the same number from each side of an equation, the two sides remain equal.

supplementary angles Two angles are supplementary if the sum of their measures is 180°.

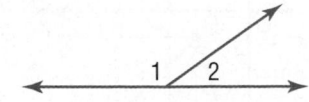

∠1 and ∠2 are supplementary angles.

pendiente Razón de cambio entre cualquier par de puntos en una recta. Es la razón del cambio vertical al cambio horizontal. La pendiente indica el grado de inclinación de la recta.

forma pendiente-intersección Ecuación escrita en la forma $y = mx + b$ donde m es la pendiente y b es la intersección y.

cuadrado Producto de un número por sí mismo. 36 es el cuadrado de 6.

cuadrado Paralelogramo con cuatro ángulos rectos y cuatro lados congruentes.

al cuadrado El producto de un número por sí mismo. 36 es el cuadrado de 6.

raíz cuadrada Factores multiplicados para formar cuadrados perfectos.

forma estándar Números escritos sin exponentes.

tallo Los dígitos del mayor valor de posición de los datos en un diagrama de tallo y hojas.

diagrama de tallo y hojas Sistema donde los datos se organizan de menor a mayor. Por lo general, los dígitos de los valores de posición menores forman las hojas y los valores de posición más altos forman los tallos.

Tallo	Hojas
1	2 4 5
2	
3	1 2 3 3 9
4	0 4 6 7

$4 \mid 7 = 47$

ángulo llano Ángulo que mide exactamente 180°.

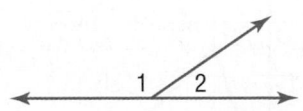

propiedad de sustracción de la igualdad Si restas el mismo número de ambos lados de una ecuación, los dos lados permanecen iguales.

ángulos suplementarios Dos ángulos son suplementarios si la suma de sus medidas es 180°.

∠1 y ∠2 son suplementarios.

surface area   The sum of the areas of all the surfaces (faces) of a three-dimensional figure.

área de superficie   La suma de las áreas de todas las superficies (caras) de una figura tridimensional.

survey   A question or set of questions designed to collect data about a specific group of people, or population.

encuesta   Pregunta o conjunto de preguntas diseñadas para recoger datos sobre un grupo específico de personas o población.

systematic random sample   A sample where the items or people are selected according to a specific time or item interval.

muestra aleatoria sistemática   Muestra en que los elementos o personas se eligen según un intervalo de tiempo o elemento específico.

Tt

term   Each number in a sequence.

término   Cada número en una sucesión.

terminating decimals   A decimal whose digits end. Every terminating decimal can be written as a fraction with a denominator of 10, 100, 1,000, and so on.

decimales terminales   Decimal cuyos dígitos terminan. Todo decimal terminal puede escribirse como una fracción con un denominador de 10, 100, 1,000, etc.

tessellation   Patterns formed by repeating figures that fill a plane without gaps or overlaps.

teselado   Patrón formado por figuras repetidas que llenan un plano sin traslaparse o dejar espacios entre sí.

theoretical probability   The ratio of the number of ways an event can occur to the number of possible outcomes. It is based on what *should* happen when conducting a probability experiment.

probabilidad teórica   Razón del número de maneras en que puede ocurrir un evento al número de resultados posibles. Se basa en lo que *debería* pasar cuando se conduce un experimento probabilístico.

tip   Also known as a gratuity, it is a small amount of money in return for a service.

propina   También conocida como gratificación; es una cantidad pequeña de dinero en recompensa por un servicio.

transformation   A movement of a geometric figure. It maps one figure onto another.

transformación   Desplazamiento de una figura geométrica. Aplicación de una figura sobre otra.

translation   The motion of moving a figure without turning it.

traslación   Movimiento de una figura pero sin rotarla ni voltearla.

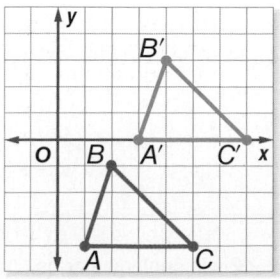

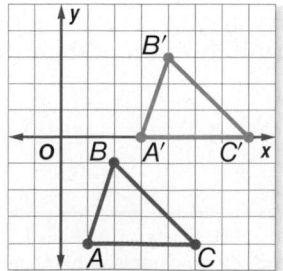

transversal The third line formed when two parallel lines are intersected.

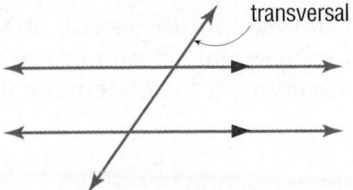

transversal

transversal Tercera recta que se forma cuando se intersecan dos rectas paralelas.

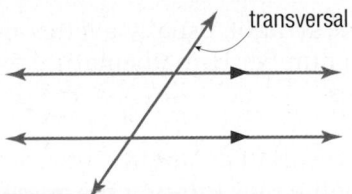

transversal

trapezoid A quadrilateral with one pair of parallel sides.

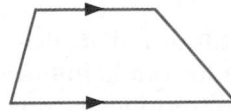

trapecio Cuadrilátero con un único par de lados paralelos.

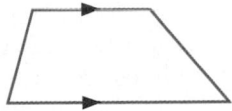

tree diagram A diagram used to show the sample space.

diagrama de árbol Diagrama que se usa para mostrar el espacio muestral.

triangle A figure with three sides and three angles.

triángulo Figura con tres lados y tres ángulos.

triangular prism A prism that has bases that are triangles.

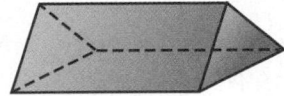

prisma triangular Prisma cuyas bases son triángulos.

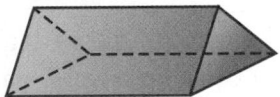

two-step equation An equation having two different operations.

ecuación de dos pasos Ecuación que contiene dos operaciones distintas.

Uu

unbiased sample A sample representative of the entire population.

muestra no sesgada Muestra que se selecciona de modo que se representativa de la población entera.

unfair game A game where there is not a chance of each player being equally likely to win.

juego injusto Juego donde cada jugador no tiene la misma posibilidad de ganar.

unit rate A rate that is simplified so that it has a denominator of 1 unit.

tasa unitaria Tasa simplificada para que tenga un denominador igual a 1.

unit ratio A unit rate where the denominator is one unit.

razón unitaria Tasa unitaria en que el denominador es la unidad.

unlike fractions Fractions with different denominators.

fracciones con distinto denominador Fracciones cuyos denominadores son diferentes.

upper quartile The median of the upper half of a set of data.

cuartil superior Mediana de la mitad superior de un conjunto de datos.

Vv

variable   A symbol, usually a letter, used to represent a number in mathematical expressions or sentences.

variable   Símbolo, por lo general una letra, que se usa para representar un número en expresiones o enunciados matemáticos.

vertex   A vertex of an angle is the common endpoint of the rays forming the angle.

vértice   El vértice de un ángulo es el extremo común de los rayos que lo forman.

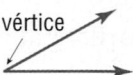

vertical angles   Opposite angles formed by the intersection of two lines. Vertical angles are congruent.

∠1 and ∠2 are vertical angles.

ángulos opuestos por el vértice   Ángulos opuestos formados por la intersección de dos rectas. Los ángulos opuestos por el vértice son congruentes.

∠1 y ∠2 son ángulos opuestos por el vértice.

volume   The number of cubic units needed to fill the space occupied by a solid.

volumen   Número de unidades cúbicas que se requieren para llenar el espacio que ocupa un sólido.

voluntary response sample   A sample which involves only those who want to participate in the sampling.

muestra de respuesta voluntaria   Muestra que involucra sólo aquellos que quieren participar en el muestreo.

Xx

x-axis   The horizontal number line in a coordinate plane.

eje x   La recta numérica horizontal en el plano de coordenadas.

x-coordinate   The first number of an ordered pair. It corresponds to a number on the x-axis.

coordenada x   El primer número de un par ordenado. Corresponde a un número en el eje x.

x-intercept   Where a graph crosses the x-axis.

intersección x   Lugar donde una gráfica cruza el eje x.

Yy

y-axis   The vertical number line in a coordinate plane.

eje y   La recta numérica vertical en el plano de coordenadas.

y-coordinate   The second number of an ordered pair. It corresponds to a number on the y-axis.

coordenada y   El segundo número de un par ordenado. Corresponde a un número en el eje y.

y-intercept   Where a graph crosses the y-axis.

intersección y   Lugar donde una gráfica cruza el eje y.

zero pair The result when one positive counter is paired with one negative counter. The value of a zero pair is 0.

par nulo Resultado de hacer coordinar una ficha positiva con una negativa. El valor de un par nulo es 0.

Index

Dd

Ee

Ff

Ll

Mm